BALL REDBOOK

16TH EDITION

VIC BALL
EDITOR

Ball Publishing

Batavia, Illinois USA

Ball Publishing
335 North River Street
Batavia, Illinois 60510 USA

Library of Congress Cataloging-in-Publication Data

Ball redbook / Vic Ball, editor.—16th ed.
    p.   cm.
  Earlier edition published under title: The Ball red book.
  Includes index.
  ISBN 1-883052-15-7
  1. Floriculture. 2. Plants, Ornamental. 3. Ornamental plant
industry. I. Ball, Vic. II. Ball red book.
SB405.B254   1997
635.9—dc21

97-6832
CIP

Cover photo from Goldsmith Seeds, Inc.; ***back cover photo*** of Vic Ball and Jack Pearlstein by Marvin N. Miller.

DEDICATION

This second printing of the 16th edition of the *Ball Redbook* is dedicated to Vic Ball, whose vision and energy helped lead *GrowerTalks* magazine and the floriculture industry for more than 60 years.

CONTENTS

CROP CULTURE BY CROP

REDBOOK AUTHORS

Ron Adams
Robert G. Anderson
Douglas A. Bailey
Warren Banner
Chris Beytes
Thomas H. Boyle
Mark P. Bridgen
Lois Carney
Charles A. Conover
Bob Cornell
Brian Corr
Simon Crawford
Bob Croft
Paul Cummiskey
Gaby and Micha Danziger
A.A. De Hertogh
Dick Devries
Paul Ecke III
Drew Effron
Orlo Ehart
Lelion D. Elledge Jr.
Willie Faber
Ron Ferguson
Arnold W. Fischer
Paul Fisher
William C. Fonteno
Ralph Freeman
Jeff Fries
Phil Gardenier
Elizabeth C. Gerritsen
Hans A. Gerritsen
Andrew Greenstein

Garry Grueber
P. Allen Hammer
Debbie Hamrick
Daijiro Harada
David Hartley
Will Healy
Royal Heins
Heliconia Society International
Gary Hennen
Kerry Herndon
Edward Higgins
Kenneth Horst
Michael Klopmeyer
Dean M. Krauskopf
Linda Laughner
Roy A. Larson
Miriam Levy
Jim Locke
Jim Long
Ed Markham
Jeff McGrew
Bob Michael
Marvin N. Miller
Robert O. Miller
Rob Miller
William B. Miller
Jim Nau
Terril Nell
John Nelson
Paul V. Nelson
David W. Niklas
Bob Oglevee

Gary Pellet
John C. Peterson
Grace Price
John S. Rader
Dennis Reynolds
Andrew C. Stavrou
Rudolf Sterkel
Jack S. Sweet
Darryl D. Warncke
C. Anne Whealy
Gary J. Wilfret
Harold F. Wilkins
Jack Williams
Keith Zary
Al Zylstra

A SPECIAL THANK-YOU TO:

Bart Bernacchi
Laura Beytes
Jeff Deal
George Dean
Tom Doak
George Griffith
Jan Hall
Margaret Kelly
Ronald K. Jones
Robert Lando
J.W. Moyer
Ronald Oetting
Gene Parsons
Hans Pein
Jay Stout
John Walters
Randy Whitesides

INTRODUCTION

The ***Ball RedBook*** is designed for two groups of readers: growers—both new ones and those with some experience—and students in floriculture production. It first covers the subjects every grower needs to know about greenhouse equipment and general plant culture, highlighting innovations and the latest in automation in flower production. Then come individual how-to-grow guides to over 120 greenhouse crops: bedding plants, pot plants, cut flowers, and a wide variety of other crops, including, of course, perennials.

You'll find a heavy emphasis on the whole field of automation and mechanization, including plugs, Dutch trays, transplanting machines (as well as automatic pot plant transplanting), computers in the greenhouse, tissue culture, and DIF (cool nights, short plants). There have been many developments and changes in these fields, as there have been with insecticides, biological insect control, and crop irrigation. We have tried to cover them all.

I'd like to credit my many excellent co-authors—the men and women listed here, all industry experts—who have contributed their knowledge and time through writing chapters and sections of this book to help make this 16th edition of the ***Ball RedBook*** the best yet. Much credit also goes to John Martens, president, and Debbie Hamrick, editorial director, from Ball Publishing, and certainly to Sylvia Hemmer, my secretary, who has organized and typed the pages of this book. Finally, many thanks to Liza Sutherland, our book editor, and Peter Nelson, who painstakingly edited and polished the pages of this book; Diane McCarthy, who carefully guided the book through its many stages of production; and Fran Manos, who diligently entered copy, made editing corrections, and proofread the contents.

I hope this book will be useful and helpful to the many of you who grow flowers. It's written for you!

Vic Ball

Vic Ball

SECTION 1

GREENHOUSE OUTFITTING

CHAPTER 1

GREENHOUSES FOR 2000 AND BEYOND

by Vic Ball

So much has changed in greenhouse structures since the *Ball RedBook's* 15th edition (only six years ago). Polyethylene is still king, but the rigid structured sheets have improved greatly—and they are successfully challenging glass and even poly. Also, the shift to hot water heat continues.

GREENHOUSE COVERINGS

There were 487,863,000 square feet (11,199 acres, 4,666 ha) of total greenhouse area in 1995, according to the USDA *Floriculture Crops Report*. Of existing greenhouses, 63% are covered with polyethylene; 23% with structured sheets—rigid plastic, polycarbonate, acrylic, or fiberglass; and 14% with glass. However, for the real story, ask half a dozen greenhouse builders, "What roofs and sides are being used on new greenhouses?" Here's the consensus:

POLY STILL NUMBER 1

Poly is used on the majority of new greenhouses built today (fig.1). It's nearly always a double sheet blown 3 to 9 inches (7 to 23 cm) apart by a "squirrel cage," a small blower fan. That air space provides great insulation—a good, tight, inflated poly greenhouse will burn one-third less fuel than a single glass roof—but the two milky sheets do cut down some on precious winter light. Most greenhouse roofs use 6-mil-thick plastic on the outside and on the inner sheet.

Poly greenhouses are very low-cost to build. In acre volume in a single sheet, good 6-mil poly costs about $.06 per square foot ($.63/m²), $.12 for a double sheet. A gutter-connected house covered with double poly (the skin only) costs about $1.50 for material only (no ends, no heat, no benches).

A "budget" hoop house—a free-standing single house, no heat, no fan-jet, 16 to 20 feet (4.8 to 6 m)

FIG. 1   Polyethylene greenhouses, like this gutter-connected range, make up the majority of greenhouses in the United States. The grower: Peter's Wholesale Greenhouse, Walnut Springs, Texas.

wide—can be had for around $.65 per square foot ($6.82/m²), materials only. Source: Agra Tech, phone 510/432-3399.

Al Reilly of Rough Brothers notes, "A complete first-class, inflated, double-poly greenhouse with pad and fan cooling, benches, overhead unit heaters, and HAF fans will cost about $9 to $10 per square foot ($94.50 to $105/m²) covered, all labor and material. Very roughly, that's $4 for structure with sides and ends, $3 to $3.50 for benches (rolling), and $2.50 for unit heaters with HAF fans."

The good news: the new polys of the 1990s will last three summers, sometimes even four. The greenhouse must then be re-covered because poly is finally destroyed by ultraviolet light, of which there's a lot more in southern areas than in the North. Growers in subtropical areas, such as Florida, often use one layer of poly—fuel expense is lower there, so there's less reason for a double-layer roof.

Poly hoop houses easily accommodate overhead unit heaters; not so with Venlo houses. Again, it's a low-initial-cost approach to growing. Poly houses with unit heaters plus fan-jets for heat distribution are frequently used worldwide.

Skip Smith of X.S. Smith, Inc. has a few observations on poly and crop used: "Inflated poly is widely used for spring bedding production, often combined with fall poinsettias in the North or South. Light limitations are not a problem for either crop. Many growers use poly for other pot plants and cut flowers even in northern areas, although this is more challenging to the grower, especially in low-light areas, such as Cleveland and Seattle. HID light is used some to offset this." Poly generally does not allow enough winter light to do quality cut roses in the North.

Huge areas of winter vegetables are grown around the Mediterranean, nearly always under single sheets of poly. There are tens of thousands of acres of such production in places like Spain.

An inflated, double-poly hoop house (or any greenhouse) can be built as a single greenhouse or gutter-connected. Initial cost is higher per square foot covered, but fuel cost is much lower for gutter-connected houses—there's much less exposed roof area to radiate heat. Also, managing a gutter-connected range is better—you can see what your help is doing, and it's easier to move materials back and forth from the head house.

Gutter-connected, double-poly hoop houses are now being built up to 30 feet (9.1 m) wide, clear span, two 15-foot (4.5 m) roofs supported by a truss. Most of all, though, the 10,000 American bedding plant growers—from single 100-foot (30 m) greenhouse "start-up growers" to 20-acre (8.3 ha) ranges—work mostly under inflated double poly.

There is a Gothic-design poly Quonset house available. It is a free-standing, single house with a rounded deck roof. It's made in 24-, 27-, and 30-foot (7.3-, 8.2-, and 9.1-m) widths and is used especially in areas with heavy snows.

Polyethylene coverings are continually improving. No longer straight polyethylene, they are a mixture of polymers. Today's three-year sheets last the full three years, largely due to UV inhibitors. Manufacturers continue to look for ways to produce a longer-lasting film.

All these films are manufactured as tubes—that's how the plastic is extruded by machines. Growers can then buy film as a tube, for double layers, or as a sheet, which is a slit tube. Tubes are not available in all sizes; there are still a lot of single sheets out there. Plus, big, heavy rolls of double sheets are difficult to get up onto a roof and handle. The folding of film is important because creases are often where rips occur. Poly manufacturers are striving to send out film with "gentler folds" to eliminate creases. How the film is folded also helps or hinders covering the house.

Additives in films address drips, sunlight (UV radiation), and heat retention. For drips the films create a sheeting effect, holding onto the condensation tighter, so it rolls down the film to the greenhouse sides. Products include CT Film's 703 DC and Armin's Tufflite III Drip-Less.

Harsh sunlight and heat retention are dealt with together in infrared (IR) films, including CT Film's Cloud Nine, Armin's Tufflite Infrared, and AT Plastic Dura-Therm. These products diffuse the sunlight, giving the house more available light, and keep the heat (infrared energy) in. Plants use visible light for photosynthesis, so this diffusion helps plant growth. Remember the light spectrum of ROY G BIV, with infrared heat beyond the red end and ultraviolet radiation beyond the violet end. Now you can picture how you can let light in and not let heat out. The plants absorb the sunlight and reradiate it as heat. Surfaces in the greenhouse do this, too. The IR film then holds that heat in.

A new blue film, called Kool Lite, is available from Klerk's Plastics. In view of the light spectrum needed for plant growth, this film is intended to refine the light available to the plant. The manufacturer claims it can reduce *Botrytis,* but it's too early to say how well this film fares. We have seen other colored films, copper and pink, come and go.

The most widely used film type is a clear, nothing-added, three-year product. Additive films do come with a higher price tag, but longer film life is something growers will probably pay for because of the tangible labor savings in less frequent re-covering. Manufacturers are striving for more affordable longer-life product.

Long-term growers need to be aware of disposal concerns with these products. Acres of used polyethylene is a lot of garbage. There is some recycling of such films, but it is by no means common.

GLASS

Glass is probably Number 3 today, passed by structured sheets in the early 1990s. Glass was here first, though, so let's talk glass before we talk structured sheets. Glass is still the Cadillac of coverings.

Glass is the choice in these situations:
- Where winter light is critical, like with northern cut roses and often by quality-oriented, northern, holiday pot growers. Glass provides substantially better winter light than double poly.
- Where the grower is well capitalized. The initial cost of glass is much higher than that of other coverings. Over the 20-year haul, however, glass costs less per square foot per year, and you don't have to replace it unless it breaks!
- Where fuel cost is less critical. Fuel cost per square foot of glass is higher than double poly. If energy curtains are used under both, poly still costs less to heat.

Glass tends to grow the best crop. It allows maximum light (fig. 2). It "breathes," the glass laps between panes allowing air to enter. Poly and structured sheets tend to be airtight and can result in excessive humidity if not carefully managed. Some

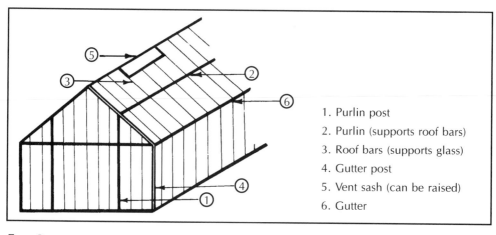

1. Purlin post
2. Purlin (supports roof bars)
3. Roof bars (supports glass)
4. Gutter post
5. Vent sash (can be raised)
6. Gutter

FIG. 2 This is a typical purlin-type glass structure. Modern glass houses have clear, open spans without posts, making it possible to crop nearly all of the greenhouse area.

greenhouse manufacturers are now using glass that's not lapped to get a tighter house, but it also requires watchful management.

Which will stand hail better? Poly, by far (unless the glass is tempered). High winds? Probably a draw if both are well designed. Snow load? Probably glass is better, again depending on snow and wind load design. A 20-pound-per-square foot (95 kg/m²) roof will hold better than a 15-pound-per-square-foot (71 kg/m²) roof, of course. However, most U.S. greenhouses today (of any roof material) are built to a 15-pound-per-square-foot design load. Normally, a safety factor of two is built in, so a 15-pound-per-square-foot roof will carry a 30-pound-per-square-foot (143 kg/m²) load safely. The ultimate design load is 30 pounds. Some houses in low-snow areas overseas are designed with only a 1.25 safety factor—thus the 15-pound roof will carry only 19 pounds safely.

We suggest that growers contact the National Greenhouse Manufacturers Association for a copy of greenhouse design load standards. For these technical specifications, write NGMA, 7800 South Elati #113, Littleton, CO 80120, phone 303/798-1338, 800/792-6462, fax 303/798-1332.

Wood bars, used until recently to support glass, require costly painting. Most new glass bars are aluminum—zero maintenance. Trusses are galvanized, so again no maintenance.

Several firms are building new Dutch-style glass houses in the United States and Canada, predominantly Venlo types (fig. 3). They permit a very high percentage of winter light to reach the crop due to narrow bars and wide glass panels. Combined with an energy curtain, they are reasonably fuel-efficient.

Venlo structures are widely used in Holland

FIG. 3 Glass and traditional Venlo design is the structure favored by many growers, especially in the North. The grower: Red Oak Greenhouses, Red Oak, Iowa.

today, including with new construction. They provide an answer to Holland's very low winter light. They are less fuel-efficient than inflated double poly, but Holland doesn't have the extreme cold of the northern United States and Canada.

Most glass used today is DSB (double strength B, 3 mm, about one-eighth inch) grade glass. This glass alone costs about $.60 per square foot ($6.30/m²). A stronger, more expensive type called "triple strength" is 4 mm (about five-thirtysecond inch) thick. There is also a tempered triple strength glass. Furthermore, there's a low-iron glass that permits more and better light transmission. The question here is the amount of photosynthetically active radiation (PAR)—the part of the light spectrum that makes plants grow—that passes through the glass.

Jeff Warschauer at Nexus, greenhouse builders, says he sees increased interest in and use of glass since the company has been using the stronger 5-mm tempered glass, which is less susceptible to hail, and sealed glass—no more lapped glass to let insects and dirt in and heat out.

In the fast-changing world of flower production, is it wise to invest in a 20-year glass structure? The grower can recover his investment in a poly greenhouse in several good years. The banker would like that!

STRUCTURED SHEETS

Structured sheets have moved up in popularity. They are expensive, but they offer compelling advantages.

Polycarbonates. Probably the most widely used structured sheet today is polycarbonate. At one time it was short-lived and discolored rapidly, but new technology has put an ultraviolet inhibitor coating onto—or sometimes blended into—the sheets. The result: much of today's polycarbonate is sold with a 10-year guarantee. Some suppliers even guarantee against hail loss, within prescribed limits, and labor costs to re-cover are reasonable. Sheets can be "point fastened"—nails or screws can be driven into the material (single or double) to secure it to a roof bar. This is not possible with acrylics, which tend to shatter. Also, newer polycarbonates are dripless, helping to control the condensation problems that often arise in such tight houses.

Polycarbonate is manufactured two ways: (1) As a "twin wall," two very thin, rigid, flat sheets of polycarbonate are held together by polycarbonate ribs. The total thickness of the two sheets is 8 mm (five-sixteenth inch). These sheets are great for insulation. The cost for sheets with necessary hardware: $1.30 to $1.50 per square foot ($13.65 to $15.75/m²). (2) As a single corrugated sheet, which costs about $.75 to $1 per square foot ($7.87 to $10.50/m²). It's a good glazing material for retail greenhouse gables and ends, especially, because it allows a good view of what's in the greenhouse.

Both types offer good light penetration, although the single sheet is better. Acrylic allows about 86% of sunlight to reach crops, polycarbonate about 80%, and glass about 88 to 89% (figures based on new material; on all of these, subtract light loss from interior structures, pipes, shade cloth, etc.).

A key selling point for polycarbonate is that it resists hail well. It's widely used in the Colorado-Texas hail belt for this reason. Also, it's used on sides and ends of glass and poly ranges everywhere to minimize vandalism damage. Polycarbonate is also fire-resistant.

Polycarbonate, even double sheets, can be bent around the shape of a hoop greenhouse. That's a real advantage, whereas acrylic bends only slightly. Polycarbonate, easily installed, is builder-friendly.

Polycarbonate is appearing on more and more new pot plant, bedding plant, and cut flower ranges these days, both new and retrofit. John Pound of Agra Tech, Inc., Pittsburg, California, says, "We've built several polycarbonate ranges in California for roses, carnations, etc., as well as a 750,000-square-foot (7 ha) polycarbonate range in Boise, Idaho, originally for fish but now growing tomatoes."

Trade names for polycarbonates are confusing! Several of the more common ones are DynaGlass (corrugated), Lexan Thermoclear, and Polygal.

Acrylics. Acrylics give some more light penetration than polycarbonates; some growers say growth is definitely better. They are long lasting, some manufacturers guaranteeing them up to eight to 10 years for light transmission, but not guaranteed against breakage. Yet, World War II fighter plane windshields were made of acrylics, and they're still okay.

An acrylic makes a fine house for such winter-light-sensitive crops as cut roses, although it costs a bit more than polycarbonate. It's normally used as a twin wall, two flat sheets one-half to three-quarter inch or so apart. The air space between the two sheets provides insulation, so it's very fuel-efficient. Acrylic may not be point-fastened—no screws or nails can be driven through the sheet—so, needing gaskets, it costs more to secure it to a roof. Also, less impact-resistant than polycarbonate, acrylic can shatter. Severe hail can cause crazing or even holes. Acrylic is also more prone to burning, about like fiberglass.

I see acrylics a lot more on northern rose ranges and generally in New England (fig. 4). Research greenhouses also use them, probably mostly for better light penetration. Ultraviolet-inhibiting polycarbonate seems to be more widely used, for both retrofit and new ranges.

One trade name under which acrylics are sold is Exolite; also, Exolite No-Drip causes condensation to run down to gutters rather than dripping onto crops. A new double-wall acrylic product, Alltop, features 91% light transmission (up from Exolite's 87%) with the same no-drip coating.

FIG. 4 Glazing made of acrylic structured sheets is chosen by many northern growers trying to maximize winter light levels.

Fiberglass. Widely used in the 1960s and 1970s, fiberglass has dropped in popularity. Corrugated fiberglass is a fairly low-cost sheet, varying from $.90 per square foot ($9.45/m²) to $1.25 per square foot ($13.12/m²) for top quality, Tedlar-coated fiberglass. Light penetration through new sheets is good. A case is made for a more diffused light reaching the crop, better than with most other roofs. Hail is less of a problem than with glass, but some hail will penetrate or craze it. There is some concern about recent quality.

The main problem with fiberglass is that it burns easier than other covers (some severe fires have occurred). It does discolor after seven to 10 years, depending on ultraviolet radiation levels; faster darkening occurs in southern areas. It is not an especially fuel-efficient roof, one reason it has been widely used in the Deep South and California, where fuel is less of an expense.

TRADE SHOWS

An interesting point: anyone considering building or retrofitting a greenhouse would get a lot of good background on roofing materials and structures by attending one of the major U.S. trade show-seminars. The GrowerExpo is held in January; the Ohio International Floral Short Course in July; the Far West Show in Portland, Oregon, and the TAN/MISSLARK show in Texas, both in August; the Professional Plant Growers Association conference in October; and the Canadian Greenhouse Conference, Guelph, Ontario, also in October. There are exhibits by principal manufacturers and distributors, with samples of all these materials available for inspection and qualified people there to talk about them. Watch the trade press for announcements of these meetings.

GREENHOUSE STRUCTURES

COST ESTIMATES

The following cost analysis figures from several greenhouse manufacturers are for some common greenhouses today:

♦ Inflated double-poly hoop house with unit heaters, pad fan cooling, 20-pound (95 kg/m²) no-load roof, no benches: $6 to $6.50 per square foot ($63 to $68.25/m²) of ground covered, labor and materials both included.

♦ A very low-cost, small hoop house with single poly, no heat, no fans or benches: to $.65 per square foot ($6.82/m²), materials only.

♦ A good, 2- to 3-acre (0.8- to 1.2-ha) Dutch Venlo range (glass), 21-ft-wide (6.4 m), gutter-connected greenhouse: $6 to $7 per square foot (($63 to $73.50/m²) of ground covered, all labor and materials, including a hot water

WHAT'S NEW

For many growers retractable-roof greenhouses are the best of both worlds, providing extra space at low cost during the spring months, when they need it most! (See fig. 5.) Retractable-roof greenhouses are designed for crops that would prefer to be outdoors but can't take the rain and cold, such as bedding plants, perennials, and field-grown cut flowers. Many growers use the new greenhouses as extra growing areas, mainly in the spring for their bedding crops and through the fall for poinsettias. They may leave the houses empty January through March; this is especially true in northern regions. Southern growers marketing early pansies and perennials, on the other hand, are likely to keep their retractable-roof houses running all year.

FIG. 5 Retractable-roof greenhouses provide extra space during the spring when growers need it most. During inclement weather, the polyethylene roof (shown here stored at the purlins) is extended to protect the crop.

Three or four manufacturers make retractable-roof greenhouses. The Cravo Retract-a-Roof Greenhouse, by Cravo Equipment of Brantford, Ontario, is a steel, A-frame-style greenhouse with gutters and trusses. Sections of woven polyethylene, traveling from truss to truss, can be extended or withdrawn. The woven polyethylene is six times stronger than regular poly. The sections come in standard widths of 21, 24, and 30 feet (6.6, 7.3, and 9.1 m).

A 40,000-square-foot (3,700-m²) roof is operated by a single electric motor. Three hundred feet (91 m) of greenhouse will open and close in less than three minutes. The crop is exposed to 90% sunshine and fresh air when the roof is retracted. It's designed for 10, 15, or 20 pounds per square foot of snow load (49, 73, or 98 kg/m²), as well as an 80-mph (129-kph) wind load, with the roof closed (extended). There is an unlimited snow load when the roof is retracted.

The manufacturer projects a five- to seven-year life, depending on the part of the country. The cost to replace the roof alone is about $.36 per square foot ($3.87/m²). The cost for an entire 1-acre (0.4-ha) structure, including labor and materials, runs about $3.50 to $4 per square foot ($37.66 to $43.04/m²). Cravo Equipment, Ltd. can be contacted at White Swan Road, RR #1, Brantford, Ontario,

Canada; N3T 5L4, or phone 519/759-8226, fax 519/752-0082. For more on the Cravo and other retractable-roof houses, see the August 1994 issue of *GrowerTalks* magazine.

It might be interesting to compare this Cravo house to the American-style roll-out system, where a crop is rolled out of the end of the greenhouse and put outdoors during the spring. Growers report that the Cravo house is cheaper per usable unit of area than the regular greenhouse with rollout and the necessary cement walks.

heating system, but not boiler, pads or fans, or benches. (These figures are from V&V Noordland.)

◆ Another figure on a 20-ft-wide (6 m) Venlo greenhouse, glass roof, polycarbonate side vents, without labor, with heat, no benches: $9 per square foot ($94.50/m²). (Rough Brothers figure.)

◆ Nexus/Vail (the company's most popular hardcover greenhouse), a single 30 × 90 ft (9.1 × 27.4 m) house, no heat or benches: $2.90 per square foot ($30.45/m²). Nexus' long-popular Teton, a double-poly house, no equipment: $2.10 to $2.20 per square foot ($22.05 to $23.10/m²).

BUDGET BUILDING

An interesting way to ease the problem of raising money for new greenhouses is budget building. John Pound of Agra Tech, a major California greenhouse builder, counsels his grower customers to "take it a step at a time." In other words, build a good, sound, basic structure, as much as you can handle financially at the time.

As your cash flow position gets stronger, you can go back and add some of the fringes. For example, take environmental controls. Start out with lower-cost, solid state controls. Later, add a mainframe computer, which really controls your whole business—accounting, dollars per square foot, yield data, crop planning—not just the environmental factors. Energy curtains and black cloth can save a lot of fuel, but they're expensive. Start off without them, then add them later as you can. All sorts of interior automation can be also added gradually.

COLD FRAMES OF THE 1990S

A generation ago a cold frame was sort of an outdoor growing bed covered with a sash of glass or poly. Today, the term is applied to a very low-cost, budget, poly-covered hoop house. Cold frames are great for spring bedding plants even in the North, if a bit of heat and a vent fan are provided. Throughout the 1990s they have been very widely used! Agra Tech alone sells 2 million square feet (19 ha) per year.

A good example is Agra Tech's 20-ft-wide (6 m) house. It's a free-standing, separate house, its roof supported by a metal hoop extending from ground to ground. It was originally designed for a single layer of poly and only for spring use, but in fact many growers are putting double poly on it, adding a little heat and a fan, and using it for spring bedding plants.

Cold frames are low cost. For example, for a 20-by-100-foot (6-by-30-m) house, about 2,000 square feet, the pipe hoops alone are under $.50 per square foot ($5.25/m²), or about $100. If you add a double-poly roof to this, a Modine heater, and one 36-inch fan, it now goes to $1.25 per square foot ($13.12/m²), materials only, or about $2,500 for a 20-by-100-foot (6-by-30-m) greenhouse. Again, growers are now using such structures for year-round grow-

FIG. 6   Cold frames like this one at Martin Viette Nurseries, Long Island, New York, are the mainstay for thousands of American bedding plant and perennial production growers. This unheated cold frame was used to overwinter perennials. Bedding plant growers typically have a unit heater in each cold frame.

ing, even in the North. Cold frames are great for early crops of pansies, carnations, snaps, and other semihardy annuals (fig. 6).

For the record, such a house has limitations. First, the framework is lightweight. Second, more complete greenhouses are better ventilated—and spring plants don't like high temperatures. It is a free-standing house—not as efficient as a gutter-connected range—and, of course, unit heaters are definitely less fuel-efficient than hot water heat, which costs a lot more initially.

To REMODEL OR REBUILD

Tom Doak, president of Yoder Brothers, Barberton, Ohio, brings much experience to this question from his long involvement in Yoder operations. Tom comments, "The first consideration should be whether the facility, when you are finished, will provide the environment to produce a quality product. What you plan to produce is a critical factor, as well as how long you plan to keep producing it. Many older facilities have low gutters, which offer limited opportunity for the use of material-handling equipment and energy-saving heat or shade blankets.

"If you do go to the expense of building a new range, be certain that when you are through, you will have important savings in both energy costs and labor costs to show for your investment. Example: double poly, we find, does save an honest 25 to 30% against a single glass roof. Also, energy curtains (gutter-to-gutter applied at night) yield between 25 to 40% savings, in our experience.

"You do have to watch snow load here, though. When snow accumulates to dangerous weight loads, you want the air outside of the roof to be warm—to melt that snow off. That means no energy curtains. Automated energy curtains also mean automated shade for mums, etc.

"We figure fuel cost alone is about \$.95 per square foot per year (\$9.97/m²) for 60F (16C) in the North today. So, if you can save one-third of your fuel bill, your return can be significant. How many years will it take to pay for your new energy curtains or lap seal?

"Certainly crop turnover has to be watched, too. It is simply that more crops—more dollars per bench per year—increase your ability to carry overhead costs, such as fuel and labor.

"Moving benches, however you do it, eliminate most of that walk space. You not only cut the fuel bill 15 to 20%, but you can build 20% less greenhouse area and turn out the same amount of crops.

"Decisions on rebuilding or remodeling also have to be looked at from a tax point of view. Talk to your tax accountant. You may be able to charge off remodeling costs in the year incurred.

"Another point: it takes a lot of money to rebuild—and even for extensive modifications, in some cases, it may be a question of whether capital is available for this purpose. To many bankers, greenhouses are not the best collateral."

Tom has also talked about industrial revenue bonds, which in effect make capital available for improvements or rebuilding at "tax-free bond" rates. In some cases, you can get 100% financing. The usual length is seven to 10 years, and they are done in cooperation with the local municipality. One requirement is they must create new jobs or save existing ones.

FIG. 7   Before you begin construction, check with your local building inspector to see what building codes you must meet.

Finally, the question of rebuilding or remodeling boils down to this: which route will enable you to recover your investment sooner while producing a quality product?

BUILDING CODES

Before you even consider building or retrofitting, check with your local building inspector to see which building codes you must meet (fig. 7). Many a grower has proceeded with construction plans without up-front clearance, only to have the plans delayed because of the local building inspector. The National Greenhouse Manufacturers Association has worked hard to incorporate considerations for greenhouse structures on a national level into the major codes, like the Uniform Building Code. If you run into snags or want to be prepared in advance with a copy of NGMA's *Standards,* call NGMA at 303/798-1338, 800/792-6462, or fax 303/798-1332.

CHAPTER 2

CONTROLLING THE GREENHOUSE ENVIRONMENT

by Vic Ball

COMPUTERS DO THE JOB

Significant changes are taking place with environmental controls! Each year you get more greenhouse control for less money. Just a few years ago, only large growers could afford computerized controls. Not so today; many growers are putting them in. We can see that within the next five years, most growers with 1 acre (0.5 ha) or more of production will have a computerized control system.

Greenhouses are becoming more and more complex. Managing the climate calls for control not only of temperature—heating and cooling—but also of shade screens, mist, irrigation (ebb and flood!), fog, CO_2, and more. How you control these systems makes a significant difference in the quality of your product and your cost of production.

ENVIRONMENTAL CONTROLS

There are four types of environmental controls (table 1). Following is a comparison by function and cost per climate zone.

Thermostats. There are two basic types of thermostats: on-off and proportioning. On-off thermostats simply switch (heat or vent) with a change in temperature. Proportioning thermostats provide continuously variable resistance that changes with temperature. On-off thermostats can control fans, heaters, and other equipment. Proportional thermostats work as sensors for electronic controllers that operate equipment.

TABLE 1   ENVIRONMENTAL CONTROLS

Type	Cost ($)[a]	Suppliers
Thermostats and timers	100–500	Many
Staged temperature controllers and timers	800–1,800	Acme, Hired-Hand, Wadsworth
Computer zone controllers	800–2,500	Acme, Hired-Hand, MicroGrow, Q-Com, Wadsworth
Computerized environmental management systems	1,000–2,000[b]	Argus, Oglevee, Priva, Q-Com, Wadsworth

[a] Cost per greenhouse climate or operation zone.
[b] These figures per additional zone follow an initial outlay of $5,500–7,000 for the base system and control of one zone.

Thermostats have many advantages, such as low initial cost, flexibility, and simple installation. Their disadvantages are many, however. They offer limited control, little accuracy, no coordination of equipment, and poor energy efficiency. Thermostats are most at home in temporary Quonsets, where you may want only simple zones with heat on-off and one or two cooling stages.

Analog controls. Using proportioning thermostats or electronic sensors, analog controls gather temperature information to drive amplifiers and electronic logic (decision-making) circuitry. These controls cost more than thermostats but are more versatile and offer significantly better performance. They also integrate heating and cooling equipment so that equipment is sequenced on and off.

Again, analog controls are less expensive than computer controls. They can coordinate multiple pieces of equipment, offering a wide range of control possibilities. If you're looking to put analog controls in a large number of zones, their cost advantages may be lost. You would not be able to expand an analog when you add additional equipment to the analog control. If you're a small grower or plan to start small, however, and you want more temperature control functions than a thermostat can provide, analogs may be for you.

Computer controls. Computers replace the amplifiers and logic circuits of analog controls with a microprocessor. The computer combines information from a variety of sensors (measuring temperature, relative humidity, sunlight, and potentially much more) to make complex judgments about how to control the environment (fig. 1). Computers can control and coordinate up to 20 devices in each zone. Most computer zone controllers offer more control capability than analogs. They are relatively simple to use, and they can be purchased to fit specific zone requirements.

Again, though, be careful when buying computer controls for multiple zones; the cost advantage may be lost. While computer controls are better than analogs, they are limited to simple feedback control and have limited expandability. Also, most computer controls have limited functions. For example, irrigation computers are dedicated to irrigation only.

FIG. 1 Bart Bernacchi, Bernacchi Greenhouses, La Porte, Indiana, uses Q-Com computerized zone controls for managing greenhouse temperature, humidity, sunlight, plus potentially much more.

Computer controls are best used by the grower with from one to three zones. The medium-sized grower who has modest expansion plans, or the small grower who wants greater control options, will find computer zone controllers ideal.

Computerized environmental management. While these are the most expensive, computerized environmental management systems also offer the most opportunities and greatest flexibility. Coordinating all aspects of environmental control allows you to tie all your pieces of equipment together and offers an unlimited array of control options for each piece. Curtains can open based on light levels and temperature, while at the same time vents open or close, and fans switch on and off. Actions are based on information sent by zone sensors to the computer, which decides on changes to satisfy the parameters you provide it.

Each zone can have different requirements, all of which are engineered into the system from the start. Such a system achieves the greatest cost savings, for as the grower expands, each additional zone lowers the average cost per zone. A computer-managed system allows you to integrate weather information into the decision-making process, too.

The biggest disadvantage is that the initial system cost is very high, especially if you plan to use it in only one or two zones. Also, a computerized environmental management system is more complex than a simple computerized controller.

Larger growers with multiple zones, especially complex zones like the plug range or the propagation house, nearly always go this route. Growers who are planning to expand by adding growing zones buy these systems, as do specialized growers for whom precise environmental control is imperative.

According to Al Zylstra of Growth Zone Systems, when you computerize, you can expect to save in many areas:

Electricity	20% reduction
Fuel consumption	30% reduction
Water	30 to 75% reduction
Fertilizer	30 to 50% reduction
Pesticide	20 to 40% reduction
Labor productivity	15% gain

Let the computer take care of it all, so you can concentrate on managing the business!

PUTTING A SYSTEM TOGETHER

Often the grower who is short of capital when needing to build or expanding a greenhouse builds the basic structure first, then worries about environmental controls. While that may have worked just a few years ago, it is not a good approach to take today. Plan your control system while you are planning the structure. Putting an analog controller in a modern, state-of-the-art facility—with overhead curtains, roof vents, side vents, and pad-and-fan cooling, and 10 mini-boilers delivering hot water heat—would make zero sense. So would a decision to use a sophisticated computerized environmental management system to control the environment in Quonsets that you use only two months out of the year for bedding plant production.

Think about opportunity costs in considering environmental controls. How much time will you spend walking around the greenhouse making sure the heat is on or off and the vents are open or closed? What else should you be doing that will not get done because you're busy worrying about temperature control? Also, the more precisely you control your growing environment, the more money you save.

When you're adding up the costs and finalizing your plans, consider these costs in addition to the out-of-pocket expense of the control system. And don't forget your peace of mind! You will live better when you turn over busywork like temperature control to computers. You'll be able to sleep better on cold nights, too. A computer will simply ring you or your grower at home to tell you if the boiler goes out!

WHICH WAY FOR WHAT CONTROL FUNCTION?

Let's take a look at the many greenhouse environmental control functions and comment on different ways each might get done.

Controlling temperature. Analog controls can control greenhouse temperature and are widely used, but computers do it more accurately! They also provide better equipment integration. A computerized temperature control system gets readings from many sensors (versus just one for an analog system). Computers also anticipate heat load—at winter sunset, for example, they can start the boiler earlier and gradually feed heat into the houses. This means greater boiler efficiency and more even temperature in the houses. Research proves that a greenhouse averaging 1F (0.6C) degree above set point over a one-year period can add 5% to your fuel bill. Increase that to 3F (1.7C), and you have a 15% higher fuel bill.

Analogs turn heat on and off. The temperature rises, the heat goes off; the greenhouse cools, the heat goes on. Computers modulate temperatures, maintaining them very close to the set point. A computer will fire just the right number of mini-boilers and only as they are needed. The computer will even maintain different water temperatures in each mini-boiler, one reason most new modern ranges are hot water-heated today.

For a sample system comparison, say you want to raise the temperature 4F (2C) at sunset. With simple analog controls, the system will go on to target the temperature of 70F (21C). The boiler will be turned on full force as long as greenhouse controls call for more heat. With computerized controls, the rise is done in steps, taking maybe 30 minutes. The temperature may rise 1 or 2F every 15 minutes. The results: No sudden load on boilers, and no sudden temperature changes on crops. Instead, the computer ramped the temperature, reducing stress on plants and systems.

A computer will also convert from day to night temperature at the correct time each day of the year. Remember that sunset varies from 4 to 9 P.M. through the year in the North. Will you get around to resetting an analog every day?

The grower can save a lot of fuel dollars with a computer. When commercial computer control firms go to large office buildings, they guarantee 15 to 20% fuel savings, and they deliver!

Humidity. Maintaining a proper humidity level is a common problem. On a typical late-winter afternoon, the sun goes down, and greenhouse temperatures drop rapidly—falling quickly to the dew point. Moisture appears on leaves, which is perfect for disease spores.

Getting relative humidity (RH) right is critical. Advanced growers consider precise RH management as *the* way to control crop disease. Powdery mildew, black spot, and *Botrytis* are all humidity related. If RH is controlled well (keeping foliage dry) and environmental temperatures are unfriendly to disease problems, the need for fungicides is minimal, and losses from disease are reduced. All pathogens have rather narrow ranges of temperature and humidity through which they can and do spread. Avoid them, and you control disease.

While analog controls can be set to begin heating once relative humidity exceeds a set point, the computer can heat and ventilate simultaneously and with great precision. Importantly, the computer takes outside humidity and temperature, as well as greenhouse conditions, into consideration.

Many of the more sophisticated plug growers gauge vapor pressure deficit (VPD) in monitoring relative humidity levels. This is the most precise way to measure a greenhouse's relative humidity, but you cannot use VPD without an environmental computer.

Irrigation. More and more growers are relying on computer-controlled irrigation systems—ebb and flood (for benches and floors!), trough, watering booms, trickle-drip, and more. And as environmental concerns regarding runoff continue to mount, more and more growers are looking for ways to control irrigation water by recirculating it.

Enter the irrigation computers that can judge when crops need water, based on accumulated light levels and temperature and media moisture sensors. Once conditions meet set parameters, crops are watered—automatically!

Soil tensiometers are coming on, too. Some growers and researchers are working to develop soil tensiometer techniques for flower crops. The cut flower grower, especially, can use soil tensiometers to know exactly when to water. Here's how they work. Several tensiometers are set into the ground, with their sensors placed at varying depths. Once soil moisture levels decrease to preset amounts, irrigation is turned on. When the water reaches another preset limit—this time to a prescribed depth— the irrigation is turned off. It works, and it saves water and fertilizers.

Irrigation computers can also manage fertilizer levels in your irrigation water. For example, you dial in desired ratios, and the computer will take solution from various stock tanks to automatically adjust water to the right ppm N-P-K and proper pH. Growers recycling their water find this very helpful.

More and more aspects of irrigation will be integrated into the environmental control computer, rather than being controlled by a stand-alone computer.

Energy and shade curtains and black cloth. Energy and shade curtains and black cloth for (short days) can be controlled by analogs, set to open and close at prescribed times each day. This may be fine for black cloth control, but it is generally inadequate for shade, light, and temperature control.

Computers can also control energy and shade curtains and black cloth. A computer, however, will open and close curtains based on environmental conditions. For example, it is not unusual to be standing in a greenhouse on a spring or summer day and have the shade curtain adjust itself automatically four or five times in an hour, depending on the light levels that are coming to the crop and the interior temperature. Analogs simply cannot do this.

CO_2. While an analog control will trigger CO_2 injection during preset hours each day, a computer control will trigger CO_2 injection based on the changing concentration of CO_2 in the greenhouse, thus making sure crops get the right amount all the time.

Other controls. Supplemental lights, alarms, and many other greenhouse controls can all be controlled by either analog controls or computers. Again, the main difference is that the analog control will turn equipment on and off based simply on preset thresholds or time of day, while a computer will take information from sensors and continually adjust conditions to the ideal using all equipment and controls.

MORE ADVANCED COMPUTER APPLICATIONS

Many of the emerging growing techniques are best done using computerized environmental controls.

Cool day–warm night (DIF) height control. Michigan State University's technology for controlling crop height by temperature is best managed using an environmental computer. The technique, basically providing cooler day than night temperatures, slows down plant stretch.

Southern growers, especially, just don't have cool days when they need them. For example, much Easter lily growth occurs during March, when there are many warm days in the South. How can you have warmer nights than days when the day is 80F (27C)?

MSU has learned through research that a plant "senses" its day temperature mainly during the first two hours after sunrise. The same is true for poinsettias. If

---

QUESTIONS TO ASK *BEFORE* YOU BUY AN ENVIRONMENTAL CONTROL SYSTEM

- ◆ How long has the system been used in greenhouses? (Should be a minimum of 10 years. Also, request a list of production greenhouses as references and call them!)
- ◆ Will the system control any type of equipment, or only certain brands?
- ◆ Is special programming required?
- ◆ Do you need electrical or computer skills to troubleshoot the system?
- ◆ Can components be replaced by you or your staff?
- ◆ What is the company's reputation for customer support? Do they charge? How much? Is help available at odd hours?
- ◆ Is all or most of the company's business in horticulture?
- ◆ Is system configuration included in the price?
- ◆ How much will adding an additional zone cost? An additional control function?
- ◆ How much will upgrades cost? Are they required or optional?

Source: Al Zylstra, Growth Zone Systems, Mount Vernon, Washington.

you can open vents and turn on cooling pads and fans at sunrise, you will in effect be providing nearly an entire day's cool-day impact in just those two hours. While the grower may complain about getting up at 5 A.M. to do this, the computer will not!

Storm settings. When a storm hits without a computer controlling the greenhouse environment, you may be scrambling around to get vents closed on the windward side and to crack vents on the leeward side. The computer will do this automatically when it senses rain and wind. Once the storm passes, the computer will adjust equipment back to normal.

THE BOTTOM LINE

If you're a grower with an acre or more of greenhouse production, you should be using an environmental control computer!

Editor's note: Many thanks to Randy Whitesides, formerly with Q-Com, Irvine, California, and Al Zylstra, Growth Zone Systems, Mount Vernon, Washington, for their input and review of this chapter. Special thanks go to George Dean, Wadsworth Control Systems Inc., Arvada, Colorado, for providing much of the background and who pioneered environmental control for U.S. flower growers.

CHAPTER 3

THE PHYSICAL PLANT: ENERGY AND SPACE

by Vic Ball

WHICH FUEL TO BURN?

Annual fuel costs for 60F (16C) nights for the northern states year-round grower, as of the early 1990s, were around $1 to $1.15 per square foot ($10.76 to $12.37/m²) of ground covered. That's 10 to 15% of total production costs. (Figure $.60 to $.70 per square foot, or $6.46 to $7.53/m², for the grower who carries some houses much cooler in January and February.)

All too often, there are neglected ways the grower could reduce this cost. The big three fuel cost cuts are: a fuel-efficient roof, a high-efficiency boiler with well-insulated mains, and fuel selected with the lowest cost per 100,000 BTUs for one's situation.

HOW TO SELECT THE LOWEST COST FUEL

To appraise actual BTU cost for various fuels, you must first reduce the available alternatives to a common language. You must compare apples with apples! The most practical language: appraise each fuel according to cost per British thermal unit (BTU), a basic measure of heat energy. Don't worry about the scientific details. Just know, for example, that if your grade of oil at your cost delivers 100,000 BTUs for $.70, and gas per 100,000 BTUs is $.40—well, switch to gas! This comparison is so easy to do; the accompanying chart (fig. 1) is self-explanatory. Remember, you're not buying coal or oil or gas—you're buying heat, so buy it as cheaply as you can.

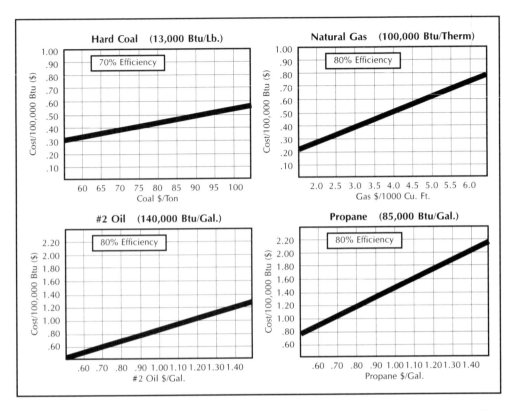

FIG. 1 Fuel cost comparisons. These handy graphs are an easy way to convert the cost of a fuel to BTUs. For example, if coal is selling for $70/ton, 100,000 BTUs costs about $0.42.

Fuel cost comparisons. A given fuel's price can vary from supplier to supplier and from season to season, making it more or less economical than other fuels per 100,000 BTUs of energy produced. Assume you can purchase hard coal (fig. 1, top left) at $70 per ton. From the price range at graph bottom, follow the $70 vertical to intercept the bold diagonal. From this point go straight left to the BTU cost range. Coal at your price would produce 100,000 BTUs for only about $.42. To receive a comparable value from natural gas, you'd need to be able to purchase it at about $3.20 per 1,000 cubic feet (28 m³).

Natural gas. By far the most used fuel for greenhouse production in the United States (and most of Europe) is gas. Why? It is clean, convenient, and is generally competitive cost-wise. Equipment for burning gas is far less expensive than for coal and even less than for oil. There are no storage tanks, as with oil, and no coal or ash handling.

Gas is used where overhead unit heaters are installed in greenhouses. Other fuels are difficult to handle in these small burners (except propane, which is expensive).

Natural gas runs about $.20 to $.40 per 100 cubic feet ($.07 to $.14/m³), plus about $.06 to $.07 delivery. Total: $.26 per 100 cubic feet (this 1996–97 rate applies to greenhouse use). Before committing to gas, though, just be sure you couldn't heat your range for 25% less if you burned oil or coal, even allowing for the added cost of burners and handling the fuel.

Fuel oil. It's more costly than gas, but some growers do use No. 2 light oil, especially for smaller ranges. A few large growers, with 3 to 5 acres (1.2 to 2 ha) or more, burn No. 6 heavy oil. This must be preheated before burning, but the BTU cost is lower than for light oil. Again, it's a question of cost per 100,000 BTUs versus gas or coal. As of 1996, No. 2 oil for greenhouse heating cost $.55 to $.62 per gallon ($.14 to $.16/l) (these are greenhouse quotes for truckload lots).

There are practical oil-burning overhead unit heaters available. Oil burners are costly, however, and require maintenance. Oil storage tanks are also expensive, and today's government agencies are tough on possible leaching of oil into the groundwater, especially from underground storage tanks. There have been costly incidents where growers had to pay big bucks to have leaking underground fuel tanks removed and the dirt surrounding them hauled away. There is also the problem of possible wide price fluctuations, especially if a sustained severe winter creates oil shortages. Some large growers invest in large aboveground storage tanks and fill them in the summer—thus hedging oil prices going into winter.

Coal. There was a strong trend toward coal, especially on larger ranges, in the early 1980s. It was triggered, of course, by skyrocketing oil prices. Whenever the country experiences a 30 to 40% increase in cost of fuel oil and propane, it sends growers, larger ones especially, back to shopping for coal and stokers. Many large growers maintain the flexibility of using both coal and oil to fire their boilers. Large northern growers who choose coal often install highly automated coal-handling and -burning equipment. With modern handling and burning, labor and handling costs become minimal.

There's another problem with coal: environmental contamination. More each year, government agencies are tightening down on the amount of sulfur dioxide that can be emitted into the atmosphere by coal boilers. Of course, hard coal is less of a problem than soft coal, but it doesn't get off free. Regulations and their interpretation seem to vary even from city to city. If you can make peace with the EPA and other agencies on a reasonable basis, this barrier is out of the way. Often it's a politicking job.

Yet, on a BTU basis in some cases, coal can save big bucks. Hard coal is about $50 per ton ($55/1,000 kg). Soft coal from Kentucky costs about $35 per ton ($38.58/1,000 kg). Illinois soft coal is about $27 to $30 per ton ($29.76 to $33.07/1,000 kg). Add delivery charges and tax. (This information was provided by

Lafayette Coal Co. of Burr Ridge, Illinois, for 1996–97.) When doing your calculations on fuel cost, include depreciation of the stoker, as well as labor and facilities for unloading, storing, and moving coal. For the record, also, coal deteriorates in heat value over long-term storage.

Propane. It's clean and convenient, and propane can be used by growers in a rural area where gas lines are not available and where other fuels are hard to bring in. Another advantage: It can be used with overhead unit heaters. Many small, rural bedding plant growers do just that, heating their houses only in late February and March.

The big problem, of course, is that propane is expensive. Prices vary widely. 1996 quotes average from $.60 to $.70 per gallon ($.16 to $.18/l) and as low as $.35 ($.09/l) in the summer and $.50 ($.13/l) in the winter. One case quoted $.44 a gallon ($.12/l) of propane for a 100,000-square-foot (9,300-m²) range. The higher rates are for smaller growers.

For bedding plant crops, especially in southern areas where there is only a short-term fuel need and no severe cold, propane is rather widely used in spite of its cost. Incidentally, there does seem to be a lot of price dealing in propane, especially for larger users. The cost for storage tanks, piping, and burners is modest. Be sure to shop for it. There are parts of the country and competitive times where propane can beat other fuels.

Sawdust and logs. Occasionally, I come across growers who heat large ranges with sawdust, wood chips, or rough logs. The Houweling Brothers in Vancouver, British Columbia, located in the midst of major timber cutting, still get sawdust practically free. They've been heating 80 acres (33 ha) of greenhouse bedding plants, pot plants, and vegetables for some years with sawdust—with no problems. Their equipment comes from Kara Engineering Co., Almelo, Holland, fax 011-31-546 870525. They comment, "Sawdust is practical for ranges of 3 acres (1.2 ha) or more, not for the small grower."

There are a few growers in the Southeast who heat ranges with low-cost rough logs. Again, the fuel cost is very low. There is some labor in handling, but on balance it is a major economy for these growers.

A GOOD EXAMPLE OF PICKING THE BEST FUEL

Hans Pein, an Urbana, Illinois, bedding plant grower, operates 100,000 square feet (9,300 m²). As of 1996, he is using propane, which he buys for a surprising $.44 per gallon ($.12/l)! Look at the cost: now about $.60 per 100,000 BTUs. Oil prices fluctuate widely, often going up in summer.

No. 2 oil, on the other hand, costs Hans about $.55 to $.65 per 100,000 BTUs. Practical Hans realizes that oil is less efficient than propane. The burner cost for oil is higher, and you must include the cost of maintaining the oil burner. Hans uses a new, superefficient Modine unit heater, which actually burns the propane at about 83% efficiency.

Gas would cost about $.38 per 100,000 BTUs, but the gas line is 6 miles (10 km) away, and the gas company declined to pipe the gas in for free. The cost would be about $40,000.

Coal, the other competitor, is $50 per ton ($55/1,000 kg) for hard coal—that comes out to about $.30 per 100,000 BTUs. Coal has a whole array of problems, however: costly stokers, cost to handle the coal and ash, and environmental contamination. This last factor comes with strict rules imposed by various government agencies, especially rules on the amount of sulfur dioxide in soft coal. Hans tried to make peace with the government but finally gave up on cheap coal.

Table 1 shows the fuel options available to Hans, circa 1996. Taking all the factors together, Hans believes he is doing best with $.44-per-gallon propane.

TABLE 1   FUEL OPTIONS AVAILABLE TO HANS PEIN

Fuel	Cost[a]	Cost per 100,000 BTUs[a]	Added costs
No. 2 oil	.55–.65/gal	.60	Burner cost and maintenance, price fluctuations
Propane	.60–.70/gal	.60	
Natural gas	.20–.40/100 ft³	.40	$40,000 to pipe the gas in from 6 miles away
Hard coal	$50/ton	.30	Stoker, coal, and ash handling; compliance environmental regulations

[a] Dollars in 1996–97.

ROLLING BENCHES—BIG SPACE SAVERS

Rolling benches may be rolled several feet sideways either right or left. A typical house has four benches and normally four walks, but with rolling benches it will now have only one walk (plus a gutter walk). That one walk can be positioned anywhere the grower wishes, simply by rolling the benches sideways. Normally, the rolling is accomplished by turning a crank that's at the head of the bench or simply pushing the bench to the side. It's not hard to do. As of the early 1990s, hundreds of greenhouses have converted to rolling benches across the United States, Canada, and Europe. They are standard equipment on modern ranges. If you're converting to rolling benches, realize that if purlin posts are installed in the greenhouse, it complicates the use of rolling benches.

The obvious advantage is that with rolling benches you now use perhaps 86 to 88% of the available space for crops, versus 62 to 66% with the old-style fixed walks. Stated a different way, that means perhaps 35% lower fuel costs and 35% less of all overhead costs (e.g., depreciation, salary, taxes, insurance) per square foot of growing area per year. That is a major cost reduction, and the grower still has access to the crop. You produce one-third more crop per year on your range!

Inevitably, we find ourselves comparing rolling benches with the tray mechanization concept. With tray mechanization, the benches (or trays) can be moved back and forth to the headhouse. Both plans offer much higher percentages of space utilization than fixed benches. In fact, the trays use more greenhouse space than rolling benches. Both provide access to the crop—trays can be spread a foot (30 cm) or so apart anywhere for this purpose. Of course, the major feature of the tray approach is fast, low-cost movement of the crop from headhouse to bench and back to headhouse. Tray conversions are more costly, though. It's an entirely different way of managing a greenhouse!

Another interesting point: a few cut flower growers are growing crops on rolling raised benches. This provides the same one-third increase in production per acre that the pot plant and bedding plant grower gets.

One American manufacturer of rolling benches, Rough Brothers, Inc., phone 513/242-0310, fax 513/242-0816, offers a commercial version of the rolling bench. Figure about $2 per square foot ($21.50/m²) of bench cost (just for the bench).

CURTAINS—NOT JUST FOR SAVING ENERGY AND SHORT DAYS ANYMORE!

Here are more new curtain technologies and important products: UV-stable energy curtains, curtains with support wires (for the greenhouse roof!), and fire retardant curtains. Many of today's greenhouse curtains serve both as thermal sheets, reducing fuel loss on winter nights, and as shades, reducing summer sun and heat. One product is effective for frost protection for outdoor areas. Insect screening is becoming commonplace.

These various curtains are widely used by bedding plant, pot plant, and foliage growers all over the world (fig. 2). The trend for such sheets is up sharply because they are now standard in modern construction.

ENERGY AND SHADE CURTAINS

Installed gutter to gutter or truss to truss, perhaps 8 to 12 feet (2.5 to 3.6 m) above the ground, curtains serve two basic functions, and both are important. First of all, they reduce excessive sunlight. Many crops just can't tolerate full summer sun. The capability of restricting sunlight to a set level (often done with computers) is a major assist to the grower striving for quality crops. Also, reducing summer sun, especially in southern areas, substantially reduces water requirements. Temperatures under shade curtains can be 8 or 10F (5 or 6C) cooler than in full sun. Also, because plant demand for good water is down sharply, that means a lot less fertilizer use.

FIG. 2   Energy curtains are an integral part of nearly every modern greenhouse. Growers in the North rely on them to insulate houses at night during the winter months, while growers in other regions use them for shading crops. They're opened and closed by computer control based on predetermined temperature and/or light level set points.

Equally important, by extending curtains daily sunset to sunrise in winter, major fuel economies can be realized—typically 25 to 35%. Now, instead of keeping summer's light radiation from entering, you're keeping winter's energy from reradiating and conducting out.

Much higher gutters—often 10 to 12 feet (3 to 3.5 m) high or more—are being built in the early 1990s. The reasons: better air circulation and more even temperatures. Also important: Higher gutters mean cooler summer temperatures at the bench level. Furthermore, with high gutters, fan-jet tubes can be placed below energy curtains.

A completely automated system, including all labor and material, of such a new, improved energy sheet will cost $.80 to $1.50 per square foot (about $8.60 to $16.10/m²) of ground covered. The grower can compute the number of years needed to pay back the investment in such equipment. Example: Given a $1 per square foot fuel cost and 35% fuel savings, that's $.35 per square foot per year saved. Divide the $1.50 cost for the thermal sheet by $.35, and in 4.3 years you'll recover your investment in fuel savings alone.

Products. An important supplier of such curtains is LS Americas (Ludvig Svensson, Inc., a Swedish manufacturer with U.S. offices in Charlotte, North Carolina, phone 704/357-0457, fax 704/357-0460). Here are some of the Svensson energy curtains, all of which are combinations of aluminum and various plastic materials.

ULS15 is the primary cloth used by North American growers today. It provides 55% energy savings over the course of the year and 55% shade. By pulling the energy curtain closed at sunset and opening it at sunrise in the winter, there is a significant reduction in heat loss. This curtain has strips of aluminum sewn into it, which reflect excess sunlight away, keeping the greenhouse cooler as well as shading it in the summer. The ULS cloth is available in shade values ranging from 20 to 85%. This material is UV-stable, coming with a five-year guarantee against UV degradation; life expectancy is eight to 12 years. Cost for the sheet alone is about $.45 per square foot ($4.84/m²). Ludvig Svensson reports typical costs for mechanized installation in a greenhouse as about $.85 to $1.50 per square foot ($9.15 to $16.14/m²), depending on the size of the installation, including all labor and materials.

The ULS-F series is a heat reduction curtain that provides shade (available from 45 to 85% shade) but no heat retention. These curtains have a more open weave, allowing air to flow through them. They are used in houses with roof vents. All these curtains go up in price as the percentage of shade increases—you're paying for more aluminum.

All the above sheets are lightweight enough to be adaptable to flat surface mechanical applications, typically gutter to gutter. Mechanical installations of such sheets are available from Van Wingerden, Inc., Fletcher, North Carolina, phone 704/891-4116, fax 704/891-8581; Van Rijn, Toronto, phone 905/945-8863, fax 905/945-9294; Wadsworth Controls, Arvada, Colorado, phone 303/424-4461, fax 303/424-6012; Conley's Greenhouse Manufacturing, Pomona, California, phone 909/627-0981, fax 909/628-3774; and many other firms. Richard Vollebregt at Cravo Equipment, Ltd., Brantford, Ontario, phone 519/759-0082, fax 519/752-0082, is a specialist in various shade and thermal sheet mechanical installations. The company's experience has positioned it to be a leader in the development of retractable-roof greenhouses.

SHORT-DAY CURTAINS

The purpose for the short-day curtain is to reduce day length. It's the same as the black cloth idea—giving near-zero light for at least 12½ to 13 hours per night to trigger flowering in crops such as chrysanthemum, kalanchoe, and poinsettia. (Actually, it's the long nights that do it.)

Among materials used today, black polypropylene can be prepared with tape and grommets for installation. It's low cost, about $.11 to $.28 per square foot ($1.18 to $3.01/m²). Polypropylene is short-lived, however, especially if abused.

ULS Obscura A+B is a two-piece blackout curtain. It's top layer is 100% aluminum, and the bottom layer of black polyethylene strips blocks any light that is not reflected away by the top layer. You can get high energy savings (75%) from this curtain because of all the aluminum. It can be used with automated installations gutter to gutter. ULS Obscura A+B reflects more sun than LS Americas' older curtains, like LS100, so there will be less heat buildup. Sateen tends to absorb more heat.

SHADE CLOTH

A different application for light reduction, shade cloth is a plastic mesh of normally about 15 to 43 threads per square inch (2.3 to 6.7/cm²). It is available in several different shade factors, ranging from 20 to 95%. Shade cloth is used mainly in southern and western mild climates; it is widely used in Florida by foliage growers, chrysanthemum cutting producers, and outdoor pompon growers. A cloth house is normally semipermanently installed overhead to facilitate easy dismantling in case of a tropical storm or hurricane. Depending on the quality of construction and installation, such materials will last several years. These are too heavy to be used in greenhouse light reduction or energy curtain applications, however.

Shade cloth comes in polypropylene as a woven plastic. Among brand names are Lumite, Pak, and V.J. Weathamaster. Knitted mesh is polyethylene plastic. It's more tear-resistant than polypropylene. One brand is V.J. Premium Knit, from V.J. Growers Supply, Apopka, Florida, phone 407/886-5555, fax 407/880-2791.

AUTOMATED OUTDOOR CURTAINS

A new Ludvig Svensson product (PLS series) can be applied overhead to large areas and be mechanically extended or retracted. Growers use it for frost protection as well as shade and protection against rain, hail, etc. It will also provide from 16 to 80% (even blackout) light reduction. It's 90% nonpermeable, keeping out most of a heavy rain.

Retractable-roof greenhouses are the latest application of these curtains. Billy Powell of Troup, Texas, has 6 acres (2.4 ha) of this material (slightly modified) for pansy plants outdoors in March, which would otherwise be subject to late freezes. Billy installed polyethylene sides and maintains 50F (10C) with some portable heat. Tom Van Wingerden, Charlotte, North Carolina, has 8 acres (3.3 ha) of Cravo retractable-roof greenhouse structures, which replaced 4 acres (1.6 ha) of roll-out. Why? "Going in and out [with the roll-out benches] was too much work." Tom cites the advantages of the new house: it has a hip roof, which handles water runoff better. Plus, it gives better protection from wind, frost, and freezes than roll-out. He also expects to get 10% more production from the same area because they can manage the crop better, and suffer less damage from the elements. It's all at a higher comfort level for the growers and Tom.

PLS-ABRI is an energy curtain with its own support wires! LS offers its PLS curtains and OLS curtains (more open outdoor curtains) with a 2½-mm (one-tenth inch) polyester wire every 16 inches (41 cm) for the full length of the sheet. The frame is suspended on stainless steel wires with hooks and optional rollers, upon which the curtain can slide. Interest in these wired sheets was on the increase in the mid-1990s.

With the QLS series, from LS for flat roof greenhouses, the QLS sheet is the roof itself! It's made of a PLS-10-ABRI sheet laminated together with a sheet of poly-ethylene. The manufacturer says it can be extended 100 feet (30 m) in one direction at 3% pitch and still have adequate water runoff. It has applications as a retractable roof. (See *GrowerTalks*, August 1994.)

For more on all automatic curtains see *GrowerTalks*, June 1995.

OTHER CURTAINS

Side curtains (ILS) can help retain heat and keep light out. This is essential for the poinsettia grower in sight of highway traffic. Already big in Europe, they can also be used as dividers between temperature zones in a greenhouse. They can be rolled up and down and range from 30% shade to total blackout.

Fire retardant sheets (ULS-Revolux) are the same as ULS series in function and look but are made of fire-retardant materials. Their cost is double that of ULS sheets. Use them around heaters or electrical outlets—places fires can start. You can use one for every third curtain as a firebreak. There is a lot of interest in them by retail green-houses and home centers and for making commercial growers fire safe.

INSECT SCREENS

Screening is no longer something just on your home to keep houseflies out. Now tiny thrips are the target. There is an increasing number of installations of insect screens. Interest in them is tremendous because of fewer pesticides being available and insect resistance. The idea: Don't let the insects in to begin with! NGMA recently put together some guidelines on these screens. To request a copy, contact NGMA, 7800 S. Elati, #113, Littleton, CO 80120, phone 303/798-1338, fax 303/798-1332.

Why are insect screens slow to be adopted into practice? Anything placed over the vents reduces airflow—the fans can't work right; the house gets hot. Screens have to be put in with eyes open; size the openings and fans right! If insect screens are added after fans have been installed, and they were not considered in fan calculations, you will increase greenhouse temperatures, use more electricity, and reduce fan motor life. If you're going to install screening, revisit your fan and pad equipment at the same time.

Sources for insect screens include LS Americas, phone 704/357-0457, fax 704/357-0460, and Green-Tek, Edgerton, Wisconsin, phone 608/884-9454, fax 608/884-9459. See *GrowerTalks*, April 1995.

CHAPTER 4

MORE EFFICIENT GREENHOUSE HEATING

HOT WATER, HOT AIR HEATING
by Vic Ball

INFRARED HEAT
by Al Zylstra
Growth Zone Systems
Mount Vernon, Washington

Since the 15th *Ball RedBook*, we have seen a trend in greenhouses toward root zone heating. The "mini" boilers continue to replace steam, and they're available in units up to 2 million BTUs. Now, with the introduction of pressurized combustion, we see greater efficiency—85% in the mid-1990s, and up to 94% efficiency! These boilers can even be in the greenhouse environment, sitting inside the range.

Greenhouse heating for the late 1990s is mainly hot water systems with root zone heating. There is also a good amount of overhead unit heaters and even infrared heating systems. We are seeing almost no new steam systems. There are also many fewer overhead heating pipes, which cast shadows. Some pipes are on walls and ends, but more heat each year is with hot water, with pipes below the bench or in a porous concrete or gravel floor. Heat the roots and plants, not the air above the plants!

LET'S TALK HOT WATER

The huge, old 20- to 30-foot (6- to 9-m) steam generators are being replaced by very efficient small hot water boilers. They do the job well, and with only 2 by 3 feet or

35

6 by 8 feet (61 by 91 or 183 by 243 cm) of floor space. It's the end of the boiler shed—and of big brick chimneys (how will seed salespeople find their customers?!).

Hot water's advantages over steam

Why is hot water taking over from steam? Steam traps are an endless maintenance problem (I still think of hammering sticky traps on a zero-degree midnight with George J. Ball).

Hot water systems deliver heat more evenly, with less fluctuation. Modern hot water systems gradually modulate water temperature up and down to control greenhouse temperature. If you turn steam into a steam pipe, on the other hand, it roars down and heats one end first. Most of all, there are only two steam pipe temperatures—hot and cold.

Steam does pasteurize benches of soil, so it's especially important for cut flower crops. Chemicals do most soil sterilizing now, but with restrictions on methyl bromide, steam is making a comeback. See the *GrowerTalks* June 1995 issue for more on methyl bromide substitutes. Most pot and bedding plants are grown in commercial mixes that growers usually don't pasteurize anyway, so this is not a great plus for steam.

How greenhouse temperatures are maintained

The simplest hot water system is based on a boiler, generally gas-fired, that's set to maintain a given water temperature. This boiler is allowed to fire only when there is a call for heat in the greenhouse zone(s). The temperature is maintained within 2 to 4F (1 to 2C) of the set point throughout each zone, depending on the firing system. Zone temperature is monitored by either a single thermostat per zone or an inexpensive multiple zone electronic controller.

With the more advanced generation of temperature controls, the temperature of the water pumped through the crops is controlled by mixing more or less cooler water (coming back from the greenhouse pipes) with hot water from the boiler. As the crop needs more heat, for example, the pipes get a warmer water mix, and pipe surfaces warm. All this is constantly adjusted, typically by solid-state computerized controls. Properly controlled, this system delivers even temperatures—only a maximum 1 to 2F difference end to end, with minimum fluctuation. Systems like this with mixing valves are typically used only for large setups—an acre (0.4 ha) or more—and are generally required on any steel or cast iron boiler system operating low-temperature (140F, 60C) hot water heating. Another application would be a combination system, with both high and low temperatures.

TYPES OF BOILER-FIRING SYSTEMS

On-off boiler. When a call for heat is made, an on-off boiler will fire at the full input given on the rating plate. This is the least expensive firing system, but it has a tendency to over- and undershoot the set point. Less efficient operation is the result if it cycles too rapidly. It's best used only on single- or dual-zone systems (a zone is a greenhouse area kept at the same temperature—perhaps 5,000 square feet, 465 m², or maybe an acre, 0.4 ha, or more).

Two-stage system. With the ability to fire at either 50 or 100% of the input on the rating plate, a two-stage system is less expensive than full modulation and provides more flexibility in matching the actual heat load. It's best used on any system up to three zones. Four-stage firing with today's highly efficient pressurized-combustion boilers make full modulation unnecessary. There is no drop in efficiency when cycling a pressurized system.

Full modulation. A boiler with full modulation has the ability to fire down to 20% (10% at an input over 1.7 million BTUs) of rating plate input. The boiler water temperature is maintained within 2F (1C) of the set point continuously, and very little cycling of the burner occurs. There is generally a slight loss in efficiency that is more than offset by no cycling losses.

BOILER TYPES

There's a whole new family of efficient hot water boilers that need perhaps only an eighth or even a tenth the floor space of old steam boilers (and with no coal bin or stoker). In these examples small and large, the areas are very rough approximations.

A 45,000-BTU, copper-tube hot water boiler with 84% efficiency will heat roughly 500 square feet (46 m²) of northern greenhouse. Cost, including piping to the bench and controls, is about $1,900. It uses a gas burner. Floor space requirement is less than 2 by 2 feet (61 by 61 cm).

A 270,000-BTU hot water boiler will heat roughly 3,000 square feet (279 m²) of northern greenhouse and cost about $4,200. Again, there's a very small floor space requirement, about 2 by 3 feet (61 by 91 cm).

A 1 million-BTU boiler occupies only 2 by 3 feet (61 by 91 cm) of floor space and can even be installed outdoors! Fuel efficiency: 88%; cost: about $16,000. Such a boiler will heat roughly 10,000 to 14,000 square feet (929 to 1,301 m²) of northern greenhouse, and, of course, the grower can use multiple units. It's a good example of a high-efficiency, modern hot water boiler. This particular boiler is made by Hamilton; another option from Hamilton is a sealed-combustion model at 85% efficiency for $11,000.

For boilers up to 4 million BTUs of capacity (up to 1 acre, 0.4 ha, of northern greenhouse), there are many options. Copper tube: Raypak, Teledyne, and Hamilton. Steel tube: Kewanee, Cleaver Brooks, Johnston, and York Shipley. Cast iron: Weil Mclain and Burnham.

The Dutch also build very advanced, large hot water boilers for greenhouses. Virtually all their new ranges today are hot water heated.

Boiler shock is a thing of the past. Copper tube boilers, which are becoming the standard, are not subject to the boiler shock of cold water hitting a hot boiler. Because copper is a better conductor, such boilers can start from cold without the problems of steam and cast iron systems.

SEVERAL MINIS, OR ONE LARGE BOILER?

The grower seeking 1 million BTUs from hot water boilers has several options, ranging from installing a single, large boiler capable of delivering all that heat, to setting up 10 separate mini boilers each with a capacity of 100,000 BTUs. What are the relative merits of the various configurations?

Setup. Merely purchasing the boilers is only the first expense with your new system. A single, 1 million-BTU boiler may be an impractical choice without a backup, so let's figure on two boilers, each at 500,000 BTUs, for about $11,000. In contrast, 10 minis, each at 100,000 BTUs, will cost roughly $18,000 to $19,000. Either way, also consider one pressurized-combustion boiler at $11,000.

It costs more to install 10 boilers. Contractors have to deliver electricity, water, gas, and computer-controlled wiring to 10 boilers in 10 greenhouses—quite a lot of work. On the other hand, with one or two large, central boilers, you do now have to pipe water out through 10 greenhouses. By the way, in terms of total floor space taken up, there is typically not a major difference between setting up one large or 10 mini boilers.

Operation. A big boiler is not necessarily more efficient than a small one. Using one big boiler for all your water-heating needs, large and small, may well mean it will be operating at only 10 to 20% of capacity some of the time, which is not efficient. The main case for the 10 small boilers is that, especially with computer control, as the load increases, more boilers will fire up automatically. Each mini will operate only at full capacity, however, and cycling on and off every few minutes is not particularly efficient.

Large modulating burners, however, can go down to as little as 25% of capacity with fair efficiency. Another good solution might be the new pressurized-combustion boiler. With four-stage firing, it does not lose efficiency.

Boilers can fail, of course, so a single, large boiler will need some kind of backup. It's almost surely best to split the load between at least two boilers at the central location, each of which can deliver hot water to all the peripheral greenhouses. If the mini serving a single greenhouse fails, you're in trouble. With a central system with two boilers, you have a backup.

A reminder on maintenance: With 10 boilers you could have up to 10 times the amount of breakdown that you'd have with one boiler. This is not to mention 10 times the routine maintenance.

A boiler strategy for the start-up grower. For the grower starting a new, small operation, where capital is always limited, here's an interesting plan. In the beginning, for each greenhouse install a unit heater or two (or as many as you need). For the time being you'll have to accept their relatively low fuel efficiency and inability to deliver heat below the crop. After two or three years, though, you may be ready to install a gas-fired hot water boiler, gaining significant fuel efficiency. Also, you can still rely on the old unit heaters for backup.

HOW IS THE WARM WATER DISTRIBUTED?

Under bench pipes. For larger areas, with acres of greenhouse crops, the most common system is aluminum, copper, or steel pipes three-quarter to 2 inches (2 to 5 cm) in diameter installed below the benches. The soil and crops are heated; the air above the crop and in the greenhouse "attic" gets less. Pipes are also installed along end and side walls of the greenhouse to prevent cold spots and often under the gutters to melt snow.

Some of these pipes are fitted with fins to deliver more intensive heat in a given area. One foot (30 cm) of 1-inch (2.5 cm) bare pipe delivers 90 BTUs per hour. A foot of fin-type pipe, with its expanded surface area, delivers 893 BTUs per hour.

Small tubes on the bench. Plastic tubes as small as one-eighth inch (3 mm) in diameter can be laid over the entire surface of a bench. With soil or flats over them, hot water is piped through, maintaining the desired soil temperature. The small tubes are ideal for a bench used for seed germination or rooting cuttings, which might be 5 by 20 feet (1.5 by 6 m). The lines run from end to end and 2 inches (5 cm) apart.

Some systems have the same layout but use five-sixteenth inch (8-mm) EPDM tubes. A synthetic rubber material that is much longer lived than PVC, EPDM withstands ultraviolet radiation exposure better than polybutylene. The same principles are also used to heat large areas, even acres of benches. For convenience, though, the tubes are laid directly under the benches.

Heat tubes on the ground. Here's a different application. The EPDM tubes are spread out on the ground, again about 2 inches (5 cm) apart from end to end of the floor or growing bay. It's low-cost, with no bench cost, and again the roots and plant canopy are kept warm, the air above cool.

Buried heat tubes.
Other growers bury smaller PVC or three-eighth inch (1 cm) EPDM pipes in a porous concrete floor 2 to 6 inches (5 to 15 cm) below the surface, with pipes spaced about 4 to 6 inches (10 to 15 cm) apart (fig. 1). Large areas—acres of pot and bedding crops, even plugs—are grown on these self-draining, heated cement floors. The results: warm soil and plants, cooler air above.

A lower-cost option is to bury the heat pipe in pea gravel instead of concrete, as is being done by some bedding

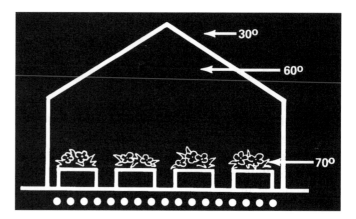

FIG. 1   Another way to save money on heating costs: Install heat *below* the plants. The heat pipes in this illustration are buried in porous concrete. Temperatures at plant level will be warm, while the attic temperature will be cooler. In this example, the attic temperature is 30 degrees warmer than the outside air temperature of 30 degrees. It is this differential between the inside and outside temperatures that most affects how much heat you need. The same house with overhead unit heaters may have 75 to 80F attic temperatures, a 45 to 50 degree difference between the outside temperature. If you can cut the differential between the inside and outside temperatures, you will save money on your fuel bill. Putting the heat at plant level helps.

plant growers. I understand, though, that burying small pipes in pea gravel is very inefficient in terms of heat getting to the crop. It's much better to put the one-fourth inch (6-mm) tubes on top of the pea gravel.

ROOT ZONE HEATING

Many researchers and growers are finding strong advantages to root zone heating, in terms of improved plant performance and substantial savings on fuel and other expenses. While there is no consensus on the value of root zone heating to growing on, it is widely accepted as the best heating method for propagating.

Plant response. Root zone heating generally keeps roots at the ideal temperature for your purposes. The aboveground plant parts are kept more or less warm to the canopy, so you don't have to be as concerned about the overhead air temperature. Bill Swanekamp of Kube-Pak Corp., Allentown, New Jersey, grows about 605,000 square feet (56,205 m²) of plugs, bedding, and potted poinsettias. He reports, "All our plugs are grown very successfully on porous concrete with bottom heat. It produces excellent uniformity of heating at the root zone, while providing cooler air temperatures to inhibit seedling stretch."

Early growth processes may need warm soil temperature, but Bill notes that bedding flats trap the heat, overheating the roots. Especially with cold weather, warm-rooted plants have a tendency to overgrow. This is wonderful if you are trying to push a plant, but it makes control difficult if you are trying to hold the plant back, Bill warns.

Savings. By heating just the soil and plants, you save energy simply because you're not heating the overhead air. Don Fleischman of Vary Industries in Grimsby, Ontario, has made a striking heat-savings comparison between root zone heating and overhead unit heaters. With root zone heating maintaining 70F (21C) at the plant canopy level and the outside air at 0F (–18C), the nighttime temperature just under the energy curtain is 40F (4C). Just above the curtain but below the glass, the temperature is 25F (–4C). On a similar cold night but with overhead unit heaters, the overhead air temperature could quite likely be 70F, as would be the difference between inside and outside. With the root zone heating and an energy curtain, the difference is only 25F, so the staggering result is a mere third as much heat loss in the root zone house.

Back in the early 1980s, Kube-Pak grew one 60,000-foot (5,574-m^2) block of poinsettias—about 35,000 6-inch (15-cm) pots—on unheated floors and without energy curtains. This crop required 28,000 gallons (105,908 l) of oil. The next year, in the same area with floor heating and thermal sheets overhead, the same size and type of poinsettias crop was grown with only about 5,000 gallons (18,925 l) of oil— a savings of more than 80%!

Jim Dickerson is a grower in Gobles, Michigan, northwest of Kalamazoo. His root zone–heated pot plant range has a one-third lower fuel cost, measured in dollars per square foot per year, than his overhead unit–heated range.

The purchase price of the root zone tube system (including the boiler) is well above that of unit heaters, $1.50 per square foot versus $.70 ($16.15 versus $7.53 per m^2). The difference is soon made up, though, by root zone's greater fuel efficiency. In addition, root zone heating wins on both counts when compared to a hot water system with pipes, which costs $2 per square foot ($21.53 per m^2).

Putting the heat tubes on the ground also eliminates bench cost, which can run $1.50 to $2 a square foot.

Tube layout. Root zone heating tubes can be placed almost anywhere close under the plant roots. Whether directly in the soil, on the bench, immediately under the bench, on the floor, or even buried in the concrete floor, heating tubes deliver heat to the root zone efficiently. Jim Dickerson's recent ground installation covers 2 acres (0.8 ha). A hard plastic mesh was spread out first, and the tubes were laid on that.

Kube-Pak installed its first root zone heating system in 1969. Now all of the range is equipped with root zone tubes buried in porous concrete. Planning on saving water as well as energy, Dickerson says, "With increasing awareness of groundwater pollution, the next time we install root zone heating, we will consider putting the heat tubes in concrete to utilize an ebb-and-flow system."

For flexibility and because of its bedding growth requirements, Kube-Pak also has overhead heating systems. In fact, all 605,000 square feet are equipped with independent bottom heating in the concrete *and* overhead heating.

Crop applications. Root zone heating is a sure winner for germinating seed, growing plugs, and rooting cuttings. The great majority of propagating is done on some form of root zone heat. In fact, most root zone installations so far have been relatively small areas to support propagation. Plants at the earliest stages of development are the most sensitive to soil temperature, so more and more the grower requires the system to maintain different set temperatures for different species of plants.

The benefits to bedding plants are not as clear. Again, Kube-Pak has experienced problems with the heat being trapped when the flats are set hard up against each other on the concrete. The plants are difficult to control, tending to become overgrown.

It is fair to say, however, that most growers find heating tubes produce high-quality pot plants. A substantial number of innovative growers even grow acres of pot plants with root zone heating, impressed with both pot quality and the large savings. Jim Dickerson has been growing poinsettias, pot mums, and other pot plants for about two years on his recently installed 2 acres (0.8 ha) of tubes on the ground. The quality that I've seen on my several visits has been excellent, and Jim is very happy with the system. Bill Swanekamp of Kube-Pak reports that the quality of pot plants on the porous concrete root zone system is fully as good as with any other system.

Most propagation is optimized with root zone heat. It is likely that more pot production will go to root zone heating, too, given the superior plant performance and the huge savings in fuel and by possibly eliminating benches altogether.

Heat loss considerations. Various greenhouse and weather factors bear on heat loss. Of course, different covering materials have different U factors. The conditions of the covering is important. Glass slippage and even the glass lap cracks can affect heat loss. A glass roof with lap cracks will freeze over on a very cold night, however, greatly reducing the heat loss.

A greenhouse with a heat distribution system below the crop—such as pipes or tubes below or on the bench or in the soil—won't require as high an air temperature. The crop stays warm from roots through canopy, which is all you really need, so the overall greenhouse heat load is reduced significantly.

Frequent high winds will increase heat loss and demand. Snow does insulate, as does freezing over of the covering.

Another point that affects heating: due to more exposed surface, heating separate hoop houses can be almost double the fuel cost of heating gutter-connected houses (fig. 2).

CALCULATING BTU REQUIREMENTS

Not that every grower should be a heating engineer, but the grower will do well to appreciate the factors influencing greenhouse heat loss and to understand how to calculate the BTU demand on a heating system. The principles and math are simple.

The heat loss of a greenhouse covering is called the U factor. The absolute amount of heat energy lost to the outside depends on the total surface area of the covering (roof, sides, and ends) and the temperature difference between the inside and outside. The British thermal unit (BTU)—the heat needed to raise 1 pound of water 1F from 58.5F to 59.5F—is the unit of measurement used to express heat loss per hour, as determined by the following equation:

BTU/hr = Area × inside temperature (F) − outside temperature (F) × U factor

To illustrate, let's assume a greenhouse with a total exposed area of 10,000 square feet. The glass has a U factor of 1.13. We want to maintain an inside temperature of 60F while it's 0F outside. How many BTUs per hour will our heaters have to provide?

10,000 ft² × (60 − 0) × 1.13 = 678,000 BTU/hr

This heat loss can be replaced either by two heaters each delivering 400,000 BTU/hr or by four 200,000-BTU/hr heaters, assuming 85% heater efficiency (800,000 BTU × .85 = 680,000 BTU).

As always, I urge you to talk with greenhouse heating engineers, equipment suppliers, and other growers. Trade shows and seminars are information gold mines. Plus they'll have handy computer programs that will calculate your heating requirements for you!

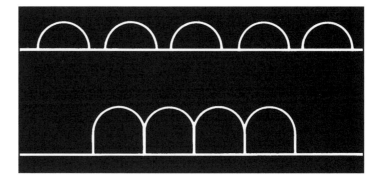

FIG. 2   Another important point on heating costs: A block of separate hoop houses *(top)* will almost double heating costs versus the same block of gutter-connected houses *(bottom)*. When houses are connected, you eliminate sidewalls, which means about half as much exposed surface.

UNIT HEATERS

Many small growers and start-ups have gotten off the ground with double-poly hoop houses and low-cost unit heaters.

A lot of unit heaters, or hot air heaters, made by companies such as Modine and Reznor, are being installed these days (fig. 3). Unit heaters can be classified by type of air mover, by physical configuration of air delivery, or by energy source.

THE SYSTEMS

Air movers. Propeller fan unit heaters are classified as zero-static pressure types and cannot be used with air discharge nozzles or in connection with ductwork. Such a unit has fixed air delivery, with a temperature rise usually in the 48 to 52F (27 to 29C) range. It is the most widely used of all unit heaters primarily because of its broad application versatility, lower cost, and compact size.

Blower fan unit heaters are designed for use with ductwork or discharge air nozzles. Since the operation of this type is generally

FIG. 3   Millions of square feet of double-poly bedding plant greenhouses are equipped with overhead unit heaters such as these Modine models.

quieter than a propeller-type unit heater having the same air delivery, it is frequently installed where lower sound levels are desirable. Also, this unit has a variable-pitch motor sheave, which permits an adjustable air volume range, accommodating typical unit heater temperature rises and corresponding rises in air volume.

Propeller unit heaters cost about one-third less than blower-style heaters. It is not possible to use an overhead poly tube to carry the heat down the length of the greenhouse with a propeller heater—you must use a fan-jet. Alternately, however, you may use horizontal airflow (HAF) blowers instead of fan-jets. The heater just blows heat toward the fans, and the fans pick up the heat and blow it down the length of the house and back up the other side—a circular trip around the greenhouse (fig. 4). All of this is a lot less expensive and simpler, and it is being done quite a bit these days by growers.

Blower heaters are blower-driven. They do involve more moving parts—but now the grower can use an overhead poly tube hooked directly to the heater without a fan-jet unit. This means one less piece of equipment, but you lose the ventilating capability of fan-jets, which is important.

Steam or hot water (hydronic) unit heaters are constructed with copper tubes and aluminum fins that provide corrosion-free heat exchangers in a humid greenhouse environment. Since the boiler is elsewhere, the heaters do not use any air from the greenhouse to support combustion, unlike the burner on a gas unit heater. The greenhouse can thus be tightly

Fig. 4   Many growers distribute the heat from overhead unit heaters using fan-jet tubes such as this one. Warm air is blown through the tube and flows down to the crop through holes in the side of the tube. Growers can also ventilate using the fan-jet by simply turning on the fan that inflates the tube.

sealed for minimum heat loss. The hydronic unit has a motor as the only electrical component, so maintenance is minimal. A hydronic unit heater will have a long, trouble-free life in the harsh greenhouse environment.

Physical air delivery configuration. Unit heaters have two basic types of air delivery: horizontal and vertical. Each has its own distinct heat throw and heat characteristics.

Horizontal delivery, separated combustion unit heaters are the best choice in such difficult applications as greenhouses using chemicals or with high humidity or negative pressure. With the separated-combustion concept, all the air needed for combustion comes from outside the area to be heated, either outside air or air from an adjacent room. This eliminates the need to use indoor air, which may be laden with chemicals, for combustion purposes.

Horizontal delivery unit heaters are widely used in many nonhostile greenhouse applications. They are also chosen for cost purposes, after the grower realizes that the separated-combustion heater is more expensive.

Skip Smith of X.S. Smith reports very wide use of HAF fans with both hot water and unit heaters. They move air around the perimeter of the greenhouse, which avoids stratifying warm air up high, cool air below. Other greenhouse builders report wide use of HAF fans. Typically, a HAF fan is 16 to 20 inches (40 to 51 cm) in diameter and moves 2,000 cubic feet (56 m^3) per minute.

A vertical delivery unit heater, due to its downward air discharge, is desirable for heating an area with a high ceiling. Optional air distribution devices that attach to the air discharge openings can provide distinctly different heat throw patterns.

Energy sources. With a natural or propane gas burner, airfoil-shaped tubes (aluminized or stainless steel) are vertically positioned over a gas burner. Flames produced by the burner burn within the tubes of the heat exchanger. Heat transferred from the flames and conducted to exterior surfaces of the tubes is removed by the airstream produced by a fan or blower. By a large margin, the most popular type of unit heater is gas fired. The most popular gas by far is natural gas, but propane may be easily used. Propane tends to be a bit more expensive.

There is generally no significant difference in the construction of a steam or hot water unit heater. Either uses a condenser, consisting of fins attached to tubes, through which steam or hot water produced by a remotely located boiler is circulated. Air moved through the condenser by a fan removes heat from the fins and discharges it to the space being heated.

Electrical heating elements consist of spiral-wound fins on steel tubes that encase electrical resistance wire. Heat generated by the electrical current is transferred through the tubes to the fin surfaces, where it is removed by the air movement produced by a fan.

No. 1 or 2 fuel oil is atomized under pressure by a gun-type burner and mixed with high-velocity combustion air to burn in a solid cone flame in an aluminized steel heat exchanger. Then, as with gas unit heaters, the heat is removed by the fan-produced airstream passing over the surfaces of the heat exchanger.

UNIT HEATERS VERSUS HOT WATER

The pros and cons of unit heaters and the new mini hot water boilers are important, especially for 20,000- to 44,000-square feet (1,858- to 4,088-m²) growers, among whom both are widely used.

Advantages. Unit heaters have lower initial cost than hot water systems. This is especially important to smaller and start-up growers. A hot water system (boiler, pipes, pump) will cost about $1.50 per square foot ($16.15 per m²) covered (for a small range) versus $.50 per square foot ($5.38 per m²) or less for a unit heater. That's a huge difference!

If you add a house or two, you can add unit heaters to accommodate the new area. Not so with a central hot water boiler. There are smaller boilers, one per greenhouse, but there are disadvantages to going this way.

Combined with fan-jets or HAF fans, heat from unit heaters can be efficiently, evenly distributed down the length of a greenhouse, so a hot water system doesn't necessarily have a huge advantage in this regard.

Many growers using bench-top root zone heat feel more secure with some overhead unit heaters as backup. They're mainly turned on for an occasional extremely cold, windy night. These same growers often report that with only small-tube, bench-top heat, the crops are warm and fine, but the air above them is uncomfortably cool for the workers. They install a few unit heaters to keep the staff comfortable.

Down side. Overhead unit heaters cost a lot less initially than hot water. On an annual basis, though, fuel costs tend to be a good bit higher than with hot water boilers. With unit heaters, you're heating the air above the crop and losing much more heat through the roof and side walls. I have seen growers install unit heaters in ways that still permit delivering under-bench heat. It can be done, but it's awkward space-wise. Root zone with hot water heats the soil and the crop more conveniently and efficiently.

Some hot water boilers modulate with the load. Unit heaters are strictly on-off, which is inherently less efficient.

Some growers will tell you that hot water heat will grow better crops than overhead heaters—"more uniform heat, less on-off cycling, less cold soil." Again, it's a judgment call. Probably, tropical plants would be more at home with hot water; cool crops, such as many bedding plants, might do better with overhead heat.

HEAT DISTRIBUTION AND EXHAUST FROM UNIT HEATERS

Distributing heat from overhead unit heaters—this is where the fan-jet shines! In an ingenious combination with an overhead unit heater (table 1), the fan-jet forces heat down the length of the house through an 18-inch (46-cm) poly tube. Holes in the tube sides permit heat distribution down the house. It works! I've seen 2 to 3F (1 to 2C) or less temperature difference end to end in houses with overhead unit heat distributed this way.

TABLE 1   GAS-FIRED UNIT HEATERS

Model	Combustion chamber	Air Mover	Vent
PAE	Regular	Propeller	Gravity
BAE	Regular	Blower	Gravity
PV	Regular	Propeller	Power exhausted
BV	Regular	Blower	Power exhausted
PSH	Separated	Propeller	Power exhausted
BSH	Separated	Blower	Power exhausted

The fan-jets have another strong advantage. On sunny days in winter, greenhouses often require some ventilation to control humidity and avoid excessive temperatures (even though it's cold outside). Fan-jets can draw in a limited amount of outside fresh cold air, mix it with air from inside the greenhouse, and deliver it down the overhead tube throughout the house. This avoids a blast of cold air (bad for crops) that would result from the use of normal vents and fans.

It is important to note that when gas- and oil-burning appliances use chemical-laden air for combustion, acids are formed. These acids can cause premature equipment failure. This phenomenon is inherent in all gas-fired and oil-fired heating equipment. Under normal conditions (with clean air) the products of combustion are vented out of the building. However, if chlorinated or fluorinated contaminants (found in most herbicides and insecticides) are present in the combustion air, it is a different story. With the high flame temperature at the burner, chlorinated and fluorinated compounds break apart to form hydrogen chloride and fluoride. The water vapor produced in the combustion process then combines with these gases to form hydrochloric and hydrofluoric acids. Both of the acids are highly corrosive and can destroy equipment and components. The intent of the separated-combustion concept is to make this phenomenon a nonfactor.

It is also important to note that many people understand the potential problems of nonseparated-combustion unit heaters in hostile environments but for economical reasons still opt to use them. Also, nonseparated-combustion gas unit heaters are perfect for greenhouses that do not use chemicals or for those careful in how the chemicals are applied.

Here's an alternate ventilating plan in lieu of fan-jets. First, open a small shutter at the end of the greenhouse. At the opposite end of the house, turn on a large 36-inch (91-cm) cooling fan. At the same time, turn on the HAF blowers. This results in drawing a small volume of cold air from outside into the house and mixing and distributing it throughout the growing area. The large cooling fan is also exhausting humid air from the greenhouse. Typically, a grower will dehumidify this way in short, 1-minute "bursts" to let the cold, dry air warm up. This dehumidification cycle is usually done late in the afternoon, partly to aid in drip elimination.

Part of what has brought on this new system of distributing air from heaters is the problem of the overhead tubes from fan-jets interfering with energy and short day curtains.

Acme Engineering describes its fan-jet system as designed to both heat and ventilate in fall, winter, and spring. When high summer temperatures arrive, however, larger fans and usually a fan-pad cooling system are needed. Rough costs for the fan-jet system—for the fan, the shutter, the poly tube, tube hangers, tube support systems, and accessories, not including the heater—in a 150-foot (45.7-m) gutter-connected greenhouse, is about $800 FOB factory.

INFRARED HEAT

Infrared systems heat greenhouses like the sun heats the earth. Solar energy travels to earth and is absorbed by the ground, buildings, and trees. These in turn convert the energy into heat, and transfer that heat to other objects and the air around them (as with the heat you feel from a warm sidewalk on a summer night). An infrared system has a 4-inch (10-cm), round tube with a gas-fired burner inside. The tube, with an aluminum reflector mounted over it, is mounted in the peak of the greenhouse, and the energy is reflected directly down to the crop below. The plants, soil, benches, and floor then absorb the energy and transfer it to the plants and air space around them. Although the plants and soil are kept at the desired temperature, the air in the greenhouse peak is kept much cooler, saving energy and reducing condensate.

Infrared heat offers unique benefits. Growers report low operating costs due to high fuel efficiency and low electrical requirements. Perhaps the most valuable benefit reported is the lower incidence of foliar diseases such as *Botrytis* and mildew because foliage and soil are kept warmer than the surrounding greenhouse air.

As with any greenhouse heating system, it is important that an infrared heat system be carefully engineered. Well-designed systems uniformly heat the entire greenhouse floor and bench area and provide sufficient heat for hanging baskets, as well. Infrared is out of the way, in the peaks of most greenhouse structures, which helps maximize the growing area. A disadvantage is that it is often not cost-effective in low-profile (e.g., 30-feet wide and 10-feet-tall, 9 by 3 m) greenhouses.

First used in greenhouses in the early 1970s, infrared now heats several million square feet of greenhouses in every climate region. Most of the development of these systems for greenhouses is due to the trials and research of Jim Youngsman, owner of Skagit Gardens and Growth Zone Systems in the state of Washington.

OTHER HELPFUL DATA

Tables 2, 3, and 4 contain useful data for growers. Table 2 lists the U factors of various greenhouse roofing materials. Then, to give an idea of which months use the most fuel, table 3 shows the heating requirements by month at State College, Pennsylvania, and table 4 reveals how much you can save by lowering greenhouse temperatures.

TABLE 2   GREENHOUSE ROOFING U FACTORS

Material	Layers or thickness	U factor[a]
Acrylic	Double sheet	.56
Asbestos board	¼ inch (6 mm)	1.10
Concrete block	8 inches (20 cm)	.51
Fiberglass		1.00
Glass	Single layer	1.13
	Double layer	.65
Polycarbonate	Double sheet	.62
Polyester fabric, spun bound	Reemy 2016	1.20
Polyethylene	Single film	1.15
	Double film, inflated	.70
Polyurethane foam, applied at site	1 inch (2.5 cm)	.14
Sateen, black		.65

[a] The lower the U factor, the more fuel efficient the roof.

TABLE 3   MONTHLY HEATING REQUIREMENTS AT STATE COLLEGE, PENNSYLVANIA

Month	Degree days per heating season	% of heating season
July	0	0
August	12	0
September	83	1
October	439	7
November	766	13
December	1,130	19
January	1,401	24
February	933	16
March	608	10
April	379	7
May	139	2
June	44	1
Total	5,934	100

Note: Averages for typical heating season.

Source: Adapted from Department of Horticulture, Pennsylvania State University, University Park.

TABLE 4 LOWERING GREENHOUSE TEMPERATURE SAVES FUEL

Outside temp (avg)		Fuel use reduction (%)[a]					
		From 65F, lower to		From 60F, lower to		From 55F, lower to	
F	C	60F	55F	55F	50F	50F	45F
20	−7	11	22	12	24	14	28
24	−4	12	24	14	28	16	32
28	−2	13	26	16	32	19	38
32	0	15	30	18	36	22	44
36	2	17	34	21	42	26	52
40	4	20	40	25	50	33	66
44	7	24	48				
48	9	29	58				

[a] Fahrenheit to Celsius equivalents: 45F = 7C, 50F =10C, 55F = 13C, 60F = 16C, 65F =18C

Source: Adapted from Department of Horticulture, Pennsylvania State University, University Park

UNVENTED HEATER PROBLEMS

Most cases of plant injury from unvented heaters happen in polyethylene structures, which are almost airtight. When open-flame unit heaters are installed right in the greenhouse and are vented poorly or not at all, the ethylene generated enters the greenhouse atmosphere and causes epinasty—plants wilt and generally just look unhappy. Steam unit heaters are never a problem, simply because there is no open flame.

The problem can occur with open-flame gas or oil-fired heaters. The "smoke" from the fire must be vented from the greenhouse. What would happen if you tried to operate a regular coal boiler in your greenhouse without a chimney? There would be no place for the smoke and gases from the fire to escape, so they would all pour into the greenhouse. Some growers who have had severe damage from unvented heaters report that the gases from the fire were so dense that they would make your eyes burn on cold nights. Obviously, there should be no noticeable smoke and gases when a greenhouse heater is properly vented, even on a cold night.

Ethylene and other problems can also occur where there is no provision for an air inlet to the burner. When carbon-containing fuels are burned completely, the end products are carbon dioxide and water vapor. If incomplete combustion occurs because of oxygen deficiency, however, reduced substances—such as ethylene, carbon monoxide, and formaldehyde—may be formed instead. Because large quantities of oxygen are needed for combustion during severely cold weather, and because the tight construction of plastic houses (particularly double-layer, air-inflated houses) prevents normal air infiltration or exchange with the outside atmosphere, inside oxygen levels can become too low for complete fuel combustion, with subsequent production of harmful air pollutants.

In regular glass greenhouse structures, there are usually two or more complete air changes per hour (depending upon outside conditions) by infiltration between laps of the glass, which maintains an adequate supply of oxygen inside the structure for fuel combustion. During extremely cold periods, though, the cracks might be filled with frozen condensate.

You must provide a supply of fresh outside air to each burner unit by installing a duct through the wall. Allow a minimum cross-sectional duct area of 1 square inch (6.5 cm²) for every 2,000 BTUs of output rating of the heating device. For example, if you have a burner rated to put out 100,000 BTUs, you need an air inlet pipe whose cross-sectional area is at least 50 square inches (323 cm²). Assuming the pipe is round, work backward from the formula for the area of the circle:

$$A = \pi \, (radius^2)$$
$$A/\pi = r^2$$
$$50/3.1416 = 15.9 = r^2$$

Therefore, the radius of the pipe should be at least the square root of 15.9, or about 4 inches: the diameter should be at least 8 inches.

What are some of the indications that you might be having self-induced air pollution problems? If you find that your heating units have gone out during the night during severely cold weather, there might have been insufficient oxygen to support the fuel combustion necessary to meet the large heat demands. If you detect unusual odors or smells emanating from your heater when you enter the house from the outside, you should also be suspicious. Both carbon monoxide and ethylene gases are basically odorless, but under combustion conditions where either might be produced, odorous substances, such as aldehydes, are also often produced. If you find that your heating unit is suddenly producing great quantities of soot, you should also have it checked, because this is another indication of poor combustion.

REMEDY

When making a new heater installation or trying to troubleshoot the problems of an old one, call upon engineers and service personnel of the heater supplier for consultation and assistance. They are anxious to help you use their equipment so that it performs properly and successfully for you, so you will come back year after year with additional business.

It would be almost certain that any unit heater installed by a reputable firm would have an adequate-sized built-in vent. Be sure that the vent is carried out through the roof so that the gas can escape. To ensure a proper draft, the outside exhaust stack must be taller than the roof peaks of nearby structures. It is also a good idea to have a proper cap on its top to prevent downdrafts, so exhaust gases will not be carried back into your greenhouse by gusting winds. If exhaust fans for greenhouse ventilation go on while the burner is still operating or just after it has shut off, this may also pull exhaust gases back into the greenhouse. Many growers have interconnected controls to prevent such occurrences. Birds' nests built in flues during the summer, when heating units were not in use, have also been found, in a few cases, to prevent escape of exhaust gases.

Still one other point needs to be watched. Each unit heater needs to be disassembled regularly—at least once a year before the major heating season begins—so that the heat exchanger panel inside can be inspected for possible cracks or holes caused by rust. The flames and heated exhaust gases pass through the inside of the heat exchanger panel on their way up the exhaust stack. Air on the other side of the panel is thus heated, then blown out into the growing area. Obviously, if this panel develops cracks, then flue gases will get into the growing area, again causing major problems. Since nearly all unit heaters are completely enclosed in an outer sheet metal cover, it means that you have to dismantle them to make your inspection possible, but it is an essential type of preventative maintenance that needs to be carried out each fall without fail. It is possible to purchase unit heaters in which the heat exchanger panel is constructed of stainless steel, but these are considerably more expensive than the stock models. If you lose a greenhouse full of salable plants from self-induced air pollution, however, steel might have been an inexpensive form of insurance.

Finally, be sure to have properly sized air intakes for each gas- or oil-fired heater unit. On the coldest nights, there is a great temptation to stuff an oil sack in them to keep all that cold air from coming into the greenhouse unit, but that is the time when air is needed most for combustion. If you try to save a few pennies during the night by doing this, you may have plant injuries worth hundreds of dollars the next morning.

Editor's note: Many thanks to Jeff Fries, Modine Manufacturing Co., Racine, Wisconsin; Jeff Deal, Hamilton Engineering, Livonia, Michigan; and Al Zylstra, Growth Zone Systems, Mount Vernon, Washington, for assistance in preparing this chapter.

CHAPTER 5

GREENHOUSE COOLING AND VENT SYSTEMS

by Vic Ball

Nearly all serious United States and Canadian production facilities for year-round pot plants and major cut flower crops are equipped with fan-pad cooling. The only exception would be cool-summer areas, such as Washington state and the Vancouver area. A pot chrysanthemum crop without cooling, even in the Midwest or the East, is a good bet for major losses from heat stall in summer. Farther south or in the Southwest, cooling is an absolute must. Surprisingly, there are major greenhouse ranges even in Florida with fan-pad cooling. In spite of Florida's relatively high humidity, there is enough cooling to make the technique economical. For the record, there is also substantial use of high-pressure fog or mist for evaporative cooling, especially in propagating areas and for cut roses.

Though most production of pot plants and cut flowers in North America is fan-pad cooled, growers are considering natural ventilation anywhere it's a viable option. The rapid rise of electricity costs has caused growers in some areas to reconsider rows of giant fans and pads as a cooling system. A big example: Aldershot Greenhouses near Toronto, with many acres of year-round pot mums, is making a serious try at relying on 50-inch (127-cm) roof ventilators to do the cooling job. The climate there (almost between two of the Great Lakes) takes a lot of the hot weather pressure away.

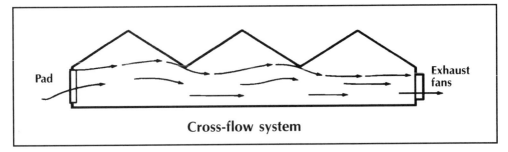

FIG. 1 This drawing illustrates how air moves through a greenhouse. The greenhouse is a ridge and furrow structure. Note how the gutters keep cool air down near the crop.

EVAPORATIVE COOLING

Drawing air through a wet pad evaporates water and cools the greenhouse air. The air released outside has absorbed 8,100 BTUs of heat energy for each gallon evaporated. The cool air comes out of the pad and into the greenhouse (fig. 1).

A wet wall—consisting of pads, a water distribution system to wet the pads, a water pump, and a sump—is erected continuously along one side or end wall of the greenhouse. The pads must be kept wet to facilitate the evaporative cooling process. On the opposite wall, fans properly sized for the greenhouse size and location are placed to provide smooth airflow across the greenhouse. Size your pads to provide the most efficient, economical system possible. Cellulose pads 4 inches (10 cm) thick operate best at an air speed of 250 feet (76 m) per minute through the pad. Aspen pads 2 inches (5 cm) thick work best at 150 feet (46 m) per minute. For 6-inch (15-cm) cellulose pads, use 400 feet (122 m) per minute.

HOW COOL?

The amount of cooling achievable by evaporative cooling varies with the dryness of the air—the differential between the wet bulb and dry bulb temperatures. This differential varies not only with location and season but also during each day. Although the dry bulb could vary as much as 25F (14C) in one day, the wet bulb varies only approximately a third as much. Therefore, cooling can be accomplished even in normally high-humidity areas in the middle of the day when it is really needed.

A well-designed evaporative cooling system should be able to reduce the dry bulb temperature inside the wet wall to approximately 85% of the difference between the outside dry bulb and wet bulb temperature. Expect a temperature rise of about 7F (4C) from the pad to the exhaust fans.

Please do not cheat on the size of the intake air openings—the larger the better. A smaller size than called for creates static resistance, which greatly reduces the efficiency of the fans and causes increased electricity usage. Use continuous vents whenever possible.

HOW MANY FANS, AND HOW BIG A PAD?

The calculations are simple—the problem is that the need varies by locality. Some areas have consistently high (90%) humidity on summer days (Dallas, Texas, for example). Others, like Chicago, will experience 90% humidity for many days in summer but not all. Generally, fan-and-pad systems work best in dry climates. Systems must be "oversized" in high-humidity areas like Texas and Florida.

To give a rough idea of the equipment needed for cooling an acre of conventional gutter-connected greenhouses, here's the way it's figured: Length of house × width × 8 cubic feet per minute (CFM) per square foot (adjusted for elevation, light intensity, etc.). This gives cubic feet per minute required for air movement. To figure the pad area required, divide that total CFM by 250 for 4-inch (10-cm) thick pads, and by 400 if using 6-inch (15-cm) pads. Then divide that total by the number of linear feet of wall the pad vent will cover. This last number will tell you how tall the system will have to be. In humid climates, oversize fans and pads by 20%. For growers interested in calculating pad and fan requirements, contact Acme Engineering & Manufacturing Corp., P.O. Box 978, Muskogee, OK 74402, fax 918/682-0134. Copies of Acme's manual *Controlled Environment Equipment for Greenhouses* and the manual *The Greenhouse Climate Control Handbook* are available at no charge.

A caution on pad installations: Pads cause resistance to airflow, so be sure that all air passes through the pads; all other openings (with less resistance) need to be closed. Even an open fan-jet shutter will greatly reduce pad efficiency.

SAMPLE CALCULATIONS

Example: 14 bays, each 21 feet (6.4 m) wide by 144 feet (44 m) long, have a total area of 42,336 square feet. With an 8-foot (2.5-m) gutter height, this greenhouse section—for proper fan-pad cooling—requires moving 338,688 CFM. After determining CFM requirements, select the fans. Then, based on their total CFM performance, size pads to match the fans. For example, to select fans, we must move 338,688 CFM. A fan drawing air through a 4-inch (10-cm) pad will move air at 250 feet per minute and can move 21,090 CFM per fan (see Acme's bulletins for fan capacity specifications). So we divide 338,688 by the 21,090 CFM capacity of each fan. Answer: We need 17 fans.

Assuming use of a pad that will move 250 CFM per square foot of pad area, we divide the 338,688 by 250 and will need 1,355 square feet of pad area. Again, a 4-inch (10-cm) thick paper pad moves air at 250 CFM and a 6-inch (15-cm) thick pad at 400 CFM. Assuming the pads are mounted to gables, we have 294 feet (89.6 m) available to mount the system (14 bays by 21 feet, 6.4 cm). The square feet required divided by the available linear feet of mounting space means the pads will have to be at least 4.61 feet (140 cm) tall to do the job. Always figure pads up to the next even foot, not down. In this example, then, 5-foot (150-cm) tall pads should be used.

When the house temperature becomes too high and cooling is needed, the equipment controller or cooling thermostat simultaneously opens the motorized shutter at the opposite end of the greenhouse and energizes the exhaust fans. When the desired temperature has been reached, the equipment controller closes the inlet shutter and turns off the exhaust fan, thereby shutting off the outside air supply. Generally, this is done in stages.

When the relative humidity in the house exceeds the desired level, it can be reduced by bringing in cooler outside air with less water vapor content or by heating the air, which increases its capacity to hold moisture. The humid house air is expelled by the exhaust fan, while the drier incoming air is heated and mixed with the house air.

FAN MAINTENANCE

For fans to work at peak efficiency with minimal electricity usage, certain maintenance guidelines should be followed. Keep all the belts very tight to eliminate slipping, which will not only reduce the air output of the fans but also increase the motor's electrical use (automatic belt tighteners are available). Be sure that all air leaks in the greenhouse are kept plugged in order to get top efficiency from your pads. Air leaks in the house can increase temperatures in the greenhouse 4 to 10F (2 to 6C) because air intake is going through the leaks and not the higher-resistance pads themselves. Be sure the pad area is kept completely wet without any dry spots. Control algae with chemical algaecides or even common bleach, but be sure bleach will not be toxic to the crop you are growing. During winter months, check for air intrusion through fan shutters or inlet shutters to reduce heat loss. All these simple steps will help keep your greenhouse cooler in the summer and warmer in winter.

FOG (MIST) COOLING AND HUMIDIFYING

Using high pressure for cooling and humidifying greenhouses for ornamental, vegetable, nursery stock, and cut flower production is steadily increasing due to its ability to provide even temperature reductions and levels of humidity. Fog offers these advantages in arid climates as well as the humid climate typically found in the southern United States. Gene Parsons, vice president of Baumac International—manufacturer of the MicroMist Fog System, phone 909/794-7631, fax 909/794-5795—reports that cooling with fog has steadily increased through the years.

Growers are becoming aware of how the greenhouse environment affects their overall quality and quantity of production. The principles of cooling with fog and with pads are primarily the same. The difference is how efficiently the water is used for cooling. In pad cooling, water saturates a pad, concentrating all the cooling potential in one location in the greenhouse, achieving 85% efficiency. Fog systems distribute the water in a vaporized form evenly throughout the greenhouse, through selected placement of the atomization lines. Because of the flash evaporation characteristics of fog, 100% efficiency is achieved.

From air inlet to exhaust fans, temperature differentials with fog cooling can be as low as 1 degree. In a typical fan-pad installation, temperature differentials can be 7 to 10F (4 to 6C) from air inlet to exhaust fans. A fog system can be installed in any house configuration, such as with natural or mechanical ventilation, and lends itself to housing when insect screens are used to protect air inlets.

Installation costs for fog are comparable to or lower than for pad cooling systems. Only pump filters and pump oil need to be replaced in a well-engineered fog system, so maintenance is low.

As many cut flower growers move into such high-tech growing supplies as rock wool, clay pellets, or coconut fiber media, the need for a fog or mist system becomes crucial. These new media have eliminated the natural humidity that would be added to the environment from the surrounding soil beds. In propagation chambers and houses, fog or mist systems play a crucial role by maintaining even and consistent humidity.

Fog systems from several firms are used by dozens of growers in the Southwest and West and in the eastern United States and Canada. They are often used for propagating, for germinating seed and plugs, for rooting vegetative material, and, importantly, for cut rose production.

Fog systems are designed for each grower, taking into consideration ambient weather conditions, greenhouse configurations, ventilation methods, crops to be grown, and desired temperature reductions or humidity levels. Once temperature objectives are known, for example, a psychrometric chart is used to define the amount of water (gpm or l pm) required to meet the cooling objective. Simple division of the nozzle flow rate into the required gpm will tell how many nozzles need to be in the greenhouse.

CHAPTER 6

EQUIPMENT AND AUTOMATION

by Debbie Hamrick
Ball Publishing
Batavia, Illinois

RETROFITS: EXPECTATION VS. REALITY

by Chris Beytes
Ball Publishing
Batavia, Illinois

Over the past five years or so, there's been a quiet revolution in the greenhouse business. If you've priced turnkey, totally outfitted structures lately, you've probably noticed it. Growers looking to build new construction today often find that the greenhouse itself is not as much of an issue as everything inside. As a matter of fact, benches, overhead curtains, cooling systems, fin heating systems, transplanters, robots, and other pieces of equipment have become the focus of the "hard" side of the greenhouse. Often it's "everything else" that goes with the greenhouse on which growers invest the most, and for good reason: efficiency. Why build another structure when you can gain an extra 10 or 15% in production by investing in equipment and bench and transport systems. You can also see significant savings in labor expense through buying equipment. Experts recommend mechanizing the most tedious, labor-intensive tasks first, which is why labor savings is the reason many growers turn to mechanization (table 1).

Efficient and flexible

After *efficiency,* the word most closely associated with greenhouses in the late 1990s is *flexibility.* Growers in today's volatile markets know they may not always produce the same crop mix from year to year. They also realize that container sizes change. Today's most modern production facilities take advantage of equipment and systems, but unlike many state-of-the-art structures from the 1980s, they also allow greater flexibility in the crops that can be produced.

Equipment and automation are the favorite topics whenever two or more growers get together. Equipment manufacturers are the busiest companies at trade shows. Fifteen years ago, the range of technology available to the grower was limited to just soil mixers, flat and pot fillers, and conveyor belts. The first larger machines were developed for the nursery industry—for containerized tree and shrub production— and adapted for greenhouse use. Today, the range of equipment is extensive, from pot and flat dispensers to transplanters and boom irrigators, from bar code labeling machines to robots to take plants into and out of the greenhouse. Greenhouse production is the main market of equipment companies today.

But mechanization and automation are different for each grower. The choices of what equipment to buy are based on the kinds of crops you grow, your location, budget, and mind-set. Some growers maintain that a set amount of money or percentage of their sales should be invested each year, while others use the same equipment until it breaks from being worn out.

Labor savings

Cut flowers require the highest labor costs of any type of floral crop. Cut flower labor as a percentage of sales is frequently 35% or higher. That, plus the ability of cut stems to be packed densely in boxes for air transportation, is probably one of the biggest reasons that production was so easily moved out of this country. Serious cut flower growers employ equipment to reduce labor whenever possible: grading and sorting machines, sleeving machines, automatic stacking and palleting. The flower is still harvested manually, though. For the cut flower grower seriously studying mechanization and automation of greenhouse activities, a trip to Holland's NTV exhibition—held annually at the RAI Exhibition Center in Amsterdam in November—is a must. The Dutch, because of stagnant prices and escalating labor costs, have developed a number of ways to increase the efficiency of cut flower production. For one, breeding brings higher yields. Sophisticated hydroponics growing systems also show higher yields.

Bedding plants are the second most labor-intensive crop. Manual transplanting (table 2) is easily identifiable as consuming a massive quantity of labor. There is thus no other single piece of equipment undergoing so much development work today as

TABLE 1   GROWING TASK OUTPUT RATES

Task	Output per man-hour
Mixing media	
Mortar mixer	1.5–2 yd³ (1.1–1.5 m³)
Transit mixer	3–5 yd³ (2.3–3.8 m³)
Shredder-tractor with bucket loader	15–20 yd³ (11.5–15.3 m³)
Drum mixer, feeder bins	15–20 yd³ (11.5–15.3 m³)
Filling flats	
Hand	60–100 flats
Machine	150–300 flats
Transplanting	
Hand	8–15 flats
Plugs with transplanter conveyor	30–50 flats
Machine	100–250 flats
Potting, 6-inch (15-cm) pots	
Hand	100–200 pots
Potting machine, 3,000 pots per hr (4 workers needed)	750 pots

Source: Adapted from John W. Bartok, Jr., paper presented at Greenhouse Systems: Automation, Culture, and Environment conference, July 1994.

TABLE 2   EXPENSE OF HAND-TRANSPLANTING BEDDING

Flat cells	Flat transplant cost ($)[a]			Flats transplanted for $100,000
	Lowest	**Highest**	**Average**	
24	0.039	0.250	0.143	699,301
32	0.052	0.270	0.161	621,118
36	0.058	0.450	0.208	480,769
48	0.078	0.600	0.264	378,788

[a] Costs of hand-transplanting bedding plants at 13 Midwest greenhouses in 1994.

Source: Adapted from P. Allen Hammer, "Investing in automation is no longer an option," *GrowerTalks*, 1995 March.

TABLE 3   PLANT HANDLING SYSTEMS

System	Investment time/flat($)[a]	Average speed		Cost ($)
		Per 1,000 flats (sec)	Per 60,000 flats (hr)[b]	
Carrying by hand	0	42	11.7	8,400
Trolley conveyor	4,400	38	10.5	7,560
Roller conveyor	4,000	36	10.0	7,200
Belt conveyor	10,000	20	5.5	3,900

[a] Assumes a 24,000-ft² (2,230-m²) greenhouse with 10 bays, each measuring 25 by 96 ft (7.6 by 29.3 m) and attached to a central headhouse.

[b] Labor at $6/hr. Flats moved in and then out of greenhouse.

Source: Adapted from John W. Bartok, Jr., paper presented at Greenhouse Systems: Automation, Culture, and Environment conference, July 1994.

the automatic seedling transplanter, which is discussed in depth later. The next most labor-intensive area of bedding plant production is moving transplanted trays into and out of the greenhouse (table 3). In some highly inefficient greenhouses, workers carry two or three trays at a time. Other growers use conveyor belts (fig. 1) or, more commonly, wheeled carts. Order assembly and packing can also require a large amount of

FIG. 1   Jim Dickerson, Dickerson's Greenhouse, Gobles, Michigan, uses conveyor belts to bring plants into and out of the greenhouses.

labor when an operation is poorly organized. Standard pieces of equipment already used by most bedding plant growers include a seeder for sowing into plugs, a flat filler and a watering tunnel for plug trays and bedding flats, a tagger to stick tags in packs, and a dibbler to dibble holes for planting plugs.

Pot plant production using palletized (Dutch tray) bench systems—developed in Denmark or the Netherlands—can be so efficient that only one or two persons are required per acre of production area. Propagation—taking cuttings, sticking cuttings—and transplanting, as well as sleeving and packing boxes, are the last areas of pot plant production to become automated. (Figs. 2, 3 and 4 show more advanced pot plant production automation in propagation and transplanting.)

Related considerations. Labor savings for one activity in the production process may be the impetus to look at a piece of production equipment, but it is only one part of the evaluation. Labor shortages are another factor pushing growers to automate. Fewer people seem to want to work at "dirty" jobs like most of those available in the greenhouse. Growers offering wages significantly higher than minimum wage are more successful at drawing and keeping high quality labor. But many growers turn to potting machines and transplanters that run without complaint 24 hours a day, 365 days a year, if needed.

A simple payback of a new machine or system is only one aspect to study when evaluating such a capital expenditure. While labor savings often drives growers to mechanization initially, once they are there, their economics change. Mechanizing one operation will often spotlight other inefficiencies. Ask any grower who buys a piece of equipment how that piece of equipment changed the operation, and it's likely that labor savings will have ended up only third or fourth on a list of five or six.

More and more, growers ask themselves how much total investment is required, rather than simply buying equipment and mechanization. Overcapitalization, especially using borrowed funds, is a competitive disadvantage today. Many times a machine enables the grower to produce more, which in turn increases production. Increased production is great in a market with

FIG. 2 At J. de Vries Potplantencultures, Aalsmeer, the Netherlands, workers can keep track of how many pots they've handled (on overhead video monitors) as they twist off rhipsalidopsis cuttings.

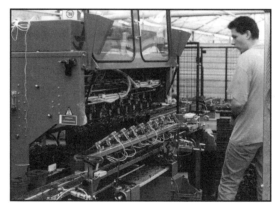

FIG. 3 Mechanically transplanting bedding plant plugs is becoming commonplace in many of America's leading greenhouse ranges. But transplanting pot plants is taking longer to catch on. This transplanter is used by the Dutch *Saintpaulia* grower Appelboom, Rijsenhout, to plant liners.

short supply, but if you are in a tight market and are looking to mechanize simply to lower costs to increase your profit margins, you may not wish to increase production volume.

Uniformity is one of the key benefits of mechanizing and automating. Machines perform repetitive tasks exactly the same way. A bedding plant crop that has been transplanted automatically is completely uniform, while a manually transplanted crop will have plugs placed in various quadrants of the

FIG. 4   Transplanter lifts plants from propagation trays and plants them into pots using a corkscrew-like mechanism.

cells and at various depths. When the uniformity of machine operation is combined with stepped-up grading before planting, a grower can significantly reduce the cull rate. Again, this results in more production. Can your market and sales staff handle it?

Probably the most important change stemming from any single piece of new equipment is a need for more equipment. To fully maximize a flat or pot filler, for example, you must have a conveyor belt coming off the end. To further take advantage of the machine, you should also add a transplanting conveyor where workers can sit comfortably while they manually plant or an automatic transplanter. All these pieces of equipment must be synchronized to work in concert.

The country is littered with equipment growers bought to solve their labor needs without thinking about exactly what the equipment would do for them. You'll find these examples out the back door of most greenhouses. Think about what you need and what you want the equipment to do before you buy.

TRAY MECHANIZATION

The most mechanized pot plant growers and plug growers use Dutch trays (fig. 5). These growing trays, known in Europe as "containers" and in the United States as "palletized benches," generally hold 100 or more pots. Most leading European pot plant growers use trays. Because of this you hear of growers with labor needs of only one person per acre for very low-labor crops, like areca palms, mini roses, or kalanchoe. Plug growers also find trays beneficial because they can easily move them from one growing-stage zone to another. This allows the growers to invest in making the growing environment as good as possible during each stage of the seedlings' development.

Inside the greenhouse bay, trays roll on rails supported by pipes, moved either manually or automatically. Generally, they are pushed into the growing bay from the main aisle. Once inside the growing bay, they can also roll horizontally, meaning each bay requires only 2 feet (60 cm) or so for a "floating" aisle.

FIG. 5 Mobile Dutch trays are supported on pipe and rails like these at Knud Jepsen's, Aarhus, Denmark. Hot water heat circulates through the U-shaped pipe.

In greenhouses with manual tray systems, center aisles in the greenhouse and the production headhouse are also fitted with rails. Trays begin their journey in the headhouse, where they are filled with newly planted pots or plug trays. A worker can then push one or several full trays along the rails into the greenhouse and subsequently into the growing bay. The key to labor savings here is that one employee is handling dozens of pots at one time. When plants or plugs are ready for spacing or to be sold, the employee simply retrieves them for return to the headhouse.

Some growers have automated the rail system so that trays are propelled down the rails on their own. Bar codes or magnetic strips enable the grower to program a tray to be placed into a designated growing bay. In the most recent construction, growers have moved away from rails in the aisles and toward robot systems (see robots at work in figs. 6 and 7). A robot collects filled trays at the headhouse and carries several at a time (six, for example) into the greenhouse, where it deposits them in the

FIG. 6 Some growers don't want to put rails and pipe down greenhouse aisles for mobile Dutch trays. Instead they use robots to take filled trays into and out of the greenhouse. This machine, in use in the Netherlands, is affectionately known as Dirk.

growing bay. The robot is guided by metal strips on the floor. Again, bar codes or magnetic strips instruct the robot into which bays to deposit the trays.

Most of the time tray systems using robots are also organized so that the crop "flows" through the greenhouse, moving progressively from potting, into the growing bay, and then out the other side to shipping. In this way the crop never moves backward, but instead in an **S** through the greenhouse, going in one side and out the other, or in a **U**.

In Denmark some growers use overhead cranes that ride on pipes attached to greenhouse gutters. A crane can lower, pick up trays, and transport them back to the headhouse.

FIG. 7   Here's a pot spacing robot hard at work. These simple machines put plants into mobile Dutch trays and take them out again. They work 24 hours a day complaint-free. Some highly mechanized growers in Europe space plants several times during the production process using robots. Growers bring tables to the robots. Pots are put onto conveyors that run into plant grading stations where they are sorted into several grades and returned to the greenhouse. The end result is more uniform quality of each bench at harvest and better total space use.

Whenever any production activity can be standardized consistently, it allows greater automation. For the 20 years or so that trays have been used in the greenhouse, they have allowed growers to just about fully automate pot plant handling. Trays also allow the grower to centralize potting, grading, and other activities in the headhouse. This centralizes employees in a clean, dry environment near washrooms and break rooms.

Tray peripherals. Trays have become widely used for pot plant production, and a number of automatic functions are now integral to their use. For example, pot spacing arms automatically take filled pots off a conveyor coming from the potting machine and place them into the trays. When plants are ready for spacing, trays are then brought back to the headhouse where one pot spacer removes them, putting them onto a conveyor belt. They may then be manually graded for size by an employee, who picks up substandard pots and places them onto a second conveyor, or by a machine using video cameras. Pots can be sorted into two, three, or more grades. Each grade has its own conveyor and its own pot spacer, which in turn puts pots back into trays. Filled trays are put back into the greenhouse and finished.

The ability to grade the crop and then space it allows the grower to even up plant quality before shipping. Large plants will obviously finish first, while smaller plants will be given extra time to attain size. In Holland and Denmark, some growers are grading and spacing two times during the production cycle to get an extra edge in space use since the process is automated and costs very little.

Trays can also be fitted with plastic or metal inserts to make them suitable for ebb-and-flood irrigation or trough irrigation (see chapter 7 on irrigation).

Costs. A monoculture pot plant grower in the North with significant production area can justify the expense of installing a mobile tray system. That is the exact case for most of the Danish and Dutch growers using trays. However, most North American pot plant growers are not monoculture. The structure of the American market forces them to produce a broad range of pot plant crops. Developing a specialized system to fully automate production of five or six crops at once is a challenge. However, parts of production can and should be automated. The second major section in this chapter, "Retrofits: Expectation vs. reality," discusses the specific experiences at Leider Horticultural Companies in Buffalo Grove, Illinois, which installed a Dutch tray system.

The cost for a 1-acre tray installation, including pipes, rails, and metal trays with expanded metal bottoms, is roughly $6 to $10 per square foot ($65 to $108 per m^2). This does not include all the other support equipment required in the headhouse and the rails or pipes or a robot or other system, to transport trays into the greenhouse.

Generally, tray systems are used by bedding plant growers only for plug production, where the added expense is offset by the value of the crop. In Europe the rule of thumb is that the closer the plants are together, the more money the grower can justify spending on equipment and mechanization. Plugs definitely meet this rule. Plug growers like trays because they can easily handle a large number of plug trays at once. Plug ranges using trays can be set up so that trays move in succession from Stage 1 to 2 to 3 and 4, which allows more precise environmental control, rather than having every part of the plug range set up to run all growing stages in succession, the plugs staying put for the whole process. (Fig. 8 shows the monorail that C. Raker & Sons uses to take plug trays off of growing trays and move them to the packing and shipping area.)

Bedding plant growers can, however, justify the expense of trays. The Van Bourgondiens on New York's Long Island are an example. There, 65,000 square feet (over 6,000 m^2) of their 2-acre range is in trays with ebb-and-flood irrigation. The Van Bourgondiens produce top-quality seed geraniums on the trays during the spring and poinsettias during the late summer and fall. Pot spacing is fully automatic. The Van Bourgondiens made the switch to trays to accommodate ebb-and-flood recirculating irrigation to eliminate runoff from their greenhouse into groundwater.

Another way to use trays for bedding plants is roll-out (fig. 9). Here the grower double-crops the bedding plant production area using trays. The first crop is on the floor, while the second crop is in the trays. During the day, trays are rolled outside, generally through greenhouse side walls onto rails running outside but adjacent to the greenhouse. The rolling out process can be manual or automatic. Using trays in this way helps the grower get more out of the same production space in the season when the space is needed the most. Developed by Jack Van de Wetering at Ivy Acres, Calverton, Long Island, New York, roll-out has become standard for most large-scale American bedding plant growers—it's just too good a way to nearly double the growing area for serious bedding growers to ignore.

The crop getting pushed outside is generally of exceptional quality because the plants are exposed to cool days and warm nights naturally (see chapter 26 on DIF). Petunias, pansies, and snapdragons are all naturals for being grown on trays in a roll-out system. Impatiens, because of their heat requirements, are ideal for the ground crop. Add hanging baskets overhead, and you can get a remarkable three crops from the same area—and all when demand is the highest. The grower can also produce crops—4-inch (10-cm) annuals or garden mums, for example—on the rolled-out trays all summer long, with much better results because of cooler day-time temperatures.

FIG. 8   Simple—but highly efficient—is one way to describe the plug tray packing area at C. Raker & Sons, Litchfield, Michigan. Finished trays are brought to and from the greenhouse using monorails like the one in the photo.

FIG. 9   Roll-out bedding has got to be one of the best ideas of the past 20 years! Developed by Jack Van de Wetering of Ivy Acres, Long Island, N.Y., this example is at Mid-American Growers, Granville, Illinois. Trays roll outside through the vents in the side of the greenhouse. Note how stocky the petunias and pansy plants are.

THE FLOOR RISES AGAIN

While Dutch trays may be the system most growers think of when it comes to a highly mechanized greenhouse, growing on the floor is the latest production trend as the industry moves to the year 2000. There's one main reason: flexibility. A grower can put any type of container on the floor. With today's newer, sophisticated ebb-and-flood systems for concrete floors, irrigation is also automated. Floor production systems tend to be simpler, requiring fewer physical materials in the form of pipes, rails, and benches, yet the grower can achieve similar results, especially for bedding plants. (For details on ebb-and-flood floor systems, see chapter 7 on irrigation.)

Equipment to place and space pot plants on the floor is becoming more and more common. One automated approach uses a forklift body with tines that picks up pots for transport and spacing out. Another approach is a robot with a flatbed conveyor that lays pots or bedding plant trays out and picks them up again.

For many bedding plant growers, using a floor system in conjunction with wheeled carts is an excellent way to increase efficiency economically. In such a setup, the grower places newly transplanted flats onto carts in the headhouse. They may be rolled into the greenhouse individually or pulled with several linked together at a time by tractor. Workers then place trays on the floor by hand, where plants remain until they are ready for sale. When plants are ready to sell, employees again use carts to take trays from the greenhouse to the order assembly area or directly to the shipping dock.

The great thing about carts is their versatility. Not only can they be used inside the greenhouse for transporting plants, but in conjunction with a lift-gate truck, they make excellent display racks at retail, where the truck driver simply rolls the carts off the truck, retrieves the empties, and goes on his way. Carts can also be used all year long, making the investment one that pays not for just a few weeks of the busy spring. Keeping track of carts at retail is difficult, however. They can be very expensive if you count on losing maybe as many as 15% of your carts. Some growers now bar code their carts for tracking during delivery.

THE ROMANCE WITH TRANSPLANTERS

Transplanters are taking off, and it's about time! Transplanting is one of the most labor-intensive, repetitive jobs in the greenhouse. The cost to transplant bedding plants by hand can run from $.25 to $.50 a flat for bare-root seedlings to $.15 to $.20 for hand-transplanted plugs. No wonder that for more than 10 years, the industry has dreamed of perfect automatic transplanting. We've gone through complicated and convoluted transplanting prototypes, for a while a different one at nearly every trade show.

Now several models have proven themselves reliable and affordable. Among them the Flier transplanter is gaining ground most rapidly. Growers using transplanters to plant bedding plants are saving $.10 or more per flat in labor costs alone.

Add to that the value of increased uniformity of transplanted flats and steady work flow, which helps to pace workers.

There are several issues to consider when taking a look at a transplanter today. Speed is among the most important. Can you live with 250 trays an hour, or do you need greater speed to meet your weekly production volumes? How flexible is the transplanter? Does it work only with one plug size or type of tray? If you can change plug sizes, how long does that take? Are extra parts involved? Can existing staff handle the operation easily?

As with seeders just five years ago, a grower often chooses to buy a couple of different transplanters, rather than rely solely on one to suit all the transplanting needs.

The increasing reliability of transplanters has put a lot of pressure on getting higher plug counts out of the plug tray to minimize skips that require patching, and this in turn has created demand for better quality seed yielding higher count trays. Some companies, such as Flier and Visser, are even developing retransplanting machines that automatically plant seedlings in no-show plug cells after just a few days.

Transplanters have also put emphasis on plug quality. Overgrown plugs do not transplant well. The machine can tangle the mass of foliage and pick multiple overgrown plugs. When this happens, operations stop. Undergrown plugs with poor root systems also do not transplant well, as the root balls tend to fall away. To automatically transplant well, the plug needs a solid root ball, which is also to the advantage of the grower in helping to minimize finishing time.

Some growers have experienced problems with some transplanters when they've reused plug trays. Trays made of lightweight plastic can be damaged or partially crushed during normal use. Damaged plug trays do not work with automatic transplanters.

Growers who have successfully installed transplanters cite many advantages. Among them is increased competitiveness from being able to refill bays quickly once shipping begins. All bedding plant growers know that replanting is the key to increasing turns. The faster you can replant, the more competitive you'll be in the market, especially if you're also using larger plug sizes as the season progresses, thus shortening your time from plant to sell. It's that last crop that adds the most to your bottom line.

Count on two or three people to run the transplanter line. You'll need one person feeding flats into the flat filler. This person will also monitor how well the filled flats are moving into the transplanter and can shut down operations if a problem arises. Another worker is needed to patch skips, unless you have perfect plug trays. A third person takes filled flats from the conveyor line and places them on carts. In a highly mechanized operation, this last function can be automated: flats can slide from a conveyor belt either into Dutch trays or onto a flatbed robot that carries them into the greenhouse.

If you're considering an automatic transplanter, it's likely that running it will take up more space than running a transplanting conveyor with workstations.

The next transplanting hurdle to clear is planting rooted cuttings or sticking unrooted cuttings. Both of these operations are highly labor intensive. Prototypes are available, but nothing to date is as reliable or economical as seedling transplanters.

For a summary of equipment costs, see table 4.

TABLE 4   EQUIPMENT COSTS

Machine	Price ($)
Cart loader	$24,000
Conveyor belt, per linear foot	$75–100
per linear meter	$246–328
Dibbler	$3,000–4,000
Flat filler	$7,000–18,000
Ink jet printer for bar-coding onto plastic	$42,000
Pot filler	$15,000–28,000
Robot to move palletized benches	$40,000
Seeder 50,000 trays min vol	$4,000–30,000
Soil mixing system, continuous	
10 yd³ (7.6 m³) / hr	$20,000
30 (22.9)	$50,000
50 (38.2)	$75,000
Tagger	$9,000
Transplant conveyor belt,	
for manual transplanting	$3,000–higher
Transplanter	$65,000–120,000
Tray/pot dispenser	$3,500–8,000
Watering tunnel	$2,800–3,500

EQUIPMENT AND MECHANIZATION TIPS

Flier USA's Robert Lando has more than 12 years' experience in mechanizing American growers. Based on this experience, he offers the following tips to growers.

Time for new equipment. Replace worn-out equipment. There's a great temptation to keep on using a paid-for, fully amortized piece of equipment until it falls apart. Be forewarned: old pieces of equipment begin to work inconsistently. A worn-out potting machine, for example, will result in crop losses. If you're a larger grower, you cannot afford the inconsistency of an old machine. A good rule of thumb is that equipment should be discarded after 10 years. There's almost always a market for used production equipment. How much will a used piece of equipment sell for? The question to ask yourself: Would you pay the price you're asking? If the answer is yes, it's likely that someone else will, too. (If you're buying a used piece of equipment, call the manufacturer and ask about that machine's history before you buy.)

Know when to mechanize. If you're producing 10,000 or more bedding plant flats, you need a flat filler. Seeders can be justified at lesser volumes, maybe only 2,000 trays, because you can buy less expensive models. To justify a high-volume drum seeder, however, you'll need about 40,000 plug trays. If you're small to medium sized, you may never mechanize soil mixing, instead choosing to buy it in premixed. Why make the investment when you don't have to? Unless you want to mechanize simply to spend money, think first about what you're doing and what you need before you buy.

Retrofit old with new. Don't hesitate to ask about integrating new pieces of equipment with your existing production equipment. Just because you're buying a transplanter doesn't mean you also must have a new flat filler. Machines can be synchronized to work together.

Desirable machine qualities. Look at your weekly volume to determine capacity. Too often growers get stuck looking at how fast a machine will operate in a given hour. It's better to look at what you need to get done by the week, then size the equipment. You can then look at how many hours a day you'll run the equipment. For peak transplanting most growers today run two shifts because it just isn't practical to gear up to transplant everything in just a few hours a day.

Think flexibility. When you're looking at any single piece of equipment, ask yourself, What else will or can it be used for? Will the flat filler work with pots? Can hanging baskets pass underneath the watering tunnel? Can the flat or pot filler operate at variable speeds to accommodate different pot or container sizes? Machines sized for one company's 512-plug tray will not necessarily work with another company's 512-plug tray.

Go for 100% soil return. When you're looking at a flat filler or pot filler, spend the extra money on the machine that returns all the soil. You'll save in the long run in labor and soil costs.

Integration into operations. Think about how to organize your production line. For example, you'll need to put a conveyor belt in front of the flat filler—15 feet (4.6 m) long or so—to assemble inserts into trays or put pots into shuttle trays. The person here will not be busy all the time and can walk the transplanting line, helping to keep production flowing. Asking this person to also remove finished trays would be too much, however.

Avoid extra steps. A lot of growers prefill flats, palletize them, and then feed them into an automatic transplanter or transplanter conveyor. However, these growers have added extra steps—having to stack filled flats and unstack filled flats—when a flat filler could be synchronized to run with the transplanter or transplanting line. There's also the issue of compacted media to consider: stacking flats on top of one another compacts media, forcing air out, which can result in problems with root disease. All plug growers should avoid stacking filled plug trays onto each other, as compaction is even more serious with tiny plug cells.

Put buffers in the system to avoid work stoppages. A system is as good as its weakest link. To avoid slowdowns, accumulate a store of flats on a conveyor belt, for example, before they're fed into a transplanter or transplanting line. This way, if the flat filler stops for a couple of minutes, the transplanter can continue to work.

Set up new equipment before the rush. Work the bugs out of equipment before you need to rely on it totally. No piece of equipment is going to come off the truck and work perfectly from the first minute. You'll also want some time to work with the equipment to test your original assumptions. Perhaps once a piece of equipment is running, you realize you need another 10 feet (3 m) of conveyor belt. Maybe you need to change insert sizes. When the manufacturer delivers the machine, be sure to have the exact products you will be running with the machine available for setup. For example, with a new transplanter make sure you run 75 or 100 plug trays in the various sizes you plan to use before the manufacturer leaves.

Making the purchase. Is leasing right for you? Just about every company offers the option to lease many pieces of equipment. Some growers prefer to lease, setting up payment terms more favorable to their cash flow cycles. However, if you're not able to get regular financing or to pay cash, you likely will not qualify to lease the equipment, either. Ask about all your financing options.

Be leery of pay-later offers. Some growers take equipment on a "try now and pay later" basis. There is a cost associated with using a piece of equipment that can't be integrated into your production line or greenhouse system: you will be tied up using that system until it is taken away. Again, rather than getting caught up by the romance of mechanization and a deal, think about what you need to do the job.

Check references. Before you write any check, call growers who have purchased the same model you are considering. Ask them not only about the machine—how long they've had it and how it is working—but their opinions of the company and of manufacturer support. Also, visit one or two growers with machines like the one you are considering—especially if you're buying an expensive machine.

Headhouse musts

Here are more helpful tips from Flier USA's Robert Lando.

The basic physical plant. Allow plenty of space. A production line can be 100 feet (30 m) long by the time conveyors going in and conveyors going out are put in front of pieces of equipment; buffer conveyors are added; patching, tagging, watering machines are added; and so on. Allow plenty of space for the line itself and for carts or trays to take planted pots or flats.

Don't forget a good roof. Machines don't like water from rain. While temperature is not a problem for the machines, you'll need to ensure worker comfort.

Put in a drain in your headhouse. Watering tunnels are standard on a seeding line and on many transplanting and pot-filling lines. You'll need to be able to handle the water runoff.

Install three-phase electricity. You can use single-phase power, but you'll need phase converters. It's best to go with three-phase, 230V as standard, even if you have to do some rewiring.

Don't skimp on the air compressor. Today's production equipment is pneumatic and has to have high-quality, dry air. It's better to oversize your compressor than to undersize it. You'll need a minimum of 15 cubic feet per minute (0.4 m³) at 100 psi (7kg/cm²). You may spend $1,500 to $2,500 for a compressor large enough to supply air to a complete transplanting line, for example.

Valuable accessories. Breathable covers are important. When the machine is not in use, cover it with a breathable material to avoid condensation and humidity buildup.

Install a phone in the headhouse. It may only be a jack for a portable, but having a telephone near the equipment is very helpful so you can troubleshoot problems or make adjustments with the manufacturer on the line.

Keep an emergency tool kit in the headhouse. Ask the manufacturer to provide you with a stash of parts that may wear out, bend, or break during normal operation, such as valves, sensors, and cylinders. The manufacturer can provide you with a list of recommended spare parts.

Seeders. Make sure your seeding room is dust free and with low or no humidity and no wind. Also, make sure the seeder has some sort of counting mechanism so you'll know how many trays of Accent Pink or Super Elfin White have been sown without stopping the seeder and counting them manually.

RETROFITS: EXPECTATION VS. REALITY

There's more to calculating the costs and benefits of a big equipment investment like Dutch trays than just figuring the up-front cash outlay. You've got to consider factors like how your investment will affect crop quality, labor savings, space use, and your ability to serve your customer.

Jim Leider, Leider Horticultural Companies, Buffalo Grove, Illinois, began a multi-phase retrofit of Dutch trays and automated plant-handling equipment at his 8-acre (3 ha) pot plant range 10 years ago. Jim's firsthand, paper-to-steel experience offers a unique before-and-after look at how retrofit expectations match reality.

REPLACE OR RETROFIT?

"I think we'd have a hard time existing today if we hadn't made this decision," Jim says of the retrofit project, which began in 1986 and is still being fine-tuned today. Leider serves Midwest grocery chains with a wide range of potted plants, especially bulbs, producing more than a million pots of crops like daffodils, tulips, and hyacinths yearly.

Most of Leider is connected by Hawe Dutch trays, and automatic handling equipment puts pots on trays and spaces them out later. Some trays use subirrigation; others have mesh bottoms and are hand- or boom-watered. Trays have kept Leider competitive in the tough, thin-margined pot plant market. "Nobody who grows this much variety is more efficient than we are," Jim says confidently. "Our system allows us to deliver quality products in the right stage and in the right quality."

Begun in the mid-1960s and expanded over a 15-year period, Leider's steel and aluminum greenhouses were in excellent condition, but after 20 years many systems needed upgrading. Plus, like most ranges built over time, the layout wasn't as efficient as the business now required. Jim had the choice either of rebuilding or of retrofitting with new systems, such as heat-retention curtains and runoff control. The high cost of replacing the structures themselves, plus their relatively good condition, helped him decide to upgrade the existing facility.

At the same time, Jim wanted to replace his existing bench system with something new: Dutch trays. Several factors led to his decision:

- ◆ Aging benches. Most benches, a collection of 1960s and 1970s wood, steel, aluminum, and wire, needed replacing. Rather than stay with inefficient rigid benches, Jim felt Dutch trays would bring his business into the twenty-first century.
- ◆ Labor costs. Labor costs were skyrocketing, with no end in sight. Leider had 85 to 90 employees on the payroll. Jim calculated that Dutch trays could save him 20% in labor while cutting out much of the "grunt work," such as loading carts and carrying trays down aisles.

- ◆ Space use. Jim was getting about 66% space use from his facilities with the existing benches. Dutch trays could bring that up to a more profitable 85 to 90%.
- ◆ Reentry restrictions and runoff control. While worker protection rules hadn't been written yet in 1986, Jim thought the speed and flexibility of a tray system might help keep employees out of the greenhouses and give him more flexibility when spraying. And the combination of subirrigation benches and concrete floors would catch all runoff, helping Leider meet any future environmental regulations.

WHAT LEIDER ACHIEVED

Dutch trays can't be beat for growing on, Jim says, at least with short-term crops that require frequent handling. Two of Leider's houses still have rigid metal benches and are used primarily for long-term, 6-inch crops (15 cm), such as mums, kalanchoes, and begonias. One house has rails down the main aisle to let workers push plants in on trays; pots have to be carried down the benches for final spacing.

The other house doesn't connect to the main facility, so plants are brought to it on carts. Jim uses this house for poinsettia stock and other crops he doesn't mind keeping isolated. While he says he'd love to retrofit both houses, he can't justify the cost.

Jim's labor-savings expectations were based on his own experience and results he saw Dutch growers getting with trays. He hoped for 20% savings; he says he's probably realized only 10 to 15%, for two main reasons. Labor costs, while high, haven't increased as much as he'd anticipated. And because Leider grows such a wide variety of crops, he can't use trays as efficiently as Dutch growers who grow only one or two varieties.

While not as efficient as if the greenhouses and tray system had been built together from the ground up, the facility has given Jim the most growing area possible. This may be the toughest part of a retrofit, as greenhouses are often of different dimensions, making tray size and flow a compromise between different widths of houses. While some of Leider's houses with 41-foot (12.5-m) bays work out perfectly with the 66- by 98-inch (1.7- by 2.5-m) trays he chose, another house with 17$^1/_2$-foot (210-inch, 5.3-m) bays left only inches between trays. Although that space is used to the utmost, the lack of aisle space reduces labor efficiency.

Concrete floors that capture runoff have been a major benefit, protecting Leider land value and preventing environmental problems down the road. Luckily, Jim didn't depend on ebb-and-flood trays alone to capture runoff; some crops don't grow as well on subirrigation and have to be grown on open-bottomed trays. All the water and runoff is recirculated through a pond.

The system also helps meet Worker Protection Standard reentry restrictions. Product can be moved quickly from greenhouse to shipping area, minimizing the time required to pull product for the next day's orders. Plus, packing and shipping are contained in a central warehouse, keeping employees out of greenhouses.

Unexpected benefits

As the retrofit progressed, more benefits that weren't initially considered became evident. As Leider increased the number and quantity of crops grown to keep up with demand, getting crops, especially bulbs, into and out of greenhouses on a timely basis had been getting more difficult. Even one day too long on the bench could mean a tulip that's too far in bloom to ship as top quality.

The tray system now lets Leider keep production precisely on schedule, moving plants into the cooler and greenhouse at exactly the right times even during peak production periods. One major benefit came through the installation of a massive, 200,000-cubic foot (5,600-m³) cooler. It's fitted with a nine-level rack and is capable of cooling a quarter of a million 6-inch (15-cm) pots at a time. Bulb crops are quickly and automatically moved in and out for cooling or holding.

Jim stresses that trays—especially ebb-and-flood trays—won't automatically improve quality. A few crops, such as calla lilies, haven't adapted to subirrigation, and it took him a while to switch from watering and fertilizing at the levels he was used to with drip irrigation.

"We've been doing things a lot better than we ever did," Jim says. "We don't make nearly as many mistakes." While total dump rate is 5%, they're aiming for 4% in 1996, and Jim thinks they'll make it.

Leider went to split shifts in 1994. Employees take orders in the morning and pull product in the afternoon. Trays can quickly be moved from greenhouses into the centralized shipping area, where evening crews can easily select, decorate, and pack plants. Grocery chains are competitive and demanding. Over time, buyers have asked for more and better service in the form of UPC codes, pot covers, tags, picks, and other decorations. Now it's easier for Leider to offer more variety per case, along with customized add-ons. "It takes a lot of time and organization to do that," Jim says. "It would be a nightmare with our old system." Finished plants move by conveyors past automated labeling equipment, which applies UPC labels, then move on to be boxed.

The downside

Improvements don't come without a price. "It cost a heck of a lot more than we thought it was going to, primarily because of the retrofit," Jim admits. "To this day we're continuing to do refinements . . . to invest money in the system to make it better and more efficient." Total investment to date is close to $2 million. Jim says that while he knew what the tray system would cost, the retrofit process—moving heating pipes and electrical boxes, pouring concrete, cutting holes in walls—wound up adding at least 20% to the total project cost.

Another cost was lost sales. "You're going to lose money the year you do it," Jim warns. "There's too much disruption to your business."

Jim initially expected payback in five to seven years, based mainly on labor and space use. In reality, "it's hard to measure what the payback really is." Jim has plenty of numbers but says, "I don't know if I believe them."

"We had to do something," Jim concludes. "We could have spent a lot less money and put conventional benches in, but we'd be stuck with a relatively inefficient old place, with 85 or 90 employees. But we wouldn't have the capital costs. So it's a trade-off. The problem is, I don't know how we would operate [without trays]. We wouldn't get the work done on a timely basis without throwing a lot more labor into it."

JIM'S SUGGESTIONS

Considering a similar project? Jim offers the following advice.

- Figure that it will cost you 20% more than you expect and that it will take longer. Be realistic about what you're going to save.
- Expect to lose money the year you do it. "If you can shut down your business without losing your customers, that's great," he says. Otherwise, expect to ruin crops, spend more on labor and make some customers mad.
- "If you have an old facility, get out of it." While his greenhouses were in good condition, he says their layout didn't give him as much efficiency as he'd like. If he could do it over again, he'd consider moving to a new location and building an all-new facility laid out for maximum efficiency. Ten years ago, however, that seemed like too big a commitment.
- Think about the future in terms of environmental impact. Use subirrigation and capture all runoff.
- Be as energy efficient as possible. Add heat curtains and modernize your heating equipment.
- Consider repair costs. Be prepared for mechanical, structural, and electrical repairs and worn-out parts. Be sure you can operate any systems manually in case of electrical failure. And don't overautomate (Jim bought a $12,000 bulb counter that wouldn't do what he needed).
- Unless you're starting from bare ground, don't go overboard. Start with the trays first, then add automation to it. But don't start too slowly—Leider's labor increased after the first phase of tray installation because the plan didn't go far enough.
- Don't install a highly automated bench system if all your other systems can't keep up. Pay special attention to: production planning, inventory and order

entry—use computer software specifically designed for greenhouses; soil mixing and potting—you can cut a lot of labor through improved material handling; packing and shipping—Jim calculates that while 25% of labor is in planting and 20% is in growing, the rest—55%—is in packing and shipping. Make sure you can ship product out as easily as you can get it in.

"There's very little margin for error left in the business," Jim says. Because of that, he predicts greenhouses of the future are going to be much more like factories: "Everyone will have to be much more efficient to survive."

COMPANY SOURCES

Berry Seeder
1231 Salem Church Road
Elizabeth City, NC 27909
Phone 800/327-3239
Fax 919/330-2227

Blackmore Company
10800 Blackmore Avenue
Belleville, MI 47111
Phone 313/483-8661
Fax 313/483-5454

Bouldin & Lawson
P.O. Box 7177
McMinnville, TN 37110-7177
Phone 615/668-4090
Fax 615/668-3209

Cherry Creek Systems
(formerly East Coast Designs)
11901 E. Palmer Divide Road
Larksburg, CO 80118
Phone 303/660-1196
Fax 303/660-1338

Flier
Zuideinde 120
Postbus 200
2990 AE Barendrecht
The Netherlands
Phone +31-180-615055
Fax +31-180-618083

Flier USA
300 Artino Street
Oberlin, OH 44074
Phone 216/774-2981
Fax 216/775-2104

FW Systems
Frank van Dijk
Leeuwenhoekweg 42
2661 CZ Bergschenhoek
The Netherlands
Phone +31-10-5215639
Fax +31-10-5212183

Gleason
(distributed by Measured Marketing Inc.)
395 N. Schuyler Avenue
Kankakee, IL 60901
Phone 815/939-9746
Fax 815/939-9751

Growing Systems
2950 N. Weil Street
Milwaukee, WI 53212
Phone 414/263-3131
Fax 414/263-2454

Hamilton
(distributed in U.S. by BFG)
14500 Kinsman Road
Burton, OH 44021
Phone 216/834-1883
Fax 216/834-1885

Hamilton
Nethercliff, Green Lane
Littlewick Green, Maidenhead
Berks SL6 3RH
England, United Kingdom
Phone +44-1628-826747
Fax +44-1628-822284

Hawe Metroplanter
(Metrolina Greenhouse)
16400 Huntersville Rd.
Huntersville, NC 28078
Phone 704/875-1371
Fax 704/875-6741

Hawe Systems Europe B.V.
Oosteindsepad 8
2661 EP Bergschenhoek
The Netherlands
Phone +31-10-5212755
Fax +31-10-5217616

Intransit
Lierweg 20
Postbus 87
2678 ZH De Lier
The Netherlands
Phone +31-174-514141
Fax +31-174-517877

Javo
Westeinde 4
Postbus 21
2210 AA Noordwijkerhout
The Netherlands
Phone +31-252-375441
Fax +31-252-377423

Javo USA
1900 Albritton Drive
Suite G&H
Kennesaw, GA 30152
Phone 770/428-4491
Fax 770/424-6635

Mayer GmbH
Postrasse 30
89522 Heidenheim
Germany
Phone +49-7321-959437
Fax +49-7321-959497

McConkey
P.O. Box 1690
Sumner, WA 98390
Phone 206/863-8111
Fax 206/863-5833

Rapid Automative Systems
2445 Port Sheldon Road
Jenison, MI 49428
Phone 616/662-0954
Fax 616/662-1007

Seed E-Z Seeder
E11290 Country Road PF
Prairie du Sac, WI 53528
Phone 608/643-4122
Fax 608/643-4289

Visser International
Beneden Havendijk 115A
Postbus 5103
's-Gravendeel, the Netherlands
Phone +31-1853-9800
Fax +31-1853-4649

Visser, U.S.
(Hove International Inc.)
44 Trowbridge Lake
Atlanta, GA 30328
Phone 404/804-0696
Fax 404/913-0026

CHAPTER 7

IRRIGATION FOR POT AND BEDDING PLANTS

by Vic Ball

Hand-held hoses have been the way to water most containerized ornamental crops since commercialized growing began. However, hand-watering is an expensive luxury!

More recently, spaghetti tubes (Chapin systems are the main kinds) have taken over irrigation of pot plants. Irrigation today can be mainly mechanical—without penalizing quality. Most growers, especially the larger ones, are moving this way for pot plants, and more and more for bedding plants, too.

At the same time, environmental pressures are pushing the grower to reduce the amount of runoff from greenhouses. The old philosophy of watering crops so "thoroughly" that ample runoff occurs is unnecessary and wasteful. Work at Michigan State University has clearly shown that most greenhouse crops can be grown with a fraction of the amount of water and fertilizer normally used. Many growers have already begun to adapt low-fertilizer and low-watering practices with great success. Often, automatic irrigation speeds the way.

AUTOMATIC WATERING CAN GROW QUALITY

Some growers say that you can't grow quality bedding plants or pot plants without hand-watering. The majority of bedding plants are more or less hand-watered, especially early in the crop. More and more, though, pot plant irrigation is fully automated. Many growers are in fact growing excellent quality without hose watering, even for bedding plants.

Pot plants are automatically irrigated with one of five techniques: ebb-and-flood benches or floors, trough irrigation, and capillary mats deliver water from the bottom of the plant; drip irrigation waters from above; and spaghetti tubes (Chapin systems) can pinpoint delivery to any part of the plant.

Bedding plants are primarily irrigated by hand-watering, but there are three important automatic ways: boom irrigation, stationary overhead sprinklers, and ebb-and-flood floors.

EBB AND FLOOD

An ebb-and-flood system is expensive, but what would it be worth to you to eliminate hand-watering and still produce top-quality plants? Hundreds of growers in Europe, the United States, and Canada have gone to ebb and flood, producing everything from plugs to 15-inch (38-cm) pots. Their crops require far fewer worker hours, suffer from fewer disease problems, and mature faster than crops grown with most other irrigation systems.

Ebb and flood is a form of subirrigation. The plants are sitting in special watertight bench (fig. 1) or floor sections (fig. 2), which are then flooded. Water is soaked up into the root masses through pot drain holes. After a few minutes, the water is drained away and piped into another bench or floor section. The foliage is kept dry, and there is no wasting of water and nutrients.

Any pot at any spacing can be irrigated with ebb and flood. Most growers who use aluminum Dutch trays use ebb and flood. A relatively new application is the ebb-and-flood floor, which we

FIG. 1 Henry Schneider, G&E Greenhouses, Elburn, Illinois, shows a few of his ebb-and-flood growing tables. The black 2-inch pipes feed water and fertilizer into the trays.

FIG. 2 Ebb-and-flood floors are in use at Plants, Inc., Huntsville, Texas.

will be discussing in detail. Many of the basic mechanisms and principles of floor flooding apply to benches, too, however.

Ebb-and-flood floors. Already widely used in the Netherlands and other parts of Europe, ebb-and-flood floors are gaining adherents here in the U.S. and Canada, reports Ratus Fischer, an ebb-and-flood floor engineer with Greenlink. He had more than 20 clients in 1996, almost half of whom are using the floors for bedding plants.

A flood floor, made of concrete with a slight pitch for drainage, can be built in sections, usually corresponding to greenhouse bay size, such as 21 or 24 by 100 feet (6 or 7 by 30 m). Ratus recommends zones anywhere from 1,000 to 3,000 square feet (93 to 279 m²). If a section were much larger, it would take too long to fill and drain.

To dry the floors quickly, you use floor heating. PVC pipes are okay, but they are not guaranteed, and they must be protected from excessively high temperatures. Polybutylene or polyethylene floor heating pipes are preferred. The floor heat will hurry the crop, but too much heat will soften the plants. To get around this problem, you need separate controls for floor heat and air heat.

To fill a section with 1 to 1¹/₂ inches (2.5 to 4 cm) of water, you'll need 2,000 to 3,000 gallons (7,570 to 11,355 l) in a storage reservoir. The section is typically filled through a 6-inch (15-cm) feed line, taking about four minutes, although pots can take up practically all the water they need in the first one or two minutes. The floor then immediately begins draining, now through an 8-inch (20-cm) drain line, which takes another four minutes—meaning a section's entire flood-and-ebb sequence takes just eight minutes.

The pots will have taken up less than 10% of the water, so the 90% is returned to the reservoir, where the balance is made up. The recirculated water can be adjusted before moving on to the next floor section. Jim Gapinski of Heartland Growers, near Indianapolis at Westfield, has a system that both filters the returning water and fine-tunes the nutrient and pH levels.

The water immediately moves to the next bay or other floor section (or bench). Jim points out that since the irrigation is automatic and rapid, one worker can oversee the irrigation of his 3¹/₂ acres of greenhouse floor in just three hours—meanwhile monitoring the crops for other concerns.

With recirculation there is no problem with water runoff or other wastage. Jim says, "We'd rather have a reused system than be told by the government [how] to handle runoff." This also means that no fertilizer is wasted, unlike with many other irrigation systems. This is a major economy!

Plants watered with ebb and flood tend to be healthier. The foliage stays dry, and the very short watering time reduces the already small risk of disease transfer in the recirculating system. With the floor heat hurrying the plants along, Jim expects that pot crops, especially, mature several days to a week earlier. Of course, with the faster crop times, Ratus notes, the grower must adjust other production factors accordingly.

Opinions vary on the suitability of ebb-and-flood floors for bedding plants and plug trays. Results with these shallow containers may depend on the quality and pitch of the floor and various growing techniques. Larry Boven of Boven's Quality Plants of Kalamazoo, Michigan, does a lot of bedding and pots. He has found that ebb and flood doesn't work with very small pots, 2 inches (5 cm) and smaller: they get too wet. His daughter, Laurie, reports that bedding plants "tend to drain slowly. It just doesn't work well for us. We use other systems for bedding."

Other growers see better results. Jim Gapinski operates 17 acres (7 ha) of greenhouse, of which 3¹/₂ acres are now in ebb and flood and 3¹/₂ more acres are slated for conversion. In the spring his whole operation is devoted to bedding. "We use our ebb-and-flood system except for the toning and finishing of the crop. Also, we occasionally hand-water to hold the crop for a few days of bad weather. But 60 to 70% of our bedding is grown with ebb and flood."

Everyone agrees, however, that an ebb-and-flood watering system works very well with larger pots. Jim's main pot crops are poinsettias, and he says his system will easily accommodate any pot size from 2¹/₂ to 15 inches (6 to 38 cm). Larry Boven grows poinsettias, geraniums, lilies, and more in his 55,000 square feet (more than 5,000 m²) of ebb-and-flood area, generally in 4-inch (10-cm) or larger pots.

Costs and savings. Of course, all of this is not cheap. Count on a minimum of $5.50 per square foot of ebb-and-flood bench, $2.50 and up for a floor ($59 and $27, respectively, per m²). Jim Gapinski figures his whole system—including concrete, tanks, plumbing, and floor heating—cost about $2.50 per square foot. Larry Boven's 10 acres of floor-heated growing area cost "less than $4 per square foot."

Nevertheless, the savings can be considerable. Besides water, you save fertilizer, both from recirculating unused water and from not leaching out nutrients already in the soil. You also save crop time and lose less plant material to disease.

Automatic irrigation means far lower labor costs. However, it does take time to learn how to control growth and adjust the operation to the warm plant roots. Also, equipment for automated handling of pots and trays on flood floors is not readily available yet, although Flier and Visser are developing just such machines. With an ebb-and-flood system installed with Dutch trays, 90% of the manual movement is eliminated.

TROUGH IRRIGATION OF POT PLANTS

With trough irrigation, pots are typically lined up in metal troughs that are 5 or 6 inches (13 or 15 m) or more wide and 1 inch (2.5 cm) or less deep. The troughs are set out across the bench and pitched slightly to one side. Water is fed in at the high end, often through a spaghetti tube, and collected in a gutter at the other end. Pots take up the water from the bottom, as with ebb and flood.

Troughs are less expensive than ebb-and-flood benches. Since the trough is in essence the bench, that is major savings. Troughs allow for excellent airflow between pots and through the foliage canopy, which ebb and flood does not allow. They can also be linked to a recirculating water system.

The main downside to troughs is that there is less flexibility in pot spacing. You can slide troughs closer together during early crop stages and farther apart at finishing, but unless you have sets of different widths of troughs, it is hard to accommodate different pot sizes.

CAPILLARY MATS

Capillary mats are especially popular with growers producing very small (2-inch, 5-cm) pots. Porous mats about one-eighth inch (3 mm) thick are spread out on top of the bench. Plants are set onto the mat, and irrigation is accomplished simply by watering the mat with a hose, spaghetti tubes, or overhead nozzles. Water soaks up into the pot by capillary action.

Mats are relatively low cost and easy to install. You don't have to waterproof benches. As with ebb and flood, there's complete flexibility on spacing and which pot sizes to use. Foliage stays dry, and there's practically no runoff.

There are drawbacks. First off, you must take extra care to establish capillary flow from the mat to the pot media. This may require wetting agents in your media. Also, mats tend to have algae problems—they can really accumulate scum. Agribrom injected at low rates into the irrigation water can help.

Saucers are used in a similar way—water from overhead irrigation drains into the saucer and is soaked up by the plant (fig. 3).

DRIP IRRIGATION

Drip irrigation is, in a way, an adaptation of the spaghetti tube system, except that there is a slow, steady drip into the pot rather than an

FIG. 3 Saucers for these poinsettias are placed in catch water from overhead irrigation at Fernlea, Delhi, Ontario, Canada. As water falls through and off the plant canopy, it drains into the center of the saucer and is soaked up by the plant. Some growers irrigate their poinsettias overhead throughout the entire crop cycle using saucers.

open line of flowing water. Drip irrigation tends to use less water and fertilizer than spaghetti irrigation. A system generally delivers one-half gallon (1.9 l) of water per hour per nozzle. With careful management, you won't have water and fertilizer running out the bottom of the pot! A system can be turned on for perhaps 10 or 15 minutes out of every two hours—automatically by computer.

Plant leaves and stems are kept dry, reducing aerial disease problems. Many hanging baskets are watered by drip systems, especially baskets grown overhead because the systems practically eliminate runoff from dripping down on crops below (fig. 4).

SPAGHETTI TUBES FOR POT PLANTS

Spaghetti tubes are widely used by pot plant growers all over the world. Chapin Watermatics Inc., Watertown, New York, pioneered this system, so most people just call them Chapin systems, although there are other manufacturers now. The plastic tubes, about one-eighth inch (3 mm) or so in diameter, are installed so that each pot on the bench is irrigated by its own tube (fig. 5).

FIG. 4   The Echo hanging basket system makes watering baskets a snap. Baskets rotate clockwise through a watering station where they are irrigated.

Typically, a water supply line runs down the length of the bench. Every spaghetti tube is attached to the supply line and runs to a single pot. The emitter end is generally weighted to keep it in place in the pot. Sometimes it is hooked and anchored to the top of the pot.

Norm White, based in Chesapeake, Virginia, and one of the country's finest growers, installs a cola bottle at the head of each bench. He puts a spaghetti tube into the bottle, just like the tube that goes into each pot on the bench. He waters the bench until the cola bottle is filled to a certain mark, so he knows each pot has received the right amount.

Question: Does the last pot on a 150-foot (46-m) bench get as much water as the first? Sometimes you see growers pipe water both ways from the bench's center rather than from one end.

Costs run about $.35 to $.50 per pot installed, but they can be used for several crops a year and for many years. Spaghetti tubes work best on 5-inch (13-cm) and larger pots. Costs are generally too high for smaller pots. Sometimes growers space pots pot tight just after potting and hand-water, waiting to install tubes until after the pots are spaced.

IRRIGATION BOOMS

Overhead booms have been with us for a long time, but during the 1980s a new batch arrived—computer controlled and better than ever. Clearly, the leader in this movement was ITS, Inc. Originally developed by Gary Lucas, ITS technology has been bought out by the McConkey Co., Sumner, Washington; phone 800/426-8124.

Booms irrigate from overhead (fig. 6), traveling slowly on pipes over the crop and spraying water. They're mostly used for plug crops and bedding plants. Some booms can move from bay to bay, while others stay in one bay only. Booms can also irrigate outdoor-grown crops.

The great thing about booms is that they can be programmed to water only those parts of the bench that need it: say the first 10 feet (3 m), then skip the next 20 feet, and water the final 30 feet. They can also be programmed to speed up or slow down to adjust the volume of water being applied to different parts of the crop.

Some bedding plant growers swear by booms, claiming that watering is even better than if done by hand. Generally, though, even when growers have booms, they also supplement with hand-watering. The best part is it takes a lot fewer people.

One of the biggest drawbacks to booms is that foliage gets wet. Also, if you chose booms, don't skimp on nozzles. Nozzle design is the key to watering uniformity!

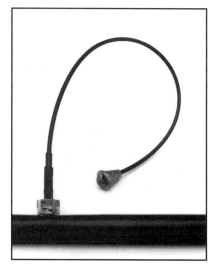

FIG. 5 Millions of pot plants in the United States are watered with drip tubes, like this one from Chapin Watermatics. An irrigation pipe runs down the center of the growing bench and feeds tubes that are placed into each pot. Note the metal weight on the end of the tube—it makes sure the tube stays in the pot.

FIG. 6 Watering booms are standard equipment for many large plug producers, such as C. Raker & Sons, Litchfield, Michigan. Many growers are installing booms to water bedding plant crops.

ITS offers several boom models. All can be moved from one greenhouse to the next. Models range in price from $1,695 for a simple boom up to $4,390 for a fully loaded version. Other boom suppliers include Growing Systems, Milwaukee, Wisconsin, and Andpro, Waterford, Ontario.

STATIONARY OVERHEAD SPRINKLERS

Fixed sprinkler nozzles are installed above the crop for irrigation in this kind of system. Many growers use overhead sprinklers for bedding plants. Most growers have them installed for emergency irrigation situations where hand-watering just isn't fast enough—those hot April and May days when there is never enough time or labor to get it all done. One major poinsettia grower even uses overhead sprinklers—without disease problems! That's done by carefully adapting the system to the threat of *Botrytis:* watering only in the morning, ventilating and heating after watering, and using fungicides.

Overhead sprinklers are low cost and can be computer controlled. One of the major disadvantages is that foliage gets wet, and they tend to waste water and fertilizer. Also, when they are not precisely engineered, they apply water unevenly. By design each nozzle delivers water in a circular pattern. To cover an entire area, the circles must overlap. Part of the crop will always be overwatered.

Mist benches are used similarly to germinate seed or to root cuttings (fig. 7).

FIG. 7   Many growers use mist benches to germinate seed or to root cuttings (note how the plastic can be pulled up or down).

SECTION 2

BASICS FOR GROWING

CHAPTER 8

MEDIA MIXES

by Vic Ball

Using the correct media mix is just as important as choosing the right variety or proper control of temperature. The media mix serves several important functions. Its physical structure should provide for air and water exchange as well as adequately support the plant. Chemically, it serves as a nutrient reservoir. Also, for the convenience of the grower, a mix should be easily reproducible and readily available.

If a mix is too tight, with too little porosity, it will hold too much water in its pore spaces, not allowing sufficient air for developing a healthy, well-branched root system (fig. 1). This will slow down growth, maybe causing a crop to miss its scheduled flower date. On the other hand, if the mix is too open, it will require too much water too often, increasing production labor costs, not to mention the quantity of water and nutrients needed to produce the crop.

While the mix serves as a reservoir for nutrient exchange, it also is critical for buffering the pH and fertilizer salts. With a media mix such as one part soil, one part

FIG. 1   A good root system is revealed by the root ball on the right. Be sure to begin with good media.

peat, and one part perlite, the optimum soil pH should be approximately 6.2; whereas with an organic mixture of bark, peat, perlite, and vermiculite, the optimum pH is

5.5 to 6.0. Also, the ability of the media to hold and release nutrients will help maximize the efficiency of the fertilizer, whether it's slow-release or water soluble.

Keep in mind that a plant in a bedding plant container requires different support than does a tall pot plant, such as a geranium. If a plant is not properly supported, then extra care will have to be taken to stake or tie it up. This means extra labor, increasing production costs. An inadequately supported plant with poor media structure or too lightweight media can also have reduced shelf life. For instance, many times plants are jostled during shipping. Plants in a lightweight mix will not be adequately supported and will be floppy and loose when they reach their destination—not of excellent quality.

You should be able to consistently reproduce a particular mix time after time, season after season. Just imagine growing on a year-round pot mum schedule and changing your mix every week. Trying to adapt to each drainage or fertilizer requirement would definitely increase your frustrations in handling that crop.

Historically, ready availability has not been a mix consideration, but today, because we're dealing with larger production areas, any media mix or its components should be readily available when you need it, be it a commercial mix from a factory or local components. For instance, I have seen seasonal outages of peat moss due to poor harvest conditions at the peat bogs. This could mean trouble if you make your own mix.

Another question to consider is where to store your mix. Is it properly stored and readily available? For instance, if you store it outside and you encounter an extremely wet and rainy period, can you get out to bring your mix into the greenhouse? In the middle of winter, is your media mix pile frozen to the point that it is unavailable? Whether it's your own mix or a commercial mix, you want to consider the availability of the material when you need it.

THE CASE FOR SPECIALIST MIXES

Commercially prepared mixes are currently receiving much attention. The mix ingredients range from straight peat moss to blends using bark, perlite, vermiculite, and peat. It is important to understand that you can grow in almost any media. But if you compare the media functions to your own conditions, there will be one mix that best fits your needs.

Several factors have created this grower shift away from making their own media mixes. Some of the benefits of commercially prepared mixes are their consistency, cost, and convenience. They also get around any problems with local scarcity of materials.

A large firm making media mixes on a year-round basis can specialize and consistently make the same mix to the same standards. Therefore, you have peace of mind when you place an order that the mix is the same as the previous batch you used.

As communities have developed around greenhouses and major metropolitan areas, good topsoil has become either unavailable or very expensive to prepare and store on site. Weed-killer contamination is a very real problem in field soils today. One clear trend of the 1990s is that outdoor field soil is rapidly vanishing as a component of growing mixes.

Other ingredients can also become scarce. Perhaps a local supplier of bark does not have his or her normal quantity. The supplier is dependent on the housing industry, which has not created a great demand for lumber at this time. Slower mills mean less bark as a by-product.

Large specialist firms making mixes are able to overcome commodity shortages and maintain larger supplies because of their economy of scale. Therefore, they can ensure a consistent supply to the marketplace.

Lately, more growers have recognized that considerable man-hours go into soil preparation, competing with the time that could be spent in production. It not only takes time to put the mixes together, but it also requires management time and organizing, planning, and resource commitment, such as inventory dollars for raw materials. With any production process, you want to tie up as few dollars as necessary in materials or labor. Buying mixes frees up this time and resource dollars so you have to inventory only finished, ready-to-use mix, and you can schedule it to arrive when you need it.

Using a commercially prepared mix allows you to concentrate your efforts on other areas of the growing operation. Let's face it—there are several things to consider in making your own mix, such as particle size, pH adjustments, soluble salt levels, and sterility. It is much more convenient to depend on an outside supplier who specializes in mixes and manufacturing than to dilute your own time and resources.

COST ESTIMATES

To give you an idea of what it costs to make mixes, Paul Nelson at North Carolina State University figured it out using a fixed-cost determination method to account for depreciation of buildings and equipment, repairs, taxes, and so on, as well as management size. He showed that if you consider the same factors that most commercial manufacturers use, the mix cost range per cubic yard was anywhere from $48 to $75 ($62 to $97/m^3), depending on your operation size. First of all, when mixing your own, you can count on about 10% shrinkage when you mix the components. Fixed costs don't drop significantly even if you're using 500 cubic yards (380 m^3), he points out. "It's why most growers buy their mixes in."

Commercial mixes are available in the marketplace from approximately $40 per cubic yard ($52/m^3), depending on where you are located, up to $70 ($91/m^3), depending on the type and quality of the mix. The fact is, growing today is done mostly in artificial mixes. All but the very large growers use a specialist-produced mix.

CHAPTER 9

SOIL TREATMENT FOR DISEASE CONTROL

by Dr. Ken Horst
Cornell University
Ithaca, New York

and Dr. Jim Locke
Floral and Nursery Plants Research Unit
United States National Arboretum
Beltsville, Maryland

Common greenhouse crop pathogens range from ones that are relatively easy to control to those that are more difficult to control. Many root pathogens fall into this latter category, especially where beds or benches are replanted with the same crop. These pathogens causing damping-off, root rot, and wilt survive in the growing media and require treatment procedures to eliminate them or at least greatly reduce their numbers.

This chapter considers the options available to treat soil and growing media to reduce unwanted pests and pathogens. Approaches can be divided into three categories: physical (application of heat), chemical, and biological (microbial agents, organic residues, or nutritional management).

STEAMING BULK SOIL

Many pot plants and bedding crops are grown today in commercially produced soilless mixes. It is generally recommended that these mixes not be steam sterilized because of detrimental effects on resident microbes. However, when field soils were commonly used in greenhouse production, sterilizing soil (usually with steam) was a requirement. Steam sterilization of greenhouse soils has been considered the best and most effective method of eliminating soilborne pathogens. In addition, weeds and other pests are killed by steam. Among other advantages of steam sterilization, there is no phytotoxicity beyond the immediate treatment area, and there are no harmful effects to plants in the same greenhouse. Little aeration time is required; steamed soil may be planted as soon as it is cool. Also, it is adaptable to many situations.

There are various methods for steaming soil (fig. 1). Boilers that heat greenhouses can be adapted to supply steam for sterilizing benches or soil bins. Boiler heating efficiency must be considered. A new boiler could have an 80% efficiency rating, whereas a 25-year-old boiler may have only a 60% rating. The heat exchange efficiency of a particular soil can have a significant effect on the number of BTUs required for steam sterilization. Heat exchange efficiency is sometimes difficult to determine. The properties that affect heat movement in soils include soil type, texture or looseness, and moisture content. As a conservative estimate, 2,500 BTUs would be required to raise a cubic foot (0.028 m³) of loose, moist soil with a heat exchange efficiency of 60% from 62 to 212F (17 to 100C).

Another option is a commercially produced soil-steaming wagon. Bouldin & Lawson (McMinville, Tennessee, 615/668-4090) builds these on a custom basis. Portable steam generators are also available. The Steam-Flo Generator (Sioux Steam Cleaner Corp., Beresford, South Dakota, 605/763-2776) is fully equipped with oil burner, complete

FIG. 1   To steam pasteurize ground beds, growers can feed steam from their boilers *(top)*, or they can use field generators to make steam *(bottom)*.

controls, and safety devices. It is on wheels and can be quickly moved into place, and steam is available in 20 to 25 minutes from a cold start. It will sterilize a bench 5 by 50 feet (1.5 by 15 m) in two hours, maintaining 8 to 10 pounds pressure while burning 4 to 4½ gallons (15 to 17 l) No. 1 fuel oil per hour with a 425,000-BTU output. Mayer IMO Soil Steamer System (Gro-May Corp., Houston, Texas, 713/561-7537) is suitable for mobile and stationary use, has standard dual-steam (up to 200C, 392F, at an operational pressure of 1 bar) or warm-water heating, providing an economic and environmentally friendly alternative to soil fumigation. With the ban of methyl bro-

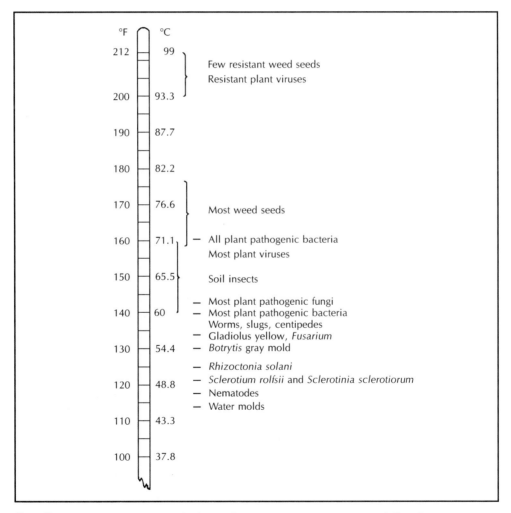

FIG. 2 This thermometer graph shows the temperatures necessary to kill pathogens and other soil-borne organisms harmful to plants. Temperatures shown are for 30-minute exposures under moist conditions.

mide use in the United States in the year 2001 under the Clean Air Act, we may well anticipate increased use of steam generators.

A guide on temperatures required for sterilizing is shown in the thermometer graph (fig. 2, from Baker, K.F., ed. 1957. "The U.C. System for Producing Healthy Container-Grown Plants." California Agriculture Experiment Station and Extension Service Manual 23). This graph indicates temperatures necessary to kill pathogens and other organisms that are harmful to plants. The indicated temperatures are for 30-minute exposures under moist soil conditions. Steam sterilization efficiency is significantly decreased in soils too moist for optimal crop planting.

Table 1 will help the grower to calculate and compare the costs of steam sterilization and other methods of soil treatment. Available BTUs per unit of fuel is a more useful common point of comparison than the price per unit, since the price changes over time and with geographic location. The formulas for cost determination include soil volume, soil heat exchange efficiency, boiler efficiency, units of fuel required, and the BTU constants of the fuel. Knowing the price of the fuel, one can then calculate the cost per measure of soil.

TABLE 1   EQUATIONS FOR CALCULATING FUEL REQUIRED TO STERILIZE SOIL

V = volume of soil in cubic feet = bench length × width × depth

E = energy in BTUs required to raise soil temperature from 62 to 212F (17 to 100C)
     E/ft³ loose, moist soil = 2,500 BTUs

S = soil heat exchange efficiency

B = boiler efficiency

U = BTUs/unit of fuel
     U No. 2 fuel oil = 140,000 BTUs/gal
     U natural gas = 100,000 BTUs/therm
     U coal (anthracite) = 13,000 BTUs/lb

F = units of fuel needed = (V × E) / (S × B × U)

Sample problem: Calculate the amount of fuel needed to sterilize a greenhouse bench with 200 ft³ of soil, using No. 2 fuel oil and a boiler of 80% efficiency.

    F = (200 ft³ × 2,500 BTUs) / (0.6 × 0.8 × 140,000 BTUs per gal)

      = 500,000 BTUs / 67,200 BTUs per gal

      = 7.44 gal fuel oil

FUNGICIDES FOR DISEASE CONTROL

Although relatively few of the fungicides that are commercially available for soil treatment have the broad effectiveness of steam, some growers not equipped for steam treatment may choose fungicides. The most widely used pesticide (fumigant), methyl bromide, will not be discussed here due to the impending ban of its use in the United States in the year 2001 under the Clean Air Act. Although other fumigant-type pesticides (chloropicrin, Basimid, metham sodium) are registered, they generally lack the broad-spectrum effectiveness of methyl bromide or steam.

A number of fungicides are registered and sold for greenhouse soil use (table 2), but results of fungicide treatment may vary, depending on such soil conditions as texture, moisture, and temperature, which require dosage adjustment. Heavy, dry, or cold soils generally require more fungicide to get acceptable results. Also, selecting most soil drench fungicides has to be based on their effectiveness in controlling specific pathogens. Always follow label directions for application rate and frequency.

TABLE 2   GREENHOUSE FUNGICIDES FOR DISEASE CONTROL

Name	Application	Organism controlled
Etridiazole	Soil drench	*Pythium* *Phytophthora*
Etridiazole + thiophanate-methyl	Soil drench or granular	*Fusarium* *Phytophthora* *Pythium* *Rhizoctonia* *Thielaviopsis*
Fosetyl-aluminum	Systemic as foliar or drench	*Pythium* Other water molds
Metalaxyl	Systemic as drench or incorporated	*Pythium* *Phytophthora*
PCNB	Soil incorporation or drench	*Rhizoctonia* *Sclerotinia*
Propamocarb hydrochloride	To seedling or transplant	*Pythium* Other water molds
Thiophanate-methyl	Soil drench or bulb treatment	*Fusarium* *Rhizoctonia* *Thielaviopsis*
Triflumizole	Soil drench	*Rhizoctonia* *Thielaviopsis* *Cylindrocladium*

ALTERNATIVE APPROACHES

Over time greenhouse production practices have changed, as dictated by production costs, consumer trends, and availability of new products, to name a few causes. Currently, concerns over worker protection, reentry intervals, and environmental protection have forced adjustments in approaches to pest and pathogen control. The impending loss of the widely used fumigant methyl bromide has given impetus to researching and developing alternative methods of controlling soilborne pests and pathogens. This section considers nonchemical soil treatment options to reduce losses caused by these root pathogens.

MICROBIAL PESTICIDES

Currently, there are four microbial biopesticide active ingredients for use on ornamentals and turf registered by the U.S. Environmental Protection Agency.

Bio-Trek and RootShield (*Trichoderma harzianum* KRL-AG2) are formulations of a selected strain of a common soil saprophytic fungus. Incorporated into growing media or used as a seed treatment, they protect against *Pythium* spp., *Rhizoctonia solani,* and *Fusarium* spp. on greenhouse ornamentals.

Galltrol A and Norbac 84C (*Agrobacterium radiobacter* K84) are bacterial cultures used as a root dip, primarily on woody ornamentals to prevent crown gall.

Mycostop (*Streptomyces griseovirdis* K61) is a powdered formulation of dried spores and mycelia of a soil actinomycete. It can be used as a seed treatment, transplant or cutting dip, soil spray, or drench or through drip irrigation systems. It is labeled for use against seed rot, root and stem rot, and wilt caused by *Fusarium, Alternaria, Phomopsis, Pythium, Phytophthora,* and *Rhizoctonia.*

SoilGard 12G (*Gliocladium virens* Gl-21) is a granular formulation of a common soil saprophytic fungus. Incorporated as a preplant soil amendment, it protects against the soilborne fungal pathogens *Rhizoctonia solani* spp. and *Pythium,* which cause damping-off and root rots of ornamental and food crop plants grown in greenhouses, nurseries, and interiorscapes.

As with all pesticides, always read and follow label directions. The one great benefit of these products is their safety for applicator, worker, and the environment. Basic protective clothing, handling precautions, and minimal reentry times may apply, as stated on the label. These microbial biopesticides are effective against specific pathogens, and, unlike with general sterilization treatments such as heat or fumigation, other pest problems (weeds, insects, nematodes) must be addressed independently.

SOLARIZATION

Solarization, capturing heat from the sun under plastic films, is a possible approach to treating greenhouse soil beds in situations where these facilities are not used for a month or more during the summer due to cropping schedules or excessive temperatures. At such times greenhouses can be closed to maximize heating, and the soil can be effectively solarized. The principal mode of action of solarization is usually direct thermal inactivation. Effectiveness is dependent on physical factors, such as solar radiation intensity and duration, air temperature, humidity at the soil surface beneath the tarp, properties of the plastics used, and soil properties (color, texture, organic matter content, and soil moisture). Many beneficial microorganisms are rapid recolonizers of solarized soil, so this provides an opportunity afterward to selectively introduce beneficial saprophytic organisms or microbial pesticides. Chemical changes can also occur in the soil, increasing mineral nutrient concentrations, so take care following solarization to avoid adding excessive levels of fertilizer.

NUTRITIONAL MANAGEMENT

Greenhouse culture lets the grower critically manage nutrients in the growing media because of the limited volume to be monitored and the uniform composition of the media compared to field soils. Calcium and nitrate-N raise soil pH, decreasing the activity of pathogens such as *Fusarium oxysporum,* which causes *Fusarium* wilt. However, ammonium-N lowers soil pH and decreases the activity of other pathogens. Nitrogen, phosphorus, potassium, and calcium have been reported to reduce damping-off, but excessive nitrogen can enhance seedling damping-off. Phosphorus and potassium are known to promote root growth and thus can help plants to overcome the challenge of root rots. Although there is much information about the effects of nutrition on disease, very few crop production systems have been extensively studied in order to correlate the positive factors into integrated production systems.

ORGANIC AMENDMENTS AND GROWTH-PROMOTING MICROBIALS

Organic matter, from either the previous crop or other sources, is often added to greenhouse beds prior to establishing the next crop. This material can have either a beneficial or a detrimental effect on root pathogens and subsequent disease. If the previous crop was diseased, do not incorporate crop residue as an amendment but remove and destroy it. Straw has been shown to suppress damping-off caused by *Rhizoctonia solani* but to increase the incidence of *Fusarium* crown and root rot. However, if organic amendments can be precolonized by nonpathogens, a suppression of pathogens can be realized. The nonpathogens produce enzymes that decompose the cell walls of pathogens. Precolonization of these materials is often achieved by composting, and extensive references are available on biological control exerted by the resident microflora of hardwood and conifer bark composts used as components of growing media. Chitin-containing substrates, like crustacean exoskeletons such as crab and shrimp shells, have been used to enhance populations of naturally occurring soil microbes. Since organic amendments have complex effects on the biological, chemical, and physical conditions of soil and the potential to suppress diseases, carefully evaluate each production situation to determine the potential benefit.

OPEN-FIELD APPLICATIONS

The increasing need for reduced maintenance in outdoor beds and interest in field production of cut flowers coupled with the pending loss of methyl bromide presents a new demand for soil treatment options for controlling pests and pathogens outside. Several options should be considered, either singly or in combination, based on the particular needs of the crop being produced and the soilborne problems anticipated.

Solarization, trapping solar radiation under a plastic cover and thus heating the soil sufficiently to inactivate organisms, can be effective in certain geographical locations in conjunction with some cropping systems. It is more feasible in some field situations than greenhouse culture because of fallow time between crops or due to excessive temperatures during the summer for crop production. Specialized solarization applications, such as with seed beds, piles of prepared container media, and cold frames, offer other opportunities. Solarization can also enhance subsequent establishment of microbial biopesticides.

Commercially available portable steam generators, such as the Steam-Flo Steam Generator and the Meyer IMO Soil Steamer System described previously, can produce steam in the field for limited areas of treatment. The size of the production operation and economic considerations would determine the feasibility of this option.

Organic amendments and plant and animal residues have been used to augment soil fertility and improve soil structure. These amendments can have marked effects on disease severity, often resulting in disease suppression either by direct or indirect action. Composted agricultural, forestry, and municipal wastes contain a residual microflora that can give biological control effects as well as cause significant chemical and physical alterations. Chitin-containing amendments have been shown to enhance populations of microorganisms that produce enzymes that destroy some pathogens, particularly nematodes. Since organic amendments have different effects on different pathogens, first consider each growing site and the known or anticipated pathogen problems.

Nutritional management and adjusting factors, such as pH, form of nitrogen, and macro- and microelements, affect both the crop and the soil microflora. Cultural systems have been worked out and put into practice that reduce disease potential for some crops. An example is the nitrate-lime-chemotherapy system for fusarium wilt control in chrysanthemum. Unfortunately, there is no general recommendation that can be made using this approach, since every crop-pathogen combination varies and must be studied independently. Nutritional management does, however, become an important component in developing alternative strategies to broad-spectrum, nonselective procedures, such as heat treatment or fumigation.

Microbial biopesticides could be utilized alone, but they probably would be more effective if combined with one or more of the above options. Most available microbial biopesticides are selections of naturally occurring soil microbes, and their effectiveness can be enhanced by practices that either reduce competition from the resident soil microflora or weaken the target pathogen.

CHAPTER 10

MODERN FERTILIZATION METHODS

by Ron Adams
Ball Horticultural Company
West Chicago, Illinois

Plant nutrition has become a more exact science as we gain more control over the growing environment. A fertilizer program should balance soil and water chemistry to provide an optimum nutrition level for the type of growth or method of application. There are two primary ways to fertilize: feeding a water-soluble fertilizer along with irrigation water through an injector; incorporating a slow-release fertilizer into the media prior to planting or topdressing the media after plants are established. Optimum growth often occurs when the grower uses both a slow-release and a constant water-soluble feed.

HOW TO CALCULATE PARTS PER MILLION

Fertilizer components in parts per million (ppm) are fairly easy to determine. Keep this standard equation in mind:

1 oz fertilizer / 100 gal water = 75 ppm

A fertilizer's formula numbers are simply the percentages of nitrogen (N), phosphorus (P), and potassium (K) in the fertilizer. Say you want to use 20-10-20 at a rate of 200 ppm N and K. Each ounce of this fertilizer has 0.20 ounce of N and K and 0.10 ounce of P. The existing concentrations of N, P, and K in diluted fertilizer are proportional.

$$N = 0.20 \times 75 \text{ ppm} = 15 \text{ ppm}$$
$$P = 0.10 \times 75 \text{ ppm} = 7.5 \text{ ppm}$$
$$K = 0.20 \times 75 \text{ ppm} = 15 \text{ ppm}$$

You want a concentration giving 200 ppm N and K, though, so you must add more than 1 ounce of fertilizer per 100 gallons.

200 ppm desired / 15 ppm per oz = 13.3 oz needed

Using 13.3 ounces of fertilizer per 100 gallons of water will give you the desired concentrations (200 ppm) of N and K. (For a listing of fertilizer salts, their analyses, percentages of N, P, and K, and the effects on pH, see table 1. For a listing of various fertilizer combinations, their analyses, and the ounces of each needed to give 200 ppm N and K, see table 2.)

TABLE 1 FERTILIZER SALTS

Compound	Formula	Analysis	Percent			Effect on pH
			N	P	K	
Ammonium chloride	NH_4Cl	25-0-0	25	0	0	Acid
Ammonium nitrate	NH_4NO_3	33.5-0-0	33.5	0	0	Acid
Monammonium phosphate	$NH_4H_2PO_4$	12-62-0	12	.62	0	Acid
Diammonium phosphate	$(NH_4)H_2PO_4$	21-53-0	21	53	0	Acid
Ammonium sulfate	$(NH_4)_2SO_4$	21-0-0	21	0	0	Very acid
Calcium nitrate	$Ca(NO_3)_2$	15-0-0	15	0	0	Basic
Sodium nitrate	$NaNO_3$	16-0-0	16.5	0	0	Basic
Urea	$CO(NH_2)_2$	45-0-0	45	0	0	Acid
Superphosphate	$Ca(H_2PO_4)_2+CaSO_4$	0-20-0	0	8.7	0	Neutral
Treble superphosphate	$Ca(H_2PO_4)_2$	0-42-0	0	18.3	0	Neutral
Phosphoric acid	H_3PO_4	0-52-0	0	22.7	0	Very acid
Potassium chloride (muriate of potash)	KCl	0-0-60	0	0	51	Neutral
Potassium nitrate (saltpeter)	KNO_3	13-0-44	13	0	36.5	Basic
Potassium sulfate	K_2SO_4	0-0-51	0	0	44	Neutral
Magnesium nitrate	$Mg(NO_3)_26H_2O$	11-0-0	11	0	0	Neutral
Monopotassium phosphorus	KH_2PO_4	0-53-35	0	23.1	28.2	Basic
Dipotassium phosphate	K_2HPO_4	0-41-54	0	17.9	44.8	Basic

Note: The analysis of compounds may vary from listed values as a result of differences in processing among chemical companies.

TABLE 2   FERTILIZER MATERIAL COMBINATIONS SUPPLYING 200 PPM N AND K

Combination	Analysis	Oz per 100 gal water
Ammonium sulfate	20-0-0	13.3
+ muriate of potash	0-0-62	5.6
Ammonium nitrate	33.5-0-0	8
+ muriate of potash	0-0-62	5.6
Urea	45-0-0	6
+ muriate of potash	0-0-62	5.6
Calcium nitrate	15-0-0	10.6
+ potassium nitrate	13-0-44	7.5
Calcium nitrate	15-0-0	16.7
+ muriate of potash	0-0-62	5.6
Sodium nitrate	16-0-0	10.6
+ potassium nitrate	13-0-44	7.5
Ammonium sulfate	20-0-0	8.5
+ potassium nitrate	13-0-44	7.5
Ammonium nitrate	33.5-0-0	5.1
+ potassium nitrate	13-0-44	7.5
Urea	45-0-0	3.8
+ potassium nitrate	13-0-44	7.5
Ammonium nitrate	33.5-0-0	6.1
+ calcium nitrate	15-0-0	4
+ muriate of potash	0-0-62	7.8

Example: Using the first combination, 13.3 oz. ammonium sulfate per 100 gal water, applied directly to the crop, will supply 200 ppm N; 5.6 oz muriate of potash in combination in that same 100 gal will provide 200 ppm K.

FERTILIZER INJECTORS

Constant liquid feeding of greenhouse crops has made fertilizer injectors standard equipment for greenhouses. The wide range of other equipment increases the difficulty of matching injectors to an operation. There are several types of injectors available, however, adaptable to a wide variety of growing situations. Some growers use small injectors in each house, and others install a central injector system to conveniently feed all houses from one location (fig. 1). Some ultramodern production ranges use highly sophisticated, computer-controlled injectors that can pull fertilizer solution from one to five stock tanks in response to in-line pH and EC readings (fig. 2).

Some growers have used Hozon siphon proportioners for feeding and drenching for years. Hozon siphons are not preferred because they make feeding a chore. One limitation of the Hozon siphon is the requirement of a high flow rate, which results in a forceful spray from the hose, often knocking pots over.

There are many types of injectors available from many suppliers (table 3). Choosing an injector requires careful planning to ensure that the unit is a benefit to getting fertilizer applied and not a limitation to production. Among the key things to keep in mind when purchasing an injector is the water flow range in which a particular model works. For example, a Smith R-3 has a rated range of 3 to 12 gallons (11 to 45 l) per minute; an Anderson 200 series is rated at 1 to 20 gallons (4 to 76 l) per minute. Proportioning is another consideration. How many gallons (or liters) of water is the concentrate blended into? How is the concentrate blended? Is it surged into the water line, giving a dose of fertilizer followed by just water, or are the fertilizer and water blended before entering the water line? Does the system require a blending tank? Is the proportion adjustable or fixed?

Portability is often another determinant of the type of injector required. Is acid injection needed? Not every injector can handle acid. Be sure to ask the manufacturers about the acid suitability of models you are interested in. Will pesticide applications

FIG. 1 Many growers have several small fertilizer injectors throughout their ranges. Here is one of six Dosatron units used through the greenhouse range at Bernacchi Greenhouses, La Porte, Indiana.

FIG. 2 Computerized fertilizer injectors with multiple stock tanks are the norm at some high-end turnkey greenhouses. This setup, in the Westland area of the Netherlands, controls irrigation and fertilization for several acres of ebb-and-flood floors.

TABLE 3 FERTILIZER INJECTORS

Injector	Manufacturer	Models	Proportion	Cost	Comments
Anderson	H.E. Anderson Co.	201, 301, 501, 1001, 1601 in S, T & V series	1:200	T & V series: $1,244, which incl. mix tank S series models, without tank, cost less	Co. has exchange program. Units last up to 20 yrs.
		J Plus series			J Plus models now with opt. serial port com for solution control and data collection; Teflon diaphragm allows acids to be pumped undiluted
DGT Voltmatic	Danish, distr. by Midwest Trading	AMI series, computer controlled	Variable	Quotes available	Can be integrated with specific greenhouse condition; able to draw from several stock tanks; can inject acid
		MGRO injectors and irrigation controllers	Variable	Quotes available	
Dosamatic	JF Equipment	MDDP1, DP30-25, DP40-25, Mobile 30-25	1:43 to 1:500	$259 to $550	2- to 3-year life
Dosatron	Dosatron International	11 GPM, 40 GPM, 100 GPM	1:50 to 1:500	$259 to $2,060	2- to 3-year life
Gewa	German, distr. by Brush King	GE04, GE06, GE15, GE26	1:100, variable	$725 and up	Portable. Fertilizer is in rubber bladder that can leak over time, causing loss of accuracy
Hozon	Distr. through local suppliers		1:15	$10 to $15	Requires high flow rates; accuracy variable
Smith	Smith Precision	R-1, R-3, R-4, R-6, R-8	1:100	$522 and up	Co. has exchange program; set flow range
		R-6, R-8	1:200	$2,078	

be harmful to the equipment? Can the injector be easily repaired, or does it have to be replaced after its useful life? Can the old unit be traded in for a credit on a replacement model?

PROPORTION IS NOT PPM

An injector is not set for parts per million (ppm). Most have a variable proportion, which means that with a given volume of water going through the metering device, a certain amount of concentrate will be drawn or injected into the water system. The rate at which you mix the concentrate determines ppm. Table 4 shows common proportions available from standard equipment.

CHECKING A FERTILIZER INJECTOR

There are two primary methods of checking an injector for accuracy. You can check the blended fertilizer-water solution as it is applied to the crop, using a solubridge to measure electrical conductivity (EC), otherwise, measure uptake of the fertilizer concentrate by a known amount of water.

Checking with a solubridge is the easier method because it does not require collecting a large amount of water and sampling can be done anywhere water can be collected. To perform a conductivity check, you need a reliable conductivity meter that has been calibrated with a standard solution before testing. First check the untreated water and obtain a salt reading. Then test a sample of the fertilized water and subtract the raw water reading from the combined reading. Refer to a fertilizer chart to see how the measured reading compares with the standards. Constantly checking your conductivity levels alerts you to improper fertilizer mixing or a malfunctioning injector.

The second method is to physically measure the amount of fertilizer water collected from the injected solution. To do a physical check, you need to collect a known volume of water, such as 50 gallons (about 190 l) of solution, at the hose just before plant application. The concentrate solution has to be measured before and after the collection to measure the exact amount siphoned into the injector and blended with the water. A 1:100-ratio injector would draw a half gallon of concentrate for every 50 gallons of dilute solution. This test can be done with larger or smaller samples, but the sample should be large enough to even out sample variation and ensure that the equipment is fully functioning and working at a normal rate when feeding.

SLOW-RELEASE FERTILIZERS

Slow-release fertilizers for quality bedding plants and pot plants are not as popular as in the past. Their decline is attributed to growers' increased use of soluble fertilizers through injectors. Water-soluble fertilizers are better understood than slow-release

TABLE 4   INJECTION RATIOS AND NITROGEN CONCENTRATIONS FOR
CONSTANT FEEDING

Ratio	Oz fertilizer per gal concentrate			Ratio	Oz fertilizer per gal concentrate		
	100 ppm N	150 ppm N	200 ppm N		100 ppm N	150 ppm N	200 ppm N
	30% N formula[a]				20% N formula[c]		
1:300	13.5	20.2	27.0	1:300	20.2	30.3	40.5
1:200	9.0	13.5	18.0	1:200	13.5	20.2	27.0
1:150	6.7	10.1	13.5	1:150	10.1	15.1	20.2
1:128	5.7	8.6	11.5	1:128	8.6	12.9	17.2
1:100	4.5	6.7	9.0	1:100	6.7	10.1	13.5
1:50	2.2	3.3	4.5	1:50	3.3	5.0	6.7
1:30	1.3	2.0	2.7	1:30	2.0	3.0	4.0
1:24	1.0	1.6	2.1	1:24	1.6	2.4	3.2
1:15	0.67	1.0	1.3	1:15	1.0	1.5	2.0
	25% N formula[b]				15% N formula[d]		
1:300	16.5	24.7	33.0	1:300	27.0	40.5	54.0
1:200	11.0	16.5	22.0	1:200	18.0	27.0	36.0
1:150	8.2	12.3	16.5	1:150	13.5	20.2	27.0
1:128	7.0	10.5	14.0	1:128	11.5	17.2	23.0
1:100	5.5	8.2	11.0	1:100	9.0	13.5	18.0
1:50	2.7	4.1	5.5	1:50	4.5	6.7	9.0
1:30	1.6	2.4	3.3	1:30	2.7	4.0	5.4
1:24	1.3	1.9	2.6	1:24	2.1	3.2	4.3
1:15	0.82	1.2	1.6	1:15	1.3	2.0	2.7

[a] E.g., 30-10-10
[b] E.g., 25-5-20, 25-10-10, 25-0-25
[c] E.g., 20-20-20, 20-5-30, 21-7-7
[d] E.g., 15-15-15, 15-30-15, 16-4-12

fertilizers and allow the grower better control, although several growers have reliably supplied all the nutrients necessary to produce an excellent crop with slow-release fertilizers.

The advantages of using a slow-release fertilizer include its being a constant fertilizer source that can be safely incorporated into the soil mix prior to planting. It reduces the need to constantly mix and proportion (water-solubles) or apply (dry-solubles). A slow-release fertilizer maintains a more constant supply of nutrients between waterings than a water-soluble fertilizer when waterings are not as frequent. You can use a capillary watering system minimizing algae buildup on mats. Slow-release fertilizers also increase plant shelf life by providing nutrients after plants leave the greenhouse. Hanging baskets, foliage plants, and pot plants, such as saintpaulia, can use the additional fertilizer, which sustains them longer.

Slow-release fertilizers have proven themselves over a number of years, but there are some disadvantages when using them. They cannot easily change the salt level, especially to lower it, when plants are growing at a slower rate. A slow-release fertilizer should be incorporated throughout the soil mix prior to planting for the most uniform distribution of nutrients. It can provide too high a fertility level at the crop finish, making growth at sales difficult to control. This is particularly a problem in the South, as in the case of foliage plants fed at too high a level for reduced growing conditions during and after acclimatization for indoor plantings. Some top-dressed slow-release materials will float out of the container under flooding conditions, reducing the effective rate. Some slow-release fertilizers are temperature or bacterial dependent, so the release rate varies, depending on conditions.

You can use either slow-release or soluble feed to produce plants. Though the trend is more to soluble, the best growth is with a combination of slow-release and water-soluble. If you plan to use the combination, it is best to reduce the effective rates by as much as 50% to provide an optimum level. For example, if you use a slow-release at 10 pounds per cubic yard (5.9 kg per m^3) and feed at 250 ppm N, you should try 5 pounds per cubic year (3 kg per m^3) of slow-release and feed at 125 ppm N.

COATED FERTILIZERS

There are two types of complete slow-release fertilizers: coated and granulated (table 5). Coated fertilizers consist of water-soluble fertilizers coated with a plastic resin or sulfur that allows the fertilizers to become available over time. Osmocote and Nutricote use a microporous plastic resin to lock up the fertilizer until osmotic pressure initiates release. Then the fertilizer diffuses through the coating until it is depleted. Release speed can be controlled by the coating thickness. A formula can be timed for release in anywhere from 40 days to a year. Osmocote or Nutricote can be either top-dressed or incorporated into the media. They should not be mixed ahead of time by more than two to three weeks, and they cannot be steam sterilized. There are several formulas of these products available. Newer formulations add trace elements and change the nitrogen type to make more complete fertilizers for soilless mixes.

Another type of coated slow-release fertilizer is sulfur-coated urea. Prokote falls into this group. Waxed sulfur coats urea nitrogen, and through bacterial action the coating degrades to make the nitrogen available over time. Other nutrients can be coblended with this fertilizer. Since sulfur coats depend on bacterial breakdown, release is variable and not totally predictable. Also, sulfur coats cannot be steam sterilized.

GRANULATED FERTILIZERS

The most commonly used granulated slow-release is MagAmp. MagAmp is a cogranulated blend of nitrogen, phosphorus, potassium, and magnesium with a formula of 7-40-6 plus 12% magnesium. MagAmp, or magnesium ammonium phosphate, is what we consider a true slow-release fertilizer, with only 1% of the fertilizer available in a 24-hour period under saturated conditions. The MagAmp formula, high in phosphorus, could be considered a phosphorus source for mixes—with the additional benefit of nitrogen, potassium, and magnesium. MagAmp is not temperature sensitive, and soils including MagAmp can be steam sterilized prior to planting without any harmful effects.

MagAmp release rates can be changed by varying the particle size. There are two grades available. Medium grade, the most desirable for greenhouse production, lasts four to six months in the media. Coarse grade, typically suited for outdoor nursery stock and outdoor plantings, can last up to 24 months. MagAmp has to be incorporated to be effective and should not be top-dressed, but it can be mixed and stored for several months ahead of using.

TABLE 5   SLOW-RELEASE FERTILIZERS

Brand	Formulation	Nutrient release time (months)	Rate incorporated (lb/yd3)
Coated			
Osmocote	14-14-14	3–4	6
Osmocote	19-6-12	3–4	5
Osmocote	18-6-12	8–9	6
Sierra	17-6-10 + minors	8–9	7
Sierra	13-12-11 + minors		
Geranium Poinsettia	+ minors	5	9
Nutricote	18-6-8 + minors	100 days	7
Nutricote	12-12-12	100 days	8
ProKote	20-3-10 + minors	7–9	5
ProKote	20-3-11 + minors	4–5	5
Blended			
Scotts Topdress	17-3-6 + minors	3–4	—
ScottKote	20-5-10	8	13
Woodace	14-12-14 + minors	5–6	6
Granulated			
MagAmp	7-40-6 medium	3–5	10

Note: For more in-depth information on fertilizers, media, injectors, and more, get a copy of *Water, Media, and Nutrition for Greenhouse Crops: A Grower's Guide,* by David Wm. Reed, Texas A&M University, College Station, available from Ball Publishing, phone 630/208-9080, toll free 800/456-5380, fax 630/208-9350.

CHAPTER 11

PLANT NUTRITION, MEDIA AND WATER TESTING

PLANT NUTRITION
by Darryl D. Warncke and Dean M. Krauskopf
Michigan State University
East Lansing

SOIL AND WATER TESTING WITH NUTRITION GUIDELINES
by Dr. Douglas A. Bailey, Dr. Paul V. Nelson, and Dr. William C. Fonteno
North Carolina State University
Raleigh

WATER TESTING AND PARAMETERS
by Dr. John C. Peterson
Massachusetts Horticultural Society
Boston

PLANT NUTRITION

Greenhouse growth media have chemical and physical properties that make them distinctly different from field soils. Over the past decades greenhouse operators have switched from mixes containing soils to peat- or bark-based mixes containing such manufactured materials as perlite, vermiculite, and expanded polystyrene beads. The "soilless" growth media have good moisture-holding and aeration properties but limited nutrient-holding capacities. As a result, fertility management in the greenhouse is more important than ever before.

115

Being knowledgeable about physical and chemical properties is a prerequisite for good management. An analysis of a growth medium (any material in which plants are grown) provides basic information on which to build a fertility program. Prior to using any new lot of growth medium, test for pH, soluble salt content, and available nutrient levels. Even though most companies maintain quality control programs, variations in growth media properties do occur. Knowing the initial chemical properties is essential to avoiding costly plant growth problems later.

The Michigan State University Soil and Plant Nutrient Lab offers a testing program specifically designed for analyzing soilless growth media used in producing greenhouse crops. Technicians analyze greenhouse growth media using a saturated media extract (SME) procedure. Approximately 400 cubic centimeters (24 in³) of a growth medium are mixed with sufficient distilled water to just saturate the medium sample. pH is determined on this saturated mix. After one hour the saturation solution is removed with a vacuum filter. All subsequent analyses are then performed on the extracted solution.

In soilless greenhouse growth media, the concentration of essential nutrients around the roots, critical to plant growth, depends upon the medium's moisture-holding capacity. With a given amount of nutrient in a container of growth medium, the nutrient concentration around the root decreases as the moisture content increases. Since growth media vary widely in bulk density (weight per unit volume), it has been difficult to develop a single set of fertilization guidelines. With the saturated media extract procedure, it's possible to use a single set of fertilization guidelines (table 1)

SOME ABCs OF WATER, MEDIA, AND NUTRITION

Total soluble salts in a medium or water sample are expressed several ways. Millisiemens (mS) is a reading of total salts given by a solubridge. It represents the same value as millimho and is also the same as electrical conductivity, or EC. The term EC is used by commercial growers today the world over to express salt levels. Roughly, EC 0.25 is a too-low nutrient level; 2.25 is too high and will cause soluble salt injury. These values are based on a solubridge test using one part medium to two parts water, volume to volume. Other measurements include the following.

♦ ppm is parts per million—of any soluble material in a solution. To convert millisiemens to approximate parts per million, multiply mS by 700.

♦ pH is a way to express acidity. From pH 0 to pH 7 is acid; from pH 7 to 14 is alkaline; pH 7 is neutral.

For those unfamiliar with industry jargon, you may see three numbers prominently displayed on a fertilizer bag, "20-10-20," for example. They simply mean that this fertilizer mix contains 20% nitrogen, 10% phosphate, and 20% potash.

since the amount of water held at saturation is directly related to the moisture-holding characteristics of each medium.

TABLE 1   MAJOR NUTRIENT LEVELS FOR GREENHOUSE GROWTH MEDIA

Nutrient	Low	Acceptable	Optimum	High	Very high
Soluble salts	0–0.75	0.75–2	2–3.5	3.5–5	5+
Nitrate-N	0–39	40–99	100–199	200–299	300+
Phosphorus	0–2	3–5	6–9	11–18	19+
Potassium	0–59	60–149	150–249	250–349	350+
Calcium	0–79	80–99	200+	—	—
Magnesium	0–29	30–69	70+	—	—

Note: Analysis by saturated medium extract method. Soluble salts expressed as mS/cm, all others as ppm.

Desirable pH, soluble salt, and nutrient levels vary with the greenhouse crop being grown and management practices. General guidelines for the most important fertility parameters are given in table 1. Acceptable sodium and chloride levels depend upon the total soluble salt content, but they generally should not exceed 10% of the total soluble salts.

To obtain maximum crop growth, adjust growth media to optimum nutrient levels before planting. Consider the following when adjusting growth media nutrient levels.

MEDIUM pH

Growth medium pH influences the availability and plant uptake of all essential plant nutrients. In a peat-based medium the most desirable pH is 5.6 to 5.8 for most plants. Most irrigation waters in Michigan are alkaline, containing excess calcium carbonate. Watering plants with alkaline water gradually raises growth medium pH; over a three-month period, the pH may increase 0.5 to 1 pH unit. Thus, it is extremely important for growth medium pH to be properly adjusted prior to planting. Too high a pH, greater than 6.5, increases the chance of micronutrient deficiencies. Too low a pH, less than 5.3, may result in calcium or magnesium deficiency or manganese toxicity.

The amount of lime to add for pH adjustment depends on the medium's buffering capacity (ability to resist change). To bring about a 1 pH unit change (e.g., 4.5 to 5.5) in a weakly buffered growth medium may require only 2 pounds of finely ground lime per cubic yard (or about 1.2 kg per m³), whereas 5 pounds or more per cubic yard (about 3 kg per m³) may be required in a more highly buffered growth medium. Amendments such as perlite, expanded polystyrene beads, and expanded vermiculite have little or no buffering capacity. Fibrous peat and shredded bark, or wood also have limited buffering ability. Somewhat decomposed peat, muck, well-composted bark, and field soil provide a higher degree of buffering.

When adding lime to a greenhouse growth medium, remember that it is better to under lime initially than to overlime. Mix up a small batch (0.1 cubic yard or m^3) of growth medium using the lime rate judged to be correct. Moisten the medium as you would before planting; place it in a large plastic bag for two weeks, and then sample and check the pH. If the pH is between 5.5 and 6, the lime rate is acceptable. If the pH is outside this range, adjust the rate accordingly. Always use finely ground lime that can pass through a 100-mesh sieve. Coarser liming materials, such as agricultural lime, may take up to six months to fully react and bring about the desired pH change. Lime will not react in dry, stockpiled growth medium; when the growth medium is moist, however, fine lime will fully react within two weeks. Calcitic lime supplies only calcium, whereas dolomitic lime contains both calcium and magnesium. Dolomitic lime reacts more slowly than calcitic lime.

Avoid growth medium with too high a pH—lowering the pH is more difficult than raising it. Acidify high-pH growth medium by mixing in iron sulfate. A decrease of approximately 1 pH unit (say, pH 7.5 to 6.5) can be brought about with 3 pounds iron sulfate per cubic yard (or 1.8 kg per m^3). The exact change depends on the buffering nature of the mix components. The rate of change in pH is slower with iron sulfate than with lime.

Adjusting the pH in pots, benches, or flats with growing plants present is more difficult and may cause plant injury. Use limewater to neutralize acidic growth medium (raise the pH). Adjustment is not suggested if the pH is 5.4 or above. Stir 1 pound of finely ground lime or one-half pound of calcium oxide into 100 gallons of water (about 480 or 240 g, respectively, per 400 l); let settle overnight, and apply the clear solution, avoiding or filtering out any settlings. The growth medium should be quite moist at the time of application to minimize root shock and injury. Avoid getting the solution on the foliage; otherwise, wash off the foliage immediately after application. Do not apply ammonium-containing fertilizer immediately before or after a limewater application. Ammonium reacts with lime to release volatile ammonia, which may burn plant foliage.

Gradually lower the pH of alkaline growth medium by watering with an iron sulfate solution. Dissolve 2 pounds of iron sulfate in 1 gallon of water (960 g per 4 l) and inject through the watering system at 1:100.

Acidify alkaline irrigation water using phosphoric, sulfuric, or some other acid. The amount of acid to use depends on the alkalinity of the water. As a starting point, 1.5 ounces (44 ml) of 85% phosphoric acid or 0.6 ounce (18 ml) of concentrated sulfuric acid added to 100 gallons (400 l) of water will bring the alkalinity closer to the acceptable alkalinity of 100 ppm and lower the pH to about 6, depending on the bicarbonate concentration in the water. Measuring the pH and alkalinity will enable more precise determination of the amount of acid to add. Some water sources may require up to 3 ounces (89 ml) of phosphoric acid per 100 gallons (400 l). Precise adjustment requires some trial additions and monitoring the resulting water pH. Injecting acid into the irrigation system also cleans out the irrigation lines and sprayers on drip nozzles.

Exercise extreme caution and care when using acid. Be sure to keep sodium bicarbonate on hand to neutralize any acid spills.

SOLUBLE SALTS

All soluble ions or nutrients, such as nitrate, ammonium, potassium, calcium, magnesium, chloride, and sulfate, contribute to the soluble salt content of a growth medium or water. Total soluble salt content in water or a growth medium extract is determined with a solubridge (conductivity meter) and expressed in millisiemens (mS). To convert mS to approximate ppm, multiply by 700.

Greenhouse operators commonly use one part growth medium to two parts distilled water (volume:volume basis) to determine soluble salt content. A 1:5 ratio may be used if more solution is needed. The MSU Soil and Plant Nutrient Lab determines soluble salt content on the saturation extract. Guidelines for interpreting soluble salt levels for each procedure are given in table 2. Mixing fertilizer into a growth medium increases the soluble salt content. In general, each pound of soluble fertilizer mixed

TABLE 2   SOLUBLE SALT LEVELS IN GREENHOUSE GROWTH MEDIA

Solubridge reading (millisiemens, or EC)

Saturation extract[a]	1 medium to 2 water[b]	1 medium to 5 water[b]	Comments
0–0.74	0–0.25	0–0.12	Very low salt levels, indicating very low nutrient status
0.75–1.99	0.25–0.75	0.12–0.35	Suitable range for seedlings and salt-sensitive plants
2–3.49	0.75–1.25	0.35–0.65	Desirable range for most established plants Upper range may reduce growth of some sensitive plants
3.5–5	1.25–1.75	0.65–0.90	Slightly higher than desirable Loss of vigor in upper range Okay for high-nutrient–requiring plants
5–6	1.75–2.25	0.90–1.10	Reduced growth and vigor Wilting and marginal leaf burn
6+	2.25+	1.10+	Severe salt injury symptoms, with crop failure likely

[a] Method used by the Soil Testing Lab at Michigan State University
[b] Parts volume to volume

in per cubic yard (593 g per m^3) of medium increases soluble salt content in the saturation extract 1.0 mS per cm. The exact increase depends on the fertilizer used.

Minimize soluble salt buildup by watering to cause some leaching. Reduce excessively high soluble salt levels by leaching the soluble salt content down to an acceptable level. Watering the container or bed so a good amount of water drains out and then repeating this procedure one to two hours later reduces the soluble salt level sufficiently. Extremely high soluble salt levels may require repeating the procedure two or three days after the first attempts.

Nitrate-nitrogen. Plants deficient in nitrogen (N) become light green in color, beginning with the older leaves. Some nitrogen-deficient plants may also show a reddish color. Nitrogen is an important component of the chlorophyll molecule. The nitrate form of nitrogen is soluble and mobile in the growth medium, so with watering, some of the nitrate may leach out.

Optimum nitrate-N levels vary with plant age and type. Some guidelines are given in table 3. Young plants and seedlings do best with low to medium nitrate levels. Most pot and bedding plants in a growing-on stage require moderately high levels. Crops grown in ground or raised beds do well with high nitrate levels. Adjust the initial level of nitrate-N in a stock growth medium using the guidelines given in table 4.

Maintain a fairly constant level of available nitrogen by injecting additional nitrogen into the watering system. When injecting fertilizer into the irrigation water, be sure to water adequately to cause some leaching and prevent excess nitrate and soluble salt buildup.

TABLE 3 DESIRABLE NITRATE-N, PHOSPHORUS, AND POTASSIUM CONCENTRATIONS IN GREENHOUSE GROWTH MEDIA

Crop or stage of growth	ppm in saturated medium extract		
	NO$_3$-N	P	K
Seedlings	40–70	5–9	100–175
Young pot and foliage plants	50–90	na	na
Pot plants, growing on	80–160	6–10	175–250
Bedding plants	80–160	6–10	150–225
Roses, mums, snapdragons in ground or raised beds	120–200	10–15	200–275
Lettuce, tomatoes in ground beds	125–225	10–15	200–300
Azaleas	na	7–12	125–200
Celery transplants	75–125	10–15	250–300

Note: Na means figures not available.

TABLE 4   NITROGEN FERTILIZER NEEDED TO INCREASE SATURATION EXTRACT NITRATE LEVEL 10 PPM N

N carrier	N content (%)	To increase test level 10 ppm N, use:				
		oz/bu	oz/yd³	g/m³	oz/100 ft²	g/10m²
Potassium nitrate	13	0.12	2.3	85	4.6	140
Calcium nitrate	15	0.10	2.0	74	4.0	122
Ammonium nitrate	33	0.045	0.9	33	1.8	55
Urea	45	0.035	0.7	26	1.4	43

Note: Analysis by saturated medium extract method

Phosphorus (P). An adequate phosphorus supply is important for root system development, rapid growth, and flower quality in floral plants. Phosphorus-deficient plants exhibit slow root and top growth. In severe cases foliage turns purple. Phosphorus plays an important role in the photosynthetic process. Phosphorus compounds, being only slowly soluble, are generally subject to limited leaching loss, but leaching may be significant (up to 30%) in fibrous, peat-lite mixes. Sufficient superphosphate can be mixed initially into a growth medium to supply phosphorus throughout the growth period without concern for undue leaching loss or soluble salt buildup.

Crops grown in ground or raised beds require higher phosphorus levels than pot or bedding plants (table 3). Plants grown at cool temperatures sometimes develop phosphorus deficiencies, even with adequate phosphorus present, due to limited root growth and activity. Raising the temperature 5F (3C) enables plants to grow out of this condition more easily than does adding more phosphorus.

Concentrated superphosphate (0-46-0) is the phosphorus source most available to greenhouse operators, but normal superphosphate (0-20-0) is better for use in greenhouses because it contains extra calcium and sulfur. Table 5 provides guidelines for increasing available phosphorus levels in most greenhouse growth media. Mixes containing greater than 25% calcined clay or muck require about 2.5 times more phosphate fertilizer to achieve the same increase in extractable phosphorus. Do not use superphosphate for lilies because of potential fluoride toxicity. Fluoride is contained in rock phosphate, the base material for production of superphosphate. Bonemeal is a better phosphorus source for lilies.

Take care not to overfertilize with phosphate. Excessively high phosphate levels may reduce the ability of plants to take up and utilize several micronutrients. If the irrigation water pH is being adjusted with phosphoric acid, additional phosphorus is probably not necessary. Each ounce of 85% phosphoric acid added per 100 gallons

TABLE 5   FERTILIZER NEEDED TO INCREASE PHOSPHORUS LEVEL IN THE SATURATION EXTRACT 2 PPM P

P carrier	P_2O_5 content (%)	To increase test level 2 ppm P, use:				
		oz/bu	lb/yd³	g/m³	lb/100 ft³	g/10m²
Normal superphosphate	20	0.75	0.90	534	1.8	879
Concentrated superphosphate	46	0.33	0.40	237	0.8	0.8
Bonemeal	25	0.60	0.75	445	1.5	732

Note: Analysis by saturated medium extract method

of water supplies 83 ppm P_2O_5 (36 ppm P). The percent phosphate listed on the fertilizer label is as percent P_2O_5, and P_2O_5 contains only 43% actual P.

Potassium. The nutrient most often limiting in greenhouse fertility programs is potassium (K). The lower or oldest leaves of potassium-deficient plants show marginal yellowing or chlorosis. Spotting over the entire leaf is sometimes also associated with potassium deficiency. Many greenhouse plants have potassium requirements equal to or greater than their nitrogen requirements. Potassium salts are water soluble and leachable in soilless growth media with low nutrient-holding capacities. As a result, potassium levels are depleted more readily than nitrogen levels. Soluble fertilizers are commonly injected into the watering system to supply 200 ppm nitrogen, and many of these fertilizers (such as 20-20-20 or 25-0-25) contain equal amounts of N and K_2O. K_2O, however, is only 83% K, so plants are receiving only 166 ppm K in the fertilizer solution.

The demand for potassium is greatest in rapidly growing plants in the vegetative stage. Seedlings and young plants usually do better with a low to medium potassium level. Optimum potassium levels are given in table 3 for various plant categories.

Potassium nitrate has a K:N ratio of about 3:1 and is ideal for building up the potassium content of a growth medium. Since plants use both the potassium and the nitrate portions of the salt, concern over soluble salt buildup is less than with other potassium sources. Potassium sulfate is a suitable potassium source, but it is less soluble. Potassium chloride is not recommended for greenhouse use because of its high salt index.

Establishing a near-optimum potassium level in the growth medium before planting is desirable to ensure a more consistent potassium supply throughout the growth period. The quantities of potassium fertilizer to obtain the necessary buildup are given in table 6.

TABLE 6   POTASSIUM FERTILIZER NEEDED TO INCREASE POTASSIUM LEVEL IN
THE SATURATION EXTRACT 25 PPM K

K carrier	K₂O content (%)	To increase test level 25 ppm K, use:				
		oz/bu	oz/yd³	g/m³	lb/100 ft²	g/10m³
Potassium nitrate	44	0.19	3.75	139	0.46	225
Potassium sulfate	50	0.16	3.25	121	0.40	195
20-20-20	20	0.41	8.25	306	1.03	503

Note: Analysis by saturated medium extract method

Calcium. Calcium (Ca) availability for plant uptake depends on growth medium pH and levels of other cations present, especially potassium and magnesium. Calcium deficiency in plants results in abnormal growth or death of the growing tip. As a growth medium becomes more acid (lower pH), especially below pH 5, calcium becomes less available.

Available calcium levels may be marginal in soilless greenhouse media, especially those having an acid peat as the base material, unless amended with lime. To effectively change the pH and available calcium level, lime must be thoroughly mixed in, and the growth medium must be adequately moist.

Many of the calcium carriers are slowly soluble, so equilibrium won't be reached if the stockpiled growth medium is maintained dry. As a result, the calcium content in a saturation extract may not accurately reflect the available calcium content of the growth medium.

Increase calcium levels by adding lime to acid growth medium. Calcium sulfate (gypsum) and calcium nitrate can be used to add calcium to growth medium not needing pH adjustment. Appropriate quantities to add are given in table 7. Calcium sulfate is insoluble and does not water in well, but calcium nitrate is soluble and can be watered in.

Magnesium. Reactions of magnesium (Mg) in growth media are similar to those of calcium. Lower leaves of magnesium-deficient plants exhibit an interveinal chlorosis. This chlorosis may sometimes appear on the upper leaves, as well. Some soilless mixes are low in available magnesium unless the pH has been adjusted with dolomitic lime. Growth media containing vermiculite usually have adequate magnesium levels since vermiculite naturally contains magnesium. Correct low magnesium levels in acid mixes by adding finely ground dolomitic lime.

Magnesium sulfate (Epsom salts), at 4 to 8 ounces per cubic yard (148 to 297 g per m³) or per 100 gallons (120 to 240 g per 400 l, for drenching), is the best material to use in a growth medium not requiring lime. When injecting magnesium sulfate into

TABLE 7   CALCIUM CARRIER NEEDED TO INCREASE SATURATION EXTRACT CALCIUM LEVEL 25 PPM CA

Ca carrier	Ca content (%)	To increase test level 25 ppm Ca, use:				
		oz/bu	oz/yd³	g/m³	lb/100 ft²	g/10 m²
Calcitic lime	30–34	0.21	4.2	156	0.53	259
Dolomitic lime	20–24	0.30	6	222	0.75	366
Calcium sulfate	23	0.29	5.8	215	0.73	356
Calcium nitrate	19	0.35	7	260	0.88	430
Normal superphosphate	20	0.33	6.7	248	0.84	410
Concentrated superphosphate	13	0.51	10.2	378	1.28	625

Note: Analysis by saturated medium extract method

the watering system, do not mix it with any other material unless you are sure it does not contain calcium or phosphorus. Several injectors have been plugged by precipitates formed when magnesium sulfate was injected with a calcium- or phosphorus-containing fertilizer.

MICRONUTRIENTS

Micronutrients are essential nutrients required in small quantities. Many artificial mixes, especially peat-based ones, may be deficient in micronutrients unless appropriate amendments are added. For these mixes, it is essential to add a complete micronutrient mix, 3 to 4 ounces per cubic yard (111 to 148 g per m³) (table 8). This can be done by either a commercial manufacturer or the grower.

All essential micronutrients, except molybdenum, become less available as the pH increases. Hence, to prevent micronutrient deficiencies, it is important to maintain the pH below 6, as well as to add micronutrients.

TABLE 8   MICRONUTRIENT FORMULATION TO MIX INTO GROWTH STOCK MEDIA

Compound	oz/yd³	g/m³
Iron chelate (6% Fe)	1	37.1
Manganous sulfate	1	37.1
Copper sulfate	0.3	11.1
Zinc sulfate	0.2	7.4
Sodium borate (borax)	0.1	3.7
Sodium molybdate	0.03	1.1

Iron (Fe) and manganese (Mn) are the two micronutrients most likely to be deficient, especially at pH above 6.5. Total yellowing of the youngest immature leaves is a good indicator of iron deficiency, whereas mottling or striping of the youngest fully developed leaves may indicate manganese deficiency. For correction of an iron deficiency, use 4 ounces of an iron chelate per 100 gallons (120 g per 400 l) of water. Iron sulfate can be used, but it is less soluble and less effective in correcting an existing deficiency. Especially with alkaline conditions (pH above 7), iron from iron sulfate becomes tied up, whereas iron in the chelate form remains available.

To correct a manganese deficiency, use either manganese sulfate at 1 to 2 ounces per 100 gallons (30 to 60 g per 400 l) or a manganese chelate at 4 to 8 ounces per 100 gallons ((120 to 240 g per 400 l). Never use a manganese chelate in conjunction with iron sulfate. The other combinations of iron and manganese carriers are compatible.

Copper (Cu) and zinc (Zn) deficiencies occur infrequently. Both sulfate and chelate forms of these nutrients are effective. Use 1 ounce per 100 gallons (30 g per 400 l) for the sulfate form and one-fourth ounce per 100 gallons (7.5 g per 400 l) for the chelate form.

Exercise extreme care when applying boron (B)—the difference between deficiency and toxicity is very small. Uniform application is very important and is best done as a liquid solution. Use no more than one-fourth ounce borax (11% B) per cubic yard (9 g per m³) or per 100 gallons (7.5 g per 400 l) water on a one-time basis. Boron in the irrigation water is a potential source of B toxicity. Levels greater than 0.5 ppm may result in injury to sensitive crops. Know the quality of your irrigation water.

Molybdenum (Mo) deficiency is seldom seen, but it may occur when the growth medium is quite acid, near pH 5 or below. Poinsettias are more likely to develop molybdenum deficiency than other ornamental crops. The quantity required is so small that uniform application can be attained only with a liquid solution.

The presence of fluoride (F) may adversely affect the quality of lilies and some foliage plants. More than 5 ppm fluoride in the saturation extract of a growth medium is likely to result in some type of fluoride injury for sensitive plants. Excess fluoride may cause some tip burn and marginal chlorosis. Adding calcium as lime or calcium sulfate (5 lb per yd³, 3 kg per m³) will fix the fluoride in an unavailable form and usually eliminate the adverse fluoride effect.

NUTRIENT BALANCE

Potassium, calcium, and magnesium compete for similar uptake sites at plant root surfaces. Increasing the concentration of one relative to the others changes the relative availability of each. Similarly, a high sodium (Na) level may depress potassium, calcium, or magnesium uptake. Hence, the balance among the essential plant nutrients, especially potassium, calcium and magnesium, is important.

When expressed as a percentage of total soluble salts, the nutrient balance given in table 9 has been found to give the best plant growth. Although the situation given in table 9 is the most desirable, having the nutrients present at other levels but in the same proportions may also represent a nutritionally balanced growth medium. Plant growth is better with balanced nutrient levels even at low fertility. High soluble salt levels are better tolerated by plants in a balanced-nutrient situation.

TABLE 9   DESIRABLE NUTRIENT BALANCE IN SATURATION EXTRACT

Nutrient	% of total soluble salts
Nitrate-N	8–10
Ammonium-N	less than 3
Potassium	11–13
Calcium	14–16
Magnesium	4–6
Sodium	less than 10
Chloride	less than 10

Note: Analysis by saturated medium extract method

SOIL AND WATER TESTING WITH NUTRITION GUIDELINES

Testing media is crucial to avoid nutritional disorders and to ensure that the subsurface environment is properly created and maintained for plant production. These tests fall into three categories: general operations, preplant, and postplant tests. General operations testing includes examining irrigation water quality and conducting injector calibrations. Preplant tests measure media moisture content, pH, and soluble salts prior to flat filling and planting. Postplant tests involve checking delivery (rechecking injectors) and monitoring nutrients, pH, alkalinity, and soluble salts during the crop. All of these tests should be simple and performed frequently.

WHEN TO TEST

General operations testing. Every greenhouse should use water testing. If your greenhouse has a history of alkalinity problems in the irrigation water, buy an on-site test kit and test your water source at regular intervals (we will discuss alkalinity testing in more detail later).

Chemical analysis of your irrigation water is critical for formulating a fertilization program. For example, some water in North Carolina already contains ample calcium for plant production, while another source would supply very little calcium. Growers at these two locations should use different fertilization programs.

Fertilizer and acid injectors should be calibrated *monthly,* more often when you suspect a problem. Injectors are only as accurate as their last calibration, so frequent calibration is essential.

Preplant testing. Preplant testing of media solution pH and soluble salts should be done prior to pot or flat filling. You're about to place a crop into this mix; wouldn't it be nice to know what its chemical properties are before you plant?

The media solution pH will affect nutrient availability to plants, so it is crucial to ensure it's in an acceptable range. Limestone may take two days to two weeks to adjust the pH fully. If you make your mix just before you fill, the pH will be different from what it would be if you mix a few days ahead and moisten the mix. Know the rate of reaction time necessary for your mix to reach its final pH. The best way to do this is to establish a *liming curve* for your mix. The rate of reaction will change with changes in peat source and quality, type, and particle size of the limestone used.

Measure soluble salts to ensure that salt levels are below levels that could cause plant damage. Acceptable salt levels depend on the sampling method as well as the crop being grown. Some growers prefer to analyze media for nutrient content prior to use. This is a good habit, especially if soluble readings are high or if you suspect a problem in the substrate blending.

Postplant testing. A complete postplant testing program should include visual monitoring of the crop's appearance; routine media monitoring of pH, soluble salts, and nutrient concentrations; fertilizer solution analysis, including pH and soluble salts; irrigation water analysis, including pH, soluble salts, and alkalinity; and plant tissue analysis.

The frequency of monitoring depends on the crop being grown. During plug production conduct weekly media and tissue analyses. For finishing flats and pot crops, every two weeks may be sufficient. In an ideal world separate media and tissue tests would be conducted for different plant species, as individual species differ in pH and fertility requirements. Check fertilizer delivery daily, or at least at every fertilization if it is done every second or third irrigation. This can be accomplished simply by capturing some of the fertilizer water in a glass, jar, or beaker and measuring the EC.

Frequency of water analysis depends on the alkalinity content and stability of the water quality. If your water quality, especially its alkalinity, changes frequently, then weekly testing may be needed, especially for plug production. The alkalinity of a water source can change drastically with weather conditions and pumping fluctuations. We have measured alkalinity ranging from 2.8 meq/l to 5.4 meq/l in well water drawn from the same well in North Carolina during the course of one year! Municipal water in many locations is derived from different sources. Although municipalities try to maintain consistent output from water plants, it is possible to encounter alkalinity fluctuations from a municipal water source, also.

For a plug producer weekly measurements may be needed, due to the rapid effects alkalinity can have on plugs, because of the small volume of medium in each plug. Alkalinity effects on larger sized containers occur more slowly, and monthly testing may be sufficient to allow the grower to adjust for alkalinity fluctuations in the water source.

TESTING PROCEDURES

Every greenhouse range should have the capability to measure pH and electrical conductivity (EC). These parameters can change too rapidly to rely solely on lab test results, and the cost of the testing equipment is no longer prohibitory.

When selecting a pH meter, look for an accuracy of +/-0.1 pH unit and a range of 1 to 14. To be useful for fertilizer injector calibration as well as media and solution testing, EC meters should have a range of 0 to 1,990 mho $\times$ 10^5/cm and have an accuracy of +/-10 mho $\times$ 10^5/cm. Many EC meters report EC in units of \micro\S/cm (microSiemens per centimeter). The conversion between S/cm and mhos/cm is simple: 1 S/cm = 1 mhos/cm. Both pH and EC meters are available from many sources, including Cole-Parmer Instruments, 745 North Oak Park Ave., Chicago, IL 60648, phone 800/323-4340; Extech Instruments Corp., 150 Bear Hill Road, Waltham, MA 02154, phone 617/890-7440; and Myron L Co., 6231 C. Yarrow Drive, Carlsbad, CA 92009, phone 619/438-2021.

Representative sampling. Whether you are collecting a sample for in-house testing (media pH and EC) or for laboratory analysis (media nutrient concentrations or nutrient analysis of plant tissue), take a "representative sample." In a problem-free, routine sampling situation, a sample should consist of material from several locations. This will provide a sample of the entire crop or greenhouse. When investigating the cause of a problem, such as why plants look chlorotic, then a representative sample should consist only of the material from problem areas, plants, or water sources. For best results, take a sample from nonaffected areas and submit it separately at the same time to serve as a comparison.

Laboratories require a prescribed amount of material, whether it is plant tissue for foliar analysis, a water sample, or a medium sample. Submitting less than the amount required results in incomplete testing or a delay until additional material is sent. Always be aware of and send the requested sample size for laboratory analysis, and use the sample containers provided by the laboratory, if provided.

Collecting a medium sample for laboratory analysis. When collecting a medium sample, always sample more than one container and collect the sample from all levels in the pots. Draw at least 10 cores of medium, each from a different location within the crop so that many different benches and locations within a bench are included. When drawing a problem sample, make sure to sample only from the affected areas. Exclude the top one-half inch (13 mm) of medium (top one-eighth inch, 3 mm, for plug samples), since it is not representative of the material where plant roots are located and could contain high salt levels, especially if the plant has been watered with a subirrigation delivery system. If a slow-release fertilizer, such as Osmocote, is in the medium, it will be necessary to remove all the fertilizer prills prior to testing to avoid skewing nutrient readings.

Refrigerate samples until they're sent to the lab, or dry them for 24 hours at 125F (52C); don't heat higher than 125F, as nutrient loss from the sample may occur. One cup (8 fluid ounces, 237 ml) of medium is usually sufficient for most laboratories; always send the volume requested by the laboratory.

Recently, affordable Cardy meters for NO_3-N and K have become available. Cardy meters allow the grower to conduct in-house measurements for both NO_3-N and K. However, for use with media solutions, Cardy meters mean that the grower must conduct saturated paste extraction.

Preparing a medium extract to measure pH and EC in-house. Routine on-site analysis of medium pH and EC allows the grower to catch fertilization errors early and to prevent major problems. One of the obstacles to successful testing is the lack of uniformity when many workers do medium sampling and testing from location to location within a greenhouse range and at different times. The best remedy is to assign the task of sampling and testing to one worker.

Probably the easiest method for a grower to measure medium pH and EC is with a 2:1 ratio of water to medium (volume:volume). When collecting a sample for in-house testing, follow the collection procedures outlined previously, taking care to collect a representative sample and removing any slow-release fertilizer, if present. Collect an 8-fluid-ounce (236-ml) volume of medium. To this volume of substrate, add twice the volume (16 fluid ounces, 473 ml) of distilled or deionized water, which is readily available at most grocery stores. Stir the mixture, then allow it to stand for approximately 15 minutes. During this time, calibrate both the pH and the EC meters against standard solutions (these come in kits with the meters) to ensure accuracy of sample measurements. Consult the instructions that came with your meters to know whether you must filter out particulate matter with a coffee filter or cheesecloth prior to reading the pH and EC. Check the sample, using tables 10 (pH) and 11 (EC) as guidelines for interpreting the readings. Out-of-range readings warrant submission of a medium sample for laboratory analysis.

TABLE 10   COMMERCIALLY REPORTED UPPER LIMITS OF NUTRIENT AND CHEMICAL CAPACITY FACTORS FOR GREENHOUSE WATER

Factor		Upper limit for greenhouse crop production		
Name or descriptor	symbol	ppm	meq/l	other
Media pH				
pH[a]	pH			5.4–6.8
Alkalinity[b]		100 CaCO$_3$	2	
Total carbonates	TC	100 CaCO$_3$	2	
Bicarbonate	HCO$_3$-	122	2	
Hardness[c]	Ca + Mg	150	3	
Plug production			1.5	
Salinity				
Electrical conductivity	EC			
Plug production				0.75 mmho/cm
General production				2.0 mmho/cm
Total dissolved salts[d]	TDS			
Plug production		480		
General production		1,280		
Sodium absorption ratio	SAR			4
Sodium	Na	69	3	
Chloride	Cl	71	2	
Macroelements				
Nitrate[e]	N	10	0.72	
Nitrate[e]	NO$_3$-	44	0.72	
Ammonium[e]	NH$_4^+$	10	0.56	
Phosphate[f]	P	1	0.03	
Phosphate[f]	H$_2$PO$_4$-	3	0.03	
Potassium[f]	K	10	0.26	
Calcium[g]	Ca	0–120	0–6	
Magnesium[g]	Mg	0–24	0–2	
Sulfate[h]	S	20–30	0.63–0.94	
Sulfate[h]	SO$_4$	60–90	0.63–0.94	
Aluminum	Al	5		
Boron	B	0.5		
Copper	Cu	0.2		
Fluoride[i]	F-	1		
Iron[j]	Fe	0.2–4		
Manganese	Mn	1		
Molybdenum	Mo	na[k]		
Zinc	Zn	0.3		

Note: Also test for iron-fixing bacteria and other plant pathogens.

[a] Water with high pH should be analyzed for alkalinity. It can be safely used if alkalinity is neutralized. The figures given are the generally acceptable range.

[b] Water with a high level of alkalinity can be used safely if treated with acid to neutralize the bicarbonates and other ions contributing to the alkalinity. Labs differ as to how they report alkalinity: *alkalinity, TC,* and *bicarbonates* are the expressions used.

[c] Hardness is a measure of Ca and Mg content, but it can be used as an indicator of alkalinity. Hard water should be checked for high alkalinity, and it can be safely used if the alkalinity is neutralized.

TABLE 10   COMMERCIALLY REPORTED UPPER LIMITS OF NUTRIENT AND CHEMICAL CAPACITY FACTORS FOR GREENHOUSE WATER (CONTINUED)

[d] TDS readings assume a conversion factor of 1 mmho/cm EC = 640 ppm TDS.

[e] Nitrate and ammonium provide N to plants and should not cause damage at moderate levels. Nitrate and ammonium levels higher than recommended indicate that the water source may be contaminated with fertilizer or something else.

[f] P and K normally occur in very low concentrations in irrigation water. If your water contains more than the recommended levels, it may be contaminated with fertilizer, detergents, or something else.

[g] These figures represent the normal range usually found in North Carolina waters. Ca and Mg content of water should be taken into account during fertilizer programming.

[h] S is usually found at low concentrations. These figures indicate the suggested optimum range of S for most greenhouse crops.

[i] This level is safe for most crops, but it may be toxic for many members of the lily family.

[j] Although 4 ppm is the recommended maximum for plant irrigation, even as little as 0.3 ppm can lead to Fe rust stains on foliage from overhead irrigation.

[k] Molybdenum is required by plants in the smallest quantities of all the nutrients; therefore a molybdenum deficiency is rare in most crops.

Alternative in-house medium testing procedures include the pour-through exfiltrate (VTEM) method and the NCSU "squeeze" method (see the *GrowerTalks* Winter 1995 issue, page 22, for more information on the squeeze). The pour-through exfiltrate method offers the advantage of nondestructive sample collection and the potential for submitting the sample to a lab for nutrient analysis after pH and EC are measured, or after use of in-house meters (Cardy meters) to measure NO_3-N and K. However, guidelines for interpreting the pour-through exfiltrate results are not as complete as for saturated paste and 2:1 (tables 11, 12, and 13).

The squeeze method also allows the grower the option of in-house NO_3-N and K analysis or submitting the solution for nutrient analysis, but it is a destructive sample method, and some crops must be harvested during sampling. Also, as with the VTEM method, interpretive tables are still in the formulation stage of development.

Collecting a plant tissue sample for foliar analysis. Analysis of leaves is the most precise method of measuring the micronutrient and macronutrient status of a crop. Routinely sample plants to establish a baseline of nutrition readings for reference in case of a future problem. For problem solving remember to first collect samples only from problem materials and to send a second sample representing a problem-free site for comparison.

Collect leaf samples in the morning (before noon), when plants are not under water stress. Collect the appropriate number or volume of leaves, as indicated on the instruction sheet included in the tissue analysis kit from your laboratory. Leaves that best represent the crop nutrient status are those that have most recently matured; collect new,

TABLE 11   ELECTRICAL CONDUCTIVITY GUIDELINES FROM VARIOUS LABORATORIES

Extraction method

Saturated paste[a]

Soil-based		Soilless				
CU	NCSU	CU	MSU	FAS	NCSU	Interpretation
—	≤0.75	—	≤0.74	≤0.75	≤0.75	Insufficient nutrition
—	0.75 to 2.0	—	0.75 to 2.0	—	0.75 to 2.0	Low fertility unless applied with every watering
2.5	2.0	3.5	1.99	—	2.0	Maximum for seedlings or newly rooted cuttings
<3.5	2 to 4	—	2.0 to 3.5	0.76 to 2.5 (no bark) or 1.5 to 3.5 (with bark)	2 to 4	Good for most crops
<3.5	—	<5.0	2.0 to 3.5	—	—	Good for established crops
>3.5	4 to 8	—	5.0 to 6.0	>3.5	4.0 to 8.0	Danger area
—	>8.0	—	>6.0	>5.0	>8.0	Usually injurious
—	0.75 to 1.0	—	—	—	0.75 to 1.0	Range for Stage 1 & 2 plugs
—	1.0 to 1.5	—	—	—	1.0 to 1.5	Range for Stage 3 plugs
—	1.5 to 2.0	—	—	—	1.5 to 1.5	Range for Stage 4 plugs
—	1.5 to 4.0	—	—	—	1.5 to 4.0	Range for finish flats of bedding plants

1 media:2 water (v:v)[a]

Soil-based		Soilless		
UC	NCSU	MSU	NCSU	Interpretation
<50	≤25	≤24	0 to ?	Insufficient nutrition
50 to 70	26 to 50	—	? to 100	Low fertility unless applied with every watering
100 to 120	100	75	—	Maximum for seedlings or newly rooted cuttings
<150	51 to 125	75 to 125	100 to 175	Good for most crops
<200	126 to 175	125 to 175	176 to 225	Good for established crops
>200	176 to 200	175 to 225	225 to 350	Danger area
—	>200	>225	>350	Usually injurious
—	25 to 100	—	50 to 150	Range for Stage 1 & 2 plugs
—	25 to 125	—	50 to 175	Range for Stage 3 plugs
—	—	—	—	Range for Stage 4 plugs
—	50 to 175	—	100 to 225	Range for finish flats of bedding plants

TABLE 11   ELECTRICAL CONDUCTIVITY GUIDELINES FROM VARIOUS
                LABORATORIES (CONTINUED)

Extraction method

Pour-through exfiltrate

Soil-based	Soilless		
CU	CU	VTU[b]	Interpretation
—	—	—	Insufficient nutrition
—	—	<0.5	Low fertility unless applied with every watering
0.6 to 1.0	1.5	—	Maximum for seedlings or newly rooted cuttings
—	—	0.75 to 1.5	Good for most crops / Good for established crops
1.0 to 2.0	≤2.0	2.0	Danger area
—	—	—	Usually injurious
—	—	—	Range for Stage 1 & 2 plugs
—	—	—	Range for Stage 3 plugs
—	—	—	Range for Stage 4 plugs
—	—	—	Range for finish flats of bedding plants

Notes: Analysis of media solution using the saturated paste, 1:2, and pour-through extraction techniques. Laboratory abbreviations are CU = Cornell University, NCSU = North Carolina State University, MSU = Michigan State University, FAS = Fafard Analytical Services, UC = University of Connecticut and VTU = Virginia Tech University.

[a] Saturated paste and pour-through ECs are given in mmho/cm (mho × 10^{-3}/cm). The 1 substrate:2 water ECs are given in mho × 10^{-5}/cm.

[b] The Virginia Tech University (VTU) pour-through standards are for outdoor nursery production, not indoor greenhouse production. They are included as a comparison of greenhouse to outdoor culture recommendations.

fully expanded leaves. If no instructions are given for your crop species, collect at least 1 cup (8 fluid ounces, 237 ml) of leaves with the petioles attached. Rinse collected leaves in distilled or deionized water. Do not use tap water, as the water nutrient content may contaminate the foliar sample. Allow the leaves to dry prior to packing for shipment. Leaf samples often rot if enclosed in plastic bags; package in paper bags for best results. Keep samples refrigerated until shipping. Shipping via overnight or next-day delivery is helpful in ensuring that samples arrive at the lab in good shape.

Cardy meters can be used for on-site testing of plant NO$_3$-N and K concentrations, usually petiole sap concentrations. This technique has been used for many years to test the nitrogen status of tomato, pepper, and other food crops. In the future, standards for floricultural crops may allow meaningful in-house testing of crop NO$_3$-N and K concentrations.

Collecting a water or fertilizer solution sample for laboratory analysis.
When collecting a solution sample, allow the water to run long enough to flush all piping prior to collecting the sample. Use clear, nonmetallic (no metal caps, either) sample containers; plastic bottles are ideal. A 16-fluid-ounce (437-ml) sample is more than sufficient. Keep the sample refrigerated until it is submitted to the lab. Transfer samples to the laboratory quickly and avoid prolonged exposure to air.

In-house analysis of water and fertilizer solution pH and EC.
Both EC and pH can be measured in-house. However, accurately measuring water pH is difficult and may require a longer measuring time than for a fertilizer solution or a media extract. This is due to the relatively low buffering capacity of tap water.

In-house analysis of water alkalinity.
Water alkalinity is caused by the presence of carbonate, bicarbonate, hydroxides, and other dissolved salts. It is measured by titrating a water sample with an acid (usually dilute sulfuric acid) to an end-point pH of about 4.6 (it varies from 5.1 to 4.5, depending on the indicator dye used and the initial alkalinity). A pH indicator dye (usually bromocresol green plus methyl red) is added to a known volume of water (indicated in the test kit instructions; usually about 8 fluid ounces, 237 ml), and acid is added until the solution changes color. With the bromocresol green plus methyl red dye system, the color will change from green to pink.

Most water sources acceptable for greenhouse use will have alkalinity of 0 to 8 meq/l (0 to 400 ppm alkalinity expressed as $CaCO_3$). When looking for a test kit, this is the measurement range that is needed. The level of accuracy varies from kit to kit; +/-0.4 meq/l (20 ppm alkalinity expressed as $CaCO_3$) is accurate enough for most situations, but more precise kits are available. We have used Hach Alkalinity kits #24443-01 (about $30 for 100 tests) and #20637-00 (about $155 for 100 tests), and are satisfied by both (Hach Company, P.O. Box 389, Loveland, CO 80539; phone 800/227-4224.) Although the second model is more expensive, it does have twice the accuracy (+/-0.2 meq/l) and also comes with a versatile digital titrator that can be used to measure other solution parameters (using different titrants and indicators), such as water hardness, chlorine, iron, nitrite, and sulfite.

INTERPRETING TEST RESULTS
Most commercial laboratories send an interpretation along with sample results. However, since laboratories differ in procedures (saturated paste extraction vs. a 1:2 extraction), it may not be possible to use interpretative guidelines from one lab for analysis conducted in another, especially for media samples. When interpreting results of media analysis, use the interpretations that correspond with the extraction method employed by the lab (tables 11, 12, and 13).

TABLE 12   INTERPRETIVE VALUES FOR ESSENTIAL MACRONUTRIENTS IN SOILLESS MEDIA

Interpretation	Concentration in extract solution (ppm)				
	Nitrates	Phosphorus	Potassium	Calcium	Magnesium
Extremely low	0–29	0–3.9	0–74	0–99	0–29
Very low	30–39	4.0–4.9	75–99	100–149	30–49
Low	40–59	5.0–5.9	100–149	150–199	50–69
Slightly low	60–99	6.0–7.9	150–199	200–249	70–79
Optimum	100–174	8.0–13.9	175–224	250–324	80–124
Slightly high	175–199	14.0–15.9	225–249	325–349	—
High	200–249	16.0–19.9	250–299	350–399	125–134
Very high	250–274	20.0–40.0	300–349	400–499	135–175
Excessively high	275–299	40.0+	350+	500+	175+

Notes: Analysis of media solution using the saturated paste extraction method. These values and interpretations are from Ohio State University. For a comparison with the five interpretation categories from Michigan State University, see table 1 (where nutrients are listed down left side, interpretations left to right).

TABLE 13   RECOMMENDED ESSENTIAL NUTRIENT RANGES IN SOILLESS MEDIA

Substance	Saturated paste extraction method (ppm)			Pour-through exfiltrate method (ppm)	
	Cornell Univ	Michigan St Univ	Fafard An Serv[a]	Cornell Univ	Virginia Tech Univ[b]
NO3-N	23–68	75–150	40–200	23	50–100
NH4-N	<12	2–10	0–20	—	50
P	5–20	10–20	5–30	15	3–15
K	150–350	75–150	40–200	50	<100
Ca	200–400	125–175	40–200	15	40–200
Mg	70	40–60	28–80	15	10–50
S	—	75–125	—	15	75–125
Fe	—	1–2	0.3–3.0	—	0.3–3.0
Mn	—	1–2	0.1–3.0	—	0.02–3.0
Zn	—	1–2	0.1–3.0	—	0.3–3.0
Cu	—	0.1–0.5	0.1–0.3	—	0.01–0.5
B	—	0.1–0.5	0.05–0.5	—	0.5–3.0
Mo	—	0.1–0.5	0.01–0.1	—	0.0–1.0
Al	—	—	—	—	0.0–3.0
Fl	—	—	—	—	<1
Na	—	<25	—	—	<69
Cl	—	<25	—	—	<71

Note: Analysis of media solution.

[a] Fafard Analytical Services

[b] The Virginia Tech University pour-through (VTEM) standards are for outdoor nursery production, not indoor greenhouse production. They are included for comparison of greenhouse and outdoor culture recommendations.

One point of confusion for many growers is interpreting soluble salt (EC) readings. Interpreting EC requires knowledge of the testing procedure employed. The two major extraction methods, saturated paste and the 1:2 (substrate:water) method, are both reported in mho/cm, but at differing decimal places. Saturated paste EC is usually reported as mmho/cm, which is mho $\times 10^3$ per cm, while 1:2 EC is usually reported as mho^5/cm (table 12). As previously mentioned, some EC meters read in \micro\S/cm; remember that 1 \micro\S/cm = 1 \micro\mho/cm (1 mho $\times 10^6$/cm). Make sure to use the correct decimal placement when using interpretative tables. For example:

1.7 mmho (1.7 mho $\times 10^3$) = 170 mho $\times 10^5$; and 170 mho $\times 10^5$
= 1,700 \micro\mho (1,700 mho $\times 10^6$).

Foliar analysis interpretation varies from lab to lab and crop to crop, especially for macronutrients. Consult the interpretation accompanying the same analysis report for recommendations. Make changes to your fertilization program based on these interpretations. Use table 13 as a guide for interpreting water sample results along with laboratory interpretations. If micronutrient or sodium and chloride levels are out of range, then these ions could potentially lead to toxicity problems if the water source is used for irrigation. If alkalinity is too high, then you may need to acidify to neutralize the excess bicarbonates in the water for pH control.

All this testing seems like a full-time job. It should be—preventing problems is a much more efficient use of an employee's time than fixing problems after they arise.

WATER TESTING AND PARAMETERS

Lack of quality water often causes severe problems for growers. In establishing a new growing operation, an adequate supply of "grower-quality" water must be a number one consideration. You can't grow quality without it! Besides its direct effects on plant tissue, bad water seriously inhibits the results from pesticides, growth retardants, and floral preservatives.

WATER pH

Water pH affects the quality and postharvest life of cut flowers; the stability and efficiency of pesticides, growth regulators, and floral preservatives; the solubility of fertilizers in solutions; and the availability of nutrients in growing media. These specific situations, plus other possible influences of water pH on cut flower and pot plant quality and longevity, are ample reasons why all floriculture firms should be concerned about the pH of their water.

Naturally, the first step in assessing your water pH is to test it. A single test is helpful but not always dependable. Whereas the pH of a municipal water source is generally maintained at a relatively constant level, the pH of well and surface water (pond, river, and so on) can be highly variable, depending on rainfall and other environmental factors. The rule of thumb for testing well and surface water sources is to do so as often as is practical—most certainly when environmental occurrences (such as rainfall) suggest a change might have occurred in the pH level.

Measure pH at commercial and university laboratories or on-site at your firm. Test procedures are relatively simple. Purchase either a high-quality pH meter or pH papers for this purpose. A good pH meter will generally cost $150 or more; besides testing water, the meter can be used for other purposes, such as monitoring growing medium pH. Using pH paper test strips is also a fairly good procedure for monitoring water pH. Test strips, relying on color-sensitive chemical systems, can be quick and easy to use. Both pH meters and papers are available from many industry supply firms and scientific supply companies.

If the pH of your water is within a range of 5 to 7, under most circumstances you have a desirable situation; it is most desirable to be within a range of 5 to 6. The entire issue does not end here, though, and you should clearly recognize that measuring pH is only half the story. It is especially important to understand and measure alkalinity, particularly if you are interested in reducing the pH of your water.

RELATION OF WATER ALKALINITY TO pH

Alkalinity is a chemical factor that is in a manner related to pH, but in itself is a different parameter. The term *alkalinity* should not be confused with the term *alkaline,* which describes a situation where the pH level exceeds 7. Perhaps *basic* is a better, more accurate term to describe a high-pH condition; a low-pH situation should be described as *acidic.*

Water alkalinity is a measure of a water's capacity to neutralize acids. Alkalinity is expressed chemically as milligrams per liter of calcium carbonate equivalents (mg/l $CaCO_3$). Dissolved bicarbonates, carbonates, and hydroxides comprise the major chemicals that contribute to the alkalinity of water.

The alkalinity of water is important in relation to pH because it establishes the buffering capacity of a water source. Simply stated, alkalinity affects the ability to change or modify water pH by adding acids.

An important example of the way in which alkalinity is related to pH modification is shown in the following example and in fig. 1. Grower A has water with a pH of 9.3 and alkalinity of 71 mg/l $CaCO_3$. To reduce the pH of this water to 5, it takes 1.2 ml of 0.1 normal (N) sulfuric acid per 100 ml (1.5 fl oz per gal) of water. In contrast, Grower B has water with a pH of 8.3 and an alkalinity of 310 mg/l $CaCO_3$. To reduce this water's pH to 5, it takes 6 ml of 0.1 N sulfuric acid per 100 ml (7.7 fl oz per gal) of water. Despite the fact that Grower B's water is a full pH unit lower than Grower A's, it takes five times as much acid to lower the pH to 5. The important perspective to

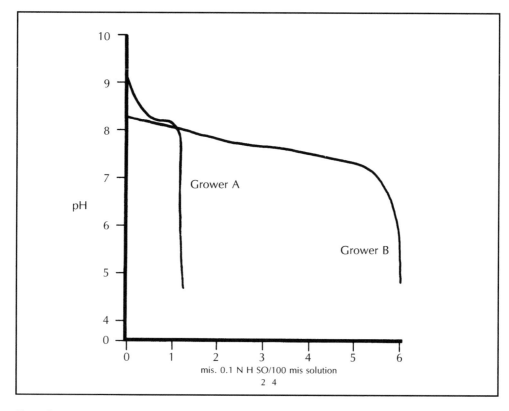

FIG. 1 Changes in pH are shown for two waters titrated with 0.1 normal sulfuric acid. Water from Grower A has pH 9.3 and alkalinity of 71 mg/l CaCO₃. Water from Grower B has pH 8.3 and alkalinity of 310 mg/l CaCO3. Neutralizing Grower B's water will require a lot more acid.

retain from this is that efforts to reduce water pH cannot and should not be based on pH measurement alone! Alkalinity must be considered.

This perspective is especially important whenever and wherever efforts are undertaken to reduce water pH levels. This includes injecting acid to lower the pH of irrigation water, adding acid to water to ensure the stability and efficacy of pesticides and growth regulator chemicals, and using floral preservatives to extend cut flower quality and longevity.

In order to be optimally effective, most floral preservatives reduce the pH of water to about 3.3 to 3.5. This pH reduction suppresses the growth of microorganisms that "clog up" the water-conducting cells in cut flower stems. Often floral preservatives contain acidifying agents, such as citric acid, to reduce the pH of water to which they are added. If the alkalinity of water is extremely high, a situation exists where the pH of the water may not be sufficiently reduced, and the floral preservative might not be optimally effective. This circumstance might account for the fact that floral preservatives might not seem to work well for you or some of your customers. The alternative is not necessarily to add more preservative to the solution, since such products contain other chemicals, which you might be adding in excess amounts. Some floral preservative companies will provide you with a special formulation containing extra acidifier, if you know for certain that the alkalinity of your water is excessively high.

The relationship between pH and alkalinity is also important for the grower interested in reducing the pH of irrigation water. High-pH irrigation water applied to a growing medium can cause the pH to rise further.

Generally, the higher the alkalinity of irrigation water, the more rapidly the medium pH will rise. As might be expected, this pH increase within a growing medium can lead to nutritional imbalances and, consequently, reduced growth and crop quality. Therefore, a program of acid injection for modifying the water might need to be employed. Both alkalinity and pH must be taken into consideration when a program is developed.

Further considerations. High pH and high alkalinity may cause the breakdown of pesticides and growth regulators. In some cases, their effectiveness can be dramatically reduced or completely eliminated. For example, the growth regulator ethephon (Florel) is optimally effective when the pH of the spray solution drops to about 4 after the ethephon concentrate is added to water. If the pH reduction does not occur, then this material may not have the desired effect when applied to plants. Water having high alkalinity may resist adequate acidification by the ethephon concentrate. The resolution to this situation might be to reduce the alkalinity level by adding acid prior to adding ethephon.

Under certain plant production systems where the volumes of growing media are very small and the amounts of water being applied as mist or subirrigation are large, as in plug production or plant propagation, or when growing media with very low buffering capacity, such as rock wool, is being used, the impact of high-alkalinity water can be highly undesirable. Under such circumstances, manage the alkalinity levels in the water to keep them low, approximately 50 mg/l of $CaCO_3$. Large amounts of irrigation water can dramatically alter the pH of the limited volume of growing media or poorly buffered growing media, resulting in serious nutritional imbalances.

When interpreting results of an alkalinity test, we have found water with an alkalinity of 100 mg/l $CaCO_3$ or less to be most desirable. Within this range if pH modification is performed, small additions of acid will generally bring about rapid reductions in pH levels. As alkalinity levels increase above 100 mg/l $CaCO_3$, the amount of acid required to bring about a pH reduction increases. We have noticed a need for a very acute awareness about potential pH-alkalinity problems as they relate to floral preservative, pesticide, and growth regulator effectiveness; as well as problems of rapid, undesirable rises in growing medium pH when the alkalinity level of water exceeds 200 to 250 mg/l $CaCO_3$. Among the water samples we tested from floriculture firms, levels as high as 500 mg/l $CaCO_3$ were found. Levels this high may present very serious problems.

OTHER WATER DEFICIENCY AND EXCESS PROBLEMS

Occasionally, other elements appear in either excess or deficiency in grower water supplies, causing severe problems. Table 14 shows the main elements involved and the approximate upper and lower parameters desirable for most crops.

TABLE 14   DESIRABLE RANGES FOR SPECIFIC ELEMENTS IN IRRIGATION WATER

Substance	Symbol	Range (mg/l)	
		Low	High
Aluminum	Al	0	5.0
Ammonia	NH_4	Undetermined	Undetermined
Boron	B	0.2	0.8
Calcium	Ca	40	120
Chloride	Cl	0	140
Copper	Cu	0	0.2
Fluoride	Fl	0	1.0
Iron	Fe	2	5
Magnesium	Mg	6	24
Manganese	Mn	0.5	2
Molybdenum	Mo	0	0.02
Nitrate	NO_3	0	5.0
Phosphorus	P	0.005	5
Potassium	K	0.5	10
Sodium	Na	0	50
Soluble salts		0[a]	1.5[a]
Sulfate	SO_4	24	240
Zinc	Zn	1	5

[a] Only for soluble salts, these figures express mmhos.

CHAPTER 12

CARBON DIOXIDE: BUILDING BLOCK FOR PLANT GROWTH

by Ralph Freeman
Cornell University Cooperative Extension
Riverhead, New York

Thirty years ago carbon dioxide (CO_2) was one of the hottest topics in the greenhouse industry. Everyone was asking, What is it? Is it really worth using? How much does it cost? Will I obtain results?

During the late 1960s and early 1970s, many cut flower growers invested in CO_2. Their results: quicker cropping, stronger stems, stockier stems, improved quality, and higher yields. Adding CO_2 to the greenhouse climate, growers rapidly discovered they needed to step up fertility programs, increase growing temperatures, and water more frequently. As the industry shifted from cut flowers and into pot plants and bedding plants, interest in CO_2 lessened. Then, in the mid-1970s the energy crisis hit hard! Fuel prices quadrupled, resulting in only a small percentage of growers continuing to invest in CO_2.

But what about today? Should you inject CO_2? Does it benefit bedding plant growers?

WHAT IS CO_2?

Carbon dioxide is a naturally occurring gas in the atmosphere. Its approximate concentration is 300 ppm. CO_2, one of the raw products required for photosynthesis, is colorless, nonflammable, and heavier than air. The gas solidifies under atmospheric pressure at −109.3F (−79C). Solid CO_2 possesses the interesting property of passing directly into the gaseous state without going through the liquid state at atmospheric pressure (sublimation).

PHOTOSYNTHESIS AND CO_2

Photosynthesis is a biochemical process using the sun's energy to chemically combine CO_2 and water (in the presence of catalysts and chlorophyll) to yield chemical energy in a usable form. The actual products of photosynthesis are carbohydrates (sugars and starches), complex chemical compounds, water, and oxygen. A simplified explanation of the process of photosynthesis is seen in the following formula:

$$\boxed{\text{Carbon dioxide} + \text{Water}} \xrightarrow[\text{Green plants}]{\text{Light energy}} \boxed{\text{Carbohydrates} + \text{Oxygen}}$$

Carbon dioxide can be a limiting factor for proper plant growth and development in greenhouses during the fall, winter, and spring because vents are normally closed to conserve heat for extended time periods. Without adequate ventilation the CO_2 level drops below the normal 300-ppm atmospheric level, and with vents closed the CO_2 is quickly used up. With inadequate CO_2, plant growth is under stress, resulting in poorer growth. Figure 1 shows approximate CO_2 levels measured in greenhouses and outdoors during the course of daylight hours.

Over the years research has shown that adding CO_2 during the daylight hours to the greenhouse atmosphere in concentrations three, four, or five times greater than that found in the natural atmosphere will give increased yields, higher quality, and often shorter cropping times.

WHEN TO USE CO_2

Figure 1 demonstrates that a very low level of CO_2 occurs in greenhouses for a significant portion of the daylight hours. Research has clearly shown that deficient levels of CO_2 occur in greenhouses generally between 9 A.M. and 3 P.M. during the fall, winter, and spring months. Therefore, if CO_2 is limited at any time during the daylight hours, photosynthesis and, ultimately, plant growth will be limited. Add CO_2 to the greenhouse environment during these hours to help overcome this limiting factor. If HID lights are used at night, adding CO_2 has been shown to enhance plug growth and development.

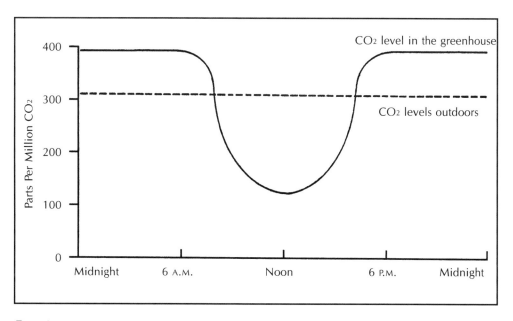

FIG. 1 Carbon dioxide levels as measured in the greenhouse and outdoors over a typical 24-hour period.

Chrysanthemums, for example, respond more favorably to CO_2 in the young plant stage (up to the time of visible flower buds) than when mature. Other plants may respond similarly. With plants showing this response pattern, a grower could stop using CO_2 during finishing.

HOW MUCH TO ADD?

Research results have been inconsistent from university to university, for various reasons, as to the exact amount of CO_2 to add to the greenhouse air for maximum plant growth and development. There are some common threads, though. Based on both research findings and commercial CO_2 use, growers should try to maintain levels between 600 and 1,500 ppm. Do not permit the CO_2 level to exceed 5,000 ppm. First, it is wasteful (costly), and second, somewhere beyond that point, greenhouse workers may experience headaches and listlessness.

SOURCES OF CO_2

There are many different sources of CO_2. One source is simply to ventilate, exchanging greenhouse air with fresh outside air. Decomposition of mulches, compressed CO_2 gas, dry ice, liquid CO_2, and combustion of various fuels (propane, natural gas, or kerosene) are other sources. Many growers remove CO_2 from flue gases after boiler fuels have been burned. To do this, special stack scrubbers must separate CO_2 from the noxious gases. The CO_2 is then collected and injected into the greenhouses. (Table 1 provides basic information on some commonly used fuels.) It is important that sufficient oxygen supplies be provided to all fossil fuel burners so that noxious gases, such as ethylene and sulfur dioxide, do not cause deleterious effects to the plants.

TABLE 1   GREENHOUSE CO_2

Source of CO_2	Generation	Comments
Organic matter	As organic matter breaks down, CO_2 is created and released.	There is little control over the level created naturally.
Solid CO_2	Dry ice is placed in special cylinders, and CO_2 is released as the dry ice sublimes.	The amount of CO_2 entering the greenhouse is regulated with gas flow meters or pressure regulator valves.
Fossil fuels[a]	Burning kerosene, propane, or natural gas results in CO_2 generation.	CO_2 concentration is controlled by the rate of burner firing or the number of units used.
Liquid CO_2	Stored in special vessels.	The CO_2 is released at a controlled rate via gas flow meters or pressure regulator valves.

[a]When purchasing fossil fuel, be sure the supplier is aware that you'll be using it for CO_2 generation. For example, request grade HD5 kerosene, a very low sulfur content fuel. If by chance the supplier has a kerosene lot with sulfur content higher than this grade, he'll choose not to deliver it.

LIGHT, TEMPERATURE, AND FERTILITY

As CO_2 can become a limiting factor reducing plant growth, likewise light, temperature, and fertility can become limiting factors. If, for example, the ideal CO_2 level for a particular crop is between 800 and 1,000 ppm and one or more of the light, temperature, or fertilizer levels were not up to par, growth would be restricted and the full potential of adding CO_2 would be lost.

Know your crop's light needs. Provide as much light as possible, yet not so much as to scorch leaves or blooms.

Higher temperatures may be needed: 62 to 65F (17 to 18C) nights and up to 80F (27C) days. Start ventilating at 80 to 82F and turn CO_2 generators off automatically or manually when top vents are open 4 inches (10 cm).

Fertilizer programs will have to be stepped up—with increased growth, more fertilizer and water are needed. Keep fertilizer levels up at all times. Use soil tests and foliar analysis. Some growers have had to nearly double the amount they formerly used when they start a CO_2 program.

Most important, again: for maximum growth, don't allow any growth factor to become a limiting factor.

CAUTIONS

As growers may have a tendency to keep adding CO_2 to reach the "upper limits," a note of caution is in order. The Mine Safety Appliance Company has reported that a concentration of 8 to 10% (80,000 to 100,000 ppm) CO_2 can be fatal to humans. CO_2 concentration of 1% (10,000 ppm) may cause headaches and listlessness. Concentrations too high make it impossible for the lungs to accomplish their function of eliminating CO_2 from the blood, and suffocation can result from prolonged breathing of such a high percentage of the gas. A small amount of CO_2 in the blood is necessary, however, to stimulate the brain centers controlling respiration. Our basic guideline may be taken from the research: most results indicate plants respond favorably to levels up to 2,000 ppm. Do not exceed this level.

Each CO_2 generator should have approved safety controls. Some of these include a flame failure safety valve, a solenoid valve that automatically switches the fuel on and off, a gas filter, a fine reading gauge showing true jet pressure, and a thermocouple that will shut off the gas flow if the pilot light is not functioning.

Use low sulfur fuels. When purchasing fuels for CO_2 generation, be sure your fuel supplier is aware that it will be used for CO_2. This will aid the supplier in avoiding delivery of batches with abnormally high sulfur content.

Unburned fuels escaping into the greenhouse air can cause serious plant damage. Leakage from the piping system or from a burner operating incorrectly has resulted in flower and fruit abortion, abnormal growth, and numerous other problems.

AVOID AIR POLLUTION PROBLEMS

Self-induced air pollution problems are caused by greenhouse burners and furnaces that are not burning properly. Failures are usually due to inadequate amounts of oxygen being supplied to the combustion process. Other causes may be dirty nozzles, off-center fires, delayed ignition, oil or gas leaks, and pulsating fuel pressure. Any of these factors and more can cause difficulties such as incomplete combustion, resulting in noxious gases contaminating the greenhouse atmosphere. When these gases are present in excessive concentration for sufficient time, damage—such as glazing of leaves, parallel veins, necrotic flecks, spotted areas between veins, or twisted and distorted growth—may occur.

DOES CO_2 BENEFIT BEDDING PLANT GROWERS?

Certainly! Bedding plants, such as geraniums, petunias, marigolds, tomatoes, lettuces, and numerous other crops, benefit from added CO_2. Expect faster seedling growth, faster crops, and improved quality. The greatest response and the most benefit from CO_2 is in the young plant stages, not near crop maturity. CO_2 is particularly beneficial immediately following germination until bud set. Propagators can make very good use of CO_2, especially during rooting of geranium cuttings, as well as in growing seedlings and plugs to the point of transplanting.

Beware: a bedding plant grower using CO_2 can expect all scheduling to change. Plants grow faster in CO_2-enriched environments. Watch your scheduling carefully.

COSTS

Even though the grower today is energy conscious, the costs of installing and operating CO_2 generators are not high. Investments in CO_2 generators amount to approximately \$.10 to \$.12 per square foot (\$1.08 to \$1.29 per m[2]) to install. That's a one-time investment! Equipment maintenance is generally minimal.

The cost of generating CO_2 from either liquid CO_2 or fuels, such as kerosene, propane, or natural gas, amounts to approximately \$.10 to \$.15 per square foot (\$1.08 to \$1.62 per m[2]) per year. Keep in mind that CO_2 will be used only from October through April, with operational hours of 9 A.M. to 3 or 4 P.M., or a total of about 1,000 hours annually in the North.

When one considers the increase in production and improved quality as a result of using CO_2, there is no question about the economics. If the grower uses it properly and manages all factors affecting plant growth, the investments in equipment, maintenance, and costs of generating CO_2 are soon paid with a good return.

MEASURING CO_2 LEVELS

Over the years a number of different CO_2 test kits have been made available to the greenhouse industry. Their purpose was to provide an indispensable aid to help the grower measure the amount of CO_2 in the greenhouse at any time.

Grower reactions to many of these CO_2 test kits have varied. Costs of the kits and supplies were often a problem area. It also seemed most of these kits soon found their way to dusty corners and were forgotten, due either to the time needed to take a test or inaccuracies. There are now, however, very accurate instruments available for measuring CO_2 concentration in air. Currently, the investment in these special sensors and meters is reasonable.

Even if you don't have CO_2-sensing equipment, you can still calculate CO_2 requirements. First, calculate the volume of the greenhouse into which CO_2 will be injected or generated.

Volume = Length × Width × Average height

Greenhouse vol. (for example) = 100 feet × 25 × 12 = 30,000 cubic feet

Second, calculate the hourly CO_2 requirements. Keep in mind that a greenhouse full of plants uses up on the average at least 400 ppm (0.04%) CO_2 per hour.

30,000 ft^3 × 0.0004 CO_2/hour = 12 ft^3 CO_2/hour.

Knowing that 12 cubic feet of CO_2 per hour are required to maintain 400 ppm, the grower can set the pressure regulator flow control valve on the liquid dry ice converter, or the gas-fired unit's pressure gauge, to deliver an adequate amount of CO_2. All reliable gas equipment should have charts to relate gauge pressure, propane, or natural gas used per hour, and the number of pounds of CO_2 generated per hour. Although the exact concentration may not be measured, combine good judgment, experience, and continued use to make necessary adjustments for daily weather changes to avoid excessive concentrations.

CHAPTER *13*

HIGH INTENSITY DISCHARGE LIGHTS

by C. Anne Whealy, Ph.D.
Proprietary Rights International
Roanoke, Texas

The application of greenhouse supplementary lighting, specifically high intensity discharge (HID) lamps, continues to increase in low light regions of North America and Europe. With proper cultural and environmental control, supplemental lighting enables greenhouse growers to increase productivity, reduce crop time, and produce consistently high-quality plants year-round, regardless of the weather or season.

BENEFITS OF SUPPLEMENTAL LIGHT

Plants produced under insufficient light levels grow slowly and have reduced vigor, stretched narrow stems, poor branching, and small leaves. Adequate light levels produce high quality plants that grow faster and flower earlier, are more compact with good branching, and have thick stems and large dark leaves.

Low light is due to the natural shortening of day length, lower angle of the sun or an increase in number of cloudy days. In the northern United States, the light intensity during November, December, and January is less than one-third of July's. Natural light levels are adequate April through August, and supplemental lighting is not economically justifiable.

The greenhouse structure, type, condition of the glazing material, and overhead equipment and hanging baskets can further reduce light levels inside the greenhouse to one-half that of outside the greenhouse. During the low light periods, light becomes the limiting factor to plant growth.

Increased profit from increased yield of high quality crops at the most favorable market price is the driving force behind installing supplemental lighting. When contemplating a supplemental light installation, consider the following: location, crop(s) to be produced, available markets, electric energy rates (present and future), and cost-benefit.

Economic considerations

Supplemental lighting costs are estimated at about $2 (U.S.) per square foot for installation or about $250 per lamp/fixture; the lamp is about 30% of this cost. Operating costs (including ballast energy) are 0.47 kw and 1.1 kw per 400- and 1,000-watt lamp, respectively. Payback time on supplemental lighting system investments is two and a half to three years.

Some power companies offer greenhouse growers reduced rates for off-peak use of electricity for lighting. Rebate incentives on lighting system purchases are also available from some utility companies.

Supplemental lighting provides supplemental heat, which reduces operating costs by decreasing heating fuel usage. Four hundred and 1,000-watt high pressure sodium (HPS) lamps/ballasts produce about 1,600 and 3,750 BTUs per hour of operation, respectively. Or, 28 1,000-watt HPS lamps operating for one hour will produce the same amount of heat as 1 gallon (4 l) of fuel oil or 140 cubic feet (4 m^3) of natural gas. It is estimated that HPS lighting can reduce heating costs by as much as 25 to 30%. The higher the light levels and longer the duration of use, the more heat is produced and the greater the reduction in heating costs. Lighting can also reduce greenhouse humidity levels. Heat given off by supplemental lighting also reduces the payback period.

LIGHT AND PLANT RESPONSES

Sunlight contains visible light of all colors in the rainbow and invisible radiation in the form of infrared and ultraviolet radiation. The radiation's wavelength, measured in nanometers (nm), determines the color of the light: violet, 400 nm; blue/green, 500 nm; yellow, 600 nm; and red, 700 nm. The visible spectrum is the radiation emitted between 380 and 780 nm. Light is composed of particles called photons that have a distinct energy level with respect to wavelength. The amount of energy increases as wavelengths become shorter. For example, blue light photons have a higher energy level than red light photons. Both visible and invisible light becomes heat when absorbed by an object, except for the energy utilized by the plant as chemical energy in the photosynthesis process.

RESPONSE TO WAVELENGTHS

Plants respond differently to different wavelengths of light. Maximum plant sensitivity for photosynthesis occurs at 675 nm. Plant leaves contain chlorophyll, a yellow-green pigment that reflects or transmits yellow-green light and absorbs blue and red light. Red light is necessary for photosynthesis and chlorophyll synthesis and promotes seed germination, seedling growth, stem elongation, flowering, and anthocyanin formation. Blue light is also necessary for photosynthesis and chlorophyll synthesis and reduces stem length and dry weight, but increases branching and stem strength. Blue light has also been shown to improve leaf and flower color. The ratio of far-red (735 nm) light to red or blue is very influential in altering plant growth. A reduction in the ratio of far-red light to blue or red light will decrease internode length and thus plant height in various flowering crops. By changing the spectral quality of the greenhouse light environment, it is possible to alter plant growth.

MANIPULATING LIGHT QUALITY

Light quality or spectral energy distribution can be manipulated by using a specific light source or a combination of light source(s) (see following section), or by using filters. Effective filters can be liquid solutions of copper sulfate in sealed double-wall glazing materials. The most promising approach is simply the use of pigmented glazing materials that absorb, reflect, or block specific wavelengths. Some coverings also reduce the heat load in the greenhouse by filtering out the longer wavelengths.

Several universities and glazing manufacturers are researching the effects of spectral distribution in the greenhouse environment. By the end of the decade, controlling plant growth by manipulating the spectral quality of the greenhouse light environment may be commonplace.

SUPPLEMENTAL LIGHT AND PLANT RESPONSES

Maximum plant densities result in the most economical use of HID lighting. Therefore, lights are most beneficial during early developmental stages. Lights should be given to bedding plants when the first true leaves are green and continue for four to six weeks. Lighting beyond six weeks is not economically worthwhile. On vegetatively produced crops, cuttings benefit from supplemental lighting that begins during propagation and continues until flowering.

Irradiation level and duration depend on crop, day length, latitude, and cultural conditions. A lower irradiance for a longer period of time is preferable to a higher level of irradiance for a shorter period of time. For example, it is better to give 300 footcandles (3 klux) for 24 hours than 600 f.c. (7 klux) for 12 hours. For most cut flowers and potted plants, light levels of 400 to 500 f.c. (4 to 5 klux) for 18 hours per day will increase plant growth, quality and profits. For plugs, 300 f.c. (3 klux) is the minimum for economic return. Light levels of 600 to 700 f.c. (7 to 8 klux) will increase growth, but are not economically justifiable. Levels higher than 700 f.c. (8 klux) may cause plant damage.

In growth chambers or growth rooms, 2,000 to 6,000 f.c. (22 to 65 klux) are used. Typically, metal halide (MH) or a combination of HPS and MH are used to balance the spectral distribution. In interiorscapes, use MH lamps that provide 100 to 350 f.c. (1 to 4 klux).

For most cut flowers and potted plants, provide lights for four to six hours on sunny days and 12 to 16 hours on cloudy days. Light most bedding plants for 18 hours. Extending lighting to 24 hours does not result in appreciably more plant growth.

TYPES OF SUPPLEMENTAL LIGHT

There is no one light source that can simulate the sun's growth effects. Most light sources emit radiation from 300 nm to 2,500 nm, but not all wavelengths are emitted by all light sources. In theory, the most effective assimilation lamp would be the one that emitted most strongly in the red range at 675 nm; however, without blue light, plants would be elongated.

Besides spectral energy distribution, efficiency, output, uniform light distribution, and the crop, are critical considerations. Efficiency—the best return on the electricity used—depends upon light source, type of reflector, and benefit to the crop. Uniform light distribution translates into uniform production. Uniformity depends upon lamp wattage, amount of area lighted, reflector design, lamp spacing, and height of lights above the crop. Uniformity of light distribution should be at least 85%.

Incandescent lights. Incandescent lights emit light primarily in the far-red range and are considered rather inefficient. Incandescent lamps convert only 6.5% of the energy consumed into light energy; the remainder is emitted as heat. Incandescent lamps have a relatively short life span (one-tenth of fluorescent lamps), but initial investment costs are low. Because of these factors, incandescent lamps are used primarily for photoperiod control and are not recommended for supplemental lighting.

Fluorescent lights. Fluorescent lighting is cooler, more efficient, has a better spectral balance, and a longer lamp life than incandescent. However, fluorescent lights are expensive and cause shading problems due to their large fixture size. Because of the lamps' bulk, they are generally only used in special situations, such as growth chambers, where they are installed in combination with incandescent lamps to balance the blue-to-red light ratio for plant growth. Fluorescent lamps are significantly less efficient than HID lamps: the growing efficiency of a fluorescent lamp is

0.83 per watt compared to 1.77 for a 1,000-watt HPS lamp. Lower efficiencies mean that more fixtures are required, adding to shading problems. Also, lamp degradation is more than three times greater for fluorescent lamps than for HPS lamps.

HID are preferable to incandescent and/or fluorescent lamps because of high efficiency, uniform light distribution, and lower amount of shading.

Metal halide. Metal halide (MH) lamps emit white light that is similar to daylight as perceived by the human eye. This white light is very effective in retail areas or interiorscapes where color and plant appearance are important. HPS lamps are not desirable in these areas because they emit a yellowish light. MH lamps provide the best spectral distribution of all lamps for plant growth, providing more reds, far-reds, and blues than HPS lamps. However, they are less efficient, less cost-effective, and shorter-lived than HPS or low pressure sodium (LPS) lamps. MH lamps have an energy conversion efficiency of 120 to 125 lumens per watt compared to an efficiency of 135 to 140 lumens per watt for HPS lamp.

Low pressure sodium. Low pressure sodium (LPS) lamps are the most efficient light source since they have the highest rating for lumens per watt and generate less heat than other lamps. The disadvantages of LPS lamps are undesirable spectral distribution, high cost, and large fixture size that causes shading problems.

High pressure sodium. High pressure sodium (HPS) or sodium vapor lamps emit mostly in the yellow-orange-red range, 550 to 700 nm. HPS lamps are considered the best supplemental light source because they provide more photosynthetically useful light per unit of electricity than other light sources, and their efficiency is almost as high as LPS lamps: 25% versus 27% conversion efficiency. HPS lamps degrade slowly and have longer useful lives than other lamps. For example, 400- or 1,000-watt sodium lamps are rated to last 24,000 hours; MH, 20,000; LPS, 18,000; and fluorescent, 15,000. HPS fixtures are more compact with less bulky reflectors and cause fewer shading problems than fluorescent or LPS lamps.

HPS lamps are available as either 400- or 1,000-watt units that provide an output of 50,000 and 140,000 lumens, respectively. The trend is moving toward the 400-watt units. One-thousand-watt lamps require a high mounting height that may not be possible in many greenhouses, and 400-watt units are more compact and less expensive to install and maintain.

Lamp reflectors. A lamp reflector is a mirror that directs the light from the fixture to the crop. The objective in reflector and fixture design is to maximize uniformity of light distribution and efficiency of power consumption. Reflectors allow increased distances between lamps and lower mounting height. Therefore, better reflectors reduce the number of fixtures required. Well-designed reflectors direct

90% of the light to the plants; poorly designed reflectors may direct less than 50% of the light.

HID units are being developed that use less electricity, operate more efficiently and coolly, are more compact to reduce shading problems, and cost less to install and maintain.

To benefit from supplemental lighting, you must provide optimum temperatures and carbon dioxide, water, and fertilizer levels to maximize plant growth. Supplemental lights will eliminate light level as a limiting factor in plant growth, but if other factors become limiting, using supplemental lights will not be worthwhile.

SECTION 3

THE BUSINESS OF GROWING

CHAPTER 14

COLLECTING YOUR BILLS

by Andrew C. Stavrou
Ball Horticultural Company
West Chicago, Illinois

Granting credit involves taking on risk. The control you administer regarding who you give credit to and the techniques you use to collect past-due amounts owed your business have a real impact on the profitability of your business.

There are two basic areas of effective credit and collection management:

Control—who you give credit to and why;

Collection tips—how to get your past-due money paid to you as quickly as possible.

CONTROLLING CREDIT

Let me emphasize one very basic, often overlooked point: No matter how good a collection technique you employ, you will only be paid when the other party has the money to pay you. Today's greenhouse grower is being called upon more and more to grant credit. That doesn't necessarily mean that it is in his or her best interest to do so. Being prepared when entering into a business relationship with *information* as the basis for making your credit decision will allow you to manage accounts receivable more effectively. Before you jump into an open credit situation, keep the following principles in mind.

KNOW YOUR CUSTOMER

I am not of the school that mandates signed credit applications from everybody. If you know the account that you have called on is solid (national chain store, long-time local merchant with a great reputation), that should be good enough. But there will always be those situations where you really don't know much about the other party. Being prepared to ask for specific information is just plain, good business sense. I recommend using a short form credit application to get the information you need (fig. 1).

As you can see, the short form credit application asks for the names of the principals, how the business is set up (corporation, partnership, proprietorship), when it started, a bank reference, and three trade references. Also on the short form is a line specifically related to the credit line the customer desires.

This single piece of paper will give you a good deal of background information and should be enough to take care of the credit relationship question. Assigning the credit reference checking to a person in your office is a good idea. If you know of any other companies that the customer in question is doing business with, have that person check with them as well. Run a check for uniformity in the way the applicant pays his or her bills; all references should show payment as being prompt, with banking relations being satisfactory and *no* NSF (not sufficient funds) checks showing up. If you see a disparity between the references that the customer supplied and the references that you supplied yourself, you can figure that trouble is brewing. What the applicant has done is to select the references he or she wanted you to contact, neglecting those that he or she has not been taking care of.

What you will find is that most creditworthy accounts will be glad to share their trade and bank reference information with you on this short form. As a group, these customers are proud of the relationships that they have built up through the years and are proud of them. By seeking out that information along with the credit line desired, you are starting your business relationship in a very healthy manner. Both sides know exactly what is expected.

INVOICING THE TRANSACTION

Properly invoice the transaction. Delays in payment are often due to reasons other than lack of ready cash. Your invoices should always contain complete information, which includes, but is not limited to, the following:

- ◆ Full name and address of the customer along with the telephone number and name of your contact
- ◆ Invoice number and date
- ◆ Date of shipment
- ◆ Full product description
- ◆ Customer purchase order number (if applicable)
- ◆ Specific terms of sale (if net 30 days, say so)
- ◆ Specific amount owed

XYZ Greenhouse			**Confidential** **Credit Application**

Firm Name	Phone	Date	
Mailing Address	City	State	Zip Code
Street Address	City	State	Zip Code
Shipping Address	City	State	Zip Code

Company Owners or Officers

Name	Title	Social Security No.	
Home Address	City	State	Zip Code
Name	Title	Social Security No.	
Home Address	City	State	Zip Code

Corporation	Partnership	Sole Proprietorship	Date Established	Sales Tax No.

Mortgage Holder/Landowner   Are you operating under the jurisdiction of any court?   ❏ Yes   ❏ No

Name	Present Balance	Monthly Payment	
Address	City	State	Zip Code

Bank Reference

Bank Name	Name of Banker		
Address	City	State	Zip Code

Checking	Account No.	Savings	Account No.	Borrowing	Non-Borrowing

Trade References

Name	Amount Owed	
Address	City	State	Zip Code
Name	Amount Owed	
Address	City	State	Zip Code
Name	Amount Owed	
Address	City	State	Zip Code

Credit Line Desired

I certify that the above information is correct.

Signature _____   Title _____   Date _____

FIG. 1   A short form credit application

◆ Latest day the prompt pay will be considered as a full payment (if allowing a prompt pay discount)
◆ Specific due date

There have been many instances, especially with larger chain stores, where if a piece of information is missing, not only are you not contacted about the missing information until you request payment from the customer, but you are also put on the bottom of their payments list once you supply the correct information. Of course, this will negatively impact your cash flow. Be your own best friend. Be sure your invoices have all the necessary information, and if you have specific customers who are requesting additional specified items, be sure to include this information on the original invoice. So much of what you control rests on documentation. Follow through on the front end; dividends will come in the form of prompt payments.

Be timely. Generate invoices immediately after the ship date—do not delay. It will only delay your payment. Remember, inventory equates to accounts receivable once it is out the door. There should be absolutely no lag time between your moving that inventory and having it become an accounts receivable entry. Generally speaking, the payment terms do not begin on the date of the shipment, but on the date the invoice is generated. Again, this is a question of being your own best friend. Get the invoice out as soon as the material ships.

COLLECTION TIPS

Assign one person to collection. If possible, give it to somebody who is not involved in sales. You will also want to give this person a title (collection supervisor, account representative, credit manager). Any of these titles would be appropriate, depending on just how important the collection function is to your business. The larger, more diverse your customer base is, the more I suggest making the title Credit Manager.

Track your past-dues at least twice per month. A letter should go out requesting any past-due money when the account has invoices that are 15 days or more beyond the due terms. On an initial letter request, especially with any institution account (school, hospital), it is a good idea to include a copy of the invoice with the letter. This should and will cut down on some of the collection time necessary for receiving payment.

Keep track of all letters you send. This should include the name of the contact and the date the letter was sent. Copies aren't really necessary as long as you know the tone of the letter, thus making a follow-up easier.

Telephone accounts that have not responded to your letters. Within 10 to 15 days after the letter is sent out, a follow-up phone call is the next step. Again, keep a log by account of exactly what activity has taken place. You should indicate who you spoke to and what was said. This will allow you to build a historical file with which you will be able to match promises with results. Therefore, you will be able to track those accounts that were able to keep their word versus those who were stalling for more time.

Make every call count. If you can't reach your contact, leave a complete message and get the name of the person you are leaving the message with. Always inquire and be sure to listen. You will find that many times you can spin off to another party and get the answer you need. For situations where this simply isn't the case, be sure to leave yourself in a position where your next phone call won't be a surprise to anyone. Also insist on a return phone call, and ask when you can expect it.

If you contact the correct person, always attempt to work out a complete payment arrangement on the entire account. It is very easy to get on the phone and work up a partial payment situation, which will necessitate an additional phone call down the road. Pushing for a comprehensive payment plan is the best idea. You will also be able to figure out whether the slow payment that you are calling about is a one-time deal, or whether you can expect the same sort of thing to happen in the future.

Follow up any commitment so that if the commitment does not come through, you will contact that party by phone within three business days after the account failed to live up to its promise. Again, it is important to be as specific as possible. If the contact tells you a check will be in the mail within a few days, ask for a specific date. This will allow for easier tracking of the money you are expecting.

Use new shipments or orders as leverage to collect old balances. It is important that the person who is in charge of your collection area interface closely with the sales and distribution areas. It is frustrating to try to collect an open account receivable balance you are concerned about when you keep shipping material to the account. You are sending a very clear signal to the customer that he can get away with pretty much anything he wants.

Keep a running tab of your aged trial balance and build up a historical file. Do this with each account as well as your accounts receivable records. It will enable you to see how much of your total receivables is current, past due by one to 30 days, etc. It is a great tool to use to weigh out your short-term borrowing requirements, if you have any, so you can have a better handle on what you can expect your receivables to do. Also use these reports for reviewing open accounts. Regular meetings between your collection person and ownership is always time well spent. In fact, monthly meetings are advised. It might take 10 minutes or three hours—either way, it's good to review what is there.

Don't state something you don't mean. This is where the interface between sales and credit areas becomes very important. Before you allow your collection person to use an order as leverage, make sure you are prepared to cancel. It is important not to bluff, as credibility will quickly slip away.

What to do if the collection looks hopeless. Plan ahead and talk to a collection agency or attorney before the next critical phone call takes place. By getting your options in line, you will know exactly what to do should payment not be made. Collection agencies are extremely competitive; most of them base their fees on strictly a contingency basis. (They get a percentage of what is collected. If nothing is collected, they get nothing.) If you plan to send something to a collection agency or attorney, do not ship anything further until the old collection situation has been resolved. This is also a great opportunity for the boss or owner to take over the account from the collection person. This adds a sense of urgency to the situation that may just get some cash rolling in.

Set credit lines. Not to be overlooked is the importance of setting credit lines and revising them as needed. As a basis for revision, use historical data (how the account has paid and kept commitments, as well as information on the short form credit application). Everyone in your account records should have a credit line (a credit limit) assigned to them. This doesn't mean that it is an out-and-out maximum limit. It also doesn't necessarily mean that it has to be exceeded before you do anything on the collection end. The invoice due date should trigger a collection response. If you see that an account is not able to take care of obligations in a timely manner and it looks like credibility is slipping, you might lower the credit line down to the existing accounts receivable balance and start holding orders.

On the other hand, you may have a situation where a customer has been paying his bills on time and your sales volume can grow appreciably. Ask yourself the questions that you would ask the customer on the short form credit application. Do you have enough information to allow you to increase the credit line and expand your sales—and still let you feel good about your specific situation? If your answer is no, then contact your customer and have him or her fill out the credit application, explaining that your credit exposure is going up and you need this additional confidential information.

Use a computer. We have talked about the need to be timely in charting out the trial balances of your accounts, both individually and in their entirety, and about the importance of getting your invoices out promptly. Of course, using a computer would make these functions easier. I recommend purchasing or leasing a computer for your aged trial balance as well as for invoicing purposes for any operation. This expense will more than pay for itself.

Do your homework. Credit and collections call for persistence. Doing your homework on the front end, knowing who your customers are, having background information on them, and keeping background information to use when you need it, makes the collection effort much easier. It also allows you to set a realistic credit line based on the risk involved. If a collection effort is necessary, make each call count, document what you are doing, and keep track of the success ratio your collection person is having. This will mean credit and collection decisions will be easier to make down the road. It also makes the credit and collection functions easier in case you have to take over from someone who has left your company—you will already have information that speaks for itself.

It is also important to remember that chronic collection problems may best be dealt with by simply not doing any additional credit business with the company involved. This, of course, assumes that you have a home for the material that otherwise would have gone to this customer. You should always strive to pool your orders toward companies that can pay within your terms. Chances are you have some good customers, and you have some customers who could be better. Being your own best friend in this scenario will make life easier for you and will allow for healthier accounts receivable records that will serve as the foundation for your business growth.

CHAPTER 15

ENVIRONMENTAL LAWS

by Orlo Ehart
Next Wave Enterprises
Belle Mead, New Jersey

Greenhouse and nursery growers, no matter how conscientious, are regulated businesses. Fortunately, not all aspects of the federal, state, or local environmental laws apply to greenhouse and nursery growers; unfortunately, however, no single comprehensive law governs growers either. Therefore, it is the grower's responsibility to determine which laws apply and to comply with them. This has become a daunting task. For example, in 1971 the *Environmental Law Reporter* summarized the national environmental laws in 33 pages; it now takes over 3,500 pages to summarize them. This does not include the volumes of regulations that interpret and implement the laws or the state and local requirements that may also exist.

Yet some growers are reluctant to consider environmental management as a part of their overall business plan. There is no question that government can have more impact on the financial bottom line than any competitor; that is, no competitor can force a business to close its doors like an inspector can.

Some laws prohibit any activity that adversely affects the environment. As a goal that may seem reasonable, but taken to extreme, it may not be practical. For example, raising crops or building a greenhouse requires purposeful disruption of the environment; under some laws and in some instances, these effects can be considered adverse to the environment. Since strict interpretation and enforcement would prohibit "farming as we know it," the adverse effects on business could be substantial too. As a result of the well-intended or unintended consequences of some laws, environmental management has become an important business practice for the agricultural industry.

While political rhetoric has recently supported more of a balance between risks or costs versus benefits, an approach that is potentially favorable to businesses, public sentiment is still pro-environmental protection and regulation. Given the public popularity of environmental issues, a significant rollback of environmental programs is unlikely. Even if laws are repealed, concerns about potential liability will continue to emphasize a business's environmental stewardship. Laws only establish minimum standards; in the wake of environmental disasters such as Bhopal, India, and the Exxon Valdez, the public expects all businesses, as good neighbors, to live up to reasonable standards. Some expectations are even more onerous than actually required by law. Most agricultural lending institutions require borrowers to submit environmental management plans to qualify for loans. Due diligence investigations that focus on environmental contamination issues are performed for most real estate transactions and also to qualify for insurance coverage. In addition, asset value is affected by the quality of a business's environmental stewardship efforts. Contaminated property loses value and cleanup costs are often prohibitively expensive. It is possible that current practices may make it impossible to operate the site as is, to modify or remodel it, to sell it or give it away.

Regardless of the desire for and possibility of regulatory relief, it is not prudent to ignore environmental issues: environmental compliance clearly must be emphasized in future business plans.

DETAILS SOMETIMES OVERLOOKED BY GROWERS

To better explain the impact of environmental laws on the grower, the term environment should be defined: it is the physical, chemical, and biological surroundings including climate, soil, water, air, and all species of plants, animals, and microorganisms. Simply put, the environment is everything around us, including the factors that affect our health, safety, and welfare. With this broad definition, many business practices can be affected by changes in environmental laws. In addition, growers should not assume that a word usage in a law is necessarily the common word usage. Growers need to carefully read law definitions, because the way a word is defined can make a huge difference in what is regulated.

The formation of agricultural policy no longer involves only the differences of opinion among agriculturists. Now, others who are far afield from agriculture also affect agricultural policy. Growers no longer have the luxury to make a profit using some of the same techniques that were acceptable just a few years ago. Growers must now be very knowledgeable about civics and willing to regularly participate in public forums, oftentimes openly discussing family business practices in order to keep reasonable compliance options available while achieving difficult environmental protection goals.

There are important distinctions between laws, statutes, ordinances, codes and rules and regulations; for sake of simplicity they are primarily referred to in this chapter as laws, except when a distinction is necessary for accuracy. In the broad scope portrayed here, laws establish, among other things, authorities, rights, penalties, procedures, mandates, prohibitions, protections, standards, and codes or concepts of conduct. Statutes are passed by elected federal and state legislative bodies; ordinances or codes are passed at the local level by municipal, county, and similar forms of government. Some principal aspects of law are historic; these common law tenets, including nuisance, negligence, liability, and torts laws, also regulate a grower's conduct. Common law joins statutes, ordinances, and codes as the body of laws that govern. Rules and regulations are promulgated by executive branch agencies to interpret, implement, and enforce laws. Except for common law, all allow for some form of public input before they are enacted. It is important to understand the differences in order to effectively participate in developing or commenting upon laws and rules. For example, it is imperative to know which legislative committees have jurisdiction or which agency has authority to act upon a request. Those participants who are well versed in procedures and responsibilities have better successes in crafting compromises that meet their needs.

DESCRIPTIONS OF LAWS

Just as there is no law that details all the grower's environmental protection requirements, there also is no single statute that covers all environmental protection issues. Instead separate laws cover individual or similar environmental issues, for example, air, water, pesticides, waste; some of them overlap in authority. While the laws of a few states repeat federal requirements verbatim, many laws spell out new authority to regulate products and/or activities without regard to other laws. Local laws can affect environmental management options. For example, zoning requirements and other land use control measures. The result is that legal requirements may be confusing and some laws, while well intended, may be difficult to comply with or even inconsistent with requirements of other laws and available technology.

Summaries and descriptions of laws are by nature incomplete and inadequate for determining compliance requirements. The following descriptions are not intended as a substitute for an in-depth review of requirements or advice of counsel and other experts. The laws are described in a way that gives the grower an awareness of the breadth of environmental control and the grower's minimum responsibility as a good neighbor and environmental steward. The actual laws will need to be consulted for details because laws are dynamic and specific citations and requirements are beyond this chapter's scope.

LICENSES, PERMITS, AND FEES

Permission is frequently required from a state or local authority to conduct business. Business and location licenses, fees, and marketing order check-off payments may be required. Some states use permits as a primary requirement for engaging in any activity. Building, well drilling, ditch digging, waste storage and wastewater, pesticide application, and environmental impact assessment permits are a few that might be needed. Overlooking the fulfillment of these requirements can result in citations just as costly as actually causing environmental damage. Once a business is cited as out of compliance, it frequently is inspected more regularly and rigorously. Therefore, check with local and state officials or counsel to be assured that the proper papers are in order. Local chambers of commerce or state marketing departments will generally assist small businesses with such requirements. It is also important to use the licensing or permitting office of regulatory agencies as a resource to determine other requirements needed to conduct business legally.

PESTS

There are plant-quarantine and -inspection, seed, and noxious weed laws that limit movement of products or require control of species so that pests do not become established elsewhere. These laws are aimed at reducing the chance of destruction of the environment and growers' crops by foreign pests, such as the gypsy moth, leafy spurge, the Mediterranean fruit fly, and the soybean cyst nematode. Certificates are necessary for movement of many plants; some may not be shipped. The United States Department of Agriculture (USDA) Animal and Plant Health Inspection Service (APHIS) has also used its plant pest authority to establish plant pest biotechnology regulations. The state agricultural agency and APHIS are the general regulatory contact points for information on the Plant Protection Act. Each state has a seed inspection service, generally as a part of an agricultural agency or a land-grant university.

PESTICIDES

Pesticide laws began as product guarantee laws in the early 1900s. Today there are so many controls placed on pesticide use that it is possible to provide only a cursory glance at the magnitude of these laws. The Environmental Protection Agency (EPA) administers the Federal Insecticide, Fungicide, and Rodenticide Act (FIFRA), while USDA provides information and administers portions of the pesticide recordkeeping requirements. The product label used to be the sole document that provided restrictions on use. As a result of the EPA placing many of the provisions for protecting agricultural workers into regulation, the label no longer serves as the only document for users to refer to in complying with federal restrictions on pesticide use.

States also have laws governing the use and control of pesticides. They can be more restrictive than federal law. State pesticide control programs, which also enforce the federal law, are generally located in the state agricultural agency, or, in approximately 15 states, they are housed in the state environmental agency. County extension agents, in conjunction with state lead agencies for pesticide control, are required to provide education to meet certain aspects of the law. There is a county extension office in nearly every county in the U.S. and a state pesticide education coordinator at most land-grant universities. The Occupational Safety and Health Administration (OSHA) also has regulations governing worker safety (see the "Right to know" section) that pertain to pesticide use and respiratory protection, other protective equipment, required education, and safety precautions. Some additional restrictions on pesticide use are discussed under subsequent sections as well.

General requirements. Nearly all aspects of pesticide registration, transportation, storage, and use (including mixing, loading, handling, and disposing) are regulated under FIFRA. All pesticide uses must be registered through the EPA by manufacturers or sellers and can only be registered if they pose no significant adverse effect on the environment. In other words, the benefits outweigh the risks. Some special provisions for exemptions from registration for emergencies and special state labels do exist. Growers, by federal definition, are private applicators, although some states have defined greenhouse growers in a special category or have lumped them into the commercial applicator category of ornamental and turf, at least for training purposes. Applicators-for-hire are commercial applicators by federal definition; additional requirements exist for them. Users must prove competency to use certain products, state or federal restricted-use pesticides. Records must be kept on many kinds of applications. Direct supervision is required for some applications; some product labels require more than one person to be present for the application.

Label directions must be followed, with few and strictly regulated exceptions. If greenhouse application is not on the label, it is necessary to consult the manufacturer to determine if use in a closed environment is allowed, regardless of whether the crop is on the label. Allowing a product to drift off target or otherwise cause damage to nontarget species is generally prohibited. Restricted entry (formerly re-entry) intervals must be adhered to. Reapplication to the same site may be prohibited, and what crops may be planted after the use of certain products may be regulated. Certain application equipment may be prohibited; equipment must be properly maintained and calibrated. Applications cannot affect water quality, unless expressly authorized for application to water sources to control a waterborne pest. This includes wastewater entering public sewer systems, unless the pesticide discharges are authorized as a part of an NPDES (wastewater program) permit.

Employers must provide safety training and information about pesticides to workers, and health concerns must be attended to promptly; i.e., a medical emergency plan must exist, and arrangements for transportation must be detailed. Workers must be provided certain kinds of protective equipment, and it must be properly maintained. Notice must be provided in some instances, and placards may need to be in place. The EPA has an entire manual devoted to the Worker Protection Standard requirements. Some personal protective equipment distributors provide copies to customers, and it is also available from the EPA, state lead agencies for pesticide control, pesticide education coordinators, and the GPO (see "Sources of information").

Recordkeeping. Records must be kept on restricted-use pesticide applications by private and commercial applicators. USDA has authority, and also has established formal agreements with the state lead agency, to inspect records. Some states have additional requirements; e.g., the grower may be required to submit records to the state agency, and the grower may be required to keep records on all pesticide applications. Commercial applicators must keep records on all applications and, in most states, are generally subject to more requirements than private applicators.

Food safety. Greenhouses and nurseries that produce plants used for human consumption are affected by the Federal Food Drug and Cosmetic Act. Only pesticides that have established residue tolerances can be used on food crops, even on seeds and seedlings. The crop and use site must be on the label, or illegal residues may result in or on the crop.

CONTAMINATION OF WATER

The EPA administers federal laws pertaining to protecting water. States also have programs and frequently administer the federal law.

Drinking water. The EPA administers the Safe Drinking Water Act (SDWA) and sets national maximum contaminant levels (MCLs) for substances; a state environmental agency or water board will generally enforce the federal and state laws. A business that has its own well for drinking water may be considered a public water supply. A public water supply is required to test water periodically for listed "priority pollutants," many of which are pesticides. Some businesses may not be required to test their water, yet be subjected to random testing by state officials. If the water for the business comes from a municipal source, the source is regulated as a public water supply, and the municipality is required to conduct the water-testing program at its facility and inform the users of the test results. If the well is in an aquifer recharge zone or a wellhead protection area, certain activities can be prohibited if they are believed to have the potential to contaminate groundwater. Some states have procedures for well abandonment that must be followed when a well has ceased to be used. Backflow prevention devices are prudent investments for any water source and are required in some states.

State agricultural chemicals programs and pesticide state management plans. Some states regulate mixing and loading sites for pesticides and fertilizers by requiring impermeable pads, secondary containment, and best management practices (BMPs), including setback distances from water sources. Under the federal pesticide law, states are required to develop management plans if a pesticide has been determined to be a "leacher," one that is presumed to be able to migrate to groundwater. Best management practices and perhaps other restrictions, developed to mitigate the potential for water contamination, are imposed, in addition to label requirements, in certain localities or for certain uses. Information on any pesticide active ingredient so regulated is available from the state lead agency for pesticide control.

Discharges to surface waters. The EPA administers the Clean Water Act (CWA). A state environmental agency may also administer a state law or may be a knowledgeable contact point for information on the regulated pollutants, wastewater pretreatment processes, and permit requirements. The municipal system is also regulated by state and federal laws; the system administrator may also provide useful information on necessary compliance practices.

Under the CWA, state and federal pollution-discharge elimination programs regulate discharges of pollutants into surface waters from point sources, such as a pipe or a business facility. The current program does not apply to farm fields, which are considered nonpoint sources. If a business discharges wastewater directly to surface water, it must apply for an NPDES (National Pollution Discharge Elimination System) permit. Some kinds of operations may require permits for stormwater runoff. It is best to consult an environmental attorney to sort out current requirements.

Point source pollution. Without a permit a business cannot discharge regulated pollutants. Discharges are limited to the substances and effluent limits set by the EPA or the state, as defined in the terms of the permit. If a facility (such as a greenhouse) discharges to a municipal system (referred to as a publicly owned treatment works, POTW), the grower is obligated to inform the municipal system of the types and amounts of discharges. Depending upon the treatment capabilities that exist within the municipal system, the business may be prohibited from discharging, or it may be charged for pretreating, or required to pretreat, the water before discharging into the municipal system. In this case, since the facility does not discharge directly to surface water, the business does not need to apply for an NPDES permit.

Discharge to a septic system is regulated if the septic system eventually discharges to surface water. Discharges from a facility (greenhouse) that would allow material to move across land, for the most part, would be regulated in the following manner. If there is a potential for any regulated pollutant to reach surface waters of the nation or state, it would require the facility to obtain a permit. It is possible, however, that in some areas a permit may not be granted, or it may require zero discharge of pesticide products or costly water pretreatment to remove the pollutants before discharging. Any discharge of regulated pollutants that would not reach surface waters of the nation or state could be considered illegal disposal, except if discharged to an authorized, regulated evaporation or holding pond or ditch.

Nonpoint source pollution. Discharge is generally not prohibited from an agricultural field (nonpoint source) unless damages occur; there may be attempts to hold a business liable, however, under general nuisance provisions of common law. Some states have programs designed to regulate nonpoint pollution. They are based primarily on implementation of voluntary best management practices (BMPs) and require specific methods of control if the BMPs don't work. The state agricultural agency is generally at least aware of these programs if it is not their administrator. USDA Natural Resources Conservation Service (NRCS) state specialists are often involved in providing technical assistance and possibly grants to implement methods to control nonpoint pollution. The federal Coastal Zone Management Act created nonpoint pollution control programs and prescribes permissible land and water uses, an authority previously left in the hands of local governments. Through this law, increased authority over pesticide and fertilizer applications and erosion control have been granted to states. The law no longer is limited to coastal protection, as it was expanded to include additional states, although few programs outside coastal states have been developed to date.

Chemigation. *Chemigation* refers to the use of irrigation water to apply an agricultural chemical. Some pesticide labels prohibit use of the product in any type of irrigation equipment; others may restrict uses to certain types of equipment. Written notification of the use of chemigation and an operational plan may be required in some states; backflow prevention and an ability to flush the system are generally required. Since the water is used for irrigation, permits may be required. Some states have specific rules for chemigation. The state lead agency for pesticide control will have information if it does not administer the program.

Nutrients. Currently, there are not many states that prescribe use practices for fertilizers or soil amendments. Growers need to use precaution to avoid nuisance or negligence suits from contamination of soil and water. The drinking water standard for nitrogen is 10 ppm (parts per million). Elevated nitrate levels can be hazardous to infants and pregnant women. Alternate water supplies may be required if drinking water is contaminated. Aesthetics is also important, as unsightly disposal of excess fertilizers or spills may cause complaints or result in environmental damage. Self-imposed BMPs are appropriate measures to add protection.

DISPOSAL OF WASTE AND PESTICIDE PRODUCTS

Disposal is regulated under the Resource Conservation and Recovery Act (RCRA) and is administered by the EPA. Solid waste includes solid materials, liquids, slurries, dry substances, and more; some solid wastes may also be considered hazardous. State environmental and some local governmental agencies administer laws.

Solid waste. Dumping, burning, burying, or any other method of getting rid of wastes is regulated, and proper disposal requirements must be followed. Recycling, reusing, and recovering wastes for appropriate other uses are encouraged; however, the material is still regulated as a waste. Zoning may also affect disposal options. Some substances are further regulated as hazardous and cannot be disposed of in a sanitary landfill.

Hazardous waste. Whenever a product regulated as a hazardous waste is used, careful management is important to minimize the production of waste. Once the material is a waste, it may be very costly to deal with it, even in limited amounts. Depending on the amount of waste, permits may be required to "generate" and to store; manifests are required to transport; few sites are approved for disposal. Proper handling is necessary to minimize legal and financial impact on the company. Regulated wastes can be accumulated only for a short time before additional burdens apply. Some sites may be required to have permits and to qualify as a collection and treatment facility. Dilution does not exempt products from regulation. Mixing two or more nonregulated wastes together may create a regulated hazardous waste. Adding even a small amount of hazardous waste to any amount of nonregulated wastes renders the new mixture hazardous.

Pesticides. Pesticide use is regulated under the federal pesticide law, FIFRA. Pesticide disposal is regulated under the federal pesticide law and RCRA; not all requirements are consistent. Labels give some direction. Some pesticide active ingredients are "listed" hazardous or toxic wastes. Wastes from commercially manufactured mixtures are regulated differently from wastes of products that contain only one active ingredient. Growers are encouraged to use leftover mixed material and collected rinsate in makeup waters for subsequent applications. They are exempt from some RCRA hazardous waste regulations if they use the makeup water on their own land and in accordance with label directions. Leftover-use dilutions and rinsates are not waste, unless the owner declares them as such (an important definitional issue that allows the grower to manage potential problems better). Commercial applicators may have additional requirements. Some states have disposal days to collect leftover, discarded, unusable, and illegal products. A few states have combined all pesticide programs, including waste, under one authority.

Empty containers must be triple-rinsed, or the equivalent, and rendered unusable (punctured or crushed). Regardless of the RCRA classification of their contents, properly rinsed containers are exempt from the hazardous waste provisions of RCRA; however, some states and localities prohibit disposal in sanitary landfills unless certain criteria are met. Container recycling programs exist in many states.

RIGHT TO KNOW

This is an area that is often overlooked or misunderstood. There is significant liability associated with noncompliance.

Community right to know and emergency planning. Even though many businesses were in place before neighbors moved in or the land was annexed by a city, right to know provisions have been enacted in many areas to mitigate "coming to a nuisance" complaints. They can be considered a compromise to zoning restrictions against business expansion. Every effort should be made to cooperate with the right-to-know concept. It can help substantially in instances where a nuisance suit is brought against the company. Several trade associations have produced training videos that are available to aid a grower's compliance.

Under SARA (Superfund Amendment and Reauthorization Act) Title III, amendments to the Comprehensive Environmental Response Compensation and Liability Act (CERCLA or the Superfund Act), businesses are required to inform fire, police, and other officials when certain substances are handled and stored. A report of stored inventories of hazardous substances is required. An emergency plan is required if certain quantities of extremely hazardous substances are stored, although in some instances farms are exempt. Any release of a hazardous substance must be reported. Employees must be trained, and buildings must be posted with specific signs. States and localities may have additional requirements.

Employee right to know and Hazard Communication. OSHA administers rules (the Hazard Communication Standard) that require employees to be informed about hazards in the workplace. Some states and localities have additional employee right-to-know requirements. Each business must have a written plan; the plan must list all the hazardous substances used in the workplace and explain how labeling and training requirements will be met. Employees must be informed of the potential for exposure to any toxic substance; at a minimum, material safety data sheets (MSDS) or equivalent information is to be available for all products used. Education of all users is required for hazardous substances used on the premise; some laws may also require that all employees be informed of hazards and trained in safety precautions; a medical emergency plan may need to be maintained.

OSHA publishes the booklet, *Hazard Communication—A Compliance Kit,* which is available through the Government Printing Office. Other OSHA publications are also available that may be useful. Manufacturers and many dealers provide MSDS for their products, and books of MSDS of common products are available. Although most growers do not manufacture hazardous substances, if they are produced in the business, the employer is responsible for providing the required safety information on these products and properly labeling them.

MISCELLANEOUS

The following laws do not affect the industry as broadly as those previously described. The individual grower, however, may be affected as severely by certain provisions.

Endangered species. Under the Endangered Species Act, the grower can be required to protect the habitat of an endangered species. Some pesticide labels also require protective practices in order to use the products in certain designated areas. The Department of Interior administers this act, and state environmental agencies generally are also involved.

Air. Under the federal Clean Air Act, administered by the EPA, air quality standards have been set to protect public health. Nonattainment areas have been designated where problems exist. Agriculture has been blamed for contributing to the nonattainment of standards in some areas. Restrictions on use of tractors, pesticides, and plowing have been suggested in some places. Such restrictions come from state or local air quality boards and could adversely affect grower activities. Burning may be prohibited or may require permits.

Laws that fill gaps between other laws. The Toxic Substances Control Act (TSCA) provided the EPA with broad authority over chemicals in commerce and gave it authority to evaluate the safety of raw materials. The act allows for data collection, evaluation and prevention of risks, and consideration of economic impacts. TSCA probably will not directly affect growers, although if fertilizers *per se* are ever regulated federally, they will likely be covered under TSCA authority. Some biotechnology products are regulated by the EPA using TSCA authority.

The Comprehensive Environmental Response Compensation and Liability Act (CERCLA), designed to fill gaps between TSCA and RCRA, creates financial responsibility for long-term maintenance of waste disposal facilities. It provides for the cleanup of old and abandoned hazardous waste disposal sites that are leaking or otherwise endangering public health. It includes a hazardous substance response plan, which establishes procedures and standards for responding to releases of hazardous substances, pollutants, and contaminants. Amendments to the act, SARA, create community right-to-know provisions (described previously). Any business contacted regarding potential involvement in a CERCLA case should consult with an environmental attorney before responding.

Storage tanks. Underground storage tanks are regulated under RCRA if they contain or have contained petroleum products, other fuels or flammable products, or hazardous substances (as listed in CERCLA). Some states prohibit some products in underground tanks; some small agricultural fuel tanks may be exempt. Notification and permits are necessary, including notice of tanks taken out of service. Minimum standards exist for tank materials, appurtenances, monitoring, containment or catchment basins, protective equipment and procedures, recordkeeping, and reporting of releases. A spill prevention plan may be necessary, and statements may need to be submitted to the state agency.

Above-ground storage tanks may have similar requirements in some states. Pesticide storage is regulated in several states; the state lead agency for pesticide control should be contacted. For other substances it is generally the environmental agency; several states have a separate petroleum authority.

Wetlands. Lands that are wet during any portion of a year may come under wetland protection provisions limiting activities in which the grower may engage. USDA NRCS, the Army Corps of Engineers, the EPA, and state environmental or agricultural agencies may have jurisdiction. Permits may be required to drain, fill, or modify wetlands. NRCS county offices should have maps of existing wetlands.

Farm bill. Various provisions of the Farm Bill establish conservation practices and programs under USDA jurisdiction that may affect growers. Some of them provide technical assistance and cost-sharing for improvements that safeguard the environment. County USDA offices can provide information.

Citizen suits and civil or criminal liabilities. Increasingly, many federal and some state environmental protection laws provide for citizens to sue agencies for nonenforcement and businesses for noncompliance. Environmental laws provide for civil and criminal penalties. Either penalty can be very costly. Business officers can be subject to jail terms for the actions of the firm.

The EPA audit policy. The EPA is encouraging businesses to voluntarily disclose the results of internal environmental audits. Although internal audits can be an excellent environmental management tool, businesses that wish to perform audits as a part of their overall environmental management efforts should discuss the EPA's audit policies with counsel before entering into this voluntary program (for more information on internal audits, see "Value of audits" in the next section).

POLLUTION AND PREVENTING PROBLEMS

This section describes some environmental issues, business activities, and legal considerations that growers should be aware of in establishing or assessing their environmental management programs. Some suggestions presented require cooperation with government to achieve. The cost of compliance and management of environmental issues has reached such proportions that preventing environmental damage is often far more effective than coping with the consequences of causing damage. Environmental management should be an integral part of a grower's business plan.

BUSINESS PLANNING AND INFORMATIONAL ISSUES

Personal business plan document. Many growers have been required to complete a farm plan to qualify for financial or governmental assistance. A broader plan encompassing regulatory issues would also be helpful to most growers. A road map of requirements—including a regulatory ledger, a compliance calendar, and an evaluation scheme tied into internal audits—is important. Growers should develop a mechanism to identify and analyze issues, set priorities, monitor activities, and evaluate results.

Legal counsel and environmental management and regulatory compliance consultants. This can be the best resource team available to the grower. It may seem as though these resources would add to the cost of doing business, but in the long run their use can result in considerable savings. It is important for the grower to spend the time necessary to locate competent and ethical assistance. Experts may not be locally available. The grower may need a separate counsel for environmental issues, as not all attorneys are expert in this specialized aspect of law. Many consulting firms exist—many with engineering capacity, fewer with specific agricultural-management or regulatory-compliance background. The grower should seek out the expertise needed and create a team between these consultants and a local counsel to provide the best representation possible.

Recordkeeping. Records are required under some laws. Proper documentation is just sound business practice as well. If a grower has any doubt about the value of recordkeeping, he or she definitely should discuss this issue with counsel before concluding that the risks associated with having records outweigh the benefits.

Value of audits. Internal audits can be an important part of a business plan. Periodic evaluation is absolutely necessary to reduce environmental risk. The audit should be conducted by someone who understands the laws and technology. It must be done with a keen eye for finding problems as they exist in every business setting.

The results must be handled appropriately. It is perhaps best to conduct internal audits under an attorney-client agreement to protect the information from discovery by opposing parties in case of litigation or some types of inspections.

Training and education. Some growers see regulatory and environmental protection training for themselves and for employees as a last resort or as a resource when a solution to an existing problem is needed. However, training should be an important part of the business plan. Without it, risk identification, problem solving, and pollution prevention will lag behind pollution production. Industry educational events, such as GrowerExpo and others, have dealt specifically with difficult and unpopular regulatory and environmental issues. Universities have compliance training courses for most laws, some given through engineering schools as well as by extension services. The Society of American Florists, the American Association of Nurserymen, and Bedding Plants, International (formerly the Professional Plant Growers Association) offer compliance materials.

There are other sources of information and training, such as chambers of commerce, the regulatory agencies, law schools, counsel, and other consultants. Some publications, such as *GrowerTalks* magazine, work to keep growers abreast of recent changes and are invaluable resources. Many workman's compensation insurance offices, other insurance companies, and lender institutions have found it useful to provide training to their customers. They are now marketing their expertise to help clients comply with certain aspects of pertinent laws.

Sources of information. Environmental laws have been around long enough that they are now the subject of many books. Information can be viewed in the shape of a pyramid with books on the bottom. They are usually easy to find and are generally accurate at the time of publication, but the information may become dated. Magazines and periodicals come next and are more timely, but they may or may not be accurate. Conferences and speeches are next; knowledgeable people are generally available. Then come the news media; they attempt to present views about an issue but frequently are not accurate. Ombudsmen, chambers of commerce, business clearinghouses, the governor's office, trade associations—although not the experts *per se*—are frequently knowledgeable about the issues, generally have access to experts, and can be helpful and current. Personally questioning experts is at the top of the pyramid: this source is the most current, but the information received depends on asking the proper questions. Whenever verbal discussions are the source of information, ask for citations and other sources. *Getting bad information is not an acceptable reason for noncompliance.*

The National Association of State Departments of Agriculture (NASDA) is the umbrella organization for many of the agricultural control officials. It is located in Washington, D.C., and its members are the agricultural secretaries, directors, and commissioners in each state. The staff will know the appropriate contact persons for many program activities affecting agriculture in each state.

There are many other information sources. The State Pesticide Education Program at land-grant universities produces educational information; some is appropriate for the greenhouse industry. The *Standard Pesticide User's Guide*, by Bert L. Bohmont, pesticide education coordinator (PEC), Colorado, published by Regents/Prentice-Hall, is a resource book that covers a wide range of pesticide issues, including some information specific to greenhouses. Similar but less comprehensive manuals are available from other states. A query of the state's PEC should provide insight into which states have quality greenhouse educational information. The National Agricultural Library at Beltsville, Maryland, has some publications used by state education programs and can help locate pertinent and current documents. Law school libraries and law students can be helpful resources in locating specific laws and rules.

Perhaps the best source for some types of information is the regulatory agency itself. Some critics say it is difficult to obtain information there as it is often bureaucratic, but contact through a staff person of an elected official generally speeds up results. A cautionary note regarding contact with regulatory agencies: On some issues it is best to discuss compliance options with counsel before contacting the agency. Even then, it may be appropriate to discuss the issue with someone outside the enforcement staff—that is, someone in the administration or in some instances an agency attorney—before discussing the issue with someone who may feel it necessary to inspect as a result of the inquiry.

The Superintendent of Documents, U.S. Government Printing Office (GPO), Washington, D.C. 20402-9328, publishes a myriad of information. For example, *Access EPA,* available from GPO, is an invaluable information source for hotlines and other locations of information and assistance. Most publications are for sale, and catalogs are available.

TECHNICAL ISSUES

IPM. Integrated pest management (IPM) principles have been effective in controlling some pests. When IPM is a component of integrated farm management, it can be an effective method for many growers. There is some desire among regulators to mandate pest management principles. IPM must be flexible and allow for judicious use of pesticides if it is to be effective, as reliance on any one form of pest control is not a sound business practice.

Resistance management. Many pests, especially insects and fungi, are capable of developing resistance to chemical and biological controls. Effective pest management strategies allow the grower to keep the tools available for the long term.

Controlling pests is only a small part of a proper pest management strategy. Resistance management must be a part of the plan.

Equipment. Maintenance of application equipment and applicator and other safety equipment is necessary for pollution and problem prevention. Include replacement of worn out or unsafe equipment in capital expenditure plans.

Application. Many products used in greenhouse and nursery operations, such as oil, gasoline, pesticides, fertilizers, and cleaning supplies, can be hazardous. No matter what type of product it is, minimizing exposure will help to reduce the risk associated with it. Eliminating off-target application—drift, overspray, volatilization, spills—should be a goal, especially with those products that are highly regulated. Regardless of the type of equipment used—backpack sprayers, granule applicators, ebb-and-flow systems, chemigation, direct injection, ground rig—careful management of pesticide application is necessary.

Pest control and losses of products from reregistration. Keeping pesticide tools available to the greenhouse and nursery industry will be difficult. Growers must understand the impact that reregistration and certain provisions of the Food Quality Protection Act can have on future business. They must also understand the limitations of special and emergency exemptions to registrations. Growers and the industry must begin planning immediately, determining contingencies for the potential loss of any principal product currently needed to make a pest control strategy work. It is imperative that growers know local company representatives; volunteer the use of their sites for research; and cooperate with university, USDA, and manufacturer representatives in the identification of pest control needs as early as possible. Do not wait until a pest is reaching economic threshold levels before acting. Growers need to work with the Minor Use offices of USDA and EPA and cooperate with the National IR-4 program, located at Rutgers University.

New products and varieties. New products, including varieties and biological pest control products may fill some future voids. Do not expect a panacea, though. Determining available and necessary pest control strategies before investing in a new crop line has always been good business but will be even more important in the future.

Research. The industry is interested in production and marketing research, those things that are perceived to contribute positively to the bottom line. Until growers realize that assessment of the state-of-the-art practices of the industry (benchmarking), potential environmental impacts, and regulatory compliance issues and problem solving options are equally worthy of research, the industry will remain behind the compliance curve. It will take a big-picture perspective to identify these additional issues; it will in fact take research to determine which ones are truly priorities for the industry.

CONCLUSION

There can be no doubt that environmental protection laws affect the ability of many growers to make a profit; however, before condemning all laws, note that many laws protect businesses as well as regulate them. It appears that objections to governmental controls would be better directed toward improving the bad laws, reducing unnecessary and unreasonable controls, and building in ways to get credit for implementing technologies and practices that mitigate existing problems.

This chapter has described some regulatory provisions of environmental laws that affect the grower. It suggested some places to get additional information and some insight into prevention tactics. It is just a skeleton of the environmental management plan and program needed to cope with the regulatory burden, though.

GREEN, NOT GREED

"The measure of character is what a person does when he believes that he will not be found out." Anon.

Many laws exist because some organized activists and government officials perceived that the agricultural industry would not act ethically and responsibly on its own. Compliance with regulations and ethical principles takes lots of thought and hard work, but growers are known for that. More thoughtful, hard work may eventually result in a positive public perception of the industry's character and appreciation of its products and its contribution to the protection of the health, safety, and welfare of those who share the environment.

CHAPTER 16

ATTENTION, YOUNG PEOPLE

by Vic Ball

The American floriculture industry—bedding plants, flowering and foliage pot plants, and cut flowers—is growing. The USDA Floriculture *Crops Summary* puts our wholesale value at $3.3 billion in 1995. Many American growers are expanding to serve the needs of mass markets and supermarkets. This expansion requires well-trained young people.

So often I hear growers talk about the critical need for good, well-trained men and women to work in the greenhouse. Most greenhouse employers are looking for employees with solid classroom training in pathology, entomology, soil science, plant physiology, and chemistry. Two years will do it; four years is better. The United States and Canada have many excellent two- and four-year colleges and universities granting degrees in horticulture.

Greenhouse employers also very much want employees with hands-on growing and people management experience. In the Netherlands, Germany, Denmark,

FIG. 1   Stephen Barlow *(left)*, Barlow Farms, Sea Girt, New Jersey, operates a busy retail growing business, and for good measure, a school! The school consists of a greenhouse, which is part of Steve's range, a teacher, and 40 high school horticulture students. They grow mostly bedding crops, from start to finish. From the left: Steve Barlow, students Scott Hurley and Jennifer Dinyouszky, and teacher Robert Lempa.

183

Sweden, England, New Zealand, and Japan, many students and other young people spend time as apprentices, learning from proven "masters" how to grow. It is a good system. Those countries' ample supply of thoroughly trained young people is a major advantage (fig. 1).

Preparation for careers in American floriculture

How can today's young Americans prepare themselves for careers as flower growers?

Formal classroom training

Get at least two years of horticultural education, if possible. Choose from one of our excellent two- or four-year schools. Univer-sities with excellent four-year programs include Colorado State University, Fort Collins; Cornell Univer-sity, Ithaca, New York; University of Florida, Gainesville; Michigan State University, East Lansing; North Carolina State Univer-sity, Raleigh; and Ohio State University, Columbus.

Another option is to attend one of the many fine local community colleges. Many students attending these schools work in local greenhouses and attend school part-time. The College of DuPage, Glen Ellyn, Illinois; New York State College, Cobleskill, New York; and Sand Hills Community College, Pinehurst, North Carolina, are but three examples.

The University of Guelph offers a horticultural correspondence course. Write to the Horticultural Correspondence Program, University of Guelph, Guelph, Ontario N1G 2W1, Canada, phone 519/767-5050.

Hands-on experience

The United States has no recognized apprenticeship program for flower growers. You can seek out a top grower in your crop preference and apply for two to three years of employment. Tell him frankly that you want hands-on experience growing crops, that you had basic classroom training, and that your goal is to be a grower of top-quality plants. Most importantly, say, "I am willing to work." Most growers will welcome such an approach.

To help students acquire such hands-on experience, my wife, Margaret, and I have established the Vic and Margaret Ball Internship with the American Floral Endowment (highlighted later in this chapter). This kind of work experience is so important!

INTERNSHIP FOR YOUR FUTURE IN FLORICULTURE:
THE VIC AND MARGARET BALL INTERNSHIP

Why an internship? An intern has the opportunity to gain a realistic view of the working world of floriculture. Practical, social, and intellectual skills are obtained by interacting with a variety of greenhouse personnel. An intern gains knowledge by applying himself or herself and taking advantage of the multitude of opportunities and experiences for learning in real floriculture production facility.

An intern is an employee of the selected business and is expected to view the experience as a learning-by-doing education. The employer is encouraged to provide as many different experiences as possible for the intern, within the limits imposed by the business.

A successful internship experience may improve a student's future employment, often with increased monetary compensation at the entry level. An internship also helps a student determine future career direction in a more specific way, as well as enhance classroom learning upon return to campus.

Eligibility. All students enrolled full-time in an accredited two- or four-year college or university with a nationally recognized floriculture program within the United States are eligible to apply if they are achieving a C or better grade point average and are making satisfactory progress toward a certificate or degree.

Internship requirements. Work experience must take place at a business not owned by the student's family and shall be performed at a production greenhouse, nursery, or botanical garden, with commercial establishments being preferred.

The student must be willing to work away from home or the school environment a minimum of three months. In this regard, a six-month training period is encouraged.

A student accepted for a three-month internship will be awarded $3,000 upon satisfactory conclusion of training, or $6,000 if the internship is for six months.

Employer responsibilities. The employer is asked to
- employ the student for a period of between three and six months at a fair wage, according to local market conditions, but not less than the minimum wage per hour.
- provide an overview of the entire company operation at the outset of the work period and subsequently provide experience with a variety of crops, tasks, and responsibilities.
- allow for periodic discussion between student and supervisors.

- allow time for a meeting with the student's floricultural faculty contact at a mutually agreeable time during the internship period.
- help locate reasonably priced housing for the student during training, although the student is responsible for payment.
- be willing to complete a questionnaire on the student's performance at the conclusion of training.

For more information. Contact the American Floral Endowment, 11 Glen-Ed Professional Park, Edwardsville, IL 62025, phone 618/692-0045, fax 618/692-4045.

VISIT GROWERS

Talk to growers and watch their crops. Major seminars and industry trade shows are other great places to learn. GrowerExpo, sponsored by *GrowerTalks* magazine each January; the International Ohio Florists Short Course, held annually in Columbus; and the Canadian Greenhouse Conference, held annually at the University of Guelph, Ontario, are three examples of fine events with excellent educational seminars and trade fairs where you can meet suppliers.

ABOUT SALARY

In past years flower production had a reputation for hard work and low pay. This has changed! Manual labor has given way to automation and computers. Compensation, especially for people with good training and on-the-job experience, is quite good. Grower supervisors who can manage several acres of crop production and the accompanying staff are in high demand. Salaries for greenhouse general managers approach $50,000 per year at the high end. Growers can expect from $18,000 starting out, up to $40,000 or more for experienced head growers. Most commercial greenhouses also offer benefits to full-time employees, such as paid vacation, health insurance, and bonus plans. Many companies have also instituted 401(k) plans and additional insurance.

Today's greenhouse professionals require top-flight technical skills when it comes to growing plants, as well as computer skills and people skills. If you are willing to make the effort to acquire the skills and to work hard, you will be ensured a long-lasting career.

SECTION 4

GENERAL CROP CULTURE

CHAPTER 17

AMERICAN FLORICULTURE: THE NUMBERS

by Debbie Hamrick
Ball Publishing
Batavia, Illinois

and Dr. Marvin N. Miller
Ball Horticultural Company
West Chicago, Illinois

American floriculture has changed in many ways over the past several years. Production processes have advanced, but perhaps the greatest changes have come in the marketplace. The days when retail florists provided the only outlet through which our products were sold are gone. Today, supermarkets, garden centers, home centers and do-it-yourself megastores, hardware stores, discount chains, landscapers, interiorscapers, and even gas stations and convenience stores sell floricultural products. These new outlets move more and more of the industry's output and are causing shifts in the foundation upon which the industry is built.

Large growers continue to expand, thus consolidating production. Smaller and middle-sized growers hone their crop assortment, often turning to direct-to-consumer sales through their own retail outlets.

Every April the United States Department of Agriculture National Agricultural Statistics Service issues the *Floriculture Crops Summary*, which covers production data for the two preceding years. The report focuses on production in the leading 36 states, providing data on units produced and wholesale sales value. Average prices for each state by crop are also provided. The *Floriculture Crops Summary* is an excellent way to get a reading on the state of the industry and to gauge production changes in response to market shifts.

STATE OF THE INDUSTRY

According to the USDA report, 1995 floriculture crop value increased less than 1% over 1994 levels, to $3.27 billion. Total greenhouse acreage rose 5% to 11,200 acres (4,670 ha).

As table 1 shows, California is the leading production state, with $672 million; however, production value there fell 4% from 1994 to 1995. Florida is the second largest floriculture production state, with 1995 output valued at $613 million, a 2% increase over 1994. Together California and Florida account for some 39% of all crop value in the states surveyed.

Of the 36 states surveyed, 23 show higher production value, while 13 show decreases. The five leading states—California, Florida, Michigan, Texas, and Ohio—produce $1.80 billion, 55% of the total value surveyed.

The largest producers, those with $100,000 in sales or greater, comprise 4,616 in number and control 92% of sales. However, the number of growers in the $100,000-and-over category dropped by 15 from 1994 to 1995. Covered greenhouse area for these growers increased 7%, to 9,720 acres, while sales were also up 1% in value. Fewer growers, more production area, and stagnant prices mean the growers in this category are competing harder than ever for large-volume-buying customers, such as mass merchandisers.

Bedding plants continue to grow, up 3.5% in 1995 compared to 1994. This crop category is produced by the largest number of growers, 3,229 in 1995. Wholesale value of all bedding plant production totaled $1.32 billion. (For the 1984 to 1995 dollar value of various bedding plants, see figs. 1–7.)

Cut flower production declined 8% in value, with 20 out of the 36 states showing lower cut flower value. Declines in hybrid tea rose production accounted for more than half of the loss of value, as imports continue to gain market share. Production of pompon chrysanthemums and standard carnations, both crops dominated by imports in the U.S. market, are relatively flat. Overall, 39 fewer growers produced cut flowers in 1995 compared to 1994. (For the 1984 to 1995 dollar value of various cut flower crops, see figs. 8–10.)

Potted flowering plants are holding their own, with a 2.5% increase in value in 1995 to $679 million. Poinsettias were up 1% in value but down 483,000 units, with production showing a small shift to smaller pot sizes. Production area for all pot plants rose 6%, while the number of pots sold increased only 3%. (For the 1984 to 1995 dollar value of various potted flowering plants, see figs. 11–15.)

TABLE 1   AMERICAN FLORICULTURE STATE BY STATE, 1995

State[a]	Area of production				Growers with $100,000 in sales					
						$ wholesale value of crops (thousands)				
	Growers	Covered[b] (1,000s ft²)[c]	Outdoors (acres)[d]	All crops[e] (millions)	Number	Cut flowers	Bedding plants	Flowering pots	Foliage plants	All crops[f]
Alabama	181	8,011	85	$48.9	76	$450	$29,547	$12,851	$1,716	$44,594
Arizona	28	1,942	206	$12.0	11	$1,458	$7,548	$738	$1,597	$11,341
Arkansas	83	2,382	53	$9.2	31	$6	$4,393	$2,203	$309	$6,947
California	952	130,116	7,882	$671.9	600	$252,985	$173,154	$126,440	$85,356	$654,279
Colorado	133	10,730	47	$69.2	75	$16,565	$37,444	$10,053	$1,993	$66,059
Connecticut	256	6,115	210	$43.4	86	$1,265	$24,525	$7,395	$2,807	$35,992
Florida	1,026	377,900	7,988	$612.6	670	$26,701	$88,353	$80,078	$309,494	$595,164
Georgia	219	7,576	115	$49.7	98	$200	$31,416	$10,072	$2,429	$44,117
Hawaii	349	24,514	648	$45.2	104	$13,711	$1,868	$9,771	$10,035	$35,480
Illinois	276	9,635	345	$70.5	121	$1,968	$35,575	$22,413	$2,691	$62,647
Indiana	255	6,589	134	$36.4	78	$3,489	$15,575	$6,601	$2,586	$28,251
Iowa	174	5,017	62	$39.4	75	$219	$22,084	$11,435	$1,583	$35,321
Kansas	72	4,013	15	$24.9	40	$72	$14,780	$6,175	$2,301	$23,328
Kentucky	162	3,907	57	$21.7	61	$351	$11,435	$5,001	$482	$17,269
Louisiana	95	3,521	59	$17.0	42	$235	$8,412	$3,187	$2,755	$14,589
Maryland	130	4,614	183	$32.3	50	$604	$21,331	$6,514	$317	$28,767
Massachusetts	419	9,392	295	$58.5	128	$6,568	$26,726	$10,184	$1,194	$44,672
Michigan	566	32,671	2,041	$184.6	300	$10,212	$131,171	$26,177	$3,767	$171,327
Minnesota	223	8,334	85	$54.6	82	$5,505	$28,233	$12,411	$1,318	$47,467
Mississippi	79	3,156	70	$9.5	27	$177	$4,632	$2,394	$233	$7,436
Missouri	186	5,762	47	$36.5	68	$615	$17,626	$10,685	$2,097	$31,023
New Jersey	362	13,087	1,824	$94.9	178	$10,264	$41,359	$29,610	$4,711	$85,945
New Mexico	50	3,665	18	$21.1	20	$3,000	$3,206	$11,925	$1,530	$19,664
New York	548	18,621	351	$126.2	215	$7,615	$68,511	$31,944	$2,165	$110,235

TABLE 1 AMERICAN FLORICULTURE STATE BY STATE, 1995 (CONTINUED)

| State[a] | Area of production | | | | Growers with $100,000 in sales | | | | | |
| | | | | | | $ wholesale value of crops (thousands) | | | | |
	Growers	Covered[b] (1,000s ft²)[c]	Outdoors (acres)[d]	All crops[e] (millions)	Number	Cut flowers	Bedding plants	Flowering pots	Foliage plants	All crops[f]
North Carolina	175	11,456	401	$82.8	81	$1,344	$38,825	$33,133	$4,461	$77,763
Ohio	515	25,867	162	$155.5	264	$4,939	$92,450	$35,965	$9,946	$143,304
Oklahoma	123	2,682	43	$13.9	38	$2	$6,886	$2,780	$874	$10,542
Oregon	215	12,805	1,020	$67.3	89	$9,283	$29,010	$18,034	$784	$61,778
Pennsylvania	665	19,800	327	$114.3	224	$7,037	$56,954	$26,460	$3,665	$94,116
South Carolina	112	2,921	400	$25.1	40	$1,124	$14,112	$4,877	$1,422	$21,535
Tennessee	162	5,760	51	$34.0	67	$768	$16,130	$11,724	$1,205	$29,827
Texas	427	33,066	230	$179.9	204	$217	$106,091	$41,127	$22,093	$170,394
Utah	90	3,853	130	$28.3	44	$2,813	$12,780	$8,581	$2,033	$26,207
Virginia	200	6,992	252	$50.1	92	$2,271	$26,840	$14,962	$1,739	$45,821
Washington	175	7,619	1,088	$72.4	85	$11,728	$41,861	$12,773	$1,458	$67,820
Wisconsin	339	8,750	232	$58.5	152	$2,911	$34,014	$12,314	$1,065	$50,319
Total	**10,022**	**842,841**	**27,156**	**$3,272.6**	**4,616**	**$408,672**	**$1,324,857**	**$678,987**	**$496,211**	**$3,021,340**

[a] The 36 leading states.
[b] Includes greenhouses, shade, and temporary areas.
[c] 1,000 ft² = 92.9 m².
[d] 1,000 acres = 405 ha.
[e] Includes all production from all growers. Values have been rounded to the nearest decimal.
[f] Also includes the wholesale value of cut greens.

Source: U.S. Department of Agriculture, National Agricultural Statistics Service, 1995, Floriculture Crops Summary.

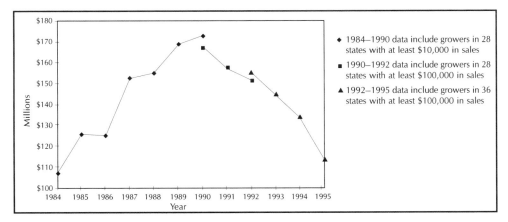

FIG. 1 Cut flowers: value of hybrid tea roses.

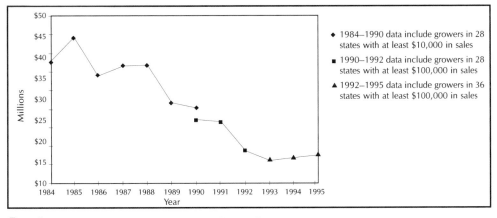

FIG. 2 Cut flowers: value of pompon chrysanthemums.

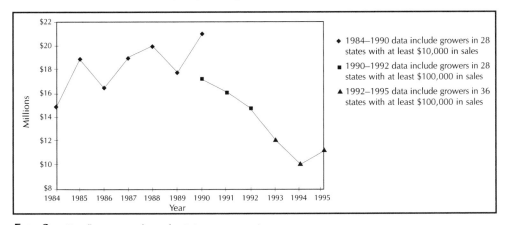

FIG. 3 Cut flowers: value of miniature carnations.

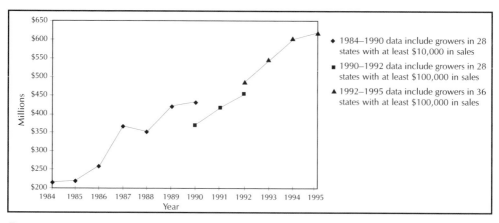

FIG. 4   Bedding plants: value of bedding-flowering flats.

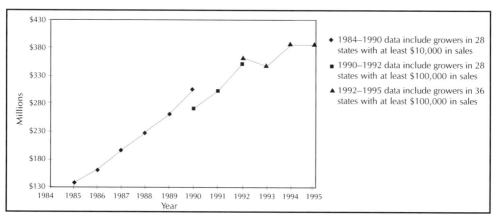

FIG. 5   Bedding plants: value of potted flowering bedding (excluding garden chrysanthemums).

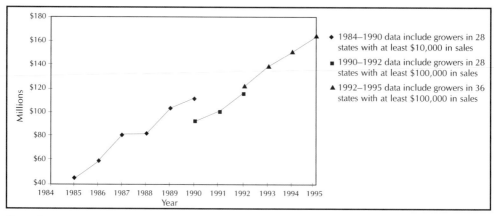

FIG. 6   Bedding plants: value of flowering hanging baskets.

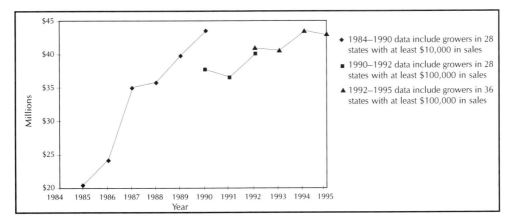

FIG. 7   Bedding plants: value of potted seed geraniums.

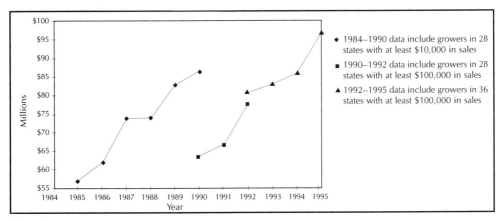

FIG. 8   Bedding plants: value of potted vegetative geraniums.

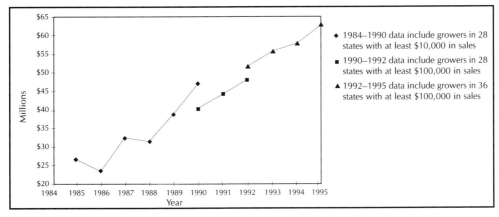

FIG. 9   Bedding plants: value of hardy garden chrysanthemums.

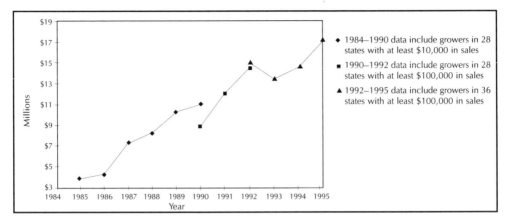

FIG. 10   Bedding plants: value of potted vegetable bedding.

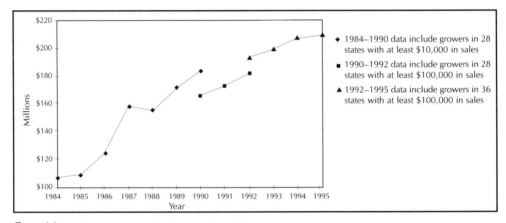

FIG. 11   Flowering potted plants: value of poinsettias.

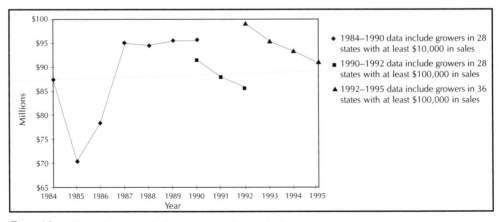

FIG. 12   Flowering potted plants: value of potted chrysanthemums.

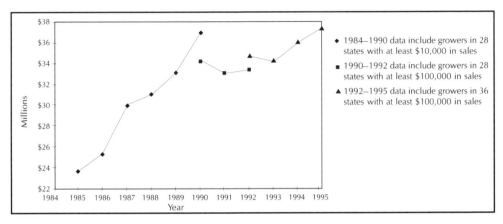

FIG. 13   Flowering potted plants: value of Easter lilies.

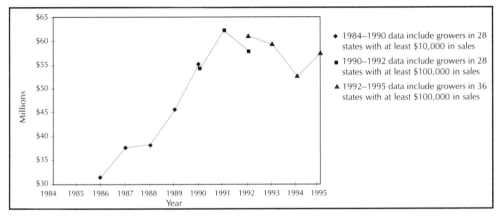

FIG. 14   Flowering potted plants: value of finished florist azaleas.

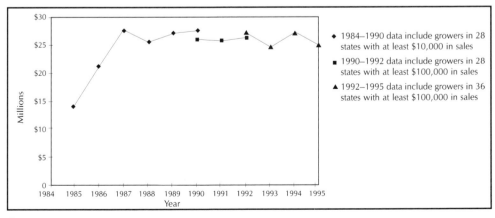

FIG. 15   Flowering potted plants: value of African violets.

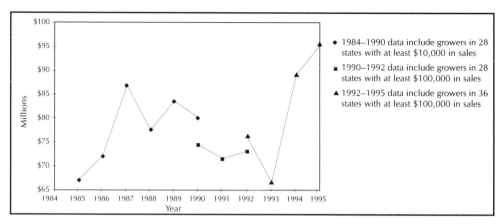

FIG. 16 Foliage plants: value of foliage hanging baskets.

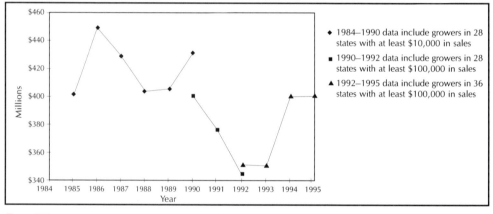

FIG. 17 Foliage plants: value of potted foliage plants.

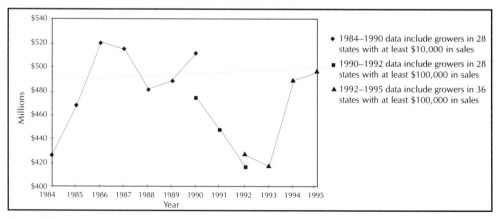

FIG. 18 Foliage plants: value of plant production (includes potted plant and hanging basket production).

Foliage plant production grew a modest 1% in 1995, but there were 126 fewer foliage plant growers. Florida continues to lead the nation in foliage plant production, as value there rose 5% to more than $309 million. (For the 1984 to 1995 dollar value of various foliage plants, see figs. 16–18.)

While these numbers point to industry consolidation and increasing competition among growers, taking the data to a different level also shows that America's floriculture is becoming increasingly reliant on fewer crops. When production statistics from 1995 are indexed for inflation and compared to 1990 production numbers, it becomes apparent that bedding plants drive the growth of today's American floriculture industry (table 2). When indexed to inflation and compared to 1990, cut flowers are down 23% in value, flowering pot plants are down 6%, and foliage plants are down 8% over the five-year period. Bedding plants are up a solid 41% over the five years. However, the rapid growth of bedding plants is slowing, as wholesale value increased only 3% in current dollars from 1994 to 1995.

TABLE 2   WHOLESALE PLANT PRODUCTION VALUE

Group	$ value (millions)		Trend (%)
	1990	**1995[a]**	
Bedding plants	$823	$1,158	up 41%
Flowering pot plants	$631	$595	down 6%
Foliage plants	$473	$435	down 8%
Cut flowers	$467	$359	down 23%
Total	**$2,394**	**$2,547**	**up 6%**

[a] Indexed to 1990.

CHAPTER 18

BEDDING PLANTS: TRENDS AND SEED BASICS

by Vic Ball

SEED STORAGE
PanAmerican Seed
West Chicago, Illinois

Bedding plants are clearly the growth part of the flower industry. Cuts are down, pots and foliage are flat, but bedding is up. Over the first five years of the 1990s, bedding plants were the only crop sector to beat inflation and still show growth. There are big ups and downs in the individual kinds of bedding plants people buy, but overall the demand is steadily rising.

BEDDING PLANT TRENDS

There are major changes in how bedding is grown and sold. The great majority of all bedding plants are now grown from plugs. Equally important, consumers are steadily moving from the traditional pack of small plants to pot annuals and various patio planters—a demand for instant color. Automation is moving in rapidly. Marketing channels are changing, with ever more bedding plants being sold to various mass outlets, such as home centers, Kmart, and Wal-Mart (fig. 1). Landscape contractors, responding to public demand for color in the landscape, are becoming major users of bedding plants. Breeders are bringing us a steady stream of really exciting new varieties—even new species.

People love the color of bedding plant flowers around their homes and offices. This creates lots of exciting potential for the grower willing to dig in and adapt. More good news: in a world dominated by giant chains and franchises, there is much opportunity in bedding plants for the start-up operations. It's not easy, but it is being done.

FIG. 1   Mass marketers are working to identify the plants they sell with brand names. Here's how Wal-Mart markets its Better Homes & Garden affiliation with bedding plant flats.

WHY SUCH CHANGE IN CONTAINERS?

Consumers. The change in the preferred sizes of bedding plant containers is the result of changes in the lifestyles of the people who buy plants. For example, most American women now work outside the home. When a woman comes back from a long day on the job, she's a lot less apt to want to go out and find tools and hoses and plants and make a garden. She'd much rather spend the money to buy some nice patio planters to set on her front porch or patio.

America's yard and garden areas are shrinking fast. Lovely 3,500-square-foot (325-m²) homes, which would have a half acre or an acre (0.2 or 0.4 ha) a decade ago, are today being built shockingly close, on small lots with 100 feet (30 m) or less of frontage. That means less space for gardens, and this means more patio color.

Chronic water shortages in some areas of the country tend to drive people away from flower beds. Patio urns and planters take a lot less water. There are ways people can irrigate flower beds with a lot less water than an overhead sprinkler, but it just doesn't seem to happen.

Growers. Containers are an important choice for bedding growers everywhere. Only eight or 10 years ago, we saw a lot of 20-inch (51-cm) flats with 72 and 96 plants. At the same time, there were very few 3- and 4-inch (8- and 10-cm) pots.

Today, it's different. Florida grows primarily 4-inch, some 3-inch, and a fair amount of 1-gallon (3.8-l) containers. California is heavily into 3-inch pots and California 17-by-17-inch (43-by-43-cm) square flats. As we move northward, it gradually becomes mainly 48 plants per flat. Through the Southeast, through the Southwest, and across the Midwest and the East—all grow predominantly 48 plants per flat. At the same time, all these growers are testing 3-inch pots or flats with six packs of six plants, and many are

CLAUDE HOPE:
AMERICAN PLANTSMAN

Claude Hope

Meet Claude Hope, if you haven't already. Claude, well known in the world of bedding plant growing, has been recognized with many national honors, including induction into the Society of American Florists Hall of Fame and receiving the first-ever Professional Plant Growers Association international award, the All-America Selections award of merit, and the Alex Laurie Award for horticulture. He was recognized as a fellow in the American Society of Horticultural Science "in recognition of his significant contributions to the development of improved open-pollinated and hybrid flowers through innovative plant genetics and breeding practices; for development of techniques and methodology for inbred maintenance and production of high-quality seed; for noteworthy leadership in the development of plant material for the bedding plant industry."

I'd like to be sure that growers today are aware of at least a few things this remarkable plantsman, breeder, and grower has done for floriculture. Claude fetched *Impatiens wallerana,* then an obscure wildflower, out of the Costa Rican jungle. He bred it shorter and with flowers on top—lots of them. It's really *the* success story of the 50-year-old bedding plant industry. To this day impatiens are the major U.S. bedding plant, often sold out by growers.

Claude first recognized the value of New Guineas. Working patiently for years, he introduced the first commercial impatiens from seed with New Guinea blood.

Claude has bred many other plants and is still working at it. His most important recent achievement is Blue Lisa, the first dwarf pot lisianthus. He developed the original Elfin impatiens and many improvements since then. Claude is responsible for the first red and the first yellow petunias. Claude also brought out all the Rainbow coleus lines that are still widely used. He is currently active in amaryllis, anthurium, and lisianthus.

Claude pioneered south-of-the-border production of F_1 seed. He was the first to realize the problem of the tedious, costly work of hand-pollinating F_1 hybrid flowers and vegetables. As a solution he set up the first F_1 production site in a low-cost labor area, creating his Linda Vista farm in Costa Rica, which now employs 900 in F_1 seed production. Claude's commitment to the education and personal welfare of the native Costa Ricans he has employed over the years is a whole separate story.

A special *Ball RedBook* hats off to a great American plantsman!

growing some 4-inch or 4¹/₂-inch (11-cm) plants. Again, lots of 1-gallon plants are grown, especially for the June and July market. With this are also a lot of color bowls and a wide assortment of window boxes and bedding plants in various patio containers. We saw 48 plants per flat at Stoffregen's in Raleigh, Norm White's in Chesapeake, the same in New Jersey, Long Island, Ohio, northern Illinois, and the Minneapolis area.

The main point is: containers are changing! The very large chains often require at least 2¹/₂- and 3-inch (6- or 8-cm) bedding. They just don't have the labor to keep plants watered, and the larger sizes do not dry out as fast. Larger containers—3 and 4 inches—are coming in, and the 48 plants per flat are steadily moving out. Some growers, even in the North, are going to 4-inch containers.

The now-famous Morgan flat is about 25 to 30% larger than the standard 10-by-20-inch (25-by-51-cm) flat. This means 25 to 30% more flats per bench or per house. Norm White at Chesapeake has converted to the Morgan flat.

Watch out, by the way, for the high cost of delivering plants to retail outlets. Bernacchi, a major La Porte, Indiana, grower reports that 10 to 12% of their total bedding cost is delivery. "We bought some new steel racks. Each holds 64 flats, 10 racks to a 40-foot (12-m) truck. Now we can load a Hertz truck in 20 minutes. It helps."

The 2-by-5-foot (0.6-by-1.5-m) carts (Cannon is one brand, shown in fig. 2) are very popular, being widely used by bedding growers. You move the cart into the greenhouse, load the flats onto it, and wheel the cart out to the waiting truck (with lift gate). When you get to the garden center or retail outlet, the truck can be unloaded very quickly.

FIG. 2   Mike Cerny, Hi-C Nursery, Sunol, California, shows a few of the Cannon carts they use for shipping bedding plants.

WAYS TO BUILD SALES

Attention garden centers and retail growers: how can you build sales? There are many ways.

ADVERTISING

Newspaper ads are clearly the Number 1 way the winners go. The great majority use newspaper ads heavily, generally more than TV and radio. The first mission is to get people to your store, and the local press seems to do it best.

In many situations, however, radio and TV do a bang-up job. My own observation is that if the proprietor takes a personal interest in radio or TV, it can really work. Many growers become celebrities among gardeners in their towns and regions by doing regular radio or TV programs each week. Also there are garden writers in radio and TV whom would be well worth your time to cultivate as friends and mutually beneficial business contacts.

Most successful retail growers and garden centers publish a price list or newsletter for direct mail outreach. Home gardeners love to read, planning warm-weather garden delights on those cold nights in February. Along with good gardening information, your message should be there. Many newsletters also include coupons good for savings when the customer comes in to shop.

ATTRACT THE PUBLIC

Open houses are often a part of a good marketing program. Lots of poinsettia growers organize a Christmas open house in December, when their greenhouses are ablaze with color. They sell lots of plants then. Some growers also have open houses in the early spring, even going so far as to organize their own indoor flower shows to jump-start sales early.

Good signs outside your establishment cost money, but they are very much a part of success in retailing. How will people know you're there and open for business if you don't tell them so?

STORE DISPLAYS

Use informative signs at point of sale. Signs can help you make several points in your display of spring bedding plants. For example, your impatiens display sign can convey the following types of information the consumer would find very helpful:

- Impatiens will flower well in shade, but not dense shade.
- What actual color can the gardener expect from Super Elfin Red, for example?
- How much space per plant should be allowed?
- How tall will each plant grow?
- What is the price per pack or pot? Nothing is more frustrating to a shopper than not being able to see the price clearly indicated.
- An 8-by-10-inch (20-by-25-cm) color photo displayed with the plant is a great sales aid.

Here and there I find a grower who makes chatty cardboard signs and puts them up among displays. They can be just inexpensive white cardboard hand-lettered with a felt-tip pen. The messages are friendly, practical pointers about the plants you offer. For impatiens: "Great in the shade." Portulaca: "Happy in the hottest, driest place you've got."

Color tags, one per pack, give color presentations of the variety in full flower. Color tags relieve much of the pressure to have annuals such as petunias sold in flower. Think about it: most other products consumers buy are marketed with colorful photos and packaging—bedding should be, too! Don't skimp on good tags.

Attractive displays are shown in figs. 3, 4, and 5.

FIG. 3   An attractive display of annuals decorates the entrance of this retail grower's garden center. The operation: Wilson Farm Inc., Lexington, Massachusetts.

MORE RETAILING TIPS

Provide ample parking. Your sales on any given day will be limited by the number of cars you can or cannot park. Provide customers with carts. I've seen all sorts of things, even children's wagons.

Your aisles should be wide and surfaced, never muddy. Have someone around who can answer questions and talk garden-

FIG. 4   Molbak's, Woodinville, Washington, makes it easy for shoppers. Notice the raised benches, easy-to-read signs with clearly marked prices, and shopping carts.

ing. Create plant displays at convenient waist height. Provide colorful garden booklets and books for people to thumb through. They get ideas on how they can use annuals around their homes, and may be encouraged to try new varieties.

Well-organized self-service is important. Build ample checkout facilities so customers who are ready to leave can give you their money. Finally, always have someone to help the customer load heavy plants and other goods into the car.

GROW QUALITY

There's no sales builder quite like good plants (figs. 6 and 7). Word spreads when you do a good job—just like with a good restaurant.

GROWING THE CROP

Some growers grow entirely for one or several of the large chains, like Wal-Mart, Kmart, and Home Depot. Many chains are famous for making low-priced deals with growers. They do order well ahead. Automation is important to grow effectively for the large discount chains.

Grocery chains are another important bedding outlet. There are many of them, and most offer bedding plants in the spring. Many growers have close tie-ins with a food chain, large or small.

The third major sales outlet for bedding is garden centers (retail growers fall into this category, too). 1996 surveys show garden centers about even with the big chains in total volume. They often price fairly high, and they offer a wide variety of bedding plants.

Regardless of the outlet, growing bedding plants is basically the same for any grower. Be sure to work closely with your seed salesperson in setting up your crop. He or she will give you valuable guidance on scheduling for your specific climate as well as on variety selection. There's a whole new technology in treating seed, especially for bedding plants. It often produces results over 90% with impatiens and even petunias. You may recognize trade names like Genesis for pansies or impatiens. There's also now the Ball Vigor Index, which is a rating of the vigor of the seed.

FIG. 5 Larry Dean and his wife Vicki operate 3 acres of wholesale bedding plants, plus a major garden center. Half of their sales go to various fundraising groups. Note the sold sign on the bench of mixed impatiens trays. The greenhouse: Dean's Greenhouse, Westlake, Ohio.

FIG. 6 David Wadsworth, Suncoast Greenhouses, Seffner, Florida, displays a great looking tray of 388 impatiens plugs.

Growers typically hand-water bedding plants or use overhead sprinklers. However, a few pioneers are doing the job with ebb-and-flood irrigation, like Jim Gapinski, Heartland Growers, near Indianapolis. See chapter 7, Irrigation for Pot and Bedding Plants.

HASTENING FLOWERING

FIG. 7 Holding a pot of pentas, Jim Pugh, manager, American Farms, Naples, Florida, meets with Rick Grossman, salesman, Ball Horticultural Company.

Since most markets want color on annuals, such as petunias and impatiens, it's worth making note that flowering of most annuals can be advanced by adjusting growing conditions.

ENVIRONMENTAL CONTROL

Petunias and most other annuals flower earlier at warmer temperatures. For example, Pink Magic petunias sown March 9 were 96% in flower when grown at 60F (16C), but only 4% in flower grown at 50F (10C). That's why some growers even grow annuals at 68F (20C) rather than 60F. Warmer temperatures stretch the plant and discourage bottom breaks, but they do flower earlier. The plant stays in flower when it's planted out. Higher temperatures are practical, and this technique is used by many growers.

Many growers would rather use growth regulators to control height than cool temperatures, which delays flowering. One of the most popular growth regulators for bedding, B-Nine, does not delay color.

On most annuals more frequent watering means earlier flowering. In Penn State trials, Comanche petunias sown March 2 flowered 70% by May 21 if watered daily, 7% if run dry.

More fertilizer means earlier flowering for most annuals, certainly for petunias. Allegro petunias sown March 2, not fed, are 33% in flower May 21; fed weekly, 90% are in flower.

What do growers do about feeding? Many initially use fertilizer injectors but often go back to clear water as the plants become established—frankly, to harden the crop and sometimes to hold it back. A few other growers use slow-release (MagAmp and Osmocote) only, often at half or less the recommended rate. In general, a pot mum crop will probably get liquid fertilizer injection from start to finish. Bedding plants generally are fed a lot less than that, some crops not at all. It's a judgment call that you get good at only with experience.

Day-length control can speed up a crop. The giant-flowered American marigolds clearly respond to short days. Just spreading a sheet of 1-mil, black poly over a bed of the old variety Moon Shot from 7 A.M. to 5 P.M. flowered the crop several weeks earlier—and at perhaps half the height. And the plant stays in flower when it's planted out. It's practical, it's real, and it's done by many growers. Note that earlier sowings made in January and February often flower just as fast without artificial short-day treatment because of natural short days at that time of year. Several of today's new American marigold varieties are less sensitive to day length. Note also that French marigolds don't respond appreciably to day-length control.

Petunias present an interesting case. On a short, nine-hour day, petunias will be bushy, free breaking, and of great quality but have delayed flowering. On the other hand, grown warm and under a 12-hour or longer day, they will flower earlier but tend to be tall and spindly. When Harry Tayama was at Ohio State, he counseled growers to "start them on long days at 60F (16C); apply B-Nine; then as flower buds are visible, stop the lights and cool them down."

Salvias have a complex day-length response. The tall, late types are naturally short-day plants, flowering only in late summer as the day length shortens. The very short, early-flowering types, like St. John's Fire, will flower short and compact in midsummer; obviously, they are long-day plants. Discuss variety selection and day length with your seed salesperson, who will be able to share the experiences of other growers in your area.

HOW GROWERS SPEED FLOWERING

High temperature, frequent watering, and fertilizer can cut crop time and give earlier flowers and a more lush plant. In contrast, a cool, dry, low-feed culture produces a sturdy, harder plant with better shelf life.

Growers can be divided into perhaps four groups. Cool-dry growers sow very early. As soon as seed germinates, they drop the crop back to 50 or even 45F (10 or 7C). It can be a 20-week crop from sow to sell. The plants at sale time will be tough, rather hard, with good shelf life, and they will perform for the customer.

A great many growers across the United States start at 60 and drop to 50F (16 and 10C). The crop is kept at 60F nights for a week or 10 days after transplanting, then dropped to 50F or even cooler to harden the plants and improve quality. There are more bottom breaks. It is probably a 14-week crop from sow to sell.

A large group of growers follow the regime of straight 60F from transplant to sell. It is now down to about 12 weeks from sow to sell. This doesn't quite result in the quality of the crop cooled down to 50F during the later part.

Straight 68 to 70F (20 to 21C) will probably cut two weeks off the time of the crop grown straight 60 to 50F. This speed is especially important when you're working for a second crop, which many growers do, and for early color on petunias.

Southern growers aim for a peak sales period of April 1, versus May 1 in the North. All my records point to perhaps a week or so less time sow to sell on their April 1 crop versus the northern May 1 crop. The reports on straight-60F (16C) crops average about 11 weeks sow to sell. Probably the warmer, more favorable spring, especially in the Deep South, is responsible for this.

Table 1 puts it all together.

TABLE 1   BEDDING GROWERS' BASIC SCHEDULES

| | Crop timing | | | Temperature | | | |
| | Sowing | Weeks | Salable | Early | | Finishing | |
Crop	date	to sell	date	F	C	F	C
Northern							
Warm-grown, early	Dec. 1[a]	13–14	March 1	68–70	20–21	70	21
Warm-grown, peak week	Jan. 25	11½	April 15	68–70	20–21	70	21
Peak week	Jan. 23	14	May 1	60	16	50	10
Slow, dry, cool	Jan. 7	16	May 1	60	16	45	−7
Many	Feb. 6	12–13	May 5	60	16	60	16
Southern							
Petunia	Jan. 15[b]	10	April 1	60	16	60	16

[a] Nov. 21–Dec. 10.

[b] Sow plugs on this date. Transplant plugs Dec. 25.

Several growers' experiences clearly confirm that exposing plug annuals, especially petunias, to HID light for even several weeks will hasten flowering.

BEDDING PLANTS IN 14 DAYS

It is possible to take only 14 days from planting plugs to a flowering flat of impatiens. Petunias and many other annuals can take 15 to 20 days. Bedding growers everywhere strive to produce more crops per greenhouse per bedding season. Producing 14-day annuals is a way to do it. (The size of the plug cells affects finishing time, shown in fig. 8.)

Van de Wetering. I was fascinated to hear that Jack Van de Wetering, a major Calverton, Long Island, New York, bedding grower, has gone over to 14-day crops. I was especially interested to hear that Jack actually produced 4.4 crops of bedding in one season (April 1 to July 1) per greenhouse. There were two crops in the house at any given time: one on the floor, and one on benches that rolled out during the day. Imagine—4.4 crops in one greenhouse in one spring season!

Jack has a reputation for pioneering in the bedding plant world. He was the first to move spring bedding crops outdoors in the day on roll-out benches. He was also first to do spring bulbs in bedding plant flats.

Jack's bedding crops are based on 48 plants per flat.

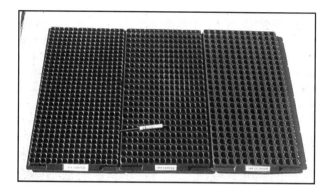

FIG. 8 Size of the plug cell affects finishing time: the larger the plug, the faster the finish. Three of Blackmore's plug trays are placed side-by-side so you can see the size difference between a 512 *(left)*, 384 *(center)*, and 288 *(right)*.

The plug used is a purchased 216, a good-sized plug. Transplanting is by an automatic transplanter that creates no transplant shock. You save 10 to 20% on crop time right there.

Jack says, "We grow them on the warm side." Also, these fast-crop impatiens and petunias are all grown on the greenhouse. "We do move some other crops outdoors (rolled out each day on benches), and they take a bit more time. . . These finished flats are typically half in flower," which seems to satisfy Jack's market.

Notes from Greiling Farms. Plug specialist Greiling does 14-day bedding plants in major volume, too. The impatiens crop has about half 14-day and half regular, 21-day timing. Greiling is currently using 288, plugs of which "many" will be in bud when planted. Greiling in Wisconsin grows 14-day plants in 606 (36 per flat) and some in 1040 (40 plants per flat). These 14-day plants are fed daily at 150 ppm. They are grown on a heated concrete floor with heating pipes at 12-inch (30-cm) intervals. Also, they are grown in 68F (20C) nights.

KEEPING PLANTS SHORT

DIF

Growing in cool day and warm night temperature (DIF) is a remarkable way to control stretch of nearly all ornamental plants, certainly all bedding plants. The technique is the result of research at Michigan State University. The term *DIF* is simply an abbreviation for the *difference* between day and night temperatures.

The principle is simple: If plants are exposed to day temperatures, say, 6 to 8F (3 to 4C) cooler than night temperature, plant elongation nearly stops. You'll notice it within several days. For the record, DIF applies to poinsettias, lilies, pot mums, and practically all ornamental crops (see chapter 23, DIF and Graphical Tracking). What a great way to control stretch of petunias and impatiens during a late, wet spring! If temperatures are just held the same during day and night, stretch will be noticeably less. Conversely, if the day temperature is perhaps 3 to 5F (2 to 3C) or more warmer than the night temperature, plant elongation will be notably accelerated.

The only important limitation of cool-day culture of bedding plants (or any crop) is simply that—in southern areas, especially—it's difficult to keep the day temperature cool enough. Remember, the night must be warmer than the day, so if the day is, say, 80 or 85F (27 or 29C), then the night temperature must be 90F (32C). The other limitation is that many bedding plant growers simply carry very low night temperatures to improve quality and save fuel. If you're going to keep the night warmer than the day, it tends to mean more of a 60 or 70F (16 or 21C) night temperature, which does add to fuel costs.

There is a quick and economical answer to these problems. Michigan researchers say that if you maintain cool-day temperatures *for just the first hour or two after sunrise*, the plant tends to accept that as the all-day temperature. In other words, if you can keep 50F (10C) for the first hour or two in the morning, the plant thinks that it has a 50F temperature all day. If the night temperature is kept at 55F (13C), then we have achieved a cooler day than night, for practical purposes. This seems to really work, and the grower can easily accomplish it by opening vents at sunrise to rapidly lower temperatures for the required length of time.

Another point about cool days: if there is a strong difference, perhaps 10F (6C) cooler days for an extended period, many plants tend to become chlorotic. Normally, a few days with warmer day temperatures will correct the problem.

Now let's talk about cool days and conventional Quonset and hoop greenhouses, which are frequently used for growing bedding plants. The fact is that most bedding plants are grown in poly hoop houses, which are typically not well ventilated. Growers tend not to heat much at night, due to saving fuel and increasing the quality of the plants. As a result, for all these years our bedding plant crops have often been getting warm days and very cool nights, which very strongly aggravates stretch. In other words, the structures we have been living with couldn't have been designed worse in terms of producing quality, compact bedding plants.

GROWTH REGULATORS FOR HEIGHT CONTROL

Growers generally use B-Nine or A-Rest as growth regulators for bedding (table 2). Many growers I know use B-Nine typically in the standard 0.25% dilution. Petunias respond beautifully to B-Nine, which keeps the plants short and stocky and doesn't seem to delay flowering appreciably. Many other annuals also respond.

Caution: Bonzi used as a growth regulator on bedding plants can retard subsequent growth in the hands of the customer. Not so with B-Nine. Bonzi can also have residual effects on subsequent crops when it is sprayed onto plastic parts in a subirrigation system or when some treated plants are left in flower beds in the landscape. If you choose to use Bonzi as a growth regulator for bedding plants, use great care. Also, growth regulators are strictly forbidden for use on any vegetable bedding plant.

TABLE 2   GROWTH RETARDANTS FOR BEDDING PLANTS

Either A-Rest or B-Nine	A-Rest only	B-Nine only
Ageratum	Catharanthus	Calendula
Aster	Celosia	Petunia
Browallia	Cleome	
Centaurea	Coleus	
Dahlia, dwarf	Dianthus	
Geranium	Snapdragon	
Impatiens		
Marigold, dwarf		
Marigold, tall		
Salvia splendens		
Verbena		
Zinnia		

SPECIFICS FROM MAJOR GROWERS

On the following pages, you'll find different approaches to growing bedding plants. Specific temperatures and schedules are given for several growers, from cool-dry to warm-faster crops, from a California grower to a Deep South operator. These are real-world growing strategies from growers in very different climates and growing for different markets.

BERNACCHI GREENHOUSES, LA PORTE, INDIANA

Bernacchi is a classic family operation. It has about four to five acres (1.6 to 2 ha) of holiday pot plants (top quality and price) plus a two- or three-acre (0.8- or 1.2-ha) bedding plant range. Bedding is nearly all done from plugs (table 3). I talked with Jerry Bernacchi and his uncle, veteran grower Bart Bernacchi.

TABLE 3   **SCHEDULE FOR BEDDING PLANTS**

Sow plugs[a]	Lot	Transplant plugs	Crop salable
Jan. 15	Begonias (slowest crop) first sowing	March 1	Apr. 20
Feb. 5	Last lot of begonias	March 5	May 1
Feb. 5	Impatiens—first lot	March 10	Apr. 15
Apr. 1	Impatiens—last lot	May 1	May 20
Feb. 1	Petunias—first lot	March 7	Apr. 15
Apr. 1	Petunias—last lot	May 1	June 1

[a]For bare-root seedlings, add one week sow to sell.

Plug growing. Temperature for plugs is 75F (24C) night and day. Plugs are grown on rolling raised benches (with expanded mesh bottoms) with steam pipes below and poly skirts on the side. Water for most of the plugs is heated to 65F.

The Bernacchis use trays with 200 cells for geraniums, 512 for begonias and most bedding plants, such as petunias, celosias, and snaps. They reuse plug trays. "It would be costly to dump them." Media is commercial peat and vermiculite material (Fafard Mix).

"Sowing is with a Hamilton here. We produce about 14,000 trays of plug sheets a year. . .We use Genesis seed on pansies and dusty miller, Vigor Index seed on all petunias and impatiens." Begonias are pelleted and double seeded. They use defuzzed tomato seed and de-tailed marigolds, too.

Watering is "a good bit by hand early in the game. Later in the spring, we use overhead mist. On a sunny day, we will often hand-water the area three to four times. The nozzles are 3 gallons (11 l) per minute."

Fertilization begins at 50 ppm at time of sowing and is increased to 75 ppm later. "We watch water acidity carefully. Our goal is pH 6 to 6.3. We use sulfuric acid, 1 fluid ounce per 100 gallons (78 ppm, or 78 ml per 1,000 l) as needed. If pH climbs, begonias get yellow." Water is tested twice a year. They inject Agribrom into the irrigation water and also use a Greenseal dip for all plug trays to control algae growth.

Culture of bedding plant flats. Flats are grown at 65 to 68F (18 to 20C) night and day. Ventilation in the day is done at 65F. Media are commercial (Fafard Mix).

Most bedding plants here are grown in 32 cells (plants) per 22-inch (56-cm) flat (eight packs of four plants). No more 72 plants per flat! "We do a few 4 inch here—mostly seed geraniums. Four-inch annuals haven't sold too well up to now." Bedding plants are grown on the ground, but they use a plastic riser that holds flats about 3 inches (8 cm) off the ground.

Bernacchi delivers to about 50 stores of one of the major supermarket chains. Once a week, out go a minimum of 200 flats, or normally 400 to 600 per trip.

Kube-Pak, Allentown, New Jersey

The plug was born at Kube-Pak, developed in the early 1970s by Fred and Bernie Swanekamp. Also, here in 1971 the first automatic seed sower ever was built, an all-stainless steel drum model that could turn out 1,200 flats per hour.

Kube-Pak is today a rather large operation, with 14 acres (5.7 ha) of greenhouse. Besides bedding plants and a major poinsettia crop, it produces 115 million plugs a year, primarily for resale. The firm grows about 350,000 flats of bedding plants, whose major distinction is that they are all grown in preformed cubes mainly of peat, the "Kube-Paks."

Plug production. Kube-Pak grows plugs in standard 512-, 288-, 162-, and 70-plug trays. Its plug production focuses on custom growing for the medium to small greenhouse, to whom it currently offers more than 2,500 different varieties. Interestingly, more than 75% of all its plugs are sold directly to plug customers, rather than through brokers. Kube-Pak developed its own software to facilitate this need.

The plugs are grown in a greenhouse roofed with a double layer of poly. In fact, a third layer of poly is in place to prevent condensation-dripping damage to the plug crop below. The 3-inch (8-cm) thick, porous-concrete floor is heated with pipes 3 inches deep and spaced 11 inches (28 cm) apart.

Kube-Pak mixes its own media to this recipe: 80% peat, 13% vermiculite (for the cubes) and 7% soil, plus lime (gypsum) and a fertilizer charge. The soil is added as a buffer, but the company is not sure it will continue with this—it's becoming too hard to get good soil. That original seed sower still works, and you can get a look at it, but it's been retired from active service. Today, Kube-Pak uses an Old Mill, Niagara, Seed-a-Matic, and two custom-built and modified B & L Drum seeders to produce its more than 250,000 plug trays annually.

The floor heating keeps the soil temperature of most crops at about 75F (24C) nights, 72F (22C) days. Begonias have an even higher soil temperature, around 76F (24C), whereas pansies and some perennials are not higher than 68F (20C).

Kube-Pak has insulated a 40-foot (12-m) truck body and equipped it with refrigeration to serve as a growth chamber for sprouting summer pansies and perennials. Pansies in particular like cool temperatures, and there are good reports on Genesis pansies here. To accommodate their temperature requirements, vinca and salvia are also germinated in the truck. There is much faster sprouting and more uniformity with many plant varieties in this growth chamber. Incidentally, the truck has an interesting air-water mixing valve to provide humidity. Mixing 40 psi of air with 3 psi of water, it makes fog at a lot lower cost than mist equipment.

Kube-Pak tests its water once a year. Plugs in greenhouse germination areas are irrigated with 26 ITS booms.

Growing on. Although growing and selling plugs comprises the great majority of Kube-Pak's business, it also finishes a good number of plants. Perennials make up 7% of its production, and 4- and 6-inch (10- and 15-cm) are another 7%.

Kube-Pak transplants on an assembly line. (Production schedule shown is in table 4.) Bill Swanekamp, Bernie's son, reports, "Our people average 45 flats per hour. Our cost is 17 cents a flat for mostly 48 plants per flat. . . . All our bedding plants are in peat moss cubes, mostly $1^1/_2$ by $1^1/_2$ by 3 inches deep (4 by 4 by 8 cm) in a flat 13 by 18 inches (33 by 46 cm). We are doing some 3-inch cubes, finding a good market for them, and will probably move more that way." That's a different form of pot annuals!

Immediately after transplanting, all annuals go to heated floors at soil temperatures of 72F (22C) nights and 68F (20C) days. The nighttime air temperature is 62F (17C), and daytime venting starts at 70F (21C). Plants grow in these temperatures for about two weeks.

After two or three weeks, the bottom heat is turned off, and overhead heat is started up. Bill notes, "Bottom heat all the way will grow plants very fast, but they will be soft and low quality." Also, with bedding plants, if soil temperature is kept at 65F (18C) and air is allowed to drop to 45F (7C), this can result in condensation drip at night and cause disease problems. Therefore, Kube-Pak recommends keeping air temperature above 55F (13C) if floor heat is used."

Kube-Pak uses the cool-days (DIF) technique to control stretch, employing it very much with poinsettias. A night temperature of 62 to 64F (17 to 18C) is followed by 47 to 52F (8 to 11C) for just the first several hours after sunrise. All the rest of the day is at 72F (22C), but the short, early-morning cool period creates a cool-day response in the plants. This tactic is known as pulsed DIF.

To spread out the availability of such annuals as petunias, Kube-Pak will grow up to 10 sowings per season. It cools earlier sowings down to 45F (7C) nights and 55F (13C) days to reduce stretching. The rest of the sowings are grown at normal temperatures and will mature quicker, but still at a later date due to their later planting.

As was transplanting, irrigation of flats is done manually. Bill reports, "We do it by hand: eight people for our spring peak weeks, the balance of the year fewer than that. We figure the cost of installing automation on our whole area would not be justified in view of our annual cost of hand-watering, and not as accurate.

"We do constant liquid feeding, but with judgment, some crops less. Example: ageratum, no feed until we see color. Same for marigolds.

"Our market wants color on annuals, unless the demand is too great—then they will accept flats without color. Peak shipping weeks here: May 1 to 15."

Do Rights Plant Growers, Oxnard, California

Dudley and Dianne Davis operate about 5 acres (2 ha) of greenhouses plus a new 1-acre (0.4-ha) "frostproof" area. Bedding plants of the highest quality are the backbone of the business. Like so many others, they do a poinsettia crop plus some follow-ups.

TABLE 4 KUBE-PAK'S PRODUCTION SCHEDULE

Crops	Sow plugs	Tranplant plugs (wks)	Crop salable (wks)	Sow to transplant	Transplant to sell
Begonias—slowest crop—Lot #1	Jan. 13	March 1	Apr. 23	7	7
Last lot	Feb. 17	April 5	June 15	6	5
Impatiens—Lot #1	Jan. 27	March 3	April 15	5	6
Last lot	March 31	May 5	May 26	5	3
Marigolds—Lot #1	Feb. 3	March 10	Apr. 15	5	5
Last lot	March 10	April 14	May 12	5	4
Petunias—Lot #1	Jan. 20	March 3	April 15	6	6
Last lot	March 3	April 7	May 12	5	5
Dahlia—Lot #1	Jan. 21	Feb. 18	April 15	4	8
Dusty miller—Lot #1	Jan. 13	March 3	April 14	7	6
Salvia—Lot #1	Jan. 27	March 3	April 15	5	6
Last lot	March 17	March 24	May 5	5	6
Verbena	Jan. 20	March 3	April 23	6	7
Vinca—Lot #1	Jan. 27	March 17	May 4	7	7
Last lot	Feb. 10	March 31	May 15	7	6
Pepper—Lot #1	Feb. 10	March 17	April 30	5	6
Last lot	March 3	April 7	May 12	5	5
Tomato—Lot #1	Feb. 24	March 24	April 21	4	4
Alyssum—Lot #1	Feb. 10	March 10	April 15	4	5
Lobelia—Lot #1	Feb. 3	March 10	April 21	5	6
Pansy—Lot #1	Jan. 1	Feb. 17	April 1	7	6
Geranium—Lot #1	Jan. 1	Feb. 5	April 15	6	10
Ageratum—Lot #1	Feb. 3	March 10	April 15	5	5
Last lot	Feb. 10	March 17	April 30	5	6
Celosia—Lot #1	Feb. 24	March 31	April 28	5	4
Last lot	March 10	April 14	May 12	5	4

It's the nature of their bedding crop that I would especially like to discuss, partly because it is typical of most California bedding growers. First and up front, you will see no 72 cell-pack plants per flat or 48 bedding plants per flat here at all. The traditional flat bedding plant is just about gone in California.

Four-inch (10-cm) annuals—mostly petunias and impatiens—make up about 10% of Do Rights' crop. They grow them 16 in a 17- by-17-inch (43-by-43-cm) flat. Four-inch annuals were once the backbone of California bedding plants, but they too are fading away now. The famous six-pack is taking over in California. The six-pack really is a sort of compromise 4-inch annual. The consumer gets 2³/4-inch (7-cm) "pot annuals" in a pack of six plants. They're always sold in flower, giving the consumer the benefit of instant color, larger soil volume, and greater ability to survive transplanting. Plus, the retail outlet has less of a problem keeping them watered than smaller pack plants. Another point in favor of the six-pack: it gives the consumer a longer flowering season than the pack annual. You don't have to wait three or four weeks for it to flower! Do Rights has also introduced a "904," with nine four-packs. Each cell is the same size as a 606, and the packs are used mainly for tomatoes and peppers. Reception is good.

Patio annuals take several different forms, but they are already an important part of California spring plants. Most importantly, there are more being grown each year. Dudley reports 5% of Do Rights' sales were in gallon (3.8-l) cans, typically planted from six-packs, using impatiens, bellis, coreopsis, statice, dwarf snaps, and more. (The Davises don't do petunias or pansies in gallons unless on special order.) "Demand (for large sizes) is increasing rapidly."

Elsewhere across California you see acres of such patio planters as color bowls, whiskey barrels, and cedar boxes. There are a wide variety and a great number of planters. People like things they can set out on their patios or front porches that will give them color now. Do Rights does a lot of color bowls in 12, 14, and 16 inches (30, 36, and 41 cm). In 1997 they introduced a new 14-inch window box for herbs or flowers. The Davises do baskets, too—mostly 10 inch (25 cm). The sales trend for baskets is generally up throughout California.

A peat, perlite, and vermiculite blend is mixed on-site and on demand. The firm has just installed a new Flier transplanter, and they are really happy with it.

Bedding plants are a year-round crop in California, with peaks in March to April, and again in October to November. Sow to sell (with plugs) takes about eight weeks for impatiens, seven weeks for petunias. In central California many annuals are kept in a 60F (16C) greenhouse for seven days after planting from plugs, then moved right outdoors. Days will be 50 to 65F (10 to 18C) in early spring. Impatiens, American marigolds, and begonias are more tender and aren't moved out until April.

The Davises germinate their plugs in a growth chamber at 75F (24C) night and day for most crops. Plug trays are 512 cells per 21-inch (53-cm) flat, all square. They have two seeding machines, a Blackmore cylinder and a KW needle seeder. They do about 60,000 trays a year—all for their own use.

DIF is used when needed at Do Rights. Convection cooling is managed by controllers with DIF setpoints programmed in.

CARL BLASIG GREENHOUSES, HIGHTSTOWN, NEW JERSEY

Carl operates about 2¹/₂ acres (1 ha) of modern greenhouses. His grand plan is to expand his place as much as possible, yet be able to do most of the work himself and with his wife and son. This means automating, especially automatic irrigation of all his crops. From seeing his operation several times, it's clear to me that he is doing a top-quality job with automatic watering and with minimal outside help. (Production schedule shown in table 5.)

TABLE 5 SCHEDULE FOR BEDDING PLANTS

Crops	Sow plugs	Transplant plugs	Crop salable	Sow to transplant (wks)	Transplant to sell (wks)
Begonias (slowest crop)–					
Lot #1	Jan. 10	Feb. 25	Apr. 25	6	8
Last lot	Feb. 20	March 25	May 15	5	9
Impatiens–Lot #1	Feb. 15	March 20	Apr. 20	5	5
Last lot	March 15	Apr. 15	May 15	4	5
Petunias–Lot #1	Feb. 5	March 10	Apr. 25	5	7
Last lot	March 15	Apr. 15	May 25	4 to 5	6
Marigolds, African–Lot #1	Feb. 15	March 20	Apr. 15	5	4
Last lot	Apr. 1	Apr. 20	May 20	3	5

Plugs. Carl maintains two temperature areas, one warmer and one cooler. Most annuals, including petunias, grow at 68 to 70F (20 to 21C) night and day. Begonias, vinca, and impatiens are grown warmer. Tomatoes and marigolds are generally in a still cooler area.

Plug trays are mainly 512. "We buy in a few of the larger 1-inch (2.5-cm) plugs for fast turnover of crops. Eighty percent of the trays we use are new. We do dip a few into LS10, but it's a lot of work at a busy time. We wonder if it pays at $10 per hour."

Plugs are grown on a solid concrete floor with buried hot water lines spaced 10 inches (25 cm) apart and 4 inches (10 cm) deep. This permits maintaining the desired soil temperature in the plug trays. Carl uses two Grower System booms to water plugs. "We do some touch-up by hand, but even begonias are mostly boom irrigated."

For seeders, Carl has one Blackmore and one Old Mill. He sows de-tailed marigold and defuzzed tomato seeds. Begonias are pelleted. He uses some of the high-tech seeds, mechanically graded for higher germination. "We do double seed a few things, especially if germination on the package is listed below 85%. All our begonias are double-seeded. We get a better, fuller tray. The same for most petunias: double-seeded. Things like portulaca and alyssum, we seed five to six seeds per cell, which we can do with either our Blackmore or Old Mill seeder. We do B-Nine most plugs—improves quality."

There is no growth chamber here. "We're thinking about it, especially for things like begonias and summer pansies." Plug sheets are set on an inverted plastic flat to keep them up off the ground.

Growing on flat annuals. Carl grows flat annuals primarily in eight packs of six plants, 48, per 21-inch (53-cm) flat. He does a considerable amount of pot annuals and various patio pots. The media for flat annuals are commercial mixes. Carl does all transplanting on a Flier automatic transplanter.

Flats in the production areas are grown on a layer of pea gravel with no soil heat. Oil-fired unit heaters are used throughout the range. The temperatures for growing on of flats of impatiens, begonias, and vincas are 65F (18C) nights, 70F (21C) days. Petunias and snapdragons grow at 55F (13C) both night and day.

Most of Carl's 2½ acres (1 ha) of flat annuals are watered almost entirely with ITS boom irrigation. He finds this model, which can be moved manually from greenhouse to greenhouse, best. "We get very even coverage," with 18 nozzles on a 10-foot (3-m) boom."

His peak week of sales of flats is April 25 to 30. "Flat sales are usually all through by May 20." Carl reports that roughly three fourths of his spring business is annuals in flats, but flat sales are steady or maybe dropping a bit, and both pot annuals and patio forms are coming up. Says Carl, "People don't seem to have time to make flower gardens with pack annuals anymore. They like the things they can set out on their patios. And if they are going to make a garden, they are more apt to buy 4-inch and a lot of 6-inch annuals to get color right now.

"We do some color bowls for Mother's Day, some 18-inch (46-cm) window boxes, and an interesting 24-inch (61-cm) long cylinder that hangs on its side—moss-lined and very colorful. Also, 30,000 8-inch hanging baskets. Baskets are big, but the trend is only steady—they just can't be hung from a patio very well.

"We do a lot of 6-inch bedding plants, lots of zinnias, some 4-inches. We do a nice 12-inch sphagnum-lined hanging basket, 40 inches (1 m) across the top when in flower: $45 wholesale."

AHRENS GREENHOUSE, OSSEO, MINNESOTA

With his three sons, Harold Ahrens turns out 70,000 flats of bedding per season. These 17-by-21-inch (43-by-53-cm) flats have 18 packs, as opposed to just 12 in the 10-by-20-inch (25-by-51-cm) standard flat. Harold says, "We have fewer flats to handle with the larger flat."

The most unique fact about Harold's operation is that he grows his bedding all from bareroot seedlings seed—no plugs! And he does an excellent job. Ahrens's petunia packs were just bristling with breaks, shoots sticking out all over the pack and ready to bloom. (Production schedule shown in table 6.)

Impatiens and begonias are grown at 60F (16C) nights, petunias and everything else at 50F (10C). There's no bottom heat, except for seed flats, which are grown at 75F (24C).

The Ahrenses use a local peaty soil and amend it with MagAmp, phosphate, and potash. Irrigation is partly automatic, with an ITS and reels of hose down the aisle for the rest.

TABLE 6 SCHEDULE FOR BEDDING PLANTS

Crop	Sow seed	Crop salable	Weeks to sell
Impatiens—Lot #1	Feb. 7	April 25	11
Last lot	March 13	May 21	10
Petunias—Lot #1	Jan. 10	April 25	—
Last lot	Feb. 8	May 20	—

Note: By the way, the impatiens are all grown with BVI (Ball Vigor Index) seed.

WAGNER GREENHOUSES, MINNEAPOLIS, MINNESOTA

Ron Wagner has almost five acres (2 ha) of plugs in the spring, and they are top quality. Wagner is a nationally recognized plug supplier (fig. 9). The trays are mostly 392 plugs each, with 648 earlier in the season.

Wagner also has a thriving retail business, including annuals. He grows impatiens (48 per flat) in 14 days from transplant to sell. Sown on March 18, the impatiens are transplanted on May 1 from 288 plugs (with a Flier transplanter). "Petunias take 21 days, transplant to sell. Also, to do 14-day impatiens, or any annual, takes good plugs!" Ron says.

Question: Are 14-day annuals good for the consumer? Ron's answer: "Yes, for impatiens. Less so for petunias and snaps. . .Without a good plug, you won't grow a good 14-days impatiens—or any annual. The plug should have buds."

Wagner has heat pipes under the Dutch trays for the 14-day impatiens. The mix is peat, perlite, and vermiculite. A constant feed at 50 to 75 ppm is used with all irrigations.

KLOOSTER GREENHOUSE, KALAMAZOO, MICHIGAN

Klooster Greenhouse is located just off Kalamazoo's Sprinkle Road, one of horticulture's best-known stretches of pavement. Steve

FIG. 9 Dave Wagner, Wagner Greenhouses, Minneapolis, Minnesota, holds a top quality impatiens plug tray. In addition to operating a substantial pot plant/bedding plant range, Wagner also operates a retail store in Minneapolis.

and Don Klooster are classic Dutch-American Kalamazoo growers, and all they grow is bedding plants.

The Kalamazoo logic. The 70 Kalamazoo growers altogether do about 4 million flats a year, two thirds of which are sold through the famous Kal-amazoo Valley Growers (KVG) Cooperative. Kalamazoo growers are as close in mind-set as they are in geography. They tend to move together on new growing and marketing concepts, and they communicate freely among themselves (fig. 10).

One unique Kalamazoo practice, for example, is that most growers carry a straight 70F (21C) on all bedding plants. To control stretch they are almost forced to use cool-day–warm-night retarding, because of the ban on chemical growth regulators on vegetables. Still, the temperature will likely average 70F, such as 67F (19C) days and 72F (22C) nights.

FIG. 10   Bedding plant production Kalamazoo, Michigan, style: a fine crop of marigolds *(foreground)* and impatiens *(back)* at Snobelt Greenhouses.

Why the warm temperature in Kalamazoo? This area, growing only bedding plants, strives for an early crop for the southern market, and warm temperatures hurry this important crop into maturity sooner. This then accommodates a major second crop, maturing in late April or early May for the big midwestern market. Since most Kalamazoo growers do bedding only and absolutely no other crops the rest of the year, the early southern crop is their key to profit.

The grower-owned marketing co-op has been especially important in developing the southern market. It has also facilitated specialization within bedding plants. One grower will do impatiens, for example, one marigolds, and another petunias.

Some of Kalamazoo's 4 million flats have 12 four-packs, or 48 plants per flat. Most KVG Cooperative bedding plants, though, are grown in 18 two- or four-packs, making 36 or 72 plants per flat. As usual, begonias and impatiens are often grown as fewer plants per flat, and each flat often brings substantially more dollars.

Klooster culture. The Kloosters, like 80% of Kalamazoo, have gone to plugs. Much more rapid transplanting is possible, especially with an assembly-line belt system. Sandwiching in two spring crops works a lot better when you can save three weeks from plant to sell on the second crop by using plugs. Then there is the greater flexibility with plugs: you can hold a tray without loss longer than a seedling flat. By the way, primed Genesis seed is used a good bit in this area.

The plug medium at Klooster Greenhouse is a commercial mix of Sunshine plus 25% vermiculite. Growing-on flats are filled with HECO, which is mainly peat with nutrients, 30% vermiculite, and 20% polystyrene beads. Steve says, "This is normal for most of Kalamazoo. We don't sterilize these mixes."

Nearly all Kalamazoo growers use root zone or some other sort of effective soil heat for germinating seed and plugs. There is very little root zone heating in the production of bedding plants here, however.

The Kloosters use HID supplemental light on seed flats. "We can sow weeks later on many crops, including impatiens, geraniums, and petunias," notes Steve. "It also speeds development of begonia seedlings—not earlier flowering, but they grow better." Light reduction—mostly whitewash on the poly roof—is needed some from mid-April on. "It does cut down the amount of watering required."

Transplanting and finishing. There's a definite trend in Kalamazoo toward automatic transplanters. Klooster is completely on mechanical transplanters.

Although the crops are grown on the ground, the Kloosters and most other Kalamazoo growers use plastic forms to keep the flats about 3 inches (8 cm) off the ground. This is a big help in keeping the occasional plant from growing through the flat, which would make a wild top growth and an uneven flat. It also keeps flats clean.

As with other good growers, these folks don't just feed bedding plants with every watering. The Kloosters mainly water their flats with clear water, adding occasional fertilizer "as needed." Don has two water lines through his range, one with constant-level fertilizer and the other clear water.

Most Kalamazoo growers are using the cool-day–warm-night (DIF) technique to control the height of bedding plants. They are also well aware of how cooling for just the first two hours after sunrise (pulsed DIF) achieves a daylong cool-day effect. How about cooling plants back for finishing? This is not done here unless the weather is holding down demand.

There are no 4-inch (10-cm) pot annuals at Klooster Greenhouse. "The Cooperative does a lot of 3$\frac{1}{2}$ inch" (or 9 cm, the A18 flat), Steve remarks. "We've tried them. It's still small for us today, and we're not gaining, except for seed geraniums."

Various forms of patio annuals, color bowls, window boxes, Dillen baskets, strawberry containers, and the like are also slow here. Among other problems with these containers, Steve reports, they are difficult to ship—and Kalamazoo does ship!

Should bedding plants be moved in color? "Our market wants color," claims Don. "You really can't sell them without it unless the market is very tight." The peak shipping weeks for the northern market are April 15 to May 15.

J & J Greenhouses, Claxton, Georgia

Judy Crosby operates J & J Greenhouses, Inc. It's a quality operation heavily into flats and hanging baskets. Almost everything is grown from plugs here, in 392-plug trays, all from specialist producers.

Temperatures for such plants as begonias and impatiens are 60F (16C) at night; vents open or fans come on when it hits 75F (24C) in the day. Petunias are often grown outdoors from the beginning, even though there may be a light frost. "We just wash it off." Plugs for petunias to be sold April 1, the peak sales time, are planted February 25 (table 7).

Vincas are kept in a separate house, with 72 to 75F (22 to 24C) nights and days if heat is needed. "Caution: Don't overwater vinca. Let them dry out before watering. Avoid wet spots—you'll lose those plants." The key to vincas is to keep them warm and not too wet.

TABLE 7 DEEP SOUTH SCHEDULE FOR BEDDING PLANTS (MOSTLY PETUNIAS)

Sow plugs	Transplant plugs	Crop salable	Weeks sow to sale
Jan. 21	Feb. 25	April 11	10

Note: April 1 is the peak week for sales in southern Georgia.

Bedding Plants International

Bedding Plants International (formerly the Professional Plant Growers Association) serves growers of all greenhouse crops and members of related industries throughout the United States, Canada, and the world. BPI membership is largely bedding plant growers. You too can benefit by joining. The association offers a monthly newsletter, an annual conference and trade show, a membership directory, and a catalog of reference materials and production and marketing aids.

BPI's affiliated organizations include the Bedding Plants Foundation, Inc., which funds scientific research selected by growers. BPI Scholarship Foundation helps support outstanding horticulture students. The FloraStar trialing program evaluates and promotes outstanding potted plants.

I urge growers everywhere to join BPI and support its affiliated organizations. For membership information, contact BPI, P.O. Box 27515, Lansing, Michigan 48909; phone 517/694-7700, fax 517/694-8560.

HANGING BASKETS

Hanging baskets are important! Baskets continue to be a major spring plant item on most U.S. and Canadian ranges (figs. 11 and 12). Growers divide baskets into two groups, hard and soft. The hard basket plants are generally vegetatively propagated, including fuchsias, New Guinea impatiens, and ivy geraniums. They are generally slower growing and tend to bring more money. Soft baskets are generally seed-propagated annuals, such as petunias, impatiens, and many others. They are colorful and widely used.

FIG. 11 Bob and Betty Koch, R.M. Koch Greenhouse, Columbia Station, Ohio, operate about an acre of bedding plants and hanging baskets. The showy basket that Betty is holding is *Streptocapella*.

HARD BASKETS

Fuchsias. Very widely used, especially for satisfying consumers in areas of cooler summers, fuchsias are long-day plants, flowering when day length is greater than 12 hours. They must be pinched. (Production schedule shown in table 8.) For late May sales, make the final pinch seven to eight weeks before the sale. If for Mother's Day, allow eight to nine weeks from last pinch to sale. Usually, two or three pinches are needed. Most growers carry them at a 68 to 70F (20 to 21C) night temperature. Cool day and warm night temperatures are very effective for retarding fuchsias.

FIG. 12 At Metrolina Greenhouses, Huntersville, North Carolina, you'll find baskets over just about every square foot of center aisle space. It's a fine example of using overhead space.

TABLE 8   SCHEDULE FOR FUCHSIA FLOWERING BASKETS

Grower	Plant date	Sale date	Remarks
Grower 1	January	April 20 to Mother's Day	Plant three 2¼ inch per 8-inch basket. Two pinches are used. Two plants don't make a quality 8-inch basket.
Grower 2	Jan. 15	April 15 to Mother's Day	Plant four 2¼ inch per 10-inch basket. May also plant a top cutting March 15.
Grower 3	Jan. 25	Color April 25, flower Mother's Day	Plant three 2¼ inch per 10-inch basket. Pinch Feb.1 and March 1. Termperature 70F night and day. Cool days and warm nights very effective.
Grower 4	Jan. 20	May 5 to Mother's Day	Plant one strong, well-branched 4-inch pot per 10-inch basket. Temperature 68F nights. Pinch Feb. 15 and March 15.

Ivy geraniums. Cuttings of ivy geraniums should be pinched two to three weeks after planting—as soon as they are well established. Apply Cycocel at 1,500 ppm when new growth starts after the pinch. Very dwarf varieties may not need Cycocel, however. (Production schedule shown in table 9.)

TABLE 9   SCHEDULE FOR IVY GERANIUM BASKETS

Grower	Plant date	Sale date	Remarks
Grower 1	Jan. 25	Mother's Day	Plant three 2¼ inch per 10-inch basket. Temperature 68F nights, 62F days to restrict growth. First pinch is March 1.
Grower 2	Jan. 15	April 15	Plant four 2¼ inch per 10-inch basket.

New Guinea impatiens. Making a spectacular basket, if fed lightly, watered regularly, and given at least fair light, New Guineas will perform all summer on patios in the Midwest and East. They do best in neither full sun nor heavy shade and become very colorful.

Growers report that New Guineas are "tough to grow," with more than an "average" proportion of failures. Why? New Guineas are damaged by being either dried out or overwatered.

Temperature is also critical—60F (16C) is too cold, but 65F (18C) is okay. If the plants are getting too tall, go to 65F days and 75F (24C) nights. (Production schedule shown in table 10.) Research at Michigan State University has shown that the optimal average daily temperature for New Guinea impatiens is 68 to 70F (20 to 21C). Faster growth occurs at 75F, and growth is extremely slow at 62F (17C).

TABLE 10   SCHEDULE FOR NEW GUINEA IMPATIENS BASKETS

Grower	Plant date	Sale date	Remarks
Grower 1	Feb. 1	Mother's Day	Plant three 2¼ inch per 10-inch basket. Plant an unrooted cutting to a cell pack on January 10. Temperature 70F night and day.
Grower 2	Feb. 1	April 1	Plant three 1-inch plugs per 10-inch basket.

Nonstop begonias. The extremely colorful, midsize-flowered versions of tuberous begonias, nonstop begonias are very free blooming, making a vivid basket. They must have some shade and good summer care to flower all summer. (Production schedule shown in table 11.)

TABLE 11   SCHEDULE FOR NONSTOP BEGONIA BASKETS

Plant date	Sale date	Remarks
Feb. 15	Mother's Day	Plant three heavy 2¼ inch.
	Mother's Day	Plant five 1½ inch per 10-inch basket.

SOFT BASKETS

Such bedding annuals as petunias, impatiens, and vincas are very widely used in baskets. They're colorful, generally easier to grow, and faster than hard baskets. Production typically takes six to eight weeks, versus the 12 to 15 weeks for the hard baskets. Most soft baskets are grown from plugs. Planted with large, 1-inch (2.5-cm) plugs, a basket can be a very fast crop. (Production schedule shown in table 12.)

TABLE 12   SCHEDULE FOR BEDDING PLANT BASKETS

Plant date	Sale date	Remarks
March 20	Late May	Plant four petunia plugs per 10-inch basket.
March 1	April 10	Four plugs (petunias or impatiens) per 10-inch basket.
April 1	Mother's Day	Plant four plugs per 10-inch basket. Temperature 70F day and night.
March 15	Early May/ Mother's Day	Plant five 200 plugs of petunias or impatiens per 10-inch basket.
		Temperature 68F nights.

Browallia. One of the few annuals where good blue flowers can be obtained, browallia is beautiful in baskets and satisfactory in shade areas. Sow seed in mid-January for mid-May flowering.

Gazania. Hardy and heat and drought tolerant, gazania is fine for full-sun patios where other plants won't survive. Sow seed late January or early February for mid-May baskets. Keep the temperature at 65F (18C) at night.

Portulaca. Making a colorful hanging basket, portulaca is great for full exposure in hot, dry situations. It grows best at a 65F minimum night temperature. Figure 16 weeks from sowing to sale.

Thunbergia. A very attractive hanging basket, thunbergia, or black-eyed Susan, drapes artistically and does well in partial shade or full sun.

Vinca. Also commonly known as periwinkle, vinca makes a satisfying basket for the consumer. Sow seed in flats and allow for a warm, 75F (24C) soil temperature. Don't overwater, or you'll lose them in the seed flat.

Tomato. There are several cherry tomato varieties adaptable to baskets. Sow in late February for May sales.

BEDDING PLANT SEED BASICS

Probably at least two-thirds of seed-grown crops today are started as plugs. You'll find much detail throughout this book on temperatures, mixes, and other conditions needed to get good germination in plugs. There are still a lot of growers, however, who sow seed in open flats for bedding plants and other crops. The following notes on seed germination are appropriate for these growers, as well as plug growers.

Germinating seed is one of the more difficult parts of the growing job, so seed is often sown and germinated by the proprietor.

QUALITY

Sow good seed. Remember, seeds are living things! Reputable seed salespersons, through dry storage and regular germination tests, generally supply good seed, although no one can guarantee a perfect stand. There are low-priced flower seed, but the occasional problems that result more than wipe out the savings.

Today, you can also purchase a wide range of enhanced seed products, like primed seed—Genesis and others—and Ball Vigor Indexed (BVI) seed. Enhanced seed is designed to increase germination percentages. Pansies, verbena, and vinca are sold frequently as primed seed. Impatiens is the main crop available as BVI, with some petunia seed, too.

MEDIA

Use the right kinds of soil media. A wide variety of commercial mixes will germinate seed. To do the job they must be loose, porous, and well-drained. The mixes should also be low in salts because high salt levels damage tender roots. You can't win if the soil is full of pathogens. The mix must either be naturally sterile or else be steamed at 180F (82C) for 30 minutes. Reasonably fine-textured soil is best; don't sow seeds into big lumps of soil.

SOIL WARMTH

Keep the soil warm. Here must be the Number 1 germination stumbling block, especially with the more "difficult" crops, like impatiens and geraniums (they really aren't difficult!). The majority of bedding annuals should have a soil temperature (day and night) of 75F (24C) to achieve prompt germination. This is not house temperature but actual temperature in the soil. This applies to petunias, geraniums, impatiens, begonias, salvias, lobelias, marigolds, and vincas. An honest 70F (21C) soil temperature will sprout most annuals most of the time, but 75F does it more promptly and generally gives a better percentage. You must have a suitable thermometer to know where you stand.

Moisture evaporation from the soil surface cools the soil 5 to 10F (3 to 6C). Temperatures in seedling flats are frequently 10F or more below the house temperature. Water applied to seed flats (as mist, etc.), often at 45 or 50F (7 or 10C), will cool the soil down sharply. Tests at Michigan State show that soil took eight hours to get back up to its original 70F temperature after an application of cold water.

There is a group of bedding annuals that do need a lower temperature for sprouting. Soil temperature of 65F (18C) is recommended for pansies, phlox, and snapdragons.

SURFACE MOISTURE

Keep the soil surface moist until sprouting. George J. Ball used to say that if a seed started to soak up moisture then was allowed to dry, it would die. Once you start, the soil surface must be kept moist.

One way to do it is to mist six seconds per 10 minutes only on sunny days. Actually, the best growers seem to prefer misting by hand. Another way is to wrap flats in poly or cover the bed with a sheet of poly. Better yet, cover with a porous material, such as Remay or Vispore. They are both white, so they reflect the light, and their porosity allows for air circulation. If flats are well watered before sowing, there is normally ample moisture to sprout the seeds. Be sure to remove the poly as soon as the seedlings come through. Beware of full sun on flats covered with poly; it's easy to burn plants!

Covering

Leave fine seed—such as of petunias and begonias—uncovered. The tiny seeds soon wash down into the mix. Nearly all growers cover the larger seeds, such as those of tomatoes, salvia, and zinnia. Many seeds need three days of darkness to germinate, among them verbena, larkspur, dusty miller, pansy, phlox, and portulaca.

After-sprouting care

Under the 75F (24C), high-moisture regime, good seed should be well up in five days or less for most annuals. Very soon these flats should be moved into a 50F (10C) house. Don't let them dry to wilting, but gradually withhold moisture. Cool and dry from here on produces well-rooted, sturdy seedlings—and a minimum of damping-off problems. As seedlings mature, they're best with full sun and lots of fresh air.

Seed storage

Seed quality is one of the most talked-about subjects among plug growers everywhere, and storage is the key to maintaining high seed quality from seed harvest to sowing. All growers have seed in storage for various periods of time. Many order their seed for the whole season at one time, then use it throughout the season. Sometimes seed is held over for the next year. The effects of storage on seed quality depend on the storage conditions provided. For large plug growers seed inventory can be a costly investment. Any seed quality loss translates into poorer seed performance in the plug.

Different types of seed have different lengths of time they can be stored effectively. (Table 13 shows relative storage life of seeds.) When seeds have been enhanced or otherwise altered, their storage life will usually be shortened. Among such seeds are pelleted, coated, primed, de-tailed, defuzzed, dewinged, and scarified seeds. Successful storage of enhanced seeds can be extended by making conditions closer to ideal.

Seed quality

Viability is the ability of a seed to germinate under optimum conditions. The ability of a seed to germinate under a range of conditions and still produce a usable seedling is called *vigor.* Viability and vigor together describe optimum seed quality. Plug growers today are demanding higher seed quality, with an emphasis on usable seedlings. What they're really asking for is seed that has not only high viability but also high vigor.

TABLE 13   RELATIVE STORAGE LIFE OF SEEDS

Short	Medium	Long
Anemone	Ageratum	Amaranthus
Asparagus fern	Alyssum	Shasta daisy
Aster	Cauliflower	Stock
Begonia	Celery	Sweet pea
Browallia	Celosia	Tomato
Delphinium	Coleus	Zinnia
Herbs	Cyclamen	
Impatiens	Dahlia	
Lettuce	Dianthus	
New Guinea impatiens	Dusty miller	
Onion	Eggplant	
Pansy	Geranium	
Pepper	Lisianthus	
Phlox	Lobelia	
Salvia	Marigold	
Vinca	Petunia	
Viola	Portulaca	
	Snapdragon	
	Verbena	
	Watermelon	

The length of time seed stays vigorous is a function of genetics, storage conditions, and time. During storage, vigor decreases before viability, meaning mere germination isn't the best indicator of how the seed is holding up in storage. Two lots of the same variety can have 90% germination in the lab, but one lot may already be declining on the vigor curve, producing fewer usable seedlings, while the other lot retains high vigor. Once a seed starts moving down the vigor or viability curve, its decline can't be reversed, only slowed down. *It's only a matter of time before all seeds eventually decline and die during storage.*

MOISTURE CONTENT AND RH MATTER

The most important factor in seed storage is moisture content; temperature is also vital. Seed moisture changes according to the relative humidity (RH) of the surrounding air. However, seeds differ in the way they adjust their moisture content to humidity. A seed's composition, especially its protein and lipid content, strongly affects its attraction to moisture.

In 1960 J.F. Harrington outlined a few rules of thumb for seed storage: Every 1% decrease in seed moisture content doubles the storage life. Every 10F (6C) decrease in seed storage temperature doubles the storage life. The sum in degrees F and % RH of good storage conditions should be less than 100.

Proper storage conditions have RH levels maintained between 20 and 40%, giving corresponding seed moisture contents of 5 to 8%, a range that's safe for many seeds. When seed moisture content drops too low, lower than 5%, storage life and seed vigor may decline. When seed moisture content rises above 8%, aging or seed deterioration can increase. Deterioration involves increases in respiration, breakdown of storage reserves, decline in nuclear structure, reduction in cell membrane integrity, and other biochemical processes, all resulting in vigor and viability loss. Seed moisture content above 12% promotes the growth of fungi and insects. Most seeds can't germinate until seed moisture rises above 25%.

Currently, most growers store seed in hermetically sealed, plastic-laminated foil packets and bags or vacuum-packed cans. These containers provide excellent barriers against moisture moving in and out. Therefore, seed must be at the proper moisture content before being packed, and the air in the containers should have an appropriately low RH.

Maintain storage temperature between 40 and 70F (4 and 21C). Pay special attention to the percent RH because humidity changes with air temperature. Store such seeds as delphinium, primula, cyclamen, pansy, and geranium at cool temperatures (42F, 6C) for best storage life. Other seeds, such as impatiens, will store longer with lower temperatures and RH.

PanAmerican Seed has compared the storability of impatiens seed. All lots were the same general age. Storage conditions consisted of different RH levels (thus different seed moisture content) and temperatures. Samples were taken out of storage at intervals up to one year and sown into plug trays. Usable seedlings were counted after 21 days under normal greenhouse germination and growing conditions. Storing impatiens seed at 15% RH appears to be too dry and reduces usable seedlings. Impatiens seed quality dramatically dropped under unfavorable temperature and RH levels. The best storage conditions for impatiens are around 42F (6C) and 25 to 30% RH. (Similar recommendations apply to pansy seed.) Storage temperature has a very strong influence on impatiens, especially with low-vigor seed lots.

The plug grower can use a frost-free refrigerator or climate-controlled room set at 42F (6C) to store all kinds of seed. Foil packets and metal cans will protect against moisture as long as they remain sealed. Once a packet or can is opened, you should prevent seeds from taking up moisture from the air. Relative humidity in a seedling area can easily exceed 80%, with all of the moisture around from filled plug trays, drum seeders, and watering tunnels. Often, open seed packets are out all day in the seeding area while the seeder operator uses them to sow the day's production. Exposed seed moisture content can easily increase 2% in two hours under such conditions. If such a process continues with the same seed lots over several weeks, seed vigor deterioration and reduction will occur, especially in sensitive crops, such as impatiens.

Simply closing a foil packet and placing it in a refrigerator won't remove the extra moisture the seed has absorbed from the air. The grower should place the seed in a water vaporproof container that can be sealed (such as Tupperware or a Mason jar), with a thin layer of silica gel desiccant in the bottom, and store it in a refrigerator. This desiccant layer should be no more than one-fourth inch (6 mm) deep because more desiccant would cause the % RH to be too low. When the desiccant's color changes from blue to pink, regenerate it by placing it in an oven at 230 to 360F (110 to 182C) until its color changes back to blue. You can also use a microwave oven.

If a refrigerator is too small to store all your seed, look into building a climate-controlled room or buying a walk-in cooler. This room should be very well insulated (R = 30) and vapor sealed, with a thermostat and humidistat for controlling conditions at 42F (6C) and 25 to 30% RH. Make sure the cooling unit fans are set on continuous operation to eliminate temperature fluctuations in the room. You can control humidity with a commercial dryer, available from Bry Air, Cargoaire, and Prime Air. Dehumidifiers for home use aren't suitable for controlling RH in a seed storage area because they can reduce RH only to 35%, and they generate heat.

TIPS ON MAINTAINING IMPATIENS SEED QUALITY FROM PANAMERICAN SEED

Impatiens seeds, like all living things, require a favorable environment to survive. Unfortunately for many seeds, they often share their living space with human beings who have different environmental expectations. Seeds often mark time in offices, in break rooms, or in a box next to the seeder. These environments are usually more suited to human comfort than seeds.

Store impatiens seeds at approximately 40F (5C) with low RH (25 to 30%). If humidity cannot be controlled, at least store the seeds in the refrigerator. Research shows impatiens seeds deteriorate at 72F (22C) *at any RH higher than 25%,* yet retain acceptable quality at 41F (5C) even up to 45% RH.

While temperature is most important, humidity must be controlled for maximum seed quality. For best success, we recommend you construct a seed storage chamber with controlled temperature and humidity. Commercially available walk-in coolers are great storage chambers, although a well-insulated room with little air exchange is also effective.

Standard refrigeration equipment will reduce humidity, but desiccant-type dehumidifiers are the only practical way to reduce humidity below 30%. These dehumidifiers use silica gel or other compounds to remove moisture from the air, then regenerate the desiccant with heated air. Do not open seed packages until you are ready to sow. Use an entire package per sowing whenever possible. If an open package must be stored, return the open packet to low temperature and dry air conditions as soon as possible. Do not reseal a seed package until the seeds have been in dry conditions for at least 24 hours to remove any moisture absorbed from the air.

Desiccant dryers are available from:

Bry Air	Cargoaire	Prime Air
Route 37W	79 Monroe Street	Hayward, CA USA
Sunbury, OH 43074 USA	Amesbury, MA 01913 USA	510/732-3400
614/965-2974	800/843-5360	

Sowing area rules

1. Remove seeds from storage only when you are ready to sow.
2. Always use open seed packages before opening new packages of the same variety.
3. Take only as much seed out of storage that can be sown within three hours, or before the next break, whichever is less.
4. Open the seed packet only when you are ready to sow that variety.
5. Open only seed packages needed for one sowing.
6. Do not leave seed on the seeder when going on break or to lunch.
7. Return seed to refrigerated storage before going on break or to lunch.
8. Do not reseal seed packages until seed has been in dehumidified, refrigerated storage for at least 24 hours.

Remember: One hour in the sowing room decreases seed vigor as much as a month at controlled storage conditions!

CHAPTER 19

PERENNIALS

by Debbie Hamrick
Ball Publishing
Batavia, Illinois

Perhaps no other crop category has undergone change as significant as perennials since the previous edition of the *Ball RedBook*. For years perennials were the neglected stepchildren of bedding plant growers who spoiled their impatiens, geraniums, and petunias and left the perennials to the care of a small but highly dedicated group of specialists driven by a love of perennials rather than the desire to mass-produce plants.

As bedding plant production grew, perennials grew alongside, but still for years as stepchildren. Then the inevitable hit: bedding plant growth began to flatten. As this happened growers turned in droves to perennials for growth in sales (figs. 1, 2, and 3). They learned that this group of several hundred species is quite versatile. Perennials can be sold before bedding plant sales begin in the spring, then alongside them during the main season, throughout the summer, and into the fall. Perennials can be pro-

FIG. 1 Bernacchi Greenhouses, La Porte, Indiana, lets their customers know the time to buy perennials is *now*.

duced in a wide range of containers and sold outside of their normal flowering seasons when special cultural conditions are met. These factors, combined with a public seeking new and novel plants, means that demand for perennials is set to boom through the end of this century and into the next.

However, deciding to include perennials in the bedding plant crop mix should not be taken lightly. Perennials are much more difficult than annuals. Perennial species grown from seed often have multiple dormancy factors that must be overcome before the seed germinates. For example, some species must have periods of moist chilling or warm stratification before they will germinate. Other seeds need to be scarified. Some seeds will not germinate when they are too young, while seeds of other species will not germinate when they are too old.

Vegetatively propagated species can in some cases be multiplied from simple division or tip cuttings. For every easy-to-propagate plant, however, there are more with difficult rooting requirements taking weeks or even months, with high losses in the propagation bed.

Once plants are potted, again perennials are more difficult. The flowering mechanisms are fully understood for only a few species, making plants difficult to force into flower for spring sales the first year. Even when flowering mechanisms are understood, providing the special

FIG. 2   *Phlox subulata* is displayed for retail sales. Well-done signs can help shoppers learn about the plants on display.

FIG. 3   Here's a portion of the containerized perennial production at T&Z Nursery, Winfield, Illinois. T&Z operates an upscale retail garden center that caters to the suburban gardener.

conditions to get them into flower is much more trouble than flowering a flat of impatiens. Because there are hundreds of species of perennials, overcoming these difficulties for all the ones you grow will be time-consuming. You can also count on losing more plants.

Breeding of perennials has also taken a back seat to breeding of other floricultural crops. For years larger hybrid seed companies would not work with perennials in their programs, claiming that because the seed was not hybrid, its price was not justified in the overall breeding programs. However, this attitude is finally changing. Several breakthroughs—such as White Knight Shasta daisy (*Leucanthemum* × *superbum*), Summer Pastels achillea, or the magnificent and reliable coreopsis Early Sunrise—have demonstrated to seed companies and growers alike that perennials from seed can be a commercial success. The Jellitto Seed company in Schwarmstedt, Germany (fax: +49-0571-4088), and the Kieft Seed Co., Venhuizen, the Netherlands (fax: +31-228-543440), maintain the industry's most extensive lists of perennial seeds and are important suppliers to growers the world over. In addition, national and regional seed companies maintain good perennial seed inventories.

Even though perennial hybrid breeding is coming up, most perennial breeding is still left in the hands of dedicated enthusiasts who make selections of superior-performing plants and propagate them vegetatively. Even though the varieties of perennials available from seed are improving, plants propagated vegetatively are more uniform and true to type than plants propagated by seed. Such spectacular perennials as Autumn Joy sedum, Moonbeam achillea, and Goldsturm rudbeckia must be propagated vegetatively.

Because of the difficulty in propagating perennials, most growers choose to buy in rooted cuttings, liners, or plug seedlings (128-size cells are very popular) to simplify production. Many times specialist propagators offer these as vernalized plants, meaning they have already been exposed to the cold treatment many perennials require for flowering. For the bedding plant grower adding perennials to the crop mix, bringing in is the simplest way to get started. Plants are received in winter, potted into 4-inch (10-cm) or larger containers, and forced into flower for spring sales.

Other growers making perennials a more important part of their production may choose to buy in plants in the summer and fall, pot them up in 1-gallon (or 4-l) containers, and overwinter them in cold frames. These plants will be much larger, better branched, and have more flowers because of the extra time to root in.

FORCING PERENNIALS INTO FLOWER

The most exciting work with perennials today is coming from the industry's floriculture researchers. Because the perennial group encompasses so many different species, it's a big challenge to put them all into the same production sequence as bedding plants. But for perennials to become as important to growers, retailers, and consumers as bedding plants, that is the direction they must go.

Several colleges are devoting research time and resources to this issue. Among them are the University of Georgia, the University of Minnesota, and the school with the best known perennial research, Michigan State University. The information about flowering perennials that follows has been gleaned from presentations given primarily by Art Cameron from Michigan State University, as well as from numerous magazine articles he and others have written for *GrowerTalks* and *Greenhouse Grower* magazines.

Light is just one factor in flowering perennials. Some species have multiple flowering factors, such as cool treatment and critical day-length requirements. Still others also have juvenility requirements.

LIGHT

Light, specifically day length, is one of the most important factors in flowering perennials. Some plants flower under short days, while others flower under long days.

One of the first steps to successful perennial production is to understand which species require long days. For example, some perennials require long days to flower (*Coreopsis verticillata* Moonbeam). Others are obligate long-day plants, meaning they do flower under short days but flower faster under long days (*Salvia superba* Blue Queen). Some, researchers have discovered, are day-length neutral (*Veronica longifolia* Sunny Border Blue).

Michigan State University researchers have not seen a definitive critical day length. They noted plants requiring long days tend to remain vegetative with less than 12 hours of light, while 14 to 15 hours of light speeds flowering. Saxifraga is an exception, requiring short days to flower. Generally, lighting can begin about Christmas for March sales and in mid- to late January for April sales.

Incandescent lights (mum lighting) can be used to extend day length. If you already have HID lights, they will work as well. Allan M. Armitage, University of Georgia, recommends that if you're going to install lamps for lighting perennials, purchase metal halide lamps, as opposed to sodium lamps, for day-length control, unless of course you already have high-pressure sodium lamps.

Breaking the night from 11 P.M. to 2 A.M. is generally sufficient; the goal is to provide 16 hours of light. Research at Michigan State University has shown that a light intensity of 5 footcandles (54 lux) is adequate; however, 10 f.c. (1,076 lux) is better to help compensate for uneven lighting of the crop. With incandescent lights, researchers recommend the lights be hung 5 feet (1.5 m) above pots. Use 60-watt bulbs spaced 4 feet (1.2 m) apart for 4-feet-wide beds, 100-watt bulbs spaced 6 feet (1.8 m) apart for two beds, or 150-watt bulbs spaced 6 feet apart for three beds.

For a detailed discussion on day length and how it affects plant flowering, see *Dendranthema* in the crop culture section of this book.

JUVENILITY

Some perennials *must* pass through a juvenility period before they will flower. Crops such as papaver, limonium, some asters, aquilegia, and heuchera show no flowering response to various cold or day-length treatments. Researchers at Michigan State University discovered that some plants simply have to be a certain age before they will flower, even when all other conditions are perfect.

Leaf count is a good indication of plant development. For example, heuchera plants with 16 leaves flower 100% of the time. For perennials that require a juvenility period to flower, attain necessary leaf counts *before* starting cold treatment. Here are the numbers of leaves needed for several varieties:

Aquilegia, some spp.	15+	*Heuchera sanguinea*	16
Aster alpinus	15	*Lavandula angustifolia*	40–50
Chrysanthemum coccineum	15	Lobelia Compliment Scarlet	6–7
Coreopsis Sunray	16	*Physostegia virginiana*	10
Delphinium grandiflorum	4–5	*Rudbeckia fulgida* Goldsturm	10
Echinacea purpurea	4	*Veronica spicata*	6–8

COLD TREATMENT

Traditional perennial production has involved propagating plants in the summer, potting them in the fall, and overwintering containers in a cold greenhouse or heeled into ground beds that are insulated with straw and a light-colored, breathable covering. Such plants emerge from the winter with well-developed roots and make uniform, compact, well-branched plants—the traditional 1-gallon (about 4-l) perennial sold all over the country in the spring.

Through research we now know that some species must have a cold period before they will flower, others flower better with a cold treatment, and still others are unaffected by cold in their flowering response. A quick tabulation of some of the results of Michigan State University research on flowering response and cold treatment appears in table 1.

FORCING TEMPERATURES

Once plants are in the greenhouse after vernalization, there are no specific guidelines as there are with poinsettias and other flowering plants, but nights of 60 to 65F (16 to 18C) are sufficient for the weeks preceding sale. Negative DIF, using reversed day and night temperatures (also known as cool days, warm nights), works well with perennials in maintaining plant height and increasing plant quality. As with most

TABLE 1   VALUE OF COLD TREATMENT AND LONG DAYS TO PERENNIALS

Variety	Value of long days
Perennials needing or benefiting from vernalization	
Achillea filipendulina Cloth of Gold	Required
Armeria latifolia	None
Armeria × *hybrida* Dwarf Ornamental Mix	None
Asclepias tuberosa (if dormant after short days)	Required
Aster alpinus Goliath	None
Astilbe arendsii	Beneficial
Aquilegia × *hybrida* (varieties requiring cold)	None
Chrysanthemum coccineum James Kelway	Required
Coreopsis grandiflora Sunray	Beneficial
Delphinium elatum Blue Mirror	None
Dianthus deltoides Zing Rose	None
Echinacea purpurea Bravado	Beneficial
Gaillardia grandiflora Goblin	Beneficial
Gypsophila paniculata Double Snowflake	Required
Heuchera sanguinea Bressingham Hybrids	None
Iberis sempervirens Snowflake	None
Lavandula angustifolia Hidcote Blue	Required
Lavandula angustifolia Munstead Dwarf	Beneficial
Lewisia cotyledon	None
Linum perenne Sapphire	None
Lobelia × *speciosa* Compliment Scarlet	Beneficial
Oenothera missouriensis	Required
Physostegia virginiana Alba	Required
Platycodon grandiflorus Sentimental Blue	Beneficial
Rudbeckia fulgida Goldsturm	Required
Salvia superba Blue Queen	Beneficial
Scabiosa caucasica Butterfly Blue	None
Veronica longifolia Sunny Border Blue	None
Veronica spicata Blue	None
Perennials not benefiting from cold	
Aquilegia × *hybrida* Songbird series	None
Asclepias tuberosa[a]	Required
Campanula carpatica Blue Clips	Required
Coreopsis grandiflora Early Sunrise	Required
Coreopsis verticillata Moonbeam	Required
Hibiscus × *hybrida* Disco Belle Mixeda	Required
Leucanthemum × *superbum* Snow Lady	Beneficial
Perovskia atriplicifolia	No information
Primula veris Pacific Giants	None

[a] If not first exposed to short days.

Source: Compiled from articles and seminar presentations by researchers from Michigan State University, East Lansing.

bedding plants, running high temperatures creates lush, weak growth. Plants grown cool over a longer period tend to be of higher quality, although flowering may take longer.

GROWTH REGULATORS

B-Nine (3,000 to 5,000 ppm) is the most important growth regulator. Aquilegia, iberis, myosotis, and salvia respond well to B-Nine sprayed as flower stems start to stretch. Platycodon, dianthus, and bellis respond well to Cycocel at 750 to 1,500 ppm. A-Rest, Bonzi, and Sumagic may also be used. Consult labels for recommendations.

FERTILIZER

When plants are not actively growing, such as during cold treatment and early winter, use nitrate-based fertilizers as a source of nitrogen. A constant liquid feed of 50 to 100 ppm N is satisfactory. During the forcing stage, as you raise temperatures and plants begin to actively grow, you can switch to ammonium-based fertilizers and increase the rate to 100 to 200 ppm constant liquid feed. When plants flower, switch to potassium nitrate at 100 ppm.

Michigan State University research so far has developed "cookbook" guidelines for a few perennial species. As researchers there and at other universities continue to devote time to perennials, the grower can only look forward to better information about how to grow and flower perennials.

Editor's note: An excellent source of additional information on perennials is the *Ball Perennial Manual: Propagation and Production* by Jim Nau, available from Ball Publishing, P.O. Box 9, Batavia, IL 60510-0009 USA; phone 630/208-9080, fax 630/208-9350; e-mail: growertalk@aol.com or gtalks@xnet.com.

OVERVIEW

A summary of perennials appears in table 2.

TABLE 2   PERENNIALS—AN OVERVIEW

Plant names: Horticultural and common	Uses[a]	Height in inches (cm)	Flowering		
			Colors	Season (northern U.S.)	Flowering month for 4-inch planted out in May[b]
Achillea filipendulina Coronation Gold Fernleaf yarrow	Rock, garden, cut	24–48 (61–122)	Yellow flower; gray-green foliage	Late spring to midsummer	July–Aug.
Aquilegia × hybrida Columbine	Garden, pot	18–30 (46–76)	Red, pink, yellow, blue, white, purple	Spring to early summer	May; needs vernalization
Arabis caucasica syn: *A. albida* Rock cress	Rock, wall	8–12 (20–30)	White, pink	Spring	May; needs vernalization
Artemisia schmidtiana Silver mound	Garden, rock	12–15 (30–38)	Yellow flower; silvery/white foliage	Grown for foliage effect	Foliage, not flowers
Astilbe × arendsii False spirea	Garden, pot, cut	24–36 (61–91)	Pink, red, white, rose, carmine, lavender	Early to midsummer	June–July; container better; needs vernalization
Aubrieta deltoidea False rock cress	Rock, wall	6–8 (15–20)	Rose, blue, purple lavender	Early to midspring	May
Aurinia saxatilis Basket of Gold Perennial alyssum	Garden, rock, wall	9–12 (23–30) prostrate	Yellow flower; gray-green foliage	Early to midspring	—
Bellis perennis English daisy	Pot, garden, rock	6–8 (15–20)	White, pink, red	Early to late spring	May
Campanula carpatica Carpathian harebell	Pot, garden, rock	8–15 (20–38)	Blue, white	Early to midspring	July
Campanula persicifolia Peach-leaved bellflower	Cut, garden	24–36 (61–91)	Blue, white	Late spring to early summer	June
Coreopsis grandiflora Tickseed	Cut, garden	18–24 (46–61)	Yellow	Early to late summer	Depends on variety; July–Aug. for Early Sunrise
Coreopsis verticillata Threadleaf coreopsis	Cut, garden	10–30 (25–76)	Yellow	Early to late summer	June–Aug.

[a] Uses include in regular garden, in rock garden, in wall, as edging, as pot plant, for cut flowers, and for dried flowers.

[b] Perennials are predominantly sold in 4-inch (10-cm) pots, but there is also an important trade in 1-gallon (4-l) and larger plants. Also, this column and the next often specify 6-inch (15-cm) pots as preferred size.

[c] FS = full sun, PS = partial shade, SH = full shade.

[d] Hemerocallis divisions made in spring seldom flower profusely the same season. Some varieties, like Stella D'Oro, will flower in June and July from a plant divided in late winter and potted one fan per 6-inch pot.

Month to sow or divide to achieve this flowering	Hardiness		Cultural preferences		Other
	Zones (USDA)	Persistence past 2–3 years	Sun[c]	Soil	
Early March, divisions	3–8	Yes	FS	Well drained, ordinary	
Sept.–Oct.; seed	3–8	Not usually	PS	Moist, rich with good drainage	
Sept.–Oct.; seed	3–7	Yes (in northern zones)	FS	Requires excellent drainage	Will melt out in South
Early to mid-March, divisions	3–7	Yes (in northern zones)	FS	Well drained, ordinary	Will melt out in South
Late winter–early spring; precooled divisions; allow 12–16 wks for 6-inch pot	4–8	Yes	PS–SH	Moist, organic	
Oct., seed	4–7	Yes (in northern zones)	FS	Excellent drainage	Cut back after flowering
—	3–9	Yes	FS	Well drained	Melts out in South; cut back 1/3 after flowering
Seed; allow 16–20 weeks to flower	4–8	Yes	PS	Cool, moist, organic	
Seed; allow 15–18 weeks for 1 plant per 4-inch pot	3–8	Yes	FS–PS	Well drained	
Seed early summer; overwinter to flower following year	4–7	Yes	FS–PS	Well drained	
Seed early to mid-Jan. depending on variety	4–9	Yes in North; no in South	FS	Well drained	Requires deadheading to prolong bloom period
Plant No. 1 transplant 6–8 weeks before sale; 1 plant per 4-inch pot	3–9	Yes, and does well in South	FS	Well drained	Drought resistant

TABLE 2 PERENNIALS—AN OVERVIEW (CONTINUED)

Plant names: Horticultural and common	Uses[a]	Height in inches (cm)	Flowering		
			Colors	Season (northern U.S.)	Flowering month for 4-inch planted out in May[b]
Delphinium elatum Delphinium	Cut, garden	36–72 (91–183)	Blue, red, pink white, violet, yellow	Early to midsummer	July–Aug.
Dianthus barbatus Sweet William	Cut, garden, rock, wall	12–20 (30–51)	Red, pink, purple white, bicolor	Late spring to early summer	In flower until June; needs vernalization
Dianthus deltoides Maiden pinks	Rock, wall, edging	6–12 (15–30)	White, pink to deep red	Late spring to early summer	In flower until June; needs vernalization
Dianthus gratianopolitanus Cheddar pinks	Rock, wall, edging	9–12 (23–30)	White, pink, rose, red	Midspring to early summer	—
Dicentra eximia Fringed bleeding heart	Garden	15 (38)	Pink flower; gray-green foliage	Early to late summer	6-inch in flower until early June
Dicentra spectabilis Bleeding heart	Pot, garden	24–30 (61–76)	Pink, white	Mid to late spring	6-inch in flower until early June
Digitalis purpurea Common foxglove	Garden	48–60 (122–152)	Purple, white, pink	Midspring to early summer	—
Doronicum Leopard's bane	Cut, garden	24 (61)	Yellow	Early to midspring	Needs vernalization
Echinacea purpurea Purple coneflower	Cut, garden	30–36 (76–91)	Purple, white	Summer	Aug.
Echinops ritro Globe thistle	Cut, dried, garden	24–36 (61–91)	Blue	Mid to late summer	July–Aug.
Gaillardia × grandiflora Blanket flower	Pot, cut, garden	12–30 (30–76)	Red, yellow	All summer	June–July
Geranium sanguineum Bloody cranesbill	Rock, garden	9–12 (23–30)	Magenta	Late spring to early summer	—

[a] Uses include in regular garden, in rock garden, in wall, as edging, as pot plant, for cut flowers, and for dried flowers.

[b] Perennials are predominantly sold in 4-inch (10-cm) pots, but there is also an important trade in 1-gallon (4-l) and larger plants. Also, this column and the next often specify 6-inch (15-cm) pots as preferred size.

[c] FS = full sun, PS = partial shade, SH = full shade.

[d] Hemerocallis divisions made in spring seldom flower profusely the same season. Some varieties, like Stella D'Oro, will flower in June and July from a plant divided in late winter and potted one fan per 6-inch pot.

Month to sow or divide to achieve this flowering	Hardiness		Cultural preferences		
	Zones (USDA)	Persistence past 2–3 years	Sun[c]	Soil	Other
January, seed	4–7	No	FS	Well drained, organic, basic pH	Benefits from fertilizer
Seed early summer; overwinter to flower following year	4–9	Divide every 2 yrs or let self-seed	FS	Well drained, organic	
Seed early summer; overwinter to flower following year	3–9	—	FS–PS	Well drained, slightly alkaline	Shear after flowering
—	3–9	—	FS	Well drained, slightly alkaline	Shear after flowering
No. 1 transplants mid-March; 1 plant per 6-inch pot	3–9	Yes	PS–SH	Well drained, moist, organic	
No. 1 transplants mid-March; 1 plant per 6-inch pot	3–9	Yes	PS–SH	Well drained, moist, organic	Usually goes dormant in summer
—	4–8	Biennial, self-seeds	PS	Well drained, moist, organic, slightly acid	
Seed early summer; overwinter to flower following year	4–7	No	FS–PS	Well drained, cool, moist	Usually goes dormant in summer
January, seed	3–9	Yes	FS	Well drained, ordinary	Drought tolerant
Seed early summer; overwinter to flower following year	4–9	Yes	FS	Well drained	Very drought and heat tolerant
January, seed	3–9	No	FS	Light, well drained	
—	3–8	Yes	PS	Well drained, moist, humusy	

TABLE 2 PERENNIALS—AN OVERVIEW (CONTINUED)

Plant names: Horticultural and common	Uses[a]	Height in inches (cm)	Flowering		
			Colors	Season (northern U.S.)	Flowering month for 4-inch planted out in May[b]
Gypsophila paniculata Perennial baby's breath	Cut, dried, garden	36 (91)	White, pink	Early to midsummer	July–Aug.
Hemerocallis Daylily	Garden	24–48 (61–122)	Wide range	Varies with cultivar	[d]
Heuchera sanguinea Coral bells	Pot, garden	12–20 (30–51)	Red, pink, white	Late spring to early summer	June–July; needs vernalization
Hibiscus moscheutos Rose mallow	Garden	48–60 (122–152)	Red, pink, white	Midsummer to frost	July–Aug.
Hosta (many varieties) Hosta	Garden	6–72 (15–183)	Lilac, purple, white flowers; green, yellow, blue foliage	Spring	July–Sept., many in Aug.; depends on variety
Iberis sempervirens Candytuft	Garden, rock, wall	9–12 (23–30)	White	Spring	Will be sold in flower; needs vernalization
Iris hybrids Bearded or German iris	Garden, rock	8–48 (20–122)	Wide range	Late spring to early summer	May–early June
Kniphofia uvaria Red-hot poker	Cut, garden	36–48 (91–122)	Red, orange, yellow	Summer	—
Lavandula angustifolia English lavender	Cut, garden, rock	12–24 (30–61)	Purple, pink, lavender	Summer	June–Aug.
Leucanthemum × superbum Shasta daisy	Cut, garden	10–48 (25–122)	White with yellow center	Early summer to frost	Depends on variety; Aug.–Sept. for Alaska
Liatris spicata Spike gayfeather	Cut, garden	18–36 (46–91)	Purple, mauve, cream	Midsummer to autumn	June–July

[a] Uses include in regular garden, in rock garden, in wall, as edging, as pot plant, for cut flowers, and for dried flowers.

[b] Perennials are predominantly sold in 4-inch (10-cm) pots, but there is also an important trade in 1-gallon (4-l) and larger plants. Also, this column and the next often specify 6-inch (15-cm) pots as preferred size.

[c] FS = full sun, PS = partial shade, SH = full shade.

[d] Hemerocallis divisions made in spring seldom flower profusely the same season. Some varieties, like Stella D'Oro, will flower in June and July from a plant divided in late winter and potted one fan per 6-inch pot.

Month to sow or divide to achieve this flowering	Hardiness		Cultural preferences		Other
	Zones (USDA)	Persistence past 2–3 years	Sun[c]	Soil	
Divisions; allow 6 to 9 wks for 1 plant per 4-inch pot to be salable	4–8	Yes	FS	Well drained, neutral-alkaline	May need staking
[d]	4–9, varies with cultivar	Yes	FS–PS	Well drained	
Seed early summer; overwinter to flower following year	3–8	Yes	FS–PS	Well drained, moist, organic	Mulch in winter to prevent heaving
Mid-Feb., seed	4–9	Yes	FS–PS	Moist, organic	
Depends on variety; for fast growing variety, allow 6–8 wks, 1 plant per 4-inch pot	3–8	Yes	PS–SH	Well drained, moist, organic	
Seed early summer; over-winter to flower following year	3–9	Yes	FS	Well drained	Evergreen; shear back after flowering
Take divisions July–Aug. and overwinter; 1–2 fans per 6-inch pot	4–8	Yes	FS	Well drained, neutral	
—	5–9	Yes	FS	Well drained; will not tolerate wet winter soils	
Seed early summer; overwinter to flower following year	5–9	Yes	FS	Well drained	
Depends on variety; if seed, Feb.	5–8	Yes	FS	Well drained	Many cultivars available
Early March, corms; 1 per 6-inch pot for May sale	3–8	Yes	FS	Well drained	

TABLE 2   PERENNIALS—AN OVERVIEW (CONTINUED)

Plant names: Horticultural and common	Uses[a]	Height in inches (cm)	Flowering		
			Colors	Season (northern U.S.)	Flowering month for 4-inch planted out in May[b]
Limonium latifolium Perennial statice, sea lavender	Cut, dried, garden	12–36 (30–91)	Blue, violet	Mid to late summer	—
Myosotis sylvatica Garden forget-me-not	Garden	8–12 (20–30)	Blue, pink, white	Spring	In flower when sold; needs vernalization
Paeonia hybrids Peony	Garden, cut	30–36 (76–91)	Red, pink, white, yellow	Late spring to early summer	—
Papavar orientale Oriental poppy	Garden	24–36 (61–91)	Red, orange, pink, salmon	Late spring to early summer	May–June
Perovskia atriplicifolia Russian sage	Garden	48–60 (122–152)	Blue flowers; gray foliage	Mid to late summer	—
Phlox paniculata Perennial phlox	Garden	24–36 (61–91)	White, pink, red, blue, purple	Summer to early autumn	Late Aug.–Sept.
Phlox subulata Moss pink Creeping phlox	Garden, rock	4–6 (10–15)	Blue, white, pink	Early to midspring	—
Physostegia virginiana Obedient plant	Cut, garden	36–48 (91–122)	Rose-purple, white	Late summer to autumn	Aug.
Primula × polyantha Polyanthus primrose	Pot, garden	6–12 (15–30)	Wide range	Spring	Sold in color
Rudbeckia fulgida var. *sullivantii* Goldsturm Black-eyed Susan Orange coneflower	Garden	30–36 (76–91)	Yellow, orange	Midsummer to frost	July–Sept.
Salvia × superba Perennial salvia	Cut, garden	15–24 (38–61)	Purple	Early to late summer	July
Scabiosa caucasica Pincushion flower	Cut, garden	24–30 (61–76)	Light blue, white, lavender	Late summer to late autumn	July–Aug.

[a] Uses include in regular garden, in rock garden, in wall, as edging, as pot plant, for cut flowers, and for dried flowers.

[b] Perennials are predominantly sold in 4-inch (10-cm) pots, but there is also an important trade in 1-gallon (4-l) and larger plants. Also, this column and the next often specify 6-inch (15-cm) pots as preferred size.

[c] FS = full sun, PS = partial shade, SH = full shade.

[d] Hemerocallis divisions made in spring seldom flower profusely the same season. Some varieties, like Stella D'Oro, will flower in June and July from a plant divided in late winter and potted one fan per 6-inch pot.

Month to sow or divide to achieve this flowering	Hardiness		Cultural preferences		Other
	Zones (USDA)	Persistence past 2–3 years	Sun[c]	Soil	
—	4–9	Yes	FS	Well drained, moist	
Nov.–Dec.	5–8	No, but self-seeds	PS	Well drained, moist	
—	2–7	Yes	FS	Well drained, moist, organic	
Seed early summer; over-winter to flower following year	3–7	Yes	FS–PS	Well drained; will not tolerate wet winter soil	Goes dormant in summer
—	5–9	Yes	FS	Well drained; will not tolerate wet winter soil	
Early Feb., March, No. 1 divisions; salable early May	4–8	Yes	FS	Well drained	Many cultivars available; susceptible to powdery mildew
—	2–9	Yes	FS	Well drained	Shear back after flowering
Jan., seed	2–9	Yes	FS		Can be invasive
Seed; allow 24–28 weeks	3–8	No	PS	Cool, moist, organic	
Seed early summer; overwinter to flower following year	3–9	Yes	FS–PS		Drought tolerant
Feb., seed	4–8	Yes	FS	Well drained	Drought tolerant
Mid-Jan., seed	4–7	Yes	FS	Well drained, organic, neutral	

TABLE 2 PERENNIALS—AN OVERVIEW (CONTINUED)

Plant names: Horticultural and common	Uses[a]	Height in inches (cm)	Flowering		
			Colors	Season (northern U.S.)	Flowering month for 4-inch planted out in May[b]
Sedum Autumn Joy Sedum	Garden, rock	18–24 (46–61)	Rosy-red flower; gray-green foliage	Late autumn	June
Tanacetum coccineum syn: *Pyrethrum roseum* Painted daisy	Garden	18–24 (46–61)	Red, pink, lilac, white	Late spring to midsummer	July–Aug.
Veronica spicata Spike speedwell	Cut, garden, rock	12–36 (30–91)	Blue, pink, white	Late spring to midsummer	July

[a] Uses include in regular garden, in rock garden, in wall, as edging, as pot plant, for cut flowers, and for dried flowers.

[b] Perennials are predominantly sold in 4-inch (10-cm) pots, but there is also an important trade in 1-gallon (4-l) and larger plants. Also, this column and the next often specify 6-inch (15-cm) pots as preferred size.

[c] FS = full sun, PS = partial shade, SH = full shade.

[d] Hemerocallis divisions made in spring seldom flower profusely the same season. Some varieties, like Stella D'Oro, will flower in June and July from a plant divided in late winter and potted one fan per 6-inch pot.

Month to sow or divide to achieve this flowering	Hardiness		Cultural preferences		Other
	Zones (USDA)	Persistence past 2–3 years	Sun[c]	Soil	
Sow early summer and overwinter to flower following year	3–10	Yes	FS	Well drained	
Dec.–Jan., seed; allow 15–18 weeks	4–9	Must be divided every 2–3 yrs	FS	Well drained	Winter mulch required to prevent frost heave
Mid-Jan., seed	4–8	Yes	FS	Well drained, slightly acidic	

CHAPTER 20

HERBS

by Jim Long
Long Creek Farm
Oak Grove, Arkansas

Herbs have become one of the fastest-growing categories in plant production in the 1990s. Plant-based businesses that stocked only parsley, sage, rosemary, mint, chives, basil, and thyme a few years ago have expanded and are now offering dozens of varieties (fig. 1). Customer demand has been the governing force behind the change, and surveys around the country reveal the trend will continue.

Several factors have contributed to this increased interest in herbs. The buying public, which is not necessarily the gardening public, has access

FIG. 1   An attractive, well-marked display of potted herbs is featured at Martin Viette Nurseries, East Norwich, New York.

to hundreds of books about herbs that were not in print five years ago. Authors have taken herbs out of folklore and brought them into today's home and kitchen.

Today there are an increased awareness of healthier eating and more interest in cooking at home to control the diet. Herbs fit into this lifestyle change because the flavors of fresh herbs make reduced-salt and lower-fat diets more palatable.

Restaurants that did not utilize herbs five years ago have changed their cooking style. Chefs that were once trained to use only dried, packaged herbs now use fresh herbs.

Ethnic influences have caused drastic market changes in America's eating habits. Catsup, traditionally the Number 1 condiment, has been replaced today in many cases by salsa, a condiment that gets its primary flavor from tomatoes, peppers, tomatillos, and the herb cilantro. Cilantro wouldn't have been found in most plant businesses 10 years ago, but now some growers are reporting it is one of their top-selling herbs.

Requirements for home landscapes are changing. Shrubs that do nothing beyond offering a perpetually green geometric shape next to the house are being replaced with herbs or other multipurpose plantings. Such plantings add fragrance, edibility, color, and usefulness to the space formerly occupied by shrubs.

THE 10 TOP-SELLING HERBS

A random survey of several wholesale and retail growers across the United States reveals these top sellers.

1. *Petroselinum* **spp. (parsley).** This includes the flat-leaf Italian and curled varieties combined. Most people know parsley from the token sprigs on their plates in restaurants. First-time herb gardeners usually choose parsley, even if they do not use it for anything more than decoration.

2. *Ocimum* **spp. (basil).** Sweet basil has the largest volume, followed by lemon and spicy globe basils. Once customers begin growing sweet basil, they often come back to try the more exotic varieties.

3. *Rosmarinus* **spp. (rosemary).** Mixed varieties—including Arp, standard rosemary, and prostrate types—are all included under the heading of rosemary.

4. *Thymus* **spp. (thyme).** The big sellers are primarily *T. vulgaris*, *T. serpyllum*, and lemon thyme, *T.* × *citriodorus*. Silver thyme (*T. vulgaris* Argenteus), caraway (*T. herba-barona*), orange balsam (*Thymus* Orange Balsam), and woolly (*T. pseudolanuginosus*) are among the other thymes listed as popular.

5. *Artemisia dracunculus* **var.** *sativa* **(tarragon, French tarragon).** Several growers surveyed included a note of caution regarding the plant industry's continued selling of Russian tarragon (*Artemisia dracunculus* subsp. *dracunculoides*) in place of French tarragon. The plants are not the same, even though some seed companies continue to list Russian tarragon as edible. Reputable growers consider this misleading advertising that is damaging to the entire herb industry. There is some evidence that ingesting Russian tarragon in any quantity can be harmful. Informed, reputable plant sellers will

offer their customers only true French tarragon and will move the invasive Russian perennial to the landscape department, where it belongs.

6. *Coriandrum sativum* (cilantro). Several growers listed cilantro as equal in sales to parsley and basil, but since this survey was taken from all across the United States, it averaged out at number six. As the popularity of Mexican, Thai, Vietnamese, and other ethnic foods increases, cilantro will work its way up the list of top sellers. The leaves, called "cilantro," are the primary flavoring for salsas and other ethnic dishes. The seeds, called "coriander," are grown for a seasoning ingredient for cookies, cakes, and pickles. Several varieties of cilantro have come onto the market recently. Leaf coriander, sometimes listed as Slow-Bolt or by similar seed company names, maintains leaf production longer, bolting to set seed later in the season. All are grown before hot summer temperatures elevate, and plantings every two weeks are recommended for extending leaf production.

Substitute cilantro varieties are gaining in popularity where summer is too hot or other conditions do not allow continued growing of *Coriandrum sativum* varieties. Mexican coriander (*Eryngium foetidum*), also called thorny coriander or ngo gai, is a native of Central America. This plant is gaining wider acceptance primarily because the leaves dry well, unlike other corianders, which lose their leaf flavor when dried. Vietnamese coriander (*Polygonum odoratum*), also called rau ram, can be grown as an indoor or outdoor perennial; it has excellent leaf flavor.

7. *Mentha* spp. (mint). All of the mints were lumped together in the survey, including peppermint (*M. piperita vulgaris*), spearmint (*M. spicata*), apple (*M. suaveolens*), and others. A new mint named *Mentha* Hillary's Sweet Lemon, offered to the wholesale trade for the first time in 1995, is the first mint to be patented in many years. A hybrid cross between apple and lime mints, it was bred by Jim Westerfield in Illinois. As it becomes available, it is gaining wide acceptance for the pleasant lemon flavor and fragrance.

8. *Origanum* spp. (oregano). The most common oregano grown for sale is *Origanum vulgare hirtum*, the wild oregano collected from seed in Greece. This plant is listed as number two, right after the wrong tarragon, for disappointing plant customers most often. While it is quite hardy, this oregano spreads rapidly from roots, often becoming invasive, and the flavor has been described as resembling something between diesel fuel and pencil shavings! According to sources in Greece, it is the flowers rather than the leaves that are commonly used from this plant. However, for American palates, other varieties offer a more "pizza-like" flavor, and growers are beginning to offer these better-quality varieties to their customers. Among those are *Origanum × majoricum,* with a flavor between oregano and sweet marjoram; *Origanum*

Kaliteri, which has a high oil content and is grown commercially in Greece for a high-quality oregano; Cretan oregano, *O. onites*, which also has an oregano-marjoram flavor more to the liking of pizza lovers.

9. *Anethum graveolens* (**dill**). Varieties listed included standard old-fashioned dill as well as newer varieties of leaf dill, like Dukat. Dill is a cool-season plant, best seeded in early spring or late summer for fall growing.

10. *Allium schoenoprasum* (**chives**). Little work has been done on introducing new cultivars of chives. The plant remains a good seller for plant nurseries and is easily grown from seed. Landscapers have begun using chives as an easy landscape herb that performs well under a wide variety of growing conditions.

THE TOP FIVE SELLING EVERLASTINGS

Not only have culinary herbs increased in sales volume, but so have the everlasting herbs used for crafts and medicinal herbs.

1. *Limonium sinuatum* (**statice**). Statice is used in dried arrangements, wreaths, and other craft projects.

2. *Artemisia annua* (**sweet Annie**). Sweet Annie is used primarily as wreath base material and for background bulk for swags and everlasting arrangements. Loved by some, about a third of the people who use this herb for crafts cannot tolerate the strong aroma.

3. *Helichrysum* (**strawflower**). This is primarily used on wire stems for arrangements or as glued-on flowers for crafts projects.

4. *Gomphrena globosa* (**globe amaranth**). This plant has gained wide popularity for arrangements and crafts due to the several colors available, including white, purple, light pink, red, and others. Easily grown in a wide variety of conditions, it can be used as a bedding plant in the landscape.

5. *Limonium latifolia* (**German statice**). This hardy perennial has delicate lavender to white flower clusters used as airy filler for floral work. *S. tarticum*, with white flowers, is also popular.

THE TOP FOUR MEDICINAL HERBS

1. *Echinacea* **spp.** (**purple coneflower**). The top-selling varieties are *E. purpurea, E. angustifolia,* and *E. pallida.* It is listed in respected herb references as an "immune system booster," due to its effect of helping the body ward off colds and flu, so customers want to grow this plant. Most landscape nurseries

offer this tough perennial as a good summer-blooming flower, not recognizing it as a popular medicinal plant sought after by the bedding plant customers.

2. *Symphytum officinale* (**comfrey**). This is another important medicinal herb that your customers will look for when they are starting their own medicinal gardens.

3. *Valeriana officinalis* (**valerian**). Another example of a popular herb that gardeners want to have in their gardens, valerian is also available in capsule form in health food stores as a mild muscle relaxant.

4. *Hydrastis canadensis* (**goldenseal**). Goldenseal is an example of a fad-driven plant. Some years back, misinformation was circulated that the root of goldenseal could possibly mask illegal drug use on drug tests. Even though there is no positive evidence supporting that idea, the plant name came into popularity. Legitimate uses of the plant include in sore throat gargle and for fever blisters. Because it has a kind of aura of mysticism, people want to grow it just to have a popular medicinal in their herb borders. It is a useful herb and perfectly legal to grow, sell, and use. The powdered root is found in many health food store herb supplements.

HERB PROPAGATION

Anise hyssop (*Agastache foeniculum*). An anise-scented perennial used as a bedding landscape plant as well as in herbal tea, this is also a good butterfly attractor. Grow from seed or cuttings. Sow in flats, barely covering seed with soil. Expect about 70,000 seeds per ounce, with viability around 70%. Keep the soil temperature at 70F (21C), and germination should occur in five to six days. The plants will be ready for transplanting in 20 to 22 days. Grow in average garden soil or a raised bed. Flowering will be continuous July through frost if flowers are kept picked off, discouraging seed setting. Keep regularly watered during the hottest part of summer.

Basil, sweet basil (*Ocimum basilicum*). There are approximately 16,000 to 17,000 seeds per ounce. Sow seed in flats indoors. Germinate at 65F (18C) for four to six days. Ready for transplanting in about 18 days, basil will be ready for sale about two weeks after transplanting. For sowing in the garden, sow seed after the last frost date, covering with soil about one-eighth inch (3 mm) deep. Thin to 15 to 24 inches (38 to 61 cm).

The most popular basil varieties besides sweet basil include the following six. Lemon basil (*Ocimum americanum*) has a fresh, lemony flavor. Keep flowering spikes clipped to keep the plant producing leaves. (Otherwise, the plant will drop its leaves as soon as seed sets, and the plant will die.) Anise basil (*O. basilicum*), also called Asian or Thai basil, has a sweet, anise flavor and fragrance. Cinnamon basil (*O. basilicum*) has a distinctive cinnamon flavor and fragrance. Genovese basil (*O. basilicum* sp.) has extra-large (2-inch, 5-cm), dark green, smooth leaves. Plant grows to about 24 inches (61 cm) and is slow to bolt to seed, and has a heavy, spicy basil flavor. Spicy globe basil (*O. basilicum minimum* Spicy Globe) keeps its globe shape, with tiny leaves. Holding well in pots for sales, it is one of the best basils for container growing and for landscaping. It is ready to sell five to eight weeks after sowing in seed flats.

Dark opal basil (*O. basilicum* Dark Opal) has dark purplish foliage. This is the one sought by customers who are making herb vinegars, as it lends the ruby-purple color to the liquid. Plant extra of this, as seed stocks often have lots of green-leafed basils upon germination, and you may have to select out the darker colors for transplanting. Several purple-leafed basils are available, some ruffled, others not. Red Rubin is an improved variety of Dark Opal with plain-edged, dark red leaves that are attractive in the annual border or on the plate. Propagation is by seed or cuttings. The plant requires full sun and is easily damaged by frost. Plant in rows in a seed flat without plastic covering. Germination should take place in about five days at 70F (21C), and plants will be ready for transplanting to pots in 18 to 20 days. Cuttings root easily under mist.

Bay (*Laurus nobilis*). While it is not one of the top 10 sellers, the demand for bay continues to be strong. Plants can be moved indoors in the North; in lower Zone 7 and Zone 8, bay is grown outdoors. Propagation is by cuttings. To propagate bay, it is important to choose the current season's stems. Choose stems that are not fully hardened but that have lost the glossiness of new growth and are a duller green, with leaves that have also lost the light green luster. Take cuttings with a sharp knife, not by using scissors or pruners. Cut 4-inch (10-cm) pieces, removing the lower leaves from the stem, leaving three or four leaves on the top. Pinch out the growing tip, then scrape the opposite sides of the bottom portion of the stem, for about one-half inch (1.3 cm) on each side, with the knife until you expose the white under-layer. This step is important in order for the stem to take up more water, as well as to encourage the stem to begin root production.

Rooting hormones in rooting powder hinder root development of bay, so avoid using the compound on this plant. Instead, be sure that the soil temperature remains warm, 65 to 70F (18 to 21C), and keep the cuttings under a gentle, repetitive mist. Rooting can also be accomplished by placing the cuttings in a shaded place in the greenhouse, such as under a bench but not directly on the ground.

Cilantro (*Coriandrum sativum*). This is also listed as "coriander." Coriander is the seed of the cilantro plant. Germinate seeds in total darkness for 10 to 14 days at

65F (18C). Transplant in about two weeks, but be aware that there may be some losses of transplants, as cilantro does not transplant as readily as some other seedlings. Growing plugs may be better.

Cilantro is a cool-season plant, like dill. When the weather warms and the days lengthen, cilantro will bolt into flowering, which changes the flavor of the leaves. Cutting back can slow the bolting slightly, but the plant will keep trying to bloom and set seed. Newer varieties, like Slow-bolt and Leisure, will lengthen the growing and selling seasons somewhat. Cilantro can be grown in the garden in springtime and fall, but it will die out in hot weather. Seed repeatedly every two weeks for a continuing supply for early spring sales.

Chamomile, German (*Matricaria chamomilla*).

Both spellings, "chamomile" and "camomile," are considered correct. An annual plant that remains popular in sales, German chamomile is grown for the aromatic flowers, used in tea after meals for relaxing and digestion. This plant is not to be confused with Roman (also sometimes called English) chamomile, which is *Anthemis nobilis*, a perennial ground cover of no use as a tea herb.

Bertha Reppert of The Rosemary House in Mechanicsburg, Pennsylvania, says that she uses chamomile tea to water seedlings in her greenhouse when she is having trouble with damping-off. This use dates back several centuries, and one ancient name for chamomile was "the plant's physician" because of the positive effects it often had on other plants near it.

Propagated by sowing seed in uncovered seed flats, the seeds are sown directly on top of the soil without being covered. Kept at 70F (21C), the seed will germinate in four to five days and can be transplanted individually to pots in about 21 days. Several growers recommend germinating the seed at lower temperatures, 55 to 65F (13 to 18C), for 15 to 20 days. Plants are ready for sales in an additional two to three weeks. Note: Do not cut back plants when small and still in pots. Stokes Seed says that cutting back in this stage of growth will deter the plant's flowering as it matures.

Peat pots are recommended for chamomile to avoid transplant shock, as it can be set back considerably when the roots are disturbed after the seedling stage. Growing seedlings as plugs may also help.

Chives (*Allium schoenoprasum*).

Seed will germinate in six to eight days at 70F (21C). Seedlings are clump-transplanted, meaning that you do not transplant individual seedlings. Mary Peddie of Rutlands of Kentucky gives the tip that chive seed can be gathered from seed stock by collecting the entire seed head and keeping it in the freezer until ready to plant. Planting one seed head per pot, upside down, gives a full, salable 4-inch (10-cm) pot, saving considerable transplanting time. An ounce of seed is about 22,000 seeds.

Dill (*Anethum graveolens*). A cool-season annual that can be grown in the garden in spring and fall, dill (shown in fig. 2) bolts to seed in hot weather. Varieties that are more compact and slower to bolt include Dukat, Fernleaf, and Bouquet.

Seeds will germinate in the greenhouse in about five days at 70F (21C) and can be transplanted in two weeks or less. Sow on top of soil, pressing down slightly but not covering over with soil. Direct seeding in pots or in the garden soil can be done, too, but seed germination averages about 60%, and thinning is often needed. Repeat planting about every two weeks, to have a continuous supply for sales in early spring, is recommended. Plants do not hold well in pots once the weather gets hot and are best kept in the cooler part of the greenhouse. When dill plants begin to bolt into flower, cut them back to slow the process a bit. Consider bolting as a sign that the weather has warmed and sales for dill plants are nearing the end.

FIG. 2   Dill is a traditional herb favorite.

Echinacea, coneflower (*Echinacea* spp.). A hardy perennial, this plant is sought for the immune-system-building components of the roots, according to the *Peterson Field Guide to Eastern/Central Medicinal Plants* by Steven Foster. Used to ward off colds and flu, this much-researched plant is of interest to landscapers and homeowners who grow it for the attractive flowers, as well as to the herb collector. Native varieties include purple coneflower (*E. purpurea*), pale purple coneflower (*E. pallida*), and narrow-leafed purple coneflower (*E. angustifolia*). Hybrid varieties used for landscaping but not for medicinal gardens include Bravado, with rosy purple, 4-inch (10-cm) flowers, and White Swan, with pure white, fragrant flowers. Widely adaptable to many soil conditions, echinacea is often found in the wild growing in poor soils made up primarily of clay, shale, and limestone.

Echinacea is grown primarily from seed, which needs to be cold-treated, because these seeds overwinter on the stalk in nature. Place echinacea seed in moist peat inside a plastic bag, and keep it in the refrigerator until ready to sow in spring. For a larger plant for spring sales, seed can be planted in the late summer and fall, starting in August or September, then wintered over in a cool greenhouse or even outside in milder climates. Otherwise, start seed in germination flats around January for April sales, transplanting to 4-inch (10-cm) pots. Echinacea will hold well through the sales season, and any leftover plants can be moved up to gallon-sized pots for landscape sales. Echinacea generally flowers the second year, although plants started early can sometimes flower a bit the first year, varying with variety.

Lavender (*Lavandula* spp.). Cuttings are the standard way to propagate named varieties. Cuttings should be taken eight to 16 weeks before the sales date. The potting soil needs to be well-drained and loose. Too much moisture can damage growth.

For seed propagation, chill seed for two weeks at 40F (4C). There are 25,000 seeds per ounce. Sow seed on the surface of the soil, barely covering. Seed should germinate in 10 to 14 days at 65F (18C).

Recommended varieties include Lavender Lady (*L. angustifolia* Lavender Lady), the 1994 AAS award winner, which is easily grown from seed, has good lavender blue flowers, and will flower the first year from seed. Hidcote (*L. angustifolia* Hidcote) is a slower growing variety with dark purple blooms; the most popular of the English varieties, it is grown from seed or cuttings. Munstead (*L. angustifolia* munstead) is named for Gertrude Jekyll's garden. Munstead is the earliest blooming English lavender, with lavender-blue flowers, and it can be grown from seed or cuttings. Other varieties that are popular, mainly with those who want unusual varieties of lavender to add to their collections, are less hardy and less tolerant of varying growing conditions: fernleaf (*L. stoechas*), Provence (*L. × intermedia*, often listed as *L. hortensis*), Spanish (*L. dentata*), and woolly (*L. lanata*). Seeds are available, but to be sure of variety, you may want to keep stock plants and take cuttings.

Marjoram, sweet marjoram (*Origanum majorana*). Grown as an annual throughout most of the United States, marjoram is easily grown from cuttings or seed. Cuttings are less likely to develop stem rot, if that has been a problem. Cuttings are taken about five weeks before the sales date for small pots. With an average of 125,000 seeds per ounce, germination is about 50% in four to seven days at 70F (21C). Soak the seeds for 24 hours before sowing. Sow one-fourth inch (6 mm) deep and transplant in 12 to 14 days. Pinch plant tops slightly at 10 days after transplanting to encourage branching if plants are getting tall. In the garden keep flowers and older growth cut back to encourage continuing new growth.

Mint (*Mentha* spp.). Mints are grown primarily by cutting and division. Even though peppermint seed will be seen listed in some catalogs, true peppermint has sterile flowers that do not set seed. To obtain the best flavor and fragrance in mints, keep stock plants and take root divisions or stem cuttings.

Especially good varieties, besides spearmint, *M. spicata* (which does grow well from seed), and peppermint, *M. piperita vulgaris*, which is propagated by cutting, include apple mint, *M. suaveolens*, a favorite for tea and jelly. Corsican, *M. requienii*, is a carpet-forming mint with tiny leaves that smell more like pennyroyal than mint. It is not hardy in colder areas of the United States but makes an interesting ground cover around stepping stones in warmer regions or treated as an annual.

Hillary's Sweet Lemon, *Menthus* Hillary's Sweet Lemon, is a cross between apple and lime mints, with a pronounced sweet citrus flavor and the background fragrance of spearmint. It was one of the first mints to be patented in many years. Lime, *M. piperita citrata*, is a good mint with a lime fragrance. Orange mint, *M. piperita citrata*, is also called eau de cologne mint and bergamot mint. The oil is an ingredient in chartreuse liqueur and perfumes, and your customers will like the fragrance in potpourri.

Spearmint can be grown by pressing the seed (about 304,000 seeds per ounce) into the surface of the seeding media; do not cover. It germinates in about 10 days in 65F (18C), with germination rates around 70%. Sow in February and March for April or May sales.

All mints produce dense root masses with underground runners that can be divided. Every root piece with a joint will root and can be ready for sale in two to four weeks. Stock mint beds should be replaced about every three years by running a tiller through the bed and adding compost and humus. Established plants need weekly watering, while seedlings or new cuttings will require two or three waterings per week.

Frequent cuttings can be taken during the growing season to propagate new plants for sale. Four- to five-inch (10- to 13-cm) cuttings, with approximately three leaves and the growing tip end removed, kept under a mist, will root in four to six days and be ready for transplanting in another seven days. Cuttings can also be rooted in water. Allow about 27 days from the time the cutting is stuck to the sales date.

Oregano (*Origanum* spp.). Aside from the *Thymus* species, this is probably the most misunderstood and mislabeled herb grown for sale. Many plant sellers, unsure of their plant varieties will simply label the plants "oregano" or "thyme," which is a disservice to the customer. The oregano variety commonly available from seed, *Origanum vulgare hirtum*, is the poorest in flavor. This plant is generally listed in the catalogs simply as "oregano," and unless you taste it yourself, you will not know you are disappointing nearly every customer who buys the plant from you. This is a plant that is best moved to the dried everlasting flower section of your garden center, where it is a worthwhile addition to dried flower crafts. The flowers dry well and come in white, pink, rose, and lavender colors. This common oregano plant spreads by roots and can quickly become invasive, taking over the space of other herbs in the garden. If you do not choose your varieties well, your customers will grow to distrust other plants in your inventory, based on their experience with a poor oregano.

Richters Herbs recommends a variety called Greek or Kaliteri, *O. kaliteri*, which has high oil content in the leaves and is the commercial oregano in Greece. They also recommend one they call Italian (*O. × majoricum*), with a pleasant flavor similar to sweet marjoram and Sicilian or Cretan oregano. Another oregano, *O. onites,* has strong flavor but is better than the wild, common oregano that spreads so rapidly by roots.

Sal Gilbertie of Gilbertie's Herbs recommends one marketed as Oregano-True. This one is also sometimes called dark oregano and can be propagated by cuttings only. Two others that Gilbertie recommends are Oregano-Maru (*O. maru aegypticum*), described as a grey-green oregano with excellent oregano flavor. This perennial is reliably hardy, but it needs some winter protection in the most northern climates. Another variety worthy of note, and which has only moderate flavor and fragrance but makes a very good ground cover in rock gardens, is compact oregano, *O. vulgare* Humile. It is reliably hardy into Zone 6. This one is easily propagated by root divisions in early spring, four to eight weeks before the sales date.

For general oregano propagation, maintain a bed of stock plants that have a good pizza fragrance and flavor. Take cuttings eight to 10 weeks before the sales date. Oregano can also be propagated by layering and root divisions (especially with compact oregano).

Seed propagation of *O. vulgare* for everlasting flowers can be done in about four days at 70F (21C) and transplanted in 14 to 20 days. Seed germination is about 50% in most cases, and there are about 350,000 seeds to the ounce. Note: The most recommended, best-flavored oregano varieties are all propagated by cuttings or divisions, not seed, according to most growers surveyed.

Parsley (*Petroselinum* spp.). Parsley is a biennial grown as an annual (curly-leafed parsley shown in fig. 3). In the second year, the plant bolts to a flowering center, and the flavor changes from mild and sweet to bitter or tasteless.

Outside in the garden, seeds need to be covered about one-fourth inch (6 mm) deep with soil. Gardeners often cover the outdoor seed bed with a board to hold moisture evenly at the seed level. Germination outdoors can take four to five weeks, but in the greenhouse in ideal conditions at 70F (21C), germination can be as quick as eight days. Cover the seeds one-fourth inch deep and keep evenly moist. Plants can be clump-transplanted in 12 days

FIG. 3   No grower's herb assortment is complete without a good selection of curly-leafed parsley. The variety: Green River.

after coming up, with the extras then thinned out as they grow. They can be moved to the garden about 14 days after transplanting or held for a few weeks for sales in the greenhouse. Germination of parsley seed is about 60%, with about 15,000 seeds per ounce. Seed at two-week intervals to ensure a continuing product to sell. When cutting back parsley, cut off the leaves on the outside of the leaf whorl, rather than shear off the entire plant, since the leaves unfold like a rose as the plant opens.

In his book *Growing Herbs from Seed, Cutting and Root*, author Thomas DeBaggio cautions that germination inhibitors naturally occur on parsley seed, which accounts for the length and variation of germination time. He suggests putting the seed in a cloth bag (or "an old piece of pantyhose," specifically) and soaking that for about two days in an oxygenated aquarium. This method works when hurrying direct seeding in the soil outdoors, although he admits that this step is not so necessary since germination improves as the soil temperature warms naturally.

The best parsley varieties for decorative planting include the double-curled ones. The best flavored, and one that customers will ask for once they have tried it, is flat-leafed Italian parsley (*Petroselinum crispum neapolitanum*). With darker green leaves and sweet flavor, this is the best parsley for culinary use.

Rosemary (*Rosmarinus officinalis*).
An important culinary herb, rosemary is also sought for topiary work. Seed will germinate in about three to four weeks, germinated at 65F (18C) in sand or Jiffy mix. Sow one-fourth inch (6 mm) deep.

Generally, however, rosemary is propagated by cuttings or layering, as those are the fastest and most productive ways for the market. Cuttings taken in December or January will be ready for 4-inch (10-cm) pots by April. Some growers indicate that they continue to take cuttings throughout the year in order to have a continuous supply, the hot summertime being the only time that the plant does not root well. Take 4-inch cuttings from plants throughout most of the year, with the exception of the hottest part of summer. Strip away the leaves from the bottom 2 inches (5 cm), and dip the cut end in rooting powder. Keep cuttings in flats under greenhouse benches or stick in the rooting bench under occasional mist.

Tom DeBaggio contends that rosemary cuttings that are treated with rooting compound produce better, bushier roots than those that are rooted without the compound. Bertha Reppert of The Rosemary House uses my method of leaving several limbs of willow (either weeping or native willow species) in water for several days. She then soaks the stem ends of rosemary cuttings in the water overnight before sticking them in potting soil to root. Willow contains a natural rooting compound that is useful for rooting several woody herbs.

Herb cuttings can be stuck in plastic greenhouse flats (11 by 21 by 2½ inches, 28 by 53 by 6 cm). You can root from 100 to 300 cuttings per flat, depending upon the size and bushiness of the cuttings. The flats can be placed under mist or kept in the shade under greenhouse benches. Cuttings rooted this way can be moved to the transplanting table easily, and the roots stay in the soil until pulling at the time of potting. This eliminates some of the air-burn that can occur when lots of cuttings are pulled at once and many remain lying out in the air for an hour or more. Roots of newly rooted cuttings are tender and should be protected from air and direct sunlight during the transplanting process.

Rosemary cuttings, as well as most other cuttings, are ready when the roots are from one-fourth to one-half inch (6 to 13 mm) long. As soon as roots have attained that size, transplant them. Delaying the transplanting time can allow disease, which can multiply in warm, humid conditions, to rapidly spread through the cuttings.

Rosemary prefers warm, well-drained soil with some lime. Neither allow plants to dry out completely nor allow them to stay extremely wet for extended periods. Grown in most parts of the United States as a tender perennial, rosemary is hardy in the south and many coastal areas.

Benenden Blue (*R. officinalis* Benenden Blue) has medium blue flowers and narrow leaves. Lockwood (*R. officinalis* Lockwood) has a trailing, prostrate habit and bright blue flowers.

Hill Hardy (*R. officinalis* Hill Hardy) is a superior rosemary hardy to 0F (–18C). The leaves are a dark green with a pine-like fragrance. Propagate by cuttings. Arp (*R. officinalis* Arp) was discovered on a trip to Texas in the early 1970s by Madeline Hill. The plant was in bloom on January 18 with the temperature at 18F (–8C)! This is an excellent variety to offer your customers. Propagate by cuttings.

R. officinalis, the common rosemary, while a good performer, is less hardy than the above varieties (hardy to mid-Zone 7). An advantage to this rosemary is that propagation can be by seed or cutting. Ingram rosemary (*R. officinalis* var. Ingram) has lovely bright blue flowers, is a strong grower, and possesses intense rosemary flavor. It is grown from cuttings only. Prostrate rosemary (*R. officinalis* Prostratus) is called Romero (the pilgrim's flower) in Spain. This rosemary is used in the same way as the more upright ones, but it grows as a trailing or hanging plant, at its best grown over the side of a terrace or low wall. Propagate by cuttings.

Sage (*Salvia officinalis*). A hardy perennial and a popular garden plant, sage is propagated by cuttings, layering, or seed. Sage is best grown in raised beds in the garden, and it will do fairly well as a container plant. Mature plants often die back in the center as the wood gets dense. The mature stock plant should be renewed by burying a limb or two and rooting by layering, then removing the older plant. Otherwise, simply replant new sage plants about every three to four years. Customers often assume there is something wrong with their sage when it dies at that age, whereas it is natural for the plant to do so.

There are approximately 3,400 seeds per ounce. Viability, based on data from Johnny's Selected Seed and Park Seed Co., averages about 60%. Sow in seed flats, covering seed slightly with soil about one-fourth inch (6 mm) deep. Held at the ideal soil temperature of 70F (21C), the seeds should sprout in nine to 10 days and be ready to transplant in 18 to 21 days.

Sweet Annie (*Artemisia annua*). This plant is primarily used by florists and crafters as a wreath and swag base. Customers will especially want it if there are strong herb groups in the area. Plants can become invasive but are easily controlled.

Propagation is by seeding two to three weeks before sales date. Seed again every two weeks for continued supply. Seed germination averages 70% and takes six to 10 days at 70 F (21C). Sown in flats, it will be ready for transplanting in about 10 days.

Statice, annual (*Limonium sinuatum*). This is probably the best known ever-lasting flower for fresh-cut floral work and as dried material. Statice is available in many clear colors and pastels. Seed four to six weeks before sales date and transplant into well-drained, rich soil. Expect 225,000 seeds to the ounce. Sow in the green-house in February or March for flowering in the garden in late July to early August. In the warm parts of the south and west, it can be sown in late July outdoors for January cutting. Another variety, often called Russian statice (*Limonium suworowii*), is also worth growing, being useful for fresh-cut and dried material. It blooms slightly earlier in the garden than some other annual varieties.

Statice, German (*Limonium latifolia*). Used in floral and craft work, primarily as dried material, plants are hardy perennials, producing lavender to light lavender flowers the second year. Propagation is by seed, but germination is often very poor. Harlan Hamernik of Bluebird Nursery says that his method for getting old seed to germinate is to seed it into flats of damp peat, leaving it for two weeks, then putting it in cold storage at 38F (3C) for four to six weeks, and then returning it to 70F (21C) for sprouting. Otherwise, if the seed is very fresh, he says, seed started in flats at 70F will pop up in about five days.

Limonium tartaricum is also a perennial that is popular in everlasting arrangements. *Tartaricum* offers stiffer masses of white flowers and is grown much like *latifolia*. Both grow well in raised beds, in average soil with full sun, spaced 18 to 24 inches (46 to 61 cm) apart. Pine needles work well as mulch between the plants, deterring weeds while not holding excess moisture at the roots. Hardy to at least −10F (−23C), it blooms reliably the second year after planting.

Strawflower (*Helichrysum bracteatum*). Used for dried floral work, these flowers appear to be dry even when picked fresh from the plant. A native of Australia, strawflower can stand high heat and arid conditions in the garden, although best production is on plants that are given regular soil moisture. Continuous harvesting of the flowers increases yields. Sow in seed flats four to six weeks before sales date, and transplant about 10 days after germination into well-drained media. Grow in well-drained, average garden soil. Mulching with straw keeps beds of helichrysum weed-free and producing from midsummer to frost.

Tarragon, French (*Artemisia drancunculus* var. *sativa*). A perennial hardy to below 0F (–18C), this plant is propagated only by cuttings or division. If your supplier offers you seed for tarragon, find a more reputable seed source: French tarragon does not set seed. Russian tarragon, *A. drancunculus*, does set seed, but the plant is not safely edible and tastes nothing like French tarragon. Russian tarragon can become invasive and will disappoint every customer who buys the plant thinking it to be French tarragon.

Cuttings of French tarragon should be taken eight to 14 weeks before sales date. Pinch growing tips to form a more compact plant in sales pots. Average greenhouse temperatures for growing tarragon from cuttings are 55 to 60F (–13 to 16C). The best time for cuttings is in early spring, taking cuttings of early spring growth. Cuttings are susceptible to root fungus while rooting, so the more sterile the medium is, the better.

Thyme, English; also called German or winter thyme (*Thymus vulgaris*). Propagation is generally by cuttings or layering. Also grown by seed, germination in the greenhouse can be quick, four to seven days or longer, at 70F (21C).

Notable varieties include creeping thyme, *Thymus serpyllum*, a hardy perennial that grows 3 to 5 inches (8 to 10 cm) tall. *T. vulgaris*, narrow-leaf French, has about 50% germination in four to five days at 70F (21C). There are about 100,000 seed per ounce. Lemon thyme, *T. × citriodorus*, has dark green leaves and a fresh, pleasantly lemon flavor. It is hardier than golden lemon or Doone Valley lemon thymes, both of which need protection in winter north of Zone 7.

Nutmeg thyme, *T. praecox articus*, is a low, creeping variety with a mild, nutmeg scent. Caraway thyme, *T. herba-barona*, is an ideal rock garden plant around rocks and over low raised-bed walls. With medium to dark green leaves, its sweet caraway flavor and scent are good in cheese and tomato dishes. Woolly thyme, *T. pseudolanuginosus*, is a good carpeting thyme around stepping stones. Its grey, woolly leaves have little flavor or scent. When growing inside, avoid excess water on leaves to avoid damage. Bottom watering in sales pots is best. Grow in a sandy mix. Move to an outdoors sales area, where air circulation will be better, as soon as the weather allows.

SOURCES

Harlan Hamernik
Bluebird Nursery, Inc.
519 Bryan St.
Clarkson, NE 68629
402/892-3457
Fax 402/892-3738

Offering more than 1,400 plants and herbs to the wholesale trade, including six varieties of echinacea, German statice. Ships throughout the United States.

Dean Pailler
Flowery Branch
P.O. Box 1330
Flowery Branch, GA 30542

One of the most diversified seed companies, offering everlastings by separate colors as well as several rare herbs and flowers not available elsewhere. Retail and wholesale seeds.

Sal Gilbertie
Gilbertie's Herb Gardens, Inc.
P.O. Box 118
Easton, CT 06612
203/452-0913; 800/US-HERBS
Fax 203/459-0142

Wholesale and retail herb plants in 4-inch, 6-inch pots, and 1-gallon containers (10-cm, 15-cm, 3.8 l). Large selection of culinary, medicinal, and everlasting herbs—including Hill Hardy, Arp, and Ingram varieties of rosemary; Maru, True, and Greek oregano; and Lady Lavender—for the wholesale trade.

David and Jodie Gilson
Gilson Farms
P.O. Box 74
Groton, MA 01450
508/448-2971; 800/720-4372
Fax 508/448-6079

Wholesale and retail herb plants, primarily 4-inch (10-cm) pots. Eighteen varieties of basil, 14 of monarda, many unusual herbs for the wholesale grower.

Ron Zimmerman
The Herbfarm
32804 Issaquah-Fall City Rd.
Fall City, WA 98024
206/784-2222

Retail seller of more than 450 varieties of herbs in 4-inch pots, including many unusual varieties. Sales at shop only. Remarkable because lunch and dinner are available by reservation (three to six months in advance). Five- to nine-course meal includes intensive workshop about herb culture.

Steve Hershfield
Hillcrest Nursery
21029 Gunpowder Rd.
Millers, MD 21107
800/452-4032
Fax 410/ 239-4396

Wholesale dealer of herb plugs in cell packs, shipping to the United States and Canada. Listing includes five kinds of oregano, nine varieties of rosemary, and 11 of thyme.

Johnny's Selected Seeds
Foss Hill Rd.
Box 2580, Dept. 598
Albion, ME 04910-9988
207/437-9294
Fax 800/437-4290

Wholesale seeds, including more than 35 medicinal and 80 culinary herbs. Guarantees correct botanical species and varieties, along with providing growing information for those crops.

George or Lynda Pealer
Millcreek Gardens
15088 Smart Cole Rd.
Ostander, OH 43061
800/946-1234

Wholesale herb plants in 3-inch (8-cm) pots. Delivery to 200-mile radius of Columbus, Ohio. Includes good selection of everlastings and medicinal and culinary herbs.

Parks Seed Wholesale
Cokesbury Rd.
Greenwood, SC 29647
800/845-3366
Fax 800/209-0360

Bulk seed for greenhouses, including statice varieties, several celosias, gomphrena. Lavender varieties include Lavender Lady. General selection of herb seeds. Publisher of *Parks Success with Seeds* and *Parks Success with Herbs.*

Conrad Richter
Richters Herbs
357 Highway 47
Goodwood, Ontario
Canada LOC 1A
905/640-6677
Fax 905/640-6641

Retail plant and seed source listing hundreds of herbs. The primary distributor for Hillary's Sweet Lemon mint, developed and patented by Jim Westerfield.

Mary Peddie
Rutlands of Kentucky
P.O. Box 182
Washington, KY 41096
606/759-7815
Fax 606/759-5745

Contract grower specializing in plugs for many varieties. Wide selection of scented geraniums. Founding member of International Herb Association and Kentucky Herb Business Association.

Jim and Theresa Mieseler
Shady Acres Herb Farm
7815 Highway 212
Chaska, MN 55318
612/466-3391

Retail herb plants in 3^1/$_2$-inch (9-cm) pots, listing 200 herb varieties.

Stokes Seed, Inc.
Box 548
Buffalo, NY 14240
905/684-6106

Retail seed, including good selection of herbs, everlastings, and flowers. Notable is the useful germination information on seed packets. Several small growers mentioned relying upon this better-than-average information.

Jim and Marilyn Westerfield
The Westerfield House
8059 Jefferson Rd.
Freeburg, IL 62243
618/539-5643

Retail sales to their herbal bed and breakfast customers. Developed Hillary's Sweet Lemon mint. Recipes available with this particular mint.

BOOKS AND PUBLICATIONS

Growing Herbs from Seed, Cutting and Root by Thomas DeBaggio
For wholesale or retail information:
Interweave Press
201 East Fourth
Loveland, CO 80537
970/669-7672

Propagation Handbook—Basic Techniques for Gardeners
by Geoff Bryant
Stackpole Books
5067 Ritter Rd.
Mechanicsburg, PA 17055

Parks Success with Herbs
Parks Success with Seeds
Reference books or for retail to your customers
Parks Seed Wholesale
Cokesbury Rd.
Greenwood, SC 29647
800/845-3366
Fax 800/209-0360

Herb Gardening at Its Best by Sal Gilbertie
With detail on the culture of the basic 15 herbs, with general information
about 150 varieties
Gilbertie's Herb Gardens
P.O. Box 118
Easton, CT 06612
203/452-0913

The Herb Companion
The largest-distribution magazine on herbs in the United States.
To order wholesale copies to sell in your book and gift section:
The Herb Companion
201 East Fourth
Loveland, CO 80537
970/669-7672

GrowerTalks
A commercial trade magazine for greenhouse producers
P.O. Box 9
335 North River St.
Batavia, IL 60510-0009
630/208-9080
http://www.growertalks.com

To purchase books by Jim Long (wholesale price list available on request with business card; retail catalog $2, refundable with order):
Long Creek Herbs
Rte. 4, Box 730
Oak Grove, AK 72660
417/779-5450
Fax 417/779-5450

PROFESSIONAL HERB ASSOCIATIONS

Contact these groups for more information. Herb business people are known for being helpful. Annual conferences and herb trade shows can be useful sources for expanding your knowledge about the industry.

David Gilson, President
International Herb Association
P.O. Box 317
Mundelein, IL 60060
847/949-4372
Fax 847/949-5896

Maureen Rogers
Herb Growing & Marketing Network
P.O. Box 245
Silver Spring, PA 17575
717/393-3295
Fax 717/393-9261
e-mail HERBWORLD@AOL.COM

About the author: Jim Long's farm, Long Creek Herbs, has been featured in Southern Living, Better Homes & Gardens, Gourmet, U.S. News & World Report, and several other magazines. He is the author of 12 books on herbs and marketing.

CHAPTER 21

FOLIAGE PLANTS

by Charles A. Conover
University of Florida
Apopka

PRODUCTION AND MARKETING

Although the United States foliage industry can be traced back to the early 1900s, it is only within the past 30 years that it has become a major industry. Production initially developed outdoors under slat sheds in southern parts of Florida, Texas, and California. In recent years production has moved into polypropylene shadehouses and plastic, fiberglass, and glass greenhouses. In northern areas production has been centered in Ohio, Pennsylvania, New York, and nearby states in glass and double-layer polyethylene greenhouses. More recently, Hawaii has developed a thriving foliage plant industry, mainly in production of stock plants for cane and cuttings. A number of companies have also developed stock production units in the Caribbean and Central America to ship propagative units to the United States and Europe. It was estimated that at least $30 million worth of propagative material was shipped into Florida in 1988 for finishing.

Major production areas within the United States are listed in table 1. Concentration of the foliage industry in Florida, California, Texas, and Hawaii is primarily because of reduced production costs associated with moderate winter temperatures and high light intensity year-round. Production in northern areas is facilitated by access to local markets and a product mix that stresses rapid turnover of easy-to-grow foliage crops in small pots and hanging baskets (fig. 1).

Table 1   State foliage plant production

State	Wholesale value ($ millions)
Florida	302
California	77
Texas	23
Hawaii	12
32 other states	73

Note: Numbers for 1994 from USDA.

Foliage crops were not economically significant in relation to other floriculture crops until the late 1960s. As late as 1970, foliage accounted for only $15 million at wholesale in Florida. The most recent USDA data (1995) indicate that the total U.S. foliage market at wholesale was nearly $487 million in 1994. Thus, foliage crops have become of major economic importance in a relatively short span of time.

Fig. 1   Many growers offer mixed trays of assorted potted foliage, especially to the supermarket and florist trade.

Detailed information on the national foliage plant product mix is unavailable, but it has been compiled for Florida (table 2), which alone accounts for 62% of U.S. production. Extrapolated to the national level, the Florida breakdown is probably low in hanging basket and terrarium plants, which are often grown for local markets in northern areas. Also, larger foliage types are listed at higher percentages here than they actually are nationally because Florida is responsible for more than 75% of plants grown in 10-inch (25-cm) or larger containers.

TABLE 2 FLORIDA FOLIAGE PLANT PRODUCT MIX

Product	% of total
Aglaonema spp.	4.0
Brassaia actinophylla	2.2
Dieffenbachia spp.	5.9
Dracaena spp.	9.4
Epipremnum spp.	9.1
Ferns	1.2
Ficus spp.	6.5
Ivy	3.1
Palms	5.8
Philodendron scandens oxycardium	3.0
Philodendron spp. (other)	1.7
Schefflera arboricola	1.3
Spathiphyllum spp.	3.3
Syngonium spp.	2.2
Combinations	1.0
Other	40.3

Note: The latest data available, 1988.

STOCK PLANT CULTURE

FIELD STOCK PRODUCTION

Producing foliage stock plants in the field (outdoors) is restricted to southern Florida, California, Hawaii, and the tropics. For the most part, stock plants are grown in full sun or under polypropylene shade cloth that provides the required light levels.

Land selected for stock production should have good internal drainage, as well as sufficient slope to allow surface water to drain off rapidly when excessive rainfall occurs. Temperatures should be between 65F (18C) minimum at night and 95F (35C) maximum day for best quality and yield. Infrequent lows of 50F (10C) and highs up to 105F (41C) will not damage plants but will reduce yields. Consider farm location in relation to wind speed and frequency, since winds may influence the design of protective structures, as well as the types of crops you grow. Wind-induced tip burn and foliar abrasion reduce crop quality and salability.

Structures can be constructed of treated lumber, concrete, or steel, with cable stringers to support the shade cloth. The height and size of the structure are important because they influence temperature at plant height. Because heat rises and air movement is slow through shade cloth, it is wise to provide a minimum of 8, but preferably 10 feet (2.4 and 3 m) of clearance. Erecting 1- to 5-acre (0.4- to 2-ha) units with spaces between units will help prevent excessive temperature buildup in the center of the structure complex.

Foliage stock plants grown outdoors or under shade cloth are usually watered with impulse or spinning sprinklers, which are also used for fertilization. It is very important to have a properly engineered system to obtain good coverage. Using low-angle-trajectory sprinklers are necessary to prevent water from contacting the shade cloth. Normally, foliage plants grown under shade cloth require 1 to 2 inches (2.5 to 5 cm) of water a week.

Native soils in tropical and subtropical areas are rarely satisfactory for foliage stock production. Sandy soils usually require amending with organic components, such as peat moss, to improve water- and nutrient-holding capacities, while heavy soils require peat moss, bark, coarse sawdust, or rice hulls to improve internal aeration.

GREENHOUSE STOCK PRODUCTION

Foliage stock plants can be grown in any type of greenhouse that provides sufficient light and required temperatures. Grow stock plants in raised benches or pots that provide sufficient media volume for good root growth, drainage, and aeration.

Growing media selection depends on local availability, cost, crop type, and personal preference. Some excellent media mixes for foliage stock plants include (1) 50% peat moss and 50% pine bark; (2) 75% peat moss and 25% pine bark; (3) 75% peat moss and 25% perlite; and (4) 50% peat moss, 25% vermiculite, and 25% perlite.

Selecting watering systems for stock plants grown in greenhouses is important since foliar diseases are reduced when foliage is kept dry. Besides the economic advantage of reducing disease and thus chemical usage, the propagative units will not have unsightly foliar residues. Use irrigation systems directed at wetting the media and not foliage. (A wood post and Saran structure, commonly used in Florida, as well as sprinkler system, are shown in fig. 2.)

FIG. 2   Most Florida-grown foliage is produced in wood post and Saran structures such as these being used to grow spathiphyllum. Note the sprinkler system for irrigation.

For maximum yield, maintain temperatures of 65F (18C) minimum and 95F (35C) maximum. Maintaining the minimum temperature can be expensive during winter months. If heat conservation is a problem, it is better to keep soil temperatures at 65F minimum and allow the air temperature to drop slightly lower.

Producing foliage stock plants in greenhouses is limited only by economics. Bench space in northern greenhouses costs between $12 and $18 a year per square foot ($129 and $194 m²) while in warmer areas costs may range between $3 and $10. Plant size in relation to yield of cuttings per square foot, as well as the need of specific crops to be grown under cover for pest protection, governs the decision for stock plant production. Only a few fast-growing foliage plants yield sufficient cuttings to make them profitable where heating costs are high. Therefore, before establishing a stock production area, compare purchase costs per cutting versus costs expected from a stock production area. Valuable, costly space devoted to stock plant production may be better used for potted plant production and propagation of purchased cuttings. Many growers use cuttings from plants being grown for sale (for example, *Epipremnum* and *Hedera*). These plants are sometimes trimmed one or two times to increase quality, and the cuttings can be used as propagation units for the next crop.

PROPAGATION

Foliage plants can be propagated by cuttings, seed, air layers, spores, division, and tissue culture.

Cuttings, one of the most popular propagation methods, can be tip, single- and double-eye leaf bud, leaf, or cane. Selecting a specific method depends on plant form (such as upright or vining) and availability of propagative material. The intended market date also matters since some types of propagation take longer than others. Numerous media have been used for foliage plant propagation. Sphagnum peat moss is the most common, either alone or amended with perlite, polystyrene foam, pine bark, or other organic components.

SEED

Propagation by seed is increasing in popularity because costs are lower than for vegetative propagation. However, seeds of many foliage plants are not available, and plant types are not stable from all kinds of seed. Some of the more popular foliage plants grown from seed include *Araucaria, Brassaia, Dizygotheca, Podocarpus,* and nearly all of the palms. Plant the seed of tropical foliage plants soon after harvest because germination percentage decreases rapidly with increased time between harvest and planting.

Vegetative

Air layering is decreasing in importance as a propagation method because of high labor costs and the need for large stock plant areas. Plants most commonly air-layered include *Brassaia* and *Ficus.* One of the problems with air layers is that their large size makes the propagative units difficult to ship without mechanical damage.

Division is the only method of propagation for a crop, such as *Sansevieria.* This is a labor-intensive method, however, and it presents problems of carrying disease, insect, or nematode pests to new plantings.

Spores are used to propagate some fern genera, although most ferns are grown from tissue culture, divisions, or offsets.

Tissue culture is an important system of propagation for foliage producers. Rapidly multiplying new varieties is an important advantage of tissue culture, but some old varieties, such as Boston fern, are also commonly propagated by this system. Two major advantages of tissue culture are improvement in plant form and disease control. Genera such as *Anthurium, Dieffenbachia, Nephrolepis, Spathiphyllum,* and *Syngonium* produce multiple crown breaks when grown from tissue culture and yield plants with more compact form and fuller appearance. Tissue culture successfully reduces disease problems with several genera since disease-free stocks can often be maintained when grown in enclosed greenhouses. Two major foliage crops, *Dieffenbachia* and *Spathiphyllum,* are grown from disease-free stock and yield high-quality potted plants.

In the past the usual propagation system was a mist bed where cuttings were misted for 15 to 30 seconds every 30 to 60 minutes. Cuttings were stuck in the bed and then rooted, pulled, and potted. During the past 10 years, many producers have shifted to direct-stick propagation, in which cuttings are placed directly in the growing pot, rooted, and finished without being moved. This system is especially adapted to vining plants; 12 or more cuttings may be placed in each pot. The frequency of misting depends on light intensity and temperature, and it should be set to keep some moisture on foliage at most times. In cooler climates, growers often use tents over the propagation bench to provide 100% humidity without misting. Mist is very useful in lowering temperatures in summer months, but it can prolong rooting during periods when the growing media temperature drops below 65F (18C). For this reason, the media in propagation beds or benches should be heated, if necessary, to maintain 70 to 75F (21 to 24C) at all times.

Cultural factors

The needs of foliage plants vary from one type to the next, and developing a detailed crop production guide here is not practical. However, by understanding the basic cultural factors influencing most foliage crops, you can have a strong foundation for making logical decisions.

POTTING AND SPACING

The potting media for growing foliage plants can range from 100% organic to approximately 50% organic and 50% inorganic. Key factors to consider in selecting potting media include aeration (measured as capillary and noncapillary pore space), moisture retention (water-holding capacity), and nutrient retention (cation exchange capacity). Also consider their consistency, availability, weight, and cost. Examples of components in commercial foliage media mixes, along with some of the commonly used mixtures, are shown in table 3. Commercially prepared media sold premixed by suppliers are the most common choices.

Normally, pH is adjusted at the time the potting media are developed. The best range for most foliage plants is between 4.5 and 6.5, but several genera, including *Maranta* and most ferns, grow best with a range between 4.5 and 5.5. Dolomite is suggested to correct pH, but any calcium-containing material can be used. When growing fluoride-sensitive crops, the best pH range is 6 to 6.5. Do not incorporate superphosphate into your potting media unless they are intended for foliage plants not sensitive to fluoride. Micronutrients are normally included in the fertilizer program, although they may also be included in media.

Potting systems fall into two main categories: hand potting and automatic potting machines. Potting machines are used by medium and larger producers because

TABLE 3   FOLIAGE PLANT MEDIA CHARACTERISTICS

Medium	Aeration	Water-holding capacity	Cation exchange capacity	Weight
Composted pine bark	H	M	M	M
Perlite	H	L	L	L
Polystyrene foam	H	L	L	L
Sand	M	L	L	H
Sphagnum peat moss	M (V)	H	H	L
Vermiculite	M	M	M	L
Peat:bark (1:1)	H	H	H	L
Peat:bark:polystyrene foam (2:1:1)	H	M	H	L
Peat:bark:vermiculite (2:1:1)	H	H	H	L
Peat:perlite (2:1)	H	M	M	L
Peat:sand (4:1)	M	H	H	H

Note: H, high; M, average; L, low; and V, variable.

of speed, uniformity, and cost. Hand potting is still used by producers with small operations. Systems vary, but usually the growing media are delivered to a central site for potting. Potted plants are then moved to growing areas. While most pot-filling machines are stationary, pot fillers that are smaller and portable are also used.

The spacing of foliage plants directly controls final plant quality. Crowding plants reduces light reaching the lower foliage and may cause it to abscise, or it may cause plants to grow tall without proportionate spread, which reduces value. Plants that will be finished within three months are usually placed at their final spacing when set on the bench or ground; spacing distance varies from zero (pot to pot) to three times pot diameter. Plants grown in containers 6 inches (15 cm) or larger can take six months to two years to reach maturity. Such plants are often placed pot to pot until they become crowded. They are then moved to their final spacing, which, depending on plant form, varies from one to six times the container diameter.

Growing environment

Light intensity is one of the most important factors to consider because it influences internode length, foliage color, carbohydrate level, growth rate, and acclimatization. Production light levels have been included in table 4 for many foliage crops. Light green foliage or faded colors, for example in Codiaeum, are indicative of excessive light and reduced chlorophyll levels. This can be corrected, in most cases, by increasing fertilization or reducing light intensity. However, increases in fertilizer often cause excessive soluble salts levels and may not be environmentally acceptable.

Fertilization directly influences growth rate and thus profitability. Because fertilizer levels also influence longevity at the retail level and for the consumer, it is important not to use excessive amounts. Maximum growth rate of acclimatized foliage plants can be obtained with moderate levels of soluble, organic, or slow-release fertilizers applied constantly or periodically. Suggested levels of fertilizer (at recommended light intensities) in table 4 are for production of high-quality acclimatized plants.

Fertilizer ratios for foliage plants need to be approximately 3:1:2 ($N-P_2O_5-K_2O$) when the growing media listed in table 3 are utilized. A ratio of 1:1:1 is also acceptable, and it is in fact desired for some flowering foliage plants. Because nitrogen is the key element in the growth of foliage plants, however, this ratio will result is higher fertilizer costs to obtain the desired nitrogen level, and it is more likely to result in groundwater contamination due to excess phosphorus and potassium. Rates listed can be calculated periodically and applied weekly or every other week. Fertilizing less frequently than every two weeks often results in reduced growth rate or quality. For constant-feed programs, use 150 ppm N, 25 ppm P, and 100 ppm K at each irrigation. Potting media used for foliage plant production (table 3) are normally very low in micronutrients; thus, most fertilizers should contain at least the minimal micronutrient levels suggested in table 5.

TABLE 4   LIGHT AND NUTRITIONAL LEVELS FOR PRODUCTION OF SOME POTTED, ACCLIMATIZED FOLIAGE PLANTS

	Light intensity (f.c.)[a]	Fertilizer requirement[b] (lbs per 1,000 ft^2 per year)[c]		
		N	P$_2$O$_5$	K$_2$O
Aeschynanthus pulcher	1,500–3,000	34	11	23
Aglaonema spp.	1,000–2,500	34	11	23
Anthurium spp.	1,000–2,000	20	20	20[d]
Aphelandra squarrosa	1,000–1,500	41	14	27
Araucaria heterophylla	4,000–8,000	34	11	23
Asparagus spp.	2,500–4,500	28	9	19
Brassaia spp.	3,000–5,000	48	16	32
Calathea spp.	1,000–2,000	34	11	23
Chamaedorea elegans	1,500–3,000	34	11	23
Chamaedorea erumpens	3,000–6,000	41	14	27
Chlorophytum comosum	1,000–2,500	34	11	23
Chrysalidocarpus lutescens	4,000–6,000	41	14	27
Cissus rhombifolia	1,500–2,500	41	14	27
Codiaeum variegatum	3,000–8,000	48	16	32
Cordyline terminalis	1,500–3,500	34	11	23
Dizygotheca elegantissima	2,000–4,000	34	11	23
Dieffenbachia spp.	1,500–2,500	34	11	23
Dracaena deremensis (cultivars)	2,000–4,000	34	11	23
Dracaena fragrans (cultivars)	2,000–4,000	34	11	23
Dracaena marginata	3,000–6,000	48	16	32
Dracaena (other species)	1,500–3,500	41	14	27
Epipremnum aureum	1,500–3,000	34	11	23
Ficus benjamina	4,000–6,000	48	16	32
Ficus elastica (cultivars)	4,000–8,000	48	16	32
Ficus lyrata	2,000–6,000	48	16	32
Fittonia verschaffeltii	1,000–2,500	28	9	19
Hedera helix	1,500–2,500	34	11	23
Hoya carnosa	1,500–3,000	34	11	23
Maranta spp.	1,000–3,500	28	9	19
Monstera deliciosa	2,000–4,000	41	14	27
Nephrolepis exaltata	1,500–3,000	34	11	23
Peperomia spp.	1,500–3,000	28	9	19
Philodendron scandens oxycardium	1,500–3,000	34	11	23
Philodendron selloum	3,000–6,000	41	14	27
Philodendron spp.	1,500–3,500	34	11	23
Pilea spp.	1,500–2,500	28	9	19
Polyscias spp.	1,500–4,500	41	14	27
Sansevieria spp.	1,500–6,000	20	7	13
Schefflera arboricola	1,500–3,000	34	11	23
Spathiphyllum spp.	1,500–2,500	34	11	23
Syngonium podophyllum	1,500–3,000	34	11	23
Yucca elephantipes	3,000–5,000	41	14	27

[a] To convert footcandles to lux, multiply by 10.76. For example, 1,500 f.c. × 10.76 = 16,140 lux, or 16.14 klux.

[b] Based on a 3:1:2-ratio fertilizer source. If the growing medium is known to fix phosphorous and potassium, they should be added at the same rate as nitrogen (i.e., use a 1:1:1 fertilizer source).

[c] One lb/ft^2 = 4.882 kg/m^2.

[d] Based on a 1:1:1-ratio fertilizer source (Conover & Henny 1995).

TABLE 5   MICRONUTRIENT LEVELS FOR FOLIAGE CROPS

Element	Spray application (lb per 100 gal)[a]	Soil drench (oz per 1,000 ft²)[b]	Soil incorporated (oz per yd³)[c]
B	0.01	0.03	0.010
Cu	0.10	0.30	0.100
Fe	1.00	3.00	1.000
Mn	0.50	1.50	0.500
Mo	0.01	0.01	0.001
Zn	0.30	1.00	0.300

[a] One oz of a material in 100 gal is approximately 75 ppm (growers' common rule of thumb). One lb per 100 gal would be about 1,200 ppm.

[b] One oz/ft² = 28.35 g per 0.0929 m² = 305.16 g/m². Therefore, 1 oz per 1,000 ft² = 305.16 kg per 1,000 m².

[c] One oz/yd³ = 28.35 g per 0.7646 m³ = 37.08 g/m³.

Notes: Average levels required for many crops; individual requirements may differ. One application is often enough for short-term crops, whereas reapplications are usually necessary for crops grown for six months or longer.

Acclimatizing foliage plants for interior use is necessary to ensure that they perform well indoors. Acclimatization is the adaptation of a plant to a new environment, such as preparing it to grow in a building with low light and humidity. One of the most important aspects of acclimatization is to develop shade foliage, which is characterized by large, thin leaves with high chlorophyll levels. A second important factor is the nutritional level, which should be as low as possible while still producing a quality plant. The major objective of acclimatization is to produce plants with low-light compensation points. Such plants are able to make the transition from production to interior environments without serious quality loss. The most highly acclimatized plants are those grown under recommended light and fertilizer programs for their entire production cycle (table 4), leaves of all ages being acclimatized. Also, plants grown under recommended light levels for full term have a more open appearance, an adaptation to reduced light intensity.

Such plants as *Brassaia, Chrysalidocarpus,* and *Ficus* are often grown in full sun, then acclimatized by being placed under suggested shade and lowered fertilizer levels for three to 12 months. Although chloroplasts and grana are capable of reorienting within sun-grown foliage, the leaf anatomy—small size and thick cross-section—prevents them from being as efficient after acclimatization as shade leaves. Therefore, such plants are less tolerant of low or medium light levels than plants acclimatized during the entire production period.

Temperature control is very important since most foliage plants are tropical and require minimal night temperatures near 65F (18C). Soil temperatures are also important. If soil temperatures can be maintained at 65 to 70F (18 to 21C), the air temperature may drop as low as 60F (16C) at night without significant loss in plant growth. The best temperature ranges for production of a wide variety of foliage plants is 65 to 80F (18 to 27C) at night and 75 to 95F (24 to 35C) during the day.

Irrigate plants enough so they remain turgid at all times. Watering during winter, when temperatures are low and growth rate is slowed, may be required less than once a week; in spring or summer, daily irrigation may be necessary. Irrigation systems that deliver water to the growing media without wetting foliage are desirable because this reduces foliar disease and residue problems. Some efficient watering systems include ebb-and-flow irrigation, capillary mats, troughs, and leader tubes.

Carbon dioxide application to foliage plants is uncommon, although plants benefit, showing up to a 25% increase in growth. However, it appears that temperature is the key to increased growth during periods when greenhouses are closed. Therefore, unless a range of 65 to 75F (18 to 24C) minimum is maintained, injecting CO_2 will probably not be beneficial.

Humidity requirements of foliage plants during production are not verified by research, but maintenance of 50% or higher relative humidity appears to be desirable. In areas where humidity falls below 25%, install mist lines or raise the humidity some other way.

PESTS AND DISEASES

INSECTS AND MITES

Factors affecting pest populations include their access to the structure, temperature, humidity, irrigation method, and growing media. When a particular pest, such as scale, becomes a problem, check cultural procedures carefully to see if stock plants are infested and if crawlers or adults have been carried through propagation to potted plant production areas.

Mites present the most problems in spring, summer, and fall when temperatures are high, while caterpillars and thrips are heaviest in spring and fall. Temperatures above 80F (27C) are conducive to rapid increases in mite populations, especially if humidity is low. Pest problems in unheated production areas are reduced by cooler temperatures, while in climate-controlled greenhouses pests present a year-round problem. Fungus gnats are more of a problem in greenhouses than outdoors, especially when organic growing media are kept too wet.

It is imperative to keep stock plants as free of pests as possible to decrease the need for spraying of potted crops for sale. Frequent, continued spraying is not only expensive, but it increases potential for phytotoxicity and causes hard-to-remove residues on foliage.

Control. Biological pest controls, such as predators, diseases, and nematodes, are becoming more important in production of quality foliage plants. The balance needed between pesticides and beneficial predators is important—many pesticides will kill the predators as well as the pests. Research has yet to provide a complete biological control program. However, at least 5% of foliage plant growers use biological control agents in some portion of their production, and this is increasing as more sources become available for purchasing them. This trend will continue as more and more pesticides are lost due to lack of labeling for foliage.

Insect and mite control outdoors and under shade cloth, where pest movement is unrestricted, requires spraying or drenching with pesticides as the primary method of control. Using high-pressure sprayers to carefully apply spray to both sides of foliage provides the best control. Air-blast sprayers, while providing high pressure and wide dispersal, often do not properly coat both sides of foliage with pesticides.

Insect and mite control in greenhouses is somewhat easier, since producers can use many of the chemicals used outdoors or under shade structures as well as indoor smoke bombs or thermal fogs and biological control agents. However, smoke bombs or thermal fogs are only effective when greenhouses can be entirely closed for several hours or more.

Table 6 lists the major pests of foliage plants and some of the hosts upon which they most commonly feed. Chemicals registered for control of specific pests are subject to change. For up-to-date recommendations on chemical pest control procedures on foliage plants, check with your cooperative agricultural extension agent.

TABLE 6 FOLIAGE PLANT PESTS

Pest	Hosts
Aphids	*Aphelandra, Brassaia, Hoya, Dieffenbachia, Schefflera*
Broad mites	*Hedera, Aphelandra, Pilea*
Caterpillars	*Philodendron, Dracaena, Brassaia, Maranta, Aglaonema,* ferns
Fungus gnats	*Schlumbergera,* palms, *Peperomia*
Mealybugs	*Aphelandra, Ardisia, Dieffenbachia, Asparagus, Maranta, Dracaena, Dizygotheca, Hoya*
Scales	*Aphelandra,* bromeliads, *Ficus,* palms
Spider mites	*Brassaia, Codiaeum,* palms, *Cordyline, Calathea, Dieffenbachia, Maranta*
Thrips	*Brassaia, Ficus, Philodendron, Ctenanthe, Syngonium*

FUNGI, BACTERIA

Fungal and bacterial diseases are more troublesome when wet foliage is combined with high temperatures and humidity. Therefore, these diseases are most prevalent in tropical and subtropical areas where rainfall is heavy. Even in these areas, however, growing plants under cover and irrigating without wetting the foliage will nearly prevent their occurrence. Soilborne fungal diseases become more severe with poor-quality growing media (without good aeration and drainage) or when plants are constantly overwatered. Several bacterial diseases become more severe when plants are grown with excessive nitrogen or are stressed because of high soluble salts, excessive temperature, or high light intensity. Virus diseases are disseminated by insects, but since some foliage plants already contain at least one known virus, insect exclusion may not remedy a particular situation.

Control. Disease control outdoors and under shade cloth in areas with heavy rainfall is difficult to achieve except with preventive spray programs. During periods of frequent rainfall, and depending on disease pressure, it may be necessary to spray weekly or more often; however, during dry seasons little, if any, pesticide application may be necessary. Irrigation applied to the soil, or overhead irrigation only during the middle of the day when rapid drying can occur, combined with proper plant spacing to increase airflow, thus receding humidity, will aid in reducing disease pressure. Directed high-pressure sprays, rather than air-blast sprayers, provide best control of disease pests and are usually worth the extra cost of application.

Greenhouse control of foliar fungi and bacteria is easiest if foliage can be kept dry. When this is impossible, fairly good control can be obtained with chemical sprays to the foliage. Soil drenches for control of most soilborne diseases are fairly successful in raised benches and in containers off the ground. No control procedures are presently recommended for plants with viruses except to rogue infected plants and use virus-free stock. Table 7 provides a listing of some of the major disease pests and primary hosts. As with insecticides and miticides, the most up-to-date microbe control information can be obtained from your cooperative extension agent.

DISTRIBUTION

The two main packaging systems are boxing and shipping loose in specially constructed racks and trucks. With the boxing system, the cartons must meet interstate shipping regulations concerning weight of corrugated cardboard, and they must be either waxed or moisture-resistant to prevent their deterioration in transit because of moisture in containers. Plants in containers up to 6 inches (15 cm) are usually placed in a waxed tray. The tray is slid into a box of the proper height. Such boxes usually do not contain dividers or other restraints, although some producers have designed boxes with features for holding the pots in position and the growing medium in the container. Plants in 6-inch or larger pots are usually sleeved and placed directly in cartons of the proper height or shipped loose in specially designed trucks. Some producers and

TABLE 7   FOLIAGE PLANT DISEASE ORGANISMS

Organism	Plant area affected	Common hosts
Fungal pathogen		
Alternaria	Leaves, stems	*Brassaia, Schefflera, Polyscias, Fatsia, Calathea*
Fusarium	Leaves	*Dracaena*
Fusarium	Stems, leaves, roots	*Dieffenbachia*
Myrothecium	Leaves, stems	*Aglaonema, Dieffenbachia, Spathiphyllum, Begonia, Syngonium, Episcia, Aphelandra, Peperomia, Aeschynanthus*
Drechslera, Bipolaris	Leaves	Palms, *Maranta, Calathea*
Colletotrichum	Leaves	*Dieffenbachia, Hedera*
Rhizoctonia	Leaves, stems, roots	Many plants, especially ferns, *Hedera, Philodendron, Epipremnum*
Pythium	Stems, roots	Many plants, especially ferns, *Epipremnum, Philodendron, Peperomia*
Phytophthora	Leaves, stems, roots	Many plants, especially ferns, *Epipremnum, Philodendron, Saintpaulia*
Sclerotium	Stems, roots	Most plants grown on the ground are susceptible
Corynespora	Leaves	*Ficus, Aeschynanthus, Aphelandra*
Cylindrocladium	Stems, leaves	*Spathiphyllum*
Bacterial pathogen		
Erwinia	Stems, leaves	*Dieffenbachia, Aglaonema, Philodendron, Syngonium, Dracaena, Saintpaulia,* and many others
Pseudomonas	Leaves, stems	*Caryota, Philodendron, Syngonium,* ferns, *Dracaena, Schefflera*
Xanthomonas	Leaves	*Philodendron, Dieffenbachia, Hedera, Aglaonema*

shippers use adjustable racks that are wheeled directly into trucks. These racks have been designed to accommodate various-sized plants. Usually, no dividers or restraints are necessary with this system, and physical damage is minimal.

STORAGE

Storing foliage plants is not a normal practice, but with increased sales in mass market outlets that use central distribution points, there is increased interest in how long plants can be held without significant decrease in quality. Research has shown that many foliage plants can be stored and shipped for two to four weeks in darkness, provided they are acclimatized and they receive proper environmental conditions during storing and shipping. In actuality, storage and shipping are very similar in nature, and environmental conditions for both are similar. However, preparing plants for storage or shipping begins during the production period.

Preshipment environmental factors. Preshipment light levels have a strong effect on postshipment foliage plant quality. Plants grown with high light will not ship well because they are not properly acclimatized. High-light production results in excessive leaf drop or other quality loss either during shipment or after placement in an interior environment. Also, plants grown in higher light are not as likely to tolerate low shipping temperatures without sustaining chilling injury.

Preshipment fertilizer levels affect acclimatization and, in turn, subsequent interior quality. Foliage plants grown with higher than recommended nutritional regimes will not be well acclimatized and will ultimately lose more leaves and be of lower quality than those properly acclimatized. Serious reduction in plant quality during shipping, however, will be noticeable only if the fertilizer level is more than twice the recommended level.

Foliage plants grown during high-light and high-temperature periods are less tolerant of the shipping environment and are more likely to decline in quality during shipment. This may be due to lower levels of acclimatization or inability to tolerate lowered shipping temperatures. It appears that plants grown in summer and shipped then or in early fall require warmer shipping temperatures than plants grown and shipped in winter or spring.

Shipping environmental factors. Although foliage plants would tolerate shipping for longer periods if light could be supplied, this is not possible with present equipment. Therefore, all research has been conducted on plants shipped or stored in darkness.

Controlling the temperature at a level specified for the crop can be a major factor in maintaining plant quality. Shipping at low temperatures, 45 to 60F (7 to 16C) versus 65 to 70F (18 to 21C), can be beneficial in maintaining the quality of some plants, provided there is no chilling damage. However, the best shipping temperature for specific plants changes with the season, being lower in winter and higher in summer. Exposure duration to a specific temperature also strongly affects plant quality; a temperature that might be optimum for a two-week shipment might cause damage when plants are exposed for three to four weeks (table 8).

Plants should have adequate media moisture at the time of shipping—not too wet and not too dry. Plants shipped wet or with saturated media often have increased leaf drop, whereas low soil moisture during shipping periods increases plant tissue desiccation and raises the potential for leaf drop or injury. In general, water plants the day before they are to be packaged and shipped so pots can drain and foliage can dry.

TABLE 8   SHIPPING AND STORAGE TEMPERATURES FOR ACCLIMATIZED FOLIAGE PLANTS

Plant	1–14 days[a]	15–28 days
Aglaonema Silver Queen	60–65F	60–65F
Aphelandra squarrosa	55–60	55–60[b]
Araucaria heterophylla	50–65	50–65
Aspidistra elatior	50–55	50–55
Asplenium nidus	50–65	50–65
Beucarnea recurvata	55–60	55–60
Brassaia actinophylla	50–55	50–55
Cereus peruvianus	55–60	55–60
Chamaedorea seifrizii	55–60	55–60
Chrysalidocarpus lutescens	55–65	60–65[c]
Codiaeum variegatum Norma	60–65	60–65
Cordyline terminalis Baby Doll	55–60	50–55[c]
Crassula argentea	50–65	50–65
Dieffenbachia Tropic Snow	55–65	55–65[c]
Dizygotheca elegantissima	55–60	55–60
Dracaena deremensis Janet Craig	60–65	n.a.[d]
Dracaena fragrans Massangeana	60–65	60–65
Dracaena marginata	55–65	60–65[c]
Epipremnum aureum	55–60	55–60[b]
Ficus benjamina	55–60	55–60
Ficus elastica Burgundy	50–60	50–55
Ficus lyrata	55–60	55–60
Hedera helix Eva	50–60	50–55
Howea forsterana	50–65	50–65
Hoya carnosa Tricolor	55–65	55–65
Maranta leuconeura	50–55	50–55[b]
Nephrolepis exaltata Bostoniensis	55–60	55–60[b]
Philodendron scandens oxycardium	55–60	55–60[b]
Philodendron selloum	55–60	55–60
Phoenix roebelenii	55–60	55–60
Pilea Silver Tree	55–60	55–60[b]
Plectranthus australis	55–60	55–60[b]
Schefflera arboricola	50–55	50–55
Spathiphyllum Mauna Loa	50–55	55–60
Syngonium White Butterfly	55–60	55–60[b]
Yucca elephantipes	50–55	50–55

[a] Plants shipped or stored for one to seven days should be held at the higher end of their temperature range.

[b] Observed to have severe loss in quality beyond two weeks.

[c] Observed to have a loss in quality of about 25% per week beyond two weeks.

[d] Data not available.

Note: Temperature equivalents: 50F = 10C, 55F = 13C, 60F = 16C, 65F = 18C.

Although all plants need to have adequate soil moisture during shipment, this will not prevent them from desiccating if humidity is too low. Relative humidity of 85 to 90% is necessary to maintain foliage plants stored or shipped for long durations. This level can be obtained by putting plants in boxes or by setting the air exchange controls on the shipping container to the closed position with sleeved or unboxed plants.

Research on foliage plants has shown that ethylene is probably the only contaminant that may occur in containers during shipping, unless there is pollution from an outside source. However, foliage plants require fairly high levels of ethylene (1 to 2 ppm) and relatively high temperatures (65F, 18C or higher) for long durations before any damage will occur. Experience has shown that ethylene does not seem to be a major problem when foliage plants are shipped at cooler (65F or lower) temperatures. In research on low-oxygen storage, no benefit was observed over foliage plants shipped in a normal atmosphere.

MARKET LEADERS

An even dozen major foliage plant genera or groups comprise the majority (60%) of the foliage plants sold in the United States. Industry dominance by a single plant, as was the case with philodendron in the mid-1950s, no longer exists. The plants and groups that follow are not listed alphabetically, but by decreasing market share of the industry.

DRACAENA

There are more than 40 species of *Dracaena* (*Dracaena deremensis* Janet Craig shown in fig. 3) native to tropical regions of Asia and Africa and almost twice that number of selected cultivars. The most commonly sold species include *D. deremensis, D. fragrans, D. marginata, D. reflexa, D. sanderana,* and *D. surculosa.* The popularity of this genus is related to its tolerance of interior environments, its range of sizes from small plants in 4-inch (10-cm) pots to trees in 200-gallon (757-l) containers, and its diversity of forms and coloration.

Most commonly grown dracaenas are propagated by tip cuttings or cane (*Dracaena*

FIG. 3   A fine crop of *Dracaena deremensis* Janet Craig.

fragrans Massangeana cane shown in fig. 4). Best acclimatized plants are grown under light levels of 3,000 to 6,000 footcandles (32 to 65 klux) with moderate nutrition. Disease pests of major importance include *Erwinia, Fusarium,* and *Phyllosticta,* while major insect pests include mites, mealybugs, and thrips.

EPIPREMNUM (POTHOS)

About 10 species, all native to southeast Asia, exist within this genus. Popularity of this group has increased considerably within recent times. The most commonly sold cultivar is *E. aureum* Golden Pothos, although not all Golden Pothos look the same, because of selection of "high color" strains by nurserymen. Other significant cultivars of *E. aureum* include Jade and Marble Queen.

FIG. 4   *Dracaena fragrans* Massangeana cane is ready to be rooted.

Essentially, all pothos are propagated from single-eye cuttings. Since bright coloration is very important with this crop, select cuttings from stock plants with the desired coloration. The best-quality plants are grown under 2,000 to 4,000 f.c. (22 to 43 klux) (lower light levels often reduce coloration) with moderate fertilizer levels. Disease pests of importance include *Erwinia, Pythium,* and *Rhizoctonia,* while major insect pests include mealybugs, root mealybugs, and thrips.

FICUS

Nearly 800 species of *Ficus* exist, including trees, shrubs, and even woody vines native to tropical areas. Although the potential for finding new species of *Ficus* is good, mainly four species—*F. benjamina, F. elastica, F. lyrata,* and *F. retusa,* with their dozens of cultivars—are most often utilized in the industry. Many Ficus are utilized as indoor trees in large plantings for their large size (10 to 30 feet, 3 to 9 m), but they are also suitable for the home in sizes from 3 to 8 feet (0.9 to 2.4 m) tall. Smaller specimens can even be very attractive. (Figure 5 shows a ficus braided tree.)

Most kinds of *Ficus* are produced from cuttings or air-layers, but *F. lyrata* is often grown from tissue culture, which yields an improved plant form. *Ficus* can be grown successfully under 4,000 to 6,000 f.c. (43 to 65 klux) or in full sun and then acclimatized under the recommended acclimatization light levels for four to 12 months; the larger the tree, the longer the acclimatization period required. Nutritional regime is moderate to high, but it should be reduced prior to sale. Disease pests of importance include *Glomerella, Botrytis, Cercospora,* and *Corynespora,* while major insect pests include scale, mealybugs, and whitefly.

DIEFFENBACHIA

About 30 species of this genus, native to Central and South America, have been identified, but most plants sold in the industry are *D. maculata, D. amoena*, and hybrids between species. Some of the most important cultivars include *D. maculata* Camille, *D. maculata* Exotica Compacta, and *D. amoena* Tropic Snow. Some new introductions include *D.* Paradise, *D.* Triumph, *D.* Victory. Sales of dieffenbachias have increased with the introduction of cultivars with new colors and forms.

Dieffenbachias are propagated mainly by cuttings and tissue culture. Best color is obtained when grown under 1,500 to 3,000 f.c. (16 to 32 klux) at moderate nutritional levels. Low light intensities or low nutrition will reduce basal breaks. There are many disease pests of significance, with *Erwinia* and *Pythium* among the most important, while major insect pests include mites and mealybugs.

FIG. 5 *Ficus benjamina* grown as a braided tree.

PALMS

The importance of palms has decreased slightly in recent years, even though the number of genera used in the industry has increased. Overall, there are hundreds of palm species, with nearly 25 common in the industry. Some of the best palms for interior use include *Caryota mitis, Chamaedorea* spp., *Chrysalidocarpus lutescens, Howea forsterana, Phoenix roebelenii,* and *Rhapis excelsa*. Many other good palm species are also available.

Palms are propagated from seed, which should be planted as soon after harvest as possible and germinated at 75 to 80F (24 to 27C). Different palm genera require varying light intensities for growth, but a range of 3,000 to 6,000 f.c. (32 to 65 klux) is best for most species when grown on moderate nutrition. Numerous diseases affect palms, with *Colletotrichum, Phytophthora,* and *Helminthosporium* among the more important, while major insect pests include scale, mites, and mealybugs.

AGLAONEMA

Approximately 30 species of this hardy foliage plant are known from the wild, but only a few species and cultivars are commonly grown. *Aglaonema* Silver Queen is grown in greatest volume, while other good cultivars and species include *A.* Maria, *A.*

Romana, *A. commutatum* Emerald Beauty, and *A. crispum.* These are excellent plants for low-light interiors, and they have increased in popularity during the past 10 years.

Most aglaonemas are grown from cuttings or root clumps, although some are from tissue culture. Aglaonemas grow best in light levels of 1,500 to 2,500 f.c. (16 to 27 klux) and with moderate nutrition. Leaves oriented vertically or bleached leaves indicate excessive light. Disease pests of importance include *Xanthomonas, Erwinia, Pseudomonas,* and *Pythium*; the most serious insect pests are scale and mealybugs.

SPATHIPHYLLUM

This genus contains about 35 tropical species, but the industry sells mostly hybrids. Some of the most commonly available hybrids include *S.* Lynise, *S.* Mauna Loa, *S.* Petite, *S.* Starlight, *S.* Tasson, and *S.* Viscount, which include both small and large cultivars suitable for 6- to 17-inch (15- to 43-cm) containers. This low-light-tolerant plant has had a considerable increase in popularity during the past 10 years.

Spathiphullum is now usually grown from tissue culture, although a few less-important species are seed grown. These plants require low light levels (about 1,500 to 2,500 f.c., 16 to 27 klux) and moderate nutrition to achieve best quality. Generally, spathiphyllums are easy to grow except for the disease *Cylindrocladium,* a severe root disease; other diseases of importance include *Pseudomonas* and *Phytophthora.* Major insect pests include mealybugs and scale.

HEDERA (IVY)

Although ivy has been used as a foliage plant for years, its popularity has increased in recent times. *Hedera* contains about 15 species native to southern Europe, Asia, and North Africa, but there are several hundred named cultivars within the single species *helix.* Some popular cultivars include *H.h.* California, *H.h.* Curly, *H.h.* Glacier, *H.h.* Gold Dust, *H.h.* Needlepoint, and *H.h.* Sweetheart. The popularity of ivy is related to its tolerance of cool locations and the wide diversity of leaf colors and shapes.

Ivy is propagated by cuttings and grown in cool greenhouses. Although it will tolerate high light, ivy grows best at 1,000 to 3,000 f.c. (11 to 32 klux) and at low to moderate fertilizer levels. Ivy grows poorly at temperatures above 85F (29C) but will tolerate night temperatures of 55 to 60F (13 to 16C) without reduction in growth. Major disease pests include *Xanthomonas, Pythium,* and *Rhizoctonia,* while insect pests of importance include broad mites, spider mites, and scale.

PHILODENDRON

Philodendron scandens oxycardium was once the major plant of the foliage industry, but it now accounts for only 3% of the industry's sales, versus 34% in 1956. There are almost 200 species native to tropical America, but, except for the species just named and *P. selloum,* mostly hybrids are sold in the industry. Most of these hybrids—*P.* Black Cardinal, *P.* Emerald Duke, *P.* Majesty, *P.* Red Princess, *P.* Royal

Queen, and others—were bred by Robert McColley, who was responsible for all the major breeding work on *Philodendron.* Many of the hybrids are "self-heading," meaning they stand up without support. Those with a vining habit, however, require a pole or slab for support or are sold as hanging baskets or in small pots or dish gardens.

Propagation is by cuttings and tissue culture, although a few species are air-layered. Best light intensities for production are from 2,500 to 3,500 f.c. (27 to 38 klux), depending on cultivar. Fertilization levels in the moderate range provide best growth without excessive softness. Numerous diseases attack *Philodendron,* and the bacterial diseases—*Erwinia, Pseudomonas,* and *Xanthomonas*—are the most troublesome. Insect pests include mealybugs, mites, scales, and thrips.

SCHEFFLERA AND BRASSAIA

These genera are often confused since the names have been interchanged, but they are closely related. *Brassaia* (commonly called schefflera), a native to Asia, is usually sold in larger sizes for its dramatic leaf form, although plants in sizes as small as 6-inch (15-cm) pots are also available. *Schefflera arboricola* (commonly called arboricola) is somewhat like *Brassaia,* but it is smaller, and the leaflets making up the compound leaves are much shorter. *B. actinophulla* and the cultivar Amate are the only plants available in this genus, while there are many *S. arboricola* cultivars, including *S. a.* Covette, *S. a.* Gold Capella, and *S. a.* Trinette. The *S. arboricola* cultivars are especially tolerant of low-light interior locations.

Brassaia actinophylla is propagated from seed, while *B. a.* Amate is produced from tissue culture. Propagation of *Schefflera arboricola* is by seed or cuttings, while the named cultivars are by cuttage or tissue culture. (Figure 6 shows schefflera cuttings being harvested.) *Brassaia* can be grown in full sun, but a light level of 4,000 to 6,000 f.c. (43 to 65 klux) is better, while *Schefflera* is grown best with 3,000 to 5,000 f.c. (32 to 54 klux). Both genera grow best under reduced light at moderate nutritional levels. Major disease pests include *Cercospora, Pythium,* and *Rhizoctonia,* and insect pests of importance include mites, scales, and thrips.

SYNGONIUM

About 20 species of *Syngonium* are native to South America, but only one, *S. podophyllum* and its cultivars, is grown in the industry. This

FIG. 6 Variegated schefflera cuttings are being harvested in El Salvador.

vining plant is excellent in hanging baskets and is also sold in smaller pot sizes (4 to 6 inches, 10 to 15 cm). The most important *Syngonium* cultivar is White Butterfly, but new cultivars appear all the time and include Jenny, Lemon Lime, May Red, Robust, and Pink Allusion, as well as others.

Propagation of *Syngonium* is almost exclusively by tissue culture because it produces the types of plant desired by consumers. Excellent plants can be produced under light levels of 2,000 to 3,500 f.c. (22 to 38 klux) with low to moderate nutrition. Because of several disease pests, it is imperative that sanitary practices be maintained for this plant. The major disease pest is *Xanthomonas,* although *Pseudomonas* and *Pythium* may also present problems. Insect pests of importance include mealybugs and thrips.

FERNS

The majority of ferns grown in the foliage industry are cultivars of *Nephrolepis exaltata*. A small number of ferns from the genera of *Asplenium, Davallia, Pteris,* and *Sphaeropteris* are also grown. *Nephrolepis* is considered a sword fern; Bostoniensis is the oldest commonly grown cultivar. Other important cultivars include Bostoniensis Compacta, Dallas, Florida Ruffle, and Maasii. Ferns are sold mainly in hanging baskets, but 4- to 8-inch (10- to 20-cm) pots are also used.

Most *Nephrolepis* ferns are propagated from tissue culture, although runners or offsets are still utilized to some extent. Other ferns are still grown from spores but are also produced from tissue culture. The best-quality ferns are grown at 2,500 to 3,500 f.c. (27 to 38 klux) at moderate nutritional levels. The primary disease problem is *Rhizoctonia,* although *Cylindrocladium* may also occur. The major insect pests are caterpillars (worms) and mealybugs.

REFERENCES

Chase, A.R. 1987. *Compendium of ornamental foliage plant diseases.* St. Paul, Minn.: American Phytopathological Society. 92 pp.

Conover, C.A., and R.J. Henny. "N and K Rates Affect Anthurium Growth and Flowering." *Proc. Fla. State Hort. Soc.*

Conover, C.A., and R.T. Poole. 1984. "Acclimatization of Indoor Foliage Plants." *Horticultural Reviews* 6:119–154.

Conover, C.A., and R.T. Poole. 1990. "Light and Fertilizer Recommendations for Production of Acclimatized Potted Foliage Plants." *Nursery Digest* 24(10):34–36, 58–59.

Daughtrey, Margery, and A.R. Chase. 1992. *Ball field guide to diseases of greenhouse ornamentals.* Batavia, Ill.: Ball Publishing. 218 pp.

Joiner, J.N. 1981. *Foliage plant production.* Englewood Cliffs, N.J.: Prentice-Hall.

Powell, Charles C., and Richard K. Lindquist. 1997. *Ball pest & disease manual: Disease, insect, and mite control on flower and foliage crops.* 2nd. ed. Batavia, Ill.: Ball Publishing. 332 pp.

CHAPTER 22

BULB CROPS

by Vic Ball

Spring bulbs are especially valuable as flowering potted plants. Aesthetically and marketwise, they are colorful and delightful—the very essence of spring. People love them! They are a fresh breath of spring either during or at the end of a long winter. In the grower's year-round space plan, they offer a profitable crop to fill January and February benches, making a great Valentine's crop, both cut and pot. Bulbs are not really difficult to force—just follow the rules.

Plan well. Before you order, plan! Calculate how many tulips (or whatever) you need, the cultivars, pot sizes, bulbs per pot, flowering dates, and selling prices. Prepare a complete growing schedule. Use *The Holland Bulb Forcers Guide* by Gus De Hertogh. It is excellent! It is available from *GrowerTalks* Bookshelf: call 800/456-5380 or 630/208-9089 for details.

POTTED TULIPS: THE HEART OF IT

Bachman's, Minneapolis's leading floral retailer and a grower as well, relies heavily on potted bulbs for sales during winter through spring. Tulips in pots are a mainstay. Flowered from January through Easter, they are Number 1 in Bachman's spring bulb crop.

ORDERING AND SHIPPING

Bachman's does its annual ordering in February and March, mainly from Holland. "Watch currency exchange gyrations—I would rather order in dollars than in Dutch guilders. I know my costs," Todd Bachman advises.

Shipping needs attention. Bachman's uses mainly ocean freight. Shipping dates must relate to planting dates—again in Groups I (January to February flowering) and II (March to April flowering). Heating during transit delays flowering; also beware of high humidity. Order that temperature recording instruments be packed with the bulbs to be sure consistent temperatures have been maintained during transit.

Upon their arrival, open and ventilate the bulbs. Ethylene can build up and cause flower bud abortion (blasting). "We always check our bulbs to be sure everything is okay. Check for a sour smell as an indication of *Fusarium.* We store our bulbs upon arrival at 60 to 65F (16 to 18C)—with good air circulation and ventilation. Cut flower bulbs are immediately put into 55F (13C) storage to start the cooling process."

Check bulbs for flower bud development. Slice a few bulbs open. Look for complete flower parts, such as anthers. If they are present, the bulb has reached "G-Stage" and is ready to go. Do not plant bulbs until G-Stage is confirmed (*The Holland Bulb Forcers Guide*).

POTTING

Bachman's grows a number of tulip pot sizes: Three bulbs per 4^1/$_2$-inch (11 cm) pot, seven bulbs per 7-inch (18 cm) bulb pan, nine bulbs per 8-inch (20 cm) pan, and 12 or 15 bulbs per 10-inch (25 cm) pan. All pots must drain well!

Bachman's uses a media mix that's a bit heavier than its media for pot mums. The mix includes sandy loam and hypnum peat. With too light a mix, the bulbs' vigorous root systems tend to push them up out of the soil. Pots are filled up halfway, tulips set in (flat side to the outside of the pot), and the remaining medium is added to fill the pot. Do not press bulbs into media—that compacts the soil. There should be at least a quarter inch of soil over the bulb.

Potting machines can work well for bulb crops. Set the equipment to fill the pot half full on the first pass. Set the bulbs, then finish filling the pot on the second pass.

Label each bulb lot carefully. Include planting date, date to remove for forcing, flowering date, and supplier name. If you get two lots of the same variety, mark them "Lot A" and "Lot B." They may not flower simultaneously.

STORAGE

Temperature. When bulbs are first placed in the cooler, the temperature should be at 48F (9C). Within four to five weeks after potting, the bulbs should be rooted. Keep the soil moist. On November 10, drop the temperature to 41F (5C), a more ideal cooling temperature—for several weeks at this rooted stage.

Early January tulips. For a great after-poinsettia crop, tulips can be flowered nicely about January 5 to 15. To do this, the bulbs must be precooled. Immediately upon arrival, they need six weeks at 48F (9C). Bachman's uses a rooting room for this. Then root at 48F (9C) followed by 41F (5C). Move to the forcing house four weeks before the flowering date and after 15 continuous cold weeks.

Group I, the early ones. Group I bulbs should normally be potted and put in the storage room in mid-September. Pot varieties like Paul Richter will need 15 to 16 weeks in the cooler, plus three to four weeks to force—a total of 18 weeks from planting to flower.

Group I goes into Rooting Room 1—Group II goes into Rooting Room 2, so temperatures can be varied (table 1). Bachman's storage rooms have concrete floors, refrigeration, and the capability of drawing in cool outside air during winter to minimize costly refrigeration. Bulbs are stored on pallets, 50 pots per pallet, stacked pot on pot, 12 high. They are moved by forklift and arranged in the rooting room in proper sequence for easy removal for forcing. Be sure there is air circulation in the rooting room!

TABLE 1 ROOTING ROOM TEMPERATURE SEQUENCE FOR POTTED TULIPS

Rooting room	Seasonal varieties	Temperature and time
Room 1	Early to mid-season varieties flowering through Feb. 14	48F to Nov. 10, 41F to Jan. 5, 35F to finish
Room 2	Mid-season to late varieties flowering Feb. 14 to end of season	48F to Dec. 5, 41F to Jan. 5, 35F to finish

Group II, the later ones. Order Group II tulips to arrive about October 1. Pot them immediately and place in the Group II rooting room (table 1). They are kept at 48F (9C) until December 10, then dropped to 41F (5C) until they are ready for forcing. (Table 3 lists bulb-forcing schedules.)

On all pot tulips, as soon as root shoots are noticeable, Bachman's applies Banrot at 8 ounces per 100 gallons (236 ml/400 l)—as a soil drench. Termil can be burned as a *Botrytis* preventative before removing the pots from the rooting room.

Forcing. Tulips need a normally light house and reasonable humidity. It's really very simple—mainly 60F (16C), or 5F (–15C) more or less, to time the crop. Later spring crops flower faster; the earlier tulips will take three or four weeks to force. Moisture is important at this stage. Never let them dry. Watering in the morning minimizes *Botrytis*. Again, avoid humidity extremes. Fertilization, long felt to be unnecessary, is required to provide both calcium and nitrate, which tulips require in higher amounts than the base soil provides. One or two feedings of $CaNO_3$ at 140 ppm (140 mg/l) during forcing produces a better tulip.

Harvest. Potted tulips should go to the retail display when they are in the "green bud" stage. Bachman's likes slight color at the base of the flower. For maximum consumer life, tulips must be sold at the right time.

Varieties for pot forcing. Due to an increased demand in the marketplace for 4-inch (10 cm), 4½-inch (11 cm), and 5½-inch (14 cm) pot sizes for tulips, varieties fall into two groups, standard and short. Short varieties for January and February flowering include Christmas Marvel, Capri, Prominence, Kareol, Abra, and Flair. Tall varieties for the same period include Paul Richter (table 2), Merry Christmas, Golden Melody, Christmas Marvel, and Kees Nelis.

Post-Valentine's Day short varieties include Arma, Plaisir, and Red Riding Hood. Tall varieties for the same period: Attila, Yellow Present, Red Present, Princess Irene, and Couleur Cardinal.

Cut tulips

These flowers are rather important at Bachman's. The Valentine's Day cut tulip crop is especially big there. Says Todd, "Good cut flowers tend to be scarce for Valentine's Day. We rely heavily on cut tulips."

Table 2   Tulip schedules for Valentine's Day

Paul Richter red pot tulip schedule for forcing

October 4: Plant date
November 10: Strong visible root
Shoots developed: Apply Chipco 26019 spray
January 5: Drop temperature to 35F (2C)
January 18: Move to 60F (16C) forcing area
February 10: Bud stage, ready for sale!

Temperatures: Oct./Nov., 48F (9C); Dec./Jan., 41F (5C); Feb., 35F (2C)
Optimum cold: 15 weeks
Forcing time: 60F (16C) for 23 days

Paul Richter red cut tulip schedule

September 14: Plant date
October 28: Strong visible root, apply Banrot
Shoots developed: Apply Chipco 26019 spray
January 5: Drop temperature to 35F (2C)
January 18: Move to 60F (16C) forcing area
February 10: Bud stage, ready to cut!

Temperatures: Sept./Oct., 48F (9C); Nov./Dec., 41F (5C); Jan., 35F (2C); Feb., 60F (16C)
Optimum cold: 17 to 18 weeks
Forcing time: 60F (16C) for 23 days.

All are grown in 6-inch (15 cm) deep plastic curvers 12 by 18 by 3¹/₂ inches (31 cm by 46 cm by 9 cm). Bachman's plants 12 rows of nine bulbs per row—108 bulbs.

In general, the same procedures used for potted tulips are used with cut tulips. Cut tulips are harvested with slight color, a bit more color than pot tulips. Cut tulips can be stored two or three days at 33F (0.6C). Bachman's stores them horizontally. Some forcers pull tulips out bulb and all, but Todd feels Bachman's gets about the same amount of life by cutting them. Get them to the point of sale fast. Give the customer the joy of watching them open.

The main varieties for Valentine's Day at Bachman's are Paul Richter (table 2), Red Gander, and Trance. For late-flowering cut tulips, Bachman's uses Golden Melody, Attila, Oxford, and Golden Oxford. (Tulips are shown in figs. 1 and 2.)

FIG. 1   For millions of consumers the world over, spring is tulips! Be sure you have pots of forced tulips on hand to meet the demand for a breath of springtime in February and March.

FIG. 2   Fresh cut tulips are a springtime treat! This crop was at the Tulip Company, Terre Haute, Indiana.

CROCUS

Crocuses are bright and colorful in pots for sales during January and February. The principal varieties are Remembrance, Purpurea, and Flower Record.

A rough schedule used at Bachman's: For early flowering (January to February), bulbs are potted, five per 4¹/₂-inch (12 cm) pot, and put into a cooler on September 10. Forcing time is short, often under one week and can be done on carts in any lighted area (table 3).

For late spring (March to April), bulbs are potted and put in a cooler on October 10. Again, just a week or less is needed to force.

Bachman's moves the crocuses into the retail shop as soon as the buds' sheaths (not the individual buds) are visible. Again, the retailer's philosophy is to let the retail customer enjoy watching them develop. Crocuses do not have long lives.

POTTED HYACINTHS

The norm at Bachman's is one bulb per 4½-inch (12 cm) geranium pot or three bulbs per 5½-inch (14 cm) azalea pot, for the mass market. For retail shops, five bulbs go in each 7-inch (18 cm) pot.

For forcing hyacinths, Bachman's just takes them out of the cooler and puts them in a warm (60F), sunny area on carts, saving greenhouse space (table 3). Hyacinths will normally be ready to sell within a week after removal from the cooler. As with other bulbs, keep them watered and ventilated.

Anna Marie, Carnegie, Delft Blue, and Ostara Blue are the main varieties of hyacinths used at Bachman's. Sell hyacinths with just a little color showing, well before the florets open.

POTTED DAFFODILS

There are two main pot sizes at Bachman's for daffodils: three bulbs per 4½-inch (12 cm) pot for the mass market, four bulbs per 7-inch (18 cm) pot for retail shops.

For early flowering, daffodils at Bachman's are potted and put in the cooler on September 10. For later flowering, bulbs are potted and put into the cooler on October 10. Daffodils need almost the same time to force as tulips, normally three or four weeks. The temperature should be maintained at 60F (16C). Daffodils go to the retail shop when buds are vertical (the pencil stage)—before they tip. There should be a slight bit of color showing.

Carlton is the main daffodil variety for 6-inch (15 cm) pots. Jack Snipe and Tete-a-Tete make excellent 4½-inch (12 cm) pots. (Flats of bulbs are shown in fig. 3.)

FIG. 3   Bulbs as bedding plants—why not? These flats were at Ivy Acres, Calverton, New York.

TABLE 3   BULB-FORCING SCHEDULES AT A GLANCE

Crop and salable dates	Dates bulbs should arrive	Potting dates[a], temperature after potting	Temperature, Phase II	Weeks needed to force at 60F
Tulips, potted				
Jan. 5 to 15	Sept. 1	Pot on arrival (Sept. 1 provided you're at "G stage"). First 6 weeks at 48F.	After 6 weeks move to 41F.	4 weeks
Group I: Early, to flower Jan.–Feb.	Sept. 1	Pot Sept. 1 to Oct. 1. Then 48F until root formation, approx. 4 to 6 weeks, then to 41F.	For optimal cooling, drop to 41F on Nov. 10.	3 to 4 weeks for early crops
Group II: Later flowering, March–Easter	Oct. 1	Pot on arrival, put in 48F cooler.	Drop to 41F Dec. 10 until forced.	Faster than early crops
Tulips, cut[b]				
Valentine's Day	Sept. 1	Sept. 15		3 to 4 weeks
Easter	Oct. 1	Approx. 20 to 22 weeks prior to Easter depending on variety.		3 to 4 weeks
Crocus, potted				
Jan.–Feb. flowering	Sept. 1	Pot Sept. 10, move to cooler temperature (48F) 4 weeks, then to 41F.		1 to 2 weeks
March–April	Oct. 1	Pot Oct. 10, move to cooler temperature (48F) 4 weeks, then to 41F.		1 to 2 weeks
Hyacinth, potted				
Valentine's Day	Sept. 1	Sept. 15		1 week
Easter	Oct. 1	18 to 20 weeks prior to Easter		1 week

[a] The above dates assume no precooling. If bulbs have been precooled at 48F for six weeks they must obviously be planted six weeks after the dates that appear here. Precooling is used by growers who do not have refrigerated bulb storage.
[b] Store all cut tulips at 55F immediately upon arrival.

PLANT GROWTH REGULATORS

A-Rest (ancymidol), Bonzi (paclobutrazol), Sumagic (uniconizole), and Florel (ethephon), also marketed as Pistil, are four plant growth regulators that are EPA-approved for use on some flower bulbs. When using any of these, follow all label instructions and adhere to WPS reentry times. A-Rest and Bonzi can be used for tulips, while Florel and Pistil are used for hyacinths and daffodils.

Research has demonstrated that the composition and pH of the planting medium can influence growth regulator action when applied as a drench. The preferred planting medium for potted flowering bulbs should contain equal parts of sterilized soil, peat, and perlite or sand. If bark or peat is used, be certain the pH is near 7. Otherwise, the growth regulator requirement may need to be increased.

The moisture status of the planting medium at the time of application of the growth regulator is also critical. The planting medium must be thoroughly watered 12 to 24 hours prior to the soil drench application. A practical procedure is to thoroughly water the pots late in the afternoon, then apply the soil drench in the morning (12 to 24 hours later).

FORCING PROBLEMS

Too tall or too short tulips could be corrected with a few more or less cold days—more days in the cooler means taller tulips. Plant one week later if they are too tall. Also, of course, variety selection is important.

Bends or crinkles on tulips are often the result of overheating during the shipping of bulbs. More than 80 or 85F (27 to 29C) often causes this problem.

In hyacinth "splitting," the flower stalk separates from the plant. Most often this is the result of freezing of bulbs. This can happen after they are potted if they are outdoors and not covered well enough.

Editor's note: Many thanks to the people at Bachman's, Minneapolis, Minnesota, for sharing their bulb-forcing techniques and to Gus De Hertogh, North Carolina State University, for his input and review.

CHAPTER 23

DIF and Graphical Tracking

by Dr. Paul Fisher
University of New Hampshire, Durham

and Dr. Royal Heins
Michigan State University, East Lansing

Managing DIF, the difference between day and night temperatures, to control plant height is a common practice for many growers. Plants grown with a positive DIF (warm days and cool nights) will be taller than those grown with a negative DIF (cool days and warm nights) (fig. 1). This chapter describes DIF and how you can use it in your greenhouse to regulate stem elongation and final plant height.

Graphical tracking, a technique that helps you know whether your crops are growing too slowly or quickly to achieve your market specifications, is also used widely now. Graphical tracking means monitoring plant height over time and plotting the height on a graph that already shows a

FIG. 1   Easter lilies grown under positive or negative DIF conditions. The shorter plant at left was grown with negative DIF, and the taller plant at right with positive DIF.

target height curve. The tool helps you decide whether to apply growth retardants or change DIF temperatures, based on whether the plant height is developing above or below the target curve.

The increasing use of computers in floriculture brings exciting possibilities for delivering research to your fingertips in the form of computerized decision-support systems. The Greenhouse CARE System, the first of this kind of tool, is a computer program that helps with your crop recording and monitoring, automates graphical tracking, and suggests a second opinion on your height control decisions. We describe the Greenhouse CARE System in the last part of this chapter.

DIF (DAY/NIGHT TEMPERATURE)

DIF RESEARCH AND BIOLOGY

The DIF concept came out of research conducted in the 1980s by John Erwin, Meriam Karlsson, and Rob Berghage under the direction of Dr. Royal D. Heins at Michigan State University. These researchers learned that stem elongation is related to the relationship between day and night temperatures. Warm day and cool night temperatures (a positive DIF) promote stem elongation, whereas cool day and warm night temperatures (a negative DIF) limit stem elongation.

The term *DIF* simply means the arithmetic difference of day minus night temperatures. For example, a 70F (21C) day and a 60F (16C) night equals a positive DIF of 10F (5C), while a 60F day and a 70F night equals a negative DIF of 10F. Plants grown under a positive DIF are taller than those grown under a negative DIF. The more positive the value of DIF, the greater the stem elongation and ultimately plant height (fig. 2). You can select a DIF somewhat like you choose the best growth retardant rate—it is not all or nothing. Figure 2 shows the effect of DIF on elongation of poinsettia internodes. Note that moving from positive to zero DIF has a greater effect on reducing internode length and final plant height than moving from zero to negative DIF.

Since the original research, many studies by a number of scientists have been undertaken to understand the biology of DIF's

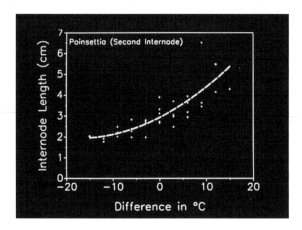

FIG. 2   Graph shows the effect of DIF on elongation of poinsettia internodes.

effects and improve its application in commercial greenhouses. We now know the following:

1. Negative DIF affects stem length by reducing internode length, not number. At the cell level, DIF affects cell length, not number.
2. For a given 24-hour average temperature, changing the DIF does not affect time to flower.
3. Under continued strongly negative DIF-conditions, leaves of some species (especially Easter lilies) tend to be more downward oriented, and most species appear chlorotic (pale and yellow) because of reduced chlorophyll content. Leaf chlorosis effects under negative DIF can be reversed by returning plants to positive-DIF conditions. Leaf orientation of Easter lilies will become more upright on return to positive DIF if the leaves have not matured fully.
4. Some species respond almost as well to a two-hour temperature dip (or drop) at sunrise as to a low temperature all day.
5. DIF has the greatest effect on final height when plants are elongating most rapidly. For determinate plants, such as chrysanthemum and poinsettia, rapid elongation occurs during the middle of the production season, rather than during the initial slow-growth period or near flowering.
6. Negative-DIF effects on stem elongation generally decrease as photoperiod increases (i.e., negative DIF is most effective under shorter days).
7. Negative-DIF temperatures late in the crop sometimes can reduce postharvest quality, particularly for poinsettia—negative DIF late in this crop may increase cyathia abscission.
8. A large number of plant species respond to DIF. Table 1 summarizes published plant responses to DIF and a temperature dip (drop) at sunrise.

PROS AND CONS OF USING DIF IN COMMERCIAL PRODUCTION

A major advantage of using DIF for height control is the ability to increase or decrease the stem elongation rate on a daily basis. Plants respond to the DIF they are exposed to each day. If elongation is too fast, it can be slowed immediately with negative DIF, while if it is too slow, it can be increased immediately with positive DIF. There is no "drag" similar to that one might find if a growth retardant had been applied and the plant had not grown out of the application.

Another advantage of DIF for height control is that it does not rely on chemicals. Pesticide use is becoming increasingly restricted, and alternatives to growth retardants are needed. DIF sometimes can be used where growth retardants may reduce flower and crop quality, for example, after visible bud on an Easter lily crop or during early bract color for poinsettias.

DIF affects all plants within a greenhouse heating zone, which is desirable if all need the same temperatures. If there are different crops with different DIF requirements

TABLE 1   EFFECT OF DIF AND TEMPERATURE DROP ON PLANT ELONGATION OF SELECTED SPECIES

Species	Strong or medium effect		Little or no effect	
	DIF	Drop[a]	DIF	Drop[a]
Aster (*Aster novi-belgii* L.)			✔	
Begonia	✔			
Campanula	✔			✔
Cyclamen	✔			
Dahlia	✔			
Dendranthema (chrysanthemum)	✔			✔
Dianthus	✔			
Euphorbia (poinsettia)	✔	✔		
Foliage (many species)			✔	
Fuchsia	✔			✔
Impatiens	✔			✔
Kalanchoe			✔	✔
Lilium (Easter lily)	✔	✔		
Lycopersicon (tomato)	✔			
Maize	✔			
Pea, bean			✔	
Pelargonium (cutting geranium)			✔	
Pelargonium (seed geranium)	✔			✔
Petunia	✔			✔
Rosa (rose)			✔	✔
Salvia	✔			✔
Viola	✔			

[a] Where there is no entry for a response to a morning temperature, it is because research has not been undertaken or results are contradictory.

Source: Adapted from Myster and Moe (1995). *Scientia Hort.*

in the zone, though, it is often necessary to select a compromise DIF temperature and spot-spray or drench tall plants with growth retardants. Also, if the greenhouse does not have good air circulation and there is a temperature gradient along the house, you may need to move plants or selectively apply growth retardants.

Controlling DIF is not always feasible. Growers in warm regions have limited ability to reduce daytime temperatures, and negative DIF is out of the question in midsummer for most locations. However, a temperature dip in the morning to achieve at least a partial negative-DIF effect is sometimes feasible in locations that have warm day temperatures.

Unfortunately, DIF is not without environmental cost—the use of energy to heat at night or cool during the day must be considered. As with all management tools, you must consider possible side effects, especially when running negative DIF with a high night temperature. For example, heat delay of poinsettia flowering will occur if the night temperature is above 72F (22C) at flower initiation.

APPLYING DIF IN YOUR GREENHOUSE RANGE

Most growers who have tried DIF believe it is a valuable production tool for controlling plant height. We believe temperatures should be a supplement to your tools for height control, not a total substitute for growth regulators. One grower we spoke with plans to use a low level of growth regulator on a regular basis as insurance against the loss of greenhouse temperature control. If you are considering trying negative DIF for the first time, here are a few suggestions:

1. Start with a zero DIF. If it does not provide adequate height control, implement a negative DIF in 2F (1C) increments. Plant response is greater when you go from positive to zero DIF than from zero to negative DIF. A commitment to trying to keep the day temperature cooler will go a long way toward controlling plant height.

2. Because warm air rises, plants grown in baskets toward the top of a greenhouse will not receive the same DIF as those grown on the benches below. Fans such as circulating or horizontal-airflow (HAF) models can maintain a more even temperature throughout the greenhouse.

3. When trying to drop the temperature in the morning, be aware that different heating systems retain heat for varying lengths of time. Hot water systems hold heat longer than unit heaters or steam pipes. Overhead systems dissipate heat faster than in-floor systems. Most growers find it most economical to lower the heat setting before sunrise so that the greenhouse cools to near the desired day or DIF temperature by sunrise. Little heat is wasted that way, and air and plant temperatures are near the desired temperature for longer periods.

4. For flowering potted crops, avoid large negative-DIF settings near flowering to avoid possible postharvest problems. Currently, our best advice is to finish a flowering potted crop at zero or slightly positive (2 to 4F, 1 to 2C) DIF. We have not observed any postharvest problems with bedding plants grown with negative DIF.

5. If you have problems with DIF, remember: you can always go back to your old production methods, including using growth retardants.

You do not need an environmental control computer to use DIF. Manual temperature control is possible, but you must monitor greenhouse temperatures closely day and night. Some growers have installed two thermostats with a time clock to regulate day and night temperatures. However, manual control or thermostats should not be necessary, since relatively inexpensive controllers that can be set to control DIF temperature are now marketed.

GRAPHICAL TRACKING: A ROAD MAP FOR GROWTH

Regardless of what height control methods you use—DIF, growth retardants, choice of cultivar, timing, or plant spacing—it can be a stressful experience growing "by the seat of your pants" without knowing whether your crop will finish at the desired height and be in flower on time.

Graphical tracking is a technique designed as a road map to show you at any point in the season whether your crops are too tall or short. Actual plant height or development is compared with desired plant height or development. A target curve is developed based on your market specifications (shipping dates and height specifications). By plotting your actual plant growth values onto a graph, you can make crop management decisions based on the comparison between the target curve and actual height.

Growers who have used graphical tracking have found the technique has increased quality, consistency, and profitability. The tool also makes them monitor the crop closely and ensures that they walk through the entire greenhouse at least twice a week inspecting height and other aspects of the crop. Finally, many growers mention peace of mind in having a guideline to help make decisions that affect the profitability of the crop. Should you apply a growth retardant? Should you change temperatures? Will the crop make the shipping date? A frequent comment we hear is "Now I can sleep at night."

A GRAPHICAL TRACK EXAMPLE

Figure 3 shows a sample graphical tracking curve for poinsettia. We will use this example to show basic principles and then go through each step needed to develop and use a graphical tracking curve. The example is taken from the Greenhouse CARE System computer program, described later, but it would be the same if tracked on paper.

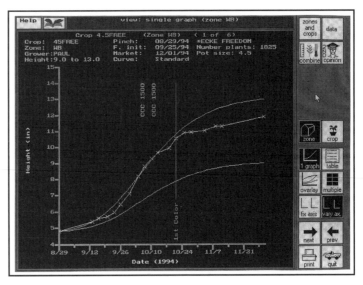

FIG. 3   Example of a graphical track for poinsettia using the Greenhouse CARE System computer program.

In fig. 3 the grower's goal was to finish Eckespoint Freedom between 9 and 13 inches on December 1. The crop was pinched on August 29, at which time the total height (pot and pinched plant) was 4.8 inches. Using this information, maximum and minimum graphical track curves were calculated and plotted on the graph at the start of the season. Twice weekly, the grower measured five plants and entered the average height onto the graph, connecting the new point with the last measurement (crosses and lines in fig. 3).

If the crop height is within the window between the maximum and minimum lines and you do not expect the crop to grow too quickly or slowly, then the crop is on track, and no changes in management are necessary. However, corrective actions are required when the crop is above or below the maximum or minimum line, respectively.

The grower applied two growth retardant sprays of Cycocel at 1,500 ppm in early October for the example crop. After first bract color, indicated by the vertical line, the grower used negative-DIF temperatures to reduce elongation rate and to keep plant height in the target window.

HOW DO YOU MAKE A GRAPHICAL TRACK CURVE?

There are at least three ways you can obtain a graphical track curve setup for your own crops. The first is to contact the Department of Horticulture at Michigan State University. For a small charge, a graphical track will be printed and faxed to you. The second way is to purchase the Greenhouse CARE System, a computer program that allows you to print as many graphical track curves as you want from your own computer. Again, contact the Department of Horticulture at MSU for information. Alternatively, you can create your own curves manually by using the following instructions.

POINSETTIAS AND CHRYSANTHEMUMS

Graphically tracking poinsettias and chrysanthemums is similar. Five pieces of information are necessary to graphically track these two crops properly: pinching date; flowering date; initial height, which includes the height of the pot plus the height of the mother (main) shoot; desired minimum and maximum height at flower; and stem-elongation curve.

The first four items are known by most growers at the start of the season. The stem-elongation curve for poinsettias and mums, however, is more complex. Lateral shoot development on poinsettias and mums follows a similar pattern after pinching (fig. 4). Growth is initially slow (lag phase) as apical dominance is overcome. The growth rate then increases during the rapid elongation phase in the middle part of the growth period. As flowers develop (the plateau phase) the growth rate declines, and elongation eventually stops on the mature plant.

Creating a graph for poinsettias and mums is easy with a computer program but is an involved process when one attempts to do it by hand. The worksheet at the end of this chapter shows the steps to making a graphical track curve by hand.

EASTER LILIES

Five pieces of information are necessary to graphically track Easter lilies: plant emergence date, visible bud date, flowering date, desired minimum and maximum height at flower, and stem-elongation curve. In con-

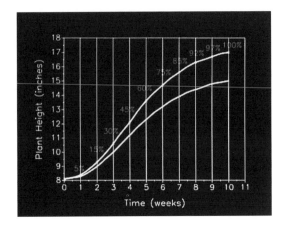

FIG. 4   Relative growth curve for poinsettia and chrysanthemum follows a similar pattern.

trast to the stem-elongation curves for poinsettias and mums, that for Easter lilies is simple. The easiest way to develop the stem-elongation curve for Easter lilies is to assume the plants double in height from visible bud to flower, which means the plant height at visible bud, not including the pot, is half the height at flowering (again, not including the pot; see fig. 5).

Considering a target total height of 22 inches at flower, the flowering plant height will be 16 inches (22 inches minus 6 inches for the pot height). Half of 16 inches is 8 inches. Assuming an 8-inch increase in height from visible bud to flower, then the plant height can be only 8 inches at visible bud (8 inches plus 6 inches for the pot equals 14 inches total).

We now know the start-ing height at emergence (6

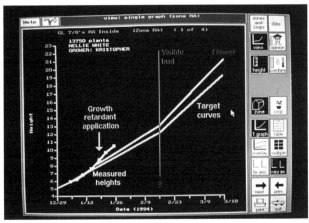

FIG. 5   Develop a stem-elongation curve for Easter lilies by assuming the plants double in height from visible bud to flower.

inches) and the total height at visible bud (14 inches) and flower (22 inches). We also know that visible bud should occur 30 to 35 days before flowering if we are to bring the plants into bloom on time. Plot these three points (height at emergence, visible bud, and flowering) and connect with two straight lines. Two lines normally are plotted to reflect the desired minimum and maximum final plant height (20 and 22 inches, respectively, in fig. 5).

HOW DO YOU MEASURE PLANT HEIGHT?

We recommend measuring at least five plants per crop twice per week to obtain an accurate average. If the same person measures each week, errors in measurement will be avoided. Choose sample plants throughout the crop that are representative of the height of all plants. If the sample plants grow slower or faster than the rest of the crop, change to more representative plants. Note that plants elongate more slowly when touched, so take care when measuring. If your crop is highly variable overall and the average height is no longer useful, you may need to split the crop in midseason into two separate graphs and physically move plants to separate parts of the greenhouse.

You can use a yardstick or make a measuring tool, following the plan in fig. 6. We developed this simple measuring tool at MSU to provide rapid and accurate measurements without touching or damaging the plants. All parts can be obtained from standard hardware stores. The metal crossbar, which can be made out of a carpenter's square or a piece of aluminum, reaches across the plant canopy. Screw or tape a carpenter's level to the crossbar to help make your measurements consistent. Screw two aluminum strips to the vertical sides of the crossbar and leave enough space to slip in a metal ruler. Screw an aluminum plate onto the aluminum strips to create a slot that allows the entire crossbar to slide over the ruler. If you affix adhesive

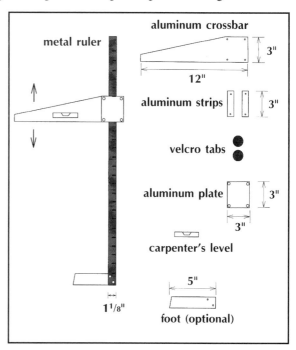

FIG. 6 Plans for a tool to measure plant height. This tool will provide a rapid and accurate measurement without touching or damaging plants.

Velcro tabs to the aluminum plate, inside the slot between the plate and the ruler, wear and friction will be reduced. You may decide to attach a small foot to the bottom of the ruler—we do this because we find it most accurate to measure from the pot rim, but the foot will get in the way if you are measuring from the bench upward.

THE GREENHOUSE CARE SYSTEM

As graphical tracking became more popular, we wanted to provide growers with the ability to generate their own curves and build crop history databases. We also knew that although graphical tracking determined whether the crop was too tall or short, it did not answer the question, What do you do when you are off track?

We decided to develop a computer program, called the Greenhouse CARE System (CARE), that would help suggest appropriate control methods based on the graphical track graph. We will use poinsettia as an example, but we also have developed modules for Easter lily and chrysanthemum. The program, which in 1995 was used by about 70 growers, is the first of what we expect will be an increasing number of computer decision-support tools available to the greenhouse grower.

One goal of the CARE project is to allow you to do all your graphical tracking on computer. Crops are organized into greenhouse temperature and light zones, and a personal crop history database can be built over several years. Crop data (heights, leaf counts, growth stages, and growth retardants) are typed into the program. Temperatures can be downloaded automatically from several environmental computer systems. CARE draws the graphical track curve, and you can either print the graphs or enter information directly into the computer.

Once data are entered into the program database, you can display crop information in a variety of ways to help make comparisons. For example, fig. 7 shows two Eckespoint Celebrate 2 crops grown over two different years on the same graph. Graphs from previous years are stored in the crop history database and can be retrieved easily for you to learn from past successes or challenges.

Temperatures can be entered manually or downloaded from the environmental computer to show you average and DIF (day minus

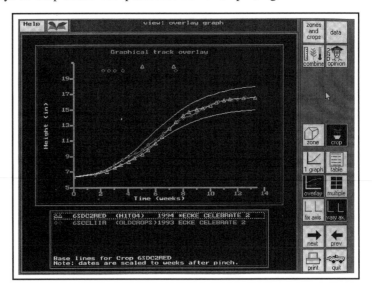

FIG. 7 Graph overlay of two Eckespoint Celebrate 2 poinsettia crops grown over two different years. Overlay of the two crops allows you to compare them easily.

night) temperatures on the same screen as plant height (fig. 8). The top graph in fig. 8 shows average temperature, the middle graph shows DIF, and the bottom graph shows plant height. This screen summarizes all the information related to growing this crop. Two growth retardant applications were made (indicated by boxes above the height curve), and the grower

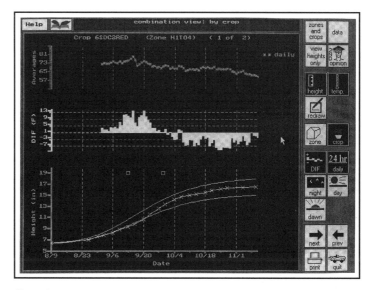

FIG. 8 Temperature and height graphs can be displayed simultaneously to help you decide the best temperature for your crop.

changed from positive- to negative-DIF temperatures (probably as outside temperatures cooled, allowing negative DIF) to reduce elongation rate later in the crop. Average temperatures were also cooled to maximize crop quality near the shipping date. Not only can you summarize and store temperature information in a useful format, but the temperature graphs help in deciding the best temperature settings for your crop.

SUGGESTED HEIGHT-CONTROL ACTIONS: A SECOND OPINION

To answer the question, What are the best growth retardant concentrations and temperatures based on a graphical tracking situation, we developed the Second Opinion modules in CARE. The goal of Second Opinion is to predict short-term elongation based on current growing conditions and suggest control actions in a consistent manner.

Height-control options for poinsettia, and their limitations, include applying chemical growth retardants, manipulating DIF, and plant spacing. Growth retardant applications are useful early in the crop to reduce elongation rate, but they can reduce bract size and crop value when applied late in the crop. DIF temperature can be changed to promote or decrease elongation rate. Negative DIF is not always feasible, especially in warmer regions or seasons. Near the end of the crop, negative DIF also can lead to premature cyathia drop. Respacing plants can have a dramatic effect on elongation. However, spacing plants widely to reduce elongation rate is labor intensive, and greenhouse space is expensive, so it is a last measure (which may not even be possible if greenhouse space is not available).

The best control actions depend on knowing the height of the crop. Is it too tall, too short, or in the target graphical tracking window? The expected growth rate and the current crop development stage (e.g., whether bract color is showing on upper leaves) are important. The production history of the crop (e.g., when the last growth retardant was applied) can also present crucial variables.

CARE Second Opinion takes these factors into account when recommending day-night temperatures and growth retardant rates. The program quantifies current height and expected elongation rate in a "worry index"—are you worried enough to promote or reduce elongation rate? Based on the worry index and crop development stage, CARE makes a recommendation, also taking into account the production history.

Figure 9 shows a sample Second Opinion screen for poinsettia. In this example there are five crops of different cultivars growing in a single greenhouse zone. The name of each crop is at the top of the graph, for example, "6" V-14 Pink Hse. 10." The number of plants in each crop is shown immediately below the crop name (e.g., 900). Arrows pointing downward indicate growth retardant applications.

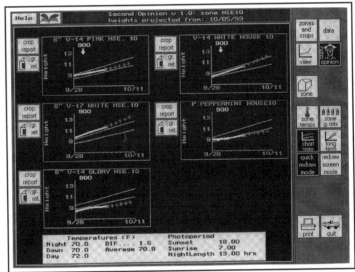

FIG. 9 A sample Second Opinion screen for poinsettia from the Greenhouse CARE System computer program. With Second Opinion, a biological model projects stem elongation over the next five days.

In the Second Opinion, a biological model projects stem elongation over the next five days. The model considers DIF temperature, growth retardants, plant spacing, and cultivar. One set of lines in fig. 9 represents actual plant height, a second set represents a portion of the target curve, and a third line represents projected plant heights based on the model.

You can try what-if situations in the Second Opinion module. Suppose you expect warm temperatures over the next week. You can click the mouse on the white temperature box at the bottom of the screen to change the temperatures from the current 72F

(22C) day and 70F (21C) night to something you expect, such as an 80F (27C) day. The model automatically reruns the projected heights and updates the graphs. Similarly, you can click on the "gr.ret" icon by a graph to see the effects of different growth retardant concentrations—projected heights are recalculated automatically.

The "crop report" icons in fig. 9 will display a report that suggests the optimum DIF and growth retardant concentration for a particular crop based on current height, expected future growth, and growth stage. The short crop report can be viewed on the screen. The "zone temps" icon at the right of fig. 9 considers all crops in the greenhouse when suggesting the best compromise day and night temperatures.

REFERENCES

Erwin, J.E., and R.D. Heins. 1995. Thermomorphogenic responses in stem and leaf development. *HortScience.*

Myster, J., and R. Moe. 1995. Effect of diurnal temperature alternations on plant morphology in some greenhouse crops—A mini review. *Scientia Hort.* 62:205–215.

WORKSHEET FOR MAKING A GRAPHICAL TRACK CURVE

A graphical track curve projects plant height from pinch to flower, with values plotted at 10% time intervals. You can overlay actual averages from your own poinsettia or chrysanthemum crop and compare its progress to the projected curve.

At the left of table 2 are columns pertaining to time; at the right, to height. The relative time (RT) column is divided into 0.1 increments, each representing 10% of the total time—whatever it may be—from pinch (0) to flower (1.0).

The next three columns give the corresponding number of days at each RT interval for crops of three different lengths.

13-week crop at RT 0.1 = 91 days × 0.1 = 9.1 days after pinch

Same crop at halfway point = 91 × 0.5 = 45.5 days

Similar simple calculations apply to periods of any total length you may project.

Anytime you schedule a crop, you will want to convert such RT-day intervals to actual dates. The next column gives calendar dates for a 91-day crop pinched on August 29. RT 0.1 is 9.1 days later, or September 7. The halfway point will be October 13.

Height projections are also easy to calculate. Plants grow at different rates at different periods between pinch and flower, so table 2 provides the rate for each RT 0.1 increment in the "Relative height" column. For example, from RT 0 to 0.1, immediately following a pinch, the new (lateral) shoot will grow only 5% of its ultimate length; from RT 0.2 to 0.3, the growth rate is three times this. Of course, the relative lateral shoot length needs to be applied to the final shoot length. This is derived from the overall plant height at flowering. Plant height consists of pot, mother shoot, and lateral shoot. The mother shoot is pinched at RT 0 to produce lateral shoots. The mother shoot will stay the same height from pinch to flower; therefore, plant growth on our curves means growth of the lateral shoots.

It's easier (on both you and the plant) to measure the whole plant, though. Lateral shoot length (Lat L) equals overall plant height minus pot and mother. Assume pot-mother height of 5 inches and ultimate overall plant height of 13 inches (plugging in cm instead follows similar equations):

Lateral shoot length = 13 − 5 = 8 inches

Lat L at RT 0.1 = 8 × 0.049 = 0.39 inches

Lat L at halfway = 8 × 0.618 = 4.94 inches

To find the plant height values between pinch and flower, simply add the lateral length at a given RT to the initial pot-mother height:

Plant height at RT 0.1 = 5.0 + 0.39 = 5.4 inches

Plant ht at halfway = 5.0 + 4.94 = 9.9 inches

Those hypothetical figures are in keeping with the curve in fig. 3 for a maximum height of 13 inches. Calculating the numbers for fig. 3's minimum height curve—or for a curve between any plant pinch height and any projected flowering height—would go along similar lines.

TABLE 2   MATH FOR AN EXAMPLE OF A POINSETTIA OR CHRYSANTHEMUM GRAPHICAL TRACK CURVE

Relative time	Crop time (days)			Dates (for 91-day crop)	Relative height	Height (inches)
	63 (9 wks)	77 (11 wks)	91 (13 wks)	Pinch date: 8/93		Start: 5 Final target: 13
0	0	0	0	8/29	0	5.0
0.1	6.3	7.7	9.1	9/7	0.049	5.4
0.2	12.6	15.4	18.2	9/16	0.144	6.2
0.3	18.9	23.1	27.3	9/25	0.291	7.3
0.4	25.2	30.8	36.4	10/4	0.456	8.6
0.5	31.5	38.5	45.5	10/13	0.618	9.9
0.6	37.8	46.2	54.6	10/22	0.749	11.0
0.7	44.1	53.9	63.7	10/31	0.849	11.8
0.8	50.4	61.6	72.8	11/9	0.921	12.4
0.9	56.7	69.3	81.9	11/18	0.972	12.8
1	63	77	91	11/28	1	13.0

CHAPTER 24

TISSUE CULTURE

by Warren Banner
Ball FloraPlant
West Chicago, Illinois

In the 1980s tissue culture (micropropagation) of ornamental crops became an important source for providing select crops to the grower. There are several crops today that are produced almost exclusively by micropropagation, such as syngonium and spathiphyllum. The early 1990s have been a period of consolidation of commercial laboratories and a shifting of some laboratories to low labor-cost countries, such as Costa Rica and India. Although this production technology has greatly improved the quality and performance of several crops, it has yet to replace many crops produced by traditional propagation methods. The next decade will probably see many more crops commercially produced through some form of tissue culture production.

NEW PRODUCTION METHODS

The exciting opportunities related to tissue culture in the late 1990s center around new production methods that will include many other crops being produced in vitro (growing in an artificial and sterile environment in culture). Three promising methods are being pursued in labs around the world.

SOMATIC EMBRYOGENESIS

This method produces many viable somatic (seed) embryos from plant tissue. Here an embryo (a seed less the protein and the coat) is "grown" by tissue culture. It is then "germinated," much as a normal seed is. This technology could be used to cost-effectively reproduce crops that are normally vegetatively propagated. Scientists envision seeding plugs with these embryos and providing a germinated plug seedling.

MICROPROPAGATION

Micropropagation (commercial propagation of crops using tissue culture) has been very labor intensive and scientists are seeking methods of automation to reduce this cost. The automation may include the possible use of robots and other forms of mechanization. The recent migration trend of micropropagation labs to low labor-cost countries will likely dominate until automation increases.

LIQUID BATCH CULTURE

This method is being used to increase bulb crops such as lilies. Each liquid culture produces hundreds of thousands of disease-free bulblets. The bulblets are then grown for one to two more years in the field before going to the commercial greenhouse grower.

These and other technologies are certain to make this decade a very dynamic one, especially in supplying our industry with superior young plants.

OTHER TECHNOLOGIES

Tissue culture has also become a valuable tool in several other important fields of horticulture. Many crops produced by traditional vegetative propagation now depend on obtaining nucleus stock free of viruses, fungi, and bacteria and to certify clean plant production. For many years, this practice has been used in carnations, mums, and geraniums; it is now being expanded to include crops like New Guinea impatiens and hiemalis begonias.

Breeders are integrating aspects of tissue culture to speed up new product introductions and expand their product lines. Some of these applications include increasing variability in horticultural traits to identify potential new varieties, rescuing embryos from seeds of select crosses for new hybrids that normally could not be obtained, selecting varieties under specified selection pressures, and preserving varieties or selections in vitro for long periods. This storage method is called *in vitro germplasm preservation.*

Breeders also rapidly scale up new varieties for fast introduction using micropropagation. These and other applications for the pathologist and breeder will continue to expand.

TISSUE CULTURE BENEFITS CROP SEGMENTS

There are five major horticultural crop segments that have benefited from micropropagation: foliage, pot plants, cut flowers, landscape ornamentals, and agricultural crops. The first four of these segments are discussed here, along with a partial list of commercially produced crops.

FOLIAGE

The foliage industry is very dependent on micropropagation as a source of microcuttings. Micropropagation provides a disease-free cutting that is vigorous and, in most cases, exhibits good basal branching. The elimination of soft rot bacterial diseases and the formation of superior basal habits in tissue culture produce superior crops compared to the plantlets produced by traditional propagation. This is especially true for syngonium, anthurium, and spathiphyllum. The basal branching is a carry-over effect from the growth regulators (cytokinins) used in the culture media. In addition, new varieties can be quickly scaled up for rapid and broad distribution throughout the world. Some varieties commercially produced by tissue culture include:

Alocasia	Cordyline	Ficus
Anthurium	Dieffenbachia	Philodendron
Banana	Dracaena	Spathiphyllum
Calathea	Ferns	Syngonium

POT PLANTS

The use of tissue culture is less prevalent for pot plants than for foliage. There is no clear reason for this trend, other than the lack of cost-efficient technology. Many pot plants could benefit from disease-free micropropagation and the uniformity possible from microcuttings. The 1990s may be the decade for tissue-cultured pot plants if some of the emerging technologies come on-line. One crop particularly helped by micropropagation is gerbera. A key selling feature here is the ability to produce plants with an identical flower color. Since the plants are clones, uniformity is also apparent within a selection. Pot plants commonly produced from tissue culture include:

African violets (limited)	Gerbera daisy	Pot roses
Aloe vera	Lilies	Rex begonia
Anthurium	Orchids	

CUT FLOWERS

Several significant cut flower crops benefit from micropropagation. Gypsophila can now be produced free of crown gall, and cut-type lilies are produced free of viruses that previously decreased quality. Disease-free micropropagation of lily bulblets produces flowers with true colors, larger petals, and increased plant vigor. Cut gerbera benefit from being reproduced with genetic purity, which ensures the unique color pattern for each variety. Several commercial success stories include:

Alstroemeria	Gypsophila
Anthurium andraeanum	Lilies
Cut roses	Orchids
Delphinium (some of the new hybrids)	Perennial asters
Gerbera	Statice

LANDSCAPE ORNAMENTALS

This classification of plants includes woody ornamentals and perennials. Many of the woody ornamentals are very difficult to propagate by traditional means and tissue culture has made this task much easier. Due to the slow growth of woody plants compared to many pot plants or annuals, tissue culture allows for a quicker buildup of new varieties. This speeds new products to market and cuts years out of the product development cycle. Perennials such as daylilies have benefited from tissue culture because many rare and beautiful varieties—previously very difficult to reproduce and seen only by a few—can now be mass-produced. Landscape ornamentals propagated through tissue culture include:

Amelanchier	Kalmia (Mountain laurel)
Birch	Lilac
Hemerocallis	Liriope
Hosta	Rhododendron

CHAPTER 25

POSTHARVEST HANDLING OF FLOWERING POT PLANTS

by Dr. Terril A. Nell
University of Florida
Gainesville

Consumers worldwide are becoming discriminating, bargain-conscious shoppers, often drawn to items having trademarks, logos, or trendy labels that represent superior quality or current popularity. Flower buyers are no different! Although purchasing habits are often related to specific marketing strategies or advertising promotions, an overriding factor in purchase decisions is quality, in terms of durability, reliability, appearance, and performance. Advertisements telling us that the "quality goes in before the name goes on" or "we will sell no wine before its time" sell products. Above all, such slogans suggest that consumers will do well to buy quality.

Selling quality products and offering service to consumers will characterize business success into the next century. The floriculture industry is no exception. Flowering plants are in direct competition with a multitude of other discretionary-purchase items, and if floral sales are to increase, consumers must be satisfied with the flowers they buy. Growers must focus on producing plants that are durable, resilient, and long lasting, as well as beautiful.

Quality is defined in different ways. One definition emphasizes consumer satisfaction, based on the consumer's comparison of product or service with his or her requirements, stated or unstated, conscious or merely sensed, technically operational or entirely subjective. In a competitive market, consumer expectations continually change. In effect, quality is a moving target.

In other segments of agriculture, quality is related to size or weight. Eggs, for example, are graded by weight, and date coding assures the customer that the product is fresh. The U.S. government has developed grades and standards for meat products that ensure quality and safety at each purchase. And most stores now date code their meats to ensure freshness to the consumer.

Using date coding on fresh meats and eggs is possible since proper temperatures are maintained from the time products are graded and processed until consumers remove the packages from store displays. Date coding is not often used on cut flowers, potted foliage, or flowering potted plants, primarily because growers cannot control the handling practices and environmental conditions to which flowers will be subjected by wholesalers and retailers after leaving greenhouse production and packaging areas.

The flower industry often refers to quality as having different levels: poor, average, good, or excellent. Phillip Crosby, a noted quality control engineer, states, "Quality does not have different levels. Products either are quality or they are not quality. Flowering plant quality and increased longevity are achievable by establishing the factors and conditions affecting quality and incorporating these specific criteria into production protocols. Achieving quality does not necessarily mean increasing cost."

One serious problem in establishing quality in the flower industry is that performance of flowers and plants is strongly influenced by production practices, shipping and handling procedures, retail display conditions, and home and office environments. Too much emphasis is placed on one criterion (appearance of the plant) at a single point in time (at the time of marketing), often because the related dimensions are easy to document. In potted foliage plants and flowering plants, sales are made by plant size or size of the container in which the plant is to be marketed. These standards do not provide any indication of quality or longevity, however.

Generally speaking, a quality plant can be defined as one having proper size and form, proper number of flowers at a marketable stage, and foliage free of nutritional disorders, insects, diseases, and mechanical injury. Most importantly, though, a quality plant withstands transit and retail display conditions, still providing enjoyment for consumers for an acceptable period of time.

Quality, as related to consumer satisfaction, can be dealt with at all levels of the industry only after it is understood that quality is simply conformance to requirements—requirements to produce, ship, and display plants and flowers according to specifications designed to achieve a quality plant meeting uniform physical dimension and appearance grades and standards. Quality begins when the cutting or seed is planted and continues until the plant is purchased by the consumer.

BUY AND SELL BY VARIETY

There are numerous production and handling methods that can increase flowering plant longevity. Growers can build quality into the plant from the time the variety is selected until the plant is shipped from the greenhouse to the retail store. Proper cultivar selection is the first place to begin. Growers and buyers should understand and be aware of the handling characteristics of various cultivars. Cultivars that do not ship well should be marketed only for local sales.

The best option today to ensure good quality plants is to buy and sell by cultivar. Information to guide cultivar selection is currently available for the major marketed crops and can easily be incorporated into existing buying and selling practices. For example, longevity differs among the most popular chrysanthemum cultivars, varying from two to four weeks (table 1). Iridon and Yellow Mandalay, both yellow decorative flowers, lasted 28 and 17 days, respectively. Similar differences in cultivar longevity have been observed for white, bronze, and light bronze chrysanthemum cultivars. Other varieties, such as Tara, are known to exhibit yellow leaves within three to 10 days indoors.

TABLE 1 POTTED CHRYSANTHEMUM FLOWERING LONGEVITY

Cultivar	Wks	Cultivar	Wks	Cultivar	Wks
Bronze		*White*		*Yellow*	
Glowing Mandalay	2	Free Spirit	2	Dark Yellow Paragon	2
Torch	2	Paragon	2	Sunburst Spirit	2
Favor	3	Power House	2	Tip	2
Mandarin	3	Mountain Snow	3	Yellow Mandalay	2
Light Bronze		Spirit	3	Mountain Peak	3
Mandalay	3	Surf	3	Sunny Mandalay	3
24-Karat	3	Puritan	3	Bright Golden Anne	4
Red				Iridon	4
Red Torch	2				

Note: Plants were placed in a simulated interior environment providing 75 footcandles (800 lux) 12 hours daily, at 68F (20C) and 50% relative humidity.

Poinsettia cultivars also differ with regard to epinasty (a plant part's outward or downward bending due to disproportionate growth of one surface of the part), leaf drop, bract edge burn, lateral stem breakage, leaf yellowing, fading of bract color, and transit durability. Poinsettia production problems of the past (leaf and bract drop and epinasty) have been largely overcome through aggressive breeding programs that have produced varieties offering outstanding interior performance. However, two new problems, bract edge burn and stem breakage, have emerged to challenge poinsettia growers. New varieties are more sensitive to transport conditions as well as changes in production conditions. Grower efforts to produce the "perfect" poinsettia in the shortest possible time have contributed to the occurrence of these problems, as well. It is surprising that the most durable, high-quality cultivar is not more commonly adopted by producers and buyers in Europe and the United States.

Since plants may be subjected to less than optimal handling procedures, durable cultivars will be sought by knowledgeable buyers (table 2). Retailers who offer high-performance cultivars can maximize the economic advantage of these crops by marketing the plants by name to their customers, thus creating a stronger market for repeat sales.

The day will come when plant buyers and consumers will select plants by cultivar name with confidence, just as they purchase cars, clothes, and candy in today's retail market.

TABLE 2   POINSETTIA HANDLING TOLERANCE AND LONGEVITY

Variety name	Shipping & handling	Keeping quality
Annette Hegg	Ships well	Excellent: very long lasting
Eckespoint Celebrate	Ships very well	Very good
Eckespoint Freedom	Good	Good
Eckespoint Lilo	Excellent; rapid recovery after shipping	Excellent: the longest-lasting variety
Gutbier V-10 Amy	Ships well	Fair; susceptible to premature leaf drop
Gutbier V-14 Glory	Ships well; sensitive to bruising	Very good

Note: Except for Annette Hegg, none of these plants exhibits epinasty following sleeving and shipping.

PRODUCTION PRACTICES DETERMINE LONGEVITY AND QUALITY

Production practices have long been oriented toward production of plants sized for designated markets. Product size will continue to be important. For several flowering plants, the Society of American Florists and the Floral Marketing Association have developed physical-dimension and flower-development criteria that are useful to growers and flower buyers. In addition, growers should simultaneously modify environmental conditions and production practices to increase longevity.

Each decision about variety selection, cultural practices, production environment, storage, transport conditions, and retail environment can have a major impact on plant longevity. Growers manipulate production temperature and light levels to maximize root development, increase branching, and—with the use of DIF (relationship of day and night temperatures)—control plant height, thus ensuring good plant quality. Growers can also use these conditions to increase postproduction quality and to extend longevity. Modifying environmental conditions to improve interior longevity does not necessarily mean incurring additional production costs.

TEMPERATURE

Flower development and stem elongation are controlled to a large extent by temperature. Temperatures from planting to bud or bract color should be conducive to good plant and flower development. Reducing temperatures 2 to 3F (4 to 5C) during the final two to three weeks of production is designed to intensify flower and bract colors— no information is available to suggest that flower longevity will be increased. (Petals on some white varieties of chrysanthemums and other flowering crops may turn pale pink with reduced temperatures.) Growers using DIF for height control of chrysanthemums should switch to a positive or zero DIF at disbudding and to lower temperatures as suggested above. Using a positive DIF during the final three weeks of production should have little effect on plant height, since stem elongation is small during this period. (High night temperatures during the final weeks of a poinsettia crop may contribute to bract edge burn, since these temperatures promote rapid bract enlargement.)

LIGHT LEVEL

Production light level can also be an important factor in extending longevity of flowering potted plants. Unlike potted foliage plants, where a low light acclimatization period reduces leaf drop, work in the United States and Europe has demonstrated that *high* light levels reduce bud drop and increase interior performance of chrysanthemums, Christmas begonias, poinsettias, and other flowering plants.

Symptoms of low production-light levels include more flower and bud drop, premature cyathia drop, and shortened interior longevity. In the northern United States, high-intensity lighting to optimize longevity during the winter months may be essential to obtain plant size as well as to develop a plant with good longevity.

FERTILIZATION

Fertilization has the greatest effect on longevity, through such aspects as fertilizer concentration, nitrogen source, and duration of fertilization during the crop period. Fertilizer application method (liquid and controlled release) will not affect postproduction longevity, provided plants are grown with fertilizers meeting the specifications described below. High fertilizer levels increase leaf size and thickness and intensify leaf color of most floral crops, but they also decrease plant and flower longevity.

For most growers, reducing fertilizer levels will increase longevity without affecting marketability of the flowering potted plant. In many cases, high fertilizer levels are necessitated by the amount of irrigation water used at each watering. The amount of fertilizer can be reduced just by lowering the amount of leachate. For example, lowering fertilizer concentration from 450 to 150 ppm at every watering increased longevity by 15 days without affecting production quality. The increase in longevity depends on fertilizer concentration, growing medium, and variety. High fertilizer levels have been related to the incidence of bract edge burn in poinsettias. Reducing fertilizer level is difficult for some growers to accept initially, but a number of growers who have lowered fertilizer levels have increased postproduction longevity without detrimental effects on crop marketability.

Stopping fertilizer applications during the final stages of production increases chrysanthemum longevity and reduces the incidence of bract edge burn of poinsettias. Terminating fertilizer on Mountain Peak chrysanthemums at disbud increased longevity by seven to 11 days. Fertilizer termination is not beneficial on Easter lilies, so it is best to continue fertilizer application until this crop is marketed to avoid premature leaf yellowing. The success of fertilizer termination for each grower will be determined by the fertilizer levels being used, growing media, and crop. Growers currently using high fertilizer levels throughout the entire production period will experience greater benefits than growers who already use optimum fertilizer levels or reduce the fertilizer levels at the end of the crop.

Basically, terminating fertilizer reduces the soluble salt level in the growing medium and prevents excessive elemental (salt) buildup in the plant. The sources of nitrogen and potassium used in fertilizer programs can also affect longevity. Longevity of most flowering potted plants is greatest when approximately 60 to 70% of the nitrogen is from nitrate sources and the remainder is from either ammonium or urea sources. Also, some growers have altered the nitrogen-to-potassium ratio at the end of the crop in order to extend longevity. There does not appear to be any benefit from this procedure, but it may be beneficial to switch from fertilizers containing ammonium nitrogen and low levels of calcium to a combination of calcium and potassium nitrate during the final two to three weeks of production as a means of extending longevity.

THE RIGHT MEDIUM

The growing medium also influences postproduction performance of flowering potted plants. The medium should provide good aeration and nutrient-holding capacity during production and maximum water-holding capacity during the postproduction period in order to minimize drying out during the retail and consumer phases. Research results have shown that applying wetting agents at the time of marketing or planting in the landscape delays time to wilt by allowing for uniform watering during the postproduction period. Growers and interiorscapers have applied wetting agents with considerable success.

Water-absorbing gels and antitranspirants have received little acceptance at the present time, due to variability in benefits obtained with these products and to phytotoxicity problems on some flowers, such as hydrangea and senecio. Also, water-absorbing gels become less effective when used in conjunction with fertilizer salts or with irrigation sources containing high salt levels.

MARKETING FACTORS

The stage of development at which a plant is marketed affects longevity. Unopened flower buds will not open with the same color intensity indoors as they would if allowed to open under higher greenhouse light conditions. However, open flowers are more easily damaged during transit than buds showing color.

Most flowering potted plants should be marketed when good flower color has developed and buds are showing a considerable amount of color. For instance, chrysanthemums should be marketed when flowers are 50% open. Poinsettia bracts should be fully colored, and cyathia (buds) should be just beginning to open. Azaleas, on the other hand, are best when eight to 12 flowers are open and a large percentage of the buds are showing color. The Society of American Florists and the Produce Marketing Association have developed a series of color photographs depicting the correct marketing stage for quite a few flowering potted plants to give growers and flower buyers specific information on each crop.

SHIP PLANTS PROPERLY TO MAINTAIN PLANT QUALITY

Shipping flowering potted plants (and other floriculture crops) was not necessary 30 to 40 years ago, when floriculture production was within a one- to two-hour drive of major metropolitan areas. In many cases, large producers have migrated, and the industry has developed in areas providing favorable year-round production climates without the high energy costs associated with production in many of the major metropolitan areas. The floriculture industry has changed dramatically over the years, and shipping of flowering potted plants has become commonplace. It appears that long-distance shipping will continue as mass-market buyers, in need of large quantities of plants at a single time, exert a greater influence on the floriculture market.

Of course, long-distance shipping of plants is not unique to the United States. In Europe, Dutch- and Danish-produced potted plants are shipped long distances to England, Italy, and southern France—even as far as Turkey.

Improper temperature, poor air movement, exposure to harmful gases, lack of light, and vibration are unfavorable conditions encountered in the transport of flowering potted plants. These conditions can lead to deterioration of even the highest quality plants. The environmental and physical stresses imposed on plants during transit is worsened if plants are improperly produced, mishandled, or not handled properly upon receipt at the retail outlet. It must be emphasized, however, that storage of plants prior to sale and transport exposes plants to the same stresses, except vibration, as placing a box of plants on a truck for a transcontinental journey from California to Virginia. Long-distance shipping of some potted plants surely contributes to the 15 to 25% estimated shrinkage of potted plants and flowers from the production area to the consumer. Clearly, the best situation is not to ship flowering plants, thus avoiding the inherent stresses in this procedure, but this option is not feasible.

The real problem in transport of flowering potted plants is maintaining the plants in conditions equal to those on the greenhouse bench. In most cases, flowering potted plants are enjoyed by consumers for two to four weeks, and there is no attempt to reflower these plants. So, the goal in designing and maintaining proper shipping and handling procedures is to provide a plant of equal longevity and quality to the consumer even if the plant has been shipped. The value, appearance, and quality of plants will not improve during transport, but proper handling and storage and shipping conditions will guarantee that the product quality will be good at the time plants are unboxed and sold to the consumer, provided proper retail and consumer care is exercised.

Many of the effects of transit stresses on flowering potted plants had not been investigated until the last five to eight years. Test results show that some of the problems that may be caused by transit stress, other than chilling injury and exposure to harmful gases, may not be apparent until one to two weeks after the plants are unboxed. In these cases, plants may exhibit delayed flower drop, leaf yellowing, or flower fading, which would not normally be associated with the stresses encountered during or immediately following the shipping and storage period. Thus, it is vital that proper conditions be provided from the time the flowering plants are removed from the greenhouse bench until they are sold to the consumer.

PRODUCTION FACTORS AND SHIPPING

For many years production practices were assumed to have little, if any, relationship to the problems associated with shipping flowering potted plants. Research results have demonstrated that production practices affect plant tolerance of shipping conditions. Chrysanthemums receiving high levels of fertilizer until marketability did not withstand shipping conditions or last as long under interior conditions as plants that had no fertilizer during the final three weeks of the crop. Also, plants receiving high fertilizer levels are most sensitive to diseases such as *Botrytis* and mildew. Plants produced under luxuriant nutritional and watering conditions are more likely to decline during shipping. Most important, plants that have been acclimatized during the final two to three weeks of production last longer.

PACKAGING

Plants need to be properly sleeved and boxed to prevent bruising and mechanical damage to leaves, flowers, and bracts during shipping. Sleeves should extend 2 to 3 inches (5 to 8 cm) above the top of the plant canopy to provide good protection to the leaves and flowers (or bracts). Paper, plastic, and fiber sleeves are available, and many growers select sleeve type based on personal preference, visual appearance, and costs. Some problems may be encountered if plants are watered immediately prior to placing a plastic sleeve on the plant, in that the plastic traps the moisture within the plant canopy. Moisture buildup is reduced when plastic sleeves have holes punched in them; fiber and paper sleeves allow for rapid movement of water, and moisture is not trapped within them. Many growers are finding that fiber sleeves reduce disease problems during transport. Of course, if moist, sleeved plants are boxed immediately, moisture buildup problems may not be due to the particular type of sleeve, since moisture will be trapped in the box regardless of the sleeve type used. While growers are encouraged to pack plants moist, moisture problems can nevertheless be minimized if plants are watered at least six hours prior to sleeving.

TABLE 3   INDUSTRY STANDARDS AND TECHNICAL SPECIFICATIONS

Pot diameter (inches)	Pack size	Box standards
3	28	Minimum 250# test carton
4	15	33# medium cardboard
4$\frac{1}{2}$	15	C flute construction
5	10	Waterproof adhesive
5$\frac{1}{2}$	8	Dividers for long distance or
6	6	high humidity
6$\frac{1}{2}$	6	Moisture-resistant tray
7	4	Cutout hand grips for handling ease
7$\frac{1}{2}$	4	
8	4	
8$\frac{1}{2}$	3	
9	4	
10	1	

Currently, the most common system for shipping plants in the United States is to place sleeved plants into strong cardboard boxes that can be stacked six to eight high without collapsing and damaging the plants. After boxes are filled, they are closed and sealed with tape or staples prior to shipment. Boxing standards have been established by the Society of American Florists and the Floral Marketing Association (table 3). Boxes should fit a standard 48- by 40-inch (1.2- by 1-m) palette once they are prepared for shipping. Closed shipping boxes are unique to the U.S. floral industry. In Europe plants are sold from auctions in trays that are delivered on movable carts. This system is not perfect, but air movement is better than with closed boxes. It may be necessary for our industry to consider other shipping containers and systems in the future in order to minimize the stresses incurred during shipping.

SHIPPING AND STORAGE CONDITIONS

Shipping and storage pose problems to flowering potted plants due to the exclusion of light in the closed containers and sleeves, presence of ethylene, and sensitivity of plants to temperature extremes. These factors may at first glance seem to have little influence on flowering potted plants. However, we must understand that potted plants are dependent on stored carbohydrates (sugars) in the plant at the time of marketability (flowering) to provide sufficient energy to live in the interior environment. Stored carbohydrates (those sugars and starch in the plants at time of marketability) are required for flowers to continue opening, for leaves to remain green, and to maximize the time the plant and flowers will maintain a good display indoors for the consumer. We are unable to provide supplemental sugars (as floral preservatives) to potted plants during the flowering period, as is done with cut flowers. Plants produce carbohydrates naturally through photosynthesis in the greenhouse, but light levels are so low that photosynthesis does not take place under most interior conditions, thus depleting stored reserves in the plant.

Carbohydrates can be rapidly depleted through respiration during shipping. The rate of depletion is directly proportional to shipping temperature—higher temperatures increase respiration rate and carbohydrate depletion. Thus, plants should be shipped at the lowest possible temperatures in order to minimize respiration. Chilling-sensitive plants should be stored and shipped at temperatures of 52 to 60F (11 to 16C) to avoid chilling injury—however, they cannot be shipped at low temperatures of 35 to 50F (2 to 10C) without injury. Chilling injury includes development of small black spots on crossandra leaves; bluing or development of white bracts on red poinsettias; leaf, bud, and bract drop on bougainvillea and clerodendrum; bud drop on hibiscus; or death of the entire plant with *Sinningia* and *Saintpaulia*. However, shipping of any of the flowering potted plants at temperatures higher than 65F (18C) is detrimental, also, since leaf yellowing and bud drop will

TABLE 4   RECOMMENDED SHIPPING TEMPERATURES FOR FLOWERING POTTED PLANTS

35–40F (2–4C)		50–50F (10–16C)	
Amaryllis	*Lilium longiflorum*	*Begonia (eliator)*	*Gloxinia*
Calceolaria	*Muscari*	*Bougainvillea*	*Hibiscus*
Crocus	*Narcissus*	*Browallia*	*Rhipsalidopsis*
Cyclamen	*Oxalis*	*Clerodendrum*	*Saintpaulia*
Dendranthema	*Pelargonium domesticum*	*Crossandra*	*Schlumbergera*
Freesia	*Rhododendron*	*Cymbidium*	*Sinningia*
Hippeastrum	*Rosa*	*Euphorbia pulcherrima*	*Streptocarpus*
Hyacinth	*Senecio*	*Exacum*	
Kalanchoe	*Tulipa*		

occur rapidly at these high shipping temperatures. Recommended shipping and holding temperatures for more than 30 crops are presented in table 4.

Another detrimental shipping factor is ethylene, which the plants can generate themselves or otherwise be exposed to during the shipping and holding periods. Ethylene is an odorless, colorless gas that can cause numerous undesirable effects on flowering plants. Some plants have not been shown to be affected by ethylene, but on sensitive flowering plants ethylene is most commonly associated with leaf and bud drop, premature aging, and leaf yellowing; other disorders have also been identified (table 5). Ethylene exposure prior to, during, or following shipping can result in these disorders, but the extent of an injury is dependent on ethylene concentration, temperature during exposure, and the duration of the exposure.

Ethylene is most damaging as temperature is increased during the exposure period, regardless of the concentration. Of course, injury is worse with higher concentrations and longer exposure periods. As an example, open dianthus are 1,000 times more sensitive to ethylene when temperatures increase from 36 to 70F (2 to 21C). So, one of the most effective means to minimize ethylene injury during shipping is to reduce temperatures, being cautious not to ship chilling-sensitive plants at temperatures low enough to cause injury.

Also, growers, shippers, and retailers should be aware that open flowers are generally more sensitive to ethylene than buds. However, it may be difficult to ship flowering potted plants in the bud stage, since open flowers are desired during the marketing period, and flowers developed under low light conditions, such as retail or consumer conditions, are generally light colored and smaller than flowers that open under greenhouse conditions.

TABLE 5   RESPONSE OF FLOWERING POTTED PLANTS TO ETHYLENE

Crop	Symptoms
Achimenes	Flower, bud drop
Begonia (elatior)	Flower drop
Bougainvillea	Flower, bract drop
Browallia	Flower, bud drop
Calceolaria	Flower, bud drop
Clerodendrum	Flower, bract drop, leaf drop
Crossandra	Flower drop
Cyclamen	Flower drop, flower wilting
Cymbidium	Wilting of the sepal
Dianthus	Failure of flower to open
Euphorbia pulcherrima	Petiole droop[a]
Exacum	Flower wilting
Gardenia	Flower, bud drop
Hibiscus	Flower, bud drop
Kalanchoe	Failure of flowers to open, petal drying
Pachystachys	Petal wilting, bud blasting, leaf yellowing
Pelargonium	Floret drop
Rhododendron	Leaf drop
Saintpaulia	Flower wilting
Sinningia	Flower drop
Streptocarpus	Flower drop

[a] Petiole droop (epinasty) of *Euphorbia pulcherrima* is caused by an upward bending of leaf and bract petioles during sleeving.

Note: The degree of sensitivity to ethylene varies with plant species, variety, ethylene concentration, temperature during exposure, and duration of exposure.

Foliar sprays of silver thiosulfate (STS) can minimize the detrimental effects of ethylene by blocking ethylene action in the plant. STS application effectively reduces the problems associated with ethylene exposure on a number of crops. STS is *not* labeled for use on flowering potted plants, but it has been shown to be effective with the following (list adapted from Pokon & Chrysal, B.V., Naarden, the Netherlands):

Begonia	*Clerodendrum*	*Gardenia*	*Rhododendron*
Bougainvillea	*Crossandra*	*Hibiscus*	*Saintpaulia*
Browallia	*Cyclamen*	*Ixora*	*Schlumbergera*
Calceolaria	*Epiphyllum*	*Pelargonium*	*Senecio*
Campanula	*Exacum*	*Rhipsalidopsis*	*Streptocarpus*
Capsicum	*Fuchsia*		

Some variability has been noted in the effectiveness of STS on flowering potted crops, and it appears to be related to improper mixing of the chemical, use of the wrong concentration, or incorrect time of application. In most instances, for best results STS is applied to flowering potted plants just as buds begin to show color. Application rates have not been established for a wide range of flowering potted plants. Additional antiethylene chemicals labeled for flowering potted plants should be introduced to the U.S. market in the next three years. These chemicals will offer protection against internally produced and external ethylene, thus minimizing the detrimental effects of ethylene.

BOTTOM LINE IS VALUE

As we approach the 21st century, the floriculture industry has the opportunity to convince consumers that flowers and flowering pot plants significantly enhance their quality of life and that flowering plants should appear on their list of necessary routine purchases along with groceries, newspapers, and toiletries. The best way to lure new customers and encourage their return business is to consistently offer *value:* high-quality plants that perform well in an interior environment.

The flower industry offers a product that enhances the quality of life. Performance and service will guarantee success with consumers. Information is now available to increase the longevity of floral crops by 30% or more and to dramatically reduce the shrinking associated with their transport and handling. The time has come for longevity of floral crops to become as important to the grower as any other aspect of production.

CROP CULTURE BY CROP

ACHILLEA (YARROW)

by Jim Nau
Ball Horticultural Company
West Chicago, Illinois

❖ *Perennial (for various species and their seed counts, see below).*
Germinates in 10 to 15 days at 65 to 70F (18 to 21C). Do not allow the
germination temperature to rise above 75F (24C). Seed should be left
exposed to light during germination.

Commonly called yarrow, there are a number of *Achillea* species and varieties on the market today. *Achillea filipendulina* (fernleaf yarrow; 200,000 seeds/oz, 7,000 seeds/g), is a rank-growing variety reaching 2 or 4 feet (60 to 120 cm) in the garden when grown from seed (fig. 1). Sown in February and transplanted within 24 days after sowing, plants can be sold green in May when grown at 55 to 60F (13 to 16C). This group of yarrows will flower only marginally at best during the summer of the first year after sowing. Flowers are golden yellow in color, blooming platelike (with flat tops) in June and July. Plants flower better when sown the year before and overwintered in quart or gallon containers. Seed varieties include Gold Plate and Cloth of Gold (Parker's Variety).

For vegetatively propagated material, consider *Achillea* × Coronation Gold, which is a related variety, though a hybrid. The flowers are golden yellow on plants no more than 3 to 3½ feet (90 to 100 cm) tall.

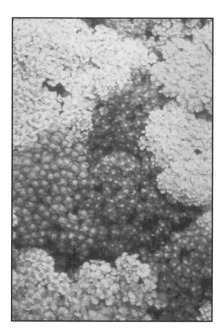

FIG. 1. *Achillea* is a fine perennial for home gardeners and an excellent fresh or dried cut flower for growers. The variety shown: vegetatively propagated Debutante.

Coronation Gold is a variety that can be used for landscaping or in the perennial or cut flower garden. The foliage is gray-green and aromatic.

Achillea millefolium (common yarrow, milfoil; 140,000 seeds/oz, 5,000/g), the most common of the yarrows on the market, comes in white or rose red flowers. The cultivars, however, range in color from lavender to buff yellow, cherry, and other pastel shades. When seed is sown in February, transplanted within three weeks, and grown on between 65 and 70F (18 to 21C), the plants will be salable green in the cell pack in May. Plants will flower more dependably (70% or better) than *Achillea filipendulina* during the first summer after sowing. Unfortunately, all the flower colors except white will readily shade to an off-color, most often to a dull or off-white, especially under high heat and humidity. They make excellent fresh cut flowers, though the flower color fades as they dry. The best blooms are taken when the buds first open. Plants can get 2½ to 3 feet (76 to 90 cm) tall.

As for varieties, those propagated from seed include Cerise Queen, which has rose red flowers that shade to pink and eventually off-white. It performs equal in habit and flower color to Rosy Red. However, there is one mixture that deserves special merit for its overall performance, especially in uniformity of both habit and height, as well as for flower colors. Summer Pastels, grown from seed, is the overall best *A. millefolium* variety from seed. It is a well-blended color mix on plants to 16 inches (40 cm) tall that flower profusely in July and August from a mid- to late-February sowing. Flowers are either salmon, cream, blush pink, or yellow.

Among vegetative varieties, be sure to try some of the Galaxy varieties. These are the result of a cross of two *Achillea* species and are related to *A. millefolium*. Varieties are sold under individual names like Hoffnang, Heidi, and Paprika, to name just a few. Though the flower colors will shade, as in the seeded types, the varieties come in gold, rose, lavender, white, and several other pastel colors.

Achillea × Moonshine, an interspecific cross, is vegetatively propagated. Moonshine's sulfur yellow flowers and fernlike, green foliage make it a staple for growers and gardeners. Plants grow to 2 feet (60 cm).

Both *A. filipendulina* and *A. millefolium* are hardy to USDA Zones 3 to 8, and flower from June to August in the garden. Instruct gardeners to trim back the plants after flowering for rebloom about one month later.

AFRICAN VIOLET (SEE *SAINTPAULIA*)

AGERATUM (FLOSS FLOWER)

by Jim Nau
Ball Horticultural Company
West Chicago, Illinois

❖ *Annual* **(Ageratum houstonianum***). 200,000 seeds/oz (7,000/g).*
Germinates in eight to 10 days at 78 to 82F (26 to 28C). Seed should be
left exposed to light during germination.

Ageratum is best used in containers, as a border or edging, and especially in mass plantings in landscape situations (fig. 1). Once sown, seedlings can be transplanted in 15 to 20 days and grown on at 60 to 65F (15 to 18C). When grown in cell packs, plants are salable green in eight to nine weeks, or in flower 10 to 11 weeks, after sowing when grown under long days. If growing for 4-inch (10-cm) pot sales, plants are salable in bloom 12 to 13 weeks after seeding (60F, 16C for one to two weeks, then 50F, 10C).

In the southern United States, allow nine to 10 weeks for flowering pack sales and 11 to 12 weeks for flowering 4-inch (10-cm) pot sales of dwarf varieties. Begin sales once all danger of frost has passed.

Among varieties, Hawaii Blue is the leading mid-blue-flowered, F₁ hybrid variety on the market. The Hawaii series is also available in White and Royal Blue (dark blue). Blue Puffs, Blue Blazer, and Neptune Blue are also available. White-flowering varieties, because they are more difficult to produce, are generally available in limited supply.

Most bedding ageratums tend to brown out in the heat of the summer, but the new Summit ageratums are perpetual bloomers, offering a show all summer long.

For something a little unusual on the market, there is also a cut flower variety, Blue Horizon, that grows to 3½ feet (1 m) and makes an excellent cut-flower or background planting.

FIG. 1. A minor bedding plant staple, *Ageratum* in full flower makes a fine bedding flat. The variety shown: Royal Hawaii.

Planted 12 inches (30 cm) apart, the variety fills in readily and flowers from June until frost. The flowers can be used fresh or dried in arrangements. Blue Horizon also makes an excellent landscape presentation.

ALCEA (HOLLYHOCK)

by Jim Nau
Ball Horticultural Company
West Chicago, Illinois

❖ *Annual or short-lived perennial* (**Alcea rosea,** *formerly* **Athaea rosea**). *3,000 to 6,000 seeds/oz (105 to 210/g). Germinates in five to 10 days at 70 to 73F (21 to 23C). Lightly cover seed for germination.*

This old-fashioned garden favorite has been greatly improved and modernized with double flowers and an array of rich colors. Hollyhocks are annuals or short-lived perennials that are often termed *biennial*. The plants can live as true perennials for several years if protected or covered with mulch once they are planted to the garden. In unprotected garden locations in areas of the country where frost can penetrate deeply, the plants often fail to perform and can die out readily. Their best performance comes from plantings made close to foundations or up against barriers to the winter winds.

Crop time for annual or perennial hollyhocks is the same: allow 10 to 12 weeks for salable 4-inch (10-cm) pots when seed is sown in mid- to late winter. Quarts require up to 13 weeks for green plant sales. Perennial hollyhocks may or may not flower their first year. For both annual and perennial hollyhocks, do not count on having plants in flower for sale. For flowering perennial plants, sow seed in November or December the year before, transplant to 1 gallon (3.7 l) containers, and move to a 50-to-55F (10-to-3C) cold frame for sales in April or May.

Hollyhock varieties are stately companions in the home perennial garden. Both single- and double-flowering forms are available, but the double-flowering varieties have the strongest acceptance. The perennial hollyhocks—Chater's Double, Powderpuffs Mix, and others—will flower sporadically the first season when sown from seed in March or after. Midwinter sowings, transplanted to quart or gallon containers when ready, will flower more freely during the summer than their late-seeded counterparts. Powderpuffs Mix is a 4- to 5-foot (1.2- to 1.5-m) variety with 3- to 3½-inch (7.6- to 9-cm) blooms in soft pastel shades, while Chater's Double grows taller, to 7 or 8 feet (2 to 2.5 m). Chater's can be found on the market available in separate colors, but the mixture is the most popular. Both Chater's and Powderpuffs are short-lived perennials.

Many times the annual varieties will flower shorter than their perennial counterparts. Often growing 3 feet (1 m) or so, these plants will flower profusely in July when planted to the garden from 4-inch (10-cm) pots from seed sown in February. The single-flowering varieties are more popular in the annual form since they often reseed if not pulled too early from the garden in the fall. Try Indian Summer Mix, which comes primarily in yellow and white shades. Summer Carnival Mix has double flowers in rose, pink, white, yellow, and carmine.

ALSTROEMERIA (INCA LILY)

by Mark P. Bridgen
University of Connecticut
Storrs, Connecticut

❖ *Annual (***Alstroemeria** *hybrids). Vegetatively propagated by division of tissue culture. Grown as a cut flower, primarily; also as a flowering pot plant.*

Alstroemeria, also known as the lily-of-the-Incas, Peruvian lily, or Inca lily, has been grown in the United States since the 1970s, mainly as a cut flower crop (fig. 1). Recently, it has been grown as a garden flower and as a potted flowering plant. The plants produce beautiful, large inflorescences in many different colors, including purple, lavender, red, pink, yellow, orange, white, and bicolors. In addition to their showy colors, the cut flowers have a long postharvest life, up to two to three weeks. These valuable characteristics have made alstroemeria one of the 10 most popular cut flowers at the Dutch flower auction.

Alstroemeria is also very popular with growers since the plants are versatile and easy to cultivate with cool temperatures. The plants produce high yields and possess an everblooming habit after flower

FIG. 1. *Alstroemeria*, with its bright colors, has become a very important cut flower over the past 10 years, both domestically and abroad. Now breeders have developed fine varieties for pot plant production as well.

initiation has occurred. Flowers may be harvested any time during the flowering season, since flowering stems will continue to develop from the underground rhizomes.

Native species of *Alstroemeria* exist in Chile, Brazil, Bolivia, Peru, Paraguay, Venezuela, and Argentina. The species are indigenous to a variety of habitats, including the snowline of the Andes Mountains, the coastlines of oceans, highland forests, and deserts. Although very diverse, the species are of little commercial value because many of the plants go dormant for part of the year and the flowers are often small and insignificant.

New varieties are the result of years of inter- and intraspecific breeding and irradiation to induce mutations. New hybrid varieties have originated in the United Kingdom by the Parigo Company and in the Netherlands by the van Staavern, Wülfinghoff, Könst, van Zanten, and Phytonova companies. In the United States, varieties have been produced by Monjonnier Enterprises of Encinitas, California; Coast Alpine Nursery of Lummi Island, Washington; and the University of Connecticut at Storrs. Plants may respond differently, depending on the cultivar and the region of the United States in which they are grown; these differences are due to the extreme variation in growth habits, the uniqueness of the plants, and the complexity of the hybrids. Objectives for breeders include vigorous growth forms, continuous, year-round flowering, and new flower colors. The first fragrant alstroemeria has been recently developed and released by the University of Connecticut. Most cultivars are protected by patents; plants are normally leased from the breeder, with a yearly royalty assessed on the square footage in production.

PLANT CHARACTERISTICS

Alstroemeria are herbaceous plants that produce two types of shoots: floral and vegetative. Shoots are initiated on white, subterranean rhizomes and can grow as long as 4 feet (122 cm) or as short as 6 inches (15 cm). Leaves on the shoots are called *resupinate* since the leaf base twists 180 degrees, the top (adaxial) surface facing down. Normally, shoots that have unfolded more than 30 leaves are vegetative and will not flower. A fibrous root system develops from the rhizome; the roots can become thickened storage roots as the plant develops, giving the incorrect impression of tuberous roots.

Flowers are irregularly shaped with inferior ovaries; flowers arise in a terminal-bracted umbel of cymes. Each cyme bears one to five flower buds. The perianth consists of six uniform tepals that open simultaneously. Mature blooms are funnel-shaped, with their tepal tips curling outward.

INDUCING FLOWERS

Controlling flowering is a process with a primary cold temperature requirement and a secondary long photoperiod requirement. The cool temperature requirement must be fulfilled prior to the long photoperiod. Once flowering begins, the plants will continue to produce flowering shoots indefinitely until the soil temperature rises above 65 to 70F (18 to 21C) for extended periods. Each variety has a unique requirement for cool temperature, varying from 50 to 63F (10 to 17C), and the amount of time exposed to that temperature. Follow the breeder's recommendations for best growth and flowering procedures.

Alstroemeria are often divided into two main classifications that generalize their flowering habits: orchid and butterfly types. However, due to intense hybridization, these categories are not always distinct. Orchid-type alstroemeria are a group with three to five months of major flower production in the spring, with little or no flowering during the remainder of the year. These varieties have tall growth habits (8 to 9 feet, 2.5 to 3 m), remain vegetative until spring, produce a large number of flowers in a short period, have a thermoperiodic cycle for flower initiation, and show little or no photoperiodic responsiveness.

The butterfly type group will flower for nine to 12 months each year, depending upon the variety and environmental conditions. These varieties have shorter growth habits and larger, more open flowers. Cool night temperatures of 55 to 63F (13 to 17C) and long photoperiods will induce and maintain flowering in this group. The cool temperatures are required, but the long photoperiods will only hasten flowering.

PROPAGATION

Alstroemeria is vegetatively propagated by rhizome division or micropropagation. Asexual propagation allows plants to grow true to type and quickly. Divide benched plants at least every third or fourth year, depending on the variety and growth characteristics. Plants that are being propagated in pots can be divided every eight to 12 weeks, depending on the time of year. Usually, three or four good divisions can be obtained for a 1-gallon pot. About one to two weeks prior to dividing, plants should be severely pruned, leaving only the youngest, 6- to 8-inch (15- to 20-cm) shoots. Take care when making divisions to dig deep to get the growing point because the rhizome can grow 12 to 14 inches (30 to 36 cm) deep. Each new division should consist of a single rhizome with an undamaged, blunt growing point, some new aerial shoots, and, most important, some large fleshy storage roots.

FIG. 2. *Alstroemeria* grows from underground rhizomes. Note the new shoots developing along the rhizome top. Some shoots will flower, and some, known as blinds, will not. *Alstroemeria* develops fleshy storage roots for storing carbohydrates, as shown in the photo.

It is essential that the pots or ground beds be ready for planting before the divisions are received or the plants are divided, since the rhizomes should be planted immediately. Extra plants may be potted up to replace plants that die or are not as vigorous. A fungicide drench is recommended at the time of planting, and again a month later if vigorous root growth is not observed. Excess watering will quickly rot the rhizomes. After the initial watering with the fungicide drench, spot-water plants as they dry. Grow the plants at 65 to 70F (18 to 21C) night temperatures until they become well established (four to eight weeks), prior to lowering temperatures.

As new growth commences, numerous shoots will form (fig. 2). Remove some of the weak vegetative shoots to encourage growth of lateral rhizomes. A shoot should be pulled, not cut, with a quick upward pull to cleanly remove it from the rhizome. Exercise care with young or poorly rooted plants because the rhizome may be uprooted or torn loose from the soil by careless stem removal.

Alstroemeria can be started from seed, but seed sources are few, and genetic variability exists. Fresh seed can be quickly germinated. First scarify the seed by pouring boiling water over them, then allow them to imbibe the water for eight hours or overnight. It is best to replace the water at least once during the imbibition process. After the seeds have been soaked, plant them into well-drained media, keep moist and grow at 75F (24C). If the seeds have not germinated in four weeks, they should be placed in a refrigerator for another four weeks or until they begin to germinate. After the cold treatment, seedlings and any ungerminated seeds should be returned to 75F.

THE CUT FLOWER CROP

Growth and culture of plants. Under normal greenhouse conditions in most parts of the United States, plants flower from January until August. As a result, transplants are preferably planted in the summer months through early fall, although transplants can be planted at any time of the year. For cut flower production, ground beds, high-sided raised beds, or large pots may be used. Any loose, well-drained, organic medium is suitable for alstroemeria plants. The beds should be 6 to 8 inches (15 to 20 cm) deep to allow the roots to grow during the three- to four-year production cycle.

Alstroemeria plants are normally spaced 18 to 24 inches (45 to 60 cm) on center in beds, depending on the variety and the number of years expected in those beds. After planting, support lines with 8- by 7- or 8- by 8-inch (20- by 17-cm or 20- by 20-cm) openings should be set up immediately, then raised later as needed.

Proper watering is the key to success. Alstroemeria beds are often watered with perimeter watering or trickle tubes down the center of the bed. Do not overwater newly established plants, or rhizomes may rot. However, established plants need abundant water and should never by allowed to remain dry for very long. Alstroemeria plants are heavy feeders, with high nutrient levels required once the plants are established. Regular fertilization with 400 ppm N each week is important for good growth. The number of flowers and the number of florets per flower will increase as nitrogen is increased to 400 ppm. Avoid ammoniacal forms of nitrogen fertilizer, because ammonia is not readily converted to nitrate under cool growing temperatures. High soluble salt levels (greater than 1.2 to 1.5 mS/cm) should also be avoided because they reduce flower production and quality.

Optimum greenhouse growing temperatures for alstroemeria are air temperatures of approximately 50 to 61F (10 to 16C) nights and 65 to 70F (18 to 21C) days. Prolonged temperatures over 75F (24C) may decrease or stop flowering. The air temperatures are not as crucial, however, as the rhizome temperatures; media temperatures should be kept between 55 to 61F (13 to 16C) to induce flowering. In warm months the flowering period may be extended by soil cooling. This is done by placing small-diameter tubes on the surface or slightly below the surface of the growing media, spaced 8 to 12 inches (20 to 30 cm) apart. Cool water is circulated through the tubes to produce lower media temperatures.

Alstroemeria respond well to high light; they can even be grown outdoors in full sun if the soil surface is mulched and adequate water is provided. Depending on the variety, the length of the flowering stems may be shortened by high light intensity. If plants are being grown in areas of low light, supplemental light will hasten flowering and increase flower production. Approximately 600 footcandles of light supplied at canopy height (approximately 5 to 6.5 feet, 1.5 to 2 m, above the ground) with an increased photoperiod of 16 hours light is sufficient. Plants exposed to supplemental light in the fall will flower, up to 12 weeks earlier, and produce 30% more flowers than plants grown under natural days. This response to additional light will not occur if the plants are not also induced with cool temperatures.

If alstroemeria cannot receive supplemental light from high intensity discharge (HID) lights, there is still some advantage to supplying long days with incandescent lights. Plants that receive 16 hours of light per day will flower faster than plants receiving fewer than 13 hours. Day length extension can occur by night-interruption or day-continuation treatment. Lighting should not occur during the first 45 to 60 days after new plantings, however. Lighting may be used on established plantings from about September 1 to April 1 at the northern latitudes.

Thinning procedures are used on alstroemerias to increase the light intensity getting to the interior of the plants, remove weak vegetative stems, stimulate rhizome branching, and stagger flowering. During the low-light winter months, 15 to 25% of the vegetative shoots are removed until flowering begins. Thinning is usually accomplished by pulling out stems, rather than by cutting.

Alstroemeria plants are relatively disease- and pest-free. Snails or slugs can be a problem in field culture, as can aphids, caterpillars, and whiteflies. The biggest pest problem is thrips because they are very difficult to control and may transmit viruses. Plants that are virus-infected should be rogued from the beds and destroyed.

Botrytis and root rots can be problems during periods of low light intensity. *Botrytis* is the most prevalent disease and can be avoided with good air circulation, removal of infected plant parts, and preventive fungicides. Root rots can be avoided by sterilizing benches, pasteurizing media, using well-drained media, avoiding overwatering, and applying fungicides at the time of planting.

As a pot plant

Alstroemeria has traditionally been grown for the cut flowers. However, since about 1985, it has been grown as a garden flower and as a potted plant. Alstroemerias make beautiful, unique potted plants that can be placed outdoors to adorn the deck or patio or planted in the garden. Interest in short alstroemeria as a potted crop and as garden plants has increased dramatically since 1990. Traits, such as their ability to be used as long-lasting cut flowers, their everblooming characteristic once induced to flower, preference for cool temperatures, and the potential for new flower colors—including white, pink, red, purple, lavender, yellow-orange, and bicolors—have made them plants in demand.

There are now at least three major breeding programs for compact alstroemeria in the United States, and prefinished liners are becoming more available. At this time, Erwin Mojonnier Enterprises of Encinitas, California, and Coast Alpine Nursery of Lummi Island, Washington, are the largest suppliers of seed-propagated liners. Vegetatively propagated clones of the patented Constitution series are available from the University of Connecticut. These include Redcoat (red flowers), Liberty (purple/pink-striped), Patriot (purple), Freedom (pink), and Sweet Laura (yellow, fragrant).

Greenhouse culture. Compact-growing varieties can be grown in pots 6 inches (15 cm) or larger to produce nice, full plants. Due to the extensive root systems that develop, standard pots are recommended over azalea pots. One 2½-inch (7-cm) liner can be planted per 15-cm pot. Liners should be planted in any well-drained media directly in the final pot. Rhizomes should be planted shallow, with the growing points 1 inch (2.5 to 3 cm) below the surface; do not expose the growing points on the surface. Shallow planting allows for earlier flowering and bushier plants. The large storage roots that sometimes accompany the rhizomes may be planted at any depth. Vegetative shoots that arise from the rhizomes of transplants should not be cut back when planting liners, since this would delay flowering two to three weeks. Remove only damaged shoots.

Plants that are to be sold in pots during the spring can be potted in the fall or 90 to 100 days before marketing. Pots planted in the fall allow for root growth with minimal care other than occasional watering. After liners are established, temperatures can be lowered to as low as 33 to 38F (1 to 3C) in refrigerators, cold frames, or cold greenhouses, and the roots will continue to grow and fill the pots. Approximately 90 to 100 days before the pots are to be marketed, the foliage can be cut back completely, and the pots can be moved to warmer greenhouses of 55 to 63F (13 to 17C). Removing the foliage will help keep the plants short. However, liners planted 90 to 100 days before marketing should not have the foliage cut back.

Good cultural practices in the greenhouse are still the best method to control the height of potted alstroemeria. Plants can remain short by choosing compact varieties, growing at cool temperatures, not keeping the plants wet, and increasing light intensity with adequate spacing. When plants are first potted, they can be grown pot to pot until the foliage starts to touch that of the next pot. The plants should then be adequately spaced to maximize light exposure. Pots 6 inches (15 cm) in diameter should be spaced 15 by 15 inches (38 by 38 cm).

Alstroemeria plants respond to pruning for height control; the more they are pruned, the more compact they will grow. Every two to three weeks, plants should be "shaped up" by pulling out dead and unsightly stems. This process will encourage new shoots to grow shorter. Pulling out the stems, instead of cutting them, will encourage more lateral breaks of the rhizome. However, if the roots are not well established, the whole plant could be pulled from the pot. These plants are monocotyledons, so once a shoot is pinched back, it will eventually turn yellow and die. No axillary shoots will be produced from a pinched shoot; instead, new shoots will arise from the rhizome. There is no commercial growth regulator registered for the control of alstroemeria height.

Full details on watering, fertilization, lighting, and requirements for flowering are found earlier in this chapter.

Growing alstroemeria in the garden. Alstroemeria are grown in the garden as herbaceous perennials in regions warmer than U.S. Zone 6. In colder regions they are treated as annuals or tender perennials. The degree of winter hardiness is dependent on the hybrid, where the hybrid was developed, the zone of growth, and the type of overwintering protection. If the plants are planted in "pansy time" after the ground is workable, they will start to flower when frosts have ended and continue all summer and fall until frost. In warmer parts of the southern United States, alstroemeria are perennial. They start to flower in February and end during the heat of summer; they rebloom in September until as late as December. Along the West Coast (fig. 3), plants will flower all summer.

In the garden, alstroemeria can grow in full sun or partial shade. Height of the plants will vary from 1 to 3 feet (30 to 91 cm), depending on the variety, the amount of light the plant receives, and plant culture. Flowering stems are shorter on plants that are grown in full sun. Plants will also remain shorter during the growing season if flowering stems are removed after they bloom. Alstroemeria plants that are grown in the garden respond positively to the application of mulch, watering, and fertilizer applications.

Rhizomes can be stored for several months if they are kept cool (34 to 37F, 1 to 3C) and not allowed to dry out. Storage in peat moss, thick-grade vermiculite, or some other light, well-drained compound is optimal. Rhizomes wrapped in plastic must be periodically checked to guarantee that they do not become too wet for long periods.

FIG. 3. Alstroemeria can be grown outdoors with minimal cover in some geographic regions. This fine field of alstroemeria is in Watsonville, California.

ALYSSUM, SWEET (SEE *LOBULARIA*)

AMARYLLIS (SEE *HIPPEASTRUM*)

ANEMONE (WINDFLOWER)

by Simon Crawford
Colegrave Seeds Ltd.
West Adderbury, England

and Grace Price
PanAmerican Seed Co.
West Chicago, Illinois

❖ *Annual* (**Anemone coronaria**). *56,700 seeds/oz (2,000/g; clean seed). Germinates in seven to 14 days at 60F (16C). Can also be produced from corms.*

Anemone coronaria, the most widely grown anemone species, originated in southern Europe around the Mediterranean. Anemones are widely grown in Europe and Japan as greenhouse cut flowers and as garden plants.

Until recently, all cultivars were grown from corms produced by specialist growers. Examples of these are the de Caen and St. Brigid (double and semidouble) selections. Although still widely grown as cut flowers, they suffer from having a poor color range, small flowers, and short stems. Another corm-raised cultivar, St. Piran, has larger flowers and longer stems, but it still has a limited color range. Recent breeding work in Israel has resulted in new corm-raised anemones called Jerusalem hybrids. This high quality series has long stems and medium flower size.

More recently, the F_1 hybrid anemone Mona Lisa has been introduced; it is grown from seed and has marked improvements in plant quality. Mona Lisa is grown predominantly for cut flower production. The colors include wine, pink, white, blue, orchid, red, and a bicolor red and white. The benefits of Mona Lisa are strong 17-inch (43-cm) stems, 4- to 5-inch (10- to 13-cm) blooms, high productivity, a vase life of 10 to 14 days, and fewer disease problems than plants grown from corms. With a good growing regime and using a planting density of 12 to 15 plants per square yard (14 to 18/m²), at least 125 stems per square yard (150/m²) can be harvested. Anemones are energy-efficient, being grown in cool houses. Compared to many other cool cut flower crops, such as carnations, they are less labor-intensive, as they do not require staking, stringing, or disbudding.

SOWING SEED, AND EARLY TRANSPLANTING

Sow Mona Lisa seed in mid-March to mid-April in well-drained, peat-based media. If seed flats are used, sow seed one-half inch (1.3 cm) apart. Seeds of

Mona Lisa are "defluffed," which makes them much easier to sow than those of other anemones. Cover the seed lightly (one-tenth inch, 22 mm) with a soilless medium and use a Banrot drench to water the seeds in. Mona Lisa should be germinated at 60F (16C). Higher temperatures reduce germination percentage.

Transplant eight to nine weeks after sowing, when the seedlings are about 1 inch (2.5 cm) long. Transplant into cells that are 1 inch in diameter. When seeds are sown in plugs, use 288 size plug trays. Keep plug trays in a cool chamber (60F) with high humidity until emergence, then move them to a greenhouse bench, where ambient temperatures should be approximately 70F (21C). Plugs should be kept constantly moist. A coarse vermiculite cover should help prevent water stress and algae buildup. Applying Agribrom or Physan 20 at the recommended rate should control algae. Plugs should be kept within a pH range of 5.8 to 6.5. With higher pH, chlorosis can develop. Young plants benefit greatly from a constant or alternate feeding using 20-10-20 peat lite special at 75 to 100 ppm. Seedlings can be transplanted to size 72 plugs four to five weeks after sowing (after the second true leaf appears).

Many cut flower growers have raised young anemone plants from seed, but the worldwide trend is to buy in plugs of Mona Lisa, thereby streamlining production.

PREPARING FOR YOUR TRANSPLANTS

The kind of growing medium influences plant spacing and bed setup. A raised bed will be of most benefit with a heavy soil, while flat beds will be more adequate for a well-drained soil.

Planting schemes may vary, as described below. The aim is to obtain 12 to 15 plants per square yard (14 to 18/m²). If a soil-based medium is used, it should be sterilized. Mulch to keep the soil temperature down.

Before transplanting, make a soil test to determine specific fertilization needs. It is necessary that the pH of soil-based media remain neutral (pH 6.8 to 7) throughout the growing season. The pH of soilless media should be approximately 6.2 to 6.5. Add 2 to 2½ pounds of superphosphate per cubic yard of medium (about 1.2 to 1.5 kg/m³).

Spacing. Several bench spacing plans are used in Europe. Most European crops are grown in ground beds. Often the walk is dug out, and the growing beds are raised 4 or 5 inches (10 to 13 cm) above the walk level. The goal is to ensure good drainage.

Scheme 1: Two-row beds widely spaced. Two rows 12 inches (30.5 cm) apart, a 30-inch (76-cm) space, and two more rows 12 inches apart. Plants are spaced 6 to 7 inches (15 to 17 cm) apart in the rows. Harvesting is easy, and there is good air circulation around the plants.

Scheme 2: Four-row bed. Four rows spaced 10 inches (25 cm) apart, with plants spaced 6 to 7 inches (15 to 18 cm) apart. A narrow, 16- to 18-inch (41- to 46-cm) walk separates the beds. This plan is used on ground beds with, again, soil mounded above the walk level. This spacing is also used on raised benches.

TRANSPLANTING

When transplanting, take care to prevent injury to the delicate root systems. Initial root growth can be stimulated by adding 150 ppm N of a starter solution (9-45-15), or 150 ppm N of calcium nitrate if superphosphate has been added to the growing media. After about two weeks, drench with a broad spectrum fungicide, such as Subdue, to prevent root rot.

To grow Mona Lisa in a pot, we have found that the best performance comes from using a 6- to 8-inch (15- to 20-cm) pot, one plant per pot. On the bench, place pot to pot, or space pots 2 inches (5 cm) apart at the rim to provide better air circulation. Pot-grown Mona Lisa may take longer to flower than bed-grown anemones due to the higher soil temperature in the pots.

CULTURE

Water before noon to allow the foliage to dry completely before sundown. Take precautions to keep water off the foliage. If possible, it is better to use a ground-level watering system.

Base fertilizer use on media and water tests. We recommend a fertilizer low in ammonium or urea N, as acid-forming fertilizers may decrease media pH when irrigation water has a low buffering capacity. Try 20-10-20 Peat-Lite Special at 150 to 200 ppm N. If no superphosphate has been added to the media, it may be best to use 15-16-17 or 15-17-17 (no sodium) Peat-Lite at 150 to 200 ppm N. It may be advisable to leach with clear water occasionally to decrease media soluble salts. Do not shock Mona Lisa with high fertilizer rates or irregular water schedules, as they can cause cracking of the flower stems. For stronger stems fertilize occasionally with calcium nitrate at 200 ppm.

Anemones grow best in cool, shady conditions. In areas of extremely high temperatures or high light intensity, heavy shading may be required. Shading should be removed in areas where cooler temperatures and cloudy weather persist throughout the growing season. During warm weather when cooler night temperatures are not possible, plants do better when grown in houses equipped with fan-and-pad cooling.

The most common diseases and insects affecting anemone are listed in table 1. If in doubt about treatments for these problems, please seek advice from a specialist.

TABLE 1   DISEASES AND INSECTS OF MONA LISA ANEMONE

Disease or pest problem	Symptoms
Botrytis cinerea	Yellowing of older leaves; soft rot at base of plant; gray mold.
Rhizoctonia	Yellowing foliage; softening and blackening of crown at soil level. Associated with wilting.
Pythium	Bluish color on foliage; wilting; a definite blackening in the crown.
Downy mildew	White, powdery fungus growing on leaves.
Colletotrichum (leaf curl/anthracnose)	Stunted and gnarled leaves; irregular margins; flower deformity or discoloration (in severe cases). High temperature and humidity promote the disease.
Aphids, whiteflies, thrips	Mottled tracks on leaves; distortion on flower.

FLOWER HARVESTING AND STORAGE

Mona Lisa flowers close naturally at night. Flowers are best cut as early as possible—before greenhouse temperatures rise and flowers begin to open. Cut the flower close to the crown, but use a clean, sharp knife, as this can become a site of disease infection. Sterilize the knife frequently.

Flowers can be cut at the closed-bud stage and stored for several days in water containing an antibacterial agent, such as Floralife. This will produce a medium-sized flower. For maximum flower size, allow the flower to open and close once before cutting, bearing in mind that this will reduce the vase life.

Flowers can be kept in cold storage prior to shipping, but we recommend that they be held at 34F (1C) for no longer than 14 days. Otherwise, stored blooms can be cooled to 40F (4C) when they are in the flower shop.

Stems need not be shipped in water if markets are close by. *Do not* ship stems lying down in a box, since "crooking" of the necks can occur. Always demand that boxes be shipped with flowers standing upright.

ANTHURIUM

by Gary Hennen
Oglesby Plant Laboratories, Inc.
Altha, Florida

❖ *Tropical plant* (**Anthurium ssp.**). *Vegetatively propagated, commonly by tissue culture. Growers typically purchase tissue culture micro-cuttings or liners.*

Anthuriums have been cultivated for many decades for cut flower production. Since the mid-1980s, anthurium's popularity as a flowering pot plant has increased dramatically, and it has become a popular addition to many foliage growers' product lines. Anthuriums are relatively easy to grow, have attractive foliage, and under the proper environment produce long-lasting flowers year-round. Currently, numerous varieties with different flower sizes, shapes, colors, and some with delicate fragrances are available. Commercially, anthuriums are grown throughout the world, with the heaviest concentrations in the United States (Hawaii and Florida) and the Netherlands (fig. 1).

Anthuriums can be divided into four basic groups: *A. andraeanum* cultivars, interspecific hybrids between *A. andraeanum* cultivars and dwarf species, *A. scherzeranum* hybrids, and foliage anthuriums. The interspecific hybrids are sometimes referred to as Lady Jane types (referring to the first widely available cultivar) or with a newer term, Andreacola types.

FIG. 1. Millions of consumers think of anthuriums when they think of tropical flowers. As a cut flower, *Anthurium* is produced in the Caribbean, Central and South America, and Hawaii for export to the United States. Florida foliage growers have made *Anthurium* a year-round flowering pot plant that consumers can buy from their supermarkets and garden centers.

Anthurium andraeanum, a generally large, somewhat open-structured plant with large flowers, is commonly grown for cut flower production and sometimes adaptable to pot culture. New andraeanum cultivars specifically selected for pot culture are more compact. *A. andraeanum*'s primary flower colors are white, pink, red, red-orange, and green. Andreacola cultivars are small to intermediate in overall size, fuller, and more compact, and generally produce smaller but more numerous flowers than andraeanum cultivars. Andreacola cultivars tend to have thicker, dark green leaves and many times show resistance to the more aggressive anthurium diseases. Primary flower colors are white, pink, red, and lavender. *A. scherzeranum*, the first widely cultivated anthurium pot plant, is small and compact. Primary flower colors are white (sometimes with polka dots), pink, and red. Foliage anthuriums come in numerous shapes and sizes and represent a minor portion of the total anthurium pot market. However, it should be noted that most foliage anthuriums are durable plants and offer the consumer distinct forms.

Culture

Most anthurium species are native to tropical rain forests and are primarily epiphytic in nature. Thus, in their natural habitat, they receive ample, frequent water with good drainage. In cultivation, anthuriums prefer evenly moist media, especially when actively growing. Overall, it is better to slightly underwater than overwater. Drying out may cause tip burn, root damage, and reduced growth rates, while overwatering can cause root damage and sudden yellowing of older leaves. Anthurium will not tolerate saturated, poorly drained media. Best results are achieved with a 1:1:1 ratio of Canadian peat, composted pine bark (watch for particle size, with not too much dust), and perlite or airlite. Avoid vermiculite, except in 4-inch (10-cm) containers. In long-term crops—that is, 6 inches (15 cm) and up—vermiculite compacts and will waterlog. Soil pH should be maintained between 5.5 and 6.5.

Young plants are primarily propagated by tissue culture and available commercially either as microcuttings or as 72- or 98-cell liner trays. Depending on the cultivar's inherent branching and flowering habit, young plant producers use one to three plants (microcuttings) per liner cell. Cultural conditions, especially light intensity, are very important for young plant production. Finish growers should avoid using young plants grown under low light conditions.

Crop time. Most pot anthuriums are sold in 6- and 8-inch (15- and 20-cm) containers, with a smaller percentage in 4 and 10 inches (10 and 25 cm). Crop finish times will vary, depending on cultivar, pot size, and cultural environment. Except in the case of *A. scherzeranum,* growers should consider anthuriums a long-term floral crop. Under the subtropical climate of Florida, most 6-inch container crops are finished in eight to 10 months, using 72- or 98-cell tray young plants. Scherzeranum is usually grown in 3½- to 6-inch (9- to 15-cm) containers and will finish in four to seven months. A young-plant supplier will be able to give recommendations on the optimum container size and finish times for each individual cultivar.

Nutrition. Moderate but consistent levels of a complete fertilizer are important. Magnesium requirements in anthurium plant tissue are higher than for most foliage crops, especially in warmer climates. Because of the long-term nature of the crop, special attention must be paid to ensure continued availability of magnesium. Per cubic yard of media, incorporate 10 pounds (4.5 kg) of dolomite and 3½ pounds (1.6 kg) of Hi-Cal lime to balance the calcium and magnesium ratio. Regular foliar applications of magnesium sources (such as Epsom salts or magnesium nitrate) will help prevent magnesium deficiencies. After 24 to 26 weeks, a topdressing of dolomite (3 tbsp per 10-inch pot) or another magnesium source will help ensure continued availability of magnesium. Topdressings of Epsom salts are beneficial but short-lived.

Generally, a 1:1:1 ratio fertilizer is recommended. Avoid high nutrient levels, especially after planting young plants. Liquid fertilizer on a constant-feed program should not exceed 250 ppm N. On actively growing mature plants, occasional rates as high as 400 ppm N are acceptable, but such feedings must be alternated with clear irrigations. Tests have shown that plants given frequent doses of 300 to 400 ppm N actually grow slower, have lighter flower colors, and produce thick, deformed leaves. When using an overhead irrigation system to dispense liquid fertilizer, a quick rinse with pure water is beneficial because liquid fertilizer left on foliage can damage leaves, causing grayish, corky scars. With dry fertilizer applications, it is very important to water frequently to reduce salt buildup. When using time-release fertilizers, carefully consider crop times; if necessary, reapply to avoid deficiencies.

Temperature. Anthuriums grow best with day temperatures of 78 to 90F (26 to 32C) and night temperatures of 70 to 75F (21 to 24C). Temperatures above 90F may cause foliar burning, faded flower color and reduced flower life. Night temperatures between 40 to 50F (4 to 10C) can result in slow growth and yellowing of lower leaves. Scherzeranum cultivars require lower temperatures, in the range of 68 to 80F (20 to 27C) daytime and 60 to 70F (16 to 21C) nights. Anthuriums will not tolerate frost or freezing conditions.

Light. Anthuriums grow under a wide range of light intensities, but their actual performance is dependent on the cultivar, elevation, temperature, and nutrition. Generally, most anthuriums grow well at light intensities ranging from 1,500 to 2,500 footcandles (16 to 27 klux). Light intensities higher than 2,500 f.c. (27 klux) can improve branching habits (i.e., fullness) but can result in faded flower and leaf color. Some growers use light intensities between 3,600 and 5,000 f.c. (39 and 54 klux) during the early crop stages to improve branching, then move the crop to lower light intensities for finishing. Scherzeranum cultivars are best grown at light intensities between 1,000 and 1,500 f.c. (11 to 16 klux).

Pest control. Preventive maintenance programs for mites, snails, slugs, worms, thrips, and whiteflies are important. Whiteflies are especially attracted to the new growth and, once established, are difficult to eradicate. A number of chemicals are effective for pest management; however, cultural conditions and cultivars will determine what you can safely use. Many growers have experienced phytotoxicity on numerous anthurium cultivars from certain pesticides. Never apply pesticides while plants are under any form of stress, such as moisture or hot temperatures.

Anthurium andraeanum cultivars are generally susceptible to bacterial blight, *Xanthomonas campestris* pv *dieffenbachiae*. This very aggressive disease starts as foliar necrosis, eventually leading to a systemic infection. *Xanthomonas* is spread by excessive water, splashing, and contact with infected plants or tools. Susceptible cultivars may perform best under hard cover with drip irrigation and very well-drained media. There are no effective chemical controls for blight.

Andreacola cultivars are generally resistant to *Xanthomonas*; however, they are somewhat susceptible to the waterborne fungi *Phytophthora*, *Rhizoctonia*, and *Pythium*. Although there are a number of effective fungicides for these diseases, the best approach is prevention via cultural practices. Keep plants off the ground, provide good ventilation, and avoid overhead irrigation during late afternoon or evening hours. As a matter of caution, every new pesticide should be used in a controlled test on a small percentage of each cultivar grown. Always allow four weeks for phytotoxic symptoms to appear.

ANTIRRHINUM (SNAPDRAGON)

by Brian Corr and Linda Laughner
PanAmerican Seed Co.
West Chicago and Elburn, Illinois

❖ *Annual* (**Antirrhinum majus**). *180,000 seeds/oz (6,350 per g).*
  Germinates in one to two weeks at 70 to 75F (21 to 24C).

Interest in greenhouse forcing of snapdragons (F_1 hybrid *Antirrhinum*) continues to increase at both the grower and floral designer levels. Many growers use snapdragons as a profitable component of their cut flower programs (fig. 1).

CUT FLOWER CULTURE AND SCHEDULING

Seedlings. Stage 1 lasts from sowing to radicle emergence (six to eight days). Maintain soil temperatures at 65 to 75F (18 to 24C). Keep soil evenly moist but not saturated. Do not cover seed; light is not necessary for germination until the radicle emerges. Keep soil at pH 5.5 to 5.8 and soluble salts (EC) less than 0.75 mmhos/cm (2:1 extraction). Snapdragons are very sensitive to a high starter charge in the mix. Keep ammonium levels less than 5 ppm.

With Stage 2 stem and cotyledon emergence occurs. Maintain soil temperatures at 65 to 70F (18 to 21C). Reduce moisture levels once radicle emergence occurs. Keep soil evenly moist but not saturated for best rooting. Light levels should be between 450 to 1,500 footcandles (5 to 16 klux). Keep soil pH at 5.5 to 5.8, and EC less than 0.75 mmhos/cm. Maintain water alkalinity at 60 to 100 ppm. Begin fertilizing with 50 to 75 ppm N from calcium nitrate- and potassium nitrate-based fertilizer once the cotyledons are fully expanded. Snapdragons are very sensitive to high salts and high ammonium levels. Liquid fertilization may not be necessary at this stage if sufficient nutrition was incorporated in the growing medium before planting. Irrigate early in the day so the foliage is dry by nightfall to prevent diseases.

FIG. 1. Colorful snapdragons make excellent cut flowers or bedding plants. Featured in the photo are cut flower trials being conducted in the Netherlands by PanAmerican Seed.

Stage 3 consists of the growth and development of true leaves. Maintain soil temperatures at 62 to 65F (17 to 18C). Allow the soil to dry thoroughly between waterings but avoid wilting. This will produce the best root growth. Increase light levels to 1,000 to 2,500 f.c. (11 to 27 klux). Maintain soil pH at 5.5 to 5.8 and EC less than 1.0 mmhos/cm. Increase feed to 100 to 150 ppm N from 20-10-20, alternating with 14-0-14 or another calcium nitrate and potassium nitrate fertilizer. Supplement with magnesium one to two times during this stage, using Epsom salts or magnesium nitrate at 1 pound per 100 gallons (479 g per 400 l). Do not mix magnesium sulfate with calcium nitrate because a precipitate would form. Occasional leaching with clear water is helpful to reduce soluble salts. Attempt to maintain a ratio of approximately 3 potassium to 2 calcium to 1 magnesium in the medium. Avoid ammonium-based fertilizer if growing below 65F (below 18C). Apply fungicides at the lowest recommended rate as needed to control *Pythium, Rhizoctonia,* and *Thielaviopsis.*

At Stage 4 the plants are ready for transplanting or shipping. Maintain soil temperatures of 60 to 62F (16 to 17C). Allow the soil to dry thoroughly between waterings. Maintain soil pH of 5.5 to 5.8. EC should be less than 1.0 mmhos/cm for transplanting, less than 0.75 mmhos/cm for shipping. Fertilize with calcium nitrate- and potassium nitrate-based fertilizer as needed. Do not use ammonium-based fertilizer.

Transplanting and growing on. Snapdragons are commonly grown in field soil or soilless media directly in the ground or in raised beds or benches. Snapdragons grow best in a growing medium that allows adequate aeration to the roots yet holds a steady supply of moisture. The greater the aeration of the medium, the more forgiving the medium is to overwatering. Media with high aeration will require frequent irrigation. Growing media in benches must be better aerated than media used to grow snapdragons directly in the ground because the bench bottom creates a "perched water table" that limits water drainage.

Ground beds in locations with sandy loam soils may be suitable for growing snapdragons without any amendments. Heavy soils should be improved prior to planting by tilling in organic material, such as peat moss, rice hulls, compost, and decomposed manure. The growing media for raised benches should consist of less than 50% field soil, with the remaining percentage consisting of a mixture of more than one of the following: vermiculite, perlite, peat moss, composted bark, rice hulls, and so on.

Whether using soil or a soilless medium, it must be free of disease. Some media, such as perlite, vermiculite, peat moss, or composted bark, are naturally free of disease organisms and may be used without treatment. However, they must be treated or replaced if diseases do develop. Most growers disinfect their growing media before they plant. Pasteurization with high temperatures is common.

Test the growing media prior to planting. Fertility should be moderate, between 1.0 and 1.75 mmhos EC, with less than 10 ppm ammonium nitrogen. Medium pH should be between 5.5 and 6.5. The more mineral soil (field soil) included in the mix, the higher the optimum pH. Organic media should have a pH at the lower end of this range. Amend the soil to adjust the pH several weeks prior to planting. Water the media and retest before planting to determine if the desired changes have occurred.

Snapdragons are often described as "light feeders," yet no crop can grow well with inadequate nutrition. Phosphorus and calcium are usually incorporated into growing media prior to planting, and the other nutrients are supplied with a soluble fertilizer during growth. Superphosphate incorporated at 5 pounds per 100 square feet (240 g/m²) should supply sufficient phosphorous for the entire crop, except in very porous media. If media tests show calcium is low, incorporate either limestone (if the pH is too low) or gypsum (if pH is acceptable), at 5 pounds per 100 square feet. If phosphoric acid is used to modify water alkalinity, superphosphate application may not be needed.

Snapdragon plugs are generally ready to transplant four to five weeks after sowing. When the second true leaves unfold, transplant plugs on a spacing of 10 to 12 plants per square foot (100 to 130/m²), decreasing to eight plants per square foot (85 to 90/m²) in seasons with low light. When buying in seedlings or plugs, allow seedlings 24 hours to acclimate to greenhouse conditions, then transplant promptly. Delayed flowering and loss of final product quality occur when seedlings are kept too long in plug trays. If holding is unavoidable, store plugs at 36 to 39F (2 to 4C) under fluorescent lights at 250 f.c. (2.7 klux) 14 hours per day. Treat with fungicide prior to storage to prevent *Botrytis*.

Irrigate seedlings with clear water after transplanting. Begin fertilizing at the next watering, using a well-balanced, low-ammonium (less than 40%) fertilizer at a rate of 150 to 200 ppm N. Constant fertilization, with occasional clear water leaching, can be used until the flower buds are visible. Once they are visible, switch to a fertilizer with more potassium than nitrogen.

Excessive side shoots are an indication of high moisture or fertility levels or just improper variety selection. Maintain a moderate to low media EC (less than 2.5 mmhos/cm) to avoid excessive side shoots. Irrigate with clear water, if necessary, to lower media EC. Light, porous media are less prone to excessive nutrient and moisture levels, resulting in fewer side shoots. It is also important to choose snapdragon varieties from the correct response groups. Subjected to long days and high temperatures, Groups 1 and 2 varieties tend to increase their side branching. If side shoots persist on edge rows, it is best to trim them off to increase light and air circulation reaching in the center of the bed.

Snapdragons need support during production. Two support nets are the minimum, but three are preferred. Mesh sizes of 4 by 4 to 6 by 6 inches (10 by 10 to 15 by 15 cm) are most commonly used and provide adequate support for the stems. Place the first level at 4 to 6 inches (10 to 15 cm) above the soil and the second level 6 inches above the first level. Raise the upper level of the support nets as the stems lengthen. Keep the net below the first flower.

Snapdragon growth and flowering response depend on the interaction of light quality, light duration, temperature, CO_2 levels, humidity, and soil type, as well as other environmental factors. Snapdragons may be grown under various light intensities, provided appropriate varieties are used. The best quality is usually achieved with the highest light levels. Shading may be necessary in some climates, though, for temperature control. While temperature affects overall growth rate, day length and quality of light are the most important factors influencing flower initiation. Initiation in young plants occurs when they have five to 10 pairs of leaves, depending on the response group and individual variety. Unusual environmental conditions during this critical stage (e.g., a long stretch of overcast weather) can greatly affect crop time.

Once flower initiation has occurred, night temperature has the greatest influence on flowering time and final quality. The ideal night growing temperature depends on the response group. For the highest quality snapdragons, optimum night temperatures by variety are as follows:

Group 1: 45 to 50F (7 to 10C)

Group 2: 50 to 55F (10 to 13C)

Group 3: 55 to 60F (13 to 16C)

Group 4: above 60F (16C)

Generally, the lower temperatures in the ranges give the best quality, but at the expense of a longer crop time. The lower temperature is advisable during extended periods of low light.

The highest quality snapdragons can be grown with supplemental HID lights. In this production method, Groups 3 and 4 snapdragons can be grown all year by lighting the plants when natural day lengths are less than 12 hours. Groups 1 and 2 are not recommended for HID culture because they initiate flowers too quickly, which causes short, weak stems. A light duration of 12 to 14 hours per day and 350 to 400 f.c. (3.8 to 4.4 klux) of supplemental light are essential. Optimize conditions by increasing fertilizer to 300 to 500 ppm N and by adding CO_2 at 800 to 1,200 ppm.

Postharvest handling. The best-quality flowers for the consumer are those cut with a minimum of five to seven open florets. Premature harvesting leads to poor color development and reduced flower size as flowers continue to open. This is especially critical on dark colors, such as rose and royal purple.

For maximum vase life, place snapdragon stems in water as soon as possible after cutting. Remove the foliage from the lower third of the stems, then grade and bunch. To condition for immediate use or shipping, place the flowers in warm water (70 to 75F, 21 to 25C) containing floral preservatives and hold at 45 to 50F (7 to 10C) at least six to eight hours or overnight. Select a preservative that contains sucrose as well as 8-HQC (8-hydroxyquinloine citrate) or another bactericide to facilitate water uptake and inhibit stem plugging. Color development is enhanced by holding the stems in the light (200 f.c., or 2 klux). Shattering in response to ethylene can be a problem with some snapdragons. Many shatter-tolerant varieties exist, so the problem can be avoided with careful variety selection. Use a floral preservative containing an ethylene inhibitor such as silver thiosulfate at the recommended rate. Avoid natural sources of ethylene, such as ripening fruit. Ventilate and reduce temperatures to slow ethylene buildup.

Snapdragons should be stored and shipped upright to prevent curvature of the spikes. Place cut stems vertically as soon as possible after harvest; stems placed horizontally may begin to bend upward in as little as 30 minutes. To maintain flower quality, it is important to sleeve the upper portion of the snapdragon bunches and use tall, upright hampers for shipping.

Snapdragons can be stored for three to four days, dry or in water, at 40F (4C). If stored dry, rehydrate and condition in the same manner as for freshly cut snapdragons. For longer term storage (five to 10 days), select only the highest quality stems, wrap each bunch in plastic to prevent desiccation, and hold the stem in a preservative at air temperatures of 32 to 40F (0 to 4C).

Special problems

Medium. High medium pH symptoms include stunted or uneven seedling growth or poor root development. Very high pH may result in iron or boron deficiency. Solution: Maintain the medium pH between 5.5 and 5.8. Test new media before sowing.

Iron deficiency may result from high medium pH or cool, wet medium, which reduces iron uptake. Iron deficiency is indicated by yellow interveinal chlorosis of upper leaves. Solution: Identify the cause of the deficiency. If it's high pH, you can lower the pH with one application of iron sulfate at 33 ounces per 100 gallons (0.25g/l) applied to saturate the medium. If the deficiency is the result of cool, wet conditions, drying out the soil and warming the air temperature will improve growth in three to four days.

Insects and disease. Snapdragons are relatively pest-free in comparison to many crops. Aphids, mites, and thrips are the most common pests on mature plants, while seedlings are damaged by fungus gnats and shorefly larvae. Frequent scouting of the growing area, to find infestations before they become severe, is essential. Many pesticides are available to control pests. Rotate between classes of insecticides after three consecutive applications. Always read the label and check local regulations regarding use of a pesticide before application. Nicotine sulfate, vapona, and malathion have been reported to cause phytotoxicity of snapdragons and should be avoided. Chlorpyrifos and some other insecticides can damage budded spikes.

Fungal diseases that affect snapdragons include downy mildew, powdery mildew, *Pythium, Botrytis,* rust, *Phyllosticta* blight, and anthracnose. **Downy mildew** affects snapdragons as seedlings with stunting chlorosis and downward curling of leaves. On more mature plants, the undersurfaces of the leaves are covered with white, fuzzy growth. The infection checks terminal growth, resulting in a cluster of stunted flower buds. Cool, moist conditions favor development of this disease. Pasteurized media, careful watering, heat, and ventilation help prevent downy mildew.

Powdery mildew affects seedlings as well as mature plants. White, powdery growth occurs on both leaf surfaces but usually starts on lower leaves. Left untreated, the fungus destroys the lower leaves and spreads to the upper plant parts, eventually causing a white, circular blemish on the flower petals.

Pythium is best known as one of the fungi that can cause damping-off in seedling trays. This soil fungus can also exist in snapdragon beds, especially where media are unsterilized or kept overly moist. At low levels, *Pythium* does not kill snapdragon plants, but it does decrease their vigor, resulting in uneven stem lengths, weaker flower stems, and poor-quality, shorter flower spikes.

Botrytis affects mainly stems and flower parts of snapdragons. Tan-colored stem lesions can encircle stems, causing wilting of the upper plant parts. With severe infections, older blossoms and infected areas show gray or light brown spore masses. *Botrytis* infections can lead to significant loss of product quality and decreased vase life, causing light gray or brown spots within the center of the flower or along petal edges. Sanitation is essential in controlling *Botrytis*, since the fungus persists in plant debris.

Snapdragon rust is primarily a disease of field-produced snapdragons. Faint yellow spots on upper leaf surfaces correspond to rusty-brown circular pustules on the lower surfaces. Keep water off the foliage as much as possible to limit infection.

Phyllosticta is generally a problem only on snapdragons grown in hot, humid areas. The disease begins as brown or black foliar spots, which enlarge to form well-defined, light brown lesions dotted with small, black fungal-fruiting bodies. Lesions can also occur on stems, cracking or girdling the stem, which may cause the plant or branch to wilt.

Anthracnose causes leaf or stem spots on snapdragons that are grayish white, sunken, and form oblong areas with dark narrow borders. Usually, infected leaves die and drop. Another fungal disease that is favored by high humidity, it is more common in the fall, when the day-night temperature differences cause condensation on the foliage.

Tomato spotted wilt virus (TSWV-L) and impatiens necrotic spot virus (INSV, previously called TSWV-1) also affect snapdragons and can be especially devastating. In young plants, it is expressed as tan leaf spots. Often these are dark bull's-eye patterns within large spots. However, it is not unusual for snapdragons infected with these viruses to remain symptomless until just before bloom. At this stage, TSWV and INSV are characterized by brown or black lesions along the length of the stem, followed by collapse of the stem in this region. The best method of prevention is to monitor and control western flower thrips, which transmit this virus, especially during the seedling and young plant stages.

The best protection against insects and diseases is prevention. Immediately discard dead or diseased plant material because it harbors insects and diseases. It is also important to keep the area surrounding and including your growing area weed-free.

VARIETY SELECTION

Snapdragons can be produced year-round in most climates. The varieties are separated into four groups, based on their optimal growing conditions (table 1).

Table 1   Variety groupings

Category	Day conditions		Night temps	
	Length	**Light**	**F**	**C**
Group 1	Short	Low	45–50	7–10
Group 2	Short (not as short as Group 1)	Moderate	50–55	10–13
Group 3	Medium to long	Moderate to high	55–60	13–16
Group 4	Long	High	60+	16+

Knowing the relationship between flowering times of varieties allows you to fine-tune crop scheduling. This is especially important in two situations: targeting a key holiday and scheduling a smooth transition between groups. If a white variety scheduled for Christmas harvest consistently blooms too short or too early, try a later-blooming white variety or sow the early white variety slightly later than recommended for that group, assuming all other factors are constant. Sowing and harvest dates presented in table 2 are purposely given in a range to account for varietal differences and regional environmental differences.

Normal weather variations from year to year can still complicate the most well-planned schedule. The fewer environmental controls available (e.g., heat, fans), the more buffers that must be added to guarantee a successful crop, such as using more than one variety or multiple sowing dates of a favorite variety.

Growers often cite the fall transition from Group 3 to Group 2 as the most difficult time to schedule a continuous succession of quality snapdragons. Excessively warm temperatures and high light at the young plant stage (late sum-

Table 2   North American snapdragon scheduling

Group	North[a]			South[a]		
	Sow (weeks)	**Transplant (weeks)**	**Flower (weeks)**	**Sow (weeks)**	**Transplant (weeks)**	**Flower (weeks)**
1	33–35	37–39	50–7	n.a.[b]	n.a.	n.a.
2	37–49	40–1	8–19	34–51	38–4	49–17
	30–32	34–38	44–49			
3	50–11	2–15	20–26	28–33	32–37	40–48
	25–28	28–33	37–43	2–10	5–14	18–24
4	13–23	16–27	27–36	11–26	15–31	25–39

Note: Times given are general guidelines only. Conditions in certain areas may warrant deviations from these ranges.

[a] The North and the South are separated by the 38th parallel, running from about San Francisco on the West Coast through Colorado Springs, Kansas City, St. Louis, Louisville, and to Washington, D.C. in the East.

[b] Not applicable.

mer) can make Group 2 snapdragons bloom too early and too short. On the other hand, unusually cool nights, even after flowers have initiated, can drastically lengthen the crop time of Group 3 varieties. Intermediate varieties are excellent choices for harvest during this period. Alternatively, use the descriptions to choose varieties that help connect Group 2 to Group 3. The logical progression as daylight decreases is Group 3, early Group 3, late Group 2, Group 2.

VARIETIES

Forcing snapdragons are classified into the four groups according to their flowering response to a combination of environmental factors. There is some overlapping of groups and varieties, and some varieties do well in more than one situation. For instance, a variety designated Group 1, 2 will perform well throughout fall Group 2, winter Group 1, and spring Group 2 harvest periods.

GROUP 1

VARIETY	COLOR	RESPONSE
Maryland Flame	Orange yellow	M
Maryland Light Bronze	Lt. bronze	E
Winter Euro Pink	Pink	E
Maryland True Pink	Pink	E
Maryland Yosemite Pink	Pink	M
Winter Pink	Lt. pink	L
Maryland Royal	Purple	M
Maryland Red	Red	E
Maryland Flamingo	Salmon rose	M
Maryland Lavender	Lavender	L
Winter Euro Rose	Deep rose	M
Winter White	White	E
Winter Euro White	White	L
Maryland Ivory	Ivory white	E
Winter Euro Yellow	Deep yellow	M
Winter Yellow	Deep yellow	M

GROUP 2

VARIETY	COLOR	RESPONSE
Maryland Appleblossom	Pink and white bicolor	L
Maryland Plumblossom	White and purple bicolor	M
Maryland Flame	Orange yellow	E
Maryland Dark Orange	Deep bronze	M
Maryland Light Bronze	Lt. bronze	E

Maryland Lavender	Lavender	M
Winter Pink	Lt. pink	ML
Maryland True Pink	Pink	E
Maryland Yosemite Pink	Pink	M
Maryland Royal	Purple	E
Maryland Red	Red	E
Monaco Red	Deep wine red	L
Maryland Flamingo	Salmon rose	M
Monaco Baltimore Rose	Deep rose	L
Monaco Violet	Purple	L
Monaco Rose	Deep rose	L
Winter Euro Rose	Deep rose	M
Maryland Ivory White	Ivory white	E
Monaco White	White	L
Winter White	White	E
Maryland White	White	M
Winter Euro White	White	M
Winter Euro Yellow	Yellow	M
Maryland Bright Yellow	Yellow	L
Winter Yellow	Deep yellow	ME
Apollo Ivory	Ivory white	L
Apollo Purple	Purple	L

GROUP 3

VARIETY	COLOR	RESPONSE
Potomac Appleblossom	White and pink bicolor	M
Potomac Plumblossom	White and purple bicolor	ME
Potomac Early Orange	Lt. bronze	E
Potomac Dark Orange	Deep bronze	M
Potomac Orange	Orange	M
Potomac Early Pink	Pink	ME
Potomac Light Rose	Pink	M
Potomac Pink	Pink	M
Potomac Royal	Purple	L
Potomac Red	Red	L
Monaco Red	Deep wine red	VE
Monaco Baltimore Rose	Deep rose	VE
Monaco Rose	Deep rose	VE
Monaco Violet	Purple	VE
Potomac Rose	Rose	ML
Potomac Ivory White	Ivory white	M
Monaco White	White	VE
Potomac Early White	White	ME
Potomac White	White	M
Potomac Soft Yellow	Lt. yellow	ME
Potomac Yellow	Yellow	M
Apollo Ivory	Ivory white	VE
Apollo Purple	Purple	VE

GROUP 4

VARIETY	COLOR	RESPONSE
Potomac Appleblossom	White and rose bicolor	M
Potomac Plumblossom	White and purple bicolor	E
Potomac Orange	Orange	ME
Potomac Light Rose	Pink	ME
Potomac Pink	Pink	M
Potomac Royal	Purple	ML
Potomac Rose	Rose	ML
Potomac White	White	M
Potomac Ivory White	Ivory white	M
Potomac Soft Yellow	Lt. yellow	E
Potomac Yellow	Yellow	M

VE: Very Early; E: Early; ME: Medium Early; M: Medium, ML: Medium Late; L: Late

RECENTLY RENAMED CULTIVARS

PREVIOUS NAME	CURRENT NAME
Axis Rose	Monaco Rose
Baltimore	Monaco Baltimore Rose
Bismarck	Winter Pink
Columbia	Potomac Early Pink
El Dorado	Maryland Dark Orange
Jackpot	Winter Euro Rose
Kansas	Potomac Early Orange
Montezuma	Winter Euro Yellow
New Mexico	Potomac Dark Orange
Oakland	Winter White
Oregon	Maryland Lavender
PanAmerican Winter Pink	Winter Euro Pink
Peoria	Winter Yellow
Rainier	Winter Euro White
San Francisco	Potomac Early White
Tampico	Potomac Soft Yellow
Winchester	Potomac Light Rose
Yosemite	Maryland Yosemite Pink

FUTURE OPPORTUNITIES

Today's demands for forcing snapdragons are being met by a variety of different types of growers and through several different distribution channels. Snapdragons are grown throughout North America, Central and South American, Japan, and Europe. Demand is also being met by producers in highland tropical and subtropical areas. A grower's choice of distribution channels strongly affects the quantity produced, the maturity harvested, and the price received.

Of the many opportunities available in the cut flower market, we see two specific areas of promise. The first opportunity is that of a traditional cut flower or bedding plant grower to supply a local market with a more mature product. The second opportunity is for outdoor summer production in a nontraditional cut flower production area for sale at a local supermarket or roadside stand. Both of these methods of supplying have led to increased demand for cut flowers and have given the grower a profitable return.

With increased worldwide competition for traditional mainline cut flowers, many growers have discovered forcing snapdragons to be a welcome addition in their move toward diversification. Year-round product availability and ever-increasing color range are two key factors that have increased their use in the floral design industry. Snapdragons are becoming a standard component in the summer flower bouquet.

SNAPDRAGONS FOR BEDDING

Most bedding plant growers and retailers list two or three types of garden snaps on their sales sheet. The well-known F_1 Floral Carpet or Tahiti series are the dwarf strains so widely used for edging and mass plantings. Growing 6 to 8 inches (15 to 20 cm) tall, they flower well in packs and 4-inch (10-cm) pots after 10 weeks from sowing in the South, about 14 weeks in the North.

Chill seed for several days before sowing to improve germination. When the dwarf varieties show breaks, move them to a cold frame or cold greenhouse and grow on at 45 to 50F (7 to 10C) to produce well-branched plants. Allowing 10 weeks, sell medium and tall varieties green.

Also in the dwarf snapdragon class are the "butterfly," or open-flowered, Bells series and the Chimes series, both excellent for pack production.

Growers looking for taller snaps—to grow in pots for the landscape trade, especially—will like Liberty or Sonnet. Both are taller growing, to 24 inches (61 cm), but do not get as tall as Rocket. Sonnet and Liberty can be used as garden cut flowers.

On the tall end of the scale are the F_1 Rocket and Madame Butterfly, usually grown for cut flower purposes and reaching a height of 30 to 36 inches (76 to 91 cm). In some areas the Rockets are grown for commercial cut flower purposes. Planted in the North by June 1, Rocket can yield two harvests as a cut.

All snapdragons perform best at cool temperatures (45 to 50F, 7 to 10C).

AQUILEGIA (COLUMBINE)

by Jim Nau
Ball Horticultural Company
West Chicago, Illinois

❖ *Perennial* (**Aquilegia × hybrida**). *15,000 to 22,000 seeds/oz (525 to 770/g). Germinates in 10 to 20 days when using fresh seed. Seed stored in a refrigerator may take 21 to 28 days to germinate. Maintain temperatures of 70 to 75F (21 to 24C). Seed should be left exposed to light during germination.*

One of the most requested of the spring perennials due to its unusual flower form, columbine is an excellent plant for the perennial border, as a pot plant, and as a cut flower (fig. 1). Due to the intercrossing between species and varieties, there are a number of flower colors available: carmine, blue, lavender, yellow, pink, rose, and white. The flowers are often two-toned or bicolor—petals are one color and the sepals another—though from seed, pure yellow, white, and violet-blue are also available. Most varieties on the market are single flowering, but double-flowering types are also available. The flowers are of an unusual form, having appendages at the base of the bloom, called spurs, which can be short or long. These spurs give the flower its unusual character. The plants flower in May and June and are hardy to USDA Zones 3 to 8.

Sowings are most often made the summer prior to the spring selling season for flowering pot sales. Germination can be irregular especially on old seed. To improve uniformity, the seed should be moist chilled two to three weeks at 40F (4C) before sowing, especially if holding seed over from year to year. Sown in July, transplanted 30 to 40 days

FIG. 1. *Aquilegia's* unusual flower form makes it one of the most requested perennials by home gardeners. The Songbird series (in photo) has a tidy plant habit and makes an excellent display in pots.

later to packs, and transplanted to quart or gallon containers when ready, plants can be overwintered in cold greenhouses or cold frames for sales next spring. Flowering will occur naturally in April in Chicago within the cold frame. Plants can become quite tall in the gallon containers (2 feet, 61 cm, and more) if not treated with any growth regulator. For green pack sales in the spring, allow 15 to 20 weeks, though plants will not flower the same season from seed.

Late season sowings (August to September) overwintered at 40F (4C) will flower in April. Use one plant per 6-inch (15-cm) pot. For later flowering, drop temperatures to 30 to 33F (–1 to 2C) once plants are rooted in.

The most common variety is McKanas Giants (30 inches, 76 cm), which is a mixture of a number of colors. This is one of the earliest columbines to flower.

The Songbird (F_1 hybrid) series includes a number of separate colors that are sold under such names as Bunting, Robin, Goldfinch, and Dove, to name a few. Check catalog descriptions carefully on this series, since some colors perform better as pot plants, while others are taller, which makes them excellent cut flower candidates. Music Mixture, also an F_1 hybrid, is available in a number of colors, blooming the same time as McKanas hybrids.

Among dwarf flowering varieties, try Biedermeier, which is a mixture of several different colors on plants to 8 to 10 inches (20 to 25 cm) tall in 6-inch (15-cm) pots. Flowers are held upright and measure no more than 2 inches across.

A note to cut flower growers: *Aquilegia* shatters within a day after cutting. Treat with an STS-substitute product after cutting to increase postharvest life.

ARABIS (ROCK CRESS, WALL CRESS)

by Jim Nau
Ball Horticultural Company
West Chicago, Illinois

❖ *Perennial (Arabis caucasica). 70,000 seeds/oz (2,450/g). Germinates in six to 12 days at 65 to 70F (18 to 21C). Seeds should be left exposed to light or very lightly covered during germination.*

Arabis is commonly called rock cress or wall cress due to its low growing habit and excellent footing for planting into rock gardens or rock walls. *Arabis* has mostly white flowers, though sometimes they are rose pink in cultivated varieties. Plants are compact and rosetting and grow to no more than 9 inches (23 cm). Flowers are small, to one-half inch (1 cm), and fragrant. Double- and single-flowering varieties are available, as well as variegated leaf types. Hardy in USDA Zones 3 to 7, *Arabis* will flower in April and May.

Seed is the most common method of propagation, though division and cuttings are popular, too, especially for double-flowering and variegated varieties. Divide plants in the spring or fall, or take cuttings in the late spring after flowering. Seed sown in winter or early spring will flower sporadically at best during the same year as sown. For green pack sales in the spring, allow 12 to 15 weeks. For flowering pot sales in the spring, sow in July of the previous year and overwinter in quart containers. Grow on at 50F (10C).

Snow Cap, the most popular seed variety of *Arabis* on the market, is sold both as seed and in plugs. Flowers are white, single, and held on stems to no taller than 6 inches (15 cm). Compinkie is a rose pink-flowering variety that is available from seed or plugs. Plants grow to 4 to 6 inches (10 to 15 cm) tall. In the garden Compinkie is weaker than white varieties.

The related *A. blepharophylla* Spring Charm is a single-flowering, rose-colored variety that is treated as a short-lived perennial. These plants do best in mild-wintered areas; they usually need winter protection in the home garden to survive even in this type of environment. Sowings made in early October flower profusely in 4-inch (10-cm) pots by Valentine's Day when grown at 55F (13 C) nights. Allow approximately 20 to 22 weeks for winter flowering.

ARGERANTHEMUM (MARGUERITE DAISY)

❖ *Tender perennial* (**Argeranthemum frutescens**). *Vegetatively propagated. Growers produce pots from purchased rooted or unrooted cuttings.*

Marguerites, with their traditional daisy flowers, are loved by gardeners everywhere. Easy to produce for growers, marguerites are increasing in production numbers. The cultural information that follows is provided by Proven Winners.

Plant in January to March for spring finish. Use three to four liners per 8- to 10-inch (20- to 25-cm) pot, one liner per 4- to 6-inch (10- to 15-cm) pot. Allow 12 to 14 weeks to finish. Use a well-drained, peat-perlite mix media with a slightly acid pH of 6.0 to 6.5. Pinch plants once two weeks after planting.

Marguerites require very high light, 5,000 to 9,000 footcandles (54 to 94 klux). Grow at day temperatures of 65 to 75F (18 to 24C), night temperatures of

45 to 55F (7 to 13C). Plants will tolerate low temperatures to 28 to 30F (−2.2 to −1.1C) , and high temperatures to 85 to 90F (30 to 32C). Keep moist but not consistently wet. Avoid wilting. If you use constant-feed liquid, provide 200 to 250 ppm N, 65 to 75 ppm P, and 125 to 165 ppm K. For periodic feeding, use 300 to 400 ppm N, 100 to 150 ppm P, and 200 to 300 ppm K. Growth regulators are optional. If needed, use Cycocel or B-Nine two to three weeks after pinch.

During growing, watch for aphids, thrips, and whiteflies. Marguerites show no apparent chemical phytotoxicity. Drench with broad-spectrum fungicides at planting.

ASCLEPIAS (BUTTERFLY WEED)

by Jim Nau
Ball Horticultural Company
West Chicago, Illinois

❖ *Perennial* (**Asclepias tuberosa**)*. 3,500 seeds/oz (1,225/g). Germinates in 21 to 28 days at 70 to 75F (21 to 24C). Moist chilling may increase germination for lots showing low emergence. Cover or leave seed exposed during germination.*

A plant native to much of the United States, this perennial is becoming more popular as a cut flower. Flowers are bright orange in the wild, although red- and yellow-flowered types are also available.

Propagation is by seed or cuttings. If seed purchased in the late fall or winter is irregular in germination, try moist-chilling it by placing seed on moist sand or in a paper wrap and placing it in the refrigerator (36 to 40F, 4 to 5C) for eight to nine weeks. Plants will flower sporadically their first year from seed sown in the winter or spring. For more reliable flowering, sow the previous year and overwinter plants, or purchase crowns from commercial propagators.

Asclepias may also be vegetatively propagated. Take terminal stem cuttings in the spring, or cut the taproot into 2- to 3-inch (5- to 8-cm) sections and root.

Flowering is more rapid under long days, four to six hours of light interruption during the night, or 14 to 16 hours continuous days. Plants must also have passed through a cooling period (about 14 weeks at 38F, 3C) before they will flower. Once established, asclepias is long lived and may be enjoyed as a perennial or cut flower for years in the same bed.

Gay Butterflies is available from seed and comes in a mixture of orange, yellow, and red on plants that grow to 3 feet (91 cm) tall.

Asiatic/Oriental lily (see *Lilium*)

Asparagus

❖ *Annual and perennial (Asparagus spp.). 500 to 900 seeds/oz (18 to 32/g). Germinates in three to six weeks at 85F (29C) days and 75F (24C) nights.*

Sprengeri is the most widely used species for bedding. It is also used as cut greens in flower arrangements, but is more widely used in hanging baskets, urns, and other patio containers, frequently combined with flowering annuals, such as petunias and geraniums. *Asparagus meyeri, A. falcatus,* and *A. pyramidalis* are other species sometimes used as potted houseplants.

Asparagus pseudoscaber is suitable for cut foliage. A hardy perennial, its dark green spikes have a long vase life. *A. setaceus,* with its fine, lacy foliage, is also used as a cut green.

New crop seed is usually harvested in January or February, and best results can be obtained by using fresh seed. Since germination is strung out over several weeks, the seed flats must be watched closely. A night temperature of 60F to 65F (16 to 18C) promotes good growth. Asparagus prefers light shade and regular fertilization after becoming established. A 2¼-inch (5.7-cm) size requires 14 to 16 weeks of growing time from seed sowing. A finished 4-incher takes about 24 weeks. Today, many growers start with liners that are 10 to 12 weeks old. These are produced by experienced specialists.

Aster (perennial)

PERENNIAL SECTION	CUT FLOWER SECTION
by Dr. Willie Faber	by Jeff McGrew
Yoder Brothers, Inc.	Horticulture Products and Services
Barberton, Ohio	Mt. Vernon, Washington

❖ *Perennial (Aster novi-belgii). Vegetatively propagated. Most growers buy in rooted or unrooted cuttings for pot plant production.*

❖ A FALL-BLOOMING PERENNIAL

Asters can be produced in greenhouses year-round using photoperiod control. They can also be easily grown as natural-season fall-pinched or fast crops. They are perfect for overwintering in standard perennial programs. Aster cultural practices are similar to those of garden mums, so no special growing practices are needed.

Asters are the ideal fall-blooming perennial to complement garden mums. The aster's delicate daisy flowers come in white and various shades of blue (a color not often seen in garden plants, including mums), lavender, pink, and raspberry. For gardeners, asters are very hardy and make a beautiful addition to perennial gardens around the home (fig. 1). When mature, asters can range from 2 to 4 feet (61 to 122 cm) in height, making them desirable for planting between mums and tall-growing shrubs. Garden asters also do well alone as accent plants in flower gardens, barrels, urns, and other large containers.

Asters originated in eastern North America; however, their development as garden plants occurred in Europe. Common names for these asters include perennial aster, Michaelmas daisy, New York aster, garden aster, and autumn aster. Generally, the species involved are *Aster novi-belgii*, *Aster ericoides,* and hybrids of these species. Recently, asters have regained popularity across North America. New varieties bear only a slight resemblance to their ancestors, which naturally flower in late September along banks, roadsides, and fields in North America.

FIG. 1. Asters, which were originally featured as flowering pot plants from Denmark, have become very popular as cut flowers used primarily for bouquet filler. American growers are now finding that asters help lengthen sales by adding to growers' garden mum and pansy fall programs.

GENERAL CULTURE

Any well-drained root media can be used. Peat moss in the media is recommended for better moisture retention. During production, use a complete N-P-K fertilizer that has the majority of N as the nitrate form and contains extra micronutrients, as in the Peat-Lite Specials. Maximum growth is attained when water and fertilizer are not limited, similar to garden mums. Mum fertilization programs have proven satisfactory for asters (200 ppm N in soil mixes; 300 ppm N in soilless mixes). However, asters are more sensitive to excess soluble salts. Leach with clear water or lower fertilization levels to avoid high soluble salts. Use only clear water once the flowers begin to open.

Soil temperatures of 65 to 68F (18 to 20C) are beneficial to root development in the starting area. During growing on, night temperatures should run 62 to 65F (17 to 18C), with day temperatures 5 to 10F (-15 to -12C) warmer. Lower night temperatures of 58 to 60F (14 to 16C) will help intensify colors during the last two weeks of production. In greenhouses, take care to reduce humidity by heating and ventilating moist air before lowering the temperatures to prevent favorable conditions for powdery mildew. Flowers on outdoor aster crops will tolerate a light frost.

Specific spacing for given pot sizes is described later in tables 2 to 5. Asters can be spaced pot to pot until the pinches are given. However, it is important to place asters at final spacing soon after the last pinch. Otherwise, lower leaves may yellow and turn brown from lack of light or foliage diseases that occur because of reduced air circulation and increased humidity from crowding. Prior to final spacing, the leaves should not touch or overlap.

Height control. The majority of height control on asters is accomplished by proper pinching. Medium or tall cultivars may need B-Nine for added height control (fig. 2). One to two applications of 3,750 to 5,000 ppm B-Nine should be sufficient. The first application should be after the last pinch when new shoots are about 1 inch (2.5 cm) long. Apply an additional application 10 to 14 days after the first, if needed. Do not use B-Nine after buds show color, to avoid clubby flower spray formations. Bonzi sprays of 5 to 10 ppm can also be effective in height control. Apply at similar stages of growth as with B-Nine, but remember that Bonzi sprays must be directed toward plant stems rather than to leaves, as with B-Nine.

PINCHING

Pinches create fuller pots. In a greenhouse pot aster program, multiple cuttings with two pinches are generally used. In a natural-season garden aster program, usually only one cutting is used, and multiple pinches are given. In general, the fewer pinches planned, the more cuttings needed to create a full pot.

The first pinch should be given when roots are well developed to the sides and bottom of the finishing container. This generally occurs 10 to 14 days after planting a rooted cutting. Pinch off enough tip so that only four to six leaves remain after pinching. This may be either a soft pinch, less than 1 inch (2.5 cm), or a hard pinch, greater than 1 inch, depending on the height of the original cutting. Rooted cuttings and unrooted cuttings rooted in cell packs for later transplanting naturally become taller during propagation, compared to using unrooted cuttings in direct-stick programs. The first pinch on rooted cuttings will be much harder than for direct-stick unrooted ones.

Subsequent pinches should allow three to four leaves to remain after pinching. Remove a minimum of one-half inch (1.3 cm) of new growth. More growth can be removed as long as three to four leaves remain after pinching. These pinches can be done by hand or with shears. Generally, pinches can be done every two to three weeks.

FIG. 2. This aster from Arhus grower Ib Elmer of Denmark shows how effective B-Nine, combined with the right cultural regime, can be in controlling plant height.

Take care to pinch all the shoots, or else there will be uneven branch height and flowering. Avoid deep pinching. Severe pinching or cutting back of asters can lead to vegetative shoots arising from the base of the plant. These shoots often become taller and flower much later than the upper branches, resulting in an uneven-flowering pot. Follow the above guidelines on leaf numbers remaining after the pinch to avoid this problem.

The last pinch date for natural-season crops should be between July 25 (northern latitudes) and August 10 (southern latitudes) to avoid any flowering delay. Later pinches may be used to delay flowering if desired, but plants must be of adequate size, since little regrowth will occur after very late pinches. Use late pinches on a trial basis only. Do not pinch after late August.

Florel has been trialed on asters to replace pinching (as done for garden mums). It appears that Florel is effective if use begins early in crop production. Only use Florel on a trial basis at rates and methods as described for garden mums.

Flowering response. Asters are similar to garden mums in that they require long days (night lighting 10 P.M. to 2 A.M.) for vegetative growth, and short days (blackout for 10 to 12 hours, no longer) for flower bud initiation and development. Most varieties will be salable five to six weeks after short days begin.

For greenhouse-forced pots or early-shaded crops, we recommend year-round night lighting to allow for maximum vegetative growth. Artificial short days (i.e., blackout) should be used from March 15 to August 15.

Asters seem to flower best under hot and bright conditions when grown under natural-season conditions. In a normal summer, when hot July and August temperatures tend to delay garden mums, natural-season garden asters generally flower one to two weeks earlier than garden mums. However, in cool and cloudy summers, when garden mums tend to flower earlier, asters are generally delayed and could flower seven to 10 days later than garden mums.

PROBLEMS

Insects. Whiteflies and thrips are the key insect pests of this crop. Regular mum spray chemicals (e.g., Avid, Dursban, M-Pede, Marathon, Mavrik, Talstar, Tame, Thiodan) have been effective for such pests on asters.

As with other perennials, the pollen on open aster flowers attracts bees, butterflies, and other assorted insects. It can be enjoyable to watch these insects collect pollen from the flowers. However, the large number of bees that may be found on plants with many open flowers could discourage a customer from picking up a plant. Therefore, plan to sell plants when only a few open flowers are present (about one-fourth open). Use color picture stakes to help with green plant sales.

Diseases. As expected, some aster varieties are more resistant to disease than others. However, reliable information on all varieties is not available. See the notes on varieties in table 6 for specific disease cautions. If conditions are favorable for disease, all varieties are susceptible to some degree. Therefore, the key to disease control for asters involves preventative cultural and chemical programs.

Major diseases found in greenhouse asters are powdery mildew and *Botrytis.* The primary diseases found in outdoor culture are powdery mildew, rust, *Botrytis,* and *Rhizoctonia* foliage blight. Powdery mildew is characterized by a white-gray, powdery coating of fungus on leaf surfaces and stems. Mildew can stunt growth but will rarely kill plants. Rust results in yellow spots on leaf surfaces, with either bright orange fungal growth or raised, brown, scalelike pustules under the leaf. Rust can advance quickly, as it is spread by wind and splashing water. It can also kill plants quickly if untreated.

Botrytis can cause yellowed and brown lower leaves on plants that are overcrowded or grown in areas of high humidity and poor air circulation. Usually, only the lower leaves are affected. *Rhizoctonia* foliage blight is characterized by brown-black necrotic leaf and stem spots, with leaves dying back from the tips

TABLE 1   FUNGICIDE SPRAYS

Fungicide	Concentration[a] (per 100 gal)	Fungus suppressed
Strike 25DF	4 oz	Powdery mildew, rust
Chipco 26019 50WP	16 oz	*Botrytis, Rhizoctonia*
Cleary's 3336 50WP	12 oz	*Botrytis, Rhizoctonia*
Dithane M-45P	24 oz	Rust
Karathane 19.5WP (Don't use Karathane on open flowers.)	8 oz	Powdery mildew
Triflorine 18EC	15 oz	Powdery mildew, rust
Zyban 75WP	24 oz	Powdery mildew

Notes: Before using any pesticide, be sure it is registered for use in your state. Check with your local county extension agent or state university extension service. Always follow label directions.

[a] One oz per 100 gal = 29.97 g per 400 l = 7.49 g per 100 l.

in advanced stages. This disease can quickly kill plants if untreated. These aster diseases generally do not spread to an adjacent garden mum crop.

All of the diseases discussed are favored by high humidity and splashing water. Use the following cultural practices to help prevent favorable disease conditions.

◆ Use proper spacing, done on time.

◆ Avoid overhead watering late in the day.

◆ In greenhouses, heat and ventilate to remove excess humidity prior to lowering night temperatures. This is especially critical whenever cool nights follow warm, humid days.

◆ Remove and destroy any isolated plants that become severely infected with disease so they will not spread disease to other plants.

◆ Practice sanitation. Remove debris and weeds from the growing area, since they may harbor diseases.

◆ Inspect crops regularly to detect diseases early.

TABLE 2   CROP PLANNING FOR EARLY, SHADED CROPS

Activity	Timing guidelines	Example
Plant rooted cutting	Upon receipt	5/1
First pinch	When ready; about 10 to 14 days after planting	5/15
Second pinch	When ready; about 10 to 14 days after first pinch	5/30
Short days	2 weeks after second pinch	6/14
Flower	5 to 6 weeks after short days	7/17 to 7/24

Note: For 6-inch or gallon pots, using three plants per pot. Plant May 1 to June 1.

TABLE 3   CROP PLANNING FOR NATURAL-SEASON FALL CROPS

Crop	Pots size (inches)	Plants per pot	Plant date	No. of pinches	Approximate spacing (inches)
Normal, pinched	8 × 5 pan 1 to 1½gal.	1	Late May Early June	3	20 × 20
Normal, pinched	8 × 5 pan, 1 to 1½ gal.	1	Mid-June Late June	2	18 × 18
Normal, pinched	8 × 5 pan	1	Early July	1	16 × 16
Normal, pinched	8 × 5 pan	2	Mid-July	1	16 × 16
6-inch fast crop	6 to 6½	1	Mid-July, late July	None	12 × 12
4-inch fast crop	4 to 4½	1	Late July, early Aug.	None	8 × 8

Notes:
1. Plant dates are based on starting with rooted cuttings. Start one week earlier if direct-sticking unrooted cuttings. Start two to three weeks earlier if rooting in 72- to 98-cell packs for transplanting to finishing container.
2. Plant dates are based on midwest/eastern region growing conditions. In general, start two to three weeks earlier for the West Coast and one to two weeks later for southern production regions.
3. One plant per pot is sufficient in most programs if water and fertilizer are not limited as with constant fertilization/drip-tube irrigation systems. Otherwise, an extra cutting may be used if desired to create fuller pots.
4. Fast crop plants will naturally flower about one week later than normal, pinched crops.
5. The above fast crop programs could also be used with night lighting to delay flowering even more. Start lighting at planting and continue lighting until six weeks prior to desired sale date. B-Nine may be needed.
6. Normal, pinched 4- and 6-inch crops may also be produced. Plant two weeks earlier than fast-crop plants for each pinch that is planned. Such crops are usually bushy and compact and do not exhibit the flowering delay as in fast crops.

Chemical programs are also helpful in controlling aster diseases. They are most effective when used along with proper cultural practices; they cannot perform effectively alone. See table 1 for recommended fungicides. Generally, if good cultural practices are followed, greenhouse pot asters need little chemical treatment. Spray as needed for *Botrytis* or powdery mildew control.

However, with outdoor aster production, we strongly recommend a preventative chemical spray program (table 1). Approximately July 15, August 15, and September 15, apply Strike to suppress powdery mildew and rust. (Do not apply more frequently than at 30-day intervals.) Use Chipco 26019 (or Cleary's 3336) sprays between Strike sprays to suppress *Botrytis, Rhizoctonia,* and other leaf-spotting fungi. To avoid potential plant injury, do not mix Strike with other fungicides. If diseases persist, try alternative chemicals and review your cultural practices.

CONSUMER CARE

Emphasize to the customer that asters are very hardy (Zones 4 to 9) and should flourish year after year in perennial gardens. Tell them that the asters' perennial

TABLE 4 CROP PLANNING FOR GREENHOUSE POT CROPS

Activity	Timing guidelines	Example	Spacing
Plant rooted cuttings	Upon receipt	2/14	Pot-to-pot
Lights on	At planting	2/14	Pot-to-pot
First pinch	When ready, about 10 to 14 days after planting	2/28	Pot-to-pot
Second pinch	When ready, about 10 to 14 days after first pinch	3/14	Pot-to-pot
Short days	1 week after second pinch	3/21	8 × 8 inches
Flower	5 to 6 weeks after short days	4/25 to 5/2	8 × 8 inches

Notes:
1. The following schedule/spacing has worked well for 4- to 4½-inch pots, three plants per pot. Plant January 1 to June 1.
2. When direct-sticking unrooted cuttings, start one week earlier using long days.
3. For larger pots, use four cuttings per 5-inch pot (final spacing 10 by 10 inches), five cuttings per 6-inch pot (final spacing 12 by 12 inches). Also, start short days two weeks after the second pinch.
4. Desired overall pot height is 9 to 11 inches for 4- to 4½-inch pots. If height control becomes a problem, reduce the number of long days between the second pinch and the beginning of short days.
5. Crops flowered for May, June, or July can be cut back halfway after blooming and planted in the garden for reflowering in the fall. Promote this extra value to customers.

garden habit will be much taller and fuller than the plants purchased in these growing containers. Pinching will be needed to keep plants compact and bushy. Staking or other support may be necessary for some varieties. Yoder Brothers consumer care and handling sheets are available.

Crop planning. Perennial asters are suitable for most of the same production programs as garden mums (tables 2 through 5). A notable exception is that they are not recommended for field digging, since the plants don't recover well after digging. Furthermore, they are not recommended for spring flowering programs (e.g., no light–no shade; lighted-shaded), since one-cutting packs and pots do not fill out well. Asters perform satisfactorily in regular hanging baskets, but they do not fill out sides well in moss baskets or Belden-type baskets.

TABLE 5 CROP PLANNING FOR PERENNIAL CROPS

Activity	Timing	Cover	Uncover
Plant rooted cuttings	Aug. 15–Oct. 15	Mid-Dec.	March 1

Note: For overwintering in 1801 packs, quarts, or gallons

❖ AS CUT FLOWERS

Several aster species native to North America have become popular cut flowers, especially used as bouquet filler. When the first new cut flower asters were introduced in Europe in the early 1990s, they became an instant success. Later they migrated back across the Atlantic, first as imported cuts from Holland; later, domestic growers and Colombians began producing them for the U.S. market.

All asters are short-day plants, requiring day length of less than 12 to 14 hours to flower. Growers producing cut asters in the greenhouse with black cloth and mum lighting can expect three to four flushes per year. Grown in the field, expect one or perhaps two flushes.

Plants are vegetatively propagated: growers purchase rooted cuttings. Plant at a density of one and one-half plants per square foot (16/m²) in low-light areas, two plants per square foot (22/m²) in high-light areas.

Well-textured, well-drained media or soil low in soluble salts is best. While asters are actively growing during long days, provide average fertility: 150 N, 75 P, 150 K, alternating with clear water. Reduce nitrogen during bud initiation (short days) to aid bud set. Potassium nitrate is ideal during this period. Media or soil high in salts negatively affects production and stem quality. An EC level between 0.75 and 1.5 is best.

Maintain good soil moisture during vegetative growth. As short days are started, gradually begin a drying-down period to help bud set and to tone plants. Be careful not to run the soil too dry, though, especially if salt levels are high or the soil is heavy.

Run night temperatures at 50 to 60F (10 to 16C) during the long-day period. These cool temperatures help keep plants vegetative. As short days (black cloth) are begun, increase night temperatures to 62 to 65F (17 to 18C) to speed bud initiation. Also, running higher night temperatures at this time gives a better and more complete bud set.

Greenhouse forcing. Plants will require one layer of wire support, which also helps as a planting guide. As a general rule, rooted cuttings are planted and left unpinched for two to three weeks. Developing a solid root system is critical for good production. After two to three weeks, top out plants, leaving two to three sets of leaves. This pinch generates two to three top shoots, plus additional ground shoots that will emerge from the root system. When newly developed top shoots reach about 15 inches long (38 cm) (average plant height), begin short days (black cloth). The short-day cycle should include no more than 12 hours of total light for optimum flower initiation. Short days will last from five to seven weeks, provided cool, 55 to 60F (13 to 16C) night temperatures are maintained. Once stems have swollen and buds are beginning to show color, short days are completed.

TABLE 6 ASTER VARIETY CHART

Variety	Size[a] and color	Flower Bloom time Natural[b]	Response group[c]	Vigor[e]	Program suitability[f]	Other comments
White						
Butterfly White *A. pringlei* × *A. novi-belgii*	Medium white	Mid	5½	Tall Most vigorous white variety Use extra pinches and B-Nine to control	Recommend for all programs except 6- and 4-in fast crops	Most sensitive white variety to foliage diseases
Monte Casino *A. ericoides*	Small white Delicate flowers	Late	6	Tall, durable	Suitable for all growing programs	Delicate foliage Closely resembles native aster forms in the wild
White Swan *A. novi-belgii*	Large white Largest flower of white true aster varieties Somewhat poor flower-keeping quality outdoors	Early	5	Medium, upright	Suitable for all growing programs	Top white variety for greenhouse pots in Europe
Pink						
Dark Pink Star *A. ericoides* × *A. novi-belgii*	Medium pink-rose	Late	6½	Short, rounded Nice ball-shaped plant	Suitable for all growing programs	Deep green foliage Reliable performer
Melba *A. dumosus*	Large pink	Mid	n.a.[d]	Medium	Not recommended for light-shaded programs, due to poor bud setting there Relatively poor keeping quality when forced, so sell as tight as possible	Longtime favorite in perennial gardens
Painted Lady *A. novi-belgii*	Medium pale pink Fades to nearly white in high temperatures	Mid	5	Tall Useful as cut variety	Suitable for all growing programs	
Patricia Ballard *A. novi-belgii*	Large lavender-pink	Mid	5	Medium Moderate vigor	Good, all-purpose variety for all aster programs	Avoid black-clothing longer than 10–12 hours for best bud set
Red Monarch *A. pringlei* × *A. novi-belgii*	Medium pink Not red, but a pleasant pink	Mid	5½	Tall, upright Upright like Purple Monarch, but always slightly fuller	Not recommended for small pots (4-in fast crop, greenhouse pots), since height control is difficult Useful as cut variety	

TABLE 6 ASTER VARIETY CHART (CONTINUED)

Variety	Size[a] and color	Bloom time Natural[b]	Response group[c]	Vigor[e]	Program suitability[f]	Other comments
Sunrose A. hybridum	Medium rose-pink Darker rose-pink than Dark Pink Star, and better color retention in warm temperatures	Mid	6	Short	Suitable for all growing programs	Seems to be more susceptible to rust than Dark Pink Star, so use protective sprays
Raspberry						
Frida Ballard A. novi-belgii	Large raspberry Slightly darker color and slightly later than Winston Churchill	Mid	5	Medium	Good performance in all programs	Very close in performance to Winston Churchill
Jenny A. dumosus × A. novi-belgii	Large raspberry	Early	5	Short, dwarf Will stay small in the garden	Suitable for all growing programs	Popular perennial garden variety
Winston Churchill A. novi-belgii	Large raspberry	Early	5	Medium	Good, all-purpose variety	Yoder Brothers' Number One-selling variety
Lavender-blue						
Ariel A. novi-belgii	Medium lavender Named for its light, airy flower display	Mid	5½	Medium Can be somewhat vigorous, so extra pinching or B-Nine may be needed to control	Suitable for all growing programs	
Butterfly Blue A. pringlei × A. novi-belgii	Medium blue Nice light blue flowers	Mid	5½	Medium Not as vigorous as Butterfly White	Recommended for all except 4-in fast crop Usually too big for small pots outdoors	Somewhat sensitive to foliage diseases
Celeste A. novi-belgii	Medium blue Striking dark blue petals with complementary yellow disc Named for its heavenly blue color Forms outstanding floriferous mounds	Early	5½	Medium	A must variety in any program	
Prof. Kippenberg #2 A. dumosus × A. novi-belgii	Large blue-purple Popular for its color	Mid	n.a.[d]	Short, dwarf	Suitable for all natural-season programs, but not light-shaded ones Marginal size in all sizes of pots	Popular dwarf for gardens due to color and habit
Purple Monarch A. pringlei × A. novi-belgii	Medium purple Nice purple petals highlighted by a yellow disc	Mid	5	Tall, upright	Although upright, useful in all programs except 4-in fast crop	
Schone Von Dietlikon A. novi-belgii	Medium blue-violet Nice flowers Earliest to flower of all aster varieties	Early	5	Medium	Suitable for all growing programs	Seems to get mildew as flowering begins, so use protective culture and sprays

TABLE 6 ASTER VARIETY CHART (CONTINUED)

| Variety | Size[a] and color | Flower | | Vigor[e] | Program suitability[f] | Other comments |
| | | Bloom time | | | | |
		Natural[b]	Response group[c]			
Sunkid A. hybridium	Small dark lavender Hundreds of flowers	Mid	5	Short, spreading, very compact	Suitable for all growing programs Somewhat short in large pots, but perfect for small pots	

[a] These are flower (not plant) sizes. Small: H to I inch (1.3 to 1.9 cm); medium: I to 1H inches (1.9 to 3.2 cm); large: 1G to 11 inches (3.2 to 4.4 cm).

[b] A third of the flowers are open under natural day length at 40∞ N latitude in Middle Atlantic–Ohio climates (Philadelphia to Columbus). Early: Sept 1 to 10; mid: Sept 11 to 20; late: Sept 21 to 30. Precise dates in other areas may fall outside these ranges, but relative timing of varieties will remain approximately the same.

[c] Response group equals the number of weeks from the start of short days until half of the flowers are open in controlled short-day crops, such as greenhouse pot asters and early-shaded garden asters. The precise response may vary from these approximations, and it can be 1–2 weeks longer in extended periods of low light.

[d] Not applicable because variety is not grown in controlled lighted-shaded programs.

[e] See table 7, "Aster vigor," for comparison of tall, medium, and short plant characteristics and needs.

[f] All varieties are recommended for overwintering perennial programs. Suitability is based on our trials. Individuals grower success may vary, depending on cultural practices and environmental conditions.

TABLE 7   ASTER VIGOR, HEIGHT CONTROL

	Tall	Medium	Short
Habit	Upright	Average height and spread	Compact
Grower			
Pinching	May need extra pinch	Not difficult to control height with normal pinching	Easily controlled by pinching
B-Nine	Up to 2–3 sprays for height control	Usually 1 spray is sufficient, if needed	Generally avoid
Gardener			
Placement	Back of border only	Middle of border	Generally suitable for front of border
Pinching	Needed to control height	Remove ⅓–½ of height before mid-July	Remove ⅓–½ of height before Mid-July
Staking	Needed to control height	Generally won't be needed if pinched as above	Generally won't be needed if pinched as above

Asters require longer day length than chrysanthemums to maintain vegetative growth: 16 to 17 hours is recommended. This vegetative period takes from five to eight weeks to reach the 15-inch (38-cm) plant height, depending on the growing temperatures and variety.

Total crop time is 10 to 16 weeks, depending on growing temperatures, cultivar, and the flush from which plants are being forced. Generally, the grower can expect four to eight flushes per plant before they are discarded.

During harvest, apply long days again to extend day length. This ensures that ground shoots that begin to develop during harvest will maintain their vegetative growing state.

After harvest, cut the plants back totally to the ground. Remove all partially cut stems, stubble, and old wood. If old wood is left aboveground, it will produce prematurely budded stems of no value.

Growing outdoors. Asters will flower naturally from September to October (once per year), depending on the variety and growing conditions. Asters are hardy in most areas. After harvest allow the plants to overwinter. In May or early June of the following year, cut back plants totally to the ground. Ground shoots will develop as above and flower in the fall.

Problems. Powdery mildew can be a problem when temperatures are cool and moderate humidity is present. Provide good air movement and keep foliage dry to prevent the problem. Some varieties are more susceptible than others.

Verticillium dahliae is a wilt disease that may affect only part of the plant initially and then spread. There is no chemical control; prevention is the cure. Other susceptible crops that may serve as hosts include liatris, phlox, chrysanthemums, and aconitum. Rotate crops to help prevent problems.

Thrips can cause flower and foliage distortion. Spray as needed. Cyclamen and spider mites can also be problems, especially in outdoor plantings. While asters are not a major host to leaf miners, they can be attacked if leaf miner populations are high.

Harvest. Small-flowered varieties are cut when 10 to 20% of their flowers are open. Large-flowered varieties are cut when terminal flowers are open and the buds behind show color. Do not use STS. However, be sure to use a bactericide. Following good postharvest practices yields shelf life of up to 14 days.

Varieties. *Aster ericoides* Monte Casino (white) and Pink Casino have small, one-half-inch (1.30-cm) flowers, ideal for mixed bouquets. *Aster novi-belgii* hybrids include Cha Cha (purple), Lambada (pink), Samba (lilac purple), and Rumba (lilac pink). Flower size is larger, 1 to 1½ inches (2.5 to 4 cm). These varieties are well suited to single flower species.

The Master series—White Master, Blue Master and Pink Master—have medium-sized flowers, ranging from one-half to three-quarter inch (1.3 to 2 cm) in diameter. The varieties are suited for mixed bouquets or mono bouquets.

ASTER, ANNUAL (SEE *CALLISTEPHUS*)

ASTILBE (FALSE SPIREA)

by Jim Nau
Ball Horticultural Company
West Chicago, Illinois

❖ *Perennial (Astilbe × arendsii). 384,000 seeds/oz (13,440/g). Germinates in seven to 21 days at 60 to 70F (16 to 21C). Seeds should be left exposed to light during germination. Most astilbes are vegetatively propagated by division in spring or fall. For the easiest crop, buy in roots during the fall or late winter.*

Astilbe, also known as false spirea, is a premier perennial plant with excellent value either for the pot plant market or for perennial plant sales in the spring. Astilbes flower in late spring and early summer and do best in areas where they get light shade in the afternoon. Plants grow 24 to 36 inches (61 to 91 cm) tall and come in colors of red, carmine, rose, pink, white, and lavender, plus shades in between. The flowers are borne in plumes that measure from 4 to 10 inches (10 to 25 cm) long. *Astilbe* is hardy in USDA Zones 4 to 8.

Though astilbes can be propagated from seed, the boldest colors and all the hybrids come from divisions that are done in early spring or fall. Upon receipt of roots, separate the crowns or pot as is, using one plant per quart or gallon container. February and March planted divisions will be salable by late May, though flowering will not be profuse from these divisions. More flower color will be on next year's plants.

For potted plant sales in flower during late winter and spring, bring the crowns (try to have three to four eyes) in around September. Pot up into 6-inch (15-cm) pots and let the plants become established in the container. Chill the plants for 10 to 12 weeks at 35 to 40F (2 to 4C). They can be overwintered with other perennials and brought in when needed. Once the pots have been brought back into the greenhouse, place the pots out of direct light and allow them to warm up gradually over a period of several days, then increase temperatures to no less than 58F (14C) nights and allow 12 to 16 weeks to flower, depending on the cultivar. In general, the early and mid-season varieties will flower about the same time. The late season varieties tend to take longer, though they would work for sales two to four weeks after the start of the sales of the earlier flowering material. It is suggested to stay away from taller varieties to avoid too vigorous growth. A growth regulator is suggested for keeping the plants dwarf.

For those who would like to grow from seed, pretreat by placing the seed on moist peat moss for two weeks at 70F (21C). If the seed does not germinate, then give 40F (4C) for three to four weeks. This will help to break dormancy within particularly hard-to-germinate seed. However, when the seed is fresh, 70F (21C) has been all that is needed to get 70% and higher germination rates. If you have been following the procedure above, once the seed is removed from the 40F (4C) cooler, it will germinate in 14 to 21 days at 70F (21C). January sowings will not flower the same season. To get free-flowering plants from seed, allow 30 to 31 months.

All seeded varieties are mixtures of the species. While the plants perform well, the flower color is muted at best and dull in its overall appearance. This, combined with long crop times, makes *Astilbe* best grown from roots brought in during the fall or late winter. As for astilbe varieties from division, the following cultivars are suggested: Fanal is a deep red to 22 inches (56 cm); Rheinland, a pink to 24 inches (61 cm); and Deutschland, a pure white to 24 inches (61 cm). These, as well as other varieties on the market, are classified by being from early- to late-season performers; check with your favorite supplier for the seasonality of the cultivars that interest you.

In related material, *A. chinensis* var. *pumila* is an excellent variety with lavender-pink flowers. It prefers a well-drained though moist soil in the garden. This dwarf variety (10 to 12 inches, 25 to 30 cm) is recommended for rock gardens and as a border or edging plant in the perennial garden. It also makes a sharp 4-inch (10 cm) pot. However, *pumila* is a July through August–flowering astilbe in the Midwest garden, which makes it one of the latest to flower.

Astilbe simplicifolia Sprite was a Perennial Plant Association "Perennial of the Year." Shell pink flowers appear on plants 12 to 14 inches (30 to 35 cm) tall. The foliage is fernlike, and the plumes are more open than on other astilbes. Allow more time for *A. simplicifolia* varieties to finish—12 to 14 weeks at 55F (13C) nights.

AUBRIETA

by Jim Nau
Ball Horticultural Company
West Chicago, Illinois

❖ *Perennial (**Aubrieta deltoidea**). 85,000 seeds/oz (2,975/g). Germinates in 14 to 21 days at 65 to 70F (18 to 21C). Seeds should be left exposed to light during germination. Temperatures from 70 to 72F (21 to 22C) will speed germination.*

Aubrieta is a low-growing perennial sometimes confused with *Arabis*. *Aubrieta* flower colors are lavender, blue, and purple. Deep rose flowers are also available, but they often appear to have blue or lavender overtones. Flowers measure three-fourths inch (2 cm) across and are held on upright stems over the rosetting foliage. It grows to 8 inches (20 cm) tall. Plants flower in April and May and are hardy to USDA Zones 4 to 7.

Aubrieta is primarily propagated by seed, though cuttings taken after flowering in late spring, and crown division done as soon as the plants can be dug out of the ground in the spring, are other methods. Seed sown in January for May sales will not flower until the following spring. Sow in summer for sales the following year. Sowings made in early October, grown at 40 to 45F (4 to 7C) during the winter and transplanted to a 4-inch (10-cm) pot, flower in May. For green pack sales, allow 10 to 12 weeks. Growing-on temperatures are 50 to 55F (10 to 13C) nights.

In varieties, the primary seed selection is called Large-Flowered Hybrids, which is a catchall name for seed collected from the species. It is a mixture of lavender, lilac, purple, and dark rose. Plants grow to 8 inches (20 cm) tall. Purple

Gem has three-fourths-inch (2-cm) purple flowers on plants to 6 inches (15 cm) tall and is vegetatively propagated. The seeded form is similar and is sold under the name Whitewell Gem.

AZALEA (SEE *RHODODENDRON*)

BACOPA (SEE *SUTERA*)

BEGONIA

FIBROUS AND TUBEROUS BEGONIAS	HIEMALIS BEGONIAS
by Dennis Reynolds Benary Seeds Glen Ellyn, Illinois	by Dick Devries Schenck Farms and Greenhouse St. Catherines Ontario, Canada and Lois Carney Oglevee Ltd. Connellsville, Pennsylvania

❖ *Annual, fibrous (Begonia × semperflorens-cultorum). 2,500,000 seeds/oz (88,000/g). Germinates in five to 10 days at 75 to 80F (24 to 27C). Do not cover raw seed or pelleted seed. Germination rate 90%.*

❖ *Annual, tuberous (Begonia tuberhybrida). 1,000,000 seeds/oz (35,000/g). Germinates in seven to 14 days at 75 to 78F (24 to 26C). Do not cover raw seed or pelleted seed. Germination rate 80 to 85%. Used as a bedding plant or in patio containers and hanging baskets.*

❖ *Annual, Hiemalis (Begonia socotrana × B. tuberhybrida). Vegetatively propagated. Finished flowering potted plants are generally produced from purchased unrooted or rooted cuttings.*

❖ FIBROUS BEGONIAS

Popular with home gardeners and landscape professionals alike, fibrous begonia *(Begonia × semperflorens-cultorum)* bedding plants are floriferous and durable, require minimal care, and can be used in almost every type of planting imaginable. Used as a summer annual in most areas and a winter annual in the warmest climates, the fibrous begonia is a staple for all bedding plant growers.

CROP CULTURE

Seed is available raw, or pelleted for easier sowings. Seed germinates in five to 10 days at 75 to 80F (24 to 27C). Do not cover seeds or pellets. High moisture in the growing media and high humidity greater than 95% are essential for fast, uniform germination. During plug production, at Stage 2 (10 to 14 days) grow at temperatures of 70 to 72F (21 to 22C). Maintain consistent moisture. Apply one to two applications of 50 to 75 ppm nitrogen per week. In Stages 3 and 4 (four to five weeks), grow at temperatures of 65 to 70F (18 to 21C) and apply 100 ppm nitrogen weekly (nitrate forms preferred during Stages 2, 3, and 4).

Supplemental lighting in the seedling stage increases plant quality and shortens crop time. Keep daytime levels at 450 to 700 f.c. (5 to 8 klux), and continue from dusk to 2:00 A.M.

After seven to nine weeks (or after receiving your plug shipment) (fig. 1), transplant into packs or containers. Grow on at 62 to 68F (17 to 20C). Feed weekly or as needed with 150 ppm nitrogen, using formulas such as 20-10-20. Maintain pH of 5.5 to 6.2.

Time to produce finished packs in the North is 14 to 16 weeks, in the South 13 to 15 weeks. Four-inch (10-cm) pots will finish in the North in 17 to 19 weeks, 16 to 18 weeks in the South. Apply B-Nine or Cycocel during finishing as needed at label rates.

During dark, cool periods, watch for signs of *Botrytis* and *Phytopthora*. Fungus gnats and shore flies are attracted to moist begonia seedling trays, so monitor for their presence and take steps to control problems early.

VARIETIES

Several breeding companies are working hard on this class of bego-

FIG. 1. Many bedding plant growers buy in begonia plugs like those shown here.

nias to create what each feels is the best combination of flowering earliness, pack performance, and garden and landscape performance. Earliness in fibrous begonias is desirable as long as the plants size up well in packs and pots and have an excellent shelf life. Strong outdoor performance is a must.

Dwarf begonias are available in dark-leafed and green-leafed series. The leading dark-leafed begonia series is Cocktail because of its superior garden performance. Newer series on the market, such as Espresso and Rio, are earlier than Cocktail. The Vision series offers a taller plant habit with solid landscape performance.

Green-leafed begonia varieties have changed significantly in the past years. The extra-early flowering Varsity series (fig. 2) gives growers a winning pack performance over older varieties, such as Linda, Scarlanda, and Scarletta, but with somewhat less stellar garden performance. The Super Olympia series offers good pack performance on medium-height plants that really shine in the garden. Green-leafed varieties include the Ambassador series; Linda, Scarlanda, Scarletta, and Viva; the Super Olympia series; the Pizzazz series; the Prelude series; the Thousand Wonders series; the Varsity series; and the Victory series.

Bronzed-leafed varieties are found among these series: Bingo, Cocktail, Espresso, Rio, Senator, Varsity, and Vision. For the grower looking to produce a green-bronze-leafed mixture, there is Organdy mix.

Taller, larger fibrous begonias are terrific performers as bedding and landscape plants. In the garden they tend to grow like miniature hedges. Their more vigorous habit isn't showy in a pack, which puts them off-limits for mass market growers, making them an excellent niche crop for retail and landscape growers. Grown as warm-weather flowering container plants or for landscape sales, these cultivars have a lot to offer. They will give your customers tremendous value. Tall fibrous begonias varieties include the All-Round series, the Baron series, the Encore series, and the outstanding Party series.

FIG. 2. The Varsity series of begonias, introduced in the mid-1980s, showed growers that begonias as bedding plants didn't have to be late to flower. Varsity is shown here in Dutch trials.

❖ TUBEROUS BEGONIAS

Several key factors have boosted the sales of tuberous begonias *(Begonia tuberhybrida)* in recent years. Today's high-tech plug growers are equipped to grow tuberous begonias more profitably, either to finish the plants themselves or to sell as liners for other growers to finish (fig. 3). Breeding companies have improved existing varieties while developing new cultivars that offer the grower and the consumer more ways to use tuberous begonias successfully. Finally, the consumer trend is to seek out more diversity when purchasing plants for the patio and garden. Tuberous begonias are beautiful and affordable.

FIG. 3. Ace Begonias, Bethany, Connecticut, is owned by Paul Rieur *(front)* and Jon Davison. They built their 57,000-square-foot pot plant business on begonias as a year-round staple. Paul and Jon also produce a range of other flowering pot plants that are marketed to the New York City metropolitan area.

CROP CULTURE

Cultural methods, plug tray sizes, and crop times will vary from grower to grower. The crop times discussed here are for tuberous begonias grown by plug growers who can use high-intensity lights and other high-tech methods to increase plant quality and shorten crop time. If you are not equipped to grow seedlings in this manner, tuberous begonias are an item that might be more profitable if you buy in liners from a plug supplier specializing in begonias. Crop time for 4-inch (10-cm) pots is 16 to 18 weeks, for 10- or 12-inch (25- or 30-cm) hanging baskets, 20 to 22 weeks.

Tuberous begonias are available as raw or pelleted seed. Sow seed from November to January for flowering plants from April to June. Seed germinates in seven to 14 days at 75 to 78F (24 to 26C). Do not cover seeds or pellets. Maintain consistent moisture of the growing media and high humidity (greater than 95%) for fast, uniform germination. For plug production during Stage 2 (six weeks), provide temperatures of 70 to 75F (21 to 24C). Apply a constant feed at 75 ppm nitrogen with 13-2-13. Transplant seedlings to a larger plug size (e.g., 72-cell tray) at Stage 3 and 4 (four to six weeks). Maintain temperatures of 70F (21C). Apply constant feed at 75 ppm nitrogen with 13-2-13.

After nine to 11 weeks (or after receiving your plug shipment), transplant into 4-inch (10-cm) pots or hanging baskets, using three to five plants for a 10-inch (25-cm) basket or five to six plants in a 12-inch (30-cm) basket. Grow on at 70F (21C). Feed weekly with 150 ppm nitrogen formulas, such as 20-10-20 or 13-2-13. Maintain a pH of 5.5 to 6.5. Apply Cycocel at 150 ppm two weeks after the first transplant. Continue weekly, if necessary.

Young plants can burn if watered during the brightest time of the day. From emergence to late March, provide 14-hour long-day treatment (extend the day or interrupt the night). Additional supplemental lighting with HID will improve plant quality and shorten crop time.

Watch for *Botrytis* in older seedling trays.

Varieties

Tuberous begonias are typically used in partial shade for bedding or on the patio in containers and baskets. The majority of tuberous begonias are produced in the northern two thirds of the United States and in Canada. While some growers in the Deep South grow tuberous begonias, many have found that heat-related problems make this a difficult item for the grower and the consumer. Newer varieties, however, especially the Panoramas and Charismas, can overcome these obstacles.

The Nonstop series, with its 3½- to 4½-inch (9- to 11-cm) flowers, is the leading tuberous begonias series. It comes in a wide range of colors. The Ornament series is a dark-leaf Nonstop type. The Fortune series has 2- to 3-inch (5- to 8-cm) flowers, while the Galaxy series is a medium-flowered single. The Panorama series has 2- to 3-inch (6- to 8-cm) blooms and is earlier and more floriferous than Nonstop. It's also more heat tolerant, especially Panorama Scarlet. Panorama has a slightly pendulous habit.

For something unusual, try the Illumination series, with 2-inch (5-cm) flowers and a multiflora cascading habit. Pin Up is a large-flowered, single bicolor.

❖ Hiemalis begonias

The Hiemalis begonia is a cross between the winter-flowering bulbous species *Begonia socotrana* and the summer-flowering *Begonia tuberhybrida*. The result of this cross is a winter-flowering begonia with characteristics similar to tuberous begonias. This type of begonia is also commonly known as Elatior begonia. (In

1955 Otto Reiger of Germany introduced new varieties that were more floriferous and more resistant to mildew; this was the beginning of the Reiger begonia.)

Hiemalis begonias are popular pot plants in western Europe (fig. 4). They are available in a wide variety of flower colors, such as shades of red, pink, yellow, white, and orange. Flower forms include single, semidouble, and double. These characteristics, combined with a wide range of leaf colors and shapes, give Heimalis begonias strong consumer appeal. In recent years the North American market seems to be growing, with interest being shown by both the

FIG. 4. In Europe, Elatior begonias, such as this 6-inch pot, are a regular feature of every retail florist and supermarket floral department. Growers purchase unrooted cuttings and liners from specialist propagators only to ensure disease-free cuttings.

florist and mass market outlets, especially in the spring. When grown and displayed properly, Hiemalis begonias sell briskly and reward the consumer with several weeks of color.

Rooted liners are available as leaf (multistem) or stem tip cuttings from specialized propagators. Multistem leaf cuttings, consisting of three or more vegetative shoots, will produce a very full finished plant from a single liner. Stem tip cuttings are well suited for 4-inch (10-cm) pot production, but when producing 6-inch (15-cm) pots, two or three rooted liners are recommended. For growers with the equipment and know-how, unrooted begonia cuttings are also available.

CULTURE TO FINISHING

Hiemalis (Elatior) begonias prefer a light, well-drained media using 80% high-fiber peat moss and 20% perlite. The pH should be adjusted to 5.2 to 5.5, and the soluble salts kept under 1.0 mmhos. Commercial mixes are readily available. Or you can blend your own mix using the following formula per cubic yard: 80% peat, 20% perlite, five to seven pounds ground limestone, 12 pounds superphosphate, one pound each of potassium nitrate and calcium nitrate (per m^3, these weights are 2.97 to 4.15, 7.12, and 0.59 kg, respectively).

Begin feeding begonias seven to 10 days after planting. Remember: This crop is sensitive to overwatering, so allow the soil to become "dry" between irrigations.

Apply 150 to 200 ppm constant feed using 15-5-15 or 20-10-20 to keep pH above 5.0 and below 6.0. Soluble salts should be kept on the low side, less than 1.2 mmhos, preferably maintained at 0.9. Remember to apply enough feed solution to achieve a 10-to-15% leach with each irrigation. For the grower using subirrigation, maintain the fertilizer concentration in the range of about 100 to 125 ppm.

For maximum vegetative growth, the night temperature should be maintained at 66 to 68F (19 to 20C). Optimum day temperature requires heating to 70F (21C), with venting beginning at 75F (24C). Air-cooled greenhouses are recommended in warm regions during the summer months. Temperatures over 75F tend to produce soft, elongated growth. As temperatures rise above 75F, start reducing the light intensity.

Maintain 70 to 80% humidity to control powdery mildew and *Botrytis*. Provide good air movement to prevent condensation on leaves, and always enter the evening or night with dry leaves. Water first thing in the morning to ensure dry foliage by the day's end.

Begonias are a low-light crop, so shade must be applied if the light level goes above 2,000 f.c. (22 klux) at 70F (21C). When temperatures exceed 75 to 80F (24 to 27C), light levels should be reduced to 1,500 f.c. (19 klux). High light produces discolored foliage and stunted growth.

Long days are required to promote vegetative growth. In northern latitudes begin long days in September and continue until May. Use incandescent light at 10 to 20 f.c. (108 to 215 lux) to extend the day length to 14 hours. Begin long-day lighting at the time of planting; maintain it for four to six weeks or until you have obtained two-thirds of the desired final growth. Start short days after the four weeks by providing at least 13 hours of darkness for a two-week period. Return plants to the long-day treatment after flower initiation to prevent a delay in finish sales date.

Use growth regulators only if necessary or during summer and early fall to keep the plants short and compact. Apply right after short-day treatment. With Cycocel, use 1,500 ppm; spray to glisten. If A-Rest, use 25 ppm; spray to drench.

Insect pests include aphids, fungus gnats, thrips, and whiteflies. As pesticide product registrations are different from state to state, check with your cooperative extension agent or university pest specialist.

Diseases affecting Hiemalis begonias—powdery mildew and *Botrytis*—can be controlled with a combination of the proper greenhouse environment and judiciously using available fungicides. However, *Xanthomonas begoniae*, or bacterial blight, cannot be controlled by fungicides. The only way to keep from getting this fast-spreading, fatal disease is to purchase rooted or unrooted cuttings from a source that uses culture index techniques and maintains sanitary conditions.

POTTING

Use two plants per 6-inch (15-cm) pot. Plant at the same depth as the cutting rootball or slightly higher. Water in thoroughly immediately after planting, and lightly water each day for the next three days.

Pinch 10 to 14 days after potting to achieve a premium finished plant. Pinching accelerates branching and encourages adventitious shoots to develop below the soil line. This growth is critical for optimum plant form and to control height. Keep begonias pot to pot during this shoot development process to enhance shoot formation; begonias are "social plants." Space the plants when shoots are well developed but before stretching occurs. Pinching also reduces the need for plant support (bamboo stake) and can be used as a tool for crop timing. Generally, plants will be ready for sale 10 weeks after the pinch.

Alternatively, if you prefer not to pinch, you can plant three plants per 6-inch (15-cm) pot and finish in 10 weeks. For producing hanging baskets and 10-inch (25-cm) containers, use three cuttings per container. Pinch as previously described, and plants will be salable in 12 weeks.

UNROOTED BEGONIA PRODUCTION

Immediately upon their arrival, unpack begonia cuttings and place them under mist in a shaded area. Follow the same media recommendations as described previously for rooted liners. Wet the soil thoroughly before sticking the cuttings. Lightly stick cuttings one-half inch (1.3 cm) deep in the wet media, pressing only enough to hold the cutting upright.

Do not allow the foliage to dry for the first three to four days. Decrease mist 5% daily on Days 4 through 8. Decrease mist 10% daily on Days 9 through 14. By Day 14, cuttings should be rooted and will no longer need mist. When grown under conditions of low humidity or during hot weather, mist rooted cuttings two to three times daily for the first three weeks. Begin feeding at Days 9 through 12 or when the base of the cutting is callused.

BELLIS

by Jim Nau
Ball Horticultural Company
West Chicago, Illinois

❖ *Perennial grown as an annual (***Bellis perennis***). 140,000 seeds/oz
(4,900/g). Germinates in seven to 14 days at 70 to 75F (21 to 24C).
Leave seed exposed to light during germination.*

Due to extreme temperatures in the Midwest, bellis plants are often grown as annuals, though they are given as tender perennials in many listings. They are best grown for spring color in borders, window boxes, or pots. Flower colors include pink, rose, red, and white, and the flowers are commonly double in form. In coastal areas plant in full sun; in other areas plant in afternoon shade.

Allow 12 to 15 weeks for flowering packs in the spring. Grow along with pansies and treat the same, growing at 50 to 55F (10 to 13C) nights. For flowering 4-inch (10-cm) pots, allow 14 to 17 weeks.

Super Enorma is still one of the more popular of the varieties available in the U.S. trade today. Flowers are double, frilled, and blossom out at 3 inches (7 cm) across. The mixture is the most commonly grown variety. Another popular variety is the Pomponette series, which has tightly held petals and a flower size of no more than 1 inch (2.5 cm). Both varieties grow to no more than 4 inches (10 cm) tall, though the crop time of Pomponette is one to two weeks earlier than that of Super Enorma.

Bellis has potential in the South as a fall crop. Due to its heat sensitivity, southern growers should buy in plugs as opposed to sowing seed in July. *Bellis* is not suggested as a fall crop in the North.

BIRD OF PARADISE (SEE *STRELITZIA*)

BOUGAINVILLEA

by Bob Cornell and Andrew Greenstein
Wes-Cor Nurseries
Sarasota, Florida

❖ *Tender perennial (***Bougainvillea*** spp.). Vegetatively propagated. Most
growers begin pots and baskets from purchased rooted cuttings or liners.*

In the past three years, bougainvillea has become one of the most popular tropical plants. It can be grown in baskets, pots, and as landscape plants throughout most of the year in the southernmost states. With care, bougainvillea can also be grown in the northern states, outside in the summer and protected in the winter.

Bougainvillea is not a difficult plant to grow if a few simple rules are followed. Plants should be grown in well-drained soil, should not be overwatered, and should be kept in *full* sun as much as possible. Use a well-balanced fertilizer with minors, such as 20-10-20 if using soluble fertilizer and a 16-4-8 mix for top dressing. No special treatment is required to enhance blooming, except the grower should realize that plants bloom only on new growth. In other words, cut your plants back, and they will bloom.

Pests are not a real problem. Those that appear are easily controlled with Orthene as a pesticide and Daconil or Kocide as fungicides. Other pesticides are available and effective.

There are many bougainvillea varieties and cultivars. However, a few varieties stand out as plants for baskets: Barbara Karst, Raspberry Ice, Purple, Single Pink, and variegated Mary Palmer. Some of the other varieties make beautiful landscape plants but are too vigorous in their growth to make good pot plants.

A few misconceptions exist regarding bougainvillea culture. Dry conditions and stress are *not* required for flowering. Special fertilizers are not required for flowering. As previously stated, plants bloom on new growth.

Feed it, weed it, prune it, and enjoy!

BRACHYCOME (SWAN RIVER DAISY)

by Garry Grueber
Kientzler GmbH & Co. KG
Gensingen, Germany

❖ *Tender perennial (***Brachycome multifida***). Vegetatively propagated.
Most containers and baskets are started from bought-in liners.*

When wandering through the Australian Outback, one often encounters rather inconspicuous little daisies with mauve-blue flowers growing by the roadside, in ditches, nooks, and crannies. The genus *Brachycome*, almost exclusively endemic to the Australasian region, is made up of approximately 60 annual and perennial species. The name *Brachycome* (or as some botanists would have it, *Brachyscome*) loosely translates as "short hair," which probably refers to the neat habit of the plants. The annual Swan River daisy, *Brachycome iberidifolia*, has been offered on a limited scale for quite some time in seed catalogs. It is still valued as a very effective, albeit short-lived, annual for color impact in summer.

However, it was not until recently that this genus came to worldwide recognition and fame. More than 15 years ago, Prof. Wolf-Uwe von Hentig, former head of the Institute for Ornamental Plants at the Research Station, Geisenheim, Germany, started a research program to develop new floriculture crops. After he visited Australia to establish contacts with nurseries and scientists working with native Australasian flora, shipments of native plant material started arriving at the research station at irregular intervals. At the time, little was known in Europe in regard to the specific needs of Australian wildflowers in cultivation. Few growers were aware of the fact that most of the native flora is extremely finicky when it comes to soil, fertilization, pH, light, and humidity. As could be expected, many of the plants perished under central European greenhouse conditions.

In one shipment, a few cuttings of a plant labeled *Brachycome multifida*, with very soft, finely cut foliage, arrived from Australia. Despite the long transport, there were still a few single, mauve-blue flowers with yellow centers left on the plants. After having so many problems keeping the Australian plants alive in their European exile, the Institute staff decided to mimic Australian soil conditions. Cuttings were planted in a gritty, virtually nutrient-free, well-drained soil mix. While they seemed to survive, they were as scrawny and inconspicuous as they were at home.

One day a brachycome was accidentally planted into a mix for growing bedding plants. The plants had not only survived the error but thrived! Later it was discovered that this species, commonly known as the Hawkesberry River daisy, occurs naturally in the rich, moist silt alongside rivers and streams—hence the obvious preference for humus-rich, peat-based media.

Once survival of the plants was secured, the brachycome's true potential began to unfold. The plants quickly formed lush, spreading cushions of fresh green, fernlike foliage, liberally studded with short-stemmed, mauvish-blue daisies about the size of quarters, each with bright golden yellow centers. The flowers appeared from May through the first light frosts. The plants seemed to thrive on full sun and the often inclement central European weather conditions. It soon became apparent that this was truly a plant that deserved wider distribution.

The German floriculture press pounced on the plant, toting it as the new "Blue Daisy," since there were few blue-flowered bedding and window-box plants available. Consumers also accepted the quiet appeal of this cute little plant. Due to the ease of propagation by cuttings, the increasing market demand for this product could easily be satiated. The brachycome was on its way to becoming an overnight success story in Germany.

As a result, the entire floriculture industry, which had been conservative and mass-crop oriented, suddenly realized that there was quite obviously a potential to be tapped with new floricultural crops. Many research stations and liner producers started intensifying their efforts in the search for new floricultural crops, especially for summer color. Over the past decade, this has resulted in such introductions as scaevola, helichrysum, trailing petunias ("Supertunias"), bacopa, and many others. Several new improved *Brachycome* cultivars are being bred and distributed from Australia and Germany.

CULTURAL REQUIREMENTS

One of the reasons that the brachycome has attained such widespread popularity is its ease of cultivation. Given the right conditions, it will quickly and easily fill a basket, pot, or window box and flower continuously all summer long.

Propagation. In contrast to the seed-grown, annual species *B. iberidifolia*, the perennial species of *Brachycome* and their cultivars are vegetatively propagated. Suckering species, such as *B. melanocarpa,* can be progated by division, and all species can be propagated by tissue culture. However, this latter method is quite costly and is usually only used to produce pathogen-free elite material for mother stock.

By far the most widespread and practical means of propagation is through soft tip cuttings. For this purpose, mother stock is maintained under high-light, short-day (under 12 hours day length) conditions to prevent flower bud formation. For maximum cutting production and optimum cutting quality, mother stock should be maintained at temperatures between 59 and 68F (15 and 20C). Cuttings can be harvested on a weekly basis by cutting with sharp knives. Optimum cutting length depends on the cultural conditions and the variety, but as a rule of thumb, a length of about 1½ to 2 inches (4 to 5 cm) is best.

Because the plants are usually a bit thin to start, many liner producers stick two to three cuttings in each multicell unit to ensure that the resulting plant is nice and stocky. Peat-based media with low pH and salinity are best for rooting. Cuttings are best rooted under intermittent misting at temperatures around 68 to 72F (20 to 22C); bottom heat speeds up the rooting process. Rooting hormones can be used to facilitate rooting, especially with hard-to-root species, either as a powder, a dip, or as a foliar spray applied after sticking. Cuttings are usually rooted after about three weeks and can then be hardened off for about a week before shipment or potting.

Plant rooted cuttings in pots 4 to 5 inches (10 to 12 cm), or directly into 10-inch (25-cm) baskets (three liners per basket). The best time for planting is in

January through March, depending on the desired size of the finished product and when the sales season for bedding plants begins. Potting media are fairly crucial for success. A well-drained, peat-based mix with a pH of about 5.0 is best; pH levels above 6.0 will result in iron fixation, thus leading to severe chlorosis (yellowing of the young leaves) and subsequent stunting. High phosphorus levels in the media—or in the fertilizers subsequently used—can also lead to iron fixation and chlorosis. If iron chlorosis does occur, it can be alleviated by applying an iron chelate preparation as a soil drench at the recommended rates.

Maintain temperatures at about 64 to 68F (18 to 20C) for the first few weeks after potting to facilitate root development. A soft pinch about a week after planting helps induce better branching and improved plant habit. After the plants have started to take off, temperatures should be lowered to 54 to 61F (12 to 16C); otherwise, the plants will become too soft and leggy. Once the plants are accustomed to lower temperatures, they can even be exposed to temperatures down to 36F (2C) for any length of time. However, the development of the plants will come to a standstill at such low temperatures.

Starting at about three weeks after planting, weekly applications of a well-balanced liquid fertilizer low in phosphorus should be carried out. If the water that is used for irrigation is high in calcium, one might consider an acid fertilizer such as Peters Excel for hard water. *Brachycome*, though not necessarily a heavy feeder, responds well to regular fertilizer applications.

As an Australian native, *Brachycome* thrives on high light levels. Thus, it is extremely important to provide as much light as possible during cultivation and to shade only if absolutely necessary. Good air circulation in the greenhouse will help produce better quality and prevent fungal diseases.

One of the most important points in growing brachycomes is good water management. Although the plant comes from an arid country, it needs very even, continuous moisture in the soil to perform well under greenhouse and consumer conditions. If the soil dries out at any point, plants will desiccate and be unlikely to recover. On the other hand, waterlogged soil will cause root rot and lead to premature death from *Pythium*.

Growth retardants are sometimes used, especially with the leggier types. Although plant habit can be kept under control by applying DIF/cool morning techniques, brachycomes respond well to Alar, B-Nine, or Bonzi as a foliar spray.

PESTS AND DISEASES

There are several pests that can cause problems: Whitefly is a common one, and aphids and leafminers can also occur. By far the most serious problem is western flower thrips, which are attracted to the bluish flowers. Thrips infestation will result in stunted, discolored flowers that render the plants virtually unsalable. *Brachycome* can be a latent, (symptomless) carrier of TSWV (tomato spotted wilt virus) and INSV (impatiens necrotic spot virus). Thus, it is extremely important to (1) buy your liners from a reliable source that has indexed mother stock and (2) maintain a stringent prophylactic insecticide spray program to prevent thrips from establishing themselves on the plants.

There are a few diseases that can also affect brachycomes, especially in the propagation phase and under low-light conditions. *Botrytis*, *Myrothecium*, and *Rhizoctonia* can cause rotting in propagation; preventive fungicide treatments and good water management should help avoid such problems. Occasionally, problems with *Sclerotinia* and *Verticillium* rot can occur; both diseases can quickly devastate a crop. Good, solid greenhouse hygiene and buying plants from a reliable liner producer that maintains pathogen-indexed stock should prevent both diseases from becoming a problem.

BRASSICA (ornamental cabbage and kale)

by Jim Nau
Ball Horticultural Company
West Chicago, Illinois

❖ *Annual* **(Brassica oleracea).** *7,000 seeds/oz (247/g). Germinates in seven to 14 days at 68F (20C). Cover seed upon sowing to keep light out.*

Ornamental cabbage and kale are often listed under a variety of names in the seed catalogs, commonly under the "ornamental" heading or as "flowering" cabbage and kale. Regardless of the common name used, it defines a class of vegetables that are used primarily as bedding plants or as garnishes in restaurants. These plants have large leaves up to 10 inches (25 cm) across, in colors of white, red, rose, and pink (fig. 1). These colors intensify and become bolder as the temperatures become cooler. Plants grow to between 10 and 14 inches (25 and 36 cm) tall and are most common in the fall garden or commercial landscape plantings in the Midwest and South. The terms ornamental or flowering cabbage are really misnomers. All varieties sold under this heading are actually kales. The terms ornamental cabbage and ornamental kale are used more to define leaf shape, not heritage. Cabbage has rounded leaves, while

FIG. 1. You see them showing up more and more during the fall in landscape plantings and in home gardens, not to mention as garnishes at trendy restaurants nationwide: the lovely colors of ornamental cabbage and kale. This is Red Pigeon, an ornamental kale with bright, red-pink inner leaves and sage green outer leaves.

kale often has frilled or notched leaves. Both crops can handle repeated heavy frosts. The plants are edible but are mainly valued for their ornamental use.

Plants are salable green in the pack, with little or no foliage color, in five to six weeks. The primary market is in 5- and 6-inch (13- to 15-cm) pots, which are salable in 12 to 14 weeks, using one plant per pot. Once the seedlings are large enough, transplant them, planting deep enough to cover the stem up to their leaves. When ready to shift to 2¼-inch (6-cm) pots, again, plant them deep, covering part of the stem. Fertilize with a low-ammonium fertilizer to minimize stem elongation.

Stem elongation can also be controlled with B-Nine. *Do not sell plants treated with growth regulators as edible or for restaurant garnishes!* Begin weekly applications at 2,500 ppm as soon as leaves have unfolded (two to three weeks after germination). Bonzi is also effective. Growth regulators are especially helpful in height control during the summer months. Grow on at 55 to 58F (13 to 14C) once the plants are established in the container, and grow along with other cool-weather plants like pansies. Temperatures above 65F (18C) may delay leaf coloring. Keep in mind that white-foliage varieties will color up more quickly than their red-leaved counterparts, sometimes as much as two weeks earlier.

Among varieties, there are many to choose from; the key question is seed availability. The U.S. ornamental cabbage and kale market has grown faster than the seed supply, making it difficult at times to get the varieties you want. In ornamental cabbage, the Dynasty (Osaka), and Pigeon series, along with Rose Bouquet, are some of the best varieties on the market. Dynasty is available in three separate colors plus a mixture, while the Pigeon series is available in red and white separate colors only; no rose or mixture. Rose Bouquet is a mounded plant that fills out a 6-inch (15-cm) pot uniformly. Though only available in one color, this variety is an excellent choice.

B

In ornamental kale, the Peacock and Feather series have deeply cut leaves on plants that grow uniformly in the container. Both are available in both red and white variegated-leaved varieties. However, the Peacock or Feather series take about four weeks longer to produce marketable plants. The Emperor (Nagoya) and Chidori series, available in either red, pink, or white separate-leaved colors, have fringed or frilled leaf ends that are not notched like the previous two kale varieties. Regardless of which you choose, you will find they give excellent performance.

BRASSICA (BROCCOLI, BRUSSELS SPROUTS, CABBAGE, CAULIFLOWER, COLLARDS, KALE, KOHLRABI)

by Lelion D. Elledge, Jr.
Rosebrook Nursery
Oklahoma City, Oklahoma

❖ *Vegetables* **(Brassica spp.).** *6,000 to 9,000 seeds/oz (212 to 317/g). Germinates in 10 days at 70F (21C). Cover seed lightly after sowing.*

Brassicas, all known as cole crops, prefer cool temperatures. Allow six to eight weeks for cell packs and eight to nine weeks for 4-inch (10-cm) pots. The number of days from transplanting varies with both the crop and the variety grown. Broccoli takes approximately 50 to 58 days, Brussels sprouts 80 to 90 days, cabbage 40 to 90 days, cauliflower 45 to 68 days, collards 70 days, kale 45 to 55 days, and kohlrabi 45 to 55 days.

BROMELIADS

by Kerry Herndon
Kerry's Bromeliad Nursery, Inc.
Homestead, Florida

❖ *Foliage plant* **(Aechmea, Billbergia, Cryptanthus, Dyckia, Guzmania, Neoregelia, Nidularium, Tillandsia, Vriesea,** *and other genera). Propagated from seed by bromeliad specialists. Also propagated vegetatively. Most finished pots are produced from purchased seedlings, meristem cultures, or pups.*

Bromeliads, members of the Bromeliaceae family, are native to tropical and subtropical areas of South, Central, and North America (fig. 1). Texas and

FIG. 1. The striking, bright orange and green or red and green color combinations of many bromeliads make them a frequent choice of enthusiastic interior gardeners.

Florida have many native bromeliads that are protected by conservation laws. Spanish moss and ball moss are two other bromeliads found across the southern states. The pineapple is the most familiar bromeliad. There are more than 2,000 recognized bromeliad species, as well as hundreds of hybrids. Each offers something unique for the grower, retailer, and especially the interiorscaper.

Bromeliads have long been a favorite houseplant in Europe, having been grown in greenhouses there for the past 200 years. Since the early 1980s, bromeliads have become increasingly popular in North American homes and other interiors. This can be attributed to three very strong selling points bromeliads offer: bromeliads are true eyecatchers, as they are available in a wide range of colors and often sport a combination of colors; they require very little maintenance and thrive on neglect; and with reasonable care, bromeliads will bloom on the store shelf or in the home for at least three to as many as four to six months or more.

Common bromeliad genera include *Aechmea, Billbergia, Cryptanthus, Dyckia, Guzmania, Neoregelia, Nidularium, Tillandsia,* and *Vriesea.* Bromeliads range in height from several inches up to 40 feet (12 m). The most commonly sold varieties range in height from 10 to 30 inches (25 to 76 cm) and are sold in 4- to 8-inch (10- to 20-cm) pots. Color development is commonly rated as low, medium, and high, depending on the maturity of the flower bract.

A colorful, well-developed bromeliad retails for $4 to as much as $40 or more, generally determined by size and desirable characteristics. A large, rare bromeliad often retails for $75 or more. Due to the wide range of colors available with bromeliads, you can tailor specific ones to the holidays: red bromeliads for Christmas, orange and yellow for Thanksgiving, pink, blue, and peach for Easter and Mother's Day.

B

Bromeliads are found growing naturally in diverse environments, from rain forests to cool mountains to hot, dry deserts. Most bromeliads available today are tropical, although there are a few desert types available. The shape, size, and color of the flower, as well as the culture, depend on the original habitat, breeding, and species of the plant.

Most bromeliads are naturally epiphytic, clinging to tree limbs, trunks, and rocks for support. Most commercially grown bromeliads have a center rosette of leaves from which a brightly colored flower-bearing spike (bract) grows. On other varieties a portion of the leaves near the center of the plant and surrounding the small flowers at the center change color to brilliant red, blue, or purple. These brightly colored bracts attract hummingbirds as pollinators. Many bromeliads have vaselike, watertight leaves that hold water and organic material, such as leaves and bugs. As the material breaks down, the bromeliads translocate the nutrients and water through sophisticated trichomes (scales, hairs) on the leaf surfaces in the cup.

Growing bromeliads from seed is for specialist breeders, mostly in Holland and Belgium. Some varieties are also available from major commercial meristem lab companies. Most growers buy seedling or meristem liners from specialist young plant suppliers. Plants that cannot be grown from seed or meristem can often be propagated vegetatively. This requires considerable bench space and time. The best source of cultural information is your plant supplier. There is also the hobbyist Bromeliad Society, Inc., with a bimonthly bulletin. Its contact address is 2488 East 49th, Tulsa, OK 74105.

While most bromeliads are truly epiphytic in nature, most adapt readily to a terrestrial or semiterrestrial culture. The adaptable bromeliads will grow in a wide variety of media. Sufficient aeration and a slightly acid pH are the only required conditions. Some favored mixes are (1) half Canadian peat, half coarse perlite; (2) a third each of peat, bark, and perlite; and (3) half peat, half rock wool. In Europe bromeliads are grown in straight coarse peat. *Cryptanthus* and *Dyckia* are naturally terrestrial, so they will tolerate many media forms. Most gray-leafed *Tillandsia* species, however, will not adapt to terrestrial culture. These plants must be mounted on wood, cork, or tree fern slabs to replicate their original habitat.

After germinating the lettuce-size seed, grasslike seedlings should be transplanted into flats after having grown to 1½ inches (4 cm). From flats, plants can be moved to cell packs and eventually into individual containers as they grow larger. The more often a bromeliad is transplanted, the better it grows. For each transplant, provide more room on the bench, more light, and, to a small extent, more nutrients. Don't crowd young bromeliads on the bench, as bromeliads thrive on aeration. Bottom heating has shown to increase the turnover in the bench space required for commercially growing bromeliads from seed, as well as reduce heating costs.

The plant flowers once in its life cycle. After flowering, axial buds are stimulated to produce side shoots. These pups draw nutrients from the mother plant as the mother withers over a period of six months to a year. If the shoots are left on the mother plant, they will grow quite fast, but you will only get one to three new plants. If the pups are harvested at a size of around a quarter of the parent, it is possible to increase the yield to 10 or more new plants. The pups normally bloom in one or two years.

Don't pot a bromeliad too deeply—just to the base of the leaves. Don't use a pot that is too large for the plant, as the danger of overwatering increases. Usually a 4- to 6-inch (10- to 15-cm) pot is sufficient. Use stable pots or containers, as any rocking or other motion damages the tender, developing roots. Staking may be necessary until roots are well developed.

To complete a typical bromeliad life cycle, a temperature range of 65 to 85F (18 to 29C) and humidity between 60 and 80% are usually required. Tropical bromeliads prefer high humidity. Outside of a few bromeliad types that can withstand temperatures as low as 30F (–1C) with protection, for most you cannot allow night temperatures to drop below 50F (10C).

Bright, diffused light and genus-specific care improves growth and plant quality of bromeliads grown for commercial production. Direct sun will burn the leaves of most bromeliads. Generally, the stiffer-leaf varieties tolerate more light and may require very high light to bring out the full color of the foliage. Some soft-leaf varieties, on the other hand, will require 85% shade or more to grow properly. Once flower color is set, bromeliads will tolerate 50-footcandle (0.5-klux) conditions for months.

Water bromeliads like all plants: by inspection. In a greenhouse environment, the watering interval is between five and 15 days, depending on environmental conditions affecting plant water use. Water-soluble fertilizer at 100 ppm should be applied with every other watering. The watering should be long enough for the fertilizer to reach all the roots. An additional 10 minutes of clear water will dilute the fertilizer in the plant cup. Time-released fertilizer mixed in the soil or top-dressed is a good supplement to liquid feed. Copper-based fungicides are toxic to bromeliads. One popular fungicide is labeled as a herbicide for Spanish moss (*Tillandsia*), in fact.

In a home environment, bromeliads require much less water. They prefer to be on the dry side inside. Water among the lower leaves and soil. Allow the soil to dry between waterings. Watering in the center of the cup can cause fungus or bacteria to attack the plant. Overwatering will shorten the flower life.

Bromeliads can be artificially induced to flower once the plant is mature. Florel is the most common way to treat plants. Label rates are too high, however, so considerable testing must be done by the grower to determine the correct rate.

BROWALLIA

by Jim Nau
Ball Horticultural Company
West Chicago, Illinois

❖ *Annual (*Browallia speciosa*). 125,000 seeds/oz (4,409/g). Germinates in seven to 15 days at 72F (22C). Leave the seed exposed to light during germination.*

Browallia is a warm-season annual with star-shaped flowers on plants recommended for baskets or contain-

FIG. 1. Browallia fills a hanging basket for spring sales.

ers (fig. 1). If planted to the open ground, be sure to have fully developed plants and not seedlings or small transplants; these often perform poorly or not at all. Flowers are 1 inch (3 cm) across.

Seedlings will be ready for transplanting in 40 to 45 days. After transplanting allow 10 weeks for flowering 4-inch (10-cm) pots, one plant per pot, or 20 to 22 weeks for flowering 10-inch (25-cm) hanging baskets with no more than seven plants per pot. Grow at 68 to 70F (20 to 21C). No pinching is necessary to encourage branching. As a hanging basket, browallia does well in fairly heavy shade.

In the southern United States, browallias will overwinter in areas without frost. Plant outdoors from mid-September to November in the Deep South for flowering until late April. In cooler areas have plants ready for sale once the danger of frost has passed.

The primary variety selection in the U.S. trade is the Bell series, which includes four shades of blue and a white. Blue Bells is the most popular variety, due to its true-blue flower color. Marine Bells is a dark indigo color, making it the darkest variety available in this series. Sky Bells has a powdery or mid-blue flower color; Heavenly Bells is Cambridge blue, and Silver Bells has a pure white flower color. The Bell series grows to 10 to 12 inches (25 to 30 cm) in window boxes and containers.

The Starlight series, in Blue, Sky Blue, and White, has an excellent mounding habit for hanging baskets and containers.

For something different, try *Browallia viscosa,* a new browallia class. Cultivate as above, allowing nine to 11 weeks for flowering packs. Leaves are darker green, and plants are very heat tolerant, even in full sun. Flower size is about one-third the size of regular browallia—about ½ to ¾ inch (1 to 2 cm). Amethyst is a deep violet blue with a white eye, while Saphyr is a lighter blue with a white eye and bright green foliage. Both grow to 10 to 12 inches (25 to 30 cm) in the garden.

CABBAGE (SEE *BRASSICA,* VEGETABLES)

CALADIUM

by Gary J. Wilfret
University of Florida
Bradenton

❖ *Tropical* (**Caladium bicolor***). Vegetatively propagated. Growers start potted crops from tubers purchased from commercial propagators.*

Caladiums, colorful members of the aroid family, are widely grown as pot plants (fig. 1) and, during the warmer seasons, as outdoor bedding plants. Caladiums are grown outdoors in the southern states from April through

FIG. 1. Caladiums are growing as pot plants for spring sales at Van Wingerden International, Fletcher, North Carolina.

November and in the central and northern sections of the United States from June through August. In climates where night temperatures are below 65F (18C), caladiums must be started in greenhouses un-til warm weather arrives. Caladiums originated in the Amazon River basin of South America, where envi-

C

ronmental conditions exemplify the ideal growing parameters for this crop: high humidity, temperatures above 65F (18C), ample water and nutrition, and loose, well-drained soil.

Plants are grown from tubers produced on the muck soils of central Florida. Tubers are grouped into grades according to their diameter: Mammoth, 3½ inches (9 cm) and larger; Jumbo, 2½ to 3½ inches (6 to 9 cm); No. 1, 1½ to 2½ inches (4 to 6 cm); and No. 2, 1 to 1½ inches (2.5 to 4 cm). New crop tubers are usually available in late December for Valentine's Day sales. Some specialists, however, store tubers under controlled conditions of 70 to 80F (21 to 27C) and 40 to 50% humidity and can ship tubers through August, although plant vigor decreases with extended storage. Tubers are generally cured for six to eight weeks after digging and prior to shipping from Florida. Tubers that have not been stored for at least six weeks at 70F will sprout slowly. If tubers cannot be planted as soon as received, they should be unpacked and stored at 70F with ample air circulation. Allowing exposure to temperatures below 60F (16C) will cause tubers to have a rubbery texture, display slow sprouting, and produce fewer leaves.

There are several ways of starting caladium tubers. They may be planted directly out-of-doors when night temperatures remain above 65F (18C). They may be started in flowering-sized pots, or they may be started in Jiffy (peat) pots for later transplanting. Some growers start tubers in peat moss beds and transplant into containers as soon as roots are visible. Many growers plant tubers in straight peat moss and grow them at 80 to 85F (27 to 29C) with high humidity, although many of the commercial peat-vermiculite media produce similar growth and plant quality. Tubers should be planted as deep as possible in the pots (2 to 3 inches, or 5 to 8 cm, of media covering the tuber), since new roots develop on the upper surface of the tuber. Media should be moist but not saturated. Containers can be stacked in a pyramid formation and covered with a plastic tarpaulin until the first leaf sheaths are visible in 10 to 20 days. If warm greenhouse space of 80 to 85F (27 to 29C) is not available, electric heating cable may be used in a section of bench to provide the necessary soil temperature.

Caladium tubers are similar to potato tubers in that new growth develops from eyes. Depending on the cultivars, there may be from one to five prominent eyes and numerous small eyes. These larger eyes will produce the first large leaves, while the smaller eyes will form later peripheral leaves. With some cultivars it is necessary to carefully remove the large eyes, allowing the secondary eyes to sprout. Be very careful not to cut or damage the small buds surrounding the main bud or eye. This "de-eying" procedure delays marketability up to 10 days but produces a more compact plant with numerous leaves. A No. 2 tuber will fit easily into a 3-inch (8 cm) pot, and one No. 1 tuber or two No. 2 tubers are adequate for a 4-inch (10-cm) pot. A de-eyed Jumbo, three No. 1s, or five to six No. 2s are required for a 6-inch (15-cm) container, and six No. 1s or three Jumbos are needed for a 10-inch (25-cm) basket. Mammoth-grade tubers produce large exhibition plants and should be used in large containers or in the landscape.

Some growers cut tubers into pieces prior to planting to get more uniform growth, but this practice increases the chance of disease problems. Plants should be grown within a 70-to-90F (21-to-32C) temperature range and a light intensity of 2,500 to 5,000 footcandles (54 klux). Lower light intensities will cause undesirable stretching of the petioles, oversized leaves, and weak plants that become prostrate rather than upright. Higher light intensities can cause color fading and foliage burning. Plants should be kept well watered and can be grown easily with a capillary mat irrigation system. If the primary roots are lost due to inadequate moisture, the plants never regain their optimum floral display. Fertilizer, in a slow-release form, can be incorporated into the medium (e.g., 5 pounds per cubic yard, or about 3 kg/m^3, of 14-14-14) or can be applied weekly at 400 to 500 ppm N-P-K.

Caladiums destined for sale as bedding plants are best started in small containers (3 to 4 inches, 8 to 10 cm) and sold in these pots as soon as three to four leaves are mature. Some retail growers plant small tubers (No. 2) in a multicell pack and sell them as multiple plant units. Temperatures should be maintained above 65F (18C) during shipping of the finished plants and in the selling areas.

Bonzi at low rates can be effective for height control. Apply Bonzi at 0.5 to 1.0 mg active ingredient/6-inch pot as a drench 21 days after planting (leaves will not yet be emerged) to moist media. Pine bark in the media will reduce Bonzi's effectiveness, requiring the rate to be increased by 50%.

Caladiums grow best with partial shade in outdoor plantings, although some cultivars (e.g., Aaron, Candidum Jr., Carolyn Whorton, Fire Chief, Gingerland, Pink Gem, Rosalie, and White Wing) perform well in full sun. They always should be planted in a well-drained location and kept well-watered. Caladium troubles are usually caused by too dry soil or too much light (foliage has a burned look), too low temperatures (tubers grow slowly and erratically), or disease (soft, mushy tubers with brown roots).

FIG. 2. Keep your eyes on this new caladium variety, Sweetheart. Plants are compact and full with deep solid pink leaves that are edged in green—a striking combination.

CULTIVARS

There are more than 1,500 named cultivars, of which fewer than 100 are available commercially (fig. 2). Two distinct leaf types are available: fancy and lance. The fancy-leaved cultivars, with their tall, upright growth and large, fleshy, heart-shaped leaves, account for 80 to 85% of the market. The lance-leaved cultivars generally have heavier textured, leathery, narrow leaves. Plant growth of the lance type is generally more compact, with numerous leaves that do not burn or fade as easily in full sun. Lance cultivars produce smaller tubers and are priced one grade above their actual size, as compared to prices for fancy-leaved cultivars. Leaves of the lance cultivars have a good postharvest life when cut and placed in water.

Probably 30% of all caladiums used are of the white cultivars Candidum, Candidum Jr., and White Christmas.

CALCEOLARIA

❖ *Annual (***Calceolaria herbeohybrida***). 740,000 seeds/oz (265,100/g). Germinates in 10 to 16 days at 70F (21C). Leave seed exposed to light during germination.*

This colorful low-temperature crop is traditionally grown for Easter and Mother's Day. Calceolarias make excellent 4-inch (10-cm) pot plants (fig. 1). Traditional cultivars are dependent on a combination of long days and cool night temperatures to initiate flowering. Some newer cultivars bloom independent of temperature and day length, and these can be produced for Valentine's Day sales or indeed any holiday or desired bloom time.

When growing the traditional varieties, it is important to remember that the shorter the day length during bud

FIG. 1. Potted calceolaria is a popular cool-season flowering pot. The variety here, a dwarf bred for 4-inch production, is called Dainty.

initiation, the cooler the temperatures must be to initiate buds. During the Easter crop bud-initiation time (generally December and January), temperatures should be below 50F (10C) for six weeks. Almost all calceolarias grown today are F_1 hybrids.

Sow calceolaria seed on top of well-pressed germinating media; do not cover. Plants need shade as soon as germination occurs, and good air circulation is a must. Germinate the seeds between 60 and 70F (16 and 21C). Transplant the seedlings into 2- or 2½-inch (5- or 6-cm) pots or cells as soon as they are transplantable. Because calceolarias are very fine-rooted plants, it is very difficult to bypass the intermediate container stage and transplant directly into the final container, as overwatering would be hard to avoid with this procedure.

To produce an Easter crop of 5- or 6-inch (13- or 15-cm) pots, sow the seed in early September. Four-inch (10-cm) crops can be sown later because less time is required to reach proper size for flower initiation. Night temperatures after transplanting should be maintained at 60 to 63F (16 to 17C).

When plants are well established in the intermediate containers, they can be potted into the final containers. As crown rot can be a problem, the crown should be above the soil line rather than below. When plants have reached a proper size (approximately as wide as the final container), lower the temperature to 50F (10C) for six weeks. After the cool period is completed, return the temperature to 60 to 62F (16 to 17C) until the crop is finished. Crop time from end of cool period is dependent on day length. Late Easter and Mother's Day crops will develop in nine to 10 weeks. Early Easter crops may take 10 to 12 weeks. For Valentine's Day flowering, use Anytime Mixture and assume a total crop time of 18 to 20 weeks from sow to sell at 60F constant.

Calceolarias are light feeders and subject to salt buildup and chlorosis if overwatered. The pH should be 5.5 to 6.0, and 100 ppm feed is adequate. Shade all plants at least 50% during bright-light periods. In the North, shade is usually required from mid-February until mid- or late October. Aphids and whiteflies, the chief pests, should be monitored for constantly.

CALENDULA

by Jim Nau
Ball Horticultural Company
West Chicago, Illinois

❖ *Annual* (Calendula officinalis). *3,000 seeds/oz (106/g). Germinates in 10 to 14 days at 70F (21C). Seed should be covered during germination.*

Calendulas are excellent cool-weather plants that tolerate light frosts, though the flowers may burn. They are most often available in orange or yellow flower col-

C

ors; additional shades and tones include golden-yellow, yellow with dark petal tips, a light yellow often called apricot, plus others. Flowers often grow to 2½ to 3½ inches (6 to 9 cm) across on plants that range from 10 to 28 inches (25 to 71 cm) tall. Plants reseed themselves in the garden.

Allow eight to 10 weeks for green packs and up to 14 weeks for flowering packs of 32 cells per flat. For 4-inch (10-cm) pots, give 13 to 15 weeks, using one plant per pot. This culture is based on dwarf varieties such as the Bon Bon and Fiesta Gitana series.

In the southern United States, allow seven weeks for selling dwarf varieties green in packs, 10 weeks for selling them in flower. Calendulas do well under short-day conditions when planted from September to April, though their performance isn't reliable in the low light conditions of December and January.

The primary varieties on the market today are the dwarf, double-flowering plants that make excellent 4-inch (10-cm) pot plants or edging or border plants. Fiesta Gitana (Gypsy Festival) is available in orange, yellow, or as a mixture of these two, plus several other colors. The plants grow to a height of 12 to 14 inches (30 to 36 cm) and are one of the more commonly grown varieties.

The Bon Bon series—Apricot, Orange, Yellow, and a Mix—is similar to the Fiesta Gitana series in colors and overall plant performance. However, these plants grow to a height of only 8 to 10 inches (20 to 25 cm).

Provide a layer of support for cut flower calendulas. By the time tall garden-planted calendula becomes top-heavy, it is past its prime, so staking isn't needed.

A historical note: George J. Ball, Sr., established his seed business in the 1920s with his outstanding strains of Ball Orange calendula—then a rather important greenhouse cut flower worldwide.

CALLA LILY (SEE *ZANTEDESCHIA*)

CALLISTEPHUS (ANNUAL ASTER)

❖ *Annual* **(Callistephus chinensis)**. *12,000 seeds/oz (425/g). Germinates in eight to 10 days at 70F (21C). Lightly cover or leave seed exposed to light for germination.*

This versatile aster was one of the favorites of the late George J. Ball, who, more than 75 years ago, developed the forebears of some of today's finest varieties. Probably no single aster series in history was as well known or has been grown by as many florists as the Ball Florist strain. Ball Florist Mix is still sold today.

Most recently, the Matsumoto asters have revived cut asters among growers for their good disease tolerance and availability in separate colors.

Asters may also be produced as a bedding plant crop. Allow 16 to 18 weeks to finish plants from February to May sowings, and 20 to 22 weeks for sowings from November and December. Sell plants green in packs after the danger of frost is passed. For bedding plant sales, try the Pot 'n' Patio series, ready for sale in just 90 days. Plants are dwarf, 6 inches (15 cm), and require no supplemental lighting to flower.

Asters are day-length sensitive. Provide three to four weeks of long days (14 to 16 hours) to seedlings until the plants are 20 to 24 inches (51 to 61 cm) tall. The only exception is from May 15 to August 1 in the northern hemisphere, during which the day length is naturally long enough. Mum lighting, or 60-watt bulbs with reflectors spaced 5 feet (1.5 m) apart is satisfactory. Turn lights on from sundown until 10 P.M., or for two to four hours in the middle of the night. Maintain a cool night temperature—50F (10C)—during the long-day period.

A mid-July sowing flowers in January, an October 20 sowing flowers in April, and a May 20 sowing flowers in September.

Asters had rapidly declined in popularity due to their disease and insect susceptibility. Aster yellows causes yellowing and distortion of all or part of the plant. It just stops growing. Flowers are also yellowed and do not open. Plants will be more than 12 inches (30 cm) tall normally, before symptoms appear. Often one side of the plant is affected first. Destroy infected plants as soon as they are discovered.

Aster yellows is a virus that affects many weeds and garden annuals. It is transmitted via the aster leafhopper. Leafhoppers pick up the yellows virus from weeds outside the greenhouse, and then they feed on the aster foliage, transmitting the disease. Aster yellows can be eliminated if plants are grown in a cloth enclosure (22 threads per square inch, 3.4/cm^2) kept tight enough to exclude leafhoppers.

Asters are also susceptible to *Fusarium* wilt. Plants rot at the soil surface. Infected plants usually show dark lesions extending up the stem. Steam soil to 180F (82C) for 30 minutes to a depth of 8 inches (20 cm) to kill *Fusarium* before planting. If steaming beds is impossible, plant asters to beds not previously used for aster culture to control losses.

Cut flowers benefit from silver nitrate, which extends vase life.

The Matsumoto series has bright yellow centers and is available in 11 separate colors and a mix from peach and blush pink to deep red, scarlet, and blue. Plants grow to 24 inches (61 cm) and produce 2-inch (5-cm) flowers in sprays. Matsumoto shows good tolerance to *Fusarium*.

CAMPANULA

C

❖ *Perennial*
(Campanula carpat-ica). 200,000 seeds/oz (7,000/g). Germinate at 70F (21C). Seed germinates in 14 to 21 days. Cover seed lightly after sowing.

FIG. 1. A popular perennial in the United States, *Campanula* is an important spring flowering pot plant in Europe. This field of flowering campanulas is at Madsen's, Odense, Denmark. The pots are grown outdoors during the summer and allowed to go dormant in the fall and early winter. During the late winter and spring, pots are brought indoors and forced into flower for sale as flowering pot plants.

The *Campanula* genus consists of about 250 species, mostly long-lived perennials (fig. 1). They range from dwarf rock garden forms to 5-feet (1.5-m) giants that require support. Flower color is predominantly blue or lavender, but some species are white or pink. Flowers are bell-shaped, hence the Latin name campanula, meaning "little bell."

AS A POT PLANT

Campanula carpatica is a perennial that grows to 6 inches (15 cm), has a neat, mounded form, and bears blue or white flowers. The Karl Foerster (sometimes listed as Karl Forster or Karl Foster) has large blue flowers that cover the entire plant. The cultivar White Clips has smaller white flowers but is just as floriferous.

These plants are relatively easy to cultivate as flowering pot plants forced out of season. Cuttings can be rooted in the spring and pots placed outside until brought in for forcing, usually December through February. These dormant roots require 10 to 12 weeks from start of forcing to finished product, in both 4- and 6-inch (10- and 15-cm) pots. Dormant liners can also be purchased and finished in the same time frame.

A well-drained, peat-lite soil mix should be used. When potting, place plants at their original soil level. The soil should be kept moist; do not allow it to dry out! Fertilize with 150 to 200 ppm N-P-K nutrient solution during the whole forcing period, using clear water every third watering.

Spacing should be pot-tight the first three to four weeks. Then place 6-inch pots on 6-inch (15- on 15-cm) centers and 4-inch pots on 4-inch (10- on 10-cm) centers. Night temperatures should be 63 to 65F (17 to 18C) for the first two to three weeks or until foliage is full, then 55 to 58F (13 to 14C) until finish.

Campanula carpatica is a long-day plant. Daytime light can be supplemented to 16-hour days of 3,000 f.c. (32 klux), or broken night lighting can be implemented. Break nights from 10 P.M. to 2 A.M. with incandescent light (mum lighting) at 10 to 20 f.c. (108 to 215 lux) until buds are fully developed. Treatment should begin 10 to 14 days after the start of forcing.

If the plants are grown outdoors and brought in, they should be given a drench of appropriately labeled fungicides and insecticides immediately. Campanulas are not bothered by many disease or insect problems, but if overhead watering is used during forcing, *Botrytis* in the crown may occur.

Growth regulators can be applied as follows: with B-Nine, use 1 tsp (85% active ingredient)/gal water; for Cycocel, 1 oz/gal (30 g/4 l) water; and for A-Rest, 1 oz/gal (30 g/4 l) of water applied as a spray or drench. These can be applied five to six weeks after planting, usually when the plant has reached one half to three fourths of desired finished size. It is important to discontinue spraying when plants are blooming, as it would cause spotting on the open flowers.

When preparing plants for shipping, keep them moist to prevent the blooms from inverting. A silver thiosulphate (STS) drench should be applied 10 to 12 days prior to shipping at a rate of 1 oz/gal (STS concentration: 40 grams of silver nitrate plus 160 grams sodium biosulphate/gal). STS splashed on open flowers will cause spotting.

Campanulas do not take well to shipping at temperatures above 80F (28C). In fact, they will simply rot very quickly enclosed in a box during warm, humid weather. Shipping should be during the bud stage and limited either to cool weather or to nearby deliveries. Any unsold pots in the spring can easily be cut back. They will flower again in six to eight weeks for a second chance at the cash register.

AS A PERENNIAL

Campanula carpatica grown from seed fits regular bedding plant production schedules well. For pack sales, January or February sowings can be sold in May, grown at temperatures of 50 to 55F (10 to 13C) nights. For pot sales, sow seed the previous summer, up to October, and grow on at 45 to 50F (7 to 10C) nights. Pots will begin flowering in April. Raising temperatures will force flowering faster.

Seed may be sown anytime before March for plants to flower during the summer of the same year.

Blue Clips and White Clips, both hybrids, flower well the first year, although these plants are not as hardy, especially when planted away from a house foun-

C

dation. Clips's habit is petite, with small leaves and flowers, and is available from seed or vegetatively. Uniform Blue and Uniform White are more vigorous than Clips, but less vigorous than straight species plants.

C. carpatica will flower repeatedly throughout the summer, making an excellent addition to annual plantings.

CAPSICUM (ORNAMENTAL PEPPER, CHRISTMAS CHERRY, JERUSALEM CHERRY)

❖ *Annuals* **(Capsicum spp.).** *9,000 seeds/oz (317/g). Germinates in seven to 12 days at 72F (22C).*

❖ *Annual Jerusalem cherry* **(Solanum pseudocapsicum).** *12,000 seeds/oz (423/g). Germinates in 15 days at 70F (21C). Cover seed.*

FIG. 1. Ornamental peppers make a showy pot plant. The variety shown: Treasure Red.

Ornamental peppers were formerly known as Christmas peppers, and in some areas of the country, they are still referred to that way. Currently, most varieties of these colorful plants are grown for sales in September, October, and November—much more than for Christmas (fig. 1).

The majority are sold in 4- or 4½-inch (10- or 11-cm) pots. A possible variation is to pan three or four established 2-inch (5-cm) plants to a 6-inch (15-cm) pot in late summer. Sowings are usually made from April through mid-July for sales in September to Christmas.

Allow eight weeks for green pack sales. Most varieties benefit from a single or double pinch to encourage branching.

Use a standard feeding program: 200 ppm N and K at each irrigation. This should be continued until the fruit is set. Once fruit is set, reduce feed and water to tone plants.

For maximum fruit set, peppers require strong light, good air movement, and temperatures between 65 and 70F (18 and 21C). As they reach maturity, temperatures can be lowered to 60F (16C). Peppers are best grown in a greenhouse.

Red Missile F_1 produces red fruit that is 2 inches (5 cm) long and tapered.

CHRISTMAS CHERRY

No longer the common sight they once were during the Christmas holiday season, some Christmas cherry plants are still grown for the holidays. Seed is sown in February, transplanted to cells or 2¼-inch (6-cm) pots three weeks later, and then moved into 6-inch (15-cm) finished pots by late May.

During the summer, plants are grown outdoors to ensure good fruit set (flowers require wind or insects for pollination). Brought inside before frost, they are finished at 50 to 55F (10 to 13C). Plants should be pinched after they develop four nodes, and can be pinched further, as needed, until July 15.

CAPSICUM (VEGETABLE PEPPER)

by Lelion D. Elledge, Jr.
Rosebrook Nursery
Oklahoma City, Oklahoma

❖ *Vegetable (Capsicum spp.). 4,000 seeds/oz (141 seed/g). Germinates in 10 days at 72F (22C). Leave seed exposed to light after sowing.*

Peppers prefer warm weather. Both soil and night temperatures should be maintained at 60F (16C) or above. As bedding plants, peppers require six to eight weeks for finishing in cell packs and eight to 10 weeks for 4-inch (10-cm) pots. Home gardeners will be able to harvest fruit 60 to 80 days after transplanting.

There is wide diversity in pepper varieties. The greatest variables for bell types are size and color, while hot peppers range from mild varieties to the superhot Habanero. Peppers have become an increasingly popular vegetable, due in part to the recent trend in Southwestern-style cooking.

CARNATION (SEE *DIANTHUS*)

CATHARANTHUS (VINCA, MADAGASCAR PERIWINKLE)

by Will Healy
Ball Horticultural Company
West Chicago, Illinois

❖ *Annual* (**Catharanthus roseus,** *formerly* **V. rosea***). 21,000 seeds/oz (740/g). Germinates in seven to 15 days at 78 to 80F (26 to 27C). After three days drop to 75 to 78F (24 to 26C) for the remainder of germination. (Also see guidelines for plugs following.) Upon sowing, lightly cover seed and place the flats into total darkness. Move to the light once the seedlings have emerged but before they have stretched. (See germination comments following.)*

FIG. 1. In the past 10 years, bedding plant vinca has become one of the most popular flowers for full sun. Be careful of culture: They need heat during production!

Vinca is an up-and-coming bedding plant (fig. 1). It is considered difficult to grow, but it is still one of the leading plants for warm, dry locations in full sun. Vinca works in tough landscape situations where no other vegetation might grow. Its glossy green leaves also repel dust. It does not tolerate cool conditions, and it prefers well-drained locations. Vinca is drought-tolerant if it is allowed to become established before water is limited.

In regard to seed, it has been found that vinca germinates better when seed is older as opposed to fresh. If seed is left over from sowings in the winter, hang onto it. It should germinate better four to six months later. For this reason, many seed companies are bringing in seed months earlier than needed, holding onto it and allowing the seed to break dormancy, improving germination by sow date.

Vinca is a warm-weather crop but is often grown in the greenhouse in the cool, early spring. The crop does best grown in plugs, limiting transplant root damage and consequent overwatering damage and *Thielaviopsis*.

Use a sterile, soilless media to ensure a disease-free start. The number of seeds grown in each cell can affect finishing time to first flower; multiple-sown plugs finish up to 10 days after single-sown. Vinca performs well from a 92-count to an 800-count plug. The crop germinates best in darkness. Cover with vermiculite to provide a constant moisture level.

A germination temperature of 80F (27C) provides more timely and uniform emergence, usually in six to nine days. At this point, remove trays from the heat. Otherwise, continue to grow on a heated bench to fill out the trays more quickly, especially in the early season. The goal is quick germination so the plants do not languish in the high humidity required for germination.

Once seeds are germinated, provide supplemental light and fertilize at 50 to 75 ppm N. Increase fertilizer to 100 to 150 ppm during Stages 3 and 4.

During Stage 3, growth regulators, such as Sumagic or A-Rest at low rates, can help control height. Avoid growth regulator application on warm days to prevent phytotoxicity. If you cannot apply growth regulators until after transplant, allow the plants to put on some growth first. When Bonzi is used to control height, black spots may form on the older leaves.

Transplant when plants are just beginning to crowd and before they stretch. At this stage the root system is large enough to withstand pulling out the plugs. Transplant into a sterile, soilless medium with a pH of 5.5 to 6.0. Make sure the medium has a minor element package included. If it doesn't, add fritted trace elements to enhance crop vigor during the cool, early spring. At a pH above 6.9, new leaves will show iron deficiency. Vinca does not like high salts—be careful not to allow salt buildup at any time during production, and use a low fertilizer concentration.

If transplanting bare-root seedlings, transplant to an intermediate cell (six or eight per full flat) for handling ease and fewer overwatering and disease problems than if transplanting directly into larger containers. Allow flatted vinca two to four weeks to establish themselves, then pot in 4-inch (10-cm) or larger containers.

Alternatively, you can grow vinca in a large plug and move it directly to a 4-inch (10-cm) pot; however, temperature control is critical. For best results, maintain night temperature at 65F (18C). Lower temperatures increase leaf chlorosis and stunted growth, which is compounded by overwatering. This risk is even higher with small plants in large soil volumes. The majority of problems with vinca—leaf rolling, leaf chlorosis—happen because of low temperature. Once seedlings are rooted through to the bottom of their growing containers, you can lower temperatures, but before that, maintain warm temperatures.

Vinca has very few pest problems but can be seriously affected by disease. *Thielaviopsis* is a major greenhouse problem. The best control is good sanitation. Grow plugs as warmly as possible. If you must grow vinca at cooler temperatures, *keep plants dry*. The major garden disease problem is aerial *Phytopthora (Phytopthora parasitica)*. This disease, typically showing up as plants mature, results from overhead watering that splashes soil and fungus spores onto lower leaves and stems. Lesions develop down the stem to the crown and kill the plant aboveground, leaving a healthy root system. Treat with one of several fungicides. An Aliette + Fore tank mix is highly effective. Still, the best control is to avoid overhead watering; use drip irrigation instead.

In the southern United States, allow 11 to 12 weeks for flowering pack sales, and 15 to 16 weeks for flowering 4-inch (10-cm) pots, one plant per pot. In the northern United States, allow 14 to 15 weeks for flowering pack sales, and 16 to 17 weeks for 4-inch pot sales, using one plant per pot. For something unique, try a 10-inch (25-cm) hanging basket using five plants per pot. Plants are salable 18 to 20 weeks after sowing.

CAULIFLOWER (SEE *BRASSICA*, VEGETABLES)

CELOSIA

by Jim Nau
Ball Horticultural Company
West Chicago, Illinois

❖ *Annual.* **Celosia cristata,** *35,000 seeds/oz (1,235/g);* **Celosia plumosus,** *34,000 seeds/oz (1,200/g);* **Celosia spicata** *26,000 seeds/oz. (910/g). Germinates in eight to 10 days at 75F (24C). Seed should be covered during germination.*

Celosias are among the best annuals for warm-weather color in the garden or as cut flowers (fig. 1). *Celosia plumosus* is characterized by holding its flowers in featherlike plumes that measure from 4 to 12 inches (10 to 30 cm) long. It comes

FIG. 1. Performance in the landscape and bold colors make celosia a bedding plant winner. The Castle series has become a fixture with landscapers looking for smaller dwarf plants. In the photo: Pink Castle.

in colors of red, scarlet, gold, cream, and rose. *Celosia cristata* is commonly called "cockscomb" due to the tight, semirounded flower heads that measure from 4 to 15 inches (10 to 38 cm) across. *C. cristata* has the same flower colors as *C. plumosus,* and both varieties make excellent fresh and dried cut flowers. Both types have green- and bronze-leaf varieties, though bronze-leaf plants are most often associated with red-flowering celosias.

Growing celosias too cool or planting them outdoors too early will cause premature flowering and spoil later performance. Restricting celosia growth at all—whether it's premature flowering in the plug or in the cell pack—will result in plants in the garden that are short, are small-flowering, and will die prematurely. If selling in 72 cells per flat, sell the plants green or use 48 or fewer cells per flat to sell in color. Allow seven to nine weeks for green packs and 10 to 12 weeks for flowering packs. Use only dwarf varieties when planning to flower in the pack. Allow 13 weeks for flowering 4-inch (10-cm) pots, using one plant per pot. If a growth regulator is needed to keep the plants dwarf, use A-Rest, B-Nine, Bonzi, and Sumagic.

In the southern United States, allow seven to nine weeks for green pack sales and 12 to 13 weeks for 4-inch (10-cm) flowering pots. Plant after all danger of frost has passed, until early April. If established before the weather gets hot, celosias will flower until August or September. Among the dwarf varieties, the feather or plume types show good color through mid-July, while crested types continue to display well until August. The taller varieties put on a good show until the first frost.

Choose varieties carefully for specific uses. The Century series and other upright, spreading celosias, as tall as 20 inches (51 cm), are ideal for mass plantings, filling in well on 12-inch (30-cm) centers. The Centurys hold up longer, and their colors shade less, than other varieties, placing them among the top performers in the American garden. Smaller, dwarf varieties, such as Geisha (10 inches, 25 cm) and Kimono (6 inches, 15 cm), make excellent borders and 4-inch (10-cm) pot plants, though the flower colors aren't as vibrant. However, the Castle series (12 inches, 30 cm) offers a medium,

C

upright (but not tall) plant with bright, nonfading colors. Of the crested types, try the Jewel Box series for 4-inch pot and pack sales as border plants. Unlike dwarf *C. plumosus* types, crested varieties hold their color regardless of height. The All-America Selections award-winning Prestige Scarlet crested celosia, intermediate in height, is a true summer performer, refusing to burn out in the heat.

For short-stemmed cut flowers or background plants, select taller varieties that hold their color well once cut. Among the feather types, the Century, Feather, and Sparkler series are suggested for outstanding performance. Of the three, Century is the shortest, and the Feather series is the latest to flower in the garden. Crested types also make good cut flowers, among them Toreador, a dark red variety, and the

Fig. 2. For something unusual, cut flower growers may want to try Chief Mixed celosia.

Chiefs, an upright series reaching 3 feet (71 cm) tall (fig. 2). For an unusual landscape celosia, try New Look, with its deep bronze foliage and scarlet red feather plumes. Plants grow 14 to 16 inches (36 to 41 cm) outdoors.

Celosia spicata, wheat celosia, makes an excellent cut flower. It germinates at 72F (22C) in four to seven days. Sown in early April, plants will flower the first week of June, grown at 55 to 60F (13 to 16C) night temperatures. Flamingo Feather, a light pink, and Flamingo Purple, which grows to twice the size, are varieties to choose. Cut the flowers as soon as they are ready because the color fades quickly in high heat.

CENTAUREA (BACHELOR'S BUTTON, CORNFLOWER)

by Robert G. Anderson
University of Kentucky
Lexington

❖ *Annual* **(Centaurea cyanus).** *7,000 seed/oz (247/g). Germinates in 10 days at 65F (18C). Cover seed lightly upon sowing.*

Cornflower or bachelor's button, *Centaurea cyanus,* is an annual flower often used as a cut flower because of its distinctive blue color. It can be useful as a bedding plant, but it is also one of the main plants in wildflower seed mixes. It can naturalize in some areas of the country and is one of the most successful seeds for late spring and summer flowers in "wildflower" areas on highways across the nation.

The bright blue flowers are excellent secondary cut flowers in flower arrangements and bouquets. Excellent white, pink, and crimson flower colors are also available in other varieties. All have good postharvest life in arrangements, but cornflower is quite susceptible to disease problems during storage and long-distance transportation, so local and regional production of this cut flower can be successful.

C. cyanus is a cool-season plant that naturally flowers under the long days of late spring and summer. Plants grow 15 to 30 inches (38 to 76 cm) tall and do not tolerate hot, humid summers.

As a bedding plant, allow six to eight weeks for green pack sales and nine to 10 weeks for flowering plants. Dwarf varieties can be grown as 4-inch (10-cm) pots; use two to three plants per pot and allow 10 to 13 weeks to flower. Be sure to grow the plant with 50 to 55F (10 to 13C) night temperatures to establish a good root system before long days initiate flowering.

For field-grown cut flowers, sow seed directly in the field or transplant. For spring flowering in Florida and south and central California, seeds are direct-sown in September. Plants grown during short days and in cool weather (i.e., fall and winter) will have good basal branching and produce many cut stems per plant when the long days of spring promote flowering. In the East and Midwest, seed is sown in the field from April to mid-May for flowering July to September.

Typical greenhouse production involves large plants as well. *Centaurea* can be flowered throughout the winter with the use of supplemental lighting treatments. Seedlings are grown under short days for at least eight weeks to produce a plant with many basal branches. After short days, use incandescent supplemental lighting (mum lighting) to extend the day to 15 to 16 hours, or use a four-hour night break in the middle of the night. Production should be at 55F (13C) night temperature with moderate fertilizer rates of 100 to 150 ppm N. Do not overfertilize because stems would become weak with many leaves. A single layer of mesh should be sufficient to support the large plants and stems.

An alternative production system using single stem plants has been developed at the University of Kentucky. Research shows that cornflower seeds in a 288-plug tray germinate in 24 to 28 hours in a germination chamber at 72F (22C). Seedlings are moved immediately to the greenhouse, where they receive 24-hour supplemental HID lighting from high-pressure sodium lamps for 10 to 18 days. The plugs

C

are transplanted to 3-inch (7-cm) pots or 4-inch (10-cm) pots (two plants per pot) and spaced pot-to-pot for a plant density of 15 to 20 stems per square foot (161 to 215/m²). The pots are placed in a greenhouse with long day lighting, as described above. Plants are grown at 55F (13C), with moderate fertilization (100 to 150 ppm N) and a single layer of support mesh. The crop requires eight to nine weeks from seed sowing to cut flower harvest. Individual cut stems have four flowers and should be harvested at the first sign of color in the primary flower bud. Individual stems are uniformly 20 to 26 inches (51 to 66 cm) tall at harvest for easy packaging and shipping. This short-term cut flower crop could fit into many winter greenhouse production schedules, and locally grown cut stems have a significant market as a secondary flower in flower arrangements.

CINERARIA (SEE *SENECIO*, CINERARIA)

CHRISTMAS CACTUS (SEE *SCHLUMBERGERA*)

CHRYSANTHEMUM (SEE *DENDRANTHEMA*)

CHRYSANTHEMUM (MATRICARIA, FEVERFEW)

❖ *Tender perennial* (**Chrysanthemum parthenium**)*. 200,000 seeds/oz (7,000/g). Germinates in seven to 10 days at 70F (21C). Leave seed exposed to light.*

Matricaria makes an excellent bouquet filler as a cut flower and a reliable addition to the annual or perennial border for consumers. Since plants are tender perennials, most that appear the following season are from reseeding.

Sell packs green, as plants have a tendency to grow tall when they flower. Green packs are ready for sale in 10 to 12 weeks grown at 55 to 60F (13 to 16C). Matricaria flowers the first year from seed. Plants will grow to 10 to 24 inches tall (25 to 61 cm) in the garden.

Jeff McGrew, McGrew Horticultural Products and Services, offers cultural tips for growers producing matricaria as a cut flower crop. Matricaria is suitable for greenhouse or field production. Crop time ranges from eight weeks in the summer to 16 weeks in the winter. Crops planted in early spring may produce a second flush of flowers under natural conditions; however, the stems will be of lesser quality.

In the greenhouse, flowering can be controlled using temperature and day length. Matricaria is a long-day plant, requiring 14 hours or more of daylight. Plantings made from August to March require mum lighting, with a minimum intensity of 15 footcandles (161 lux) to make a 16- to 18-hour day to initiate flowers.

Under low light conditions, plant four plants per square foot (43/m²), five to six plants per square foot (54 to 65/m²) in high light conditions. Night temperatures of 55F (13C) and day temperatures of 65 to 80F (18 to 27C) are ideal. In high light areas, stem length can be increased by applying 20 to 30% shade from late spring through the summer.

Provide a well-balanced fertilizer program up until visible bud. At that point, discontinue feeding, using clear water up until harvest. Harvest when 75% of the individual flowers within the spray have matured.

Try Santana for bedding plant sales. Plants grow 8 to 10 inches (20 to 25 cm) tall, making an excellent replacement for, or complement to, ageratum in the garden.

For cut flowers, try Ball Double White or White Pompon for the traditional double-flowered look. For something different, try the Vegmo series, specially bred for cut flower production. Golden has golden yellow button flowers; Single has white daisy flowers with yellow centers; Snowball has white button flowers; Star has white anemone flowers; and Tetra is a white tetraploid.

CLARKIA (GODETIA, SATIN FLOWER)

by Robert G. Anderson
University of Kentucky
Lexington

❖ *Annual/biennial (Clarkia amoena G. whitneyi). 37,000 seeds/oz (1,305/g). Germination in six to 10 days at 70F (21C). Cover seed for germination.*

Godetia, or satin flower, has been transformed from a unique and uncommon garden plant into a dependable cut flower for greenhouse and

FIG. 1. Godetia comes in a range of colors from solid white to pastel pink to red, also with stripes and picotee coloring. Shown here is the Satin series.

field production and for pot plant production. Open-pollinated varieties, such as Cattleya, Furora, Kyohuhai, and Maidenblush, have been available for a number of years. The Grace series of hybrid cut flower godetia was introduced in 1987 with salmon, red, rose pink, shell pink, lavender, and white flower colors. The Satin series, introduced in 1992 (fig. 1), is a dwarf hybrid godetia suitable for pot production with the same flower colors as Grace, plus deep rose, lilac rose, and red and white varieties.

Godetia is a cool season crop. All *Clarkia* species are native to the west coast of North America from California to British Columbia and grow in the cool climate within a few miles of the ocean. Optimal temperatures of 55 to 65F (13 to 18C) during the day and 45 to 55F (7 to 13C) at night will produce the best quality. Hardened plants will tolerate temperatures down to 20F (–7 C) with little or no plant damage. Continued warm days and nights, however, cause the plants to deteriorate quickly.

Nutrition is also very important for godetia production. These plants prefer very low fertility. Stems will be strong and erect, with few lateral branches, when grown under low fertility, but the stems will be soft, with many lateral branches, at what most greenhouse growers consider low fertility (50 to 100 ppm constant liquid feed).

Supplemental lighting can speed winter flowering of godetia because it is a facultative long-day plant. Flowering in the Grace series of cut flower godetias can be speeded four to six weeks using incandescent (mum lighting) or HID (200 to 400 footcandles, 2 to 4 klux) supplemental lighting from 6 P.M. to midnight each day during production. Using HID supplemental lighting for 24 hours per day for four to six weeks after germination will also reduce production time significantly. The Satin series for pot plant production does not react as dramatically to long-day treatments.

Godetia is infested by aphids, thrips, whiteflies, and spider mites when these insects are a problem on other plants in greenhouse or field production. Root- and stem-rotting diseases, such as *Fusarium* and *Pythium*, can be a problem when these plants are grown in poorly drained soils or when they are overwatered. Gray mold (*Botrytis*) can cause leaf and stem damage on tightly packed, overfertilized plants in the greenhouse.

PLUG PRODUCTION

The first four weeks of godetia production for cuts, or six weeks for pots, are best completed as plugs. Seed germination is usually uniform and at high percentages. Seed is often single-sown into 288 or similar-sized plug trays for cut flower production, or larger cells for pot production.

Stage 1, Days 1 to 10. Sow seed into plug media with little or no starter charge. Lightly cover seed with media or vermiculite, maintain uniform soil moisture with intermittent mist, and maintain media temperatures of 70F (21C) for seven to 10 days, until germination is complete.

Stage 2, Days 11 to 21. Move seedlings to a cool, bright, well-ventilated greenhouse; optimum temperatures are 55 to 60F (13 to 16C). Fertilizer applications directly affect the number of lateral branches. Use almost no fertilizer (beyond a starter charge in the plug media) if the plants will be grown as single-stem cut flowers. Fertilize pot crops and multiple-stem cut flower crops one or two times during plug production to get four to eight lateral branches per plant. Apply only 50 to 100 ppm fertilizer at one time. Do not use constant liquid fertilizer applications, or the crop will be too soft. Use a well-balanced fertilizer with calcium, or consider using a balanced organic fertilizer. Supplemental HID lighting for six to 24 hours each day will greatly enhance plug growth in low light areas of the country.

Stage 3, Days 21 to 28. Maintain cool temperatures. Negative DIF or weekly applications of B-Nine (2,500 ppm) will help control plant height for pot production. Cool temperatures and wise use of water and fertilizer will make the best plugs.

Stage 4, Day 28. Plugs are ready to transplant or ship. Transplant plugs as soon as possible. Obtain plugs for cut godetia crops for Thanksgiving or Christmas from a cool climate, probably California, because seedlings do not tolerate warm temperatures in the plug stage.

FIELD CUT FLOWER PRODUCTION

Most field-grown godetias are produced on the West Coast because uniform cool temperatures are common during production

FIG. 2. If you've got to pick one crop to gamble on for the future and you're in an area with moderate to cool temperatures, select *Clarkia* (godetia). Godetia is rising fast in popularity as a cut flower. It also performs well as a flowering pot plant and bedding plant. Flowers are exceptionally long lasting, and buds continue to open either in pots or once stems are cut. In the photo: White Satin.

(fig. 2). Field production is possible in the upper Midwest, New England, and at higher elevations with some experience. California growers harvest sequential crops from May into July from fall, winter, and spring sowings. Crop time and stem length decreases as average temperature, light intensity, and day length increases from spring into summer.

Plants can be planted in rows 5 feet (1.5 m) apart with in-row spacing of 2½ to 3 feet (76 to 91 cm). Large individual plants with 20 to 50 branches are produced, and most cut stem lengths will be 14 to 20 inches (36 to 51 cm).

Cut stems will be 20 to 36 inches (51 to 91 cm) long if the plants are planted in double rows with a spacing of 9 to 12 inches (23 to 30 cm) between plants. Transplants are pinched at transplanting to remove the primary stem and leave four to eight lateral branches. These rows must be supported in the field.

Outdoor production requires more fertilizer than greenhouse production but still much less than most crops. Use low-fertility soils with good drainage. Keep the soluble salts below 0.6 mmhos. Organic fertilizers have been successful in field production in California. Be careful not to damage plugs at transplanting, and do not plant too deeply.

GREENHOUSE CUT FLOWER PRODUCTION

Godetia is a spectacular greenhouse cut flower crop with long stems and large flowers. However, the crop must be grown with almost no fertilizer through the 12 to 16 weeks the crop is in the greenhouse. Godetia can be grown in ground beds, which must be nearly free of fertilizer. Godetia planted in beds following other crops or where bedding plants or pot crops were grown will probably be a failure due to the relatively high fertilizer residue in the soil. Overfertilized plants are soft, their stems are crooked, there are many lateral branches along the stem, and plants are easily knocked down with watering and difficult to support.

Godetia is a great candidate for cut flower production in pots, though. Cut stems 24 to 40 inches (61 to 102 cm) long can be grown from single-stem plants grown as one plant per 4-inch (10-cm) pot or two to four plants per 6-inch (15-cm) pot, with a plant density of eight to 10 plants per square foot (86 to 108/m²). Do not set these pots on ground beds, where they will root into the soil below and deteriorate because of the fertilizer residue in the soil. If ground beds are used and the soils are low in fertility, pinched plants with four to eight lateral branches could be planted at a density of one to two plants per square foot (11 to 22/m²). Pinched plants will require two to three weeks more production time than single-stem plants. Plants will require one to two layers of support. Younger plants, eight weeks old to when buds are 1 inch (2.5 cm) long, are softer and easy to knock down with careless overhead watering.

Always water the soil rather than the foliage. It's best to keep the plants on the dry side; however, wilted plants will have permanently crooked stems when they recover.

Supplemental lighting will speed flowering in Grace godetias by four to 10 weeks, depending on the variety and time of year. Best winter greenhouse production will involve plugs grown under four to six weeks of HID supplemental lighting for 24 hours per day. After transplanting, plants can be grown with supplemental light from incandescent lamps (mum lighting) for six hours per day.

Scheduling depends on greenhouse temperature, supplemental lighting treatments, time of year (total light accumulated by the crop), and variety. In Kentucky, four-week-old plugs transplanted in late September flowered in mid-December (11 to 12 weeks, for Christmas); plugs transplanted in mid-October flowered in early February (15 to 16 weeks, for Valentine's Day); transplanted in early November, they flowered in mid-March (17 to 18 weeks). These results all came when supplemental incandescent lighting was used for the whole crop in a greenhouse with 50F (10C) night and 60F (16C) day temperatures.

Godetias are excellent cut flowers. Grown properly, the stems are strong and straight. Each stem will have four to 15 flower buds, depending on the overall vigor of the plant. Cut stems have a vase life of 14 to 18 days, and all flower buds open to normal size and color when the flowers receive proper care in a flower arrangement. Research has shown that floral preservatives with sucrose will damage the leaves and reduce vase life. Cut stems in tap water performed equally well or better than in preservatives in vase life trials.

POT PLANT PRODUCTION

Godetias in the Satin series are dwarfs suitable for pots. Overall plant size is reduced significantly, and the flower size is slightly reduced, compared to cut flower types. Satin godetia is well suited for 4- to 6-inch (10- to 15-cm) pots and is grown as one plant per pot.

For pot production, it is easier to transplant a larger plug (from a 1- to 1½-inch, 4-cm, cell) that is approximately six weeks old. A late December or early January sow date should result in plants for Mother's Day in most of the northeastern and midwestern states. Keep plants in a cool greenhouse at 50 to 55F (10 to 13C) night and 60 to 65F (16 to 18C) day temperatures, preferably with negative DIF, after transplanting. Although the growth retardants B-Nine, Cycocel, Bonzi, and Sumagic are effective, they are not necessary when godetia is grown cool, dry, and with very low nutrition.

Watering and fertilizing practices are critical to pot godetia production. Careless overhead watering will weaken the plants and open the plant canopy. Consider subirrigation or drip irrigation. Use very little fertilizer on the plants after transplanting. A single application of a quarter teaspoon of Osmocote 14-14-14 at transplanting was sufficient for 4-inch (10-cm) pots of Satin godetias in our trials. Consider applications of 50 to 100 ppm from a balanced liquid fertilizer every 15 to 30 days, or a balanced organic fertilizer, as you evaluate production practices for pot godetia. Be sure individual plants always have sufficient space as they develop, to prevent stretching of stems.

In spring California-grown dwarf godetias are shipped east for combination pots and other uses. Dwarf godetias are as tolerant of cold temperatures as pansies, snapdragons, and dianthus. The unique flower color patterns of godetia make it an excellent addition to early spring bedding plant sales.

Godetia may be flowered as a pot plant for Mother's Day and early summer using the Satin series. Following is a production schedule from Satin's breeder, Sakata Seed of Morgan Hill, California, and Yokohama, Japan.

Stage 1, Days 1 to 10. Single-sow into a large plug cell (128 is ideal), using a plug media with a low starter charge. The larger plug cell allows more natural light around the plant, helping reduce stretch and increase basal branching. Cover the seed lightly with either media or vermiculite and maintain a soil temperature of 70F (21C) with even soil moisture.

Stage 2, Days 11 to 21. As soon as one sees green in the plug tray (around 10 days), immediately move the tray to a cool, bright, and well-ventilated greenhouse. Supplemental lighting can benefit the plug and ensure its healthy development. Optimum temperatures are between 55 and 60F (13 and 16C). The use of a negative DIF (55F, 13C day; 65F, 18C night)—or a two-hour temperature drop of 5 to 10F (–15 to –12C) at daybreak, followed by moderate day temperatures of 60 to 65F (16 to 18C)—is ideal for godetia production. If the plug medium has no starter charge, feed the plugs lightly with 50 to 100 ppm nitrogen, preferably from a well-balanced, calcium nitrate–based fertilizer. Negative DIF combined with low media fertility will keep the plugs compact and toned. High day temperatures, high media fertility, and low light will weaken the plugs, resulting in soft, leggy growth.

Stage 3, Days 22 to 40. Maintain cool temperatures and use a negative DIF, if possible. Weekly sprays of B-Nine (5,000 ppm) will help control plant height, but temperature manipulation has proven to be the most effective tool. Feed plugs lightly, using 50 to 100 ppm N every 10 to 14 days from a well-balanced, calcium nitrate–based fertilizer. The use of high ammonium- and urea-based feeds is *strongly* discouraged.

Stage 4, Day 40. Plug trays are now ready for transplanting or shipping. They can be held in a well-lighted area at 40F (4C) to tone before shipment or until space is available for planting.

4-inch (10-cm) pot to eight weeks. Godetia plants are sensitive. Dislodge them from the plug tray by pushing up from the bottom. Avoid pulling the plants out of the tray by hand, which may damage the stem. Avoid planting the plug below the soil line, to guard against stem rot and ensure a healthy transition.

Maintain cool growing conditions of 55 to 60F (13 to 16C), preferably employing a negative DIF.

Ideally, use drip tubes or subirrigation, especially when in flower. Overhead watering with strong water pressure will weaken the plant and open up the plant canopy.

You have two fertilization choices, liquid and dry feed. With liquid feed, fertilize every 14 days with 100 ppm N from a well-balanced fertilizer, like 15-5-15 or 20-10-20. With dry feed, top-dress each pot with a quarter teaspoon of 14-14-14 Osmocote with *no additional fertilizer required.* (Note: Plants that appear undernourished during the middle of the production cycle will have stronger stems. Plants can be greened up in the last two weeks as the flowers begin to open.)

Maintain a media pH between 5.5 and 6.5. Many growers rely on the acidifying properties of liquid fertilizers to control media pH. Since little fertilizer is being used, especially with the Osmocote-only method, acidification of the irrigation water to reduce alkalinity to 100 to 120 HCO_3 may be necessary to maintain optimum media pH.

Godetia Satin is responsive to day length, and extending the day length to midnight (six hours), using ordinary mum lighting (10 f.c. [108 lux] from incandescent bulbs placed on 6-foot, 1.8-m, centers), will hasten development.

Godetia plants need to be spaced to ensure good plant quality. Acclimated plants of godetia Satin can tolerate light frosts, permitting spacing outdoors in late April or early May, when greenhouse bench space is at a premium.

For garden performance, godetia Satin does best under mild weather conditions. In areas where summer temperatures regularly exceed 80F (27C), plants will benefit and perform better if given shade during the hot afternoon.

CLEOME (SPIDER FLOWER)

by Jim Nau
Ball Horticultural Company
West Chicago, Illinois

❖ *Annual* **(Cleome hasslerana,** *formerly* **C. spinosa)**, *14,000 seeds/oz (490 seeds/g). Germinates in 10 to 12 days at alternating 80/70F (27/21C) day/night temperatures. Cover the seed lightly upon sowing.*

Cleome is one of the best old-fashioned annuals still available in the U.S. trade today. Robust plants from 3 to 5 feet (91 to 152 cm) tall and up to 4 feet (122 cm) wide, grow without branching for the first 10 to 12 inches (25 to 30 cm) and then branch off with three to five developing shoots that will all flower. The primary flower colors include white, rose, and purple, though other colors are known as well. The flower heads grow to 6 inches (15 cm) across and often wilt in the afternoon sun, though they fully recover by nightfall. Cleomes have scented foliage that may be overpowering in small, enclosed gardens. Landscapers are hard-pressed to find another flowering annual to cover so much ground and create such a large presence. Cleome is ideal for highway plantings or other areas where size and a bright color show is the goal. Plants have short, sharp spurs similar to thorns, along the stems.

Cleome can be difficult to germinate. Germination occurs in 10 to 20 days or in flushes, depending on whether seed is dormant. To germinate dormant seed, alternate day and night temperatures.

Cleomes should be sold green using one plant per 4-inch (10-cm) pot since the plants may be rather tall by the time they flower. For green pack sales, allow seven to nine weeks, and the plants will be 5 to 8 inches (13 to 20 cm) tall in the pack. Instruct home gardeners to plant out into the full sun and not to pinch the plants back. If pinched, cleomes often need staking before the season ends. In the southern United States, cleomes can be planted from summer or early fall until late December. Cleomes have not performed very successfully, however, in the very deep South, including Florida and southern Texas.

The Queen series is the only known variety on the market. Rose Queen is a soft rose-pink color, while White Queen (also called Helen Campbell) is a pure white. Both varieties grow to 4 feet (122 cm) tall, with large blossoms. Cherry, violet, and a mix are also available.

COLEUS

❖ *Annual (Coleus × hybridus). 100,000 seeds/oz (3,500/g). Germinates in 10 to 14 days at 70 to 75F (21 to 24C). Leave seed exposed to light during germination.*

The attractive foliage of coleus makes it a versatile addition to any garden, serving as a backdrop for showy flowers or adding depth and texture to a planting. It is one of the few bedding plants that thrives in low light conditions and still puts out bright colors. Foliage color is determined by light levels. The variety Wizard Sunset, for example, is bright bronze-scarlet in the shade but deep, dark red in full sun.

Allow nine to 10 weeks for salable packs; 12 to 13 weeks for 4-inch (10-cm) pots. Use five to six plants per 10-inch (25-cm) hanging basket and allow 14 to 16 weeks. Southern growers can reduce crop times by one week or more.

While older varieties required a pinch to create full, well-branched plants, more recent varieties such as Wizard do not. For height control, A-Rest, Bonzi, and Sumagic are effective.

For large, robust plants, try Rainbow or Wizard. The heart-shaped leaves of these two varieties are among the largest offered. Both grow to a height of 14 inches (36 cm) in the garden and fill in quickly. Wizard comes in separate colors, too: Golden, Jade, Pastel, Pineapple, Pink, Rose, Scarlet, Sunset, and Velvet. The bushy, compact Carefree series features small leaves with attractive, deeply curled edges on plants growing to 10 inches (25 cm) tall. Fiji Mix has large, fringed leaves.

While seed coleus has been losing ground over the past few years, vegetative coleus has been creating excitement. Bred for full-sun conditions and pot production, plants go through heat and drought without sunburn, maintaining their color no matter what.

George Griffith, Hatchett Creek Farms, Gainesville, Florida, offers the following background and cultural tips. Provide normal nutrition of 75 to 200 ppm 20-20-20 or 20-10-20. In the landscape, plants benefit from Osmocote application, so nutrition is constantly available. Provide light levels of 3,000 to 15,000 footcandles (32 to 161 klux). Coleus changes color when grown under different light intensities. At medium light levels of 3,000 f.c., leaf color is different than under full, bright, tropical sun of 12,000 to 15,000 f.c. (129 to 161 klux). On the plant, one leaf shades another leaf, thus producing still more color variation, adding to the excitement.

Several series are available in vegetative coleus. The Sunlover series comes in Rustic Orange, Red Ruffles, Cranberry Salad, Gay's Delight, Thumbellina,

C

Freckles, and Olympic Torch. The Solar series is available in Sunrise, Shadow, Storm, Eclipse, Flair, Shade, Furnace, Spectrum, and Red, while the Ducksfoot series comes in Dwarf Yellow, Dwarf Red, Midnight, and Camouflage.

Dr. Allan Armitage, University of Georgia, selected a number of sun tolerant coleus—the Sunlover series—in 1994. The plants withstood the summer's hot sun and had a tendency not to flower. When used in the landscape or home garden, seed coleus must be deadheaded frequently.

COLLARDS (SEE *BRASSICA*, VEGETABLES)

COLUMBINE (SEE *AQUILEGIA*)

CONSOLIDA (LARKSPUR)

❖ *Tender perennial* (**Consolida ambigua**)*, 8,200 seeds/oz (289/g). Germinates in 10 to 20 days at 55 to 60F (13 to 16C). Cover seed. Chill seed at 35F (2C) for seven days for best germination.*

The half-hardy nature of the larkspur can be taken advantage of by outdoor cut flower growers in the near South. If a sowing is made in the open in this area six to seven weeks before the ground freezes, plants well enough established to winter over should be produced. An exceptionally severe winter will sometimes destroy them, but usually they come through very well. Most growers find that an overwintering cover of even, coarse material tends to rot plants. With perfect drainage and some covering, a fall sowing outdoors does usually come through nicely if made late enough to avoid germination before the ground freezes. The advantage of such a sowing lies in the promptness with which larkspur germinates in early spring. Such a sowing will flower at least two to three weeks earlier than if sown out after the ground dries in the spring.

Midwestern growers can sow seed in December for harvest in the greenhouse in mid-April. For outdoor crops, sow seed direct to the field in April and May for flowering plants 10 to 13 weeks later.

In figuring seed requirements for an extensive planting, do so on the basis of 25 ounces of seed covering an acre (175 g/ha); this is figured on double 8-inch (20-cm) rows spaced 3 feet (91 cm) apart. Spacing plants in each row is not so important. They will fill out the row if spaced 10 to 14 inches (25 to 36 cm), but will do so more promptly if allowed half that distance. Some southern growers plant out March 1—sown seedlings, usually getting good results, but we believe that if the fall sowing comes through, it will be more profitable because of the cost of greenhouse plants and transplanting. If you are depending on spring sowings, two should be made two to three weeks apart. By all means, get the first one in as early as possible and use deep, fairly well-enriched soil. It will pay to irrigate during dry weather, if it can be done.

Larkspur is an important minor crop among California and Florida cut flower growers. It is also, of course, a great favorite for the homeowner's cut flower garden. For this it can be sown direct or started from quart or gallon material purchased in May or June (December or January sowing). Plants should be supported—use one to two layers of string or wire.

Try the Giant Imperial series, available as a mix or in separate colors. Plants grow to 3 to 4 feet (91 to 122 cm) and make excellent fresh cut flowers. Use an STS preservative to reduce flower drop. Store flowers standing in water.

COREOPSIS

by Jim Nau
Ball Horticultural Company
West Chicago, Illinois

❖ *Perennial (*Coreopsis grandiflora*, sometimes* **C. lanceolata***). 10,000 seeds/oz (350/g). Germinates in nine to 12 days at 65 to 75F (18 to 24C). Cover the seed lightly or leave seed exposed after sowing.*

FIG. 1. Early Sunrise, an All-America Selections award winning flower, put coreopsis on the map for many growers and home gardeners alike. It produces bright golden flowers throughout the summer and reliably returns to the garden every year.

Coreopsis offers single to double golden yellow flowers on plants that grow from 18 to 24 inches (46 to 61 cm) tall (fig. 1). It is a dependable perennial, with excellent color that remains stable throughout its flowering period. Flowers can have a slight scent and are held on upright stems that work well in cut flower arrangements. Flowers are 1¾ to 2½ inches (4 to 6 cm) across, depending on the variety. In Midwest gardens coreopsis tends to be short-lived, lasting only three to four seasons. Plants flower in June and July and are hardy to USDA Zones 4 to 9.

Seed is the most common method of propagation, though division can also be done in the spring or fall. Home gardeners should divide coreopsis as needed to increase its life as a perennial. If seed of Sunray or Sunburst is sown during winter, plants will not flower profusely the same season. In fact, if sown in early February and planted into cell packs, grown at 50F (10C) nights, plants remain vegetative all the following spring and summer. However, in the case of the newest variety, Early Sunrise, sowings made in March, transplanted to cell packs, and grown on at 50F (10C) nights flower profusely in mid-July until late August.

With most varieties (with the exception of Early Sunrise), sow seed in July or August for overwintering in gallon containers, one plant per pot. These will flower in the spring of the following year. It is suggested to grow Early Sunrise from seed sown early in the year, as opposed to trying to overwinter this variety. Grow on at 50F (10C) nights.

Early Sunrise is the most recent variety of coreopsis from seed. Early Sunrise grows to 2 feet (61 cm) tall, with semidouble blooms of golden yellow. Flowers are 2 inches (5 cm) across, and plants can bloom within 100 days of sowing. Early Sunrise is a 1989 AAS Gold Medal Winner—AAS's first Gold Medal Winner in 15 years! Sunburst is a vigorous performer to 3 feet (91 cm) tall that has a somewhat open, rank habit in the garden. Its flowers are yellow, to 2 inches (5 cm) across, with either a single or a semidouble appearance. Sunray is an excellent double-flowering variety to 2 feet (61 cm) tall. It has a tidier habit than Sunburst, with flowers to 2½ inches (6 cm) across. It has some semidouble blossoms, as well. Flower color is a rich golden yellow.

CROSSANDRA

❖ *Annual* (**Crossandra infundibuliformis**). *4,000 seeds/oz (140/g).*
Germinates in three to four weeks at 80F (27C). Maintain 70F (21C).
Cover seed after sowing.

Crossandra is an unusual, attractive pot plant that originally came from India. Introduced to the American trade by the Ball Seed Company many years ago, it had declined in popularity until the 1980s, when it started to make a comeback (fig. 1). *Crossandra* is popular today as a 4- to 5-inch (10- to 13-cm) spring- and summer-flowering pot plant. The plants have glossy, gardenia-like foliage with flower spikes of overlapping, clear salmon-orange florets.

A minimum temperature of 75 to 80F (24 to 27C) is necessary for satisfactory germination, with a minimum of 65F (18C) nights for growing on. Alternate day and night temperatures for the first three to four weeks to improve germination. Germination is slow and sporadic. *Crossandra* needs rather rich, well-drained potting media and will flower seven to eight months after sowing.

Transplanting must be done at intervals as the plants appear and when they are large enough to handle. Use care in transplanting so that the ungerminated seeds will not be disturbed. Light should be between 2,000 to 3,000 footcandles (22 to 32 klux) for an average day length. Summer growing will require about 30% shade. *Crossandra* is an average feeder, requiring about 200 ppm of N, P, and K with each irrigation. It will send up flower spikes with salmon-orange flowers and bloom over a period of time. Northern growers should allow seven months for flowering plants in the summer, nine months for flowering plants in the winter.

FIG. 1. *Crossandra* is a popular pot plant, especially in Europe.

The seed-grown variety Tropic Flame is fast growing and heat tolerant, making it a regular for landscape plantings in the South. Plants grow to 10 inches (25 cm) tall. There are four other varieties available that are propagated from cuttings and are more cold tolerant: Florida Flame, Florida Passion, Florida Summer, and Florida Sunset. They should be pinched about two weeks after transplanting. The foliage is not glossy, but it will turn glossy and dark green when treated as follows. When new breaks are 1½ inches (4 cm) long following a pinch, apply B-Nine at 5,000 ppm. A-Rest and Bonzi sprays are also effective.

CUCUMBER (SEE *CUCURBITS*)

CUCURBIT FAMILY (CUCUMBER, MELON, SQUASH, PUMPKIN, WATERMELON)

by Lelion D. Elledge Jr.
Oklahoma City, Oklahoma

❖ *Vegetables* (**Citrullus** *spp.,* **Cucumis** *spp.,* **Cucurbita** *spp.*) *180 to 900 seeds/oz (6 to 32/g). Germinates in seven to 10 days at 72 to 75F (22 to 24C). Direct-sow to final container.*

Plants in the cucurbit family all have marked similarities. They are all deep-rooted vines and do not like having their roots disturbed during transplanting. Because of this, vine crops are not often grown for the bedding plant market. Plus, germination with direct-sowing is very good, which reinforces their use as seed items by the gardening consumer.

It is possible, however, to produce vine crops as bedding plants. The best way is to sow three seeds to each Jiffy pot, then thin them down to two plants after seedlings emerge. When transplanting to the garden, the liners can be planted—pot and all—directly to the soil.

Vine crops need four to six weeks before they're ready to retail. Cucumbers and watermelons are the exception: cucumbers require only three to five weeks, while watermelons need eight weeks. Home gardeners will have to wait 48 to 70 days before cucumbers are ready for harvest, 72 to 90 days for muskmelons, 90 to 120 days for pumpkins, 49 to 85 days for squash, and 70 to 85 days for watermelon.

CYCLAMEN

by Hans A. Gerritsen
The Hortus Group
Salinas, California

❖ *Annual* (**Cyclamen persicum**). *2,500 to 3,000 seeds/oz (80 to 100/g). Germinates in 21 to 28 days at 65 to 68F (18 to 20C). Seed may be exposed to light or covered very lightly.*

The well-known cyclamen is part of the Primulaceae family (fig. 1). The name "cyclamen" is derived from the Greek word *kyclamenos,* which means "circle form," reflecting both the plant's tuber and foliage. As a native of the

Mediterranean, some of its first research goes back to Plato. Some of the first crossings were made in England by a British grower in the 1800s. From there cyclamen breeding went to Dresden, Germany, in 1870 and later to Holland and Switzerland.

Over the past 100 years, we find the key breeders in European breeding stations: Goldsmith Seeds, S & G, PanAmerican Seeds, Morel Diffusion, and De Ruiter Seeds are the main trendsetting F_1 developers. Morel is the only company that still breeds in the cyclamen's native Mediterranean environment; the other breeding companies are turning to high-tech growing areas in the Netherlands.

FIG. 1. Cyclamens are some of the most stately of the flowering pot plants with their large, elegant flowers. Many American growers have added cyclamens to their pot plant mix during the fall and winter to lessen their dependence on poinsettias.

PRODUCTION

Traditionally, cyclamen has been a winter and spring plant. However, in some parts of the world, we see growers producing this crop year-round (fig. 2). In general, it has been considered to be a cool crop, but we have seen the nicest crops in the Sacramento Valley in California, where summer temperatures can exceed 100F (38C) daily! Cyclamens perform in such conditions because of their native Mediterranean roots.

Most of the seed is sold by the 1,000 seed count, and the average germination rate is between 75 and 85%.

Stage 1. Sow the seed in a plug tray with a sterile peat mixture. The newest trend is not to cover the seed until after germination. Place trays in a dark, cool 63 to 65F (17 to 18C) environment with relatively high humidity for at least three to four weeks.

Stage 2. Place the trays in a shaded area and cover the new corm with some vermiculite. It is important to keep humidity high until the seed coats come off entirely. The tray will stay here at least two weeks.

Stage 3. The seedlings are placed in a brighter greenhouse and can be grown at 65 to 68F (18 to 20C) for the next six to eight weeks.

C

Stage 4 to potting. After total of 12 weeks, transplant to a 4-inch (10-cm) pot. After 16 to 18 weeks, depending on the plug size, transplant into a 5- to 6-inch (13- to 15-cm) pot. Use a well-draining, nutrient-rich, sterile, peat-based mixture with pH of 5.5 to 6.3. Plant the young seedlings "on the high side" to avoid corm diseases: try to keep half the corm above the soil. After transplanting, maintain the temperature at 65 to 68F (18 to 20C) for at least four to five weeks; preferably, run day and night temperatures

FIG. 2. Mini cyclamens—for 4-inch pots—are the rage for some growers and retailers. Newer breeding has produced F_1 varieties that flower consistently and together across all colors for bench production.

the same. After good root development, lower the temperature slightly and grow the cyclamens at a slow pace to develop nice foliage.

TECHNICAL NOTES

Keep a check on pH and keep the EC at around 1.0 to 1.2. Try to avoid wet foliage during the night and make sure that plants do not dry out. Ebb-and-flow systems are recommended for even watering. After eight weeks apply 100 ppm N in a well balanced, soluble fertilizer every two weeks. In the beginning of the feeding, we keep the EC at 1.0 and use a 10-7-20 fertilizer. Later in production the EC will go to 1.2; then feed with a 15-15-30 fertilizer. In general, cyclamen is daylight neutral; however, high light levels will increase flower development, and during dark winter production, extra light is recommended for proper flower development. In the bright summer months, 30% shade is recommended. Space plants when the leaves begin to touch each other. For 6-inch (15-cm) pots, we recommend spacing 15 by 15 inches (38 by 38 cm); for 5-inch (13-cm) pots, we recommend spacing 12 by 12 inches (30 by 30 cm); for spacing 4-inch (10-cm) pots, 10 by 10 inches (25 by 25 cm).

In some parts of the world, growers are convinced that applications of gibberellic acid (GA) will generate a more uniform and earlier flowering crop. They apply 10 to 25 ppm GA 45 to 60 days prior to the sales target day and see a steady flower acceleration. With today's new hybrid breeding, however, there is definitely better series uniformity, making GA treatments unnecessary.

PLUGS

The key transplanting months are July and August. Most growers buy in their plugs to avoid the high cost of growing a cyclamen transplant during their key spring production time.

For a 6-inch (15-cm) pot, we recommend purchasing at least a 72 tray. For 4- and 5-inch (10- and 13-cm) pots, there are 128 and 200 trays available. The smaller plug is extremely sensitive and will not hold up well under stressed conditions. Long shipping periods, heat, drying out, and delay of transplanting are major triggers of wilt disease in an early stage.

In addition to plugs, prefinished 4-inch cyclamen pots are sold in the fall to reduce the crop time. Prefinished cyclamens allow growers in very hot areas to avoid the summer heat and give everyone a chance to produce this great crop.

COMMON DISEASES AND INSECT PROBLEMS

While cyclamens are quite sensitive to disease, they are very rewarding when grown under strict clean conditions. *Fusarium oxysporum,* or *Fusarium* wilt, was first observed in Germany in 1930 and did not enter the United States until 1949. Major research was done in 1970. Today, we know that *Fusarium* can be present latently within the plant and appear after transplanting or other stress such as shipping, heat, or lack of moisture. *Strict sanitation,* clean soil, and clean pots are a must. *Remove any yellowing plants*. When cutting the corm of a diseased plant, a purple-red ring is visible. Many growers apply a preventative drench of Celery's or Mycostop the fifth week of growing on.

TABLE 1   COMMERCIAL F₁ CYCLAMEN SERIES

Type	Series	Container size[a] (inches)	(cm)	Crop time (weeks)
Fringed-flowering	Butterfly, New Wave	6–8	15–20	36
Large-flowering	Colorado, Concerto, Halios, Pannevis, Royal, Sierra	6–8	15–20	32–34
Mid-types	Intermezzo, Lasers, Latinia, Novella	4–5	10–13	28–30
Mini-types	Marvels, Minimate, Miracles	3–4	8–10	22–24

[a] 3, 4, 5, 6, and 8 inches equal approximately 8, 10, 13, 15, and 20 cm, respectively.

Cryptocline cyclaminis is a wilt disease that creates stunted growth in the stems and leaves after transplanting. The fungus appears black at the edges. The vascular system will have reddish discoloration in petioles and the corm. Dead plant tissue will have masses of pale orange spores. The disease will spread rapidly by splashing water. Remove the diseased plants and apply Euparen.

Erwinia chrysantemi, or soft rot, shows symptoms of sudden wilting followed by plant collapse. The corm appears slimy. Deep plantings plus warm, humid weather may trigger *Erwinia*. Cultural practices are the main problems—avoid overhead watering and high temperatures.

Botrytis cinerea is visible by its soft decay of the flowers and leaves. Flowers are spotted and have gray mold on them. Lower the humidity and increase the air circulation. Space crowded plants and avoid cold night temperatures. Some fungicides can be helpful, but they will not be needed if you make the greenhouse environment unfavorable for *Botrytis*.

Tomato spotted wilt virus (TSWV) shows yellow rings and brown streaks on the petioles. Flowers are malformed, and the plant stops growing. TSWV is transferred by thrips, and the only good cure is insect control.

Cyclamen mites create distorted leaves and curling, some discolored flowers, and twisted flower stems. Standard insect control will apply here.

DAFFODIL (SEE *NARCISSUS*)

DAHLIA

by Hans A. Gerritsen
The Hortus Group
Salinas, California

❖ *Annual (***Dahlia** × **hybrida***). 4,500 seeds/oz (159/g). Germinates in five to 10 days at 60 to 65F (16 to 18C. Cover seed lightly for germination.*

The dahlia is native in Central America and was used by the Aztecs. It came to Europe through the Spanish explorers and was named after Dahl Anders (1751–1789), who was a friend of the Swedish botanist Linnaeus.

Thousands of varieties have been developed from the hybrids of *Dahlia pinnata* and *D. coccinea*. The genus *Dahlia* has plants from 12-inch (30-cm) bedding plant dahlias to the 20-foot (6-m) tree dahlia. Flower forms and color ranges are almost unlimited; this alone gives breeders and consumers an exciting future.

DAHLIA AS A POTTED PLANT

Until recently, the traditional potted plant was grown from special-selected dwarf dahlia varieties and sold as tubers. The growers stored the tuber at 44 to 50F (7 to 10C). When they were ready to plant, they planted the tubers in 6-inch (15-cm) pots. The crop time varied between 12 and 14 weeks. Due to the high start-up cost of tubers and limited pot sizes, the first vegetative dahlias were introduced from tissue culture in 1985. These new types are sold as "Dahliettas" in standard plug trays (128, 72) and as unrooted cuttings. Dahliettas are an excellent choice for 4-inch (10-cm) and 5-inch (13-cm) pots. Crop time is reduced to six to 10 weeks, and the grower can produce more pots per square foot with greater uniformity compared to the tuber types. Flower forms and colors, however, are still limited. For large containers and novelty flower forms and colors, the dahlia tubers are still the only production alternative.

PRODUCTION

Dahlias require well-drained media with a pH of 7.0. Plant the tuber slightly higher than the soil, and cuttings or plugs slightly deeper, to create a stronger plant.

To create a compact, well-branching plant, pinch transplants after transplanting. Tubers are pinched after five to six weeks and drenched with A-Rest (ancymidol). A-Rest must be applied no later than two weeks after planting if growing from tubers. For vegetative types use 1,500 ppm B-Nine.

Dahlias require at least 14 hours of light and 2,000 footcandles (22 lux) to develop a nice plant and good flower count. Short day length will prevent flowering. Maintain temperatures of 62 to 65F (17 to 18C). Lower temperatures will delay flowering, and high temperatures (80F, 27C) will delay and abort flowering.

Feeding can begin immediately after transplanting for vegetative types and after the first shoots appear for tubers. A well-balanced fertilizer, such as 14-14-14, can be used two to three times a week, or use a liquid feed at 200 to 250 ppm with 20-10-20.

If you must hold plants prior to shipping, lower the temperatures to 50 to 55F (10 to 13C). Do not put plants in cold storage.

Any flower removal, disbudding, or pinching will add another five to seven days to the program.

AS A BEDDING PLANT

Bedding plant dahlias have average germination of between 85 to 95%. Plugs are widely available. Recently, we have seen a trend of separate colors in dahlias from the breeders. There are also some F_1 hybrids on the market. The breeding objective is to develop a uniform, double-flowering, dwarf type. The Figaro series is still a trendsetter in the industry, growing 12 to 15 inches (30 to 38 cm) tall.

Crop time for packs is around 10 to 12 weeks, for 4-inch (10-cm) pots around 13 to 14 weeks. Depending on the growing area, B-Nine is recommended. Use 750 to 1,500 ppm to maintain a uniform height. For large containers, such as 1-gallon pots and hanging baskets, use the larger-type series Rigoletto or the true hybrid Sunny series, which are both slightly taller, 15 to 18 inches (38 to 46 cm) tall. Production recommendations are similar to those for the pot types. Seed dahlias will make small tubers and can change their flower colors under high light conditions.

PEST CONTROL

These plants need to be properly spaced to avoid touching in order to prevent aerial disease. Thrips, mites, whiteflies, and aphids are potential pests. Watch for yellow ring spots as a potential indicator of TSWV. Effective insect control can be achieved by Marathon, Margosan O, Avid, Mavrik, and Decathlon. Avoid Dursban and Vydate, since they can cause phytotoxicity.

DAYLILY (SEE *HEMEROCALLIS*)

DELPHINIUM

❖ *Perennial (Delphinium elatum). 8,000 to 10,000 seeds/oz (about 282 to 353/g). Using fresh seed, germinates in 12 to 18 days at alternating 80/70F (27/21C) temperatures.*

Today's stately delphinium, an excellent garden perennial (fig. 1), has some commercial importance. While delphinium is admittedly one of the most valuable hardy perennials for outdoor cutting, it also responds well to forcing. Outdoors, it seems to be profitable from Canada to Florida, though the finest flowers are produced where summer temperatures are moderate. In Florida it is largely grown as an annual. For potted perennial sales, some Southern growers find it best to use well-established seedlings that are started farther north.

FIG. 1. Few flowers have the outright stately elegance of a well-grown *Delphinium elatum*.

DELPHINIUM AS A CUT FLOWER

Delphinium makes an elegant cut flower in electric colors of blue not available on other species. The "bees," or conspicuous centers, of delphinium blossoms, are often a contrasting white or black. Seed-propagated plants will flower their first year, although exposing plants to cold temperatures for six weeks at 35 to 40F (2 to 4C) will increase yield and quality. Fall is the best season for outdoor planting, with plants flowering the following spring. Grow at 50 to 60F (10 to 16C) nights. Mum lighting speeds flowering, lengthens stems, and improves stem quality, especially for *D. belladonna*. Use bottom irrigation to avoid diseases on flowers and stem rot after harvest begins. Expect annual yields of five to six stems from *D. elatum* plants and up to 12 shorter stems from *D. belladonna* plants. Delphiniums planted in the greenhouse in late summer will flower the following January. Use mum lighting to speed flowering, and avoid temperatures over 75F (24C).

Harvest flowers when one-fourth to one-third of the flowers are open. Use an STS preservative to avoid shattering because delphinium is hypersensitive to ethylene.

Magic Fountain is a shorter growing hybrid popular with many cut flowers. The Pacific Hybrids, growing 4 to 5 feet (1.2 to 1.5 m) tall have double flowers. In *D. belladonna* types, the variety Belladonna has blue flowers, while Bellamosum has deep blue flowers.

AS A PERENNIAL

Use only fresh delphinium seed for greatest success. Seed is harvested in late summer; the best germination will come when seed is sown in the first four to six months. Pacific Giants tend to show poorer germination than belladona types, which germinate well under a constant temperature of 65 to 70F (18 to 21C).

December to March sowings will produce plants that flower the following summer. At 20 to 28 days, transplant seedlings and grow on until they're well rooted. Place plants in a cold frame in early March when night temperatures are at least 50F (10C). Plants will flower by mid- to late April. Commercially purchased delphinium seedling liners can be finished in seven to nine weeks at 50F (10C).

Blue Fountains is a mix of blue shades and white. Flowers are single to semi-double, and plants grow 3 to 4 feet (91 to 122 cm). Plants tend to show greater hardiness than the shorter growing Magic Fountains. The Pacific Giants produce 17- to 28-inch (43- to 71-cm) flower spires on 4- to 5-feet (1.2- to 1.5-m) plants. Stake Pacific Giants. Separate colors are named after characters of King Arthur's court—Astolat, Black Knight, Galahad.

AS A FORCED FLOWERING POTTED PLANT

Tests at Pennsylvania State University have shown that delphinium can be forced as a flowering pot plant. Researchers suggest growers begin their 5- or 6-inch (13- or 15-cm) pot crop with bought-in plugs. Light intensity seems to be more important than photoperiod in forcing; use supplemental light during winter months. In March and April, natural light is generally intense enough to stimulate fast flowering. Grow plants at a night temperature of 60F (16C) and day temperatures of 70F (21C). To control stem elongation and maintain compact plant habit, keep day and night temperatures as close together as possible. Plants are responsive to A-Rest and Sumagic, though controlling plant height by temperature is the first choice.

DENDRANTHEMA (CHRYSANTHEMUM)

by Vic Ball
Ball Publishing

and Edward Higgins
Yoder Brothers, Inc.
Barberton, Ohio

❖ *Annual and perennial (**Dendranthema grandiflora***). *Vegetatively propagated. Growers purchase cuttings from commercial propagators.*

Chrysanthemums, in all of their product forms, are among the most popular flowers in the world today. Pot mums, garden mums, pompons, and standard mums can be found in stores, homes, offices, and landscapes at any time in North America and in much of the world. Consumers know, purchase, and value chrysanthemums for their wide range of colors, flower forms, flower sizes, and good keeping qualities. Growers appreciate the chrysanthemum's reliability in scheduling, consistent year-round demand, and continual cultivar improvements (fig. 1).

FIG. 1. A small section is shown here of the top quality pot mum crop at Milgro Nurseries, Oxnard, California.

Basics of controlled flowering

Controlled flowering underlies the major success of mums the past 20 or 30 years. Any mum crop can be forced to flower at any target date anywhere—if the rules are followed. Controlling the flowering is based on the fact that mums are a short-day crop. Simply stated, they will set bud and flower if exposed to short days, which reflects the way natural-season, fall mums have been grown for hundreds of years. Fall mums form buds and flower precisely because the days naturally shorten in fall. The researchers who discovered this most important phenomenon were two USDA scientists, W.W. Garner and H.A. Allard, in 1920. Since then, many other crops have been found to either be short-day responsive or, in some cases, flower under long days.

By the 1940s growers began to force mum crops into flower early by pulling black cloth over the beds, typically from 5 P.M. to 8 A.M., beginning in the early summer. The shade cloth provided an artificial short day (9 hours long), and varieties normally flowering naturally in November would now flower in September. Prices went up; growers were happy. Day-length control gradually led to flowering mums, both pot and cut, 52 weeks a year.

If short days are applied at planting a new cutting, the crop immediately goes into bud and flower, which results in plants perhaps 6 or 8 inches (15 or 20 cm) tall. Because of this, there is a need for several weeks of long days immediately upon planting the cutting to allow the plant time to develop enough stem length and substance before flowering.

The "response group" a variety belongs to refers to the number of weeks from the beginning of short-day, or black cloth, application, until the crop flowers. Typically, a crop is planted, given two or three weeks of long days, then perhaps 10 weeks of response period (short days). Total crop time: roughly 12 to 13 weeks from planting to flower.

Varieties vary in response period from as little as seven and eight weeks to as long as 14 to 15 weeks. In general, the very early responding varieties (seven and eight weeks) are the garden varieties that flower normally outdoors in September and early October. The great majority of commercial pot and cut mum varieties have a nine-week or 10-week response. These are varieties that naturally flower in early November. There is also a 13- to 14-week response group, which are varieties that naturally flower in December. This last group includes mainly pompons, formerly used for winter-flowering controlled crops in northern greenhouses. Because of certain unreliability aspects, they have been replaced by new, improved 10- and 11-week varieties as the northern winter pompon crop.

Scheduling. Once this basic knowledge was in place, propagators developed complete schedules prescribing a precise number of weeks of long days and short days for crops to flower each week of the year. Schedules were also developed for varying degrees of latitude from north to south. Typically, these schedules also recommend certain varieties for each of these crops in each latitude. Such schedules are available, in fact, to this day from major propagators, such as Yoder Brothers, Barberton, Ohio.

Typically, northern winter pompon crops might need three or even four weeks of long days to develop enough stem and substance during the low-light, slow-growing winter crop. A grower in the same northern greenhouse flowering a crop in summer or fall would reduce the number of long days to a week or two because the crop grows so much faster in the higher light intensities and the longer days of summer and fall.

Growers in the Deep South experience, in effect, high light and longer days year-round. Therefore, their schedules tend to closely follow the schedules and even selections of varieties used by northern growers in summer and early fall.

EXACTLY WHAT MAKES A MUM FLOWER?

Precisely what are the conditions of day length and other environmental factors that cause a chrysanthemum to set bud and flower, or not to flower? Since varieties differ substantially in their responses, one answer cannot be precise, but some ground rules can be established.

Twelve or fewer hours of day length will cause most varieties to set bud and flower. A few short-day varieties will flower with 12½ hours of day length. In fact, early researchers pointed out that it's the long night, not the short day, that does it, but the result is the same.

Another clear requirement to cause the mum to set bud and flower is temperature, normally stated as 60F (16C). Again, varieties differ: some will set bud and flower at 55F (13C); some require even a "warm" 60F or perhaps 62F (17C). For most commercial crops of pot plants or cut flowers, growers maintain 60–62F, at least until buds are set. Note these are night temperatures.

Some of the earliest garden varieties, naturally flowering in late August, and many of the English "early mums," which also flower outdoors naturally in late summer, seem to flower more in response to temperature than they do to day length. If grown cool at 50 or 45F (10 or 7C) nights, they will remain vegetative. If grown warm, 60 or 65F (16 or 18C), they tend to set bud and flower. In fact, some of the most responsive of these varieties will set bud no matter what you do! The Japanese also have a race of mums that are widely used for April and

May flowering—without day length control—which seem to flower more due to temperature than to day length.

How to prevent flowering. What conditions keep the mum plant vegetative? Maintaining long days, which of course occur naturally in northern summers, is the key. To prevent flowering during the short days of fall and winter months, it is necessary to supplement natural day length with artificial light. The rule to remember: Buds will not form as long as the uninterrupted darkness periods are less than seven hours long. For example, assume midwinter natural light is from 8 A.M. to 4 P.M. Lights provided from 10 P.M. to 2 A.M. will produce two six-hour periods of darkness, 4 P.M. to 10 P.M., and 2 A.M. to 8 A.M. No buds will form. Note also that most varieties of chrysanthemums will simply rosette, making short, spreading growth and no stems or flowers, if temperatures are 50F (10C) or even 40F—too cold.

Intermittent light can be used for long days and save a lot of electricity besides. Use the same hours per night as with continuous light, but during the hours of light, turn the lights on only six minutes out of each 30-minute period. In other words, six on, 24 off, then six on, 24 off, and so on. Some growers, to be sure, use 12 minutes on, 18 minutes off.

There have been cases where light "leaking" onto the crop by accident from a nearby source has delayed flowering of a crop. The effect is like heat delay. The tip-off nearly always is the area affected: plants nearest to a window of a nearby home (whose light is shining out at night), sometimes to lights used on main greenhouse walks by night men, or to streetlights.

It is obviously important in a year-round flowering situation to carefully block in light being applied to a bench of young plants. If a bench is being lighted and the benches on either side are not being lighted, then the light must be confined to the one bench by means of sateen curtains or something similar. Light that leaks to other benches would cause blindness, failure to flower.

At latitudes near the equator, winter days are longer, and fewer hours of light are needed. In fact, growers in Bogota, Colombia, very close to the equator, find that no buds occur on some varieties even with no lights at all in midwinter; the same happens in Hawaii on certain varieties. Of course, the natural day length at

TABLE 1   LIGHTING ARRANGEMENT FOR MUMS

Number 4-ft beds[a]	Bulbs Spacing (ft)	Watts
1	4	60
2	6	100
3	6	150

Note: All bulbs are 5 ft (1.5 m) above soil.
[a] Just one row of lights, no matter whether 1, 2, or 3 beds.

TABLE 2 YEARLY MUM LIGHTING CYCLE

25–30 degrees N lat[a]		35-40 degrees N lat[b]	
Time period	Night light (hrs)	Time period	Night light (hrs)
Oct. 1 to Mar. 31	4	Sept. to Mar. 31	4
Apr. 1 to Mar. 31	3	Apr. 1 to May 15	3
June 1 to July 31	2	May 15 to June 15	2
		June 15 to July 15	0
		July 15 to July 31	2
Aug. 1 to Sept. 30	3	Aug. 1 to Aug. 31	3

[a] Florida Keys to St. Augustine; Brownsville, Texas, to Houston
[b] Charlotte, North Carolina, to Philadelphia; Memphis to Indianapolis; Albuquerque to Denver; Bakersfield, California, to Reno, Nevada.

the equator is 12 hours year-round; at the North Pole it's zero hours in midwinter and 24 hours in midsummer.

Mum lighting. (See tables 1 and 2.) Light reflectors must be held up off the bulbs. For a 20-foot-wide (6 m) house, a single row of 300-watt reflector bulbs, facing down at a 45-degree angle and spaced every 10 feet (3 m), will do it. Bulbs should be staggered on alternate sides of ventilators, not directly under, because of possible rain damage to the bulbs. Figure about 1½ watts per square foot (16 W/m²) of ground covered. Use flood, not spot, bulbs. It takes only 10 footcandles (108 lux) to prevent bud formation.

On a large installation, half the area may be lighted before midnight, half after. This halves the demand cost. Using 230-volt lines reduces main sizes greatly.

Inclination to flower. Speaking of preventing mum flowering, there is one other factor. A mum's proclivity to set bud and flower increases as the stem elongates. Propagators learn from experience that stock plants (in beds to produce cuttings) allowed to grow older tend to set buds regardless of day length. Conversely, the first flush of cuttings from a healthy, soft, succulent stock plant will nearly always be free of buds. A mum stem physiologically

FIG. 2. Norm White of White's Nursery & Greenhouses, Chesapeake, Virginia, is America's premier pot mum grower.

oriented to bud and flower will first display "strap" leaves (long leaves without notches) up near the growth tip. It is an early warning sign!

ALL ABOUT POT MUMS

Pot mums are holding their own in the mid-1990s and are, in fact, the most important year-round flowering pot plant in North America. They are second only to the poinsettia in total dollars for flowering pot plants in North America.

Part of the success of the pot mum is the dramatic rise in supermarket sales, where its long shelf life, bright colors, and wide varieties of flower types make it Number 1 among supermarket flowering pot plants. Certainly, nonflorist outlets in total are selling 75% or more of U.S. pot mums today.

The rise of sales in the supermarket and other mass merchandisers is changing the nature of pot mum production. Many smaller and medium-size pot mum growers have switched to other crops. The large grower is better able to fill the needs of the chain retailers for large quantities, long-distance shipping, and volume pricing (fig. 2). Smaller growers, as well as some of the largest, do find niches open to develop certain products for retailers, florists, and garden centers, such as highest-quality pot mums, diverse product forms (e.g., large, 7- or 8-inch, 18- or 20-cm pots, bulb pan, and centerpiece mums) and upscale flower forms (among them, large disbud daisies.)

If you're targeting the supermarket trade, be sure to get a copy of *Pot Plants Grades and Standards* from the Floral Marketing Association, P.O. Box 6036, Newark, DE 19714-6036; fax 302/731-2409.

Market the crop first! Study your market before you commit to production. A good starting point is to realize the wide variety of pot sizes available to the pot mum grower, from 2 inches up to 7 and 8 inches (5 to 18 and 20/cm), including pinched and single-stem crops. Another very important variable is simply your crop spacing. Growers talk more and more about growing a crop of pot mums or poinsettias to the "specifications of the buyer." If it's a quality retail shop, a 6-inch (15-cm) mum may get up to 15 by 15 inches (38 by 38 cm) of space. For a supermarket or discount store wanting a lower price, the spacing will be more crowded, and the pot size will often go down to 5 or even 4 inches (13 or 10 cm), with fewer cuttings.

Talk to several of your most promising prospective customers or your present ones. Ask them what they are looking for and what sort of price they are willing to pay. Offer them a variety of alternatives—with your pricing.

Each year more and more growers break from the standard 6-inch (15-cm) pot mum. Some major producers are moving toward 5 inches (13 cm), also 4 and 4½ (10 and 11 cm). Europe grows all its mums at 13 cm.

D

"Grow" the crop on paper first! Simplified cost forecasting is not that big a job. The objective is to be sure that for the crop you are planning, the spacing planned, and your price, you will produce a fair profit, including packing cost, selling cost, all overhead expenses, and shrink. Try different spacings and different prices and compute a net profit on each variable before you grow the crop. A simple spreadsheet program like Excel or Lotus 1-2-3 can help you do this. *Plan* to make money.

A reminder about expansion: Often existing growers plan and build new greenhouse areas without seriously studying where all the new product will be sold. The result can be a scramble just to sell the product at distressed prices, at least at first.

The pot mum crop is potentially 52 crops a year, one each week. Poinsettias and lilies are each one crop a year; that's a big difference. The original uniform-production, 52-weeks-a-year program has given way, in most cases, to considerable peaking for major holidays. A typical year-round grower will often "bump" his or her weekly average production by three to five times for Easter, Mother's Day, and other holidays. In some cases, pot mums are produced only in season when space availability or the market makes them profitable.

Now a production plan. Having decided what you can produce, and at a profit, the next and very important step is to put your production plan in writing. How many pots do you want, what varieties, how many cuttings, and what spac-

TABLE 3 POT MUM VARIETIES

White	Yellow	Pink	Bronze	Red	Coral/Peach
Spring and fall					
White Diamond	Yellow Diamond	Charm	Dark Bronze Charm	Red Delano	Coral Charm
Boaldi	Miramar	Pomona	Pelee	Cherry Pomona	Coral Pomona
Shasta	Iridon	Regal Davis	Orange Davis	Lucido	Coral Blush
Summer					
White Diamond	Yellow Diamond	Regal Davis	Dark Bronze Charm	Red Delano	Coral Charm
Boaldi	Miramar	Dark Charm	Pelee	Lucido	Salmon Charm
Shasta	Dark Yellow Boaldi	Deep Luv	Orange Davis	Rage	
Winter					
Claro	Miramar	Delano	Pelee	Cherry Pomona	Coral Charm
Shasta	Iridon	Dark Charm	Dark Bronze Charm	Red Delano	Coral Pomona
Envy	Yellow Envy	Splendor	Orange Blush	Lucido	Coral Blush

ing for each phase of the crop? Most pot mums are grown in a starting climate (70F, 21C, and long days) for the first several weeks, then moved to final spacing. Some growers use three spacings, not two. Of course, all this spacing must be carefully coordinated with available bench area. Your plan must also specify complete schedules, pot sizes, dates to pot, pinch dates (some judgment here), short-day dates, and flowering dates.

Your production planning must be in writing a minimum of six months ahead of plant date for several reasons. The plan is the basis for ordering cuttings from suppliers. Substantial discounts are offered for early orders. Late orders are often not fulfilled with the desired varieties, especially orders for holidays. Quality, profitable growers normally have a definite commitment from their major customers well in advance of even planting the cuttings. With six months planning, you are in a much better position to plan other crops around the space needed for the pot mums.

Variety selection is a critical part of planning the crop (see table 3). It's a point that requires constant communication with your customers. A constant stream of new varieties is coming onto the market. You must evaluate most of them and put some into your program, at least on a trial basis.

Sorting varieties so that all pots in a planting are in the same response group is very helpful. Propagators classify pot mums as short, medium, or tall varieties. The short ones are those that naturally grow short; the tall ones of course tend to grow tall. As you would expect, the short varieties are generally given a week or two more of long days, especially for winter flowering, to provide adequate stem length. The tall ones are given a minimum of long days. Commercial growers try very hard to include varieties of the same response group—and if at all possible, the same height group—in a given week's flowering. Height of individual varieties can be controlled somewhat by more frequent growth regulator applications.

Schedules. Samples of typical commercial pot plant growing schedules are shown in table 4. Most are based on flowerings once a week year-round.

Note that all varieties of the "tall treatment" group are potted January 28 and handled the same. The eight-week variety flowers April 1, the nine-week variety

TABLE 4   SAMPLE COMMERCIAL POT MUM SCHEDULES

Treatment					Variety flowering date		
group	Pot	Lighting	Pinch	Shade	8-week	9-week	10-week
Tall[a]	Jan. 28	Jan. 28–Feb. 3	Feb. 11	Mar. 15	Apr. 1	Apr. 8	Apr. 15
Medium	Jan. 28	Jan. 28–Feb. 10	Feb. 11	Mar. 15	Apr. 8	Apr. 15	Apr. 22
Short	Jan. 28	Jan. 28–Feb. 17	Feb. 11	Ma.r 15	Apr. 15	Apr. 22	Apr. 29

[a] Tall treatment is explained later in this chapter.

D

(one week longer response) flowers April 8, and the 10-week April 15. These flowering dates are from a short-day date of February 3. Shading in all cases does not start until March 15 because days are naturally short enough until then.

The short treatment is for a naturally short-growing variety. Short varieties need several weeks more time to develop strength and substance, compared to a naturally tall variety. Therefore, from the same potting date (January 28), the short treatment varieties have a later short-day date (February 17 versus February 3). The extra two weeks from plant until short days allows the plant to develop more substance. Therefore, the short-treatment 10-week variety will flower two weeks later than the 10-week tall treatment (April 15 versus April 29).

When you order cuttings, the catalog will designate pot varieties as tall, medium, or short treatment.

Early crop culture. Most pot mums today are grown in commercial mixes. Some very large growers produce their own, but they have essentially the same ingredients, such as peat, pine bark, vermiculite, and perlite.

A quality crop starts with quality right from the beginning, including a good cutting. Wherever it comes from, it should be a reasonably heavy stem, free of insects and diseases, and uniform in caliper and length. Also, it is very important that it be succulent, not hard and woody. Rooted cuttings should have several ½- to 1-inch (1.3- to 2.5-cm), bristly white roots, not a mass of 2- to 3-inch (5- to 8-cm) roots. These roots should be solid, not hollow and tubular.

Years ago, almost all pot mums were started with rooted cuttings. Today, the majority of pot mums are started with unrooted cuttings that are stuck directly into the finished container. Savings are realized in planting time, labor, and cutting costs. A potential drawback is that more attention to detail is needed to successfully start the crops and ensure success. Some of the key factors to address in starting a pot mum crop from unrooted cuttings:

- Investment in a quality starting, propagation area
- Grading the unrooted cuttings for length and caliper (stem thickness), and planting comparable cuttings together
- Mist or fog available to keep cuttings turgid while rooting
- HID lighting, especially for winter-flowering crops, at 500 to 800 f.c. (5 to 9 klux) at plant level
- Carbon dioxide injected at approximately 1,000 ppm
- Fertilization during the rooting process

Any pot mum, no matter what the size or type of pot, no matter the variety, and no matter whether it starts from an unrooted or a rooted cutting, will owe almost all its success or failure to the initial environment and care in the first few weeks of production. So invest wisely and pay attention to detail in the environment right from the start.

Most U.S. and Canadian pot mums are fed and irrigated by drip tubes in each pot. The grower can water a bench by just turning a valve for a couple of minutes manually or by computer. Dutch growers often use mat irrigation with a thin sheet of perforated poly on top of the mats to prevent algae. Many use an ebb-and-flow system, where the tray is actually flooded. Pots take up water, then the excess is drained away. Ebb-and-flow systems can be set up to recirculate irrigation water.

Commercial pot mums are nearly always fed at every irrigation. The fertilizer is injected into the irrigation water line with a fertilizer injector. The normal injection level is 200 ppm N and potash, 300 ppm or more on the northern winter crop.

Hard, high-salt water detracts from the quality and flower size of a mum or any crop. Typically, the problem creeps up gradually; the grower isn't really aware of it. One interesting way to discover if you have a problem is to grow a few plants with distilled water from a grocery store on the same schedule and culture as a few grown with your tap water. The difference might surprise you. A solubridge will give you a definite reading on the salt content of your water supply.

Growers are also beginning to realize the major impact of soil temperature on all pot crops. Cold water can drop soil temperature 20F (11C) in a minute or less. As far as air temperature, use 63F night temperatures, 65F on cloudy days, and maybe 70F sunny days (17, 18, 21C). The grower who has a separate area for starting crops can maintain a warmer night temperature, often 70F (21C) for this first several weeks.

Later culture and control. Where possible, cool plants down at the finish for increasing color intensity and quality. Gradually drop temperatures off to 55–58F (13 to 14C). Also, just as buds are starting to form, run pots on the dry side for a week or so. Just let them dry at the edges a bit, not severely. You'll end up with a bit less flower size but a lot more keeping quality and substance from these toned plants.

Mums subjected to high temperatures (90F, 32C) for one to three weeks may flower late. In aggravated cases of a month or more of very high temperatures, this can result in a flowering delay of seven to 10 days or even more, plus excess height. The first sign of heat delay is simply a delay in flowering beyond the scheduled date. Also, you'll see heat delay buds; the little sepals coming up the sides of the buds are curled inward in a very distinctive way.

D

The answer, in most cases, is evaporative pad cooling. Some moderate-summer areas, such as the Northwest and New England, may not need this. Mist cooling as an alternative is appearing in some Southwest areas. Also, some varieties are clearly more tolerant of high temperature than others. If you are in a high-temperature-prone area, be sure to lean on the heat-tolerant sorts. Your propagator will know them. Breeders, such as Yoder, are increasing their efforts in developing varieties with greater tolerance of hot summer temperatures.

In northern areas, pot mum winter quality depreciates seriously. Depending on the number of cloudy days, quality can drop as early as late November and December. The late December noon sun, when you can see it at all, rises only a bit above the southern horizon. By early February, as day length increases, the sun rises higher each day, and cold, clear days occur more often; quality increases again.

What can you do about low winter light? Be sure the roof is clean and clear. Even a clean double-poly roof allows perhaps 8 or 10% less light to get to the crop than good, clean glass. Remove summer shade on time! Fiberglass ranges that have been allowed to deteriorate to a point of serious light loss are not great for winter mums in the North—or anywhere, for that matter.

Another part of the answer is to space the plants out a bit farther, more light reaching each leaf, and allow an extra seven or 10 long days to develop more plant substance. Another critical point is variety selection. Some varieties perform much better under marginal winter light than others. Some growers use supplemental light from high intensity discharge (HID) lights. They deliver on the order of 500 f.c. (5 klux) and are most often used during the first several weeks of the crop, when many plants can be influenced at low cost.

If your pot mums are too tall, try one or more of these height control tactics:

◆ Select shorter-growing varieties

◆ Use less water or fertilizer

◆ Increase plant spacing

◆ Give more light. Is the roof glass clean?

◆ Reduce the long-day period, although most growers like to allow a minimum of one week

◆ Increasing frequency of B-Nine applications, three or four times is not uncommon

On pot mums B-Nine is very widely used and effective in controlling height. For taller varieties flowered in August and September (highest light and tallest plants), growers often make several applications. For a 2,500-ppm dilution, use 0.4 ounce of 85% wettable powder per gallon of water (or 4 g/l); at 5,000 ppm, use 0.8 ounce of 85% wettable powder per gallon of water (8.1 g/l).

The most important point about B-Nine is to be sure not to water the foliage immediately after application—you'd just wash it off. Application on wet foliage is okay. The first application is made normally 10 to 15 days after pinching, or with a minimum of 1 to 1½ inches (2.5 to 4 cm) of new growth on new shoots. Applications are often repeated every three weeks, and sometimes more often than that. One of the few penalties of using B-Nine is the tendency for white varieties to turn cream.

Also, don't overlook the cool day-warm night (DIF) approach to height control. It's very effective and inexpensive.

Crowd plants carefully—for profit. The more space you give a plant, the better the quality, but the lower the profit tends to be. Space costs a lot of money; therefore, pressure is on the grower to crowd plants. The traditional 14-by-14-inch (36-by-36-cm) spacing for four or five cuttings per 6-inch (15-cm) mum has given way to 12 by 13 inches (30 by 33 cm) or even 12 by 12 inches (30 by 30 cm). The effect on profit in moving from 14 by 14 inches to 12 by 12 inches is monstrous: you get 27% more plants on a bench at 12 by 12 inches. (The same facts, by the way, apply to poinsettias.)

Partly, it's a matter of being a good grower. It takes skill to crowd plants and maintain quality—not growing a deluxe plant, but a somewhat more restricted plant with fewer flowers. Use the height control measures already discussed. Also, select varieties that tend to grow more upright, such as Torch and, somewhat, the Annes.

What if plants are too short? This is usually a much simpler problem. In general, select medium or tall varieties and add some long days to the schedule. Increase feed and water, check plant roots, and use less B-Nine.

What is the optimum height for pot mums? Many use the 12-12 rule of thumb—the plant should be 12 inches (30 cm) high from the pot rim, 12 open flowers on a plant. Supermarkets may want plants more like 11 or 10 inches (28 or 25 cm) from the pot rim. A rule of thumb on what's best for a florist-quality pot mum is 15-15-15. That's 15 inches (38 cm) high from the rim of the 6½-inch (17-cm) pot, 15 flowers (disbudded), and 15 inches diameter.

Finished pots. Traditionally, pot mums are disbudded. Each one of the eight, 10, or 12 stems is disbudded, leaving only the top bud to flower. It's costly, 25 cents or more a plant. Growers frequently put this job on a piecework basis, but that requires careful follow-up inspection of the work. With the immense pressure of labor costs, growers are looking for ways out of this expense. There seems increasing evidence that consumers who buy pot mums may not always want the disbudded plant! For example, daisy mums are virtually all not disbudded. Removal of the center bud is practiced, but not lateral disbudding.

You have to consider, too, carefully selecting varieties that present themselves best with center bud removal only and trying them on your customers. The Boaldis are a good starting point. With only center bud removal, they're a very colorful plant. The Annes are a good example of a variety that must be disbudded. Watch Yoder's Fleurette series, with tiny, semisingle flowers, and dozens of them. This is a novel, showy plant, especially three per 4½-inch (11-cm) pot or, even better, one cutting in a 2-inch (5-cm) pot. No disbudding needed. Three cuttings per 4½ inches is an innovation in the 4-inch (10-cm) pot mum field.

Some growers consistently produce a small percentage of their crop single-stem, maybe 7-inch (18-cm) pots with seven cuttings each. Each cutting is disbudded and grown up to a single flower; there is no pinch. The result, if well done, is a very showy plant. One of the problems involved is height control; inevitably, more B-Nine applications are used, and shorter varieties are often selected. The question is: will the market pay the additional premium in space and cutting required to produce such a plant and still yield a fair level of profit? An adaptation of this plan is the one cutting per 4-inch (10-cm) pot.

Centerpiece mums are pot mums grown in bulb pans with the goal of producing compact, mounded plants rather than a taller or more upright pot mum. The centerpiece mum is tailored for use on coffee tables, office desks, reception areas, and the like.

Six-, 7-, and 8-inch bulb pans are used. Unlike with traditional pot mums, plant one cutting in the center of the 7- and 8-inch (18- and 20-cm) pans to help achieve the mounded appearance. Table 5 shows the number of plants to use per pot and the targeted finished height, including the pan. Culture is the same as for pot mums, except that additional B-Nine is used to keep height in control.

As with single-stem pot mums, ask if your market will accept this premium product. Some growers find that acceptance is better at major holidays, when demand is high and consumers are often looking for an upscale product.

TABLE 5   CENTERPIECE MUM GROWING GUIDELINES

| Pan size | | | Target finished height | | Spacing range |
in	cm	Plants	in	cm	inches[a]
6	15	5	11	28	12 × 12 to 14 × 14
7	18	6	12	30	13 × 13 to 15 × 15
8	20	7	13	33	15 × 15 to 18 × 18

[a] 12, 13, 14, 15, 18 inches = 30, 33, 36, 38, 46 cm, respectively.

FIG. 3. Most cut chrysanthemums sold in the United States today are imported—primarily from Colombia. Here, bunches of pompons are being unpacked after 15 days of transport by sea in controlled atmosphere containers.

Pressures to mechanize. For several reasons, the U.S. pot mum grower is under increasing pressure to automate. Mums, one of the two most widely grown flowering pot plants, are more or less a year-round crop, not a one-day holiday plant. Mums are moved several times: from the starting area to final spacing and then to the headhouse.

Growers discussing mechanization respond that the wide variety of crops necessary on most pot ranges precludes efficient mechanization, but there are several arguments against this way of

thinking. Many growers could halve the number of crops they grow, buy in a few from other specialists, and simply not try to do everything themselves. Also a 3- or 4-acre (1.2- or 1.6-ha) pot range would probably have 1 or 2 acres (0.4 or 0.8 ha) of year-round mum production, which is ample to justify advanced automation.

POMPONS

North American production of pompon chrysanthemums is under major pressure, especially from Colombia. The Colombian crop enters the United States through major importers, mainly in Miami (fig. 3). Many Northern retail growers continue to grow pompons for their own retail shops, especially in the summer and fall. Their great advantage: the mums are strictly fresh.

FIG. 4. In the Netherlands, pompons are the most produced floricultural crop. The number one color: white.

TABLE 6 SCHEDULING A CUT MUM CROP

Plant[a]	Lighting period[b]	Approx. pinch date	Start shade[c]	Flowering
Jan. 7	Jan. 7–Feb. 25	Jan. 28	Mar. 15	April 29
Jan. 21	Jan. 21–Mar. 03	Feb. 11	Mar. 15	May 05
Jan. 28	Jan. 28–Mar. 10	Feb. 18	Mar. 15	May 12
Feb. 11	Feb. 11–Mar. 17	Mar. 4	Mar. 17	May 19
Feb. 18	Feb. 18–Mar. 24	Mar. 11	Mar. 24	May 26

Note: All these mums from response group 9.
[a] For single-stem crops, plant on pinch date.
[b] In these examples, lighting commences with planting.
[c] With pompons, stop shade when buds show color. With October through May standards, stop shade when bud is size of a nickel. With June through September standards, stop shade when bud is taken.

In Holland pompons are the second most important greenhouse crop, with sales of $371 million through the Dutch auctions in 1995, down 6% from 1994 sales. Dutch growers, too, are feeling the pressure of imports (fig. 4).

Plan ahead. Before you commit to grow pompons, make a plan to market the crop, including some commitment from buyers to handle most of the crop at a profit to you. Planning is critical. A crop planned well in advance of planting time is a lot more apt to be profitable than one thrown together the last minute. The plan, based on a marketing study, must include exactly what varieties and quantities will flower on what dates. From that, specific planting, pinch (if any), and short-day dates for each planting should be set. Certainly, the plan should include spacing the total cuttings needed.

Propagators offer substantial discounts for cuttings ordered six months or more in advance of shipment. Most important of all, you get the varieties you want in advance.

Year-round cut flower schedules are designed to flower crops at predictable dates (table 6). They are adjusted for dark winter weather versus bright summer or fall weather. Typically, pompon schedules are done on a year-round basis with successive crops that flower one week apart. Tables from cutting suppliers provide schedules for flowering each of the 52 weeks of the year. Often there are separate schedules for three or four zones across the continent.

Response groups. Most commercial poms are eight- and nine-week varieties. Summer and fall crops in the Northern greenhouse will be typically eight- and nine-week varieties. The winter crop will be nine- and ten-week varieties, often with three to five weeks more plant-to-sell time. In Florida and Southern California, the eight-, nine-, and 10-week varieties are used year-round. A few

EUROPE'S SENSATIONAL SANTINIS

Santini chrysanthemums are rapidly gaining popularity in Holland and Germany. Developed by the Netherlands breeding company Fides Holland BV, these short-stemmed, small-flowered cut mums bring average auction prices almost 10 Dutch cents higher (about $0.06) than other mums. Santini sales grew rapidly upon introduction from half a million stems in 1990 to 24 million stems in 1992.

Santinis have smaller flowers than the traditional 1½- to 4-inch- (4- to 10-cm-) spray mums—no more than 2 inches (5 cm) in diameter. They also have a different umbel, as all flower buds form in the upper level of the spray. Santinis have at least eight flowering buds in the upper 4 inches of the stem and have a slow to average growth rate.

Santini stems are only 18 to 22 inches (46 to 56 cm) long. Short stems means Santinis have about a two-week shorter growing time than traditional long-stemmed mums. The long-day treatment for Santinis is about 50% shorter than for other mums. Santinis are planted eight to nine stems per square foot (or 86 to 95 stems/m2).

Culture. Santinis grow in almost any high-quality soil. Intensively growing four to five crops per year places high demands on soil. Maintain pH at 5.5 to 6.5. Fertilize with 10 to 20 pounds of triple superphosphate and 10 to 14 pounds magnesium sulfate per 1,000 square (4.9 to 9.8 and 4.9 to 6.7 kg per 100 m2, respectively). After planting, EC should be below 1.5, with low levels of salts and sodium phosphate to avoid starting problems. Fungi have fewer opportunities to develop when plants start easily. Begin fertilizing about one week after planting, when cuttings are making new roots outside blocks. Fertilize through the irrigation system with water-soluble fertilizers.

With supplementary lighting of 279 to 325 f.c. (3 to 4 klux) from mid-October to mid-March, you can produce Santinis year-round, even during winter's low light levels. Plant density can be up to 20% higher, about eight to nine plants per square foot (86 to 94/m2), in winter with supplementary lighting.

Cuttings rooted in peat blocks work well because roots aren't damaged during planting. It is important to water correctly, though. Peat blocks that are too dry or too wet could cause Pythium or Rhizoctonia.

Bare-rooted cuttings, unlike peat block cuttings, should be watered several times a day in warm weather. During extremely hot weather, provide additional shade. Optimum temperatures are 63 to 66F (17 to 19C) during the day and 66F at night. Large temperature fluctuations can cause negative DIF reactions, resulting in stems that are too short.

When Santinis grow in day lengths longer than 16 hours, they stay vegetative until a certain number of leaves develop. This long-day leaf number, different for each variety, can be between nine and 55. Keep them vegetated until they're 7 to 8 inches (18 to 20 cm) tall in summer and 11 to 12 inches (28 to 30 cm) tall in winter.

D

During the short-day period, temperatures are important for response time, measured from the start of short days until flowering.

Use growth regulators like B-Nine to retard growth. Never spray under sunny or dry circumstances. After spraying, the foliage must stay dry for at least 12 hours. Because leaf quality is especially affected by growth inhibitors, they should not be applied until the last growing period, when stems are about 12 inches (30 cm) long. Varieties that grow vigorously may require earlier applications.

Northern winter crops are 11-week varieties. To simplify your growing operation, plan the crop to grow together various mums from the same response group, such as all eight-week varieties or all nine-week varieties (again, that's eight or nine weeks from start of short days to flowering).

Commercial growers are always pushing for faster-responding varieties that will still make quality pompons. An eight-week variety, with substantially less total time on the bench than a 10-week variety, means less overhead cost for the crop. In grouping varieties, growers also strive to combine varieties that take roughly the same number of long-day weeks. A tall, rapidly growing variety may produce a quality crop with one week less of long days than a naturally shorter-growing variety. The idea is to not mix the two and, whenever possible, to use the naturally taller variety that will make a crop in a week's less time.

Starting with a good cutting is one of the basics of producing a good crop. Good cuttings aren't cheap, but planting a crop with cuttings that are hard, uneven, and tending to bud prematurely is starting the job with a penalty. Worst of all, it is also asking for insect and disease problems.

Early growing. Pompons will grow in a variety of soil types. Excellent-quality crops are grown in the rather hard clay found in northern California field soils. Equally good crops can grow in peat-lite (peat and vermiculite) media. The media must drain reasonably well: mums will not grow in a heavy, wet, poorly drained soil. Soluble salt levels and pH must be within reasonable limits, and the soil must be free of pathogens and soilborne insects.

Irrigation is nearly always mechanized today. You sometimes see it done with overhead sprinklers, "Rainbird nozzles," that slowly rotate around a 360-degree circle. Many greenhouse crops are irrigated with drip lines around the periphery of the benches. Fertilizer is injected at each irrigation. A normal rate of application is 200 ppm N and potash. Raise rates somewhat in winter and if you use very porous nonsoil mixes.

The standard temperature for pompons is 60F (16C) nights (65F, 18C, on cloudy days) from planting cuttings until buds appear. The temperature may be, and often is, gradually lowered to 55 or 50F (13 or 10C) as color appears—both to conserve fuel and to improve quality and color. Temperatures on sunny days are perhaps 70F (21C). Nearly all commercial varieties will set buds uniformly at 60F minimum night temperature. A few prefer 62F (17C) or 63F (17C). Grown too cool, the plant will rosette, making short clusters of leaves down near the ground. Crops grown much above 60F (especially under low winter light conditions) will be drawn, thin, and of generally poor quality.

In areas where midsummer temperatures can go to 90 or 95F (32 or 35C), heat delay may be encountered. Buds will form but will fail to develop. You'll see the little heat bud, a bud surrounded by sepals or little ridges that curl up around the bud; they will be concave-shaped. The only answer is fan-and-pad cooling to moderate greenhouse temperatures.

Mums tend to "remember" temperatures. For example, high day temperatures of up to 75 to 80F (24 to 27C) may offset low night temperatures below 60F (16C). Even cuttings grown under high temperatures tend to flower better at marginally cool temperatures.

Remember that the mum is a natural-season crop, which means planting and long day periods naturally occur during midsummer, with its warm temperatures and long days. As fall approaches, day length shortens and temperatures are typically 50F (10C) until flowering. Buds set during August and September in warm temperatures and under shortening days.

There is substantial production of poms under northern winter low-light conditions. The best example is the Dutch crop, grown at 53 degrees north latitude, equal to Hudson Bay, Canada. There a winter day is short, with lots of cloudy weather.

A truly low-light crop like this demands changes in growing. First, be sure that all possible available light gets to the crop. Clean shading compound off the roof! Next, carefully select varieties—some perform better during dark, midwinter weather than others. Extend the long-day period, in effect giving the crop several weeks more total time from plant to flower. This allows time to develop a quality pom even though light is sharply reduced.

Plant cuttings on wider spacing. Carefully control crop irrigation; don't overwater under such conditions.

Later growing. Pompons are done as both pinched and single-stem crops. Typically, the greenhouse crop is single-stem. Overhead costs are so high under glass that the several weeks time saved on the crop more than offsets the added cost of cuttings for single-stem growing. On the other hand, most Central and South American crops are pinched.

Considering your overhead cost per week, the additional total crop time for a pinched crop, and the cost of the pinch, you can quickly make a decision based

D

on cost. There is also the element of quality, however. The single-stem crop is typically heavier and has better sprays.

Where crops are to be pinched, the tip pinched out several weeks after planting, the propagator's schedules normally include a pinch date. The pinch should be soft, always allowing several new leaf axils to provide breaks for the crop. A typical "bunch" of pompons for North American markets will include a minimum of five stems and weigh about 10 ounces (284 g).

Clubby sprays are short peduncles, 1 or 2 inches (2.5 or 5 cm) long, rather than the 8 or 10 inches (20 or 25 cm) that retailers like. This problem of peduncle length, partly a matter of variety, is aggravated by cold, dark winter weather. Day-length manipulation—applying a 10-day long-day treatment—can help. For example, if the normal day to start short days is February 1, you would start short days 10 days earlier (January 21), continue short days for 10 days, then use a 10-day long-day period into the crop schedule. After that (February 10), go back to short days to flower. Also, Dutch growers commonly use B-Nine on pompons. The application is made mostly on winter crops, as it improves spray formation.

If a pompon crop flowers normally, with good quality, but it is more than 3½ to 4 feet (107 to 122 cm) tall, it's a good sign that the long day period was excessive. The same crop grown next year could be cut down to correct total height by reducing long days a week or two. (Crops where bud development is delayed by excessive heat or excessive air pollution may stretch in height and perhaps never flower. That's a different story.)

Pompon crops that flower too short and are still of good quality indicate an insufficient long-day period. For the same crop next year, add another week of long days. If there are obvious quality problems in the crop, it may be due to an excessively wet, heavy soil, nematode or other soilborne pest problems, high salts, or other factors.

Spacing is one way the grower can adapt the crop to a market that wants a little less quality or, conversely, to a quality retail shop that does want better pompons. Spacing varies with the season, especially in northern areas. Single-stem winter pompons in the Midwest are typically grown 5 by 6 inches (13 by 15 cm). The same crop flowered in August or September will be grown at 4 by 5 inches (10 by 13 cm). Pinched crops are adjusted accordingly. Pinched crops are often pruned to several stems per plant, typically one extra stem on the outside row. Often growers completely omit the center row on a bench of northern winter pompons or mums, since quality problems are worse in the center. Removing this row gives the other middle rows a lot more light.

STANDARD CUT MUMS

Standards, also known as the football mum, have showy, large, 5- and 6-inch (13- and 15-cm) blooms, great for large basket designs and wearing to football games. They were a major crop for years. However, demand has been steadily down in recent years.

Growing the crop. Just as with pompons, plan the crop well at least six months in advance of planting date. Many of the cultural details are identical to those in the preceding section on pompons. The Northern winter crop, however, is tough. Standard mums, perhaps even more than pompons, suffer in quality with the very short days and cloudy weather of the northern winter.

Regional crops. In northern California it's almost entirely a greenhouse crop, often under poly, sometimes fiberglass or glass. There has been, and still is a little, shade house production of standards and pompons in summer and early fall in the West. In northern greenhouses standards are grown under glass by whole-

TABLE 7   GARDEN MUM VARIETIES

Best spring-flowering garden mums

White	Yellow	Pink	Bronze	Red	Coral/Salmon
Linda	Lisa	Debonair	Denise	Bravo	Grenadine
Nicole	Jessica	Grenadine	Remarkable	Helen	Christine
Tolima	Anna	Megan	Jennifer	Red	Stunning Lynn
			Robin	Remarkable	

Best summer-flowering garden mums for black cloth (shaded) programs

White	Yellow	Pink	Bronze	Red	Coral/Salmon
Frolic	Donna	Debonair	Remarkable	Bravo	Christine
Tracy	Lisa	Barbara	Jennifer	Minngopher	Stunning Lynn
Nicole	Jessica	Emily	Triumph	Red Remarkable	Blushing Emily

Best garden mums for fall container programs

White	Yellow	Pink	Bronze	Red	Coral/Salmon
Linda	Lisa	Debonair	Harvest Emily	Bravo	Grenadine
Tracy	Jessica	Emily	Jennifer	Helen	Christine
Tolima	Yellow Sandy	Lynn	Sandy	Raquel	Blushing Emily

Best garden mums for fall fast crop pot plants

White	Yellow	Pink	Bronze	Red	Coral/Salmon
Linda	Lisa	Lynn	Sandy	Bravo	Grenadine
Nicole	Jessica	Symphony	Jennifer	Helen	Christine
Tracy	Target	Emily	Harvest Emily	Raquel	Blushing Emily

sale specialists, sometimes retail growers or smaller or midsize growers supplying a local group of retailers directly.

It's nearly all under poly in Bogota, Colombia. Interestingly, some varieties really don't need black cloth shading. The natural day length is so near the short-day requirement that some varieties, even in summer, will make it without black cloth. The Bogota crop is shipped almost entirely to the U.S. market. Standard mums don't ship as well as pompons, and having to go through several forwarding points doesn't help their quality.

GARDEN MUMS (HARDY MUMS)

No other flowering plant is as associated with the colorful fall season as the garden mum. Its bright, colorful blossoms provide exciting sales opportunities in the fall. But garden mums are not just for fall (table 7). They are also widely grown and marketed for spring and summer sales, too. They are so adaptable that container sizes range from cell packs all the way up to 14-inch (36 cm) color bowls and even moss hanging baskets.

When growing garden mums, it is important to determine the parameters of your market. Consider your market's needs in terms of quality, plant size, and colors. Produce the best-quality plant that you can. Take advantage of the garden mum's versatility and offer customers a variety of sizes (price points), colors, and flower forms.

The term garden mum applies to mum varieties that will flower naturally in most parts of the United States early enough in the fall to be showy well before the first heavy frost. This contrasts with most commercial varieties, which bloom naturally late in October and early November and thus would be nipped long before flowering in most parts of the United States. The term hardy has been abandoned by most suppliers—not many pot flowering and cut flower varieties are indeed very hardy. The hardiness of garden mum varieties may vary significantly from one part of the country to another and from season to season.

Today, garden mums are selected for plant habit, flower color, flower form, plant flexibility, resistance to diseases like bacterial leaf spot, and flowering response in the fall (i.e., early, midseason, and late). Cushion growth is a favorite habit, the plant possessing a rounded, mounded form. Older garden mums were often upright in their growth and were not as adaptable to the wide range of production choices available today.

Garden mum flowers are typically smaller than their pot mum cousins. Flower size ranges primarily from 1 to 3 inches (2.5 to 8 cm) with a few exceptional varieties that may reach 5 inches (13 cm) in diameter. Generally, the largest flowers are not especially weather-tolerant. A range of flower forms is available—decoratives, daisies, spoons, buttons, quills, and more. Flower colors go from rich, saturated hues to soft pastels. Resistance to fading, improved flower-form retention, and better weather tolerance is found in many of today's cultivars.

Production methods. For many growers the first mum crop of a year is for spring sales. For flowering up until Mother's Day, cuttings are planted, pinched seven to 10 days after planting, and require no more than general growing care until maturity. A selection of 15 to 20 varieties will flower over a two to three week period, not all precisely on one date. This growing technique can be practiced from about January 15 planting through March 15 planting in the North. Not all varieties sold are suitable for this type of production; read variety descriptions from your cutting supplier carefully.

The classic crop is potted March 1, is given no shade-no light and 60F (16C), and flowers early May. This method is ideal for small pots (3 to 4 inches, 8 to 10 cm) and 18- or 24-cell packs. It is also ideal for the increasingly popular 6 unit handle basket. Since the day length steadily increases in the spring, this program is not reliable for flowering after early May or Mother's Day.

Garden mums are often flowered any week of the year using the light-and-shade technique, just as are any other chrysanthemums. Four hours of light in the middle of the night are provided, beginning when the plants are potted. The schedule suggested for mid-to late-winter growing depends on variety. For short-treatment varieties, give three weeks of light and pinch two weeks after potting. Medium treatment varieties take two weeks of light; pinch two weeks after potting. For tall treatment varieties, give one week of light and pinch two weeks after potting. Almost every variety accommodates this type of production. Again, read the variety descriptions.

The ultimate retail customer takes home a flowering plant to use for decoration indoors or out. This plant, when past its prime, can be planted in the garden (cut back to just its lower foliage), where it will produce a full-sized garden mum

TABLE 8 SEVEN-WEEK VARIETIES TO FLOWER AUGUST 10

| Container (inches[a]) | Number of plants | Plant | Pinching | | Start short days[c] |
			1st	2nd	
4	1	June 8	June 22	n.a.[b]	June 22
6–6½	1	May 18–25	June 1–8	June 22	June 22
7–8	1	May 18–25	June 1–8	June 22	June 22
7–8	2-3	June 8	June 1	June 22	June 22

[a] 4, 6, 7, 8 inches = 10, 15, 18, 20 cm, respectively.
[b] Not applicable.
[c] With these examples, corresponds to last pinch.

D

for natural fall flowering. The advantage to flowered sales is that the customer sees the variety purchased.

For spring sales, vegetative plants sold with color-variety labels are growing in popularity. This crop is timed to be marketed with the big push of the bedding plant season. Cuttings are planted in single plastic pots or 8s, 24s, or larger cell packs. For best results these plants should be placed where they will receive three hours of light in the middle of each night from planting until sale. They are ready for sale when breaks are ¾ inch (2 cm) or longer. If sales are slow, a second pinch can be made two to three weeks after the first. B-Nine is helpful to control growth.

FIG. 5. Bart Bernacchi, Bernacchi's, La Porte, Indiana, holds one of his 8-inch potted garden mums.

Summer and later sales. Many garden centers need fresh, quality flowering plants in July and August. By flowering garden mums in the summer, you can (1) create an extended sales period from the end of the bedding plant season until the fall crops flower naturally, (2) use empty greenhouse space during the summer, and (3) capture a niche in the early retail market, which can help pave the way for smooth sales of the natural fall crop of garden mums.

Summer crops can be timed by controlling day length and using variety response information to develop a schedule. Long days are given to start, followed by short days, provided by black cloth or black plastic. For example, a crop of seven-week, medium-treatment varieties to flower for August 10 (table 8) should start short days on June 20 (short days provide 12 or more consecutive hours of total darkness to the plants, such as 8 P.M. to 8 A.M.) The number of long days prior to that can vary by the size of the plant desired. See table 8 for guidelines and comparisons.

For decades garden mums were planted from early to late May, given two or three pinches, and sold as they flowered in September and early October. With today's improved varieties and better cultural techniques, such as drip irrigation, there is a shift to later planting. Growers have the luxury of choosing a production scheme that is best suited to their facilities and abilities.

For traditional fall crops, containers vary from 6-inch (15-cm) pots to 2- and 3-gallon (7.6- and 11.4-l) containers, with 6- and 8-inch (20-cm) pots being the most popular (fig. 5). Plant dendranthema cuttings from mid-May to early June.

Space on 18- to 24-inch (46- to 61-cm) centers for container-grown plants, 24- to 30-inch (76-cm) centers for field-grown plants.

Give the first pinch seven to 14 days after planting. Additional pinches can come every two to three weeks thereafter. The breaks should be 2 to 3 inches (5 to 8 cm) long. The last pinch should be no later than July 15 for a plant to flower naturally in September or October. In the Northwest, though, stop pinching by July 4. In southern areas pinching can go on until July 15 to 25. Mums begin flowering from September 10 to October 15, depending on variety.

Alternative crop strategies. With the newer, free-growing mums, some growers are planting late—not quite a fast crop—and gaining several benefits. By planting two cuttings two fingers apart from June 20 to July 5 in the center of an 8-inch (20-cm) or so container and pinching only *once* (approximately two weeks after planting), the grower can produce a garden mum of equal size and comparable flowering date as one started in late May or early June. Benefits to transitional multiple-cutting crops are:

- There is less stress and less work to do in May or early June, a busy time of year, anyway.
- Plants are set outside into warm temperatures, not into cool nights, which could promote premature budding.
- Less water is used, meaning lower cost and less runoff.
- Less fertilizer is used, meaning lower cost and less runoff.
- Fewer pesticide applications are needed.
- Less labor is required by planting directly into the finishing container; no time is spent with cell packs or small pots because it is too cool to plant outside.
- Less labor is needed for pinching, since you pinch each cutting only once.
- Better uniformity results from less crown budding.

Fast crop is a strategy when you desire a natural-season crop that can save greatly on labor and crop time. Fast crops are planted as rooted cuttings from July 15 to 25, one cutting per 6-inch (15-cm) pot or two per 8-inch (20-cm). *Do not pinch this crop.* See the dramatic savings in time and labor. Yet these plants flower only a few days to, at most, seven days after a typical crop planted and pinched in early June.

General culture. Here are some general garden mum cultural comments that apply to all programs. They can be summed up as follows:

D

- Use the best cultivars.
- Plant promptly.
- Use well-drained media.
- Keep media moist.
- Keep plants well fertilized.
- Provide adequate spacing at all times.

Garden mum media can be soil based or soilless. The media should be loose, well drained, and provide a solid anchor for the root systems. The pH for soil-based media should be 6.0 to 6.5. The pH for soilless media should be 5.5 to 6.0.

A key to success is adequate spacing. Even the best variety can be ruined by close spacing. Adequate spacing is necessary from Day 1 for a quality finish. As a general guide, space 6-inch (15-cm) pots on 12- to 15-inch (30- to 38-m) centers, 8-inch (20-cm) pots on 18- to 24-inch (46- to 61-cm) centers.

Always plant or stick garden mum cuttings into moist media. Planting into dry media would reduce initial growth and future potential. Water the cuttings in thoroughly immediately after planting. It is beneficial to mist or syringe the plants frequently for the first few days until the plants are fully turgid.

Liquid feeding at planting helps get the cuttings off to a good start. Use a 200- to 300-ppm fertilizer solution to water the plants in immediately after planting. Garden mums thrive with copious amounts of fertilizer concentrate and moisture. While perhaps the ideal feed program would center on providing constant liquid feed at approximately 250 to 300 ppm of a balanced fertilizer, such as 20-10-20, many garden mums are grown with slow-release fertilizers or a combination of liquid feed and slow-release fertilizer.

Proper ongoing watering is crucial to successfully produce a high-quality garden mum. Apply enough water to thoroughly soak the media. Garden mums should never be allowed to wilt in the first half of their crop life. Wilting during the first few weeks of growth could restrict branching action and overall growth. In later stages of growth, slight wilting can be beneficial in controlling height and toughening up the plant. Whenever possible, use drip tube irrigation. Overhead watering can contribute to the development of leaf-spotting foliar diseases, especially in the mid- to late summer.

Pinching. Pinch a garden mum when the plant is ready, not on a schedule. Generally, a rooted cutting is ready for a pinch about two weeks after planting. Look for approximately an inch (2.5 cm) of new growth and make sure plants have very active root growth. The pinch should be a medium-hard pinch and remove ½ to ¾ inch (1.3 to 2 cm) of growth.

Second and third pinches, if given, are made when the breaks are 3 to inches (8 to 10 cm) in length. For natural-season crops, the last pinch should be performed between July 1 and 15 in northern latitudes, up to seven to 10 days later in southern states.

Garden mums are very reproductive. They can bud prematurely with cool tem peratures, stress from low moisture or feed levels, late pinching, and other reasons. They can bud prematurely even if lighted at night. Therefore, it is all the more important to strive to keep a young garden mum plant actively growing with plenty of moisture, plenty of fertilizer, and plenty of space.

Fortunately, garden mums that have developed premature buds can develop into quality plants. If you have premature buds, pinch harder (deeper) to remove unseen but developing buds, and be sure you are employing the best cultural practices, especially with watering and fertilization.

Florel. Florel (ethephon) is simply ethylene and can be used as a tool in gar den mum production, primarily for fall crops. Florel has the potential to (1) delay the natural fall-flowering response, (2) reduce or eliminate pinching, and (3) inhibit or reduce the number of premature crown buds. The potential ben efits are exciting, yet there is still much to learn with this recently registered growth regulator.

The following guidelines for using Florel are based on a crop planted in late May or early June. The first application of Florel should be made only after plants are well established (four to seven days for rooted cuttings, 10 to 14 days for unrooted cuttings). Mix according to the labeled rate of 500 ppm for best results. *Never apply to stressed plants, that is, ones that are wilting or fatigued.* Use a spreader sticker in spray solutions and mix Florel with deion ized water or water with low alkalinity for best results. High alkalinity could diminish Florel effectiveness.

Pinch after Florel by hand when cuttings are ready for the first pinch. Trial grow some plants without any manual pinch to prepare for a decision on whether you want to manually pinch next year's crop.

The second Florel application should come approximately 10 to 14 days after the first pinch, again using a 500 ppm solution. If the second Florel application is made by late June to approximately July 5, you can stop applications and expe rience little or no delay in flowering, yet still obtain a fuller plant with more uni form bud formation than plants grown without Florel. In effect, this Florel application acts as a second pinch.

If the goal is to delay flowering, continue Florel applications every two weeks until late July or early August. This will delay flowering of some early- and midseason varieties into October. However, there are as yet no accurate guidelines for predicting flowering times by cultivar based on the number of Florel applications. This is an area where additional trialing is required to correlate final application dates and actual flower date by variety. As a guideline, you may find that mid- to late July application delays the natural flowering by one to two weeks. A final application in late July to early August may delay natural flowering by three to four weeks.

DIANTHUS (CARNATION)

by Phil Gardenier
Van Staaveren
Aalsmeer, the Netherlands

❖ *Tender perennial* **(Dianthus caryophyllus)***. Vegetatively propagated. Start from disease-indexed cuttings purchased from a reputable supplier. Produced primarily as a cut flower. Growers crop plants for as long as two years before replanting.*

After roses and dendranthema, carnations are the world's third most popular cut flower crop (fig. 1). The flower has a long history, tracing its botanic origins to the Mediterranean. In the year 300 B.C., the Greek philosopher Theophratus named this beautiful flower *dianthus*, after *dios* ("god") and *anthos* ("flower"). The divine flower was later described by Linnaeus as *Dianthus caryophyllus*.

FIG. 1. Cut carnations are among the leading cut flowers worldwide, forming the basis of many florist arrangements and mixed bouquets. The crop is easily shipped when stems are properly treated during postharvest and when long-lasting varieties are grown. Most carnation production is centered in Colombia for the United States, and in Africa and Israel for Europe. The variety Terra—a terracotta-tinged medium pink—is shown in the photo.

Carnations have many characteristics which suit today's consumers and marketers, including a wide range of colors, a pleasant, clove-scented fragrance, and a long vase life. There are many different types—standard, miniature, and single-flowering types, such as Gipsy, and they work quite well in mixed bouquets. Although the carnation has a lot of competition from other crops, these characteristics offer breeders, growers, and marketers tremendous opportunities to increase their carnation business.

Since the 1980s, production of carnations has moved from traditional consumer markets, such as the United States, Germany, and Japan, to countries where the flowers can be produced at lower costs. Currently, the main carnation production areas are Colombia, Israel, Kenya, and Spain. Flowers are shipped north from these locations to consumer areas. Because they travel well, imported carnations have built up an important position in consumer markets (nearly all North American carnations are imported, for example), although fresh, locally produced carnations are often considered to be of superior quality.

PLANTING

Soil preparation. Before planting, till soil so that it is well loosened and able to retain sufficient air and water. Such conditions will stimulate rooting. The soil must be well drained, so break up impermeable layers.

As basic fertilizer, a mixture of 50% old stable manure and 50% garden peat should be mixed into the soil before disinfection. Add 3 to 4 cubic meters per 100 square meters (4 to 5 yd^3 per 120 yd^2) of greenhouse area. Before planting and on a regular basis during growing, analyze the soil to measure its nutrient content. Following are soil nutrition target values based on a 1:2 volume extract.

	Target	**Range**
EC:	1.0 MS/cm at 25C	0.8–1.6
pH:	6.0	5.5–6.5
NH$_4$	<0.2 mmol/l	0.1–0.5
K	2.1	2.0–4.0
Na	1.8	1.0–3.5
Ca	2.0	1.7–2.8
NO$_3$	4.3	3.3–6.0
Cl	1.0	0.5–3.5
SO$_4$	1.8	1.0–3.5

Disinfect the soil before planting. You will obtain the best results from steaming the soil. Make sure the soil is well watered after the disinfection (also when steaming) in order to rinse out remaining chemicals. Have the soil analyzed. If the EC, especially the sodium and chloride content, is over 3.5 mmol/l, leach with water again.

Please note: Always disinfect the soil before planting. Virgin soil may be clean, but when reused, soil may contain many pathogens. *Fusarium,* especially, can buildup in soil over time, making some beds unsuitable for carnation cultivation even after disinfecting.

Planting cuttings. Plant only clean, healthy plant material from a reputable supplier. Cuttings should come from virus-free mother stock. When planting older generation cuttings, problems may and probably will occur during the culture. Only high-quality cuttings can produce the best-quality flowers.

Preferably, keep the base of the cutting above the soil level. Never plant well-rooted cuttings deep. Cuttings planted too deep are at high risk for root rot.

A planting density of 32 plants per net square meter (27 per yd²) is normally recommended. This equals 20 plants per square meter (17 per yd²) of greenhouse area.

Carnations can eventually build up a huge plant mass, so the support system should be very strong. The distance between the poles should be no more than 3 meters (10 feet). Use four to five layers of nets, preferably made of wire. Insufficient support will result in flowers with bent stems.

CULTURE

Environment. Carnations perform best in relatively cool climates. When beginning a new crop, the soil temperature should be about 60F (16C), and the greenhouse temperature 61 to 68F (16 to 20C). High temperatures combined with low light intensity will result in flowers with inferior quality. Do not allow the greenhouse temperature to exceed 77F (25C). Ventilation (natural or mechanical), shade cloth, whitewash, or a cooling system—such as a fog, a roof sprinkler, or evaporative pad-and-fan cooling—will be required in most regions.

The optimum greenhouse humidity for producing carnations is between 65 and 90%. When humidity is higher, plant activity drops, and they become more susceptible to diseases. When condensation occurs, there is a high risk of fungal diseases such as *Alternaria* and *Botrytis*. High light levels and too little humidity are harmful to a young crop and may cause leaf burning, weak foliage with closed stomata, and drought-related infections of parasites, such as red spider mites.

Diseases and pests. The most common insect problems are aphids and thrips, along with the red spider mite. *Fusarium*, root rot, *Rhizoctonia*, stem rot, *Alternaria* and *Botrytis* can all cause troubles. Preventive spraying is recommended. Monitor the plants well during culture in order to locate and treat infections efficiently. Be careful using chemicals and read the label to determine if a pesticide is suitable for use on carnations. When in doubt, experiment first on a small section to find out if it is phytotoxic.

Keep the plants as dry as possible during culture, especially after harvesting and pinching. Wounds should dry up well in order to avoid disease infections.

Pinching. Two to four weeks after planting, pinch the plants, leaving four to six leaf pairs. Basically, each leaf pair will give a shoot. If too many leaf pairs remain, stem quality and growing speed may be dramatically reduced. If you leave too few leaf pairs, production will be insufficient.

To spread out production, a 1½ pinch is a possibility. About four weeks after the first pinch, 50% of the shoots should be pinched a second time. To push production for a particular selling time, a double pinch might be considered. In the end, all shoots should be pinched a second time.

Light and flower induction. Carnation flowers are induced when the plant has developed seven full-grown leaf pairs, and when the day length is 13 to 14 hours of sufficient intensity. Artificial lighting of 10 to 15 watts (38 to 50 lux) per square meter (8 to 13W or 32 to 42 lux per yd²) is commonly used for flower induction in periods of naturally low light levels and short days.

Disbudding. The side buds of standard carnations must be removed (disbudded). This can be done in three phases. Buds should be taken away before they get too big, as they will negatively effect flower size and will take too much energy from the plant. Be careful when disbudding close to the main bud. This should be done just before it flowers, otherwise the flower head may become deformed. With miniature carnations, the center bud should be removed as soon as the bud shows color.

HARVEST AND CUT FLOWER CARE

Harvesting and cropping. Flowers should be harvested at the correct stage. With standard carnations, this is when they are half open and the first petal lays horizontally. With miniature carnations, it is when the first flower is open. Do not cut the flowers too deeply as this may negatively effect production during the next flush. Keep flowers in fresh, clean water as long as possible and handle them with care.

There are several possible cropping programs: 1-year, 1½-year, 2-year and crown-flower culture. The choice of culture depends on the climate (temperature), the quality perception of the flowers in the market, crop planning, and personal preference. Two-year culture is most common.

Postharvest. Your flowers may have superior quality at harvest, but if postharvest treatments and handling are sloppy, the consumer may be unhappy with your product. Again, provide fresh, clean water for your cut flowers as long as possible. Be careful with any handling or movement.

Carnations are very sensitive to ethylene. Always pretreat them with an STS-based preservative, such as Chrysal AVB or Florissant 100. Preservatives work best at temperatures between 59 to 68F (15 and 20C).

After pretreatment with a preservative, store flowers in a cooler at 41F (5C). To avoid condensation in the flowers and stem breakage, do not put the flowers in a cooler immediately after arrival from a "hot" greenhouse.

Maintain cool temperatures during transport. When shipping the flowers in boxes, make sure the insides of the boxes have been cooled thoroughly, either by precooling or by keeping the open boxes in a cooler for at least eight hours.

Transport should be as quick as possible and under cool conditions.

This article provides only basic information about the culture of carnations. For more information, contact a reputable supplier of carnation cuttings. Most companies have qualified salespeople and technicians who can provide complete cultural information for your growing climate.

DIANTHUS

(PINK; SWEET WILLIAM)

by Jim Nau
Ball Horticultural Company
West Chicago, Illinois

❖ *Annual* **(Dianthus chinensis).** *25,000 seeds/oz (872/g). Germinates in seven days at 70 to 75F (21 to 24C). Cover the seed lightly upon sowing.*

❖ *Biennial* **(Dianthus barbatus,** *also called Sweet William). 25,000 seeds/oz (872/g). Germinates in seven to 10 days at 60 to 70F (16 to 21C). Cover the seed lightly upon sowing.*

FIG. 1. Slowly but surely bedding plant dianthus is making its way into the major leagues as a strong spring bedding crop. Varieties such as the Ideal series, Princess series, and Telstar series with their improved heat tolerance, more compact plant habit, and bright range of colors are driving the dianthus class. In the photo: Ideal Cherry.

ANNUAL DIANTHUS

Dianthus is a diverse class of plants that is only briefly covered here. Bedding plant types in the United States today center around *D. chinensis* varieties. Predominantly cool-season crops, most *D. chinensis* varieties perform like pansies in the Midwest and South, showing good color for the spring and early summer and failing in the heat and humidity of August. In moderate climates, such as coastal California, dianthus puts on a vibrant flower show unrivaled by other annuals. Recent improvements in varieties have helped this class to become as popular as it was in the gardens of yesteryear (fig. 1).

Green packs are salable 10 to 11 weeks after sowing (nine to 10 weeks in the South), with flowering packs ready in 15 to 16 weeks (11 to 12 weeks in the South). Using one plant per container, 4-inch (10-cm) pots flower out in 15 to 16 weeks (11 to 12 weeks in the South). In the southern United States, sow seed in August for green pack sales in November. Plants will flower from February until May or possibly June. If the plants become leggy and are in need of a growth regulator, A-Rest, Sumagic and Bonzi are effective. Note that these plants are often robust once they come into flower when grown in the pack. Shelf life at retail is better if they are grown in pots and marketed as such.

Some of the best varieties for all-around uses in the garden today include the Ideal, Princess, and Telstar series. The Princess series is made up of six separate colors plus a mixture, while Telstar has eight separate colors plus a mixture. The Ideal series, in 13 separate colors and a mix, moved quickly into the marketplace because of its superior flower color and uniformity of production. Also now on the scene is the Floral Lace series in six separate colors. Flower petals are serrated, giving a lacy appearance. Floral Lace also has the large flower size of Ideal.

Among other varieties, consider both the Charms and Carpet series for 4-inch (10-cm) pots and other containers. Be aware, however, that these varieties do not have as good heat tolerance outside. They are best suited to use in combination pots or as flowering plants for Mother's Day, for example.

The Parfait and Rosemary series combine the dwarf pot performance of Charms and Carpet with the bicolor appearance and flower size of the old types. Both of these series are excellent for 4-inch (10-cm) pots and as bedding plants. The flowers, more than 1¼ inches (3 cm) across, are characterized by a prominent eye or spot of either purple or rose red in the center.

PERENNIAL DIANTHUS (SWEET WILLIAM)

While annual forms of this old-fashioned garden favorite are available, most varieties are biennial. Sowings made in February or March usually grow 6 to 12 inches (15 to 30 cm) tall during the summer and flower sporadically at best. Once the plants go through the winter, they will flower the following year and can be used for garden accents or as cut flowers. In bloom the plants reach up to 18 or 20 inches (46 or 51 cm) tall for cut flower types, while the carpet or prostrate varieties grow to 12 inches (30 cm) at best.

Green packs are salable 10 to 12 weeks after sowing, and the plants will flower the following May and June (a year later). Grow on at 50 to 55F (10 to 13C) nights and treat as you would pansies. If the variety you buy is touted as blooming the first year from seed, it is an annual and will not overwinter in northern areas.

In selecting varieties, note that many U.S. suppliers sell the dwarf strains of *D. barbatus* in the perennial sections of their catalogs. These varieties are recommended for quart or gallon containers for summer or fall sowing and overwintering along with other perennials for sales the following spring. There are a number of varieties available, including Indian Carpet Mix, a dwarf variety to 8 inches (20 cm) with single blooms. Cut flower growers want to be sure to look for the long-stemmed varieties, which are often not listed by most seed companies. These make excellent plants even in Midwest summers.

DORONICUM (LEOPARD'S BANE)

by Jim Nau
Ball Horticultural Company
West Chicago, Illinois

❖ *Perennial (***Doronicum caucasicum***). 26,000 seeds/oz (917/g). Germinates in 4 to 21 days at 68 to 72F (20 to 22C). Older seed germinates over a longer period. Lightly cover or leave seed exposed to light during germination.*

One of the earliest perennials to flower, *Doronicum* is a hardy plant in Midwest gardens, with golden-yellow flowers from 2 to 2½ inches (5 to 6 cm) across on plants to 2 feet (61 cm) tall. Flowers are single to semidouble, daisylike, and have no scent. Plants flower in April and May and are hardy to USDA Zones 4 to 7. The foliage is kidney-shaped and deep green in color. *Doronicum* can be used within the perennial border as an accent planting. However, large mass plantings should be avoided, due to the short duration of flowering and expected heat stall during the summer. Plants often go dormant in August under high temperatures and humidity in Midwest gardens.

Doronicum caucasicum is frequently propagated by seed and division. Most plants on the market come from seed. Divisions can be taken in either fall or spring.

Doronicum is a spring-flowering plant that will not flower from a winter sowing. If seed is sown in December and grown on in 4-inch (10-cm) pots, plants will not flower until April or May a year and five months later. Sow seed in July, transplant to quart containers in late summer, using one or two plants per pot, and overwinter in a cold frame. Growing-on temperatures should be 50 to 55F (10 to 13C).

Most often the variety sold is labeled *D. caucasicum or D. cordatum*. Magnificum is more uniform than the species, and grows to 20 inches (51 cm) tall. Spring Beauty has double-flowering yellow blossoms on plants 12 to 15 inches tall. Finesse has large, full blooms and grow to 20 inches tall.

ECHINOPS (GLOBE THISTLE)

by Jim Nau
Ball Horticultural Company
West Chicago, Illinois

❖ *Perennial* (**Echinops ritro**). *2,600 seeds/oz (92/g). Germinates in 14 to 21 days at 65 to 72F (18 to 22C). The seed should be left exposed to light during germination.*

Echinops is a metallic blue flowering perennial from 24 to 36 inches (61 to 91 cm) tall. Flowers are globe shaped and range from 2 to 2½ inches (5 to 6 cm) across. The individual florets are small, to one-eighth of an inch (32 mm) across, and the globe opens from the top down. The leaves are distinctively thistlelike, with spines at the ends; underneath, the foliage is gray-green in appearance. Plants flower July and August and are hardy to USDA Zones 4 to 9.

Plants can be used as either fresh or dried cut flowers, though they are primarily used in perennial border plantings. As a fresh cut flower, harvest when the uppermost flowers emerge. As a dried cut, harvest when about one-third of the flower has opened and before the top blooms have faded or died. To dry, hang in a warm, dry place.

Any variety sold under the name of *E. ritro* is the species. This is available as seed, while the cultivars can be propagated by division or root cuttings. The species is more vigorous than the cultivars and can get 5 feet (1.5 m) tall after three years. Dividing the plants is done in the spring every three to four years, depending on the variety and how vigorous it grows.

Sowings made in the winter and spring seldom flower the same season from seed. Sowings are more often made in July and transplanted directly to a quart container. Seedlings do not like numerous transplantings and can often die for a number of reasons. In some cases the die-off is from burying the crown too deeply. In others, root damage is the cause. For best performance, sow seeds directly into plug trays and then transplant into the final container. For green plant sales in the spring, sow in January or February, transplant directly to 3- or 4-inch (8- or 10-cm) pots and sell 11 to 13 weeks later. Growing-on temperatures are 55 to 58F (13 to 14C).

In varieties, our opinion is that the species type is a very good strain, though some variability is to be expected. However, it makes excellent cut flower plants. Tapglow Blue is the most popular cultivated variety today. Flowers are steel blue and 3 inches (8 cm) across. Tapglow Blue grows 3 to 4 feet (0.9 to 1.2 m) tall once established and is vegetatively propagated. Vetchii's Blue has a darker blue flower color than Tapglow, and it is earlier to flower in the garden. It is becoming more popular in the U.S. perennial trade. This variety is also vegetatively propagated.

EUPHORBIA (PERENNIAL)

by Jim Nau
Ball Horticultural Company
West Chicago, Illinois

❖ *Perennial (***Euphorbia polychroma***). 3,500 seeds/oz (123/g). Use fresh seed and germinate at 70F (21C). Seed germinates in 8 to 15 days. Do not cover seed after sowing.*

The most popular *Euphorbia* known on the market is the poinsettia, *E. pulcherrima*. There are other types, however, that make excellent cut flowers or perennials. *Euphorbia polychroma*, a perennial plant growing to 18 to 24 inches (46 to 61 cm), is admired more for its dense foliage than its inconspicuous yellow flowers, borne in April and May. As the weather gets cooler, the foliage turns crimson in color. *Euphorbia* often exudes a white latex or sap when the stem is cut or bruised. When taking cuttings or harvesting as a cut flower, be sure to sear the cut end with a flame before working with the plant. Wear gloves to avoid skin irritation. *E. polychroma* is hardy in USDA Zones 4 to 8.

E. polychroma germinates in 8 to 15 days at 70F (21C), though without a pretreatment, the germination percentage often falls below 30 to 40%. If time permits, chill the seed for three to four weeks at 35 to 40F (2 to 4C), then sow it on moistened sand. Twenty to 25 days after sowing, seedlings can be transplanted into cell packs, or 3- to 4-inch (8- to 10-cm) pots using one to two plants per container. Green packs are salable eight to 11 weeks after sowing, but plants will not flower well the first summer after a midwinter sowing. For the best plants, finish sowing by April, so they will develop a good root ball by fall to survive northern winters. Overwinter pots and sell the following spring. For the easiest way to produce flowering pots the first year, buy in vernalized plugs.

E. variegata (also *E. marginata*) is an annual plant commonly called snow-on-the-mountain. The foliage is green and white on plants to 2½ feet (76 cm) tall. Like the preceding species, *E. variegata* is sold for its colorful foliage rather than its white flowers, which color up during midsummer in the garden. Plants are primarily used as fresh cut foliage, though they can be used for bedding plants or as pot plants, being especially nice in combination pots.

E. variegata germinates in 14 to 20 days at 68 to 70F (20 to 21C), and the seed should be covered during germination. For cut flowers, sowings made directly to the field will be ready nine to 12 weeks later. In the Midwest the last sowing is in late June, which will produce large, robust plants suitable for cutting by Labor Day.

EUPHORBIA PULCHERRIMA (POINSETTIA)

EUPHORBIA CULTURE
by Dr. P. Allen Hammer
Purdue University
West Lafayette, Indiana

EUPHORBIA PROBLEMS
by David Hartley, Ph.D., and Jack Williams
Paul Ecke Ranch
Encinitas, California

EUPHORBIA CULTIVARS
by Paul Ecke III
Paul Ecke Ranch
Encinitas, California

❖ *Annual (**Euphorbia pulcherrima**). Vegetatively propagated. Most growers purchase cuttings for stock plants for self-propagation in spring or cuttings for finished production from specialists in midsummer. Buying prefinished pots is increasing in popularity.*

❖ EUPHORBIA CULTURE

Poinsettia production starts with planning and not with panning (potting). The first step in the planning process is to determine when you want to have salable plants. From that date you determine the timing of every other process (fig. 1).

STOCK AND PROPAGATION

Poinsettia stock can be planted in March for a three-pinch program, April for a two-pinch program, or May for a one-pinch program. Each program has advantages and disadvantages. More cuttings are produced per plant with the earlier planting; however, saving early spring space for other production is often more valuable. You can also grow single-pinched stock plants in small containers, but they certainly require more greenhouse space.

Stock plants should receive adequate light and space for pinching and for harvest-

FIG. 1. *Ball RedBook* Editor Vic Ball (*center*) with Ellen (*left*) and P.J. (*right*) Ellison during Ellison's Greenhouses' open house in Brenham, Texas.

E

ing cuttings. Stock plants must be grown in long days. Provide incandescent lighting (10 footcandles, 108 lux) from 10 P.M. to 2 A.M. from planting until mid-May. Pinch plants two weeks after potting and then every four to six weeks to build up cuttings and maintain soft vegetative growth. Leave two to four nodes on the pinched shoot. The cultivar affects the number of cuttings from a stock plant as well as the schedule for pinching and cutting removal. Seek guidelines for specific cultivars and follow them. Guidelines are available from breeders and suppliers.

Snap or cut poinsettia cuttings, 2½ to 3 inches (6 to 8 cm) long, from stock plants with a sharp, clean knife. Disinfect the knife between stock plants. Take cuttings in the morning when plants are turgid, and place them in clean plastic bags or on clean newspaper. It's neither necessary nor desirable to remove leaves from the cuttings.

Treat cuttings with a rooting hormone to improve rooting uniformity and speed (fig. 2); stick them quickly and don't allow them to wilt. Poinsettia cuttings can be rooted in peat pellets (Jiffy 7, Jiffy 9), phenolic foam (Oasis), rock wool, or cells filled with root medium. You can also direct-stick cuttings in the finish container, but this method requires a great deal of bench space with mist. Whatever rooting medium or method you select, be sure the medium is free of disease at the start and that it remains disease-free throughout propagation.

Cuttings must be rooted under intermittent mist. Apply mist from sunrise to sunset at a frequency that keeps cutting leaves uniformly moist. Mist frequency may start with a 10-second cycle every four to six minutes for the first four days. Reduce mist frequency every four to five days as plants callus and root. After you stick cuttings, it's very important to arrange leaves so cutting terminals (growing points) aren't covered.

Temperature is very important during propagation. The minimum air temperature should be 70F (21C). Bottom heat should maintain a root medium temperature of 75 to 80F (24 to 27C). Cuttings should show visible roots in 21 days and be ready for potting in 28 days. Fertilize cuttings under mist beginning 14 days after sticking.

Fig. 2. Poinsettia time at Wolfe Wholesale Florist, Waco, Texas. The plant in the photo, grown for a nearby chain, is a 10-inch with five cuttings. On the left is Ben Selanders, manager, and on the right, Ron Null, president.

SCHEDULING

You must schedule your poinsettias to provide top-quality plants for the ever-lengthening poinsettia market cycle. The present market requires flowering plants from as early as the first week in November through the Christmas season. It's impossible to provide plants in prime condition without scheduling. Manipulating growing temperature should not be used to schedule poinsettias. Low temperatures, lower than 62F (17C) nights and 70F (21C) days, will delay flowering and reduce bract size. High temperatures, greater than 65F (18C) nights and 80F (27C) days, will speed flowering but reduce plant quality and fade bract color.

The average growing temperature is very important to poinsettia quality and timing. The daily average greenhouse temperature should be approximately 67 to 70F (19 to 21C). As an example, if plants were grown at 75F (24C) for 10 hours and 65F (18C) for 14 hours, the average growing temperature would be 69.2F (20.7C). Average daily temperature is calculated with the following formula:

$$\text{Avg daily temp} = \frac{(\text{day F} \times \text{hrs}) + (\text{night F} \times \text{hrs})}{24 \text{ hours}}$$

$$\text{Example avg} = \frac{(75\text{F} \times 10 \text{ hrs}) + (65\text{F} \times 14 \text{ hrs})}{24 \text{ hours}} = \frac{750 + 910}{24} = \frac{1{,}660}{24} = 69.2\text{F}$$

Cultivars can also be selected for early or late flowering; however, your customer may complain if cultivars are changed during the marketing period. Therefore, cultivar selection, although useful in specific cases, can't always be used to spread poinsettia flowering.

All the new cultivars and their differences, along with the need for prime plants over at least a four-week period, make poinsettia scheduling much more difficult. Different growth habits among cultivars add to the complexity. Along with this, add the differences in growing conditions between northern and southern production areas.

Poinsettias are photoperiodic, which means flowering is controlled by day length (or night length). They are classified as short-day plants, with a critical day length of approximately 11.5 hours. Poinsettias will flower when the nights are longer than 11.5 hours but will not flower (will remain vegetative) when the nights are shorter than 11.5 hours. Natural short days start in the United States around September 21 to 25. Grown under natural day lengths, poinsettias remain vegetative until September 21 and initiate flowers after September 21. We can artificially alter day length by covering plants with black cloth (to provide artificial long nights) or by adding lights (to provide artificial short nights). Plants are usually covered with black cloth from 5:00

P.M. until 8:00 A.M., which provides a 15-hour night or dark period. Lighting is best done in the middle of the night from 10:00 P.M. until 2:00 A.M., with a minimum of 10 f.c. (108 lux) of light at the plant canopy. This can be accomplished with 60-watt incandescent lamps 5 feet (152 cm) apart, 3 feet (91 cm) above the plants.

To schedule poinsettias, begin with the date you want the plant to flower (fig. 3). What is 'flowering' in poinsettias? Anthesis, or visible pollen on the first cyathia, should be your guide to a salable poinsettia. Plants without pollen will never develop their full potential in the postharvest environment, and plants held in the greenhouse after anthesis, even at cool temperatures, will show reduced postharvest quality and longevity.

From the target date for visible pollen, it is simply a matter of counting backwards and following a few simple guidelines. Once you select a flowering date, the next question is: When do short days start? The response group of the specific poinsettia cultivar gives the number of weeks from the start of short days until flowering. Generally, begin lighting on September 5 to prevent initiation and make sure plants remain vegetative.

Once we determine when short days start, the next information needed is the pinch date, or the number of days of growth required between the pinch and the

POINSETTIA CROP SCHEDULE

Cultivar: ___Freedom___

Response Group: ___Eight-weeks___

Growing temperature: ___65F___ Night   ___70-75F___ Day

Daily average temperature: ___69F___

Region: ___Midwest___

Tall, medium, or short: ___Short___
(Will you need growth regulator or extra vegetative growth?)

Activity	6 1/2 inches	6 1/2 inches
Sell date	Nov. 20	Dec. 11
8 wk response group Begin short days	Sept. 26	Oct. 16
No black cloth		
26 days' growth		
Pinch date	Aug. 31	Sept. 20
Lights on	Sept. 5	Sept. 5
14 days' growth Pot date	Aug. 17	Sept. 6

FIG. 3. Scheduling a poinsettia crop.

start of short days. This varies among the cultivars, depending on growth habit and the growing conditions.

Two to three weeks of growth from panning is needed to develop a good root system before pinching. The time between pinching and the start of short days (table 1)—which can vary from two to five weeks, depending on cultivar and growing conditions—is the most critical decision for overall crop quality. This period greatly affects final height and overall plant size. The maximum potential height and overall plant size are in some respects determined at the start of short days. We can control size, but few tools exist to increase size after the start of short days.

Growing conditions also vary between the different regions in the United States (figs. 4 and 5). Each grower will need to use some judgment in deciding on growing conditions in the specific greenhouse and climate. Once the pinch date is determined, the potting date is determined by adding 10 to 14 days of vegetative growth between pinching and potting. This time requirement can also vary among greenhouses, but roots should be visible at the sides of the pot before pinching.

Scheduling of single-stem poinsettia plants is similar, except the time between short days and potting is shorter.

A couple of things to remember in scheduling poinsettias: First, black cloth must be used if short days are necessary before September 21. Black

TABLE 1   MINIMUM DAYS FOR VEGETATIVE GROWTH

Pot size (inches)	Pinch to start of short days			Potting until short days[a]		
	Tall	Medium	Short	Tall	Medium	Short
South						
3	0	0	0	0	0	5
4	5	10	10	5	10	10
5–5½	9	14	19	9	14	19
6–6½	12	17	22	12	17	22
Midwest						
3	3	3	8	3	3	8
4	7	12	12	7	12	12
5–5½	12	17	22	12	17	22
6–6½	16	21	27	16	21	26
North						
3	5	5	10	5	5	10
4	9	14	14	9	14	14
5–5½	15	20	25	15	20	25
6–6½	22	27	32	22	27	32

[a] Single-stem plants

E

cloth treatment can also be discontinued after the first couple of weeks in October, as the natural day length will then provide short days. If short days are scheduled to start after September 21, then the plants should be lighted beginning September 5 until the start of short days. I recommend that all poinsettias be lighted beginning September 5 until the start of short days so that plants are precisely timed each year.

FIG. 4. Poinsettia time at City Floral, Denver, Colorado, with partners Dave Gross (*left*) and Ron Brady (*right*). The story: Dave's more open plant in a 6½-inch brought $25 at retail!

MEDIA AND FERTILIZATION

The root medium for poinsettias should be porous and well-drained, have a moderate nutrient content and a pH of 5.8 to 6.2, be free of insects and disease pests, and be easy to manage. You also want to select media that will be appropriate for the consumer. It shouldn't be so well-drained that in the home environment it's impossible to keep moist.

Poinsettias generally have been considered a crop requiring high fertility. Although the light leaf cultivars fit into this category, the newer dark leaf cultivars require lower nutrient levels (150 to 200 ppm nitrogen). In fact, high levels of fertility (300 ppm nitrogen) can reduce crop quality in the darker-leafed cultivars. My experience would suggest that you really need to fertilize the light and dark leaf cultivars differently and not compromise somewhere between high and low levels when growing both groups in the same greenhouse.

FIG. 5. Will Weatherford (*left*), Weatherford Farms & Greenhouses, Stafford, Texas, and son Jack are pictured with one of their outstanding poinsettias.

Micronutrients are important in poinsettia production. Poinsettias have a high requirement for molybdenum, which is generally added at each watering at 0.1 ppm stock: 1 ounce of ammonium or sodium molybdate per 40 fluid ounces (24 grams per liter) of water; application: 0.15 fluid ounce of stock solution per 100 gallons (12 ml per 1,000 l) of water.

Research in poinsettia production has also shown that the calcium-to-magnesium ratio is important for adequate calcium uptake. High levels of magnesium interfere with calcium uptake, while low levels of magnesium will cause magnesium deficiency, another common problem. The best ratio of calcium to magnesium for poinsettia production appears to be 2:1. Both calcium and magnesium should be monitored with root medium analysis because they're both very important macronutrients.

SPACING

Finished poinsettia quality is greatly affected by plant spacing during all phases of production. Greater spacing results in larger, better-proportioned plants. The grower generally must compromise on a spacing that provides adequate return for a given finished plant size and quality. Table 2 below provides spacing guidelines.

TABLE 2   POINSETTIA POT SPACING

Pot size		Plants	Pot-to-pot spacing		Area per pot	
inches	cm	per pot	inches	cm	ft²	cm²
Branched plants						
4	10	1	9	23	0.5	465
5	13	1	12	30	1.0	929
6–6.5	15–17	1	13–14	33–36	1.2	1,115
6–6.5	15–17	2	15	38	1.5	1,394
7	18	2	17	43	2.0	1,858
8	20	3	19	48	2.5	2,323
Single stem plants						
4	10	1	9	23	0.5	465
5	13	2	12	30	1.0	929
6–6.5	15–17	3	15	38	1.5	1,394
6–6.5	15–17	4	17	43	2.0	1,858
7	18	7	22–23	56–58	3.5	3,252
8	20	9	25–26	64–66	4.5	4,181

Source: Adapted from *The Poinsettia Manual*, 3rd ed., 1990. Encinitas, Calif.: Paul Ecke Ranch.

HEIGHT CONTROL

Growers generally apply chemical growth retardants to poinsettias to reduce height and tone the plants. Commonly used chemicals are ancymidol (A-Rest), daminozide (B-Nine SP), paclobutrazol (Bonzi), chlormequat (Cycocel), uniconzole (Sumagic), and a B-Nine SP–Cycocel tank mix. Because different cultivars respond differently to various chemicals, growers must adjust rates in their own production. General ranges are Cycocel 1,000 to 3,000 ppm, Bonzi 10 to 30 ppm, Sumagic 2 to 10 ppm, B-Nine SP–Cycocel tank mix of 2,500 ppm B-Nine SP and 1,500 ppm Cycocel, and A-Rest 0.25 to 0.5 mg a.i. drench.

Chemical growth retardants generally shouldn't be applied after the start of short days because bract size can be significantly reduced in the northern U.S. Research has shown, however, that very low concentrations (1 to 2 ppm) of Bonzi or A-Rest can be applied as a drench (table 3) as late as early November for height control without significantly reducing bract size. This approach is useful to correct height problems late in the production cycle and to avoid late stretch.

DIF is also used to manipulate poinsettia height in areas with cool day temperatures. Positive DIF (warm days and cool nights) increases stem elongation, and negative DIF (cool days and warm nights) decreases stem elongation. Computer software is available for using DIF as a tool to control plant height. It's very important not to change average daily temperature when using DIF, and day and night temperatures need to be adjusted to maintain the same average daily temperature. Average growing temperature can affect timing and bract size.

TABLE 3   LATE DRENCHES OF GROWTH REGULATORS

Chemical	Concentration	
	ppm	fl oz/gal[b]
A-Rest	1	0.485
	2	0.970
Bonzi[a]	1	0.032
	2	0.064

Note: Use higher rates in the South, lower in the North.
[a] These ppm figures are based on a drench of 4 fluid ounces of solution per 6-inch pot. 6 inches = 15.24 cm; 1 fl oz = 29.573 milliliters.
[b] There are 128 fl oz in a gallon; 1 gal = 3.785 liters.

❖ EUPHORBIA PROBLEMS

PHYSIOLOGICAL DISORDERS

Leaf distortion. For many years leaf deformity has been seen in some stock plants and often on pot plants in greenhouses. The symptoms are extremely variable. In some cases damage has occurred only at the tip of the immature leaf, which will give the appearance of having been chopped off at a later stage of development. Where the entire margin of the leaf has been affected in earlier stages, later growth of all except the margin causes a puckered appearance, as if a drawstring around the leaf margin had been pulled tight.

On Christmas season plants, leaf distortion frequently occurs in late September and early October after the plants have been moved from propagation to the finishing area. Branches that develop after pinching may have two to three misshapen and distorted leaves. In most instances these leaves remain distorted, but green, throughout the forcing period. Leaves that expand later are usually normal and hide the damaged leaves by market time.

The causes of leaf distortion are not well understood. It seems that when cells in very young leaf tissues are ruptured or killed, the leaf becomes misshapen as it expands. Drying of tissue, burn from fertilizer, nutrient deficiency, and chemical burn have all been suspected of damaging young leaf cells. Plants under stress from bright light, extremely warm temperatures, or moving air often have more leaf distortion. It is helpful to provide shade and syringe (mist) the foliage until roots are well established and the side branches begin to develop. Leaf distortion may also result from insects, like thrips, feeding on the young leaf tissues. As leaves mature, damage from the pest becomes apparent and prevents normal expansion.

Many plants, including poinsettias, have leaf structures that include hydathodes, or vein endings opening along the edges, tips, and sometimes leaf surfaces. Under cool, humid conditions, with ample growing medium moisture supply and elevated growing medium temperature, high fluid pressure in the conducting system may occur. If a rapid rise in temperature and drop in humidity occur simultaneously, as frequently does happen in the mornings of bright days, dissolved contents will become more concentrated. Sudden use of air conditioning fans or natural movement of air from wind can cause the same effect. This concentrated solution may be strong enough to cause cell damage, and when sudden stress on the plant occurs simultaneously, the concentrated fluid may be drawn back into the vein endings and cause damage to cells in and around the area. Since the phenomenon occurs only on immature leaves still undergoing expansion, subsequent growth in areas of cell injury will be inhibited, and developing leaves will be distorted.

Control of such leaf-edge damage can best be achieved by maintaining low humidity at night and avoiding conditions of rapid drying in the morning. Syringing of foliage in the early morning may also help by slowing transpiration. A complicating factor is frequently that of infection of injured tissue by *Botrytis*.

Growers have also experienced leaf dehydration of dark leaf cultivars at various times in production. This damage is usually evident on expanded, mature foliage. Symptoms begin with darkening of the leaf tissue, followed by one edge of a leaf rolling up. The affected tissue eventually dries with no further progression of symptoms. Damage is usually isolated to a few leaves per plant, and does not continue beyond this.

This damage is most likely a result of sudden changes in the environment. Conditions typically noted prior to leaf dehydration include extended cloudy weather followed by bright, warm days. Leaf dehydration is most evident on plants that are well fertilized or where the growing medium is allowed to dry. The best prevention for this disorder is to maintain uniform soil moisture with low soluble salts. Whenever rapid changes in growing conditions are noted, mitigate the conditions by syringing plant foliage.

Bract edge burn. A condition or disorder that affects blooming poinsettias, bract edge burn (BEB), first appears as small, brown necrotic spots at the tips or along the edges of mature bracts. As the condition progresses, entire bract margins may die and turn brown, giving a burned appearance. This injured tissue is also ideal for *Botrytis* to establish on and cause further damage to the plant.

Bract edge burn is not a new problem. It has been the focus of study in the late 1970s and early 1980s by researchers at the University of Florida and more recently at other research stations and universities. This ongoing research has provided us with insight into the disorder, but no one has solved why it occurs. Nutrition and environment seem to be the most important factors contributing to the occurrence of BEB, but we also know that pesticide applications have a significant influence on the likelihood of this disorder being experienced.

Severe bract burn has also been encountered where extreme rates of fertilizer have been used. Under these conditions the leaves may show no damage. One theory is that during growth there is a diluting effect of plant-absorbed fertilizer, but at flowering, new tissue development has virtually ceased, and the fertilizer salts accumulate in the youngest mature and most sensitive tissue, the bract. This accumulation causes cell damage, usually starting on the bract edges. With slow-release fertilizers, the usage rate should be modest, and application should be early enough to ensure almost complete depletion at time of flowering.

Calcium seems to be the most important nutrient associated with bract edge burn. Plants use calcium to build strong cell walls. Without strong cell walls, plant tissues are vulnerable to damage from soluble salts, drying, sunburn, or invasion by diseases or insects. The young, developing bract margin cells are the kind of tissue most likely to be damaged under any of these stress conditions. Providing sufficient calcium levels is critical to protecting these young cells from damage. Calcium is moved from the root solution through the plant in the water stream. Any condition that restricts water movement in the plant will also limit the uptake of calcium required for strong bract cell walls.

Factors that affect the uptake of water and calcium include loss of roots to disease or burn, high relative humidity and poor air circulation, which limit transpiration, and low soil temperatures. Among soil chemistry factors are high soluble salts in the growing medium, nutrient imbalances that favor the uptake of elements other than calcium, low pH in the soil, affecting the availability of calcium for uptake, and low calcium availability in the soil as a result of fertilization programs.

Monitor calcium levels in the soil and tissue throughout the crop. If deficiencies are noted prior to or during bract formation, foliar applications of calcium can be beneficial in preventing BEB. Use of laboratory- or technical-grade calcium chloride at 300 ppm may provide the calcium required to overcome this problem.

Efforts to understand more about bract edge burn have brought to light the influence pesticide applications may have on this disorder. Although there is no one pesticide that will absolutely cause BEB, there is concern about the use of Thiodan early in the production cycle. Studies performed in Canada identified a consistent pattern of damage for plants that were treated at or near the time of flower initiation. Later applications of Thiodan did not appear to cause damage to the bracts. With the introduction of chemicals providing long-term systemic control of whiteflies, such as Marathon, growers do not have to treat crops frequently with more traditional pesticides. A reduction in the occurrence of bract edge burn has been noted as a result of fewer spray applications.

Another factor that influences the occurrence of bract edge burn is the age or maturity of the bracts. The disorder is more likely to happen on plants that are being held for sale in the greenhouse. The conditions used to hold a crop are the same conditions that limit water movement in the plant. As plants sit waiting to be sold, BEB, *Botrytis*, root rot, or cyathia abscission is likely to occur. To help avoid these conditions, schedule crops for precise timing and improved shipping quality.

To reduce the incidence of bract necrosis, the following practices should be observed.

- ◆ Avoid high fertility rates and heavy watering practices during the final four weeks of the production period. It may be advisable to discontinue fertilization and use water only, beginning two weeks before the flowering plants are to be sold.

E

- Avoid the use of fertilizers which contain 50% or more of their nitrogen in the ammoniacal form. Fertilizers which contain mostly nitrate nitrogen are readily available or are easily formulated.

- Avoid using high rates of slow-release fertilizers that maintain high fertility levels late into the production period. Second applications of slow-release fertilizers late in the production period should be especially avoided.

- Modify the greenhouse environment to reduce humidity and increase air circulation throughout production. Chemical disease-control methods may be used to minimize the spread of *Botrytis* on the bracts.

- Use calcium chloride as a foliar spray on developing poinsettia bracts whenever tissue and soil analysis indicate calcium deficiency in the crop.

- Produce cultivars that are less sensitive to bract edge burn.

Premature cyathia drop. During some Christmas flowering seasons, the true flowers, or cyathia, may drop from the center of the bract presentation before the flowers reach maturity. This may occur before the plants are ready for market, particularly in northern climates with low light conditions. Because this detracts from the appearance of the poinsettias and makes them appear to be overly mature, it also reduces their economic value.

A study at Michigan State University determined that premature cyathia drop is caused by low light levels or high forcing temperatures. Water stress exacerbates the problem. These conditions allow the food reserves of the plant to become depleted. As the food reserves become low, the plant reacts by dropping the cyathia.

Low light levels may result from dark or cloudy weather conditions or simply from spacing the plants too tightly on the bench. It is not uncommon for skies to become overcast during late October and November in northern states. With the cloudy weather comes lower light levels and cooler temperatures. These are not ideal conditions for high-quality poinsettias. With lower light levels, plants do not make as much food. Also, as poinsettia bracts develop, they shade the green leaves below and further restrict the amount of light available for photosynthesis.

If poinsettias are grown at lower than optimum temperatures during the early part of the production period, flower development may be delayed when the cloudy weather begins. To speed up the rate of flower development, it then becomes necessary to raise greenhouse temperatures. As temperatures increase, the plants food reserves are used at a faster rate.

To produce high-quality poinsettias and reduce the possibility of premature cyathia drop, it is important to follow the old adage, "Make hay while the sun shines." Especially in northern areas, it is important to maintain optimum greenhouse temperatures for rapid poinsettia development during the early part of the production period. When cloudy weather begins, it may then be possible to start lowering greenhouse temperatures and slowing the depletion of food reserves.

The Michigan State University research also demonstrated that if the growing medium is allowed to dry to the point that the poinsettias begin to wilt after the time the flower buds become visible, the chances of premature cyathia drop become greater.

To lessen the possibility of premature cyathia drop, the following cultural guidelines are suggested.

- Schedule your poinsettia program early enough so that the plants can do most of their "growing" early in the fall, while good light intensities are available.

- Do not attempt to grow a poinsettia crop at lower than optimum temperatures during the early part of the production period in an attempt to save energy and reduce fuel costs.

- Do not allow the growing medium to become excessively dry. This only hastens the start of premature cyathia drop.

- Grow your crop under a clear greenhouse cover to admit as much light as possible during October and November.

- If the crop develops properly it should be possible to reduce temperatures late in the production period, when light levels are low, thus preserving part of the plants' food supply.

Latex eruption. Plants belonging to the Euphorbia family contain latex, which is exuded upon cell injury. This became a problem in poinsettia production when the variety Paul Mikkelsen and its sports first became popular. The malady is sometimes termed *crud*. The mechanism is one of bursting cells resulting from high turgor pressure, with latex spilling over the tissue and, upon drying, creating a growth-restricting layer. When this occurs at developing stem tips, distortion or stunting of growth results. The exuding of latex has also been observed on fully expanded leaves, sometimes giving the appearance of mealybug infestation due to the white splotches scattered over the leaf surfaces.

All contributing factors have not been clearly defined, but several obvious ones include high moisture availability and high humidity, both of which result in high fluid pressure within the cells. Low temperature is an important contributing factor. Sudden lowering of temperature can trigger the reaction, but, fortunately, most varieties are not highly sensitive to this problem. Mechanical injury from rough handling or from excessively vigorous air movement may also increase injury to cells. High rates of photosynthesis may contribute by building up a high osmotic pressure in cells from carbohydrate accumulation.

Control is best attained by using a growing medium which dries out in a reasonable length of time. Also avoid extremes of high humidity, particularly during the night. Moderate shading in extremely bright weather might also be helpful.

Stem splitting. Under certain conditions poinsettias will suddenly produce stem branches at the growing tip. Careful examination will reveal that the true stem tip has stopped growth or aborted. This phenomenon is known as splitting.

Splitting is actually the first step in flower initiation. The stimulus to flower increases with the age of stems, exposure to cold temperatures, and the lengthening of nights. Even with short nights and normal growing temperatures in the 60 to 70F (16 to 21C) range, splitting can be expected if the stem is permitted to grow until 20 or more leaves are present. Some cultivars, such as Lilo, are more sensitive to splitting that results from stem maturity, splitting at leaf counts as low as 12.

Keeping stock plants pinched back on a regular basis will help prevent splitting of cuttings harvested. Stem tips that are continuously propagated carry an increasing tendency to flower. To ensure against this, lights should be supplied to stock plants until May 15. Plants propagated prior to July 15 should be grown as multiflowered or branched plants *only* with tips discarded.

Another cause of splitting can be cold temperature exposure. Lowering night temperatures to around 60F (16C) was an old trick used to help initiate flowering at the end of September with poinsettia cultivars of the 1950s and 1960s. Growers have experienced early flower initiation and splitting with today's cultivars, however, when growing temperatures become cool and heaters are not yet operating. Avoid greenhouse temperatures that drop below 60F whenever plants should be maintained in a vegetative state.

Stems heavily shaded by a canopy of higher foliage may experience such a reduction in light as to split even in periods when day length would be considered adequately long to keep apexes vegetative. If daytime conditions are unusually cloudy, plants may not receive enough total light stimulus to maintain vegetative growth. During weather patterns like this or under fluctuating day length, it is prudent to use night lighting to prevent premature bud set.

Bilateral bract spots. Often called "rabbit tracks," this condition is characterized by breakdown of the tissues between the veins located on either side of the midrib. It occurs in late November and early December during the flowering process. Although the plant is not killed, the condition can effectively lower the quality of the plant, rendering it unsalable.

Bilateral bract spots seem to occur under a wide temperature range, in various types of houses, under both gas and oil heat, and on about eight different varieties. The condition might be confined to a few plants or found throughout the greenhouse.

A study was conducted in West Germany to consider certain factors that may predispose poinsettias to bilateral bract spots. The most obviously susceptible varieties were of the Annette Hegg family, more so than most other commercial varieties. Cultural and environmental factors had their greatest influence during the bract development stage. In this study incidence of bilateral bract spots was associated with high relative humidity or changing humidity levels. High temperatures above 70F (21C), especially high night temperatures during the bract development phase, caused a greater incidence of bilateral bract spots. And high levels of nitrogen fertilization near the end of the crop, or high nitrogen content in the plants, caused a higher frequency of bilateral bract spots.

Leaf drop. The older varieties were much more prone to sudden loss of leaves than are modern varieties. There are several indirect causes of leaf drop. Under conditions of moderate to severe stress, it is not uncommon for older leaves to form an abscission layer at the juncture of the petiole and the supporting stem. It is believed this is due to loss of auxin from the leaf blade under stress conditions. Once started, the reaction is irreversible, and the leaf petiole is virtually severed from the stem. Also, when plants are kept under very low light intensity for a period of several days, lower leaves will turn yellow and drop.

Before better sanitation procedures reduced or eliminated disease problems, leaves of the older varieties would frequently drop in the greenhouse as root disease reduced the plant's ability to supply water to the top. A parallel contributing factor was the deliberate attempt by growers to keep the growing medium dry in order to restrict disease organism activity. Even with healthy roots, many of the cultivars would drop leaves within a day or two after being moved from the humid glasshouse to a warm, dry home or office. The change in environment caused more water stress than the leaves could tolerate. The moisture loss exceeded the ability of the roots to supply water.

Modern varieties are far more resistant to leaf abscission, though not completely immune. Modern methods of sanitation should make it unnecessary to impose dry growing medium conditions in the greenhouse or the home. Healthy poinsettias thrive with high moisture availability and moderate to high light intensities. Waterlogging should be avoided, however.

INSECTS

Poinsettias are subject to attack by various insect pests under greenhouse conditions. Whiteflies, fungus gnats, thrips, and spider mites are the most prevalent pests of poinsettias, although other pests may occasionally cause problems.

E

One of the first lines of defense against insect pests is prevention of their entrance into the greenhouse. With increasing regulation and restriction of chemical pesticide use, screening all vents and doors can be an effective, economical means of excluding insects from greenhouses. Sanitation and cleanliness are also of utmost importance in an effective control program. Weeds and ground covers in and around greenhouses provide favorable locations for pests, which may easily move onto cultivated plants. Growers with the fewest insect problems are usually those with the cleanest operations.

Another essential element of pest management is an effective scouting and monitoring program. Frequent and thorough inspections of plants for the location and identification of insect pests may help prevent an infestation from becoming an unmanageable epidemic. Yellow sticky insect traps are an effective tool for monitoring insect populations.

Biological control is becoming an important pest management option as an alternative to chemical pesticides. The use of parasites, predators, and diseases of insect pests is receiving more attention and appears more promising. Effective biological control of whiteflies, fungus gnats, and spider mites on poinsettias may soon become a reality.

Pesticides, although an effective and necessary part of pest management, continue to become more restricted and expensive. This situation increases the importance of sanitation, prevention, scouting, monitoring, and biocontrol in any insect management program. Highly effective, insect growth regulating pesticides and systemic pesticides introduced in the 1990s have provided growers with effective and long lasting chemical pest control options making crop health management easier to accomplish. Good chemical management and rotation should be used to ensure longevity of these tools. Pesticide registrations may vary from state to state. It is the grower's responsibility to read and follow the label rates approved in his or her state.

MICROORGANISMS

Pathogens of primary importance include fungi and bacteria. For disease to occur, the organism and the host plant must be in close proximity. Fungi infect plants through wounds, natural openings, such as stomates, and intact epidermal surfaces. Bacteria infect primarily through wounds or natural openings, including stomates, lenticels, nectaries, hydathodes, and glandular hairs. Under favorable conditions, wounded tissue is quickly covered by a suberin film, which protects against bacterial infection.

Disease control can be attained only by using clean plants, clean growing media, and complete sanitation and by providing an appropriate environment. All other procedures must be considered as suppression, not control! The use of chemicals anticipates that the control measures will not be, or have not been, properly executed.

The diseases described include pertinent information on the ecology of the pathogens. This background often provides the most important basis for planning control measures and preventing infection.

Where chemicals are to be used, limited trials should be employed before treating an entire crop, unless there has been adequate prior experience.

***Botrytis cinerea* (gray mold).** Plant symptoms: Tissue rots, frequently starting on young leaf edges or other immature tissue. Sometimes damping-off symptoms appear at or near the soil line. Red varieties develop purplish color on infected bracts. When bracts are affected, it is difficult to distinguish from edge burn due to chemicals or salts.

Organism characteristics: This fungus has airborne spores which can be assumed to be present everywhere at all times. It is not an aggressive parasite unless favored by injured, aging, or succulent tissue, moderately low temperature, and 100% humidity at the site of infection. It thrives on plant debris on the floor of greenhouse.

Control: The first line of defense is control of the environment. Avoid physical injury to plants, maintain air circulation at night, use night heat plus ventilation to lower the humidity, and keep temperatures above 60F (16C) if at all possible. Remove all dead plant material. The dense habit of multiflowered types presents a special problem of leaf and bract overlap.

Suppression: Numerous fungicides are effective as inhibitors of germination of spores and of growth of mycelia. New, developing plant tissue must be repeatedly covered to provide continuous protection. Materials that leave no residue are preferred to maintain salability. Fungicides employed include Benefit F, Chipco 26019 50 WP, copper complex (Phyton 27 21.36%), Cleary's 3336 F or 50 WP, Daconil 2787 & Ornamental F, Domain FL or 50 WP, Exotherm Termil fumigator, Fungo Flo F, Ornalin FL, or Zyban WP. Exotherm Termil programs have been widely used on poinsettias; however, special care should be taken not to apply it on the cultivar Red Sails or to mature bracts of Freedom and Pepride.

***Corynebacterium poinsettia* (bacterial canker).** Plant symptoms: Black, elongated, watersoaked streaks occur on green stems. Stem tips abort or bend over. Spots or blotches occur on leaves. In a favorable (warm, humid) environment, disease progresses rapidly, resulting in death of the stem above infection or of the entire plant. This is not a common disease except during hot, humid weather, such as found in the summer climates of the midwestern, eastern, and southern United States. It has shown up in other areas where inocula were present and the environment favorable.

Organism characteristics: This bacterium is transported in water, in soil, on contaminated tools, and on the hands of workers. Entering a plant through stomates or wounds, it spreads in the plant through thin-walled parenchyma cells.

Suppression: Severe roguing should be practiced, and avoid all overhead irrigation or syringing. Humidity should be kept as low as practical, and excessive temperatures should be avoided. Plants should be protected from wind or rain. If stock plant infection is suspected, sterile knives should be used in removing each cutting to avoid spread.

Erwinia carotovora **(bacterial soft rot).** Plant symptoms: Bacterial soft rot occurs primarily in propagation. Cuttings develop a soft, mushy rot beginning at the basal end within three to five days of sticking.

Organism characteristics: The bacterium, prevalent on dead plant material, can be carried on windblown dust, nonsterilized tools, and the hands of workers. It spreads readily in water and may be found in pond water. Wounded tissue, waterlogging of the rooting medium, high temperatures, and other factors that stress the cuttings favor this organism.

Suppression: Grow stock plants under cover or other controlled environment. Use good sanitation practices throughout the harvest and propagation of cuttings. Avoid waterlogging of the rooting medium. Keep temperatures below 90F (32C) in propagation. Avoid, as much as possible, stressing the cuttings. Spray applications of copper-containing fungicides, such as Phyton 27 21.36%, have proven beneficial when mother plants are treated in advance of the cutting harvest.

Oidium **spp. (powdery mildew).** Plant symptoms: The fungus may begin growth on the undersides of leaves, resulting in chlorotic patches on the upper surface. During early stages of infection, leaves or bracts develop spots that resemble pesticide residue. Colonies of powdery mildew grow rapidly and have a characteristic white, powdery appearance on the plant surface. Tissue infected with powdery mildew can become necrotic.

Organism characteristics: *Oidium* is a fungus whose spores are easily spread by physical movement and air. The disease can occur at any phase of poinsettia production, but it is most likely to develop in the spring or late fall, when environmental conditions are most favorable. Powdery mildew generally develops faster in cooler weather, when there is a greater fluctuation between day and night temperatures. High humidity is a necessity for development of the fungus.
Suppression: This problem can be managed through good environmental control in the greenhouse and through periodic fungicide applications. Chemicals that may be applied to suppress powdery mildew include Benefit F, Cleary's 3336 F or 50 WP, copper complex (Phyton 27 21.36%), Domain FL or 50 WP, Fungo Flo F, SysTec 1998 F or WDG, Strike 25 WP, Terraguard 50 W, or Zyban WP.

Pythium **spp. (water mold root rot).** Plant symptoms: Root tips and cortex are rotted, a condition which may advance up the stem. The plant is stunted, lower leaves yellow and drop, and the entire plant may collapse. The growing medium tends to stay wet, since roots are incapable of removing moisture, leading to the erroneous diagnosis of too much water.

Organism characteristics: It carries over in the growing medium or infected plants and is spread in water, with no airborne spores. It requires high moisture availability and is active at cool temperatures. Inactive spores may live in dry growing medium for several months.

Suppression: Rogue obviously infected plants, taking care not to spread debris to healthy plant areas. Maintain low moisture in the growing medium. Drench with such fungicides as Banol 66.5%, Banrot 40W, Chipco Aliette 80 WDG, Subdue II WSP, Terrazole 35 WP, or Truban 25 EC or 30 WP.

Phytophthora parasitica (Phytophthora crown and stem rot). Plant symptoms: This fungus is closely related to *Pythium,* but the pattern of symptoms that develops on the plant is quite different. Poinsettias infected by *Phytophthora* may have no root rot at all. A characteristic sign is a brown canker about three-quarters of an inch (2 cm) long just above the soil line. The canker often shows a black rim around it. Under more humid conditions, gray, wet lesions develop at the soil line. As the disease progresses the affected stem or the entire plant may wilt and die.

Organism characteristics: *Phytophthora* is an organism which historically has not been as prevalent or damaging as *Pythium* root rot or *Rhizoctonia* stem rot. *Phytophthora* crown and stem rot is caused by a water mold, as is pythium root and stem rot. The organism is able to invade tissue very rapidly through wounds. This fungus often attacks plants at the soil line, where optimum levels of humidity exist. A lesion or brown canker is formed just above the soil line. Additionally, a black streak may run up the stem from the canker. The shoots above the stem discoloration eventually become brown; stems may become extensively brown and shrivel. Any or all of these symptoms may be on a given plant. The disease affects the vascular system, so wilting may precede the externally visible black discoloration.

Suppression: This problem can be overcome with strict sanitation. All plants with symptoms must be discarded, and you must not handle healthy plants after touching diseased plants. Splashing during watering is very likely to spread the contamination. This organism can be carried over in soil, and contaminated soil or pots should be disinfected before reusing. Chemicals that suppress other water molds, such as pythium root rot, are also usually effective against *Phytophthora.* Apply such fungicides as Banol 66.5%, Banrot 40W, Chipco Aliette 80 WDG, Subdue II WSP, Terrazole 35 WP, or Truban 25 EC or 30 WP.

Rhizoctonia solani (stem and root rot). Plant symptoms: There is brown rot of the stem at the soil line, roots may have brown lesions, and leaves can become infected under mist propagation where they touch soil. Infected plants are stunted, with leaves yellowing from the bottom and sometimes dropping. The disease may progress up the stem as well as down into the roots. Complete plant collapse can occur under severe conditions.

Organism characteristics: This fungus carries over in the growing medium or on infected plants. It is easily spread by water, but there are no airborne spores. It is favored by moderately high available moisture, high temperature, and factors which weaken the host, such as salinity.

Suppression: Rogue infected plants and avoid scattering debris from infected plants. Drench with such fungicides as Banrot 40W, Chipco 26019 50WP, Cleary's 3336F or 50 WP, Defend 2F or 75WP, Domain FL or 50 WP, Fungo Flo F, SysTec 1998 F or WDG, Terraclor 75 WP or 400 F, or Terraguard 50W.

Rhizopus spp. (*Rhizopus* rot).

Plant symptoms: Poinsettia plants growing under high humidity, high temperatures (80 to 90F, 27 to 32C), and poor aeration are subject to a soft, wet rot of foliage and stems caused by *Rhizopus*. Cuttings in propagation during hot weather are attacked, especially when they are placed too close together. The stems, leaves, or leaf petioles become very soft, brown, and mushy. When *Rhizopus* attacks the stems of poinsettia cuttings, the resulting rot can resemble bacterial soft rot.

Organism characteristics: The spores can be carried by air currents, and the organism can live over in plant debris. It requires high temperatures (80 to 90F), high humidity, and wounded or weakened host tissue for activity. It grows rapidly, forming abundant and visible surface mycelia. The mode of attack is similar to that of bacterial soft rot, with an enzyme being released to cause cell deterioration.

Suppression: Improving environmental conditions for the plants or cuttings, such as lowering the temperature and humidity, should help control this relatively uncommon but potentially destructive disease. Suppression is best attained by sanitation, careful handling of the cuttings to avoid injury, and possibly applying a fungicide such as Zyban 75 WP.

Thielaviopsis basicola (black root rot).

Plant symptoms: Roots develop black, rotted areas. The stem may accumulate black sclerotia, which form in the pith area. Plants show lack of vigor, leaf yellowing, leaf drop, and sometimes sudden collapse, particularly after temperatures have been lowered below 60F (16C).

Organism characteristics: This fungus has a long life in the growing medium in its sclerotia resting stage. It is favored by a cool, moist environment, its growth slowed at elevated temperatures and in an acidic growing medium (pH below 5.5). There are no airborne spores.

Suppression: Rogue infected plants, avoid low temperatures, and use acid growing media and acidifying fertilizers. Drench with such fungicides as Banrot 40W, Cleary's 3336 F or 50 WP, Domain FL or 50 WP, or Fungo Flo F.

PRACTICE SANITATION

Good sanitation practices are essential for avoiding or minimizing disease problems. As a first defense against poinsettia diseases, always think clean.

Disinfect hands, knives, and other equipment before handling plants. Use copper naphthenate on all wood, metal, or composition surfaces. Presteam all soil or sand benches. Steam or fumigate all growing media or use a clean, ready-to-use, soilless mix.

Avoid inoculation from dust. Keep plants high enough off the ground or the soil mulch so that splashing water will not come in contact with pots. Keep feet off benches. Keep hose ends off the floor. Never use cuttings which have fallen on the floor. Disinfect tools which have fallen on the ground before reusing them.

Remove soil and plant debris from tools, pots, and benches before disinfecting them. Rogue diseased leaves and plants and remove them from the greenhouse. Eliminate weeds and debris both inside and outside the greenhouse, as they can harbor disease and insects.

POSTHARVEST

As a poinsettia grower, you have a responsibility for the postproduction longevity of your plants. You can't afford just to be happy to have the plant out of the greenhouse door. Happy poinsettia consumers make for repeat customers.

The grower's responsibility in postharvest care is to provide the best poinsettia possible at the proper stage of development, free of insects and diseases, and low in soluble salts (from fertilizer). Don't sell a poinsettia before its time! Research has clearly shown that poinsettias have a much longer and better display life when sold at visible pollen. Young, underdeveloped bracts will never develop good color in the postharvest environment. Pink bracts on red cultivars result when plants leave the production environment too early.

Fertilizer salts should be reduced before selling, but it's not necessary or desirable to completely eliminate fertilizer application. Simply reduce the rate to half or a quarter of the rate used during early production. Poinsettias can also receive chilling injury when exposed to 50F (10C) or lower for as little as two hours. Chilling injury can cause epinasty as well as leaf loss in the most severe cases. Transporting plants in unheated trucks in the North and uncooled trucks in the South can significantly reduce poinsettia quality. A general guideline is that poinsettias do much better when placed in a uniform, nonstressful environment at 60 to 65F (16 to 18C).

As a poinsettia producer and marketer, you can do a great deal to maintain the potential postharvest beauty and longevity built into the modern poinsettia cultivars. At the same time, you can reduce the postharvest life of the poinsettias if you mishandle them. It's extremely important that we all do our best to give the consumer the best possible poinsettia. Plant abuse anywhere in the marketing chain will ultimately show up in the final consumer setting, whether it's a hotel lobby, mall display, or home living room. I like the concept that each poinsettia plant should be handled as if it were the one you're taking home for Christmas. Every poinsettia consumer deserves a plant receiving such treatment.

GREENHOUSE GROWER INSTRUCTIONS TO RETAILERS

Upon receiving plants, unpack and unsleeve them immediately. Poinsettias left in the sleeve become droopy. This epinasty is caused by ethylene production from the sleeving process. The longer poinsettias are sleeved and the higher the temperature above 65F (18C), the greater the droopiness problem. The plants generally recover from epinasty in a couple of days when placed in a lighted area at 65 to 75F (18 to 24C) if the sleeving period was longer than a couple of days.

Place poinsettias in bright light at 60 to 65F (16 to 18C). They should not receive direct sunlight under postharvest conditions. Plants should also be out of hot or cold drafts. A heat duct or outside door shouldn't expose plants to sudden changes in temperature. At no time should poinsettias be stored in a garage area.

Poinsettias are fragile. Rough handling will bruise bracts and cause stem and leaf breakage. Poinsettia plants can't be handled like hard goods. Also, be sure to provide adequate spacing in the display area. Plants shouldn't be spaced so close together that the bracts from one plant rub against the bracts of an adjacent plant.

❖ EUPHORBIA CULTIVARS

RED CULTIVARS

Eckespoint Freedom. Eight-week response with dark green foliage. Freedom initiates flowers earlier than other cultivars and is in full flower from November 15 to 20. Freedom has dark red bracts and a short to medium growing height. Freedom, the most popular poinsettia cultivar in North America, comes in a full range of colors.

Mikkel Donner. Eight-week response with green foliage. Donner is a deep red cultivar with large bracts. More vigorous than Dasher, Donner branches well from a pinch. It does not heat

FIG. 6. Paul Ecke III, Paul Ecke Ranch, Encinitas, California, is the author of the poinsettia variety section.

delay or have a problem with splitting of the bracts, and it exhibits no epinasty when boxed for up to five days.

Brightpoints Nutcracker Red. Eight-week response with green foliage. Nutcracker is free branching, with scarlet red bracts. Nutcracker red can be flowered in warmer temperatures, which makes it a good choice for the southern United States.

Eckespoint Lilo. Eight-and-a-half-week-response with dark green foliage. Lilo is very popular in lower light areas, like the Pacific Northwest. Coming in a full range of colors, Lilo has bright ruby red bracts and a medium growing height. Lilo is an extremely long-lasting cultivar, ideal for commercial interiorscapes and mass markets.

Eckespoint Celebrate 2. Eight-and-a-half-week response with green foliage. In full flower in late November, Celebrate 2 has bright red erect bracts that are resistant to drooping and a medium growing height. Although it is free branching, it also makes an excellent single-stem plant with three to four plants per pot. Celebrate 2 also comes in white and pink.

Gross SUPJIBI. Eight-and-a-half-week response with green foliage. SUPJIBI is in full flower in late November. SUPJIBI has large, red, fleshy bracts that have a velvety appearance and a short to medium growing height. SUPJIBI is tolerant of warm temperatures that might delay flowering of other cultivars (fig. 7).

Mikkel Dasher. Eight-and-a-half-week response with green foliage. Dasher is a compact red cultivar with intermediate bract size. It does not heat delay or have a problem with splitting, and it exhibits no epinasty when boxed for up to five days. Dasher branches well.

FIG. 7. Here's SUPJIBI with its creator Eduard Gross of Establissement Horticole, Blanzak, France.

Pelfi Cortez. Eight-and-a-half-week response with dark green foliage. The dark red bracts and dark green foliage give it a colorful contrast. Cortez branches freely and is a vigorous grower.

Peter Jacobsen's Petoy. Eight-and-a-half-week response with green foliage. An improved SUPJIBI, it has smooth, horizontal bracts with a bright red color. Petoy grows more uniformly than SUPJIBI and has good postproduction performance.

Peter Jacobsen's Peterstar. Eight-and-a-half-week response with green foliage. This intense red mutation from Angelika is perhaps the most freely branching poinsettia cultivar on the market. It is earlier flowering and requires less growth regulator than Angelika V-17. Good in all climates, Peterstar is the most popular variety in Europe.

Eckespoint Red Sails. Nine-week response with dark green foliage. Red Sails is in full flower in late November to early December. Red Sails has large, dark red bracts and a medium growing height. It is suitable as either a branched plant or a single-stem plant. Its postproduction qualities are not as good as several other cultivars, however.

Gutbier V-17 Angelika. Nine-week response and green foliage. Angelika reaches maturity in late November or early December. Angelika has bright red bracts, a medium growing height, and is free branching. Its postproduction qualities are not as good as several other cultivars. Angelika V-17 comes in a complete range of colors.

Annette Hegg Dark Red. Nine-week flowering response with green foliage. Dark Red Hegg is in full flower in late November to early December. It has dark red bracts and a medium to tall growing height. First introduced in the 1970s, this cultivar is being replaced by the newer generation of cultivars and represents only a small percentage of sales in North America.

Peace Jolly Red. Nine-week response and dark green foliage. Jolly Red has rich, deep red bracts. It is free branching and performs best as a pinched plant. Cyathia size is dependent on growing temperatures; however, they tend to be small. Leaves and bracts do not droop after shipping and sleeving.

Brightpoints Dyanasty Red (Dynasty II). Nine-week response with dark green foliage. Dynasty has large, dark red bracts that are held upright. Dynasty is suitable for a wide variety of growing applications, from small packs to 14 inch (36-cm) displays, as single-stem or pinched plants.

Pelfi Bonita. Nine-week response with green foliage. Bonita has bright orange-red bracts and an upright growth habit. Bracts have medium width and may be slightly rippled.

Pelfi Picacho. Nine-week response with green foliage. Bright orange-red bracts, similar to Bonita, except Picacho's bracts are more rippled and elongated. It has upright growth with excellent branching. Picacho is suitable for hanging baskets, 4-inch (10-cm), and small 6-inch (15-cm) pinched plants.

Mikkel Red Delight. Nine-week response with green foliage. Red Delight has large, bright red bracts and a short, free-branching growth habit.

Pelfi Sonora. Nine-and-a-half-week response with dark green foliage. Among the darkest reds, Sonora requires little growth regulator. Bracts don't fade under high light, but this variety droops if finished warm.

Mikkel Yuletide. Nine-and-a-half-week response with green foliage. Yuletide does not require cooler growing conditions (below 60F, 16C) to develop a deep red bract color. It does well in warm growing areas, and the bract color holds under low light conditions.

Gutbier V-14 Glory. Nine-and-a-half-week response with green foliage. V-14 Glory is in full flower in early December. V-14 Glory has large, red bracts and a medium growing height. It responds well to the high light intensity and warmer temperatures of southern climates. V-14 Glory comes in a full range of colors.

Eckespoint Success. Nine-and-a-half-week response with green, oakleaf-shaped foliage. Good as a branched or single-stem plant, Success has elegant, bright red bracts that do not fade. A mid- to late-season cultivar that is particularly suited for the high-end florist trade, it grows best at 65 to 66F (18 to 19C) night temperatures. It makes a fine replacement for V-14 production, although it requires different cultural conditions.

Red Splendor. Ten-week response with dark green foliage. Red Splendor has bright red bracts and a thick canopy of branches. It performs well in the home.

WHITE CULTIVARS

Eckespoint Freedom White. Eight-week response with dark green foliage. Freedom White is a sport of Freedom with creamy white bracts. The flowering time, plant height, and cultural requirements are the same as for Freedom.

Brightpoints Nutcracker White. Eight-week response with green foliage. Nutcracker White performs well in warm temperatures without heat delay, and it resists splitting.

Eckespoint Lilo White. Eight-and-a-half-week response with dark green foliage. Lilo White is a sport of Lilo with creamy white bracts. The flowering time, plant height, and cultural requirements are the same as for Lilo.

Mikkel Blitzen. Eight-and-a-half-week response and green foliage. Blitzen is a compact white cultivar similar in response and habit to Dasher. It does not heat delay or have a problem with splitting of the bracts, and it exhibits no epinasty when boxed for up to five days.

Peter Jacobsen's Pearl. Eight-and-a-half-week response with green foliage. Pearl is part of the new SUPJIBI family. The first white SUPJIBI, it has thick, strong stems and smooth, narrow bracts. The beautiful Pearl white color is comparable to that of V-17 Angelika White.

Peter Jacobsen's Peterstar White. Eight-and-a-half-week response with green foliage. Peterstar White is the white mutation of Peterstar, with the same response period and cultural habits.

Eckespoint Celebrate 2 White. Nine-week response with green foliage. Celebrate 2 White is a sport of Celebrate 2 with white, erect bracts. The plant height and cultural requirements are the same as for Celebrate 2.

Gutbier V-17 Angelika White. Nine-week flowering response with green foliage. Angelika White is a sport of Angelika with clear white bracts and excellent branching. The flowering time and cultural requirements are similar to those of V-17 Angelika Red.

Annette Hegg Topwhite. Nine-week response with green foliage. An Annette Hegg sport with genuinely white bracts. The flowering time and cultural requirements are similar to those of Annette Hegg Dark Red.

Mikkel White Yuletide. Nine-week response with green foliage. White Yuletide has true white bracts and a more compact growth habit than Yuletide and Pink Yuletide, but it has a similar response time.

Gutbier V-14 White. Nine-and-a-half-week response with green foliage. V-14 White is a sport of V-14 Glory with large white bracts. The flowering time and cultural requirements are similar to those of V-14 Glory.

PINK CULTIVARS

Eckespoint Freedom Pink. Eight-week response with dark green foliage. Freedom Pink is a sport of Freedom Red. The flowering time, plant height, and cultural requirements are the same as for Freedom. Color is best when finished at cool temperatures: night temperatures 62F (17C), day temperatures less than 75F (24C).

Brightpoints Nutcracker Pink. Eight-week response with green foliage. Nutcracker Pink does well in warm temperatures without heat delay and is recommended for southern growers. It has a soft pink bract and it is a bit later than Nutcracker Red or White.

Beckmann's Altrosa Maren. Eight-and-a-half-week response with green foliage. Its timing is similar to that of Peterstar or Angelika. Maren has salmon pink bracts that resist fading. It is recommended for southern growers.

Eckespoint Lilo Pink. Eight-and-a-half-week response and dark green foliage. Lilo Pink is a sport of Lilo Red. The flowering time, plant height, and cultural requirements are the same as for Lilo.

Eckespoint Celebrate 2 Pink. Eight-and-a-half-week response and green foliage. Celebrate 2 Pink is a sport of Celebrate 2 with pink, erect bracts. The flowering time, plant height, and cultural requirements are the same as for Celebrate 2.

Gross Darlyne. Eight-and-a-half-week response and green foliage. Darlyne is an improved sport of SUPJIBI Pink and part of the new SUPJIBI family. It offers smoother, narrower bracts and a more uniform habit. Darlyne has the usual SUPJIBI strong, thick stems and a raspberry pink bract with a highlighted margin.

Peter Jacobsen's Peterstar Pink. Eight-and-a-half-week response and green foliage. Peterstar Pink is the pink mutation of Peterstar, with the same response period and cultural habits.

Gutbier V-17 Angelika Pink. Nine-week response and green foliage. Angelika Pink is a sport of Angelika with pink bracts. The flowering time and cultural requirements are similar to Angelika's.

Annette Hegg Hot Pink. Nine-week response with green foliage. This is an Annette Hegg sport with vibrant pink bracts. The flowering time and cultural requirements are similar to those of Annette Hegg Dark Red.

Pelfi Flirt. Nine-week response with green foliage. Flirt has pink bracts that are not as large as Maren's. Under warmest conditions, bracts may have green margins and veins. The color is not a "hot pink," but nevertheless a good pink for cool growing climates.

Mikkel Pink Yuletide. Nine-and-a-half-week response time with green foliage. Pink Yuletide is a deep pink mutation of Yuletide, with the same general habit and culture.

Gutbier V-14 Pink. Nine-and-a-half-week response with green foliage. V-14 Pink is a sport of V-14 Glory with large, dark pink bracts. The flowering time and cultural requirements are similar to those of V-14 Glory.

Novelty cultivars

Eckespoint Freedom Marble. Eight-week response with dark green foliage. Freedom Marble is a sport of Freedom with pink-and-white bicolored bracts. The flowering time, plant height, and cultural requirements are the same as for Freedom.

Eckespoint Freedom Jingle Bells. Eight-week flowering response with dark green foliage. Freedom Jingle Bells is a bicolored sport of Freedom with red bracts with light pink flecks. The flowering time, plant height, and cultural requirements are the same as for Freedom.

Bevelander's Marblestar. Eight-week response with green foliage. Marblestar has a nice bract color variation, a rich pink spreading to a cream margin—perhaps the best of the marbles. Excellent postharvest performance is reported.

Dahlqvist's Noblestar. Eight-week response period with green foliage. Nobelstar is an exciting new shade of coral that does not fade under high light and temperatures. Good shelf life is reported.

Peter Jacobsen's Peterstar Marble. Eight-and-a-half-week response with green foliage. Peterstar Marble is the marble mutation of Peterstar, with the same response period and cultural habits.

Eckespoint Lilo Marble. Eight-and-a-half-week response with dark green foliage. Lilo Marble is a sport of Lilo with pink-and-white bicolored bracts. The flowering time, plant height, and cultural requirements are the same as for Lilo.

Eckespoint Lemon Drop. Eight-and-a-half-week response with very dark green foliage. Lemon Drop has golden yellow bracts. It is compact and slow growing, making it ideal for smaller pot sizes. Because of its slow growth habit, it should be started two weeks earlier than other cultivars, and it may not require growth regulators.

E

Eckespoint Pink Peppermint. Nine-week response and green foliage. Pink Peppermint has pastel peach to pink bracts softly speckled with flecks of red. Pink Peppermint has a medium growing height.

Gutbier V-17 Angelika Marble. Nine-week response with green foliage. Angelika Marble is a sport of Angelika with large, pink-and-white bicolored bracts. It has more pink color than most marbled cultivars. The flowering time and cultural requirements are similar to Angelika's.

Pelfi Puebla. Nine-week response with green foliage. Puebla has pink-and-cream marbled bracts with good contrast. This plant is a semicompact, free-branching variety with good keeping and transport qualities.

Eckespoint Monet. Nine-and-a-half-week response with green foliage. Monet has unique multicolored, cream-rose-pink bracts that vary from light to dark tones. The bracts darken over time, even after leaving the greenhouse. It has large, wide bracts and a medium growing height.

Eckespoint Jingle Bells 3. Ten-week response with green foliage. The unique, bicolored bracts of Jingle Bells 3 have good color contrast between the light pink flecks and the dark red background. It has a medium growing height.

Pelfi Dark Puebla. Ten-and-a-half-week response with green foliage. This is a late-season marble with two shades of pink and cream, creating an excellent color contrast. Dark Puebla has good reported postproduction performance.

EUSTOMA (LISIANTHUS)

by Bob Croft and John Nelson
Sakata Seed America
Morgan Hill, California

❖ *Annual* (**Eustoma grandiflorum**). *624,000 seeds/oz (22,012/g).*
Germinates in 10 to 20 days at 70 to 75F (21 to 24C). Do not cover seed.
Also known as prairie gentian.

Lisianthus is native to Texas, Arizona, Colorado, and Mexico (fig. 1). Although it is found in desert areas, it is not a true desert plant. In its native habitat, lisianthus is found growing along riverbeds and low areas where it always has access to fresh water. In midsummer, when the rain is less frequent, native lisianthus plants push down deep roots into the soil to gain access to fresh water. Therefore, the root system is also the grower's key to producing lisianthus.

PLUG CULTURE (SOWING TO DAY 60)

Stage 1—Days 1 to 14. Carefully single-sow pelleted seeds in deep plug trays filled with well-drained media. Do not cover the seed and never allow it to dry out during germination. Sufficient moisture must be provided to melt the pellet. Maintain a soil temperature of 70 to 75F (21 to 24C) and keep the media moist throughout the entire germination period with a mist system. A pH between 6.0 and 6.5 is recommended to provide sufficient calcium levels. Placing the seed trays on capillary mats or plastic helps to keep the media moist and encourages very uniform emergence. (Note: 100 to 300 footcandles, 1 to 3 klux, of light are needed for germination.)

FIG. 1. Up-and-coming from all angles is the best way to describe lisianthus production today. An elegant and long-lasting cut flower, lisianthus also makes an excellent bedding plant and a fine flowering pot plant in most regions of the country.

Stage 2—Days 15 to 21. After the seedlings emerge, remove the seedling trays from the germination area and place them in a location with good air circulation. Lower the temperature to 60 to 70F (16 to 21C) and provide a light feed between 100 to 150 ppm N from a well-balanced, calcium nitrate-based fertilizer. Be careful not to allow the day temperature to exceed 75F (24C) or the night temperature to drop below 60F (15C) to avoid rosette problems, such as an induced resting stage, which is difficult to cure. To avoid rosetting, one can use a new production method. After the seed germinates, keep cool temperatures at night 59 to 63F (15 to 17C) and days between 75 to 80F (24 to 27C) until transplant. The key point is to grow cool at night for 12 hours. This cool system will help to avoid rosettes even under hot conditions.

Stage 3—Days 22 to 56. The young seedlings, very slow in growth, require extra care in avoiding high or low temperatures to prevent rosetting. Other factors to avoid are low light levels and excessive humidity, which will invite both disease and overgrowth of the seedlings. Since lisianthus is native to the alkaline soils of west Texas, Arizona, and southern Colorado, calcium-based feeds seem to help maintain stronger, healthier seedlings. Fertilize the seedlings with 150 ppm N as needed and maintain media EC levels between 0.7 to 1.0.

Stage 4—Days 57 to 60. The seedlings should have four true leaves at this stage and are now ready to transplant into cut flower beds. Lisianthus has a sensitive root system; be careful to avoid *checking* the plugs. Timely transplanting will ensure that the root system stays active and takes hold in the cut flower bed. Older plugs will have twisted root systems, and the transition into the cut flower bed will be more difficult. Also, older plugs will flower later on shorter stems, especially under long-day conditions. During short day conditions in winter, it takes extra days to grow out plugs to finished size.

CUT FLOWER PRODUCTION (DAYS 61 TO 90)

Choose a flower bed with a rich organic soil that is pest- and pathogen-free. A soil pH of 6.0 to 6.8 is recommended. Cultivate it to a depth of 18 inches (46 cm). Covering the bed with black plastic will increase soil temperature in winter and reduce crop time. For summer production, silver plastic will keep the soil temperature lower by reflecting the hot summer sun. Maintain a minimum soil temperature of 65F (18C) and a maximum of 73F (23C) for optimum results.

Transplanting. Transplant seedlings when they are young and actively growing, around the fourth true leaf stage. In order to avoid stem rot, take care not to bury the plants too deep. Setting the plugs a little "high" in the flower bed will help guard against *Rhizoctonia*. To ensure a healthy start, maintain high relative humidity for 10 days after transplanting and do not let the soil dry out.

Spacing will depend on whether one is growing a pinched crop or a single-stem crop. In general, space 4 by 6 inches (10 by 15 cm) and try to arrange the plants to enhance air movement for disease prevention.

Culture. Since lisianthus is native to low-humidity areas, *Botrytis* is a major disease problem. Using drip irrigation is best to reduce free moisture on the plants. Some growers bury the irrigation lines 2 to 3 inches (5 to 8 cm) under the soil, which imitates the natural habitat of lisianthus and helps to promote a deep, strong root system.

Lisianthus does not require high fertilizer levels like chrysanthemums. Maintain a soil EC level around 1.0 mhos (2-to-1 saturated paste method). Calcium nitrate-based fertilizers are recommended to build strong stems and reduce soft growth. Lisianthus requires higher moisture levels in the early stage of development. As the plants begin to mature and show flower buds, watering can be reduced, and the crop allowed to dry out more. Support wire is necessary to support the plants as they grow.

Flowering. During periods of high light and warm temperatures, a light shade on the greenhouse roof is recommended to avoid fading the flowers. Stems are usually harvested when one or more flowers are open. There is a longer period of time between the opening of the first and second flower than from the opening of the second and third flower. Therefore, some growers remove the first flower and sell it for small bud vases, then harvest the stems when the second and third flowers open. After cutting, place in tepid water in a shaded area for two hours to allow the stems to rehydrate. After rehydration, place in a refrigerator to keep fresh.

Since lisianthus is a long-day-response plant, using mum lighting at the sixth true leaf state from 10 P.M. to 2 A.M. during the short days of winter will reduce flowering time. Be sure to maintain at least 55F (13C) soil temperature.

For early flowering varieties from January to June, choose the Heidi Tyrol (single) and Echo (double) series. For midseason flowering from July to October, choose the Flamenco (single) series or the Mariachi (double) series.

POT PLANT PRODUCTION

Follow the guidelines for plug production (fig.2). Transplant to the final growing container (4-inch, 10-cm pot) when four to eight true leaves appear (60 to 70 days after sowing). Use a well-drained, sterile media high in organic matter, with a pH of 6.5. Fertilize with a calcium nitrate fertilizer, maintaining an EC of 0.5 to 0.8. Provide maximum light (and supplemental HID lighting in areas where needed) and good ventilation to create strong, disease-free plants.

FIG. 2. Blue Lisa lisianthus, a Florastar flowering pot plant award-winning variety, is an excellent addition to flowering pot plant programs. Or it can be sold as a potted bedding plant.

Growth regulators may be required, depending on the variety being grown. B-Nine at 2,500 ppm or A-Rest at 0.5 mg per pot drench are effective. Apply growth regulators when breaks are 2 to 3 inches (5 to 8 cm).

Seed sown from October to January can be potted from December to March for flowering from May to June.

Aphids, thrips, cutworms, *Botrytis,* and *Fusarium* may be problems. Disease problems include root rot and damping off. Preventive drenches are recommended. It is also important to start with well-drained, pasteurized media and to run the crop on the dry side.

The best way to control plant height is by variety selection. Naturally dwarf varieties include the Mermaid series and Blue Lisa.

EXACUM

by Jack S. Sweet and Paul
Cummiskey
Earl J. Small Growers, Inc.
Pinellas Park, Florida

❖ *Annual (***Exacum affine***). May be*
propagated vegetatively or from seed.
Growers generally purchase plug
seedlings or rooted cutting liners to
produce flowering potted plants.

Exacum is a beautiful blue-flowered
plant that has exploded into popularity as
a pot plant in the last few years. The
myriad of dime-sized flowers, blue with
bright yellow pollen masses in the center,

FIG. 1. With dozens of blue, lavender,
or white flowers, exacum is in demand
year-round as a flowering pot plant.

tend to cover the whole plant when grown well (fig. 1). Plant and pot size vary
with the grower and the market. Most seem to be grown as 6- or 6½-inch (15-
or 17-cm) pots, but with the advent of some of the newer, faster growing, compact varieties, we should see a lot more grown in 4- or 5-inch (10- or 13-cm)
pots. Some growers have white exacums to add a little variety.

Most exacums are purchased from specialty growers as small plants ready for
potting into 4-, 5- or 6-inch pots. These plantlets may be from cuttings or from seed,
depending on method of propagation.

Seeds are available, but they are quite tiny—even smaller than begonia seeds—
and must be handled very carefully. Exacum seeds are slow to germinate—two to
three weeks—and should be planted in a lightweight, starting media with little or no
covering. The seedlings will be ready to transplant into 2-inch (5-cm) pots in about
six weeks. About seven weeks later they can be put into the final pots for flowering.

CULTURE

The cultural procedures that follow are based on using plants, generally produced by
specialty growers, ready for final transplant from 1½- to 2-inch (4- to 5-cm) pots.

Use a very light, loose, well-drained potting soil. It should have plenty of
amendments such as perlite, calcine clay or polystyrene beads, and peat to allow
good root aeration. One plant per pot is sufficient. It should be placed deep
enough in the soil to remain stable as it grows larger. Initial watering and for the
first 10 days should be very light to encourage root action.

Exacum grows best at 60 to 65F (16 to 18C) nights and 75 to 80F (24 to 27C) days.

Plants should be grown under full sunlight in winter months. During the summer months, a light shade should be applied (4,500 to 6,000 footcandles, 48 to 65 klux). During the spring and early fall, plants should be grown in full sun, with light shade applied when plants start to flower. Shading at this time will produce darker-colored blooms. Excessive light and heat will cause flowers to be faded.

Exacums are moderate feeders. We have found they grow best by alternate use of fertilizers such as 15-16-17 Peatlite Special and calcium nitrate at the rate of 2 pounds per 100 gallons (about 959 g/400 l) every third watering. During the summer months a slow-release fertilizer should be used in addition to liquid feeding. Osmocote 14-14-14 or another slow-release fertilizer should be used at one-quarter to one-half teaspoon per 6-inch (15-cm) pot.

Some growers have experienced excessive leaf curl or crinkle. This seems to be related to excessive light and possibly a low copper level in the leaf structure. A foliar spray using Tri-Basic Copper at 1 pound per 100 gallons (479 g/400 l) applied two weeks after potting has been very successful in reducing crinkle on exacum. Soil applications of copper on exacum have not been useful.

Winter growing. Winter seems to produce more growing problems than summer. Apparently the lower light levels and shorter days make a softer plant that can be easily injured and attacked by disease. This must be compensated by lower fertilizer levels and reduced watering to make a harder plant. Starting around October, pots should thoroughly dry out between waterings. Water early in the morning so foliage is dry by late afternoon. Provide good air circulation around plants, and reduce fertilizer levels by half. Overwatering and high nutrient levels, besides promoting disease, cause delayed flowering.

Exacums require much less fertilizer and soil moisture than pot mums, lilies, or poinsettias. They respond poorly if fed with high levels of fertilizer on constant feed with every watering, such as other crops may require. *Remember:* To accelerate winter flowering of exacum, you should lower fertilizer levels and make sure plants dry out between waterings. Light at normal mum intensity will force winter plants into bud and bloom.

Production time. Timing of the crop is seasonably variable, with a marketable 6-inch (15-cm) flowering plant being produced in seven to eight weeks in the summer and up to 12 to 14 weeks in midwinter. Smaller plants, grown in 4½- or 5-inch (11- or 13-cm) pots for mass market sales, can be produced in less time and on less bench space, as exacums can be forced to flower at an early stage. No pinching is necessary, as they are self-branching plants. In case of premature budding of small exacums, larger plants can be produced by removing the earliest flowers.

PROBLEMS

The major trouble with exacum seems to be related to *Botrytis* (gray mold). Showing up as a gray lesion at the soil line or small, gray lesions at the forks in branches, it causes part or all of the plant to wilt and die. This form of *Botrytis* is actually caused by excess fertilizer, too much water, or water left on the leaves in the evening or overnight. Specific *Botrytis*-control chemicals, such as Chipco 26019 (Rovral) at 16 ounces per 100 gallons (435 g/400 l) as a foliar spray, will help control this problem, but lower fertilizer strength and less water are also needed. *Pythium* and *Phytophthora* can be controlled by a tank mix of Cleary's 3336 WP at 16 ounces per 100 gallons (435 g/400 l) and Subdue 2E at 1 to 1½ ounces per 100 gallons (43 g/400 l). Banrot at 8 ounces per 100 gallons (227 g/400 l) is also useful as a light drench right after planting. Regular use of the above chemicals can eliminate almost all diseases on exacum, but remember: they need it right after potting and repeated two to three times during production.

For best results, use a scheduled fungicide program. Immediately after potting apply Chipco 26019 at 16 ounces per 100 gallons (435 g/400 l) as a light drench, penetrating one-quarter- to one-half-inch (0.6 to 1.3 cm) of the medium. Two weeks after potting use Cleary's 3336 at 16 ounces per 100 gallons plus Subdue at 1½ ounces per 100 gallons (43 g/400 l). Four weeks after potting spray with Chipco at 16 ounces per 100 gallons.

The most damaging pests of exacum are the broad mites. Usually found on the upper parts of the plant, they cause the leaves and growing tips to become yellow and distorted, and the buds to fail to open. Broad mite can be controlled by various miticides, such as Pentac, Avid, or Sanmite. Avid 0.15 EC and Sanmite have proven quite useful in control of mites and can be used on flowering crops without flower injury. Worms can be controlled with Thuricide or Resmethrin aerosol. Thrips injury to growing tips can be controlled by Orthene, Sanmite, Naturalis, or Avid.

ADDITIONAL CARE

Exacum height can be controlled with B-Nine at regular mum strength (0.25% solution) applied one week after potting. If needed, a second application can be given two to three weeks later for plants that are being grown under lower-light conditions. Height can also be controlled by regulating the amount of water they receive. If small plants are desired, they should be allowed to dry out more between waterings.

Exacums react according to the total light energy on the leaves. Flower bud initiation is *not* affected by day length, but plant growth is increased by longer days. Therefore, supplemental lighting in winter is very beneficial. Lighting—such as HID lights or even mum-type lighting of 10 to 20 f.c., used four to six hours a night starting at dusk—can speed up production time in winter months by two or more weeks.

After leaving the greenhouse, exacum should be placed next to a window or some artificial light source for long-lasting quality. Low-light conditions will cause flowers to fade. Exacum may also be placed outside in a semishaded area or on a patio. It is hardy to 32F (0C).

VARIETIES

There are several good exacum varieties used by commercial growers today. Blue Champion is an improvement on the original that we introduced many years back. White Champion is a white form of the original. Royal Blue is a new, deeper-color form that comes in about seven to 10 days earlier. Little Champ, a smaller F_1 hybrid developed for the 4- to 4½-inch (10- to 11-cm) trade, is smaller, grows smaller leaves, is sky blue in color, and flowers 15 to 20 days earlier than Blue Champion.

FRAGARIA (STRAWBERRY)

FIG. 1. Add something that will really get your customer's attention: Grow strawberries in hanging baskets!

❖ *Strawberry (***Fragaria** × **ananassa***). 60,000 seeds/oz (2,116/g). Germinates in 21 to 28 days at 65F (18C). Temperatures above 75F (23C) are detrimental. Cover seed lightly.*

Bedding plant growers can grow strawberries alongside their regular bedding crops with Sweetheart and Fresca, two varieties produced from seed. Sweetheart is suitable for pack and jumbo production, while Fresca is also great for hanging baskets (fig. 1). Both produce medium-sized berries until frost. Fresca is more suited for automated bedding production than Sweetheart.

Seedlings should be ready to transplant from a 392-plug tray about six weeks after sowing, at the three- to four-true-leaf stage. The seedlings are slow to stretch, and transplanting can be delayed for a week or two with no damage. Fresca may be grown in various container sizes, from 48-cell flats to 8-inch (20-cm) hanging baskets. Grow at 60 to 70F (16 to 21C) days and 55 to 60F (13 to 16C) nights. Fertilize every other watering with 200 ppm N from a fertilizer such as 20-10-20 or 15-5-15.

FREESIA

by A. A. De Hertogh
North Carolina State University
Raleigh

F

❖ *Annual replacement corm* (**Freesia** × **hybrida***). **Vegetatively propagated. Growers purchase corms from commercial suppliers for forcing as cut flowers and potted plants.***

Freesias originated in South Africa. The commercial cultivars (tables 1 and 3) are hybrid products of extensive breeding efforts. They come in a wide range of colors, are often fragrant, are highly suitable for low-temperature forcing, and have good keeping quality. They can be forced as either cut flowers or potted plants.

Normally, 5/7-cm corms are used, and most corms used for forcing are produced in the Netherlands. Flowering is regulated primarily by temperature and light intensity. Because of the need to precisely control growth and development of the corms, the forcer must coordinate the cooling and planting schedule with the corm supplier. Corms for forcing must have been properly stored at 86F (30C) before being shipped to the forcer. Because the quantities are generally small and transportation time needs to be short (less than seven days), air freight from the Netherlands is absolutely necessary.

FIG. 1. One of the most fragrant of all floricultural crops, freesias are popular as cut flowers. Many growers—especially in northern California—are also producing freesias as flowering potted plants by using dwarf cultivars selected just for pot production.

CUT FLOWERS

The information here covers flower production in the greenhouse only from corms. It is also possible to produce flowers from seed (see Smith, D. 1985 [revised]. *Freesias*. England Grower Books. Fax (44) 322-667633).

Planting can begin in September in northern areas and be as late as December in southern areas. Depending on the cultivars and forcing temperatures, flowering usually starts 110 to 120 days after planting and lasts about four

weeks. The forcer who wants flowers for several months needs to stagger plantings. There are a wide range of freesia cultivars available (table 1). The forcer
should consult the supplier for cultivars in addition to those listed. There are both
double and single flowers. Many are fragrant and available in a wide range of
colors. Average plant heights are 20 to 30 inches (51 to 76 cm), but the actual
cut flowers are usually 10 to 14 inches (25 to 36 cm) long.

When corms arrive always check a sample to be certain they are free from
serious diseases or physical damage. The forcer should be prepared to plant the

TABLE 1   FORCED *FREESIA × HYBRIDA* FLOWERING CHARACTERISTICS

Cultivar	Color	Planting to flower (days)	Florets/ inflorescence	Primary flower scape Length cm	Length in[a]	Number/ corm	Secondary scapes/ primary stem
Aida[b]	Pastel blue	123	13.7	44.3	17	1.1	2.8
Apollo	Pure white	125	7.5	31.4	12	1.0	2.8
Athene	Clear white	134	10.5	37.3	15	1.0	3.4
Aurora	Light yellow	126	7.8	30.4	12	0.9	1.4
Ballerina	White, yellow	129	10.4	37.5	15	1.2	4.1
Blue Heaven	Blue	139	9.5	31.6	12	1.4	1.0
Blue Navy[b]	Blue	135	10.9	39.4	16	1.2	3.4
Carmen	Red, yellow	120	9.6	32.6	13	1.7	2.3
Czardas	Red, yellow	131	10.0	33.8	13	1.1	3.3
Diana[c]	Cream	129	9.3	36.8	14	1.3	4.0
Fantasy[c]	Soft yellow (ivory)	141	9.4	27.5	11	1.4	1.0
Golden Melody	Deep yellow	127	9.3	38.3	15	1.6	1.7
Golden Wave	Yellow	133	11.6	39.7	16	1.1	2.7
Granada	Golden yellow	128	9.6	30.5	12	1.2	2.9
Melanie[b]	Yellow, rose	135	7.4	35.8	14	0.9	1.6
Miranda	White	133	10.8	36.7	14	0.9	3.8
Moya[b]	White	139	8.3	29.4	12	0.8	4.0
Oberon	Red	133	9.8	39.7	16	1.5	4.9
Prominence	Orange/red	125	11.2	36.0	14	1.2	2.3
Red Lion	Red	121	8.3	26.7	11	1.8	2.3
Romany[c]	Violet	134	11.6	33.9	13	1.0	3.3
Rosalinde[b]	Deep pink/yellow	140	8.5	32.3	13	0.8	2.8
Rose Marie	Medium pink	136	8.8	37.6	15	0.8	4.6
Royal Blue	Blue	127	10.1	29.3	12	1.1	2.5
Safari	Soft yellow	132	9.5	39.2	15	1.2	2.6
Silvia[b]	Purple, white spots	128	11.4	35.3	14	1.4	2.1
Tosca[b]	Pink-purple	138	9.9	35.9	14	0.5	6.3
Wintergold	Deep yellow	125	10.0	40.9	16	2.1	3.0
Yellow Ballet	Yellow	135	10.9	36.0	14	1.4	3.5

Source: Dr. Terry Ferriss, research report, University of Wisconsin, River Falls, WI. 54022.
Notes: Cultivars forced, as fresh cut flowers under North American conditions. Each figure is the average for
that cultivar.
[a] Rounded to nearest whole inch.
[b] Semi-double flower
[c] Double flower

corms on arrival. If they must be stored, place them at 55F (13C) under nonventilated conditions. They can be held up to three weeks at this temperature.

The planting media should have a pH of 6.5 to 7.2, be free from fluoride-containing additives, and be sterile. Plant in either well-drained ground or raised beds 8 to 10 inches (20 to 25 cm) deep. They can also be started in special propagating trays and then transplanted. The bed or bench needs to have a mesh support system for the growing plants.

Plant corms 2 inches (5 cm) deep and use about 80 to 100 corms per square yard (97 to 120 per m^2). Under most North American conditions, planting is usually from September to December, but it is year-round in certain West Coast areas. Planting depends on prevailing soil temperatures, which should be 55 to 60F (13 to 16C). Keep the planting medium moist but not wet.

Freesias require a greenhouse with medium to high light intensity (2,500 to 5,000 footcandles, 27 to 54 klux). Use 50 to 55F (10 to 13C) night temperatures and avoid day temperatures over 63F (17C), especially during the short days of winter. During warm-temperature months, use soil cooling systems to maintain a soil temperature less than 63F.

Freesias can be forced in a greenhouse with 1,000 ppm CO_2 during the daylight hours. After plants begin to grow, use 200 ppm of 20-20-20 every other week.

Other than viruses, which come with the corms, *Fusarium* is the most common disease. The most common insect is the aphid. Freesias can exhibit flower abortion. As with Dutch iris, low light intensities or high temperatures during the period of rapid flower development can cause abortion. Leaf scorch can be induced by fluoride; thus, *do not use superphosphate* or other fluoride-containing amendments.

It is possible to harvest the corms and store them for forcing in the next season. This requires proper storage facilities and meeting specific temperature, relative humidity, and ventilation targets.

Cut flowers when the first (lowermost) floret opens. For short-term storage hold flowers dry at 32 to 35F (0 to 2C) and 95% relative humidity. For long-term storage keep flowers in water at 32 to 35F. Little or no storage is advised, and they are sensitive to ethylene in storage. Some flower preservatives will aid in the bud opening of cut freesias.

POTTED PLANTS

Potted freesias require practical experience, and forcers are advised to determine how the plants will force and market under their conditions. The objective is to produce a marketable plant in 60 to 80 days from planting, with an average total plant height of 10 to 16 inches (25 to 41 cm). At present these goals are not

TABLE 2   STANDARD DUTCH-GROWN FREESIAS FOR FORCING AS FLOWERING POTTED PLANTS

Color	Cultivar	Greenhouse days to market stage[a]	Marketable plant ht[b] in	cm	Bonzi (ppm)[c]
Blue	Amadeus[d]	65–75	14	35	100–200
	Blue Lady (Scorpios)	65–75	14	35	0–50
	Blue Navy[d]	65–75	12	30	50
	Caravelle	65–75	14	35	50–100
	Castor	75–85	14	35	50
	Sailor	65–75	12	30	100–200
Orange-red	Oberon	65–75	16	40	200
Red	Figaro	55–65	14	35	100
	Oberon	65–75	16	40	200
	Rapid Red	55–65	14	35	200
	Rossini[d]	65–75	14	35	200
	Washington	60–70	16	40	100
Pink	Bloemfontein[d]	65–75	15	35	200
	Florida	60–70	14	35	200
	Lewyna[d]	65–75	12	30	200
Rose	Diva	65–75	14	35	50
	Michelle	65–75	14	35	50–100
	Mosella	65–75	14	35	200
	Pink Glow	80–85	14	35	50
	Rossini[d]	65–75	14	35	200
	Sandra	65–75	14	35	100
White	Athene	65–75	16	40	200
	Dalba	80–85	14	35	50
	Elegance	70–80	14	35	50
	Napoli	60–70	16	40	300
	Poolzee	70–80	16	40	50–100
	Vienna[d]	65–75	16	40	300
	White Wings[d]	80–85	14	35	50
Yellow	Aladdin	70–80	12	30	50
	Desert Queen	65–75	14	35	200
	Golden Wave[d]	70–80	12	30	50
	Graced	75–85	14	35	50
	Senator	80–85	14	30	50
	Yellow Dream	60–70	14	35	200
van Staaveren dwarf selections("Easy Pots")					
Golden Yellow	Popey[e]	50–60	14	35	0 not required
Lemon Yellow	Pinokkio[e]	50–60	12	30	0 not required
	Smarty[e]	50–60	14	35	0 not required
Cream	Seagull	——	——	——	
	Suzy[e]	50–60	16	40	0 not required

Note: All standard Dutch-grown cultivars except for four van Staaveren dwarf selections ("Easy Pots"; see note "e" below).

[a] Average range

[b] Approximate, so English-metric conversions are inexact.

[c] One-hour preplant dip (see also table 3, "Bonzi dip concentrations").

[d] Double flower

[e] van Staaveren dwarf selections

always achieved with the cultivars and treatments evaluated; however, progress is being made. Recently, four dwarf cultivars called "Easy Pots" have been released. Under most forcing conditions they do not require plant growth regulators (see table 2).

When corms arrive, always check a sample to be certain they are free from serious diseases or physical damage. "Easy Pot" corms should be planted immediately upon arrival. Other cultivars (table 2) should be stored in open trays at 55F (13C) at a high relative humidity, with good air circulation but no ventilation, for 45 to 49 days. Do not return corms to 86F (30C)! The forcer should not store corms in excess of 49 days, since this can cause the corms to pupate, that is, form a new corm instead of a shoot. It is possible that the 55F treatment can be provided by the corm supplier, but if this is done, transportation time must be exceedingly short.

A preplant soak in paclobutrazol (Bonzi) can significantly reduce the flowering height of potted freesias. The concentration needed varies with each cultivar (table 2). Dipping corms should take place after the 55F (13C) storage treatment and immediately before planting. The corms should be dipped for one hour. Also, it is important that the corms be planted immediately after dipping. Do not allow them to dry out!

The guidelines for preparing a series of concentrations of Bonzi (paclobutrazol) as a dip are in table 3.

Plant corms 1 inch (2.5 cm) deep. Use four to six per 4-inch (10-cm) pot, six to ten in a 6-inch pot, or 10 to 15 corms in an 8-inch (20-cm) pot. Use a well-drained, fluoride-free, sterilized planting medium with pH of 6.5 to 7.2. After planting keep the medium moist but not wet. For plants that require staking, special rings are available from suppliers.

Freesias require a greenhouse with medium to high light intensity (2,500 f.c./54 klux). Use 55 to 60F (13 to 16C) night temperatures. Avoid temperatures above 63F (17C), especially during the short days of winter. Forcing times range from 55 to 90 days, depending on the cultivars (table 2).

TABLE 3   BONZI DIP CONCENTRATIONS

ppm	Bonzi/water oz[a]/1 gal	ml[b]/1 l
50	1.6	12.5
100	3.2	25
200	6.4	50
300	9.6	75

[a] 1 oz contains 112 mg of active ingredient.
[b] 1 ml contains 4 mg of active ingredient.

After plants begin to grow, use either 200 ppm N of 20-20-20 every other week or 14-14-14 Osmocote.

Market plants when the first floret begins to open. If plants need to be stored, place at 32 to 35F (0 to 2C), but not for a long period. Homeowners should be advised to place plants in the coolest but well lighted areas of the home in order to obtain maximum flower life.

FUCHSIA

by Will Healy
Ball Horticultural Company
West Chicago, Illinois

❖ *Annual* (**Fuchsia** × **hybrida***). 70,000 seeds/oz (2,450/g). Seed germinates in eight to 10 days at 70 to 75F (21 to 24C). Cover seed lightly for germination. Most fuchsia crops are grown as hanging baskets and started from bought-in rooted cuttings.*

Perhaps few other spring plants have the sheer grace and elegance of a well-grown fuchsia hanging basket. For some people, the fuchsia is a standard Mother's Day gift, purchased year after year (fig. 1).

While one variety, Florabelle, is available grown from seed, most fuchsias are started from rooted cuttings purchased in the early winter. Tip cuttings that are 2 to 3 inches (5 to 8 cm) long root in two to three weeks. Plant fuchsia into well-drained, aerated media with

FIG. 1. The vegetatively propagated fuchsia Dollar Princess makes a fine hanging basket.

pH of 6.0 to 6.5. Plants are sensitive to high soluble salts, so avoid water stress and use water with low alkalinity levels.

Alternate a balanced fertilizer such as 20-10-20 with 15-0-15. Excessive ammonia causes soft growth and plants that do not last well in the retail or home garden setting.

Fuchsia is a facultative long-day plant: plants require 10 to 25 long days to promote flowering. Use mum lighting to extend day length. Once induced, fuchsia will continue to flower, no matter what the day length is. When light intensity is low (500 footcandles, 5 klux), plants are day neutral; while at 1,000 f.c., plants are long day. Crops started in the low-light days of winter take longer to finish; applying HID light for 18 hours a day can shorten crop time.

Pinch plants one to two weeks after potting or once plant roots have reached the edge of the growing container. Pinch back to above the fourth or fifth set of leaves. Florel can increase lateral branching when plants are grown under short days or low light conditions. A-Rest at 25 to 75 ppm can control growth and speed flowering. Spray five to six days after pinching, and reapply every four weeks as required.

Plant one plant per 4- or 5-inch (10- or 13-cm) pot and pinch once or twice. When timing plants for Mother's Day, allow eight to nine weeks from last pinch to sale. Plants will finish in eight to 10 weeks in the winter, six to nine weeks in the spring. Use four or five plants per 10-inch (25-cm) basket and pinch once or twice. Plants finish in 16 to 18 weeks in the winter, 14 to 15 weeks in the summer.

Fuchsia is a magnet for insects and diseases. Follow good sanitation practices and use preventative sprays to avoid problems with whiteflies, aphids, and thrips. Do not allow media to stay wet for extended periods of time to avoid stem canker *(Botrytis)*. Rust may cause yellow spots on the top leaf surfaces and red spots on the undersides.

Fuchsias like cooler temperatures during production; move baskets started overhead down to benches during the spring. Plants may fail to develop flowers due to exposure to too low or too high light, too hot or too cold temperature, excessive Florel application, overfertilization under low light, or low light and overwatering. Another cause may be long days applied too late in the crop or exposure to low temperatures after flower induction.

Retail growers should advise their customers to keep plants out of full sun in a cool location easily accessible to the watering hose. Fuchsia thrives in the coastal climate of California.

Florabelle is the only fuchsia from seed. An F_1, it is well-branched and free flowering. Flower size, however, is much smaller than with vegetatively propagated varieties. Flowers are single and deep purple with deep rose sepals. Florabelle is more heat tolerant and free flowering than the vegetative varieties. Many growers produce it for 4-inch (10-cm) pots and 8-inch (20-cm) hanging baskets, reserving their 10- and 14-inch (25- and 36-cm) baskets for the larger-flowered vegetative varieties.

A number of fuchsia varieties may be grown. Among some of the more popular names are Swingtime, Dark Eyes, Dollar Princess, and Southgate.

GAILLARDIA (BLANKET FLOWER)

by Jim Nau
Ball Horticultural Company
West Chicago, Illinois

❖ *Perennial* (**Gaillardia** × **grandiflora**, *formerly* **G. aristata**)*. 7,000 seeds/oz (247/g). Germinates in 5 to 15 days at 70 to 75F (21 to 24C). Seed should be left exposed to light during germination.*

An easily grown perennial from seed, *Gaillardia* has mostly single flowers in either bright yellow or crimson and in-between bicolors. The scentless flowers are 2 to 4 inches (5 to 10 cm) across in an open, daisylike habit, on mounded plants 12 to 30 inches (30 to 76 cm) tall and 18 to 24 inches (46 to 61 cm) across. Plants have excellent weatherability and flower consistently from June until frost. Flowers are borne heavily at first, then provide better than average color the remainder of the season. *Gaillardia* is one of the few perennials to flower for the majority of the summer. *Gaillardia* is hardy to USDA Zones 3 to 9.

Seed or stem or root cuttings can be used for propagation. Cuttings can be taken from newly developing shoots in the spring or from roots while the plant is dormant. Seed can be sown directly to the final container, although plants may become lanky if they are not transplanted once.

From seed, sowings made in midwinter or spring will flower sporadically the first year after sowing. Plants put on a better show the second season. Allow 10 to 12 weeks for green pack sales in May. However, for larger and fully blooming plants, sow seed in July, move up into quart or gallon containers, and overwinter for sales next spring. Grow on at 55F (13C). More easily, buy in vernalized plugs.

Among varieties, there are a number of seeded and vegetative forms to pick from. The key difference is habit and height; the seeded forms are bigger in both aspects. Baby Cole is propagated by division only; there is no seed available. Plants have red flowers with yellow margins on plants to 10 inches (25 cm) tall. Burgundy is a rich wine red flower on 2 to 2½ foot (61 to 76 cm) plants. The seed form of Burgundy can be dull in color, and the foliage can cover flowers. Dazzler has crimson red petal ends with yellow centers on plants to 16 inches (41 cm) tall. It is a vigorous version of Goblin. The plants are commonly propagated vegetatively, though seed is available. Goblin, also a vegetatively propagated variety, grows to 12 inches (30 cm). The seeded form of this variety isn't bad, though it can get to 15 inches (38 cm) tall. The flowers can get up to 4 inches (10 cm) across on established plantings, and the flower color has a red base with yellow tips. This is the most preferred variety for its uniformity and overall performance in the garden. Golden Goblin is the pure yellow version. Monarch

Strain is an upright, vigorous variety to 30 inches (76 cm) tall. Flower colors are primarily a mixture of red and yellow combinations throughout. Excellent as a cut flower, it may require staking in the garden.

GAZANIA

by Jim Nau
Ball Horticultural Company
West Chicago, Illinois

G

❖ *Annual (***Gazania splendens***). 12,000 seeds/oz (423/g). Germinates in 10 to 12 days at 70F (21C). Seed should be covered during germination.*

Gazanias are low-growing annuals reaching 10 inches (25 cm) tall, in bright colors of white, yellow, orange, and crimson; although lavender, pink, and red shades are also available. Gazania flowers, like portulaca, close during cloudy weather and will not reopen until the sun appears. Flowers measure up to 3 inches (8 cm) across and are single in appearance (fig. 1). In warm-winter areas, the plants are perennial, but they are treated as annuals where there are killing frosts.

FIG. 1. Gazania's bright, perky flowers are a welcome addition to any full-sun garden. The variety shown: Garden Sun.

Allow 11 to 12 weeks for green packs and 12 to 13 weeks for flowering cell packs. Using one to two plants per pot, gazanias require 14 to 15 weeks to flower in a 4-inch (10-cm) pot. *Gazania* is susceptible to *Botrytis*. To avoid problems, maintain good air circulation around plants during production.

In the southern United States, allow 10 to 11 weeks for flowering pack sales, and 13 to 14 weeks for flowering 4-inch (10-cm) pots, one to two plants per pot. Gazanias can be planted out from late March to June for flowering during early summer. The plants are tolerant of salt and can be used in coastal regions very effectively.

In varieties, gazanias are divided between dwarf and vigorous types. Dwarfer varieties are best in pack and 4-inch (10-cm) pot production, sold in full bloom. Among the dwarfer varieties, Chansonette Mixture provides the earliest flowering in packs. Another good performer, the Daybreak Series has separate colors with the same dark-ringed eyes as some of the colors in Chansonette Mixture, while the Ministar Series has clear, unringed eyes. All of these grow from 8 to 10 inches (20 to 25 cm) tall with a 10-inch spread in the garden.

Newer introductions, such as Talent Mix, combine the gray-green foliage of dusty miller with the bright, sunny flowers of gazania. While flowering is a little later, the effect is worth it.

GERBERA (TRANSVAAL DAISY)

❖ *Annual* **(Gerbera jamesonii).** *7,000 to 8,500 seeds/oz (247 to 300/g). Germinates in 10 to 14 days at 68 to 72F (20 to 22C).*

Gerbera daisies have now taken a place alongside geraniums in many greenhouses as an important pot and bedding crop (fig. 1). Significant improvements in uniformity, earliness, habit, color, and flower form have been made in seed varieties since Happipot was first introduced. There are now cultivars that flower consistently double, others with dark eyes, separate colors, and bicolor petals. Among the dwarf pot types are Tempo, Festival, and Masquerade.

Tissue-cultured *Gerbera* has also been bred for production in larger, 6-inch (15-cm) pots. Some shorter cut flower varieties from Holland are adapted to pot production, and several of the cut flower breeders have developed pot series from their lines.

Breeding programs for cut gerberas in Holland are advancing at a rapid pace. The Dutch are emphasizing keeping quality and high productivity. The Dutch auctions have implemented a certification program for all

FIG. 1. Gerberas make striking pot plants and can be produced from seed or tissue-cultured liners. The seed variety shown: Festival Pink.

varieties to ensure that they meet minimum vase life standards. Dutch breeders are also developing smaller flowers more suitable for bouquets.

PROPAGATION

Seed is the method most used for propagating potted gerberas. However, tissue culture is used more often for cuts. Division is now seldom practiced commercially.

A mix of 60% perlite and 40% peat is good for germinating gerberas. Approximately 1,000 seeds may be broadcast in a 14-by-18-inch (36-by-46-cm) flat and covered with a thin layer of vermiculite. About 70% usable seedlings at transplant is an average yield. Use bottom heat to maintain a soil temperature of 68F (20C). Avoid direct sunlight, and keep the seeds moist at all times, or germination will be uneven. It is important *not* to cover the seed. Immediately after germination provide good light levels, using supplemental light on seedlings during winter months. Emergence will be in seven to 14 days. In about four weeks, when two true leaves develop, transplant seedlings to cell packs, 2¼-inch (6-cm) pots, or Jiffy Strips.

Tissue-cultured plantlets are usually sent to specialist propagators for adaptation to greenhouse conditions. At Stage 3 of their growth, they may still come in culture jars or in plastic bags. If the plants are shipped in jars, the agar medium must be thoroughly washed away with water. They should be immediately planted into cell packs, 2¼-inch (6-cm) pots, or Jiffy Strips. Soil should be sterile and well drained. Several of the soilless plug mixes do a good job and are easy to handle. After transplanting, put plantlets under 50% shade with intermittent mist and drench with Banrot, Truban, or Subdue. Mist cycles will depend on the stage of growth and individual conditions at each operation. Mist can be gradually reduced, and most plantlets will be ready for planting in six weeks. Keep the temperature at 77F (25C) during the day and a minimum of 60F (16C) at night. Feeding may begin at the rate of 100 ppm N and K in two to three weeks or when the plants become established. Since tender new growth is easily burned, soluble fertilizers should be rinsed from the foliage. Because establishing young plants from tissue culture Stage 3 is tricky, many growers prefer to buy in established liners that may immediately be planted to ground beds or rock wool cubes.

Division is a method of propagation still used in the garden. June is a good time to divide one- or two-year-old clumps, so that size is reached for flowering in the fall.

POT PRODUCTION

When five true leaves have developed, plant liners in a well-drained mix with a pH close to 6.5. Some research indicates that gerberas grow best in mixes with less than 20% bark. *Do not bury the crowns!* Be very careful about planting depth because plants will die when the crown is covered.

Use 250 ppm N from a balanced fertilizer at each watering. In some programs Osmocote 14-14-14 at 8 pounds per cubic yard (3.7 kg/m³) is the only source of fertilizer. During warm weather in Florida, 14-14-14 releases too rapidly; instead, slower 18-6-12 is used at the same rate. Magnesium and iron deficiencies are not uncommon when gerberas are grown in soilless mixes. Magnesium sulfate at 1½ pounds per 100 gallons (71 g per 400 l) or chelated iron drenches will restore vigor quickly. Foliar application of minor elements is less effective.

Space requirements are dependent upon light quality and market demands. For the first month, plants may be spaced pot to pot. Final spacing for 6-inch (15-cm) pots is typically 12 inches (30 cm). In northern areas more space may be needed in winter. Final spacing for 4-inch (10-cm) pots can be as close as 7 inches (18 cm) for dwarf types to 10 inches (25 cm) for the larger varieties.

Growth regulators are helpful in situations where pots must be spaced tightly. B-Nine may be used as a 2,500-ppm spray two weeks after potting if leaf petioles appear to be stretching. Bonzi at 1 ounce per gallon (30 g per 4 l) is effective and does not fade color or reduce flower size when applied late. A-Rest as a drench at 0.25 mg active ingredient (a.i.) per 6-inch (0.246 mg a.i. per 15-cm) pot is also used.

Gerberas grow best under maximum light intensity. Northern greenhouses should not be shaded at all during fall, winter, and spring. Temperature is not as critical as with some other crops. Day temperatures of 70 to 80F (21 to 27C) are optimum. For pot production most growers hold night temperature near 60F (16C).

Production time from 2¼-inch (6-cm) liners to flower is approximately 10 weeks. Gerberas will not bloom at one time. The first few flowers appear about eight weeks after transplanting, and the last plants begin blooming about four weeks later.

CUT FLOWER PRODUCTION

Plant rooted plants on arrival to the greenhouse. Avoid exposing young plants to high temperatures or wind, especially before planting. Pressman Gerbera of Rijsenhout, the Netherlands, provides the following cultural guidelines.

Soil preparation. The soil should have a good air-water balance and be deeply tilled—harrowed, if needed. Make sure the soil is disinfected before planting. Methyl bromide, if allowed in your area, works fine. Before treating beds, cover them with leakproof plastic for several days to germinate any weed seed. After methyl bromide treatment, leave the plastic in place for seven to 10 days, then allow the treated area to air for several days before planting. Add water slowly until field capacity is reached before planting.

Spacing. Plant in a two-row system measuring 39 to 43 inches (1 to 1.1 m) across, including the aisle. Space the rows 14 to 18 inches (36 to 46 cm) apart, and plants within the row 10 inches (25 cm) apart. For regions with high light and low humidity, the plants can be spaced closer together. Most growers mound beds to a height of 10 to 14 inches (25 to 35 cm), depending on soil type, to allow leaves to droop, improving air circulation and facilitating harvest.

Planting. Place the plants into beds, making sure that the soil line of each young plant matches the soil line of the bed. Deep planting causes rot. If daytime temperatures are greater than 86F (30C), plant in the morning or evening.

Irrigation. Use overhead irrigation in the initial stages of plant establishment only. Using sprinklers during this crop phase allows the entire bed to become wet, which increases humidity around plants, thus improving the microclimate. Water thoroughly rather than frequently during plant establishment. During the first weeks, irrigate two to three times a week, decreasing to one to two times a week later. Discontinue all overhead irrigation as soon as leaves cover the soil.

During the establishment period, provide moderate light levels and prevent direct wind from affecting the plants.

Two weeks after planting, begin to use drip irrigation, supplying about 75 to 100 cubic centimeters per plant, while you are still using the overhead irrigation. As plants become established, make sure the water column extending underneath the plant from the drip emitter is uniformly moist from top to bottom.

Nutrition. Gerberas are light feeders. Maintain an EC of 1.2 to 1.6. Supply fertilizer in a constant liquid feed through the drip irrigation system. Use an N:P:K:Mg ratio of 8:1:4:2.

Diseases. Gerberas are subject to a number of diseases: *Botrytis, Phytophthora, Pythium,* and *Fusarium.* Many growers help to minimize disease problems by running a heating pipe lengthwise between two rows of plants. This helps keep leaves dry, thus preventing *Botrytis.* For root diseases, use a preventative drench of Previcur and Topsin M.

Disease problems are one reason many growers have switched from soil culture to rock wool culture in troughs (fig. 2). Plants grown in

FIG. 2. Cut gerbera production is rapidly changing over from culture in ground beds to hydroponic systems such as this one at Ever Bloom in Carpenteria, California, where plants grow in rock wool cubes that sit in hydroponic troughs.

rock wool in troughs are raised above the ground, thus increasing air circulation. Rock wool is also a sterile medium, so crops begin with a cleaner start. Yields are also higher—by at least 20%.

Insects. Leafminers are the most troublesome insect pest. The flies feed on young leaves. Later their larvae tunnel through leaves, causing minelike markings. The best control is not to get leafminers by treating with Vertimec and Tamaron weekly. Whiteflies and thrips can also cause problems. Thrips can cause streaking on petals and, if present in high numbers, flower deformity. Because gerberas are harvested with no leaves on the stems, many growers are successfully working with biological controls for whitefly and leafminer control.

Harvest. Gerberas are harvested beginning eight to 12 weeks after planting. A flower may be harvested when two to three rows of stamens show in the center of the flower. Harvest the stems by pulling them from the plant. You can find the natural breaking point of the stem by pulling up and down while slowly pulling the flower from the plant. Depending on the variety, plants should be harvested between two and four times a week. Place the stems in water immediately after harvest. To stimulate water uptake, cut off the stem ends. Growers use a special postharvest rack that allows the flowers to stand supported upright in water, thus preventing bent stems. Most growers also use a bactericide in water (fig. 3). After stems have taken water for three to four hours, they may be packed for shipping. Gerberas are packed so that each flower is protected, either by netting each individual flower or by packing flowers flat on cardboard, before they are boxed. Ship cool, maintaining 10 to 15C (50 to 59F) to avoid condensation on flower heads.

FIG. 3. To ensure that each gerbera flower arrives damage-free at its final destination, most growers pack stems in special boxes, laying each flower flat. Flowers also spend time after harvest hanging with stems in an antibacterial soak to preserve freshness (as shown in photo).

GEUM

by Jim Nau
Ball Horticultural Company
West Chicago, Illinois

❖ *Perennial (G. quellyon, formerly G. chiloense). 10,000 seeds/oz (353/g).*
   Germinates in 8 to 15 days at 65 to 70F (18 to 21C). Cover the seed
   upon sowing.

Geum can be somewhat difficult to germinate; by alternating day and night temperatures by 8 to 10F (−13 to −12C), germination can be increased. This is especially true if using seed saved from last year's sowing. Fresh seed should yield 65 to 80% germination with no pretreatment. Flowers are semidouble to double, scentless, and measure up to 1½ to almost 2 inches (4 to 5 cm) across. The flowers are predominantly colored red, yellow, and shades between these two and appear for one to two weeks in the spring. Plants produce a mass of foliage, in which long flowering stems to 20 inches (51 cm) develop. Overall, plants can get 18 to 24 inches (46 to 61 cm) tall. *Geum* is hardy in USDA Zones 4 to 8.

   Sowings made in midwinter to midspring will be of good size and salable 10 to 11 weeks after sowing. Plants will be sold green and will not flower the same season from seed. For blooming plants in June, sow in July of the previous year, transplant to quart or gallon containers as needed, and overwinter with other perennials. More easily, buy in vernalized plugs from a commercial supplier. These can be potted, grown on at 48 to 50F (9 to 10C) nights, and sold green in the spring or in flower in June. *Geum* can also be propagated by divisions taken in the spring or fall.

   Mrs. Bradshaw is the most common of the geum varieties on the market. It has 1½-inch (4-cm) flowers of bright red held on upright stems in May. Plants range in height from 18 to 24 inches (46 to 61 cm) tall. Lady Stratheden has deep golden-yellow flowers to 1¾ inches (4.4 cm) across that are semidouble or double. Plants bloom in May.

G

GLADIOLUS

❖ *Annual (***Gladiolus** × **hybrids***). Grown as cut flowers from pur-chased corms.*

Gladiolus is the primary crop of the "spe-cialty bulbs." Botanically, gladiolus is a corm, not a bulb. Glads are important in Florida, with the largest grower having more than a thousand acres. Florida produces glads from November until June. In California they are produced year-round. Other growers are spread from Florida to Canada. The earliest Midwestern crop would be planted April 1 and flower in midsummer. Glads are also a delightful cut flower and border plant for the home gardener in the United States and Canada (fig. 1).

FIG. 1. Small-flowered gladiolus varieties are expanding the tradi-tional glad market with a new size that's suitable for bouquet work and smaller vases.

Here's a brief outline of how Manatee Farms, Bradenton, one of Florida's largest gladiolus growers, produces its crop. Man-atee does about 1,200 acres (486 ha) a year of cut glads, plus a variety of other cut flower crops.

A furrow is dug, and the corms are hand-set (upright), then covered mechanically. Rows are 6 feet (1.8 m) apart. Corms are spaced with 1 inch (2.5 cm) or more between each bulb. It's mostly sand in southern Florida, so lots of fertilizer is required—around 2,000 pounds per acre (about 367 kg/ha) per crop.

Aphids are a particular problem, not only because of the aphids them-selves, but also because they are a vector for cucumber mosaic virus. Also, thrips are widespread.

Watering is done by controlled subirrigation. In fact, the water table is main-tained under the roots to provide water as needed. Glad crops are grown out-doors in the area roughly from Bradenton to Naples.

Stems are cut as the first buds show color. Flowers are cut from November to June 1. Manatee ships them over the entire United States and Canada. They are shipped standing upright packed in boxes (gladioli are geotropic, so the tips would turn upright if the stems were laid flat).

The trend on the gladiolus market is steady. However, in Florida there are problems, such as with the government, scarcity of good land, hurricanes, and freezes.

Methyl bromide availability—to fight *Fusarium*, especially—is also becoming an issue. Manatee develops its own glad hybrids and propagates them through tissue culture, which helps to keep virus out of the crop. It then grows its own corms to use later to grow cut flowers. Keeping *Fusarium* out of corm production is critical, which makes losing methyl bromide as a soil fumigant hard.

During cut flower production, beet army worms can also be a problem, Manatee reports. To control it, there are many sprays, such as Dimilin, Durshan, or any of the *Bacillus thuringiensis* products.

As previously mentioned, the gladiolus is a delightful cut flower and border plant for the home gardener. Corms are planted in the spring after the last freeze, around April 1 or later. In temperate areas, they normally flower in about 90 days. A succession of plantings can be made; the last must be timed to flower before a killing frost.

GLOXINIA (SEE *SINNINGIA*)

GODETIA (SEE *CLARKIA*)

GOMPHRENA

by Jim Nau
Ball Horticultural Company
West Chicago, Illinois

❖ *Annual* (**Gomphrena globosa**)*. 5,000 uncleaned to 11,500 cleaned seed/oz (176 to 406/g). Germinates in 10 to 14 days at 72F (22C). Seed can be covered or left exposed to light during germination.*

Gomphrena is a heat- and drought-tolerant plant that can be sown directly to the field or transplanted to the field from cell packs for cut flower production or bedding plant use. Flower colors are pastel and come in white, pink, lavender, and bright purple. Plants will fill in quickly, and the flowers are a traditional dried cut flower.

As bedding plants, allow seven to eight weeks for green packs and nine to 10 weeks for flowering packs of the more dwarf varieties. For 4-inch (10-cm) pots allow 13 to 14 weeks and use one plant per pot.

For bedding, 4-inch pots, and landscaping, the primary choice on the market is Buddy. Buddy is a deep purple-red flower color growing to 12 inches (30 cm) tall in the garden. In some catalogs you will find Buddy White or Cissy listed. For all practical purposes these two are the same variety. Both have off-white flowers on plants to 12 inches tall. The Gnome series has pink, purple, and white, offering greater color choice. These plants have a more uniform height than Buddy, growing 10 to 12 inches (25 to 30 cm) tall.

In the taller strains, gomphrena is most often just sold as the globosa type instead of a variety name. These are available in separate colors of rose, pink, white, and purple on plants to 2 feet (61 cm) tall. The Woodcreek series is available in lavender and red, with 1½-inch (4-cm) flowers. These make excellent plants for cut flower growers or for landscapers who want a gomphrena with a little more vigor than the more dwarf strains.

GYPSOPHILA

by the staff of Danziger
The Danziger "Dan" Flower Farm
Moshave Mishmar Hashiva, Israel

❖ *Perennial (**Gypsophila paniculata**). 26,000 to 30,000 seeds/oz (917 to 1,058/g). Germinates in five to 10 days at 70 to 80F (21 to 27C). Do not cover seed.*

Gypsophila is a member of the Caryophyllacae family, of which 125 species are known. The species originate mostly in Europe and North Asia. They are annuals, biennials, and perennials. Several species are used as ornamentals. The main quality of the inflorescence is that it produces a misty, cloudy effect in rock gardens and flower arrangements. *G. elegans* is an annual that is dominant in spring gardens. *G. repens,* creeping baby's breath, is a hardy perennial. *G. paniculata* is the major species used in commercial cut flower production. The varieties Bristol Fairy and Perfecta belong to this species. Natural flowering in Europe and North America is from late spring to autumn, during which there may be two to three flowering flushes (fig. 1).

The information that follows provides an overview of cut flower production. Cultivation methods differ in different countries, from field production to heated to unheated greenhouses. To achieve high-quality production, use plants that have been vegetatively propagated.

One of the problems that accompanied this plant's growth in the past was crown gall disease, caused by *Erwinia herbiculata*. This disease caused weakening of the plants and eventually led to death. High-quality nurseries have tried to deal with this problem by producing the mother material in a clean stock program, which includes meristem propagation from carefully selected clones and varieties in a laboratory. At the end of the laboratory stage, the plantlets go through the following stages: nuclear house, foundation house, and finally a clean, young plantation of mother plants. Plants produced from such a system are healthy and vigorous.

G

CULTURE

Gypsophila is clearly affected by a few major environmental factors: day length, temperature, and light intensity. The plants need a proper combination of environmental factors in order to go through the four growth stages: vegetative stage, flower induction, elongation (bolting) and flower initiation, and flower formation and flowering.

FIG. 1. Once the backdrop for nearly every corsage, flower arrangement, or mixed bouquet sold in the United States, gypsophila sales are still strong but are being challenged by other fillers. Gypsophila production is also shifting from California and Florida to offshore locales like Ecuador. Bunches shown here were harvested from outdoor fields in Encinitas, California.

Studies of flower initiation and development have shown that factors such as radiation (light intensity) and temperature are especially important and may play major roles in the ability of the plants to respond to the long-day needs. Gypsophilas may be grown both indoors and outdoors, provided that the environmental conditions and the specific variety are coordinated. They can be grown in all kinds of soils, as long as the soil is well aerated.

The varieties Golan, Gilboa, Arbel, and Tavor respond to shorter day lengths than Perfecta and Bristol Fairy. Therefore, they can flower outdoors without additional lighting and without a structure in spring, summer, and autumn. Their stable branches are an advantage for the flower grower in picking, sorting, bunching, and bouquet-making for the flower arranger.

Gypsophilas are relatively resistant to pests and disease. Common pests are the leafminer, thrips, *Spodoptera littoralis,* and aphids. Diseases related to unaerated soils are *Pythium* and *Rhizoctonia*. It is possible to deal with these problems by controlled growing and high-quality plant material.

PLANTING DATES

Gypsophilas are usually planted in summer and autumn. The flowering flush occurs in summer and autumn. Another flush occurs in the spring. When growing in a closed structure, flowering can be programmed by lighting, and thus another flowering flush may occur in the middle of the winter. It is of course possible to plant or prune the plants in spring; consequently, the flowering flush will occur in the summer. Generally speaking, gypsophilas can be grown all year-round, as long as controlled day length and temperatures are maintained.

Harvested flowers should be treated with postharvest solution for long shelf life and high-quality opening. The solution should consist of STS 0.1 to 0.2%, sugar 5 to 7% and a bactericide, such as Floron G 0.4% or the new "Forever" gypsophila solution.

According to its growth pattern, *Gypsophila* is defined as an obligatory and quantitative long-day plant. This means that long-day conditions will enable the plant to proceed from the vegetative stage to the flowering stages. The day length required for the plant to achieve flowering reaction, at least 13 hours, varies with the clone and the variety. At a given temperature level, the longer the day, the earlier the flowering reaction. The duration of the growing process and time until flowering are also affected by temperature: the higher the temperature (above a certain minimum), the earlier the flowering.

To summarize, day length, temperature, and light intensity affect flowering at all stages and are crucial to flower induction, initiation, and formation. Rapid reaction is not necessarily a desirable characteristic from the commercial point of view. What is important is that the growth rates at the different growing stages should be coordinated with one another.

When growth is rapid (with high temperature and day length of 14 to 16 hours), the interval between planting and flowering will be short (50 to 60 days), but the quality and yield will be poor. This is likely to occur when pruning or planting is done in spring and summer.

Under short-day conditions and low temperatures, the plant will be characterized by vegetative growth with much branching. Flowering will not occur unless one of the key environmental factors (day length, temperature, or light radiation intensity) is changed. Light intensity contributes both to the growth rate and to the abundance of flowers. The higher the light radiation intensity, the larger the number of flowers on the flowering stems.

HELIANTHUS (SUNFLOWER)

by Will Healy
Ball Horticultural Company
West Chicago, Illinois

❖ *Annual* **(Helianthus annus)**. *920 seeds/oz (32/g). Germinates at 68 to 72F (20 to 22C) in five to 10 days. Cover seed for germination. Seed may be direct sown into the final growing container.*

Sunflowers grace gardens and roadsides everywhere at the height of the summer. In recent years they've also become a popular cut flower. Some growers even sell them as flowering pot plants or bedding plants (fig. 1 and 2).

Seed can be sown direct into the final container or into a large plug cell or cell pack. Use well-drained, disease-free media with a pH of 5.8 to 6.2.

Provide 100 footcandles (1,000 lux) of light during germination to improve uniformity and seedling quality. Provide 1,000 to 2,500 f.c. (11 to 27 klux) immediately after germination to avoid seedling elongation. As seedlings mature, increase light to 5,000 f.c. (54 klux).

Begin fertilization with 50 ppm 15-0-15 as radicles emerge. When cotyledons expand, increase fertilizer to 50 to 75 ppm N. Fertilizer can be increased to 100 to 150 ppm N during Stage 3 of plug production. Maintain media EC between 1.0 and 1.5 mS/cm.

To prevent branching, maintain temperatures above 60F (16C): use 65 to 68F (18 to 20C) during the night and 70 to 75F (21 to 24C) during the day. Avoid water stress on plants to prevent leaf yellowing and necrosis.

FIG. 1. Sunflowers may very well be the most popular flower today. Its image has appeared on nearly every type of consumer goods from fabric to dishware to stationery. Home gardeners have also shown an eager demand for live plants as well.

During finishing, provide a constant liquid feed of 200 to 250 ppm 15-0-15, applying 20-10-20 as needed. Fertilization can be increased during flower bud elongation to prevent leaf chlorosis. Provide 14 hours of light to accelerate flowering. From a 384 plug, 4-inch (10-cm) pots will finish in five weeks; 6-inch (15-cm) pots in six weeks.

The best way to control sunflower height is through environmental manipulation, such as withholding fertilizer (avoid ammonia fertilizers, such as 20-10-20) and water. Using a negative DIF also controls plant height. B-Nine at 2,500 to 5,000 ppm, Bonzi at 5 to 10 ppm, or A-Rest at 33 to 68 ppm is also effective. Use lower rates for the first application; higher rates if a subsequent application is needed. Do not use growth regulators after buds are visible.

Sunflowers can be attacked by a wide range of diseases. *Alternaria* causes purple or blue-black lesions on leaves. Maintain good air circulation, keep foliage dry, and lower humidity during production to prevent problems. Chipco 26019, Ornalin, or Zyban can be used as preventative sprays. *Pseudomonas* may also cause leaf spots (yellows). Again, maintain good air circulation and low humidity during production. Phyton 27 or Kocide 101 copper sprays can be used as preventative sprays. *Pythium* and *Rhizoctonia* may cause problems, as well.

FIG. 2. Sunflowers are popular as cut flowers, too. The right variety, cut at the right time, can be transported with success. The photo shows outdoor sunflower production in the Arava Valley of Israel for export to Europe.

Varieties. Big Smile (fig. 3), golden yellow, is excellent for container production as a single flower. Teddy Bear, a branched variety for container production, has double flowers. Both varieties grow 2½ to 3½ feet (76 to 107 cm) tall in the garden. Prado (gold or crimson red) and Sonja (golden orange) are larger growing varieties, to 3 to 4 feet (91 to 122 cm) in the garden. For cut flowers, try Hallo (bred for cut production, golden yellow), Double Sun (double, golden yellow), Sunrich (pollenless for easier harvesting and handling), or Sun and Moon.

FIG. 3. Varieties like Big Smile helianthus are popular with consumers.

HELICHRYSUM (STRAWFLOWER)

❖ *Tender perennial (***Helichrysum bracteatum***). 45,000 seeds/oz (1,587/g). Germinates in seven to 10 days at 70 to 75F (21 to 24C).*

Helichrysum is known as strawflower for its crisp-textured, long-lasting blooms. Flowers come in fall colors: reds, creams, yellows, and burnt reds, as well as pink.

Helichrysum can be produced as a bedding plant started from seed. However, plants should be sold green, as generally they are too tall by the time they flower. Allow six to eight weeks for green pack sales; 10 to 12 weeks for 4-inch (10-cm) pots. As a cut flower, grow *Helichrysum* in the field—sown direct or started from transplants. Plants flower from July until frost. Strawflower is an excellent dried cut (fig. 1).

For a bedding plant in packs or pots, try Bright Bikini Mix. Plants grow to 12 inches (31 cm), with flowers in a bold blend of red, yellow, dark red, hot pink, and white. The King Size series, available in four separate colors, is an excellent choice for cutting.

FIG. 1. Tom Bebout, Bebout Farm, Venetia, Pennsylvania, displays a fine helichrysum hanging basket.

While seed-grown strawflowers are the most well-known plants, it is the new vegetative *Helichrysum* that is becoming popular for combination pots and planters as well as hanging baskets.

Helichrysum Golden Beauty has a compact, spreading plant habit, with long-lasting, golden yellow flowers. The cultural information that follows comes from Proven Winners. Plants can be sold in pots or baskets in flower, too. Plant cuttings in January to March for spring sales. Use three to four liners per 8- to 10-inch (20- to 25-cm) pot; one liner per 4- to 6-inch (10 to 15-cm) pot. Allow 12 to 14 weeks to finish.

Use well-drained, peat-based media with slightly acid pH of 6.0 to 6.5.

Allow plants to fully dry out between waterings; foliage will yellow if kept too moist. Use a constant liquid feed of 200 to 250 ppm N; 65 to 75 ppm P; and 125 to 165 ppm K. Pinch plants one to two weeks after planting.

H. petiolatum Rondello is so striking in combination pots—with its mint green leaves that are marked with dark green blotches in the center—that it may become a standard part of the grower's spring plant mix.

HELICONIA

by Heliconia Society International
Miami, Florida

❖ *Tropicals* (**Heliconia** *spp.*). *Vegetatively propagated by rhizome division. Most growers buy in rhizomes from a commercial nursery for planting out.*

The genus *Heliconia* consists of more than 250 species and about 1,000 forms. Intensive but environmental-conscious collection of species in their natural habitats by some renowned botanists has brought beautiful flowering plants for use as cut flowers and container plants for the landscape (fig. 1). There is no actual breeding of varieties at the moment.

The genus *Heliconia* was named after the Greek Mt. Helikon. Heliconias are closely related to the bananas (Musaceae). They are found primarily in South and Central America (their greatest region of biodiversity) and in islands of the southwest Pacific. Major commercial production areas for cut flowers are the Caribbean, Central America (Costa Rica, El Salvador, Honduras), South America (Brazil, Colombia, Guyana, Venezuela), Hawaii, and increasingly in parts of tropical Africa, northern Australia (Queensland), and Southeast Asia.

FIG. 1. Lady Di *Heliconia psittacorum.*

These exciting, bold tropical plants have inflorescences with a wide variation of shapes, sizes, and colors. The colorful bracts can hold up to 20 small florets. Heliconia bracts can have combinations of up to six different colors, including red, yellow, pink, orange, green, and white. Their size differs from small to very large, with stems from 8 inches (20 cm) to 12 feet (4 m) or more. Inflorescences may be either erect or pendant.

PROPAGATION

Heliconias usually are propagated by division of their rhizomes. Some plants do produce viable seeds. However, seed propagation is not reliable, and it is more labor-intensive. The pulp of the blue ripe fruits needs to be cleaned off. The seeds then should be rinsed with a mild bleach or fungicide solution. The species have different germination requirements. Most seeds have a short life span and should be sown as soon as possible.

Therefore, mostly rhizomes are used as planting material, which are commercially available from a number of reputable growers. Rhizomes are prepared for storage or shipment with a 4- to 24-inch (10- to 61-cm) cut stem attached. They are cleaned thoroughly to be free from soil and diseases before shipment and are shipped bare root. Their stems might be coated with antitranspirants. Rhizomes with one or more developing eyes are preferable. Before shipment and planting, treat rhizomes with insecticides, fungicides, and bactericides to help increase storage life and sprouting. Rhizomes are best packed for shipment in damp sphagnum moss or shredded newsprint.

For smaller species, the first inflorescence may appear within eight weeks of planting, within seven months for some larger species. Some need up to two years to flower.

CULTIVATION

It is best to start rhizomes in containers (sprouting). In tropical areas with adequate observation, water, and pest control, direct planting in beds is often the most common way to start the crop (fig. 2). Heliconias are usually grown in full sun to 30% shade. Planting distances for large species are 10 by 14 feet (3 by 4.3 m), and 5 by 7 feet (1.5 by 2.1 m) for smaller species. Grow in raised beds in well-drained soils with a pH of 5.5 to 6.5. Heliconia

FIG. 2. Field production of Kaleidoscope *Heliconia psittacorum.*

plants react beneficially to mulch and manure. They are heavy feeders. Use a granular, 3:1:2-ratio fertilizer with micronutrients in a slow-release form. Apply one-fourth cup (0.06 l) every two months for a 3-gallon (11-liter) container. For soluble fertilizers, apply 21-5-20 or 7-9-5, each with micronutrients, in a dosage of one-half to 1 tablespoon per gallon of water once per month. To encourage flowering, reduce nitrogen and increase potassium and magnesium. One of the most important cultural requirements is to provide high soil moisture; good drainage and high relative humidity are also important.

Growing media. Expert growers differentiate between media for sprouting rhizomes and for growing container plants (table 1). In both cases use containers with drain holes. These pots need to be at least two times wider than the rhizomes.

Sprouting media: Canadian sphagnum peat (1) + pine bark (1) + perlite or aerolite (3).

Possible alternatives: Horticultural charcoal (1) + ProMix BX (1) or volcanic rock (1) + ProMix (1) or silica sand (1) + sphagnum peat (1), vermiculite (2), perlite (2).

Container media: Canadian sphagnum peat (1) + pine bark soil conditioner (1) + sponge rock (larger size perlite, 1), add 1 quart dolomitic lime per 60 gallons of this mix.

Alternatives: Silica sand, ProMix BX (4) + perlite (1) or vermiculite (1.5) + sphagnum peat (1.5) + perlite (2).

Problems. Strong winds can cause leaf laceration, which is unsightly. Some species cannot tolerate high winds. Rats have been reported from Miami to love young heliconia leaves. Common pests include ants, caterpillars, grasshoppers, mealybugs, mites, nematodes, and snails. Common diseases include *Alternaria*,

TABLE 1   HELICONIAS AS CONTAINER PLANTS

Container size	Varieties
6–10 inches (15–25 cm)	H. psittacorum Sassy H. stricta Dwarf Jamaican H. aurantiaca H. thomasiana H. latispatha (dwarf types)
3–10 gallons (11–38 l)	H. rostrata H. latispatha H. stricta Bucky H. stricta Sharonii H. stricta Tagami H. wagneriana H. psittacorum × H. spathocircinata
10 gallons and larger (38+ l) Landscape or high style interiorscaping	H. caribaea H. bihai H. pendula H. pastazae H. marginata H. mutisiana H. caribaea × H. bihai H. orthotricha H. champneiana H. pseudoamegdiana

Cercospora (leaf spot fungi), *Fusarium* (Panama disease), Erwinia bacterial blight, moko disease or bacterial wilt (*Pseudomonas*).

Postharvest treatment for cut flowers. Heliconias are tropical flowers and cannot tolerate extreme heat or temperatures below 58F (14C). They are more sensitive to chilling than *Strelitzia* (bird of paradise) or gingers.

Harvest cut flowers during the cool morning. Keep the flowers away from direct sun. Submerge in cool water for one hour, cut the stems again, and place them in clean water. Flowers do not continue to develop once harvested; inflorescences remain at the stage at which they are cut. Most preservatives are ineffective because heliconias have marginal water uptake once cut. Antitranspirants do extend vase life with most varieties up to 25%. Florists should recut 2 to 4 inches (5 to 10 cm) from stem ends.

TABLE 2   HELICONIA VARIETIES RANKED AS CUT FLOWERS

Species and cultivars	Height		Flowers	
	feet	meters	Held	Months
H. psittacorum Sassy, Petra, Andromeda, Lilian, Rhizo, St. Vincent Red, Winter Pink, Lizette, Kathy, Strawberry and Cream, Lady Di, and many others	1.5–9	0.5–2.7	Erect	1–12
H. psittacorum × *H. spathocircinata* Golden Torch, Golden Torch Adrian	2.5–9	0.8–2.7	Erect	1–12
H. caribaea Chartreuse, Cream, Flash, Gold, Purpurea	12–18	3.6–5.5	Erect	1–12, peak 4–10
H. caribaea × *H. bihai* Green Thumb, Grand Etang, Jacquinii, Kawanchi	12–16	3.7–5.5	Erect	1–12, peak 4–11
H. bihai Arawak, Aurea, Lobster Claw II, Chocolate Dancer	5–18	1.5–5.5	Erect	1–12, peak 12–8
H. chartacea Sexy Pink, Sexy Scarlet	6–16	1.8–5	Pendulous	1–12
H. collinsiana Collinsiana	5–16	1.5–5	Pendulous	1–12, peak 1–9
H. orthotricha She, Edge of Night, Garden of Eden, Eden Pink, Imperial	2.5–16	0.8–5	Erect	1–12, best winter
H. angusta Holiday, Orange Christmas, Yellow Christmas	4–10	1.2–3	Erect	Winter-flowering "Christmas heliconia," peak 1–7
H. stricta Carli's Sharonii, Cooper's Sharonii, Dwarf Jamaican, Fire Bird, Las Cruces, Tagami	3–5	0.9–1.5	Erect	Very seasonal, differing between cultivars
H. rostrata Previously known as Lobster Claw, there are now 12 cultivars	3–20	0.9–6	Pendulous	1–12

Optimal storage and shipping temperatures are 58 to 65F (14 to 18C) and 90 to 95% humidity. Store in original packages (plastic sleeves and moist, shredded paper) to avoid dehydration in air-conditioned storage rooms.

The most cold-resistant heliconias include *H. acuminata*, *H. episcopalis*, *H. hirsuta*, *H. latispatha*, *H. schiedeana*, *H. pastazae*, and some *H. rostrata*. The most cold-tolerant heliconia is *H. caribaea* × *H. bihai* cv. Jacquinii.

Fig. 3. *Heliconia* flowers are diverse. *Alpina purpurata* cultivars not only have a different flower structure than other *Heliconia* species, but also come in a range of colors from cream and pastel pink to medium pink and deep red.

Editor's note: This chapter was written by the Heliconia Society International. Members contributing to this section were Alan Carle, Australia; Mark Collins, Hawaii; Fred Berry, Florida; Gilbert Daniels, Indiana; and Rudolf Sterkel, Illinois.

HEMEROCALLIS (DAYLILY)

by Jim Nau
Ball Horticultural Company
West Chicago, Illinois

❖ *Perennial* (**Hemerocallis** *spp.*). *Propagated primarily by division. Tissue-cultured plants are also available.*

Daylilies are the grande dames of gardens in the United States from coast to coast. Few other perennials have the staying power and can tolerate the neglect, poor soil conditions, or drought daylilies can. While individual flowers last but a day (hence the name), flowers are produced in abundance, held high on stalks, with many varieties providing a guaranteed summer show from June through frost. Flowers are primarily orange or shades of orange, yellow, red, pink, rose, purple, maroon, and crimson. Plants are hardy from Zones 4 to 9 (fig. 1). In

FIG. 1. Daylilies are practically indestructible in the home garden or commercial landscape, a feature that makes them a flowering perennial favorite for summer.

more southern regions, some species are ever-green or semievergreen.

Daylilies form clumps at their base, with en-larged, thickened, fibrous roots. Most plants are started from divided clumps. As you divide clumps, make sure each fan has one to three root shoots. Divide clumps in the spring or fall. Pot immediately. Make sure the plants are rooted in well before temperatures drop consistently below 40F (4C). Otherwise, maintain 45F (7C) night temperatures. Bought-in roots are available from late summer through early winter. After potting, grow them at 55F (13C) nights for sales eight to 10 weeks later. Plants potted in February or March will generally be rooted in well enough for May sales as green plants.

Provided the plants are started early enough, they will flower in the garden their first year, but the best flowering comes from plants potted the prior fall and overwintered. The exceptions to this are the dwarf-flowering varieties, which often can be sold in flower in the early summer, even when potted as late as March or April.

Plants that remain vegetative after planting out, only developing a foliage canopy, are often from divisions taken from plants that were only one year old and are too young for flowering.

There are more than 20,000 different daylily varieties from which to choose! Many catalogs sell varieties by chromosome count, such as diploid and tetraploid. Diploid varieties, with two sets of chromosomes, have 3- to 6-inch (8- to 15-cm) flowers and softer colors. Tetraploid varieties with four sets of chromosomes, have larger plants and flowers. Miniature daylilies have 3-inch (8-cm) flowers.

Three varieties do stand out. Stella D'Oro produces 2½-inch (6 cm) golden yellow flowers from June to frost. Mary Todd, a tetraploid, has buff yellow flow-ers that are 6 inches (15 cm) across and lightly ruffled. Chicago Fire is red with a light band of yellow.

H

HEUCHERA (CORALBELLS, ALUMROOT)

by Jim Nau
Ball Horticultural Company
West Chicago, Illinois

❖ *Perennial (***Heuchera sanguinea** *and* **H. micrantha,** *sometimes listed as* **H. americana***). 500,000 seeds/oz (17,636/g). Germinates in 21 to 30 days at 65 to 75F (18 to 24C). Leave the seed exposed to light during germination.*

Heuchera is an excellent perennial with rounded, rosetting leaves on plants to 12 to 18 inches (30 to 46 cm) tall when flowering. *H. sanguinea* is the most common of the coralbells, with deep green leaves with pastel red or scarlet flowers, while *H. micrantha* is the deep purple- bronze-leaved type with yellow flowers. This latter variety is sold primarily for its leaf color rather than its flowers. Plants flower in June and July and sporadically after that until frost. *Heuchera* is hardy to USDA Zones 3 to 8 and does best in areas where it gets afternoon shade.

From seed, allow 10 weeks for green pack sales of *H. sanguinea* and up to 13 for *H. micrantha.* Plants will not flower the same season when sown from seed. Sow seed during July of the previous year and move up into quart or gallon containers to overwinter. These plants will flower this season. More easily, purchase vernalized plugs from a commercial supplier. Allow eight to 10 weeks for a 50-plug to root into a quart container at 55F (13C) nights.

Heuchera is also propagated by crown division in the spring or fall. Again, fall-propagated plants that are overwintered produce the highest quality plants.

H. sanguinea has several varieties of importance, including Bressingham Hybrids, a mixture of scarlets and light crimson. Plants grow upright to 18 to 22 inches (46 to 56 cm) tall. Splendens (Spitfire or Firefly) has deep red flowers on uniform plants. Chatterbox, vegetatively propagated, is a deep rose-pink, showing flowers on 18- to 20-inch (46- to 51-cm) stems. Snowstorm and Snow Angel have variegated cream-white–splashed foliage and soft pink flowers. Both are vegetatively propagated.

In *H. micrantha,* Palace Purple is the most common variety available. Plants have dark red foliage that will bronze and turn deeper with increased sunlight. Flowers are off-yellow or cream in appearance on plants to 18 to 22 inches (46 to 56 cm) tall. While Palace Purple can be propagated by seed or division, division will provide true-to-type plants; about 5% of seedlings will be off-types.

HIBISCUS

by Dr. Harold F. Wilkins
Deer Wood, Minnesota, and Anna, Illinois

❖ *Tender perennial (**Hibiscus rosa-sinensis**). Vegetatively propagated from tip cuttings. Growers generally purchase either rooted cuttings or half-grown prefinished plants to force into flower.*

The hibiscus continues to be popular in the homes of northern Europe, illustrating that it can be acclimatized to the home environment. Growers in the United States and Canada also have found it a profitable pot plant. This plant is a specialty item in the North for home use, as well as in the South, where it survives in the landscape under frost-free conditions. Regardless, this plant is tolerant of high temperatures and full sun in the South, as well as being an excellent patio plant in the North (fig. 1).

FIG. 1. The tropical flowers of hibiscus have been a consumer favorite for decades. Most growers choose to buy in hibiscus as a prefinished plant, and force them for fall and winter sales.

A member of the Malvaceae family, *Hibiscus rosa-sinensis* has been commonly grown in gardens for so long that its site of origin is lost. However, it is believed to have originated in China and Cochin China (Vietnam), and it is extremely common in the East Indies, as well. In China some varieties have been cultivated since the dawn of history, as recorded in ancient art and writings. Women have used the sap from the flowers to color their hair black, and the juice can also stain shoes black.

There are about 250 species, which are widely diffused geographically but are particularly abundant in the tropics. The name *hibiscus* probably derives from *ibis*, a bird that was believed to live off certain hibiscus plants. In fact, many species of hibiscus are naturalized in marshy localities where such birds abound.

We do not know for sure how many varieties have been developed during the past centuries or how many have been lost since *H. rosa-sinensis* was first introduced into Europe in 1731. This species is certainly the most beautiful of all the hibiscuses, and there are numerous magnificent hybrids, each more attractive than the next.

THE COMMERCIAL POT HIBISCUS

The hibiscus is a shrub whose leaves are a shiny dark green, sometimes variegated, and usually simple and palmately veined. The flower, mostly solitary in the leaf axils, consists of five petals with a bell-shaped calyx; the stamens are united into a tubular column that is frequently longer than the petals. The style usually has a five-branched stigma, which can be quite ornamental. The ovary is a five-celled structure with three or more seeds per cell. Colors of the blossoms range from vivid red to white to various shades of pink, yellow, and orange. Flowers can be single or double. Flowers last but a day; however, selections do exist whose flowers last somewhat longer. Hawaiians harvest hibiscus blossoms in the morning, place them on a table, and they remain turgid and open for 20 or so hours without water.

Propagation of hibiscus can be achieved by seed, cuttings, grafting, or layering. However, hybrids of *H. rosa-sinensis* must be propagated from cuttings because they do not come true from seed. Cuttings with two or three mature leaves are taken from stock plants every two weeks. Researcher Karl Wikesjo states that in Sweden stock plants are grown in beds or in large containers for five to seven years. When stock plants are grown in containers for nine to 12 months, they can be sold as specimen patio plants. Many producers buy unrooted cuttings or buy rooted liners. Cuttings can also be obtained from plants in production when the pinch occurs. Thus, production can be self-perpetuating.

Plants thrive in a variety of well-drained media. Propagation can occur under mist. Wikesjo reports that rooting takes place in 35 to 40 days under milk-white plastic tents in the summer or under clear plastic in the winter without mist. Little air is allowed under the tent until rooting commences; then the center is opened 4 inches (10 cm) and afterwards gradually enlarged.

Rooting hormone hastens the process and may be desirable on some cultivars. Various concentrations have been recommended: 3,000 ppm IBA in talc, or Hormodin #1. Bottom heat of 72 to 75F (22 to 24C) is beneficial and will speed rooting. Air temperatures for production are 68 to 70F (20 to 21C) nights and 72 to 74F (22 to 23C) days. Do not go above 90F (32C) days.

Cuttings are to be stuck directly into liners and later shifted into larger pots, or directly stuck into the final pot size. After rooting, cuttings can be transplanted—one, two, or three cuttings per pot, for 4-, 5-, or 6-inch (10-, 13-, or 15-cm) production, respectively. When cuttings are first stuck, a fungicidal drench combination can be used. Dead leaves should be continually removed.

Fungicidal sprays can be used on plants every second week under the tents during rooting.

Production of hibiscus may be a profitable summer fill-in crop. Spring and summer sales are brisk. Cuttings taken in December can be sold in late March; if taken in March, plants are ready to sell by early July. The average production time from cutting to sale is 15 to 16 weeks. In winter it can be 18 weeks, in spring and summer 14 weeks.

Growth and flowering time of *H. rosa-sinensis* is greatest in the summer. It has been established that flowering of *H. rosa-sinensis* is not photoperiodic. High light and long photoperiods result in maximum flowering. Maximum leaf unfolding occurs near 90F (32C) with adequate light. In Florida in the summer, 8.3 leaves can unfold weekly; in the winter the rate decreases to 5.8 leaves per week.

H. rosa-sinensis is considered a moderate feeder. Recommendations for nutrition have varied. If regular or treble superphosphate is used, frequently only the nitrogen and potassium supply must be of concern. Osmocote (17-7-12) and micromix can be used to top-dress pots. A pH of 6 to 6.5 should be maintained. Criley, working in 1:1 volcanic ash:wood shavings, incorporated 4 ounces Osmocote (14-14-14), 2 ounces treble superphosphate, and 6 ounces dolomite per cubic yard of medium (148 g, 74 g, and 222 g/m^3). Plants were also fed two times daily with 200 ppm N and K. Never allow the plants to dry.

The height of outdoor plants in China can be up to 30 feet (9 m). In Florida and California, though, they seldom reach 15 feet (4.6 m). Pinching in the greenhouse can control the growth and shape of the hibiscus plant. Von Hentig and Heimann report that when new shoots reach 1¼ to 2 inches (4 to 6 cm) in length, they should be pinched. As many pinches as deemed necessary to achieve the desired shape and form can be used. For 4-inch (10-cm) production, only one pinch is needed with Cycocel sprays.

In commerce for houseplant and patio specimens, growth regulators are commonly used. Sprays of Cycocel or chloromequat not only induce shorter plant internodes and darker green leaves but also more flowers, and sooner, during the summer months. Initial sprays are applied two or so weeks after the plants have been pinched and active axillary growth has commenced. Shanks used single sprays of Cycocel (1,000 to 4,000 ppm active ingredient), while Criley used 3,000 ppm to control growth of hedges in Hawaii. However, an application of two to three individual sprays spaced three to four weeks apart have evolved as commercial recommendations. Wikesjo recommends 200 to 300 ppm (0.02 to 0.03% a.i.). The different levels could be relative to amount applied per plant, location (Hawaii versus Sweden), and cultivars. Criley used 0.05% Tween-20 as a wetting agent. A-rest (ancymidol) has also been reported to retard growth on several cultivars, as has Bonzi.

Dwarfing of *H. rosa-sinensis* by Cycocel has also been reported in India by Bhattacharjee et al., Bose et al., Hore and Bose. Bhattacharjee et al. report that a soil drench of Cycocel at 2,500 and 5,000 ppm suppressed growth on 10 cultivars. With Cycocel drenches some cultivars produced fewer flowers; all produced larger flowers, and the effect persisted for more than 360 days for most cultivars. Reduced flower numbers and persistent activity were also reported by Criley when drenches were used.

Common insects on hibiscuses are red spider, aphid, and whitefly. Exclusion and scouting is a must to prevent major infestations.

A common disease is xanthomonas angular leaf spot. The best control for this is to keep a close watch over temperature, moisture, and humidity levels. Many times this disease cannot be controlled, and infested plants should be discarded.

Hibiscus rosa-sinensis is very beautiful and useful as a garden, terrace, or balcony plant, or as a houseplant in the North or South. Postharvest stress,

TABLE 1   GROWING PROGRAM FOR *HIBISCUS ROSA-SINENSIS*

Planting		
	Substrate	Pure peat substrate or ordinary manufactured substrate mainly containing peat Light or low humified peat Normal nutrient levels Important trace elements: Fe and Mn pH 6.0–6.5
	Propagation	Top cuttings Stick two directly into 4- or 4½-inch (10- or 11-cm) pot
	Temperature	72–74F (22–24C)
Growing on		
	Temperature	Mar.–Sept.: 68–70F (20–21C) nights 72–74F (22–24C) days Oct.–Feb.: 65F (18C) nights 68–70F days
	Fertilizing	Apr.–Aug.: nitrogen 120–140 ppm phosphorus 25–30 ppm potash 120–140 ppm minor elements
	Pest control	Exclusion and scouting are best methods Use chemicals only if other methods don't work well enough
	Pinching	Soft pinch
	Other growth control	First Cycocel spray comes when the laterals, after a pinch, reach 1–2 inches (2.5–5 cm) Use at 0.02–0.03% a.i. Repeat two or three times (200–300 ppm)

Source: Adapted from Wikesjo 1981, p. 585.
Note: Diploid cultivar type Miesiana.

ethylene, dryness, and darkness can cause flower, flower bud, and leaf abscission. STS spray reduces leaf and bud drop. Plants are injured at temperatures well above freezing, 40F (4C). Ship at 50 to 60F (10 to 16C) for not more than three days.

HIPPEASTRUM (AMARYLLIS)

by A.A. De Hertogh
North Carolina State University
Raleigh

H

❖ *Tender perennial bulb* (**Hippeastrum** *hybrids*), *vegetatively propagated. Growers purchase bulbs from commercial suppliers for cut flowers and pot plant production.*

Amaryllis (Hippeastrum) originated in South America. Flowering is regulated by bulb size, temperature, and moisture. The commercial cultivars are the products of extensive breeding efforts. The primary sources of bulbs forced in the United States and Canada are Israel, South Africa, and the Netherlands (fig. 1). The major use is for pot plant forcings, but they can also be used as cut flowers. The general marketing season is from September to May. Normally, the South African-grown cultivars are forced early, and the Israeli- and Dutch-grown cultivars medium to late. With special growing and handling, some Israeli and Dutch cultivars are suitable for November-December forcings. The objective is to market a plant that has simultaneously produced at least one floral stalk and growing leaves.

FIG. 1. Based on new breeding from the Netherlands, pot plant growers should keep their eyes on the *Amaryllis*. Fast forcing, smaller dwarf cultivars in brighter colors are on the increase.

The number of floral stalks produced is influenced by bulb size and cultivar. Examples of commercial-sized bulbs are (in circumference) 20/22, 24/26, 28/30, and 32/up cm. The number of flowers per stalk is primarily a cultivar response, but most cultivars produce four flowers per stalk. The range is two to six. Larger bulbs tend to produce two floral stalks.

After harvest the bulbs are quickly dried and cured. During this and all subsequent processes, it is critical that the old root system be kept viable. Normally, the bulbs are cured for two weeks at 73 to 77F (23 to 25C) with high ventilation rates. They are subsequently stored at 48 to 55F (9 to 13C) at 80% relative humidity for at least eight to 10 weeks. Bulbs stored for longer periods are held at 41 to 48F (5 to 9C). Bulbs should be transported at 48F (9C). In addition, they must be protected against freezing and drying out.

Forcers should be prepared to plant bulbs as soon as they arrive. If they must be stored, place them at 41 to 48F (5 to 9C). The precise temperature for preplant storage will depend on the sprouting condition of the bulbs on arrival. If they have begun to sprout, store them at 41F. If no sprouting is observed, store at 48F. Keep bulbs from drying out during preplanting storage.

Amaryllis must be planted in well-drained, sterilized planting media with pH of 6.0 to 6.5. Never use fresh manure or bark as part of the media. Sunshine Mix No. 4 is a very satisfactory commercial mix. Growing media must be capable of being firmed-in tightly around the roots.

Normally, one bulb is planted per 6-inch (15-cm) standard pot. Plant the bulb with the nose above the rim of the pot; one-third of the bulb should be out of the planting medium. Force bulbs pot to pot on the bench.

Amaryllises are tropical plants, and they can be forced over a wide range of temperatures, but 70 to 80F (21 to 27C) is preferred. Bottom heat should be used. The average forcing time to the market stage of development is three to seven weeks. It will vary with each cultivar and forcing period (see table). It is also important to note that most lots are variable. Thus, the forcing information in the table should be used only as guides to average dates of marketing and flowering.

Plants should be forced in a greenhouse with medium light intensity (2,500 to 5,000 footcandles, 27 to 54 klux). It is possible to start bulbs in a dark, temperature-controlled area before the bulbs are placed under lighted conditions. Force plants in a well-ventilated greenhouse. Do not allow the relative humidity to build up!

After planting, water the medium thoroughly. Subsequently, the medium should be kept only slightly moist. It is important not to overwater the plant in order to stimulate regrowth of the basal root system. Normally, watering once a week is satisfactory. Use tepid water and do not water over the bulb noses.

Initially, the bulbs do not need fertilization. After they are marketed, however, consumers should be advised to fertilize the plants. Use care tags when the plants are marketed.

The primary disease of *Amaryllis* is fire or red spot *(Stagonospora)*. Overwatering can sometimes promote development of *Fusarium*. In addition, it is possible to have mites, thrips, and mealybugs.

Market plants when the floral stalks are 12 inches (30 cm) tall. At marketing, it is desirable to have leaf growth of 6 to 12 inches (15 to 30 cm) and a second stalk beginning to grow. Do not cold-store the plants! If they need to be held, place them at 48F (9C). Wholesalers and retailers should use tepid water after they receive the plants.

Whenever possible, the plants should be marketed with care tags. The consumer should be informed that *Amaryllis* should be fertilized at least one or two times per month when it is growing. They should keep plants in the coolest area of the home and out of direct sunlight in order to obtain maximum life from the flowers. *Amaryllis* can be placed outside in the pot when the danger of frost has passed.

To reforce the plants, two systems are available. They can be taken into the home in the fall, allowed to dry, and stored for at least eight weeks at 50 to 60F (10 to 16C). Then the dried leaves should be cut off, the planting medium watered, and the plants placed in a warm area to start the forcing process. If one does not want to store the bulbs, the plants can be grown in the light at 50 to 60F for eight to 10 weeks, then forced into flower.

Amaryllis can also be used for cut flowers. They should be cut when the floral buds are fully colored but not open. To prevent splitting and outrolling of the cut stems, the flowers can be held in 0.125M sucrose (60 ounces of sucrose per gallon of water or 43 g/l) for 24 hours at 72F (22C) before shipping.

HOLLYHOCK (SEE *ALCEA*)

HYDRANGEA

by Dr. Robert O. Miller
Dahlstrom and Watt Bulb Farms, Inc.
Smith River, California

❖ *Tender shrub (***Hydrangea macrophylla***). Vegetatively propagated. Growers typically buy in prefinished plants and force them for Easter or Mother's Day sale.*

Native to Japan, the florist's hydrangea belongs to the Saxifragaceae family and is known botanically as *Hydrangea macrophylla*. Varieties designated as suitable for outdoor use may be bud-hardy in Zone 6, but greenhouse forcing varieties are usually bud-hardy only into the lower part of Zone 7. Outdoor plants grow vigorously even though winter-killed to the ground, and early budding varieties may still develop flowers by late summer (fig. 1).

The hydrangea, like the poinsettia, has the capability of being a long-lasting flowering plant because the showy parts are not petals that rapidly fade and fall, but instead sepals. The types with flat cymes and staminate flowers only at the outer edge are known as lace caps.

Hydrangeas grown outdoors make vegetative growth in July and August, with initiation of terminal flowers in September and October, after which the flower buds are in a resting state and resume growth after normal winter chilling and leaf shedding. Overwintered flower buds are usually in flower by late June. The usual method of greenhouse forcing simply mimics the natural sequence, with propagation in May or June and the substitution of a controlled cold period of six to eight weeks for winter chilling, followed by 12 to 14 weeks' forcing in the greenhouse. The period of

FIG. 1. This fine crop of hydrangeas was grown at Foxpoint Growers, Encinitas, California.

availability for "summer production" of hydrangeas as blooming pot plants extends from January to early June. With freezing techniques or photoperiod manipulation, this season can be extended.

The cardinal points to remember about hydrangeas are that the flowers are initiated in the late summer to early fall. They develop in cool temperatures, either outdoors or in refrigeration, during the required 1,000 to 1,200 hours of cooling. They can then be grown at warmer temperatures to allow the stem with flower buds to elongate and the buds to further develop.

The resting hydrangea bud contains five to eight sets of leaves in addition to the initiated and partially developed flower bud. The flowering shoot must unfold these leaves and the flower by flowering time; therefore, a relatively long forcing period is required.

Hydrangea production declined in the recent past as a result of their need for space. Hydrangeas, unlike lilies, require a great deal of bench space in forcing, so hydrangeas have become relatively expensive plants to produce. Plants with three flowers need nearly 1¼ square feet (0.12 m²) per plant; plants with four to five flowers need 1½ square feet per plant. Calculate about ⅓ to ½ square feet per flower as a space requirement. More recently, however, hydrangeas have appeared to be regaining popularity. The plant fulfills a need for a showy, long-lasting plant that can be accurately timed for a holiday market. A well-grown hydrangea fulfills the demand for a distinctive, high-class flowering plant for all occasions.

Today, few hydrangea growers perform all phases of growth. Specialists take care of separate operations, including cutting production and propagation, summer growing, and cold storage. The discussion that follows covers only the greenhouse phase of production, with reference to summer growth only as it is necessary to reference greenhouse practices.

GREENHOUSE FORCING SEQUENCE

Dormant plants are placed in the greenhouse at forcing temperatures of 60 to 64F (16 to 18C) immediately after removal from storage. Plants can be placed in the greenhouse prior to potting, then potted as time permits, provided potting is done quickly.

One serious problem can develop when the plants are repotted in fresh media in larger pots than used for summer growth, 4 inches into 6½ inches (10 into 17 cm), for example: failure to root into the new soil, with subsequent lack of stem elongation and flower expansion. This is especially true when plants are forced early, and especially in the North. There are several ways to overcome this problem.

- ◆ Use 6-inch (15 cm) plants instead.
- ◆ Scarify the root ball rather severely to damage and expose root tips, thus encouraging them to grow into the fresh medium.
- ◆ Use media low in fertility.
- ◆ If a lead weight watering tube is used, place it directly over the original root ball. This ensures that the original root ball does not shrink away from the new medium.
- ◆ Some experts suggest starting forcing in trays out of the pot until root growth starts.

In any event, understanding the nature of the problem is necessary to solve it and avoid further problems.

Space pots closely for the first two to four weeks to save on heat and space. Final spacing can be made after potting.

During forcing make every effort to prevent plant growth from becoming soft and subject to excessive water loss or desiccation injury on removal from the greenhouse. Maximum sunlight, adequate space, and low humidity are important. Do not wet the leaves! Tube watering is practical during the forcing period, but mat or capillary watering is acceptable. Growth will be more vigorous with a constant supply of moisture to the roots.

Growth regulators are frequently used to prevent excessive height and to reduce space requirements. Plants forced for Mother's Day, particularly, may need a growth regulator. Application of a retardant during the forcing season is usually made during the third week of forcing, when leaves are 1½ to 2 inches (4 to 5 cm) long. B-Nine at 1,250 to 2,500 ppm, or A-Rest at 25 to 50 ppm, is satisfactory as a foliar spray. Some varieties, such as Bottstein, do not need growth regulators, while Rose Supreme and Lacecaps require more. Sister Therese should have a "delayed" application for toning up plants.

Should plants show signs of insufficient cold storage—as evidenced by slow development, short internodes, small leaves, or a general rosetted appearance— consider an application of gibberellin (GA). To overcome cold storage deficiency, GA at 2 to 5 ppm is used in the forcing period. A single foliar application may be adequate, but weekly applications may be made if plants do not respond. Careful observation is the only means to determine the number of applications necessary to restore growth.

As the sepals enlarge and become pigmented, it may be necessary to reduce light intensity to prevent fading and injury to sepals from excessive transpiration. Harden the plants as the sepals approach maturity by giving cooler night temperatures and ample ventilation. If plant growth has been restricted through environmental manipulation and growth regulators, staking and tying of flower heads should not be required. Multiflowering plants usually need no support.

Mature hydrangea plants can be held in refrigeration at 35 to 40F (2 to 4C) for several weeks, if necessary.

Out-of-season forcing. Forcing out-of-season hydrangeas extends their period of availability, increases market potential, adds variety to available potted flowering plants, and reduces production costs by growing hydrangeas at other than the coldest period of the year. Some of the principles involved in early flower initiation are also of value to southern growers who lack the cool fall temperatures for flower initiation.

Forcing hydrangeas for sales later than Mother's Day has always been possible by holding plants in refrigerated cold storage. Much of the heating costs of winter forcing is avoided, but market potential must be developed for any out-of-season production. Freezing dormant plants after their cold storage requirement has been tested and offers a means of having plants at other periods of the year. Some growers routinely force hydrangeas for late January and Valentine's Day market.

DETAILED CULTURE

Root media. As have growers of other crops, hydrangea growers have changed to media containing little or no soil. Because of that, watering is easier, a more vigorous root system is possible, and where plenty of peat moss is included, the all-important water supply to the hydrangea plant is provided. Soilless mixes contain little buffering capacity relative to soil, however; therefore, attention to initial pH and pH maintenance is very important.

Field soil, where used, should not constitute more than one-third of the total bulk of media, with other ingredients being peat moss and perlite, vermiculite, pine bark, composted hardwood bark, and so on. In a soilless mix, the media should have at least one-third peat moss for its moisture-holding capacity. With peat moss becoming scarcer, other forms of organic matter, particularly composted materials, can be substituted with excellent results.

All media should have ground limestone added to attain a pH of 5.5 to 6.0 (except when plants are to be "blued"). A source of slowly available minor elements is also necessary. Media with compost usually need no further minor elements added. Steaming can eliminate weeds and pathogens in media containing field soil. In the shift to a larger size of pot at forcing, a soilless mix containing peat moss or peat moss alone can be packed around the original mutilated soil ball. Gypsum, to add calcium without affecting pH, should be added to soils for both pink and blue plants.

Fertilization and color control. Understanding the effects of fertilization on color change in pink-blue hydrangeas is essential. The sepals of hydrangeas contain a red anthocyanin pigment, which becomes blue upon reacting with certain metals, including aluminum. The relative availability of aluminum, which is abundant in most field soils, is thus the principal factor in determining the color of the florist's hydrangea. Unless steps are taken to prevent aluminum uptake, the pink sepals gradually become blue. Just as importantly, unless enough aluminum is present to react with all the anthocyanin completely, an intermediate color will be produced, instead of the desired clear blue color. Intermediate colors are not attractive in most varieties.

Aluminum becomes more available to plant roots as soil acidity increases (pH values become lower). Growers usually lime to pH 6.5 to produce pink hydrangeas and acidify to pH 5.5 or less for the production of blue flowers. *In artificial media these pH values often are one-half to one unit lower.*

Phosphorus will also render soil aluminum unavailable—high phosphorus and high nitrogen during flower development promotes clear pink sepals. Low phosphorus and nitrogen, but an abundant supply of potassium, promotes clear blue sepals when the medium contains plenty of aluminum.

Hydrangea growth during forcing requires a relatively high nitrogen ratio; a 2-1-1 or 3-1-1 ratio is adequate. Plants are not fertilized in cold storage, nor are they heavily fertilized during the late phases of summer growth. For this reason, give attention to the effects of fertilization in both the potting media and the liquid feed program in the greenhouse. Since the root ball of the dormant plant is likely low in fertility, using potting media high in added fertilizers will not promote rooting into the new media when, for example, 4-inch (10-cm) dormant plants are potted into 6½-inch (17-cm) pots. Use potting mixes low in initial soluble fertilizers. Raise fertility with liquid fertilizers or top-dressed dry fertilizers *after* root growth is initiated in the media.

If plants have been cold-stored in the pots that they are to be forced in, begin fertilization immediately. Solubridge readings of 1.25 to 1.75 mmhos × 10^{-6} with a 1:2 soil:water dilution are satisfactory.

For production of pink sepals, alkaline-residue fertilizers, such as calcium nitrate, are used. When pink color is difficult to attain, ammonium phosphate (either mono- or di-ammonium) at 700 ppm N can be used on alternate weeks after flower buds are visible. The phosphate fertilizer ties up aluminum in the soil, which will cause blue color to develop. Replace fertilization when sepals are in full color and plants are hardening.

In the production of blue-flowering plants, the dormant (summer) plant grower needs to have made summer applications of aluminum sulfate. Summer applications of aluminum sulfate alone, however, are not sufficient to produce a reliable blue color. Four or more applications are necessary in the greenhouse, in addition to planting in a soil low in lime. Aluminum sulfate at the rate of 10 pounds in 100 gallons of water (4.8 kg per 400 l) is satisfactory. This material should be applied only to moist media.

Fertilization for plants being forced for blue sepals should be lighter than for pink sepals. Use low phosphorus and high potassium levels for the clearest blue color. Additional applications of aluminum sulfate, made on several alternate weeks after flower buds are visible, should ensure complete bluing of the sepals. Additional applications may be required if the soil or water is alkaline.

Low nitrogen rates are usually used on plants being forced as blue to produce clear blue colors. For blue, use 1 to 200 ppm N; use 2 to 400 ppm N for pink varieties. White varieties are best fertilized on the pink sepal program for best plant appearance.

Temperature, photoperiod, and forcing. Photoperiod may affect the rate of development and type of growth during forcing. Plants placed in cold storage early and forced under the long nights of November, December, and January, or which have not had an adequate storage period, will benefit by a night break with 10 footcandles (108 lux) of incandescent light. Additional light will have little effect on plants forced late in the season, as these plants have had longer bud development and rest periods.

There may be a slightly greater effect from increasing the lighting period to eight hours or all night, but a light break from 10:00 P.M. to 2:00 A.M. appears to have near-maximum effects during forcing in increasing the rate of development, height, and flower size without adversely affecting quality.

Forcing temperatures regulate not only the rate of development but the ultimate height, size of cymes, intensity of sepal color, and quality of the finished plant. Basically, hydrangeas are cool-temperature plants, making their best growth at night temperatures below 60F (16C), although the rate of development will be faster at a higher temperature. Night temperatures in the mid-50sF (11 to 14C) will produce taller stems, larger leaves, and larger flower heads than growing at 62 to 65F (17 to 18C). Representative forcing periods at different night temperatures are 16 weeks at 54F (12C), 12 weeks at 60F (16C), and 10 weeks at 65F (18C). At a temperature of 60 to 62F (16 to 17C), buds are visible eight weeks before bloom; they measure ¾ inch (2 cm) at six weeks and 1½ inches (4 cm) in diameter at four weeks before flowering. The old rule of buds being pea size eight weeks before sale, nickel size six weeks before sale, and half-dollar size four weeks before sale is still valid.

HYDRANGEA VARIETIES

Most hydrangea varieties originated in Germany, France, Belgium, and Switzerland. Some have retained their original names, while others have been renamed by the introducers and are known by different names in the United States. Only one currently popular variety, Rose Supreme, is of United States origin.

The varieties most popular in the United States have quite large, distinctive flower heads, or cymes, which are usually grown with three or more flowering stems to the finished plant. Some single-stem plants are being grown; however, the danger of blind shoots deters many growers from growing and forcing single-stem plants. Many newer varieties that branch freely but have smaller cymes are grown with five or more flowering stems per plant.

Plant characteristics (table 1), such as branching, stem length, size of flower head, and especially the precise time required to force, depend upon climate and

TABLE 1   HYDRANGEA CHARACTERISTICS

| Variety | Flower | | | Plant height | Heat tolerant |
	Color	Head size	Days to flower at 60F (16C)		
Bottstein	Red	Medium	92	Medium	Not very
Jennifer	Red	Small	85	Medium	Yes
Kuhnert	Light blue	Medium	90	Medium	Okay
Mathilda Gutches	Blue	Medium	90	Medium tall	Yes
Merritt Supreme	Medium pink	Large	85	Short	Yes
Rose Supreme	Light pink	Very large	95	Very tall	Yes
Sister Therese	White	Large	78	Short	Not very

culture, so it is difficult to accurately characterize varieties. In the north early varieties are expected to force in 12 weeks, midseason varieties in 13 weeks, and late varieties in 14 weeks under usual forcing conditions. Sepal colors are always more intense when forced at cooler temperatures. In the interest of plant quality, water relations, and consumer satisfaction, use the following schedule of pot sizes for finishing plants in forcing.

Rose Supreme. At one time Rose Supreme was the most popular hydrangea variety (table 2). In the United States, Merritt Supreme, a very different plant, is now first in sales. Rose Supreme is very vigorous and produces extra-large, light pink or light blue heads. Intermediate colors are also good. It is tall, requiring B-Nine applications during summer growth and, usually, two to three B-Nine applications during forcing to reduce height. Rose Supreme is a late variety, requiring about one week longer to force than most, and two weeks longer than Merritt Supreme. It stands up well in the heat and is used extensively in Texas, Florida, and other southern areas. Rose Supreme may be difficult to bring to flower for the earliest Easters. It is always suitable for Mother's Day, though.

Merritt Supreme. Rose Supreme's replacement, Merritt Supreme is the leader because it is an early, strong-growing variety, flowering one week earlier than most and two weeks earlier than Rose Supreme. Its flowers are large but smaller than Rose Supreme's. Less bench space is needed compared to Rose Supreme, and the plant is shorter, requiring less B-Nine. It usually does not require more than one application of B-Nine at 2,500 ppm during forcing. Merritt Supreme tends to a dark rose color. It makes a very dark blue. Intermediate colors may be unattractive, though. Merritt Supreme is grown extensively in the South and seems to stand up well in the heat, but perhaps not as well as Rose Supreme.

Kuhnert. Also difficult to grow in the summer, Kuhnert is an excellent forcer if grown cool. It blues easily, giving clear light blues. Kuhnert is not particularly adapted for southern forcing, although some use it for its color. At Dahlstrom and Watt, all Kuhnerts are blued in the summer. It branches well and has smaller flowers. Kuhnert typifies the often-discussed "new European" varieties, but it has been grown in the United States for years.

TABLE 2   HYDRANGEA BLOOMS AND POT SIZES

Blooms	Rose Supreme		Other varieties	
	inches	cm	inches	cm
1	6	15	5.5	14
2	6.5	16.5	6	15
3,4	7	18	6.5	16.5
5	8	20	8	20

Note: Pot sizes for finishing plants in forcing.

FIG. 2. Lacecap hydrangea varieties are adding new interest to this floriculture favorite.

Sister Therese. Most whites are not of high quality as regards hardiness of flower and compactness of growth. However, Sister Therese is as good as any and better than some. It forces early and readily with large-size heads. Provide extra protection against sunburn during forcing. Sister Therese usually will not need B-Nine early, but it will benefit from a "toning" application later, after its height is nearly established. Sister Therese flowers 10 days ahead of Merritt Supreme. When it is received in shipment the same time as Merritt Supreme, it should be delayed by cold storage or in a cold greenhouse for 10 days before forcing.

Regula. Another white, Regula has a more symmetrical habit than Sister Therese. It also flowers later, which as a result may offer advantages. Regula is not often offered, though, because of its rank growth and extremely soft nature.

Blue Danube. With Blue Danube, it has been difficult to develop adequate plant size during summer growth, so it is not popular.

Bottstein. A short variety with red-rose sepals, Bottstein is midseason in forcing. It is very short and does not require B-Nine. Quality is excellent, with dark green foliage and rather large flowers. Bottstein is a distinctly different color and deserves consideration as a variety to supplement Merritt Supreme. However, its short height may be a problem.

Jennifer. Among the reddest hydrangeas available, Jennifer has lighter green foliage, and the sepals are greenish in the center, especially in the early stages. It makes an outstanding addition to the color selection. Growth is not especially vigorous with respect to stem diameter, and flowers are generally smaller than Merritt Supreme. Flower numbers, however, make up for flower size. Some growers use up to 10% of finished plant production in red so that they can include from at least one plant to half the case in nearly every mixed case.

Mathilda Gutches. This European variety branches profusely. Its flowers are smaller, but the show is comparable because of flower numbers. It makes an excellent blue as well as medium pink. Mathilda Gutches, a well-formed plant with excellent steel-blue flowers, is the most popular variety at present for blue. It can be tall, requiring B-Nine because of its smaller-diameter stems.

Fire Light (Leuchtfeuer). A new variety in the United States, Fire Light is dark pink and more rose-colored than Merritt Supreme. Under very cool temperatures (50 to 55F, 10 to 13C) at finishing, it can be nearly red. Fire Light breaks freely, has sturdy stems, and features pretty, large flowers. Some say it is susceptible to mildew, but that has not been everyone's experience. We believe it to force slightly faster than Merritt. It apparently performs well in the South. Fire Light is a vigorous variety and will require B-Nine.

Masja. A new variety about which we know very little, Masja is a pleasing red-pink color, forcing along with Merritt Supreme. It has good growth habit and light green, pleasing foliage, and grows well. Try it as a trial on a limited basis. Masja's vigorous growth will likely require B-Nine.

Lace caps. This large, diverse group of plants has one row of sterile flowers surrounding a cluster of fertile flowers. The fertile flowers are usually an attractive blue. The lace caps are novelties. Forcing time is about two weeks slower than Merritt Supreme. The color of the flower improves with age, the sepals being very light when they first expand. White lace caps are of great interest. Powdery mildew can be a problem with all lace caps, but Rubigan will give control.

Sara. A light pink of very good substance, Sara has a forcing time about one week slower than Merritt. It is worthy of trial.

DISEASES

Hydrangeas are not often subject to root and stem rots, provided usual sanitation and good cultural practices are followed. Propagation benches, all media, as well as pots, flats, and so on, should be disinfected by steam or other methods to eliminate pathogens and weeds. Good drainage is important, as is avoidance of overwatering. Fungicide drenches are usually not needed with good culture. Note: *Terraclor should never be used on hydrangeas.* It acts as an herbicide and can cause severe crop damage.

Botrytis. Hydrangeas may get gray mold on leaves, stems, and cymes in the greenhouse or on buds in storage. It occurs in situations with high humidity and high moisture. It frequently starts on injured or dead tissue, so injured leaves should not be left on plants. Ventilation, air circulation, and avoiding overhead watering greatly reduce the incidence of gray mold. Ventilation to reduce humidity, especially at sundown, is important as the cymes mature. *Botrytis* often starts in the center of the cyme and as such it is hard to detect. A fungicide such as Daconil can be used, if needed, taking care to consider spray residue.

Powdery mildew. The most prevalent disease on outdoor plants in the fall, powdery mildew is also a problem in the greenhouse under conditions of high humidity and crowding. Older leaves are most susceptible. A protective mildicide can be used in the fall. Humidity control is usually adequate to prevent serious infection in the greenhouse. Rubigan is a possible control.

There is a wide range of variety susceptibility to both gray mold and powdery mildew. Do not grow susceptible varieties where an environment conducive to disease cannot be avoided.

Viral agents. The hydrangea ring-spot virus has been found in most present-day commercial varieties. Typical symptoms show only during winter growth, and the effect on susceptible varieties is generally weakened or smaller growth. Roguing is difficult, and virus-free plants of commercial varieties are not currently available.

Another disease, the green sepal mycoplasma complex, has been responsible for a series of problems. Its effects may be seen in different but related groups of symptoms, according to severity of disease. With a severe infection, extreme stunting, small leaves with vein yellowing, and dwarf, green cymes are followed by death of the plant. An intermediate case presents a reduction in vegetative growth, but with normal leaf expansion and continued vein yellowing. The cymes will contain both green or bronzed sepals and normal-colored sepals. With mild virescence, stock plants gradually decline in vigor. Forcing plants retain their normal patterns of growth with green leaves, but cymes contain large, green sepals, and the reproductive parts may revert to a vegetative type of growth.

OTHER PROBLEMS

Bud failure. Failure to initiate flower buds or evidence of crippled buds during forcing may be due to poor culture during summer growth, frost injury during storage, or bud rot (gray mold) in storage or shipping. Initiation of flowers early in the summer may result in fewer than normal leaves at forcing, causing poor flower development because of the lack of leaf area. Cymes containing leaves are also associated with early initiation. Removing leaves from the cymes early in forcing usually permits the cyme to develop normally.

Iron chlorosis. Interveinal yellowing is frequently a problem during rapid forcing without adequate root development. Chlorosis can also be due to alkaline soil, overfertilization, or overwatering during summer growth. Plants grown for pink are more likely to show iron chlorosis than those being blued because of the greater availability of iron at lower pH.

Chemical burn. The young growth of hydrangea leaves and flowers at forcing is very susceptible to injury from insecticides, fungicides, and growth regulators. Use caution with any chemical spray and confine dosages to the lower recommended rate. Hydrangeas do not tolerate herbicides of any kind. Clorox use in close proximity to hydrangeas can cause severe damage.

Insects. The usual run of insects may be found on hydrangeas, but the most common problems are aphids during the forcing period and two-spotted spider mites during summer growth. Plants should undergo continuous inspection. Use appropriate insecticides or miticides to bring any infestation under control. Soil-applied systemic insecticides may not always be satisfactory, particularly during forcing after the plant has developed a woody stem. Slugs and snails may be present on plants as they are brought from storage. They can also be particularly troublesome where plants are forced on soil or on solid-bottom beds.

Author's note: Much of this discussion is adapted from articles by Dr. James B. Shanks published in previous editions of the *Ball RedBook*. Additional information on propagation, summer growing, and more on hydrangeas can be found there.

TABLE 3   USEFUL CHEMICALS FOR HYDRANGEAS

Trade name (chemical name)	Function	Registration
Growth regulation		
A-Rest (ancymidol)	Growth retardant. *Hydrangea* spray applications as for chrysanthemums have been effective	Not registered for *Hydrangea*
B-Nine (daminozide)	Growth retardant	Registered for *Hydrangea* on summer growth and at forcing
Pro-Gibb (gibberellin Ga³)	Powerful growth stimulant	Gen. stimulation of blooms
	Young growth out of cold storage: 2–5 ppm	
	Rosetted plants in full leaf: 25 ppm	
	Before defoliation and storage: 50 ppm	
Pest and disease control		
Daconil 2787 (chlorothalonil)	Spray to protect against gray mold and mildew	A number of flowering and foliage plants
Enstar (kinoprene)	Juvenile hormone to control aphids.	General use on ornamental plants
Karathane (dinocap)	Labeled as both a fungicide and miticide for control of powdery mildew and two-spotted spider mites on *Hydrangea*	
Orthene (acephate)	Controls many insects	Greenhouse and outdoor floral crops
Pentac (dienochlor)	Controls resistant two-spotted spider mites	Many ornamental plants and flowers
Rubigan (fenarimol)	Controls powdery mildew	
Truban (ethazol)	Soil drench for control of *Pythium* and *Phytophthora* root rot	Many types of flowers and ornamentals

HYPERICUM

by Jeff McGrew
McGrew Horticultural Products and Services
Mount Vernon, Washington

❖ *Perennial (***Hypericum androsaemum***). Vegetatively propagated. Grown as a cut flower.*

Hypericum species have been part of the landscaping and perennial garden for years. Most hypericums flower under long-day conditions and are considered long-day plants. Recently, *Hypericum androsaemum* and *H. inodorum* species and hypericum hybrids have been used by professional cut flower growers to produce a unique and colorful filler for flower arrangements. The most common hybrids are Autumn Blaze, which has clusters of reddish brown berries, and Excellent Flair, which is quicker to flower than Autumn Blaze. Excellent Flair has fewer berries (also brown to red in color), however, and is also more susceptible to rust than Autumn Blaze. Both perennial varieties are hardy to about Zone 5. Extreme heat and humidity are also detrimental to plant survival and quality.

Propagation of these varieties is vegetative, normally by rooted cuttings. It's important to build the young rooted plant properly so its growth habit will yield maximum stem length and number of stems. All rooted cuttings should be pinched, leaving behind three to four pairs of leaves.

Hypericums do best in well-drained, balanced soil with average fertility and a pH of 5.0 to 6.3.

Established rooted cuttings are planted out at a finished spacing of about 15 to 18 inches (38 to 46 cm) from one plant to the next, going down a row. Final spacing may vary, depending on whether the crop is grown in the field or greenhouse and how mature the crop is.

Both Excellent Flair and Autumn Blaze thrive under a moderate growing environment, with night temperatures in the 50sF (10 to 15C) and days 75 to 85F (24 to 30C). Under these conditions, and with mum lighting to extend day length, it may be possible to achieve two flower flushes per year. Dutch research shows that stem quality, length, and production are improved when plants are given a total of 14 hours of light (normal and artificial combined) per day until average stem length is about 12 to 15 inches (30 to 38 cm) long. At this time, increase total day length to 20 total hours and maintain it until flowering is completed. Then reduce day length to 14 hours until berry maturity or harvest.

Normally, though, when only natural light levels are available, the crop produces one quality flush and, consequently, only one quality berry setting per year. One-year-old plants from rooted cuttings yield two to three stems the first year; six to eight the second year, and 12 to 16 thereafter.

Good culture practices also include managing crop fertility, especially nitrogen. Before flowering occurs, minimal amounts of nitrogen encourages flowering. After the berries have fully developed, back off on excessive watering so as not to burst them.

After the berries have colored, it is time to harvest. Leave about one-half inch (1.3 cm) of stem behind after harvest. Postharvest treatment, using a bactericide designed to help products with semiwoody stems take up water, will greatly increase the vase life.

H. androsaemum and *H. inodorum* are both susceptible to mildew and rust fungi if conditions are mild and wet or humid. A proper fungicide control program should be followed. Drip irrigation versus overhead watering is also beneficial in controlling disease.

IBERIS (HARDY CANDYTUFT)

by Jim Nau
Ball Horticultural Company
West Chicago, Illinois

❖ *Perennial (Iberis sempervirens). 10,000 seeds/oz (353/g). Germinates in 14 to 21 days at 60 to 65F (16 to 18C). Leave seed exposed to light during germination.*

A low-growing, evergreen perennial that ranges from 8 to 12 inches (20 to 30 cm) tall, *Iberis* has flowers that are pure white and held in clusters in early spring. Plants are dependably hardy and are among the first returning perennials to show flower color in the spring. As they age, the base of the plants turns woody. Plants flower in April and May and are hardy to USDA Zones 3 to 9. Commercial propagators offer *Iberis* as plugs or liners from both seed or vegetative cuttings.

Seed sown anytime after the first of the year will not flower the same season. Green packs are salable 10 to 14 weeks after sowing, and plants will attain their full height the same season from seed. For flowering plants in April and May, sow seed in June and July of the previous year, transplant to cell packs by the end of August, and shift up to quart (liter) containers when ready. Overwinter and bring back into a cool greenhouse to finish for spring sales. Bare-root divisions can be potted in the fall, overwintered, and sold in flower the following spring.

Snow White is about the only *Iberis* variety sown from seed that is sold by name. Seed-grown plants produce smaller flower clusters than the vegetative material. Vegetatively propagated varieties are also more uniform than seed-sown material, and you will notice variability in both purity of flower color and foliage darkness in Snow White.

Vegetatively propagated varieties include Snowflake, Alexander's White, and October Glory, which is unusual because it blooms the first spring and again in the fall.

IMPATIENS (BEDDING PLANT)

by Brian Corr
PanAmerican Seed Co.
West Chicago, Illinois

❖ *Annual (***Impatiens wallerana***). 46,000 seeds per oz (1,623/g). Germinates (radicle emergence) in four to five days at 70 to 75F (21 to 24C). Leave seed exposed to light during germination.*

To most people "impatiens" means *Impatiens wallerana* (fig. 1). This species is native to Kenya, Tanzania, Malawi, and Mozambique but has naturalized in many places around the world, including Central and South America, Southeast Asia, and some Pacific islands. It has even become a common weed in some coffee plantations.

Impatiens wallerana was brought to England in 1896 by Dr. John Kirk, a physician who had been a member of Dr. David Livingstone's (of "Dr. Livingstone, I presume" fame) expeditions to Africa. This species was originally called *Impatiens sultanii* in honor of the sultan of Zanzibar but was renamed *Impatiens wallerana* to honor Horace Waller, a British mis-

FIG. 1. Impatiens form the base of most growers' bedding plant crops. The hundreds of flowers each plant produces over the course of the summer make it the favorite of home gardeners everywhere. Try impatiens in flats, pots, and baskets. The variety shown: Super Elfin Coral.

sionary in Africa who published the journals of Livingstone.

Up until the mid-1960s sales of impatiens were slight. In 1968 the seventh edition of the classic greenhouse textbook *Commercial Flower Forcing* by Kipplinger, Laurie, and Post mentions impatiens only once, in a paragraph labeled "Other bedding plants." By 1983, however, impatiens had become the third most popular flowering bedding plant from seed, and by the end of the 1980s, impatiens was the best-selling bedding plant in the United States.

Sales of impatiens continue to boom. In the 36 states surveyed in the United States Department of Agriculture *Floriculture Crops Summary*, impatiens accounted for 15.1 million flats with a value of $104.3 million (an average of $6.90 per flat). Potted impatiens accounted for $17.7 million from 25.0 million pots sold (fig. 2). Hanging baskets resulted in $19.2 million in sales from 3.6 million units (fig. 3).

FIG. 2. Jeff Lovell, Lovell Farms, Miami, Florida, shows off his fine impatiens pot. Lovell Farms has 175 acres and primarily produces 4-inch annuals.

Much of the commercial success of impatiens as a garden plant can be traced to the early breeding work of Claude Hope of Linda Vista, Costa Rica. Claude went to Costa Rica in 1943 and saw impatiens growing in the hills and fencerows. Some time later, after he was successful in a flower seed production company, Claude began to hybridize impatiens. The impatiens of that time was more foliage than flowers, with large leaves, and did not branch well, and Claude's goal was to produce a dwarf, symmetrical, basally-branched plant that flowered profusely.

By 1965 Claude Hope had an impatiens series of eight colors that was tested successfully in trial gardens at

FIG. 3. Pat Brister, Hickory Hill Nursery Ltd., Forest Hill, Lousiana, an excellent bedding grower, displays an impatiens basket of which she is justly proud.

Michigan State and Purdue Universities. The following year PanAmerican Seed Company grew this series, which was introduced in 1968 as Elfin. In the span of a few years, the Novette, Fantasia, and Futura series also appeared.

The hard work of numerous breeders has resulted in many series of impatiens from seed which have all the characteristics Claude Hope was looking for when he began his breeding efforts: prolific production of large flowers in many colors, compact habit, and basal branching. In addition, traits that were less important in the 1960s now receive significant attention in breeding, especially seed quality. Through diligent breeding, impatiens are now available with flowers of many shades and colors, with two-tone petals, and with double flowers. Impatiens with double flowers are available from seed, with some variability, and also from cuttings. Impatiens with variegated foliage cannot be propagated reliably true to type from seed, however, and must be propagated vegetatively.

PROPAGATION BY SEED

Plugs have revolutionized the bedding plant industry, but they have also led to higher standards for seed germination. To be most successful, plug producers must be able to achieve nearly 100% germination and near-perfect uniformity. All factors must be maximized to achieve these goals. Breeders must develop cultivars with genetic potential for vigorous seedlings, seed producers must produce and handle the seeds under ideal conditions, and distributors must store and handle the seeds properly. Once these factors are in place, the grower must provide the ideal conditions for germination. Of course, some growers buy plugs from a specialty propagator to simplify production.

Seed storage. Impatiens seeds, like all other living things, require a favorable environment to survive. Unfortunately for impatiens seeds, they often have to share their living space with human beings with different environmental expectations. Impatiens seeds often mark time in offices, in break rooms, or in a box next to the seeder. These environments are usually more suited for human comfort than for impatiens seeds.

Store impatiens seeds at approximately 40F (4C) with low humidity (25 to 30%) for longest life and best viability. If humidity cannot be controlled, at least store the seeds in the refrigerator. Research shows impatiens seeds deteriorate quickly at 72F (22C), yet they retain acceptable quality at 41F (5C), even with up to 45% relative humidity.

Temperature is most important, but for maximum impatiens seed quality, humidity must be controlled, as well. For best success, construct a seed storage chamber with controlled temperature and humidity. Desiccant-type dehumidifiers are the only practical way to reduce humidity below 30%.

Do not open impatiens seed packages until ready to sow. Whenever possible, use an entire package per sowing. If an open package must be stored, return the open packet to low temperature and humidity conditions as soon as possible. Do not reseal a seed package until the seeds have had been in dry conditions for at least 24 hours to remove any moisture absorbed from the air while the packet was open.

Germination and seedling culture. Lowell Ewart, professor of horticulture at Michigan State University, in 1985 wrote in one paragraph the essence of what is needed for uniform impatiens germination: "Improper moisture maintenance can lead to very uneven stands or much reduced stands. The seed is light requiring and usually is not a problem unless the seed has been covered too deeply. . . . Germination has been excellent when the medium temperature is 70 to 75F."

Sow seed into a well-drained, disease-free seedling medium. Use a medium with a pH of 6.0 to 6.5 and electrical conductivity less than 1.0 millimhos per centimeter with a 1:2 extraction. The seed does not need to be covered. Irrigate the plug trays after sowing, then never allow the soil surface to dry until the radicle has penetrated the medium. As the cotyledons expand, begin to allow the soil to dry, making sure the soil is never waterlogged. Once roots reach the bottom of the plug cell, reduced moisture levels will ensure a healthy root system.

Germinate and grow impatiens seedlings at 72 to 76F (22 to 24C). Impatiens germination only varies by a few percentage points between 72 and 80F (22 and 27C) yet drops off very quickly outside this optimum range (fig. 4). Seed temperatures above 80F from sowing to radicle emergence will reduce germination percentage and uniformity. Plugs may be held at 65F (18C) to tone the plants prior to transplant.

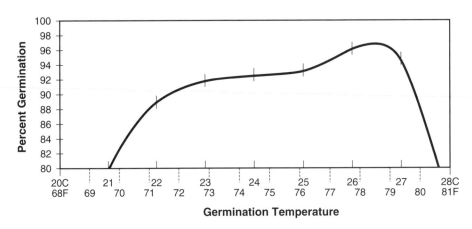

Fig. 4. Temperature and impatiens germination in eight days.

Maintain 100% relative humidity until cotyledons emerge. Low humidity from sowing to cotyledon development will lead to reduced germination and a lack of uniformity. If the walls or floor of the germination area is dry during this stage, the humidity is too low. Reduce humidity to approximately 50% as plugs mature, though.

Fertilize impatiens with 50 ppm nitrogen from 15-0-15 as soon as radicles emerge. When cotyledons expand, increase fertilization to 100 to 150 ppm nitrogen from 15-0-15. Use 20-10-20 with every other fertilization if growth slows. Maintain growing medium electrical conductivity around 1.0 mmhos/cm (1:2 extraction). Maintain pH between 6.0 and 6.5.

Light. Not long ago, when all seeds were germinated in the greenhouse with natural light, there was no need to be concerned with supplying light to improve germination. With the use of germination chambers, however, light requirements are more of an issue.

Only 10 footcandles (108 lux) of light are necessary for improved impatiens germination. This is approximately the amount of light necessary for most people to be able to read a newspaper at arm's length. Keep in mind that this is the amount of light needed at the seed, not at the edge of a plug tray at the edge of a pallet. Cool-white fluorescent lights are inexpensive to buy, economical to operate, and they can be mounted on the walls of the germination chamber to provide uniform light. Use them for eight to 12 hours per day.

Seedlings must receive more light immediately after germination to avoid elongation, so they are usually removed from the germination chamber at this point. Maintain light levels between 1,000 and 2,500 f.c. (11 and 27 klux) during Stages 2 and 3. As seedlings mature, light levels may be increased up to 5,000 f.c. (54 klux) if temperature can be controlled.

Growth regulation. Control plug growth first by environment, nutrition, and irrigation management. Temperature differential (DIF) can also be used to minimize height. Then consider chemical plant growth regulators, if necessary. Minimize ammonium-form nitrogen fertilizer (such as 20-10-20 or 20-20-20) to avoid elongation of seedlings. For chemical plant growth regulation, test Bonzi, Sumagic, or A-Rest sprays at 5 ppm, 1 ppm, or 15 ppm, respectively.

Plug storage. Best growth occurs if impatiens are transplanted as soon as the seedlings have sufficient roots to be pulled from the plug tray. Occasionally, seedlings need to be held until they can be transplanted. Impatiens seedlings may be stored at 45F (7C) for up to six weeks. Temperatures lower than 45F increase the chance of chilling injury. Light is not essential during storage but may be beneficial.

PROPAGATION BY CUTTINGS

Several of the variegated and double-flowering impatiens are vegetatively prop-agated. Harvest cuttings with two to three nodes. Be sure the nodes are vegeta-tive, indicated by a small shoot or leaves emerging from the node. Cuttings with only reproductive nodes will not branch. Reproductive nodes will have flower buds or a scar where the flowers have fallen off.

Store cuttings at 45F (7C) for up to 24 hours prior to sticking them. This removes field heat from the foliage and will improve rooting. Use a rooting hor-mone to increase rooting uniformity and speed, especially in the winter. Maintain 72F (22C) air temperature and 75F (24C) rooting medium temperature. Maintain moisture in the cuttings as they root by misting the foliage or by cov-ering with white plastic over hoops to maintain high relative humidity. Excess mist will waterlog the rooting media and delay rooting, however, so decrease mist frequency as soon as the cuttings are turgid.

Once roots begin to form, begin feeding the cuttings with 100 ppm nitrogen from a balanced fertilizer to enhance root and new shoot growth. Cuttings will root within 21 days under normal conditions. If the cuttings are stressed, rooting is significantly delayed. Temperatures below 65F (18C) will inhibit rooting.

GROWING ON

Transplant into a well-drained, disease-free, soilless media with an initial nutri-ent charge and a pH of 5.5 to 6.3. Ideal fertilization varies, depending on water quality and grower practices. A general recommendation is to fertilize every sec-ond irrigation, alternating 15-0-15 with 20-10-20 at 150 ppm nitrogen. Excessive fertilization will result in lush growth with few flowers. Maintain medium elec-trical conductivity around 1.0 mmhos/cm (using 1:2 extraction).

Impatiens can be grown within a wide range of temperatures and light levels. Most growers find night temperatures of 62 to 65F (17 to 18C) with day temper-atures of 65 to 75F (18 to 24C) to be effective. Lower temperatures result in more compact plants but increase crop time. Carbon dioxide added to the green-house air at 1,000 to 1,500 ppm speeds plant development.

Impatiens will tolerate light levels down to about 2,000 f.c. (22 klux) but grow best with more light. Impatiens will grow well at light levels up to 8,000 f.c. (86 klux) if moisture is in adequate supply and temperature is in the appro-priate range. High light levels coupled with high temperatures can result in leaf scorch. Most growers find they need to reduce light levels for temperature con-trol rather than to avoid light injury.

Effective height control of impatiens can be accomplished with environ-mental manipulation. Once plants are rooted to the sides of the containers, they can be allowed to wilt prior to irrigation to provide some height control. Height can also be controlled by withholding fertilizer, especially phosphorous and ammonium-form nitrogen. Impatiens are responsive to day-night temperature differential (DIF), being shorter with a negative DIF. Bonzi and Sumagic are effective for impatiens height control, although these compounds are not labeled for this use in all locations. Always follow label recommendations. Test sprays with Bonzi at 15 to 25 ppm or Sumagic at 5 to 10 ppm.

PROBLEMS

Seedlings. Anything that inhibits germination will increase the variability between seedlings in a plug tray. Some of the most common practices that result in seedling variability include (1) low relative humidity during Stages 1 and 2 resulting in uneven drying; (2) lack of light during Stage 1; (3) low nutrient charge in the germination medium and no soluble fertilizer applied immediately after germination; and (4) temperatures above 78F (26C) during germination. Low seed vigor will also result in uneven seedling development. Seeds which are old or have been stored improperly may have low vigor.

Occasionally, some impatiens seedlings will germinate normally up to the point of cotyledon expansion, but the growing point will abort, and the seedling will stop developing. In some cases, the growing point will not abort until after one or two leaves have developed, or it will continue to develop, but growth will be abnormal. Tip abortion is more severe on seedlings grown from low-vigor seeds. Numerous cultural factors have been implicated in impatiens tip abortion. Often, though, two growers will have what appear to be identical cultural practices, yet one will have tip abortion while the other will not. Until this mystery is fully resolved, avoid waterlogged soil and sources of ethylene. Do not use overhead irrigation if your water has high electrical conductivity. Be sure the medium and fertilizer have sufficient boron and calcium to avoid deficiencies.

Diseases. With *Rhizoctonia* stem rot, dark brown stem lesions form near the soil, often causing the plant to collapse. Use good greenhouse sanitation to avoid spreading the disease. Drench with Banrot, Cleary's 3336, or a similar product. With *Rhizoctonia* leaf and stem blight, foliage "melts" and collapses, often spreading from one location in a plug tray or flat. This is usually only a problem with temperatures above 75F (24C) with high relative humidity. If humidity is high enough, tan or dirty white cobweblike fungus growth can be seen around the infection. Control by keeping humidity low, maintaining good sanitation, and spraying with fungicides such as Cleary's 3336 or a similar product.

Botrytis blight causes tan or brown leaf or flower spots, usually followed by gray, fuzzy spores. Maintain good air circulation and reduced humidity during production. Remove dead or injured leaves or flowers. Apply preventative sprays, such as Chipco 26019, Ornalin, or Daconil, or smoke with Exotherm Termil.

Pseudomonas bacterial leaf spot symptoms are easily confused with those of *Botrytis* or tomato spotted wilt virus (TSWV) or impatiens necrotic spot virus (INSV), so the disease can only be confirmed with a laboratory test. Leaves develop black, brown, tan, or purple spots, sometimes becoming water-soaked and infecting the petiole. The disease is spread easily by splashing water or handling the plants. Sanitation is the key to control. Disinfect anything that contacts the plants. Be especially careful when patching plug trays, as this could be an excellent way to spread the disease. Workers should wash their hands frequently. Keep the foliage dry as much as possible, especially at night. Drought stress in the seedling stage may also increase susceptibility to *Pseudomonas*. Some cultivars are more susceptible than others, notably those with apricot, coral, salmon, lavender, or bicolor flowers. No chemical control is completely effective, although some growers claim to have reduced the spread of the disease by spraying with either Phyton 27 or a mixture of Protect and Kocide.

TSWV and INSV symptoms include black, brown, tan, or purple leaf spots, often in concentric rings. Portions of the stem may be affected and turn black. Plants may be infected with the virus and not show symptoms until the plant is under stress, at which time the plant will collapse rapidly. The disease is spread by thrips feeding on infected plants, then transmitting the disease to other plants. The disease can be spread in a matter of minutes. All seed-propagated impatiens plants are free of the disease at germination. Impatiens propagated vegetatively must be purchased from a reputable supplier who produces cuttings certified to be disease-free. Whether propagated vegetatively or from seed, disease-free impatiens will be infected with the disease if not kept isolated, however. Many species of plants, including common weeds, can harbor the disease. The only effective controls are to be sure no plants in the greenhouse are infected with TSWV or INSV and to exclude thrips.

Other difficulties. Impatiens are naturally floriferous and bloom freely under all photoperiods. Certain cultural conditions can lead to poor flower production, however. Excessive fertilization, especially with ammonium-form fertilizer, will result in excessive foliage and few flowers. Impatiens plants which are moderately drought-stressed will flower more profusely than those with uniform moisture levels. Plants grown in low light (less than 2,000 f.c., 22 klux) will not flower profusely. Flower buds can also abort in response to ethylene in the air or use of Florel.

Impatiens are native to tropical climates and do not tolerate low temperatures well. Temperatures below 45F (7C) can result in stunted plants and leaf yellowing.

VARIETIES

Following is a short list of *wallerana*-type impatiens, all excellent for bedding plant production. Dazzler is excellent for cell packs, 4-inch (10-cm) pots and especially for the landscape market. Expo, available only outside North America, was developed for European environmental conditions and customer preferences.

FIG. 5. Impatiens are winners in hanging baskets. This grower has put cell pack plants into 10-inch baskets on the bench.

The Accent and Super Elfin series are the standards. Both are very well known and widely used, featuring a wide color range. Also in this class are Impulse and Tempo. Showstopper and Blitz 2000 are both extra-large-flowered series, and both are available in a wide color range. Try either for hanging baskets (fig. 5) or the landscape market.

Try novelties to spice up your regular impatiens mix, especially if you're a retail grower. Carousel, a seed-grown double with a higher percentage of double flowers than other series from seed, has excellent branching and germination. Dazzler Stars and Novette Stars have the best star patterns available. Deco is the first dark-leafed impatiens series from seed. Be sure to also include some Swirl or Tempo Frosts for their unique, dark-margined flower forms.

IMPATIENS (NEW GUINEA)

by Brian Corr
PanAmerican Seed Co.
West Chicago, Illinois

❖ *Tender perennial (**Impatiens** × **New Guinea**). 11,340 to 17,010 seeds/oz (400 to 600/g). Germinates in seven to 10 days at 78F (26C). Because of the limited colors and leaf forms of varieties from seed, most are vegetatively propagated. Growers produce pots and baskets from purchased rooted and unrooted cuttings.*

Few plants have risen in popularity as quickly as New Guinea impatiens. Introduced to the gardening public in 1972, sales have climbed at an annual rate of more than 15%.

Breeders continue to make significant advances in developing varieties that not only have the large, tropical-colored flowers the plants are famous for, but also have reliable flowering, excellent branching, and garden performance (fig. 1). Bicolor flowers add interest to the variety assortment.

CUTTINGS

When growing New Guinea impatiens, begin with high-quality plant material from a reputable supplier who propagates from certified disease-free mother stock. Begin the crop with vegetative cuttings; cuttings with buds take longer to root and flower and do not branch well. Cuttings root in three to four weeks at a minimum temperature of 68F (20C). Use mist or tents to maintain high humidity until callus forms. Then mist as needed.

FIG. 1. New Guinea impatiens are the bedding plant industry's single biggest success story over the past decade. Grown mainly in pots or hanging baskets, New Guinea impatiens' large, tropical-colored flowers are traffic stoppers. Most varieties are vegetatively propagated: Growers buy in and finish rooted cuttings for spring sales.

Cuttings are susceptible to a wide range of root and stem diseases. Therefore, it is vital to begin with sterile media, certified stock, and to maintain strict sanitation.

Use one rooted cutting per 4- or 6-inch (10- to 15-cm) pot, two to four for 10-inch (25-cm) pots or baskets and patio containers (figs. 2 and 3). Allow 10 to 14 weeks for finishing; hanging baskets may require longer. Do not pinch plants if you are using free-branching varieties, as you would only increase production time.

After planting provide high light, up to 7,000 footcandles (75 klux) and temperatures above 60F (16C). Maintain moderate moisture; avoid overwatering, but do not drought-stress plants, which would cause severe stunting. Fertilize with 200 ppm N at every other irrigation. If height control is a problem, especially for growers in the South, experiment with Bonzi or Sumagic.

Tango, with its bright orange flowers, is an excellent bedding plant, especially for landscape plantings. Plants grow to 18 to 24 inches (46 to 61 cm).

NEW GUINEA IMPATIENS FROM SEED

The first commercially significant New Guinea impatiens from seed was Tango, an All-America Selections award winner in 1989. Gardeners and growers were impressed with the large, 2- to 3-inch (5- to 8-cm), bright orange flowers on vigorous plants. An improved Tango became available in 1994, with increased

branching, a more compact habit, and brighter flowers. Tango foliage is bronze-green and effectively sets off the flowers.

The second commercially significant New Guinea impatiens from seed was the Spectra series. Spectra is available as Pink Velvet, Rose Shades, Lilac-Rose Bicolor, Red Shades, Salmon Shades, Borneo mixture, and a formula mix. The cultivars are designated "shades" because flower color differs slightly within a cultivar. The colors are uniform enough for plants to be sold in the same pack or planted together in a hanging basket, however. The degree of foliage variegation differs by cultivar. Lilac Rose has the greatest amount of variegation, Rose has almost none, and Salmon and Red seldom have variegation. Excessively variegated seedlings of some cultivars are occasionally produced and may be discarded at transplanting.

FIG. 2. Bob Bedner, Bedner's Farm Market, Cecil, Pennsylvania, a 1-acre grower, also has a garden center outlet where his top quality New Guinea impatiens baskets are popular items.

Spectra New Guinea impatiens are best produced in packs (36 or fewer plants per flat) and pots with diameters of 3 or 4 inches (8 or 10 cm) when planted one plant per container. Tango requires at least a 4-inch container. Hanging baskets with multiple plants per pot are also practical. A 10-inch (25-cm) basket with three (Tango) to five (Spectra) plants works well. Spectra will grow to 10 to 14 inches (25 to 36 cm) in the garden.

Germination and seedling production.

Production of New Guinea impatiens from seed is in many ways similar to production of any bedding plant from seed. The germination and seedling phase is the most critical. Although New Guinea impatiens and garden impatiens *(Impatiens wallerana)* are both produced from seed, there are significant differences in their culture. To reach optimum germination levels, the grower must pay attention to the following details.

FIG. 3. Russ Johnson, Red Oak Greenhouses, Red Oak, Iowa, displays 8-inch patio pots of New Guinea impatiens.

Temperature has a greater influence on germination success than any other factor. Optimum germination temperature for New Guinea impatiens is 78F (26C). The grower will have little success with germination if the media temperature is below 75F (24C) or above 80F (27C). Temperature control is less critical after Stage 1 (radicle emergence). Temperatures in the 70-to-75F (21-to-24C) range are acceptable from Stage 2 to transplant.

To be successful germinating New Guinea impatiens seeds, it is important to realize that their germination timing is different from that of *Impatiens wallerana*. While *I. wallerana* seedlings may be ready to remove from Stage 1 after three or four days, New Guinea impatiens requires seven to 10 days. Maintain optimum temperature and moisture level throughout this stage. New Guinea impatiens is ready to transplant from five-eighth-inch- (1.6-cm-) diameter plug cells (392-plug tray) approximately six weeks after sowing.

Research has demonstrated that New Guinea impatiens germinates with a higher percentage and greater uniformity if light is available during germination. A light level of about 100 footcandles (1,100 lux) is sufficient. Seedlings develop quickest with moderately high levels (4,000 to 5,000 f.c., 43,000 to 54,000 lux) from germination to transplant.

Although New Guinea impatiens has been shown to germinate better in the light, covering the seed lightly with coarse vermiculite also has been shown to improve stands. Presumably, sufficient light passes through the vermiculite to enhance germination. New Guinea impatiens seeds are larger than many bedding plant seeds (400 to 600 seeds per gram for New Guinea impatiens, compared to 1,400 to 1,600 per gram for *I. wallerana*). Perhaps covering helps maintain sufficient moisture around these moderately large seeds.

Fertilize plugs of New Guinea impatiens with 50 to 100 ppm N from a fertilizer such as 15-16-17 or 20-10-20 from when the cotyledons are fully expanded until transplant.

Growing on. Transplant New Guinea impatiens seedlings as soon as they can be removed from the plug tray without damage to their root systems. Always maintain temperatures above 60F (16C). Total light level has little effect on time of first flower, but it does have an impact on plant shape and number of flowers, which may influence time to sale. Plants grown at high light levels have more branches and flowers than those grown with less light. Grow seed-propagated New Guinea impatiens with greenhouse light levels up to 7,000 f.c. (75,000 lux), provided temperatures can be maintained at a moderate level.

New Guinea impatiens from seed differs substantially from *Impatiens wallerana* in its response to drought stress. Some growers use drought stress to control the height of *I. wallerana*, but drought stress of New Guinea impatiens would severely stunt the plant and could result in yellowing of lower foliage. Strive to maintain a moderate moisture level, though, since overwatering can encourage root rot disease.

Fertilize seed-propagated New Guinea impatiens throughout the crop cycle with 200 ppm N from a fertilizer such as 15-0-15, 20-10-20, or 15-5-15 at every other irrigation. Maintain pH 5.8 to 6.5 in typical soilless media.

Spectra New Guinea impatiens is naturally compact and basally branching. Under virtually all conditions, it does not require any height control. Tango is more vigorous and may require height control under some conditions. B-Nine (daminozide) and A-Rest (ancimidol) do not have adverse effects on New Guinea impatiens when used at rates appropriate for bedding plants (B-Nine, 1,500 ppm; A-Rest, 67 ppm), but they reduce height only slightly. Cycocel (chlormequat) at rates above 750 ppm result in chlorosis on the margins of Spectra foliage. Bonzi (paclobutrazol) and Sumagic (uniconazol) are effective for height control. Appropriate concentrations for grower testing are 5 to 15 ppm from Bonzi and 2 to 5 ppm for Sumagic. The grower in a cooler climate should either try growing seed-propagated New Guinea impatiens without growth regulators or test the lower rates. The grower in a warmer climate should experiment with the higher rates. Please note, however, that the use of any growth regulator on Spectra New Guinea impatiens is usually unnecessary and may result in abnormally compact plants.

The time required to produce any greenhouse crop depends on the environment under which it is grown. Under most good greenhouse conditions, Tango

New Guinea impatiens should be in flower and salable 15 weeks after sowing for early spring sales and 12 weeks after sowing for summer sales. Time to flower of Spectra New Guinea impatiens is 12 to 13 weeks from seed when grown at an average temperature of 65F (18C). Seed-propagated New Guinea impatiens branch naturally and do not require pinching, which would delay flowering. Temperature and light levels have the greatest influence on time to sale. While low light levels do not delay flowering, plant growth is reduced, increasing the time needed to produce a plant large enough to be appropriate for the container (table 1).

Seed-propagated New Guinea impatiens is not immune to any of the diseases common to all New Guineas, but it is disease-free initially. Good cultural conditions will usually limit disease development. Control insects with insecticides as

TABLE 1   WHICH VEGETATIVE NEW GUINEA IMPATIENS IS BEST FOR WHAT CONTAINER?

Flower color	Pots		Hanging baskets	
	4 inch or smaller	6 inch or larger	8 inch or smaller	10 inch or larger
Purple (blue tones)	Antares Bora-Bora		Antares Bora-Bora	
Purple (red tones)	Syrtaki	Cel. Raspberry Rose	Syrtaki	Cel. Raspberry Rose
Lavender	Serenade	Cel. Lt. Lav. II Flamenco	Serenade	Cel. Lt. Lav. II Flamenco
White	Cel. Pure White Moorea	Moorea	Cel. Pure White Moorea	Moorea
Blush white		Samoa		Samoa
Light pink	Equinox	Equinox	Equinox	Equinox
Medium pink	Kallima	Rosetta Cel. Rose	Kallima	Rosetta Cel. Rose
Pink bicolor	Pago Pago	Pago Pago	Pago Pago	Pago Pago
Dark pink	Impulse	Impulse		Impulse
Hot pink	Bonaire	Aglia Cel. Elec. Pink	Bonaire	Aglia Cel. Elec. Pink
Cherry red, fuchsia	Anguilla	Martinique	Anguilla	Martinique
Red bicolor	Cel. Cherry Star	Ambience	Cel. Cherry Star	Ambience
Red scarlet	Prepona Lanai	Cel. Red Lanai	Prepona Lanai	Cel. Red Lanai
Dark orange		Ambrosia Timor		Ambrosia Timor
Orange	Escapade	Nebulous	Escapade	Escapade
Salmon		Cel. Salmon		Cel. Salmon
Coral		Cel. Lt. Salmon Cameo		Cel. Lt. Salmon Cameo

Note: Cel. = Celebration Series
Source: C. Anne Whealy, Proprietary Rights International, Roanoke, Texas, Copyright, 1996.

TABLE 2   VEGETATIVE NEW GUINEA IMPATIENS COMMERCIAL CULTIVARS 1996

Flower color	Mikkel Sunshine series	Pure Beauty series	Celebration series	Danziger series	Lasting Impressions series	Paradise series	Liberty, Patriot series	Bull series
Purple (blue tones)	Antares[1-2,nv-g]			Debka[1-2,nv-g] Swing		Aruba[1,nv-dg] Bora-Bora[1-2,nv-g]	Patriot Purple Lilac / Patriot Lilac	Elvira
Purple (red tones)		Apollon[3+,nv-dg]	Celebration Raspberry Rose[2-3,nv-g]	Samba[2+,nv-g] Danserra[1-2+,nv-g] Syrtaki[1-2,nv-dg]	Rhapsody[2-3,,nv-g]	Papete[1+,nv-g]	Patriot Violet	Wanda
Purple bicolor		Octavia[2-3+,nv-g]			Shadow[2,nv-dg]	Guadaloupe[2-3,,nv-g]	Liberty Lilac Red	
Lavender			Celebration Lt. Lav. II[2-3,nv-g]	Flamenco[2-3,nv-g]	Heathermist[2,nv-dg] Serenade[2,nv-dg] Tiffany	Tonga[1+,nv-db]	Liberty Lavender Patriot Lavender	Alice
White			Celebration Pure White[2,nv-g]	Ballet[2,nv-g]	Innocence[2,nv-g]	Moorea[2,nv-dg]	Patriot White	
Blush white						Samoa[2,nv-g]		
Light pink	Equinox[2,nv-b]			Blues[2,nv-dg]		Tahiti[1-2+,nv-g]	Patriot Soft Pink	Tina
Medium pink	Gemini[3+,var]	Kallima[1-2,nv-dg]	Celebration Deep Pink[2,nv-dg] Celebration Rose[2-3,nv-g]	Merengue	Rosetta[2,nv-b]		Patriot Pink	Barbara
Pink bicolor			Celebration Candy Pink[2,nv-db]			Pago Pago [2,nv-db]		
Dark pink		Dark Delias[1+,nv-g]		Dandin[2-3+,nv-dg]	Impulse[2,nv-g]			
Hot pink		Aglia[2+,var]		Danga[2,nv-g] Lambada[1-2+,nv-b]		Bonaire[1-2,nv-dg]		Doerte[2+,nv-g]
Cherry red, fuchsia	Pulsar[2+,var]		Celebration Elec. Pink[2+,nv-dg] Celebration Cherry Pink[2-3,nv-g]	Rondo[2+,nv-g] Cha-Cha-Cha[1,nv-g]	Masquerade	Martinique[2,nv-g] Anguilla[1-2,nv-g]		Anna[2,nv-g]
Red bicolor			Celebration Cherry Star[1-2,nv-db] Celebration Apple Star[2-3+,nv-g]	Danlight[2-3+,nv-g]	Ambience[2,nv-g]			

TABLE 2   VEGETATIVE NEW GUINEA IMPATIENS COMMERCIAL CULTIVARS 1996 (CONTINUED)

Flower color	Mikkel Sunshine series	Pure Beauty series	Celebration series	Danziger series	Lasting Impressions series	Paradise series	Liberty, Patriot series	Bull series
Red scarlet	Mirach[2,var]	Anaea[2,nv-g] Prepona[1-2,nv-g]	Celebration Brt. Scarlet[1-2,nv-g] Celebration Red[2,nv-g] Celebration Deep Red[2,nv-dg]	Danhill[2-3+,var]	Blazon[2+,nv*-g]	Lanai [2+,nv-g]	Patriot Dark Red Patriot Red	Karina[1-2+,nv-g]
Dark orange		Marpesia[2,nv-db]	Celebration Bonfire Orange[1-2,nv-d]		Ambrosia[2-3+,nv-db] Timor[2+,nv-g]			Susanne[2-3-,nv-g]
Orange	Nebulous[2-3+,nv-b]			Bingo	Escapade[2,nv-g]	Antigua[2+,nv-g] Tanna[1-2,nv-b]	Patriot Orange	
Dark coral, salmon			Celebration Deep Coral[2-3,var]				Liberty Salmon Orange	Flora
Coral, salmon		Melissa[2+,nv-dg]	Celebration Bright Coral[2+,nv-g] Celebration Salmon[2,var]		Charade[1-2,nv*-db]	Grenada[2,nv-db]	Patriot Pink Salmon	Rosemarie[2,dg]
Light coral, salmon			Celebration Light Salmon[2-3,var]	Danshir[1-2,nv-dg]	Cameo[2-3,nv-g] Illusion[2-3+,nv-g]			
Orange bicolor	Twilight[2+,var]			Danova[2+,nv-db]	Tempest[2,var]			

[1] Compact cultivar
[2] Moderately vigorous cultivar
[3] Very vigorous

+ Upright habit
nv Nonvariegated foliage
nv* Variegated foliage under high light

var Variegated foliage
g Medium green foliage
dg Dark green foliage

b Bronze foliage
db Dark bronze foliage

Source: C. Anne Whealy, Proprietary Rights International, Roanoke, Texas, Copyright 1996.

Note: Recommendations (italicized varieties) determined from greenhouse and field trials in California, Florida, Illinois, and Texas.

Vigor and foliage key: Vigor ratings and flower and foliage colors are averages from multiple greenhouse and field trials. They may vary with environmental conditions.

Product sources:

Mikkel Sunshine and Lasting Impressions series are products of Mikkelsens Inc., Ashtabula, Ohio.
Pure Beauty and Paradise series are products of Kientzler Jungpflanzen, Gensingen, Germany.
Celebration series is a product of Ball FloraPlant, West Chicago, Illinois.
Danziger series is a product of Dan Flower Farm, Beit Dagan, Israel.
Liberty and Patriot series are products of Dummen Young Plants Ltd., Vancouver, Washington.
Bull series is a product of Gartenbau Norbert Bull, Goennebek, Germany.

needed. Dursban (chlorpyrifos), Dithio/PlantFume 103 (sulfotepp), Kelthane (dicofol), and Vydate (oxamyl) have shown phytotoxicity in some instances, particularly on flowers.

Garden performance. Seed-propagated New Guinea impatiens performs as well in the garden as impatiens propagated vegetatively. Plants should be transplanted outdoors after all danger of frost has passed and the soil has warmed. Gardeners in most climates should plant in semishaded areas for best bloom. Tango and Spectra tolerate high light levels better in humid climates. Tango is more tolerant of high light levels than any other New Guinea impatiens.

IRIS (DUTCH IRIS)

by A.A. De Hertogh
North Carolina State University
Raleigh

❖ *Semi-hardy annual replacement bulbs* **(Iris hollandica)***, vegetatively propagated. Growers produce cut flowers from bulbs purchased from specialists.*

FORCING

Many growers, both retail and wholesale, find it profitable and not too difficult to force Dutch iris during the winter months. These bulbs are grown in Holland, Israel, and the Pacific Northwest.

Plant bulbs in deep flats (4 to 5 inches, 10 to 13 cm), ground beds, or raised benches. The bulbs should be 1 inch (2.5 cm) below the surface of the soil. They must be kept moist and never allowed to dry out. In flats, irises form a thin, matted layer of roots on the bottom; thorough watering is a must. Placing the flats too near the heating pipes can cause rapid drying out and subsequent flower bud abortion (blasting).

Spacing depends on the cultivar and size. With standard cultivars, such as Ideal and White Wedgewood, plant size 10/11 bulbs at 12 per square foot (129/m²), size 9/10 at 15 per square foot (161/m²). Special colors: Apollo, Blue Ribbon, Golden Harvest, Purple Sensation, Blue Magic—plant 10/11 bulbs at 8 per square foot (86/m²), size 9/10 at 10 per square foot (108/m²), and size 8/9 at 12 per square foot. The bulbs should be planted in a well-drained soil. Some growers mark the flat or bench at recommended spacings after thoroughly watering the soil, then plant the bulbs into the soil, covering them with an inch (2.5-cm) layer of peat, which is again watered. When placing flats on a bench, it is a good idea to allow a 1-inch air space between them.

Dutch irises need high light intensities and ample ventilation with uniform low (50 to 60F, 10 to 16C) temperatures. After the bulbs root and begin to grow, fertilize weekly with $Ca(NO_3)_2$ at 2 pounds per 100 gallons (959 g per 400 l).

Due to the excellent work by the USDA and universities, schedules are available that allow the grower to plan the crop—selecting the proper cultivar, size, and temperature treatment—to force Dutch irises for months. East Coast growers can produce from mid-December through May, but West Coast growers in areas where temperatures do not exceed 70 to 75F (61 to 24C) can produce flowers year-round.

Bulbs are usually dug in July and immediately given a 10-day curing period at 90F (32C), which accelerates formation of flower buds. Immediately upon arrival at warehouses, all Dutch irises

FIG. 1. Dutch iris in the spring are a tradition in cut flower arrangements and bouquets.

that are to be shipped to customers for planting after mid-November are put into a retarding room at 82 to 86F (28 to 30C) with 80 to 85% humidity. They are stored there until they are precooled, which is approximately four to eight weeks before planting, depending on size, cultivar, and time of year. Those bulbs that are to be planted before mid-November can be left in open storage for about 30 days until they are precooled.

Precooled bulbs force more quickly and uniformly. Bulbs must be placed in 45 to 48F (7 to 9C) precooling storage at the proper time and then removed just prior to planting. Precooled bulbs must be planted immediately. Some growers have facilities to do their own precooling and thus use regular or heat-treated bulbs, depending on the intended planting date. Generally, these are shipped six weeks prior to planting. However, the time of precooling varies from four to eight weeks, depending upon the time of the season, cultivar, and size of bulbs. If in doubt, consult firms where scheduling services are available. Trials have shown that a two-week temperature treatment of 65F (18C) either before or after the 45F (7C) treatment reduces the amount of foliage and improves flower size. Also, this treatment increased the percentage of bulbs flowering.

Widely used cultivars are Ideal, a lobelia-blue sport of Wedgewood; Blue Ribbon, a dark blue; Royal Yellow, a deep yellow; and White Wedgewood, a creamy white.

Dutch irises are measured in centimeters of circumference: 6/7, 7/8, 8/9, 9/10, 10/11, and 11 up. Depending on the temperature, the forcing time ranges from nine to 12 weeks on standard cultivars, such as Ideal and White Wedgewood, while other varieties may take 11 to 14 weeks before they flower. Regular Dutch irises cannot be timed accurately, so only growers with cooling facilities use them.

There are three principal disorders encountered in forcing Dutch iris: blindness, flower abortion (blasting), and stem topple. Blindness, or "three leaves," is the failure of a bulb to form a flower. Generally, it occurs when small bulbs are forced too early or the bulbs have not been given an adequate heat or ethylene treatment after lifting.

Flower abortion is when the flower aborts after being formed; thus, the bulb fails to produce a salable flower. It can be caused by low light intensities, high temperatures during forcing (even in the field), and moisture stress. The critical period for flower bud abortion is about 14 days before flowering, that is, when the flower stalk is rapidly elongating. Avoid stresses during this period of development.

Stem topple is caused by calcium deficiency and is similar to that of tulips.

The major diseases are *Fusarium* and *Penicillium*. Aphids, and sometimes thrips, can also be a problem.

When Dutch iris flowers are shipped long distance, they should be cut in the bud stage just as their flower color begins to show. For local selling, cut flowers when they are nearly wide open. If Dutch iris flowers need to be stored after cutting, hold them in water at 33 to 35F (1 to 2C).

THE WEST COAST IRIS CROP

There are large field-iris producers of Dutch irises in the Watsonville and Half Moon Bay areas both just down the coast from San Francisco. There are other large growers from southern to northern California, all located close to the Pacific Ocean.

Most production occurs in open field conditions in full sun. There is an increasing number of Dutch irises being grown under saran (40 to 50F, 4 to 10C), usually in the summer months and more in southern California than northern.

Bulb sizes are 8/9 and 9/10 cm for spring-summer growing, and 9/10 and 10 cm and up for fall-winter forcing. The most popular varieties are Blue Ribbon (Prof. Blaauw) and Ideal, along with Telstar, Apollo, White Wedgewood, Hildegarde, and a few others.

Growers between Watsonville and Half Moon Bay often plant nine to 10 months a year. Only the months of high rainfall (December and January) keep growers from planting and harvesting. In Watsonville, an October-November planting will usually flower in April. The same planting in southern California flowers about one month sooner. A January-February planting in Watsonville flowers in May-June, while in southern California it flowers four to six weeks earlier.

Ivy Geranium (see *Pelargonium peltatum*)

Jerusalem Cherry (see *Capsicum*, ornamental)

Kalanchoe

❖ *Annual (***Kalanchoe blossfeldiana***). Vegetatively propagated. Most growers buy in rooted or unrooted cuttings from specialists.*

The kalanchoe is a member of the family Crassulaceae, succulent herbs and pliable shrubs of the temperate and tropical regions. It is a short-day plant flowering in January or February in the temperate regions.

The kalanchoe can be flowered year-round on a scheduled program very similar to the pot mum program: long-day requirement, followed by short-day response, using a long-day period and growth regulators to control plant habit and flowering (fig. 1).

FIG. 1. This fine crop of 6-inch kalanchoe was grown using subirrigation on troughs at Van Wingerden International, Fletcher, North Carolina.

There are many new commercial varieties with bright colors, pleasing foliage, long shelf life, and customer satisfaction. Sales have been increasing for 4-inch (10-cm) pots flowered for mass-market outlets. Plants are easy to care for in mass-market displays and have good postharvest life in the home for consumers.

Propagation. While kalanchoes have been propagated from seed in the past, it is no longer practical because of long crop time. The best method is to purchase either liners or unrooted cuttings from specialist propagators. Because problems make maintaining disease-free vegetative stock plants both costly and difficult, most large growers purchase unrooted cuttings scheduled to arrive weekly or biweekly for a steady supply of flowering plants.

The best cutting is either a two- or a three-node cutting, 1½ to 2½ inches (4 to 6 cm) long, from a clean stock plant growing under long days with a minimum of 13 hours of light. These cuttings will root within 14 to 21 days with soil temperature of 68F (20C). Give 40% shade and very little mist to prevent stress or wilting. No rooting hormone is required.

Growing media. The root system is very fibrous. Media should provide for aeration but also hold about an 80% field capacity. The higher field capacity reduces the need for frequent waterings. A good soil mixture contains 50% peat, 25% soil, and 25% perlite. In a bag mix, look for a finer mix, rather than a coarse one—better water-holding capacity for the fine fibrous root system.

The pH should be between 5.8 and 6.5, with a peat-lite mix being closer to 5.8. In any medium bring calcium to good supply by the addition of dolomite. Also add superphosphate and micronutrients if a complete water-soluble fertilizer is not used.

Watering. Kalanchoe should not be watered overhead on a regular basis. Smaller sizes work very well on a mat or an ebb-and-flood system. The growing media and watering practices go hand in hand in plant culture. Kalanchoe does not have a high transpiration rate. It dries out from air circulation rather than plant transpiration, which makes it drought resistant.

The young plants, in the early stages of growth through bud initiation, should not be stressed by high temperature or lack of water. It is very important to maximize the growth of the young plants through bud initiation to size them up in relation to the pot size.

After flower initiation, overwatering can soften and stretch the flower stems. It is a good practice to tone the plants at the finishing stages.

Fertilization. Kalanchoes require less water than chrysanthemums, which reduces the frequency of watering. Since you are watering less frequently, you need to increase the fertilizer concentration. Three hundred to 400 ppm N and K is recommended through visible bud. After that stage, clear water and fertilizer can be alternated, or the fertilizer can be reduced to 150 to 200 ppm N and K. As in all fertilizer programs, regulate media pH to between 5.8 and 6.3.

Use calcium nitrate and potassium nitrate rather than ammonium nitrate, as sources of nitrogen, or apply a complete fertilizer, such as 20-10-20. The kalanchoe crop requires a consistent fertilization and watering program and will not perform at its best if neglected.

Growing temperatures. There have been many complaints that kalanchoes are difficult to schedule. This is because temperature is critical. Low night temperatures slow down the growth and make schedules and plant size difficult. The minimum night temperature through visible bud should be 65F (18C). A 65 to 68F range is optimal. Bud initiation and development is slowed down by night temperatures over 75F (24C), and heat delay can occur.

Should the crop be developing too fast, the night temperature can be reduced to below 60F (16C) after the buds are initiated. During the winter months, with low light in the north, CO_2 can be beneficial at 1,200 and 1,500 ppm.

Pinching. Kalanchoes can be grown either pinched or unpinched, depending on the size of the pot and the number of cuttings used. As a rule, with the smaller-sized single plants (2½ to 4½ inches, 6 to 11 cm), no pinch is necessary if the minimum number of long days and growth regulators are used to develop a compact plant habit. In growing a single plant in a larger pot, apply more long days and a pinch to increase plant size. If growing three plants in a larger pot, crop time can be reduced by four weeks by not pinching.

When pinching kalanchoes, it is best to take the tip with one set of leaves to get a plant with more basal breaks.

Plant spacing. Because of their compact growth habit, kalanchoes can be grown closer than most pot plants. They can be grown pot to pot until the foliage touches, usually at the beginning of the short-day period. A final spacing can be established when short days begin (table 1).

Long-day treatment. The number of long-day weeks is critical to kalanchoe vegetative development, determining plant size and plant habit. Not enough long-day weeks reduces the size and height of the plant. Too many long-day weeks stretches the plant. The long-day weeks on pinched plants should be balanced. For example, if three long-day weeks come before a pinch, three long-day weeks should follow the pinch.

If too many long-day weeks are given, the result is top-heavy plants rather than balanced, compact plants. The schedules in table 2, given good culture programs, should produce balanced plants for any size. There are varietal differences, and experience will fine-tune the results.

Long-day treatment consists of 10 footcandles (108 lux) at plant level. Use two hours per night March 1 through October 31 and four hours per night November 1 through February 28. Using this schedule takes some of the details

TABLE 1   KALANCHOE FINAL SPACING

Pot size		Spacing	
inches	cm	U.S.	metric
4	10	4–6 plants/ft^2	43–45/m^2
5	13	7 by 7 inches	18 by 18 cm
6	15	10 by 10	25 by 25
6½[a]	17	12 by 12	30 by 30

[a] This size is planted with three plants per pot; spacing is between pot centers. All other sizes have only one plant.

TABLE 2   KALANCHOE CROP SCHEDULE[A] (LONG SCHEDULE; MINIMUM TEMPERATURE 65F NIGHT)

Pot size (inches)	Plants per pot	Long-day weeks	B-Nine spray [b]	Approx. wks. plant to pinch	Short-day weeks	B-Nine spray [b]	B-Nine spray [c]	Wks. plant to sell [d]
Uprooted cuttings								
2–3	1	0–2	Week 3	No	6	Week 6	Yes	9–15
4–4½	1	3	Week 4	No	6	Week 7	Yes	12–16
5–5½	1	5	Week 4	Week 5	6	Week 8	Varietal	14–18
6–6½	1	8	Week 5	Week 6	6	Week 9	Varietal	16–20
6–6½	3	4	Week 5	Optional	6	Week 9	Yes	13–17
Liners								
2–3	1	0	Week 2	No	6	Week 5	Yes	9–13
4–4½	1	2	Week 4	Optional	6	Week 7	Yes	11–15
5–5½	1	3	Week 4	Week 2	6	Week 7	Varietal	12–16
6–6½	1	6	Week 6	Week 3	6	Week 9	Varietal	15–19
6–6½	3	3	Week 4	Optional	6	Week 7	Yes	12–16

Note: Short-day treatment is not needed in northern areas from October 1 to March 1. Days are naturally short enough to induce flowering.

[a] Schedule used from March through September in the North; Sunbelt year-round. Longer crop times may be required in winter. Lower light areas.

[b] B-Nine for height and foliage size control, 5,000 ppm, applied for example, third week after plants are potted.

[c] B-Nine for peduncle length control, 2,500 ppm, bud visible—apply only if needed. Some varieties will not need this third application.

[d] Two long-day weeks plus six short-day weeks totals eight weeks, not 11 to 15 weeks. Reason: You need not shade kalanchoe all the way to maturity. Also, why the range of weeks from 11 to 15? Reason: Varieties differ in their earliness.

out of changing schedules every month. Mum lighting is compatible with kalanchoe lighting time and light levels.

Short-day (black cloth) treatment. Kalanchoes, being photoperiod short-day
plants, require a longer night than dendranthema (chrysanthemums). At higher
night temperatures, the black cloth should be applied between 14 and 15 hours,
such as from 6:30 P.M. until 9:30 A.M.

The consistency of applying black cloth is more critical to kalanchoes than
to chrysanthemums. Since most of the larger sizes require long crop time, do not
miss any nights of black cloth application, or the crop will be delayed. Short-day
treatment should start March 1 and end October 1. The grower should also prevent any lighting spillover from adjacent areas from October 1 until March 1. Be
aware that heat delay can occur in summer production if there is high temperature buildup under the black cloth.

Minimum short-day weeks for flowering is six weeks (42 days). Why the
long-day and short-day periods? The kalanchoe flowers when exposed to short
days (normally under 12-hour day lengths). If short days are applied on cuttings
that are just planted, the finished plant will be too small. Therefore, some weeks
of long days are included in the schedule—time for the plant to build strength
before it is put into flower by application of short days.

Growth regulators. B-Nine controls internodal stretch and leaf size on more
vigorous varieties. B-Nine can be scheduled as often as every three weeks during the plant development period to produce a more compact plant. Usually no
more than two applications of 5,000 ppm are necessary; 2,500 ppm is also effective for some growers, depending on how vigorously the plants are growing.

Some markets like a very short kalanchoe with flower heads right down on
top of foliage. B-Nine can also shorten peduncle stretch on tall-growing varieties. Use 2,500 ppm when the buds are clearly visible. Not all varieties need B-
Nine to shorten peduncle stretch. Some growers have used Bonzi as a growth
regulator on kalanchoes.

Light. Kalanchoes do not usually do as well under high temperatures and high
light. Leaf temperature is everything in growing a good kalanchoe. Under high
temperatures and high light, the leaves will bleach out and harden, even developing a red color. When temperatures exceed 75F (24C), reduce the sunlight to about
3,500 or 4,500 f.c. (38 to 48 klux). Reduce the light to lower the leaf temperature.

Under low-light levels, kalanchoes will stretch and not flower as heavily, and
flowers will be thin and weak. In the winter or other low-light periods, use supplemental light (HID) to start plants.

Insects. The most common insect problem is aphids at the later stages of maturity, when buds and flowers develop. Plants should be inspected on a regular

basis for any developing insect problems. Some of the other insect problems are worms, mealybugs, thrips, and whiteflies.

Kalanchoes are very sensitive to certain spray materials. Emulsifiable oils often burn the leaves. This burning happens because many of the kalanchoe leaves cup up and hold the spray material.

The best materials for kalanchoes are wettable powders or water-soluble pesticides. Oil-based sprays should always be tested before application. If an oil-based spray is used, wash it off after a short period; don't let spray material remain on the leaves. Also, be cautioned against fog materials that have an oil base and can settle on the leaves.

Diseases. The most difficult disease problem of kalanchoes is bacterial soft stem rot, which can develop at any stage of plant development. The crop should be started from disease-free plants, and sanitation is important in culture. The benches should be clean and treated for disease with bleach. If growing on the ground, the soil should be sterilized between crops.

Spraying plants with a fungicide for mildew or bacterial wilt is almost impossible. The best way to control foliage diseases on kalanchoes is through the environment. Keep the foliage dry at all times, have good air movement in the house, and water from the bottom, not overhead. In plastic houses and where you do not have good air circulation, *Botrytis* can often become a problem. Daconil or Chipco can help in the control of *Botrytis*.

Postharvest. Kalanchoes should be about 40 to 50% open before shipping for the best shelf life under low light levels in the store or home. Shipping the plants before terminal flowers open will set the plant back, and the flowers will never develop fully. A customer with well-lighted windows or a garden exposure can handle a budded kalanchoe. Be sure the plant has mature flowers before shipping. Plants shipped at this stage will have six to eight weeks of shelf life. Breeders continue to work for newer varieties that will perform better in the marketplace and be easier for the grower to produce.

KALANCHOES FROM SEED

Kalanchoes may also be propagated from seed. Vulcan is a bright-red variety to be sown carefully in a well-drained soil. Keep the soil temperature at 70F (21C) and do not cover the seed. Transplant into 2¼-inch (6-cm) pots and grow at 62 to 65F (17 to 18C). Move the plants into 4- to 4½-inch (10- to 11-cm) pots for finishing. Seed sown in June should provide finished plants for Thanksgiving. Plants should be given short-day (black cloth) treatments from late August until October 1. For Christmas, shade from September 10 until October 1. To flower for Valentine's Day, light plants from September until November 1. Plants do not usually need a pinch or B-Nine.

KALE (SEE *BRASSICA*, VEGETABLES)

KALE, ORNAMENTAL (SEE *BRASSICA*, ORNAMENTAL)

KOHLRABI (SEE *BRASSICA*, VEGETABLES)

LANTANA

by Jim Nau
Ball Horticultural Company
West Chicago, Illinois

❖ *Tender perennial* (**Lantana camara**)*. 1,300 seeds/oz (46/g). Germinates in six to seven weeks at 65 to 75F (18 to 24C). Seed should be covered during germination. Most lantana is vegetatively propagated from cuttings. Lantana from seed is available only as a mixture or in purple; propagation by cutting is strongly advised.*

Lantana comes in a wide range of colors; many times the buds will exhibit one color and the resulting flower will show another. Then flowers often shade to a third color, in some cases. Flowers come in pink, orange, yellow, purple, cream, and shades in between, and they are held on plants that grow horizontally and flower as long as the temperatures stay warm. Lantana is treated as a perennial in some areas of the Deep South and Far West but as an annual in the Midwest. Plants grow to a height of 3 feet (91 cm) and should be grown in full sun. Space 18 to 24 inches (46 to 61 cm) apart to fill in.

The predominant culture of these plants was for growers to dig the plants in the fall, pot them up, and move to a 55F (13C) house to take cuttings. Today, commercial propagators perform this task, and the resulting plants are shipped to growers to finish off. However, if you want to take your own cuttings, use only softwood cuttings and stick them into sand or sand mixed with peat moss. Provide bottom heat of 65F (18C), and mist. Roots will develop in three to four weeks. Resultant plants will finish off in another seven to nine weeks in a 4-inch (10-cm) pot when grown with a soft pinch. Growers producing large containers of lantana prune plants regularly to maintain shape. Bonzi at low rates (less than 0.5 mg active ingredient per pot) may be helpful in height control.

If buying liners from a commercial propagator, allow six to eight weeks to finish, unless otherwise noted by the supplier, and stick directly into 4-inch

(10-cm) pots or 8- and 10-inch (20- and 25-cm) hanging baskets. Water and feed only when necessary, as plants will be shy to flower from an abundance of water or fertilizers.

New Gold lantana is a popular, prostrate form with golden yellow flowers.

LARKSPUR (SEE *CONSOLIDA*)

LATHYRUS (SWEET PEA)

by Jim Nau
Ball Horticultural Company
West Chicago, Illinois

❖ *Annual (***Lathyrus odoratus***). 350 seeds/oz (12/g). Germinates in seven to 14 days at 60 to 70F (16 to 21C). Seed should be covered for germination.*

Often in horticulture, certain crops or classes of plants go through a period of breeding and improvement, are offered to the trade, and enjoy a place of prominence for years to come. Sometimes, as in the case of cut flowers, the market changes dramatically, and the crops fall out of favor and go by the wayside. Such is the case of the sweet pea.

Sweet peas offer softly scented flowers in pastel colors atop foliage that grows and trails well. There are two classes of sweet peas: those that are summer flowering and those that are winter flowering. Summer-flowering varieties branch close to the ground and produce a number of active shoots. The winter-flowering varieties produce one vegetative shoot, which first flowers and eventually will branch along the stem. Of the two classes, the summer-flowering varieties are long-day plants and will not flower under the short days of winter.

FIG. 1. For cool climates, sweet peas are unbeatable in creating a country cottage look in the garden. Growers generally sell green plants in pots.

Therefore, since most cut flower material is sown in the summer months for winter-flowering, use only winter-flowering varieties in northern greenhouses for best performance.

In cropping, sow seed in June, July, August, and September to flower from October to March or April. Sow seed direct to the final container or production bed (cuts).

Knee High Mix is a bush sweet pea growing 24 to 30 inches (61 to 76 cm) tall. Winter Elegance mix, flowering under short days from winter to early summer, is ideal for cutting.

LAVANDULA (LAVENDER)

by Miriam Levy
Ball Horticultural Company
Carlsbad, California

❖ *Annual and perennial (Lavandula spp.). 25,000 seeds/oz (882/g). Germinates in 14 to 21 days at 65 to 75F (18 to 24C). Leave seed exposed to light.*

The genus *Lavandula* contains numerous species, many of which are increasing in popularity for use as annuals and perennials for aromatic leaves and unusual foliage colors. There are many choices, but what are the differences? How do you choose which ones to produce? Seed or vegetatively produced; gray or green foliage; deep purple, light blue, pink, or white flowers; dwarf, tall; winter-hardy or semihardy; Spanish, English, or French?

All lavender species produce seed, but the most common production method is to produce cuttings vegetatively. With the exception of Lavender Lady (fig. 1), most seed-produced types are not true hybrids, and finished plants vary considerably. Starting with

FIG. 1. Growing the All-America Selections award winning Lavender Lady lavender is one way growers can capitalize on the current trend of herbs and scented plants.

vegetatively produced plants ensures a product that is consistent in leaf color, growth habit, and bloom color.

The most common lavender is *Lavandula angustifolia*, or English lavender. This is the species from which all others were bred, the true garden fragrant type and the lavender used in perfumes and potpourri. Munstead and Hidcote are named varieties of *angustifolia*. All have gray foliage with smooth leaf margins. The major difference between Munstead and Hidcote is growth habit. Munstead, as a rule, will be taller in the garden than Hidcote, 12 to 24 inches (30 to 61 cm) versus 12 inches for Hidcote. Flower color is also different. The Hidcote type available is a Jean Davis variety with light pink or purple flowers; Munstead has deep purple blooms. The most popular Munstead type is Lavender Lady, a seed-produced variety that blooms the first year. It should be treated as an annual. True English lavender can reach 3 feet (91 cm) in the garden and has blue flowers. Bloom time is summer and fall for Hidcote and Munstead, spring and fall for *angustifolia*. All require full sun with well-drained soil.

French lavender, *Lavandula dentata*, has a different leaf shape than English lavender. French has green or gray foliage with a deep serration on the leaf margin. It's more sensitive to cold temperatures and thus less winter-hardy. Its primary use is as topiary and in indoor pots. *Lavandula dentata* will grow to 3 feet (91 cm) in the garden and has blue flowers with a bloom period of spring and fall.

Spanish lavender, *Lavandula stoechas*, has the showiest flowers of all lavenders and is the most vigorous grower. Two varieties of *stoechas* are becoming very popular, Quasti and White. Both have gray foliage. Quasti has a very showy purple flower and will reach 3 feet (91 cm) in the garden. White will only attain 2 feet (61 cm) in the garden and, as its name implies, has white flowers.

Lavandula pinnata is very tender and is used in the landscape only in very mild southern climates. It is not winter-hardy in the United States, but when treated as an annual, it will bloom from spring through the first frost. Pinnata, with gray-green foliage and deep purple blooms, will reach 2 feet (61 cm) in the garden.

Intermediate hybrids add to the confusion. These are crosses between hybrids of *angustifolia* and *latifolia*. Provence is similar to *angustifolia* in fragrance but is a larger, taller plant with a light-colored flower. Because of its vigorous growth, Provence is used in the garden primarily as a hedge. Goodwin Creek Gray is a cross of *dentata* and woolly lavender. It looks like a *dentata,* with gray foliage and deep blue flowers. In warm climates, Goodwin Creek will bloom year-round and thus is a good southern variety.

The best time to propagate lavender is late summer and early fall. Cuttings benefit from bottom heat of 65 to 75F (18 to 24C). As with most silver-leafed plants, too much mist will cause leaves to rot.

Growing lavender from seed is a slow process. Germination takes 14 to 21 days, but you may wish to wait 28 days to transplant for extra plant growth.

TABLE 1   LAVENDER VARIETY CHARACTERISTICS

| Variety | Flower | | Height (feet[a]) |
	Color	Season	
L. angustifolia	Blue	Summer, fall	3
Hidcote Blue	Deep blue	Summer, fall	1
Hidcote Pink	Pink	Summer, fall	1
Jean Davis	Light pink	Summer	1
Lavender Lady	Deep blue	Spring, late	1–2
Munstead	Purple	Summer, fall	2
L. dentata	Blue	Spring	3
Gray	Blue	Spring	3
L. stoechas	Purple	Summer	3
Provence	Deep purple	Summer, fall	2
White	White	Summer	2
L. pinnata	Dark blue	Spring	2

Note: Foliage color of these varieties is always gray, except for *L. dentata,* which is the only green in the species, and *L. pinnata,* which is either gray or green.

[a] One foot = 30.48 cm

Allow 15 to 18 weeks for green pack sales and 18 to 20 weeks for 4-inch (10-cm) pots. The variety Lavender Lady, an All-America Selections award winner, shows better germination than species *L. angustifolia* and finishes faster, as well.

Lavenders require full sun and a well-drained, porous soil during production and in the garden. They are light feeders; a 20-10-20 fertilizer every few waterings is sufficient. In southeastern and mid-Atlantic states, high relative humidity leads to their decline. They don't perform well in warm, humid areas and should be treated as annuals in these locations. Winter-hardy lavenders don't like to be mulched. This would keep moisture levels too high, and they'd rot. In northern states overwintering success is determined by the size of the root system before the plants become dormant. Most varieties benefit from a spring planting so they can establish themselves before winter temperatures drop.

LEUCANTHEMUM (SHASTA DAISY)

by Jim Nau
Ball Horticultural Company
West Chicago, Illinois

❖ *Perennial (***Leucanthemum** × **superbum,** *formerly* **Chrysanthemum** × **superbum** *and* **C. maxima***). 15,000 to 35,000 seeds/oz (525 to 1,225/g). Germinates in 9 to 12 days at 65 to 70F (19 to 21C). Cover seed lightly upon sowing.*

One of the backbones of the perennial garden, the Shasta daisy is one of the best perennials to use when white is a needed color. Large blooms, from 3 to 4 inches (8 to 10 cm), appear on plants that flower readily from seed. Plants vary in height from 10 inches (25 cm) (Snowlady) to as tall as 48 inches (122 cm) (Starburst and Alaska). The flowers, with no scent, are most often single, though semidouble-flowering varieties are also available. Double-flowering varieties grown from seed are not 100% reliable, nor are the blooms of high quality. Once a seed-grown plant produces a quality double flower of any merit, it is advised to prop-agate it vegetatively instead of relying on seed to perpetuate it in the future.

Shasta daisies are hardy to USDA Zones 5 to 8. In the Midwest they are most often treated as biennials due to their short life expectancy. Shasta daisies need well-drained soil to avoid winter kill; even then, Midwest winters are generally too severe for plants to last more than two to three seasons. Shasta daisies flower from June to August in the gar-den, and old blooms need to be removed as they fade to encourage rebloom.

FIG. 1. Shasta daisies are some of the easiest perennials for growers and home gardeners alike. Shown here are the bright white flowers of the dwarf variety White Knight.

Growing Shasta daisies from seed varies with the variety selected for pro-duction. For instance, both Alaska (single-flowering, white, 4-inch, 10-cm, blooms on plants to 24 inches, 61 cm) and Snowlady (single-flowering, white, 3-

inch, 8-cm, blooms on plants to 10 inches, 25 cm) will flower from seed when sown before March. Alaska flowers sporadically with a higher profusion of blooms the second year, while Snowlady is an excellent pot plant. Sow Snowlady in January and February for flowering pot sales in May. For green packs, sow in early March and sell in May.

Other varieties, like Silver Princess, Starburst, and any of the double-flowering varieties on the market, will not flower the same year from seed sown in January. For any of these, sow in the summer for overwintering in quart or gallon containers, and sell green next spring. Plants will flower in June. Growing temperatures should be 48 to 50F (9 to 10C) nights. Increasing temperatures to 60F (16C) will speed flowering.

For vegetative production, consider the double-flowering varieties of either G. Marconi or Diener's Double to offer to customers. Diener's Double is the dwarfer of the two, to 2 feet (61 cm) tall, with frilled white flowers to 3 inches (8 cm) across. G. Marconi is 3 feet (91 cm) tall, with pure white flowers to 4 inches (10 cm) across. Both of these varieties are still in production, though somewhat hard to find.

LIATRIS (BLAZING STAR, KANSAS GAY FEATHER)

by Jim Nau
Ball Horticultural Company
West Chicago, Illinois

❖ *Perennial* (**Liatris spicata**). *9,400 seeds/oz (332/g). Germinates in 21 to 28 days at 65 to 70F (18 to 21C). Leave seed exposed to light. Reduce night temperatures by 10F (−12C) if seedlings are slow to emerge.*

Liatris, especially *L. spicata,* has become an important perennial and cut flower in recent years. The showy 10- to 15-inch (25- to 38-cm) spikes make great accents at the back of the perennial border or in cut flower arrangements.

Although it is a native U.S. plant, hardy to Zones 3 to 8, liatris has been primarily shipped from Holland. In fact, the Dutch exports are what triggered the present interest in liatris as a cut flower.

Liatris spicata is a tall, stately perennial that blooms in July and August with long, tapered spikes of lavender or white. Flowers are small, to one quarter inch (64 mm) across, and tightly arranged on 10- to 15-inch (25- to 38-cm) spikes. Liatris grows 1½ to 3 feet (46 to 91 cm) tall.

Seed, corms, and divisions are the usual ways to propagate *L. spicata.* Divisions are taken either in the spring or autumn.

For plants that will flower in summer, sow seed during the previous summer for transplanting to the final container during August or early September. These plants are overwintered for green container sales the following spring. For seed sown during the winter for green packs in spring, allow 12 to 14 weeks at 55 to 58F (13 to 14C) nights for rooted plants. Plants will flower the same season as sold, although seed should be sown by January or February for best performance.

Commercial propagators offer *L. spicata* in small pots, liners, or plugs. Primarily available during the autumn and winter, any of these can be potted up into quarts and gallons for green, budded, or flowering pot sales in the spring, depending on how early you transplant the crop.

If one corm is planted per quart container in mid-March, the plants will be salable green in mid-May. These plants will flower in late June and July. Commercial propagators also offer liatris as bare-root transplants. These can be potted up into 3- or 4-quart containers in the winter or early spring for May sales.

Kobold (also called Goblin) grows 18 to 24 inches (46 to 61 cm) tall. Its flowers are lavender. This variety is available from seed, corms, or divisions. Those propagated by vegetative means, however, are usually darker in flower color and more uniform in appearance. Floristan is another *spicata* type, available either in white- or blue-flowering plants, the blue offered as Floristan Blue or Floristan Violet. The plants are 3 feet (9 cm) tall and produce excellent cut flowers. Floristan is propagated by seed.

CUT FLOWER PRODUCTION

A gibberellic acid soak of corms at 500 ppm after a five-week cold storage results in 100% flowering. Using long days in conjunction with cold treatment accelerates flowering and makes longer stems. Be careful—if corms receive short cooling treatment, then short days will enhance flowering. Temperature also affects flowering. At cool temperatures (55F, 13C), long days enhance flowering, while at warmer temperatures, day length has little effect. Research has shown that the greatest effects on flower acceleration come from use of long days in the first five weeks after foliage emergence.

While liatris may be planted year-round, corms planted in February and March have the highest yield per corm. Plants may be cropped for multiple years. At the University of Georgia, corms showed increasing yield from Year 1 to 2 and Year 2 to 3. After the first year, use support wire for stems.

Once harvested, remove lower stem foliage before placing in water. Use a 24- to 72-hour pulse with a 5% sucrose solution for tight flowers; a 2.5 to 5% sucrose solution for normal stems.

In addition to Floristan Blue or Floristan White, Gloriosa is a good cut flower variety.

LILIUM (ASIATIC AND ORIENTAL LILIES)

by William B. Miller
Clemson University
Clemson, South Carolina

Rob Miller and Dr. Robert O. Miller
Dahlstrom and Watt Bulb Farms, Inc.
Smith River, California

❖ *Tender perennial* (**Lilium** *hybrids). Specialist propagators multiply lilies from bulb scales. Growers buy in bulbs and produce them as flowering pot plants or cut flowers.*

Asiatic and Oriental lilies are a highly diverse group of plants that can be forced for holiday or year-round markets. The major advantages of hybrid lilies are relative ease of production, high crop value per square foot, an ever-increasing variety selection, better height control possibilities through chemicals and genetics, excellent value for the consumer, and in the case of Asiatic hybrids, low greenhouse temperature requirements.

PRECOOLING

Most Asiatic hybrids are precooled at least six weeks at 34 to 36F (1 to 2C). Orientals are cooled eight to 10 weeks at 34 to 35F (1 to 1.7C). Additional time at these temperatures can be used for short-term holding. For later plantings (early January on), bulbs are frozen-in, Asiatics at 28F (–2C) and Orientals at 30 to 31F (–1 to –0.6C). Usually, the freezing process is done by the bulb supplier rather than the grower. Orientals do not tolerate low temperatures well. Freezing allows long-term storage for near-year-round planting, while lessening danger of sprouting. To minimize water loss, bulbs are packed in moist peat moss and wrapped in polyethylene before freezing. If bulbs arrive frozen, thaw slowly (one to three days) below 55F (13C), then plant as soon as possible thereafter.

PLANTING AND MEDIA

Most hybrid lilies are heavily stem-rooted, which means that deep planting in the pot is essential (fig. 1). Deep pots (standard pots) are also important for proper rooting depth.

Some growers use "3-quart gallon," smooth-sided nursery pots for greater soil depth and volume. These pots grow excellent lilies and should be considered in your program. Use good, well-drained media with high air-filled porosity. The pH should be 6.0 to 6.5; lower pH could lead to fluoride leaf scorch. Avoid perlite and superphosphate in hybrid lily media mixes. Keep in mind that mixes con-

FIG. 1. This plant was too shallowly planted, so stem roots were not able to benefit the plant. Solution: deep planting in deep pots.

taining pine bark tie up growth regulators, thus reducing their efficacy. Typically, three bulbs (12/14 size) are planted per 6-inch (15-cm) pot, five bulbs per 8-inch (20-cm) pot. Water in very well!

Fertility programs vary widely. Easter lilies are taller when grown hungry, and hybrids probably are as well. The basic program at Clemson is to use 200 ppm N at each irrigation, with clear water applications on the weekends. This approach has never resulted in salt problems yet produces good foliage color. Since there is low-alkalinity water in South Carolina, we give at least one application of calcium/potassium nitrate at 200 ppm N per week and use 20-10-20 peat-lite for the other irrigations. The specifics of fertilizer—that is, the best nitrogen form, micronutrient levels—have not been intensively studied for hybrid lilies. We also like to let the crop run on the dry side for the first two to three weeks, or until stems are 1 to 3 inches (2 to 8 cm) tall because this improves stem root growth.

TEMPERATURES AND TIMING

Asiatics need to be grown cooler than Orientals. Grow Asiatics at 55 to 60F (13 to 16C) nights and 70F (21C) or lower days, no more than 85F (29C) days. Orientals respond better to 65 to 67F (18 to 19C) nights and 75F (24C) days. As with Easter lilies, timing is mainly controlled by 24-hour average temperature, and some level of height control can be achieved by using DIF (warmer nights than days).

Always keep in mind that the "average temperature" is not simply the average of the day and night temperatures! Calculate average temperature as follows:

[(Day temp × hours of day temp) + (night temp × hours of night temp)] ÷ 24

As a general rule, the number of days from planting to emergence is based on the length of cold storage or freezing in. Days to emergence vary from two to

TABLE 1 *LILIUM* GROWTH RATES

Cultivar[a]	Plant to emergence (days)	Plant to visible bud (days)	Visible bud to flower (days)	Plant to flower (days)
Amber	22	31	25	56
Sinai	25	37	25	62
Delicious	28	42	23	65
Sun Ray	33	49	25	74
Red Carpet	33	54	20	74
Sun Pearl	33	51	24	75
Mt Blanc	30	49	31	80
Whitebird	33	53	29	82
Montreaux	23	49	35	84
Dimples	25	51	36	87
Stargazer	27	58	48	106

Note: Table arranged from fastest to slowest to flower. Data based on spring trials planted Feb. 15 at Clemson University, Clemson, South Carolina.
[a] All cultivars are Asiatic, except the two slowest to flower, Dimples and Stargazer, which are Oriental.

three weeks for early crops to three to four days for crops planted in May and later. Beyond that, timing is highly cultivar- and season-dependent. For example, timing to visible bud may vary from two to four weeks after emergence. The number of days from visible bud to flowering is also highly variety-specific, and it is a long period for most Orientals.

Table 1 gives data for a number of cultivars trialed at Clemson University. Night temperatures averaged 63 to 65F (17 to 18C) for the trial, and daytime temperatures were frequently over 80F (26C), especially later in the crop.

As with Easter lilies, bud sticks can be used to time the crop from visible bud on to flowering. Unfortunately, bud development rates vary widely by cultivar, and bud sticks must be developed for individual cultivars. In fig. 2, bud sticks for the Oriental cultivars Stargazer, Dimples, Melody d'Amor, Mona Lisa, and Ready are given, all based on greenhouse temperatures of 65F (18C) nights and 75F (24C) days. As usual, warmer temperatures will speed bud development, and cooler temperatures will slow development.

LIGHTING

Sans Souci, an older Oriental cultivar, can be forced three to four weeks faster through the use of long days (mum lighting). Use the lighting for four hours, from 10 P.M. to 2 A.M. with at least 10 footcandles (60-watt bulbs, 4 feet, 1.2 m, apart, 4 feet above the benches). Stargazer responds in a similar manner, but

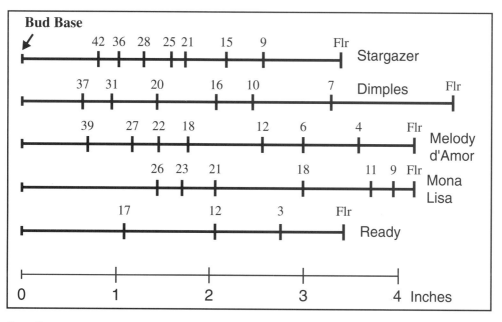

FIG. 2. Use bud sticks to time flowering during later production stages. Hold the end of the stick up to the bud and read the number of days for the bud to open at a temperature of 65F nights/75F days. Warm temperatures speed bud development; cool temperatures slow development.

extensive studies with other oriental hybrid cultivars have not been reported, although some positive timing response should be seen.

HEIGHT CONTROL

A-Rest. Drenches with A-Rest are the most common means of chemical height control. Drench rates are typically in the 0.25 to 0.5 mg per pot range. Some recommended rates are given in table 2.

A key to good chemical height control is *early application*, as soon as roots are growing (fig. 3). When multiple bulbs are used per pot, apply the

FIG. 3. Providing growth regulators at this stage of development is too late. Although the application will control height, it will not suppress early stem growth.

drench when the first two shoots are one-half inch (1 cm) tall. Don't wait for all shoots to emerge! Split applications are usually better than a single application since the first application (at one half inch) can help reduce early stretch, and the second (seven to 14 days later) is more readily absorbed due to the better mass of stem roots present. Again, pine bark mixes tie up growth regulators and reduce their effectiveness.

Sumagic. A Sumagic drench, also highly effective, has the advantage of substantial cost savings compared to A-Rest. At Clemson we trialed Sumagic at 0 to 0.3 mg per pot on 11 hybrid cultivars. *These high rates are above those recommended on the label, and the plants were growing in pine bark media!* The treatments had good height control, with no effect on days to visible bud or number of flowers, and there was no flowering delay. Data for all cultivars are given in table 2.

Work at Michigan State University has shown that preplant bulb dips with Sumagic are very effective with Stargazer. Dipping bulbs for one minute in a 10 ppm Sumagic solution gave an overall height decrease of 35%, relative to untreated controls. Most of the height effect was seen during the rapid-growth phase from 10 to 45 days after planting. Dips of 5 ppm were much less effective, and dips of 20 ppm were only marginally more effective, than 10 ppm.

SHIPPING AND STORAGE

The proper time to ship hybrid lilies is when the first bud is swollen and colored but before it opens, as shown in fig. 4. Before this stage, younger buds

TABLE 2   SUMAGIC DRENCHES AND *LILIUM* FINAL HEIGHT

Hybrid cultivar	Sumagic drench rate (mg a.i. per pot)			
	Control	0.1	0.2	0.3
Amber	24.9	23.4	23.6	22.7
Delicious	20.4	17.0	16.6	14.0
Dimples	27.0	24.8	23.7	23.8
Mt. Blanc	32.1	23.3	18.3	19.7
Montreaux	36.2	28.7	26.5	27.0
Red Carpet	20.1	18.1	15.6	16.1
Sinai	28.1	24.7	21.4	21.9
Stargazer	27.0	24.6	21.2	21.5
Sun Pearl	23.9	17.8	19.5	15.8
Sun Ray	22.4	17.6	19.5	18.3
Whitebird	21.0	16.3	18.6	13.7

Note: Height in inches, including the pot (12, 18, 24, 30, 36 inches = 30, 46, 61, 76, 91 cm, respectively). Cultivars grown in a pine bark mix.

FIG. 4. These pots have a few more days to go before shipping. Harvest pot lilies when the lowest flower bud is fully colored, but before it opens.

often fail to open properly; if harvested too late, the plant and flowers are correspondingly older, and open flowers are damaged. Maximum consumer satisfaction and repeat sales are dependent on proper postproduction handling, starting with proper harvesting time.

As with Easter lilies, there is the temptation to cold-store budded plants prior to sale. Incoming crops that need space or plain bad timing can exacerbate the need to store hybrid lilies. Oriental hybrid lilies store much less well than Easter lilies. Upon removal from the cooler, plants often drop a large portion of their leaves, and quality is drastically reduced. Recent research at Clemson University has indicated that using lights in the cooler can help with this problem. We held Stargazer plants either in darkness or under two 4-foot (1.2-m) fluorescent lamps about 4 feet above the plants in a 40F (4C) cooler for 17 days. We saw a very beneficial effect from lights, as seen in fig. 5. While this is a starting point, there clearly needs to be more work done on postharvest storage of hybrid lilies.

L

TABLE 3   FUNGICIDE DRENCH PROGRAM FOR ROOT ROT CONTROL IN HYBRID LILIES

Potting drench:	Terraclor	4 oz per 100 gal
	Subdue	½ oz per 100 gal
Monthly intervals:		Mixture of one each from Group 1 *and* Group 2 below. During alternate months, Banrot 40 WP may be used by itself at 6–12 oz per 100 gal. Banrot has activity against both *Rhizoctonia* and *Pythium*.

| **Group 1 (*Pythium* control)** | | **Group 2 (*Rhizoctonia* and *Fusarium* control)** | |
Fungicide	**Concentration**	**Fungicide**	**Concentration**
Banol 65 EC	20 oz per 100 gal	Cleary's 3336-F	1.5 pt per 100 gal
Terrazole 35 WP	3–10 oz per 100 gal		
Terrazole 25 EC	4–8 oz per 100 gal		
Truban 30 WP	3–10 oz per 100 gal		
Truban 25 EC	4–8 oz per 100 gal		

FIG. 5. Lights in the cooler may improve cold tolerance of Oriental lilies. The plants on the left had about 200 footcandles from fluorescent bulbs; the plants on the right were in the dark. Both were held at 40F for 17 days. The photo was taken seven days after they were removed from the cooler and placed in a postharvest room. Note the high number of abscised leaves from plants on the right.

TROUBLES

Caused by low-light forcing or ethylene in the greenhouse atmosphere, abscission of buds (blasting) can happen during postharvest transit and is exacerbated by ethylene and high temperatures.

Root rot is prevented through a regular fungicide program alternating between materials effective against *Pythium, Rhizoctonia,* and *Fusarium.* Within one to three days of planting, drench the crop, using the schedule in table 3.

CULTIVARS

Many cultivars are being introduced to the market each year. There are probably even more cultivars of hybrid lilies than of New Guinea impatiens! The new cultivars arise mainly from Dutch breeding efforts. To be successful in the marketplace, a new cultivar must be amenable to tissue culture, propagate and grow well in the field, and force easily. Certainly, in the past most emphasis on lily breeding was for garden and cut types. Greater awareness of the rapidly increasing pot crop market, though, has stepped up Dutch and domestic selection efforts for pot types in the last five to seven years, and ever shorter and better pot varieties are in the pipeline for the future.

LILIUM (EASTER LILY)

by Dr. Robert O. Miller
Dahlstrom and Watt Bulb Farms, Inc.
Smith River, California

❖ *Tender perennial* (**Lilum longiflorum**). *Propagation from scales by commercial propagators. Grown as a flowering pot plant.*

The Easter lily belongs to the subgenus *Eulerion,* the true lilies. The Easter lily, with its white trumpet, is the most popular lily for greenhouse pot lily production. Asiatic and Oriental hybrid lilies are also gaining in popularity. A new group, the LA hybrids (for *longiflorum*-Asiatic crosses) are creating interest. This group has foliage most similar to *longiflorum* and flowers similar to Asiatics. An improved attribute is better tolerance of storage conditions than Asiatics.

Easter lilies are a major holiday pot plant crop that has maintained its popularity over the years (fig. 1). It is a traditional plant, with a core of demand centered around its religious and traditional signifi-cance. Although the lily of reli-gious paintings and writings is the Madonna lily *(L. candidum),* longiflorum types have replaced it for practical cultural purposes. Tradition has also played a role quite apart from religious consid-erations. Probably the most impor-tant factor contributing to the con-tinued popularity of the Easter lily has been the profitable nature of the crop, for both the producer and the seller, and the perceived value by the customer. The Easter lily was one of the first plants to be sold through mass market outlets.

FIG. 1. Well-budded Easter lilies are on their way to becoming a fine crop.

Easter lilies are the most profitable major-holiday pot-plant crop produced. This is true even though the initial cost of a lily bulb is high. When combined with the cost of the pot, media, and labor of potting, growers often have 30 to 35% of the selling price invested initially. The space occupied by the crop compensates for this, so the return per square foot is high. Consider that a chrysanthemum, a poinsettia, or a hydrangea may occupy ¾ to 1½ square feet (0.07 to 0.14 m²) of bench space at finished spacing. Assuming an average of 1¼ square feet (0.12 m²) used and a price of $4 per plant, then the return per square feet of finish space is $3.20 ($34.43/m²). Lilies are grown at a density of 2 to 3½ per square foot (22 to 38/m²). If 2.4 lilies are grown per square feet (26/m²) and sold at $3.50 each, then $8.40 per square foot ($90.38/m²) is realized. This is more than two times the per-square-foot income to be realized from other holiday pot plant crops. Depending on marketing area, lilies may be even more dramatically profitable.

VARIETIES

The variety Ace first gained widespread popularity about 1953, and the variety Nellie White was slowly accepted from about 1964, replacing Ace in popularity probably in 1979. At present Nellie White is the only important variety. These varieties are "old" by present horticultural standards. Although persistent efforts to breed superior varieties continue, Ace, and especially Nellie White, have many good characteristics and remain preeminent. Great hopes were held for several newer varieties, but they have not met expectations.

A large number of varieties are being created and tested at the Easter Lily Research Foundation, Brookings, Oregon. This research effort is supported by the lily bulb growers. Additional efforts are directed towards better field growing practices and to the support of university lily research. Numerous varieties from this project have been released for testing at various universities and research stations. Most of these, while promising, are not ready to be mentioned yet.

Georgia lilies, either from the United States or from Japan, once used in significant numbers, are no longer important. Their problems included height, virus content, and particularly overall plant appearance and the lasting quality of individual flowers. Attempts have been made by the Dutch to export Easter lilies into the United States. Varieties were obtained from Oregon State Lily Research or developed by Dutch companies. While some varieties have been satisfactory in some situations, bulbs in Holland must be spring-planted. With their often wet springs, the Dutch have not been able to deliver quality bulbs consistently. While the bulbs themselves are cheaper, the grower costs are in labor, heat, overhead, and so on, and they have not felt the quality risk worthwhile.

Which cultivar to grow? Since the major varieties continue to be Ace and Nellie White, a comparison of several traits follows:

- Height. Nellie White is shorter than Ace.

- A-Rest tolerance. Nellie White is more apt to suffer leaf yellowing from late or excessive growth regulator applications than Ace.

- Bud count. Nellie White will usually have one-half to one fewer buds for a given size of bulb compared to Ace. This is why most growers force larger bulbs of Nellie White.

- Flower size. Nellie White is generally credited with having larger flowers than Ace.

- Forcing time. There is no difference. In most years Nellie White is slower to emerge than Ace, but from emergence to flowering, it is faster.

- Leaf number. Nellie White, for a given size of bulb vernalized the same way, usually has fewer leaves than Ace. In some years it's as many as 10 to 20 fewer leaves. This is why Nellie White flowers in the same time as Ace, even when it emerges later.

- General toughness. Nellie White is not generally as "tough" as Ace. Ace is more tolerant of temperature extremes, growth regulators, root rot, fertilizer excess and deficiency, and such cultural irregularities. Ace foliage does not curl down with the use of lower day temperatures than night temperatures (negative DIF) to the degree that Nellie White does.

- Precooling temperature. Nellie White is best vernalized (precooled) near 45F (7C), while the optimum for Ace is 40F (4C).

- Scorch. Nellie White is more resistant to true leaf scorch than Ace. Ace is much more resistant than the older variety Croft, which is no longer grown. Ace is not troubled by scorch if a few rules are followed (see "Fertilization" section).

- Space. Nellie White and Ace have the same space requirements.

- Fertilizer requirements. Nellie White and Ace require essentially the same fertilizer. Nellie White may show higher leaf nitrogen content than Ace under similar regimes.

- Timing from visible bud. Nellie White and Ace have the same time of development from visible flowers until flower opening. The same "bud stick" may be used.

- General plant picture. Nellie White generally is given credit for having broader foliage and being more pleasing overall. It is difficult, however, to distinguish between well-grown Ace and Nellie White.

Grower preference is the single determining factor in choosing one of these cultivars over the other. Larger growers, especially, should be familiar with the forcing characteristics of both varieties and be prepared to use either. Nellie White is more prone to summer sprouting in the field than Ace. Since summer sprouting can on occasion affect 10 to 50% of the crop, summer sprouting could affect the supply of bulbs in a given year.

It is also true that weather conditions on the West Coast can affect bulb size and the cultivars differently. There are "Ace years" and "Nellie White years." The ability to handle either variety is an advantage.

Height is perhaps the single most important consideration. The fact that Nellie White is the shorter of the two varieties is balanced, in many instances, by the fact that Ace is less likely to suffer leaf yellowing and can tolerate a higher negative DIF or more grow regulators, thus being held as short as Nellie White. The cost of growth regulators may be offset by the higher bud count obtained with Ace. Nellie White is more prone to leaf curling under excessive negative DIF conditions than Ace. Proper DIF management will eliminate problems.

FIELD CULTURE

Most Easter lilies are produced on the West Coast, around the Oregon-California border area (many fields overlook the Pacific Ocean) between Harbor, Oregon, and Smith River, California. Production requires two to four years' growth in the field, depending on size and whether scale production is used or plants are started from bulblets (small bulbs formed around the belowground stem above the bulb). From scales (modified leaves broken from "mother bulbs"), bulbs called scalets can be produced in one year. Scaling, combined with tissue culture, offers a means of more rapid buildup of desirable stock. Other factors being equal, scalets produce more uniform crops than bulblets.

Bulblets or scalets are graded and planted to produce 4- to 8-inch- (10- to 20-cm-) circumference bulbs called yearlings the first year. These produce "commercials" in the second year. Because of demand for larger-sized bulbs, some smaller commercials are replanted for still another year.

Packing and size. Bulbs are harvested in late September and October, usually being completely packed by October 20. Rain is a determining factor in some years. The rainy season can begin on the West Coast about September 15, thus delaying the harvest. Forcers should be aware of this and prepare to adjust their procedures should harvest be delayed.

New lily pack standards are presented in table 1. Changes were made from the old style cases to facilitate handling and to improve cooling of bulbs in cases.

Bulbs are packed in peat moss of a standardized moisture content. The ratio of bulbs, peat moss, and moisture is of critical importance. *Bulbs must not dry out during the vernalization period or afterwards.*

VERNALIZATION (PRECOOLING)

Easter lily bulbs of presently grown cultivars have similar, though not exact, vernalization, or precooling, requirements. *Vernalization* is the proper term to describe the cold treatment, which lasts several weeks and must precede initiation of flower buds. While *vernalization* is the proper word, the process is often referred to as precooling, cooling, chilling, cold treatment, and others. It is important to remember that not only is the cold treatment critical, but it must be given under *moist* conditions. Cold received in late October, November, and December is "remembered" by the stem to cause flowers to be initiated in January. If plants are not exposed to cold (or long days), the stem will eventually grow—perhaps indefinitely—and not initiate flowers. Stems with over 300 leaves have been recorded. Cold thus causes bulbs to cease making leaves and to form flowers.

Case-precooled bulbs are precooled in the packing case, wherever it takes place, by the bulb grower or by the finished producer.

Pot cooling is a broad term meaning that plants are precooled after potting. Pot cooling has many variations, such as cold framing, which refers to early potting and placing in a cold frame or another location that will prevent freezing but otherwise is not temperature controlled. Outdoor cooling refers to potting and placing pots outside, with perhaps a straw cover to prevent drying, protect from heating during the day, and ward off frost at night. CTF, or controlled temperature forcing, is a popular, preferred system of pot cooling. The CTF system, or variations of it, allow more definite control of vernalization. When the investment in bulbs, soil, pots, and labor is considered, we believe it prudent that all pot cooling be done under controlled conditions.

A variation in pot cooling that we have found helpful is to pot bulbs in 4½-inch (11-cm) pots for vernalization. This saves space in expensive refrigeration facilities. After vernalization (and before emergence), the knocked-out bulb and rootball are carefully placed in the bottom of an inverted pot (to pot deeply). The pot, turned right side up, is then filled with soil. This entails two pottings but saves space in coolers and allows potting in fresh, noncompacted media. Root growth is explosive after final potting.

Case cooling. Bulbs are shipped from the production area to either commercial cold storage facilities or to greenhouse growers, who place them into refrigerators in the cases in which they have been shipped. It is important that temperature and time be carefully controlled. Table 2, though reflecting some variation in the data, illustrates the effect of too little or too much vernalization or precooling.

Almost always a vernalization time of six weeks (1,000 hours) is recommended. With six weeks expect forcing times of 110 to 115 days. Longer vernalization results in faster forcing but lower bud count. This is a trade-off. Nellie White has an optimum vernalization temperature of 44 to 46F (7 to 8C), and Ace at 39 to 41F (4 to 5C). Times longer than six weeks are not suggested. Bulbs probably vary from year to year either in the amount of cold they have accumulated in the field or in the time requirement for vernalization, and perhaps also as to optimum temperatures as a result of seasonal changes. It has not been necessary to vernalize longer than six weeks or less than four weeks. For practical purposes, never vernalize less than five weeks, and do that only in the years when bulbs have received some cold in the field.

TABLE 2   CASE COOLING

Weeks of storage	Flowering	
	Days	Number
0	196	10.0
1	176	9.7
2	160	9.1
3	135	7.1
4	123	6.4
5	114	6.5
6	109	5.6
7	112	5.6
8	110	5.2
9	103	5.0
10	100	4.9
11	98	4.4
14	103	4.5

CTF cooling. Bulbs are potted immediately after being received in October. After potting, many schedules call for three weeks of 63F (17C) for root growth, and six weeks at 40 to 45F (4 to 7C) for vernalization. This is a total of 63 days. When bulbs are received early and Easter is relatively late, this is no problem. When Easter is early or bulbs are received late, there is just not time to accommodate the entire CTF nine-week schedule. The six weeks of vernalization is most important. More on this later.

Many texts place the forcing time for Nellie White and Ace at 120 days. Earlier editions of the *Ball RedBook*, for example, say "timing of lilies for Easter centers around a basic rule that the bulb requires approximately 120 days from potting to flowering." Table 3 shows (after six weeks' vernalization) 109 days to flower. The point is, *at least* 110 days should be allowed for bringing the pot into the heated greenhouse until shipping. The 120-day figure is a good one because of slower forcing. Usually a third of the crop is shipped up to two weeks before Easter. Many growers in recent years have not allowed enough time for forcing.

Consider the forcing times shown in table 3 for a crop that is to be shipped *seven days before* Easter:

TABLE 3   FORCING TIMES TO EASTER

Easter date	Forcing days from		
	Dec 1	Dec 8	Dec 14
March 26	109	101	95
April 7	121	113	107
April 14	128	120	114
April 21	135	127	121

For the two earliest Easter dates, plants must be brought into the greenhouse by December 1 to 8 or earlier to allow time for forcing. On the latest Easter dates, enough time is available from a December 15 to 21 date.

Most practically, whatever the date of Easter, we strongly suggest that plants be in the greenhouse no later than December 15. On a very early date, December 1 is much preferred. This allows time for the plant to develop. If December 15 is used as a date to begin forcing (no matter what the Easter date), then either the schedule before forcing or the forcing time, or both must be adjusted.

Note that for the earliest Easter, a vernalization date of October 26 is suggested (table 4). Earlier is better. If bulbs are shipped from the West Coast October 10, then plan five days for transit and three days to get potted—to October 18 or thereabouts. At most, there is only seven to eight days for 63F (17C) rooting treatment. The movement to the greenhouse at the proper time and a full vernalization treatment are more important than three weeks at 63F; therefore, the rooting period should be cut short. Note also that in many cases, weather and other factors prevent early bulb shipments. Also, as more and more forcers elect pot cooling, more and more early shipments are requested, so it is obvious that all shipments cannot be made at once. Further, bulbs are not always out of the ground to honor all early shipment requests.

In summarizing pot cooling techniques, remember that Easter lilies need 110 to 120 days from the start of forcing to flowering. Many troubles in forcing result from bringing pots into the greenhouses too late and "starting from behind." Pot cooling is recommended for those who can use and understand it. The 63F (17C) rooting period should be adjusted (eliminated, if necessary) to allow for a full six weeks of vernalization and getting pots into the forcing greenhouse in time. In many cases, a slower start in the greenhouse at relatively low temperatures helps early root growth. It is also critical that forcers understand that they must keep pots moist to ensure that bulbs can perceive the proper cold temperatures. Many problems are blamed on the bulbs when, in fact, the problem was caused by drying during pot cooling.

With today's earlier poinsettia-shipping schedules, lilies can be brought into the greenhouse at the proper time. While this may require special management,

TABLE 4   CONTROLLED POT-COOLING SCHEDULE

	Vernalization		Forcing	
Easter date	Days	Start	Start	Days to Easter
March 26	42	Oct 26	Dec 7	102
April 7	42	Nov 3	Dec 15	114
April 14	42	Nov 3	Dec 15	121
April 21	42	Nov 3	Dec 15	128

the extra trouble will be more than repaid by the crop quality that results. Raising the temperature of the storage after precooling, prior to moving the pots to the greenhouse, is a method to start forcing at the proper time if some problem prevents moving pots to the greenhouse on schedule and if sprouting has not started. Growth in storage is probably slower than in the greenhouse because there is no solar radiation. There is also danger of sprouting in dark storage, so use care.

Modifications of pot and case cooling. In the past some growers have requested that bulbs be shipped to them early, prior to the finish of case cooling, in order to be potted in late November. These growers then run cool temperatures, 50 to 55F (10 to 13C), until late December, when they raise temperatures to 60 to 65F (16 to 18C) to start forcing. This is a workable system. The cool temperatures during December allow for some rooting and some vernalization. The four weeks at 50F equates to nearly two weeks at 40F (4C). This is a system that was widespread before the long-lasting poinsettia varieties became common.

In some warm areas, such as the South and the inland valleys of California, a modified pot cooling program works, although controlled conditions are preferred. Some forcers allow bulbs to have two to four weeks of vernalization in the case. Then they pot them up and allow the balance of the vernalization to proceed under natural conditions. This procedure, too, allows some rooting to occur.

Lighting. Long days can substitute for vernalization on a day-for-day basis. Long days also have the same effect on reducing bud count as does increased vernalization. From a practical view, lights can be very beneficial if combined with sorting. Light of 10 footcandles (108 lux) (mum lighting) is used four hours nightly. If a new lighting is installed just for lilies, these costs can be reduced by using intermittent lighting. Begin lighting immediately on emergence (have lights on one to two days prior to emergence) and continue to light for the number of days desired.

There is one potential dilemma: early-emerging plants could receive more long days than they need, and late emerging plants not enough. Observations indicate, however, that slow-emerging plants often have fewer leaves than early-emerging ones, thus flowering in nearly the same time with slightly less lighting, eliminating the worry. Remember: Lighting for too long reduces bud count.

Bulb drying and vernalization. Vernalization is a process that takes place under cool, moist conditions. Drying during vernalization, either in the case or in the pot, can prevent the bulb from receiving cold temperatures and result in uneven or partially vernalized bulbs. If vernalization is not complete, exposure to temperatures of near 70F (21C) or higher can erase the cold treatment and can cause growth anomalies. If vernalization has been completed (that is, by having six weeks at temperatures near 40F, 4C, under moist conditions), temperatures at or slightly above 70F (21C) will not cause devernalization.

BUD COUNT AND BULB SIZE

Bud count is controlled by bulb size, vernalization, and growing factors. Bulb size affects bud count. Table 5 is idealized, but it gives a rough idea of the number of flowers to be expected from a given size of bulb treated properly. Note that pot cooling can increase bud count. Bud count is probably controlled by the meristem area at the start of the flower initiation period (January 7 through February 7). Thus, larger bulbs, which have larger meristems, have more flowers. Similarly, anything that promotes vigorous growth of the new stem can increase flower count. The most recognized of these growth factors is temperature. Other factors, such as sunlight, good fertilizer, high CO_2 levels, good roots, and proper watering, all have definite effects on bud count.

Forcers who use smaller bulbs, such as 6½- to 7-inch (17- to 18-cm) Ace or 7- to 8-inch (18- to 20-cm) Nellie White, must bear in mind that should any difficulties arise during the forcing period, where they lose a bud or two, they may end up with an undesirable, nonsalable plant.

By reducing the night temperature during the flower initiation period, growth slows and the meristem apparently expands, allowing more flowers to form. Reducing temperatures to 55 to 58F (13 to 14C) for seven to 14 days can increase bud count appreciably. *Perhaps more importantly, it must be emphasized that raising the temperature during this period can cause a severe loss of flowers.* In no instance is it suggested that the grower use a temperature dip unless the leaf count method of timing is being used to monitor crop development.

The bad effects of oververnalization have already been covered. The most beneficial effect of pot cooling is that roots are established when flower initiation occurs. This allows lots of water and nutrients to be absorbed to promote vigorous growth. Much as a pot mum cutting fattens after planting, a lily stem also expands. The better the growth, the higher the bud count. A further effect of rooting prior to stem emergence is apparently control of leaf elongation by the

L

TABLE 5   EASTER LILY BULB SIZE AND FLOWER NUMBER

Bulb circumference		Case-vernalized		Pot-vernalized	
Inches	Cm	Ace	Nellie White	Ace	Nellie White
6½–7	16–19	3–4	2–3	4–5	3–4
7–8	18–20	4–5	3–4	5–6	4–5
8–9	20–23	5–6	4–5	6–7	5–6
9–10	23–25	6–7	5–6	7–8	6–7
10–11	25–28	7–8	6–7	8–9	7–8

Note: Figures are average number of flowers per plant, not maximum or minimum. Pot cooling, in many of its variations, will usually produce one or more buds more than indicated in each category here.

root system. Plants that emerge prior to rooting will have shorter lower leaves than those well rooted prior to emergence.

POTTING

Bulbs should be potted either upon receipt or completion of cooling. Delay in potting is a serious potential problem. Plant bulbs deep in standard pots (6 by 6 inches, 15 by 15 cm, for example) to protect against early emergence in controlled storage and to allow adequate room for the development of stem roots above the bulb. One inch (2 cm) of media in the bottom of the pot is adequate, and 2 inches (5 cm) of media over the bulb is preferred.

A few bulbs may sprout in the case. Sprouted bulbs are not hurt. The critical factor is to bury the entire etiolated (white) stem below media. Sometimes planting the bulb on its side can accomplish this. If the entire stem is covered, growth will be normal on emergence. If a portion of etiolated stem remains aboveground, leaves will not elongate, and small stem bulblets will form. If this is 1 inch (2 cm) or less, it will not be noticed at flowering.

Some forcers still prefer gravel in the bottom of pots, mostly for weight to prevent tipping. Gravel serves no drainage purpose; in fact, because it reduces the medium's water column and contributes to a wetter root area, it is a practice that should be discarded.

With more height specifications by large buyers, more three-quarter pots (6 by 5 inches, 15 by 13 cm) are being seen. The 1 inch (2 cm) shorter pot could make a difference between three and four layers in a truck. However, plant height is best controlled by growth regulators, negative DIF, spacing, and so on, rather than by using short pots, because deep potting is beneficial and should be a goal.

Media. A good lily medium not only has high water-holding capacity but also good drainage and especially good fertilizer-holding capacity. It is now thought unlikely that the keeping quality of lilies grown in lightweight peat mixes is as good as that of lilies grown in heavier soil or compost-based mixes. Further, lily media should have enough weight to prevent tipping should lilies get taller than desired.

Depending on physical characteristics, 17 to 50% of the mix can be soil. Vermiculite is excellent to increase nutrient exchange capacity. Peat moss at 25% or more and bark (be sure to add nitrogen to correct for bark decomposition) are suitable. Bark can be a problem if A-Rest drenching is used because of absorption. Remember that on a volume basis, peat moss has very little fertilizer-holding capacity. A useful mix is 33% soil, 33% peat moss, 17% vermiculite, and 17% perlite. You could replace half the peat with bark, giving 17% of each in the mix.

Adjust pH to 6.2 to 6.5 with calcium carbonate, using 2 pounds per cubic yard (1.2 kg/m^3) as a basis for soil-based mixes, but nearly 0.5 pH unit lower for artificial mixes. Certain limestone deposits have apparently high fluoride content. Known concentrated sources of this material should be avoided. For this reason, perlite should not be used in lily mixes. Calcine clay is an excellent aggregate where available. Styrofoam has been used, but it is not in favor because of the litter problem.

Fertilization. A complete fertilizer, 12-12-12, for example, can be added at the rate of 1 pound per cubic yard (0.6 kg/m^3). *Do not use superphosphate* (read about leaf scorch under "Problems"). If bark is added to the mix, we suggest an extra half pound of urea formaldehyde fertilizer per cubic yard (0.3 kg/m^3) be added to compensate for nitrogen tie-up in bark decomposition. If media is to be stored, organic nitrogen should not be used. Trace elements are probably best added in liquid form.

Easter lilies need fertilizing early in their development to produce vigorous stem expansion and a large leaf canopy. For this reason, media should contain adequate fertility. Good nutrition begins with a soil of high initial fertility but not exceptionally high in total soluble salts. High soluble salts could cause erratic sprouting and in extreme cases could prevent sprouting completely.

Ace is more susceptible to leaf scorch than Nellie White, but Nellie White can show scorch symptoms. Leaf scorch is a serious disorder, and many younger growers have not seen the extreme loss that can result. Base fertilization of lilies helps to eliminate leaf scorch as a problem. Leaf scorch can be controlled—whatever its true cause—by high calcium levels and low phosphorus. This is the reason why lime should be used for pH control, and why no superphosphate should be used in the initial mix. After potting, calcium nitrate at 200 to 750 ppm actual nitrogen, coupled with soluble trace elements and potassium at 200 ppm, is satisfactory, depending on soil mixes and irrigation schedules. Calcium nitrate can be applied at 200 to 400 ppm regularly, and at up to 750 ppm to boost fertility. Phosphorus at 20 ppm in irrigation water is not likely to induce leaf scorch and certainly provides adequate phosphorus.

One of the most important cultural considerations is to provide adequate nitrogen to prevent lower leaf yellowing and subsequent leaf loss. Experiments that removed the lower third of leaves induced tall lilies. Lower-leaf loss caused by nitrogen deficiency most likely starts at bud initiation. A high nitrogen requirement at this time plus a possible deficiency because of organic matter decay can result in an incipient nitrogen deficiency. Often at this time, it is difficult to make an adequate number of liquid fertilizer applications because of constantly wet soils due to dark weather and other liquid applications. Frequently, the yellowing of a few lower leaves is attributed to root loss or drying, and nothing is done. Nitrogen deficiency then becomes progressively more pronounced until it is too late to correct it.

The best method to prevent nitrogen deficiency and to ensure a dark green, shiny, healthy leaf surface is to top-dress with a dry, long-lasting nitrogen source, such as urea formaldehyde nitrogen (Nitroform, for example), Osmocote, or other slow-release nitrogen fertilizers. One-half heaping teaspoon of Nitroform per 6-inch (15-cm) pot at mid-January and another February 15 is adequate. If urea-formaldehyde is used, extreme care must be used not to "dump" the whole half spoon in one spot, but to spread it around the pot. Application costs are more than repaid. Soluble salts should be kept below 2.0 micromhos per square centimeters using a 1:2 soil water dilution.

For best keeping quality, plants should be well fertilized during the growing season, so that the fertilizer level can be reduced in the greenhouse during the last week. Clear water applications will leach out any excess salts. Lilies, like foliage plants, use much less fertilizer and are more susceptible to fertilizer injury when they are moved from the growing environment to a nongrowing environment. This is especially true when lilies are boxed and cold-stored prior to shipping.

IRRIGATION

In most cases, lilies should not be automatically watered by mat or drip tube systems. Excess height almost always results, unless special conditions prevail. Overhead sprinkling, however, is satisfactory. As with all crops, water management is tied to media management. Heavier, soil-based media are harder to manage with respect to water relations than lighter media, but the extra customer satisfaction resulting from soil-based mixes makes the extra effort worthwhile.

Excessive watering encourages root rot problems. It is probable that currently available fungicide drenches have allowed growers much more latitude in watering. However, with the loss of certain fungicides, watering may become as critical as in the past. Maintain careful watch over root systems should rot appear. Carefully assess water levels, soluble salts, and fungicide drench timing.

Early in the life of the crop, in the dark days of December and January, fungicide drenches, liquid fertilizer applications, and liquid growth regulator applications are difficult to time. Our theory has been that "wet is wet," so that needed fertilizer or growth regulator applications are made mostly on schedule.

CROP TIMING

A great amount of research work in recent years has made the lily crop more predictable. Timing and height control are two areas that were highly unpredictable as recently as 10 years ago. In spite of the progress made in recent years, lilies are a difficult crop to grow to perfection. There are several reasons for this:

- ◆ Easter dates vary widely at a time of the year when the environment rapidly changes.

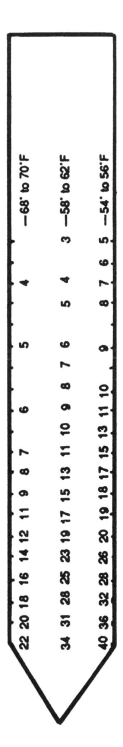

♦ Bulbs are field grown and exposed to differing environmental conditions yearly.

♦ Only one crop a year is grown.

♦ Vernalization, as well as season, has a great effect on the number of leaves that are initiated prior to flower initiation, thus influencing the time to force.

♦ Specifications by large buyers have been recently adopted. These criteria are perhaps overly restrictive given the nature of the crop.

Until A.N. Roberts, Oregon State University, devised the leaf counting method of timing lilies, most growers relied on height to time the crop until flower buds were visible. From visible flower buds, timing was aided by "bud sticks." (See fig. 2.) The pattern can be traced onto a pot label to make a "bud stick." The pointed end is sharpened and then aligned with the base of the small developing bud (where the peduncle ends and the petals and sepals begin). The tip of the flower is then aligned with three numbers that show the number of days at three temperatures required to bring that bud into flower. Note that a bud nearly one-half inch (1 cm) long—about the visible stage—can be brought to flower in as little as 20 days by "hard forcing" at near 70F (21C) night temperature (day temperature 10 to 20F, 6 to 11C, higher) or as many as 36 days at 54F (12C).

FIG. 2. This is the famous lily bud stick. How to use: If you grow Easter lilies at 68 to 70F, place the bud stick tip where the flower joins the peduncle (1). Next locate the corresponding number at the bud tip (2) to find the number of days at 68 to 70F until the bud will open. In this example the answer is about seven days. Note: A lot of bright sun will hasten the process.

TABLE 6   INFLEXIBLE, OLD-STYLE EASTER LILY DEVELOPMENT SCHEDULE

Weeks to Easter	Days	Suggested stage of development
17	Dec. 3	Potted, 60F (16C)
16	Dec. 10	Making roots
15	Dec. 17	Making roots
14	Dec. 24	Making roots
13	Dec. 31	Growth coming through
12	Jan. 7	Growth 2–3 inches (5–7 cm)
11	Jan. 14	Growth 4–6 inches (10–15 cm)
10	Jan. 21	Growth 6–8 inches (15–20cm)
9	Jan. 28	Growth 10 inches (25cm)
8	Feb. 4	Growth 15 inches (38 cm)
7	Feb. 11	You can feel the buds
6	Feb. 18	You can see the buds
5	Feb. 25	Buds ½–1 inches (1–2 cm) long
4	Mar. 4	Buds 2–3 inches (5–7 cm) long; few bending down
3	Mar. 11	Buds 3–5 inches (7–13 cm) long
2	Mar. 18	Buds fully developed
1	Mar. 25	Buds whitish; cooled
½	Mar. 28	Some opening; cooled
0	Apr. 2	Easter

Time of emergence and height are still useful guides for early development. Such detailed schedules were published by many colleges in the past; few are now. In table 6, these criteria are presented in an old schedule based on an April 2 Easter.

The essential problem of such a height schedule is that it completely fails to take growth factors that influence height into account. It is rigidly fixed in allowing six weeks for development from visible bud, and also relatively rigid upon emergence. The only flexibility is in the middle of the schedule. Since it doesn't allow any method to determine how fast development should be in the middle of the schedule, it really is quite useless.

Lily leaf counting. If buds are not seen by a desired time, temperatures can be raised. Obviously, it would be advantageous if an earlier measure was available. Such a system was devised by Oregon State University Research and has been widely publicized and refined by Harold Wilkins, retired, University of Minnesota.

Using the leaf counting procedure, a grower can determine by mid- to late January exactly how many leaves the crop has (remember, the number of leaves varies with the amount of cooling). Since the rate of leaf unfolding is determined by temperature, a grower can figure the best forcing temperature as early as the leaf count is made. By monitoring the rate of leaf unfolding, development can be continuously monitored. The number of leaves the crop has must be determined after flower initiation has occurred. To count leaves, use the following procedure:

1. January 15 to 20, select three to five representative plants in each major lot to be monitored.
2. With a felt tip pen or by notching a leaf, select the uppermost "unfolded" leaf.
3. Start at the bottom and count all the leaves that have unfolded, up to the notched or marked leaf (in step 2). Write this number down.
4. Start with the notched or marked leaf and remove and count leaves toward the growing point. This is easy until the leaves get to be one-quarter to one-half inch (0.6 to 1 cm) long. At this point, a mounted hand lens and needle will be necessary. Count the leaves right into the growing point; buds should be visible. Write down the number of leaves. As an example, the numbers written would appear as follows:

Leaves unfolded	50
Leaves not unfolded	<u>45</u>
Total leaves	95

5. Average the number of leaves for three to five plants. If, for example, the count is made on January 15, and 45 leaves are left to unfold, compute the number of leaves per day that must unfold to make your schedule:
    a. Counting back from Easter, compute the number of days before Easter that buds should be visible. Usually six weeks is used but less can suffice. So, if Easter is on April 19, then six weeks earlier would be March 8.
    b. There are 52 days from January 15 to March 8. Divide 45 leaves not unfolded by 52 days. The answer is that 0.87 leaf must unfold per day in order to see buds on March 8.

Now the question is, how many leaves per day can be unfolded? After the leaves per day are determined, the data in fig. 3 can be used as a starting point to pick a forcing temperature.

It's nice to read in a book that for a "large bulb," 60F (16C) will cause one leaf per day to unfold. It's even better to actually measure the average number of leaves unfolding per day after the indicated temperature change has been made. To accomplish this, follow this procedure.

1. Select another three to five plants from each major lot. Put a label in each pot and use a tall, flagged stake to mark it so that it will be easy to find.
2. Write the date on the label and notch or mark a leaf in the same relative position as indicated in step 2 in the previous procedure.
3. Wait four to five days and again mark a leaf in the same relative position as in step 2.
4. Then count the number of leaves that have actually unfolded between the most recently marked leaf and the earlier marked leaf. Write this on the label.

L

5. Compute the leaves per day as follows: days/leaves unfolded = number of leaves per day. For example, five unfolded leaves in four days means 1.25 leaves per day.

6. The computation in step 5 of the previous procedure indicated that 0.87 leaves must unfold per day. This current computation shows that 1.25 leaves per day are actually unfolding. Therefore, temperatures need to be slowly lowered to decrease the rate of unfolding to 0.9 to 1.0 leaf per day.

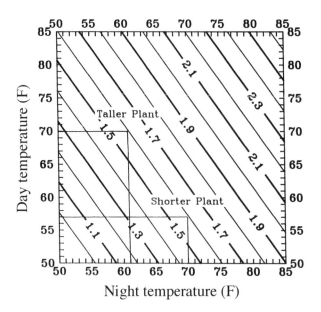

FIG. 3: Lily leaf unfolding rate at various day and night temperatures.

By counting leaves every four to five days, crop progress can be followed. This is a workable system. All lily growers should use it. *Remember:* Leaf counting allows timing to start January 15 rather than when buds are visible. Growers have four to six weeks longer to manipulate temperature using this system.

Leaf counting also allows more reasoned decisions regarding temperatures, which can result in fuel savings. When coupled with graphical tracking (to be discussed later), keeping consistent records of the rate of leaf unfolding along with major cultural events—periods of high or low light intensity, irrigation and fertilizations, CO_2, temperature changes, root loss, fungicide applications—makes it possible to determine the effects of these changes on growth rate. It builds a body of knowledge that can be used to judge the effects of future cultural changes.

HEIGHT CONTROL

Chemical growth regulators. At this writing, A-Rest and Sumagic are the only practical chemicals to reduce the height of pot lilies. While expensive, their use often is necessary, and probably will continue.

Growth regulators may be applied as a drench, as a spray, or as a combination. If roots are present, application to media may be made early, before the leaf surface is expanded to absorb a spray. The treatment may have a more lasting effect (but only if a root system is present). Shredded pine bark absorbs A-Rest, thus reducing its effectiveness; fir bark is less absorptive. More material must be used to counter either loss. The greater expense incurred may offset the advantage of soil application. Rates of A-Rest near 0.5 mg per pot as a media application and 33 ppm as a spray (see A-Rest label) are effective. Growth regulators are less effective during periods of low light. In higher light areas, or if bark is not used (these rates presume use of bark media), the lower rates probably should be used. Spraying is gaining favor over media drenches because of flexibility, especially in view of negative DIF.

Growth regulators should be avoided, if possible. The known effect of rapid senescence (leaf yellowing from the bottom to the top) has been associated with A-Rest applications, particularly on Nellie White. Most yellowing, though, has occurred under conditions of carbohydrate shortage. Prolonged high-temperature forcing, long shipping distances, long storage of boxed plants, and late applications have been most detrimental. Some thought persists that A-Rest reduces flower count. However, it is unlikely that this is true unless it is applied too early at excessive rates.

Growth regulators should be applied in relation to leaf number in order to have more predictable response from year to year. First spray applications can be made when 25 to 30 leaves have unfolded. Sprays any earlier have less dwarfing effect, apparently because of lack of leaf surface. Usually two or more applications are necessary, the second seven to 15 days after the first. There may be some advantage to a first spray application, which gives a rapid effect, followed by a drench, which seems slower to take effect but lasts longer. The spray is done before extensive root growth, and the drench after some root growth.

Negative DIF. Work at Michigan State University by Dr. Royal Heins and his graduate students has had an impact on height control of lilies. Studies have shown that reversing the normal pattern of day and night temperatures has a profound effect on stem elongation. The normal pattern is a higher day temperature than night temperature. By changing to a higher night temperature than day temperature, shorter plants result. Low day temperatures are most effective during the first hours of light each day. This is fortunate because it is sometimes impossible to maintain low day temperatures in high sunlight situations or when outside temperatures are high.

When night temperatures are higher than day temperatures, it is referred to as negative DIF situation. Conversely, with higher day temperatures than night temperatures, a positive DIF results. The greater the DIF, positive or negative, the greater the height difference will be, taller or shorter. As a practical matter, growers have had good results with negative DIFs on the order of 5 to 15F (3 to 8C). Excessive negative DIF can cause growth changes, especially downward leaf curling on the variety Nellie White. Unless the negative DIF has been too excessive, a few days of positive DIF will correct the problem.

When using DIF to control height, it is critical to remember that the total amount of heat a lily receives controls flowering. In the past, lilies often were exposed to day temperatures considerably higher than today and thus were forced faster.

If it is determined that in order to flower the crop on time, what is needed is an average 24-hour temperature of 65F (18C) and a 10F (5.5C) negative DIF, then the following computations can be made. It assumes (for example) a nine-hour light period and a 15-hour dark period.

24 hours × 65 degrees (18C)	=	1,560 degree-hours	(432)
15 hours × 69 degrees (20C)	=	1,035 degree-hours	(300)
degree-hours remaining	=	525	(132)

Dividing 525 (132) degree-hours by a nine-hour day gives a day temperature of 58F (14C), about a 10F (5.5C) negative DIF, and an average temperature of 65F (18C).

Graphical tracking. The tools to control height are largely available. The challenge is to apply them. Royal Heins has devised a method to track height changes on a graph on a regular basis, so that A-Rest or DIF can be used to change the rate of stem elongation.

The graph shown in fig. 4 illustrates tracking a crop for an April 7 Easter, with planned flowering one week earlier. The graphical tracking method is based on the assumption that the height of a lily will double from the top of the pot to the top of the plant from buds visible to flowering. (Under conditions favoring tall lilies—high day temperatures, low light, no growth retardants, too much water, etc.—height can more than double from the *bottom* of the pot to the top of the plant.)

To follow the graphical tracking procedure, record the date of emergence on a graph. This is the starting point. The next step is to determine the desired final height—for example, 20 to 22 inches (51 to 56 cm). Third, specify the date buds should be visible. In the case illustrated, this is February 28, 42 days before Easter. Assuming a height of one-half final height at buds visible, a buds-visible range of height should be 10 to 11 inches (25 to 27 cm). Lines connecting the emergence date at the top of the potato buds visible, then to final height, defines the envelope of height over time that is the goal.

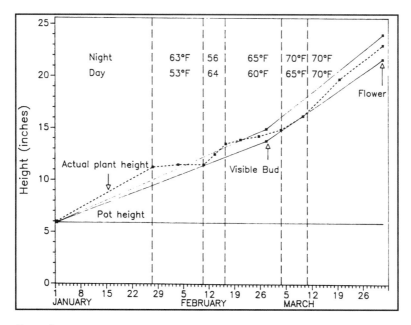

FIG. 4: Crop tracking for an April 7 Easter

Height can be monitored closely by measuring test plants at least twice per week. The graphs should be filled in while in the greenhouse so that any modifying actions, such as temperature changes or A-Rest applications, can be entered at the same time, thus creating a record for future crops. A potential problem of graphical tracking is false information that could result from handling the plants to get measurements. More than one uniform plant should be selected at the beginning of the tracking. This allows switching plants, should stunting from handling occur.

Graphical tracking is a practical method to follow crop progress. Graphical tracking is not, however, a substitute for watching other aspects of culture. It is interesting to note that while great progress has been made in methods to control and monitor growth, buyers' specifications have become more demanding. Growing the crop has therefore remained difficult.

PROBLEMS

Diseases. Root rots are probably the most serious disease of Easter lilies. Well-drained media and attention to good irrigation practices are important. A wide array of soil drenches are available for control. This may change as chemicals become harder to obtain. *Pythium, Rhizoctonia,* and to a lesser extent *Fusarium* appear to cause the most problems. The drench schedule in table 7 has proven effective.

TABLE 7   EASTER LILY FUNGICIDE DRENCHES

Fungicide	Concentration (oz per 100 gal)	Time		
		Potting	Jan 15	Mar 1
Cleary's	4–8	No	Yes	Yes
Subdue	½	Yes	No	Yes
Banrot	4–8	No	Yes	No
Terraclor	4	Yes	No	No

Note: This schedule exactly applies to either pot-cooled or case-cooled bulbs.

Roots should be monitored and the drench schedule adjusted as needed. One Terraclor application is important because of its effectiveness in *Rhizoctonia* control. *Rhizoctonia* infections may be associated with bulb mite infestations, and Terraclor is most effective in preventing this.

Two major viruses affect lilies. When cucumber mosaic virus and lily symptomless virus are at high levels, fleck symptoms appear. This virus can be crippling. The fleck spots on the leaves and sometimes distorted plants can cause economic loss. All lilies in commercial production have lily symptomless virus. Clones made free of this virus grow much taller than infected clones; but flower count and other characteristics are apparently not much affected.

Poor growing conditions can magnify virus symptoms. Low temperature at starting, low humidity, and other conditions can cause more symptom expression. Due to roguing and aphid control in the field, viruses are usually not a problem, however.

Under low-light, high-humidity conditions, *Botrytis* can be a problem on unopened buds and flowers. The best control is to add heat while venting to reduce relative humidity. A Daconi spray over the bulbs is also used at 1 pound wettable powder per 100 gallons (479 g per 400 l) with a recommended spreader. *Botrytis* is seldom a problem on foliage in greenhouses, although it is the most important disease in field production. Under severe humidity and free-water situations, *Botrytis* can occur on leaves.

Insects. There has been much publicity about bulb mites. Mites have long been associated with bulbs and are responsible for consuming the sloughed-off outer scales of the bulbs as they grow from the inside out. Mites will tunnel into stems, apparently entering between the bulb and the medium surface. It is debated whether the mites are primary, causing the lesion themselves, or secondary, entering a lesion caused perhaps by *Rhizoctonia*. To protect against the latter, Terraclor is specified as a drench at potting. Most severe bulb mite-type damage has been seen in pot-cooled lilies. Plants are often bent because the lesions on one side of the stem stop growth while the opposite side elongates. Dwarfed, stunted growth with thickened leaves and crippled growing points has been attributed to mites, but this has not been established.

Bulb dips in Kelthane can be effective to eliminate mite populations. Rates are 1⅓ pounds of Kelthane 35% wettable powder per 100 gallons of water (639 g per 400 l) with bulbs immersed for 30 minutes. Plant immediately to prevent drying.

Aphids transmit viruses in lilies and should be rigorously controlled in greenhouses. A few aphids early in the crop can cause considerable damage. Distorted foliage, honeydew secretions with subsequent black mold, as well as virus buildup can result from aphid infestations. Controls that have been effective include Orthene W.P. at 1 pound per 100 gallons (479 g per 400 l), or Diazinon AG 500 at 1 pint per 100 gallons (a 1:800 ratio).

Preventive sprays are not suggested, but beginning in early January, carefully monitor the crop. Early spot treatments can sometimes eliminate the need for general sprays. Aphids can build rapidly near the end of the crop, however, due to warmer temperatures.

Fungus gnats can build up on lilies that are kept especially moist or when much algae is present. It is not known if direct damage can result; however, their presence is not a good sign. Granular Diazinon and Vecto-Bac have been used for control.

No-shows. Lily bulbs are a field crop, and by the very nature of production, all bulbs are not perfect, so that each year a percentage of the crop may not emerge. There are several causes of this phenomenon, although some nonemergence cannot be explained. Among the most common causes are broken sprouts. If lily bulbs summer-sprout in the field—if the stem that normally emerges in the greenhouse emerges in the field—the sprout may be broken off during the harvest operation. Bulbs are inspected prior to packing, but the broken sprout may not be detected. A bulb with a broken sprout will eventually make good roots and sprout, but the plant will be off cycle and will not bloom for Easter.

Phytophthora, and sometimes other soilborne diseases, can cause a dying back of the flowering stem in the bulb. This may occur on a prematurely sprouted bulb in the field or in a sprouting bulb after potting. Such bulbs are very difficult to detect at packing. At other times, the dieback occurs after packing or potting. Such bulbs eventually make good roots and sprout, but they too will be off cycle.

During the harvesting process some bulbs are broken. This is evidenced by bulbs that do not emerge and, when dug up and scaled down, show excessive amounts of bulblet formation.

In addition to these determinable causes, there are still unexplained nonemergences. Some bulbs make roots but do not sprout. This is seldom a problem, but it can occur. Other bulbs do not sprout or make roots. Nonemerging bulbs of this kind can occur in the field, during vernalization, or in the greenhouse. Severe drying, anaerobic conditions from media high in readily decomposable organic matter or water, and probably other factors, can cause rest in the bulb. That is, even though the current environment is favorable, internal factors prevent sprouting and growth.

Bud abortion and bud blasting. There are three phenomenon which pertain to loss of flower buds. The first is buds that have blasted. The first sign of blasting is bud growth stoppage, followed by shriveling which begins at the base of the bud. This is followed by further browning of the bud. Most often the cause is high-temperature forcing, especially if low humidity has been allowed. Lack of water is also critical. This situation may be made worse by root rot at the time.

The second phenomenon is bud loss by abortion, which occurs just after bud initiation. A grower will see signs that buds were present but then were lost. Again, high temperatures just after the initial bud set period are most often responsible. Drying or any other growth factor can affect this. Small scars, small "pimples," and bractlike leaves (which always are present below a bud) are all signs that buds were there, or were potentially there, but were lost.

The third main bud problem is loss of potential. In severe situations, buds may be lost but leave no telltale signs. This phenomenon is best observed by comparing bulbs from the same case which were grown under different conditions, such as pot and case cooling. Again, high temperatures during the bud set period is most frequently responsible, but poor growing conditions from potting though the bud initiation period can also be the fault.

In most cases, excessively low bud count is caused by greenhouse forcing problems. So many factors affect bud count that simple answers are often not possible. Most problems, however, relate to temperatures during the bud initiation period.

Leaf yellowing. Leaf yellowing from the bottom up is of two types. The first type, catastrophic yellowing that occurs in a very short time, is usually associated with a carbohydrate shortage in combination with, or as a result of, late growth regulator applications, negative DIF, high-temperature forcing, or long-term storage. All of these can cause nearly complete yellowing of the plant. The second type, gradual yellowing at the bottom of the plant, usually starts at bud initiation and often progresses to an unsightly appearance at sale time.

Avoiding late growth regulator applications, late high-temperature forcing, long box storage, and long delivery times are essential in controlling catastrophic yellowing. Some research shows higher phosphorus levels near shipping helps. There is also evidence to suggest gibberellin applications prior to shipping can help. We have used 200 ppm Pro-Gib, and some improvement has been noted.

Gradual bottom leaf yellowing is a complex problem. We believe that the primary control of leaf yellowing is by preventing nitrogen deficiency in the lower leaves. This usually requires extra applications of dry fertilizer or high rates of liquid fertilizer. Stress of any kind is also likely to be important. Research has shown that high-density planting, more than 2½ plants per square foot ($27/m^2$), is contributory. In addition, excessive drying, root rot, and high-temperature forcing are all very critical. Dry fertilizer applications to supplement liquid feeding are very important, especially when it is difficult to sequence liquid feeding with fungicide or growth regulator drenching.

Other problems with mysterious causes. Leaf scorch symptoms are very characteristic: half-moon–shaped areas, often with concentric rings of varying colors of brown. These scorched areas are almost never located at the tip of the leaf but are usually one-quarter inch (0.6 cm) or more from the tip. Uniformly brown "dieback" of leaves is not true leaf scorch, and while occasionally seen, the cause of this problem is not known. Often such symptoms are noticed after a period of clear weather that followed a long period of dark weather. For control, refer to the "Fertilization" section.

The cause of greenhouse twist is debated. Some believe an organism—perhaps a bacterium—is the cause. Whatever its cause, twist can be serious in limited situations. Symptoms are circle-shaped leaves appearing at the growing point, often with brown necrotic areas bordering the inside of the circle. The plant may produce only one or two such leaves or several, and then may outgrow the problem and be salable.

Shipping

Lilies are sold through retail florists and mass merchants. Most retail florists receive plants unboxed, while most mass market lilies are delivered in boxes. In many instances, plants are packed when they reach the "white, puffy" stage and are placed in a cooler. This is the proper stage for packing.

However, *excessively early packing is being abused.* Packing too early can reduce plant quality for stores and ultimately, for consumers. Good quality means good future retail sales. Efforts should be made to reduce box time in all instances. Lack of keeping quality in stores and homes is potentially the most severe problem in the industry. Long box time, high-temperature forcing from late potting, media with no soil base, high growth regulator rates, and other factors may also be responsible for this problem.

LIMONIUM (STATICE, SEA LAVENDER)

by Daijiro Harada
Miyoshi & Co., Ltd.
Yamanashi, Japan

❖ *Annual or perennial (**Limonium** spp.). 13,000 to 28,000 seeds/oz (459 to 988/g). Germinates in 15 to 20 days at 70F (21C). Do not cover seed. Much cut flower limonium is started from tissue-cultured liners.*

Perennial limoniums are in the family Plumbaginaceae. Botanical species originate in the vast regions of the northern hemisphere. Limonium's unique branches are essential as filler for bouquets, corsages, baskets, and other flower arrangements (fig. 1).

Plants for commercial production originate from tissue culture produced in Japan and Taiwan. The advantages of tissue-cultured plants are uniform growth, longer stem length, better flower color, and longer vase life. Other advantages include disease-free plant material and much higher crop yield.

In 1984 we began hybridizing limonium from wild species collected from 112 different sources. While different limoniums look very similar, growers should recognize that there are differences in plants bred from different genera or parents. Growing different limonium varieties under the same cultural conditions and crop management may cause problems.

Basically, limonium can be divided into

FIG. 1. This field of statice was grown outdoors in southern California by Mellano & Co., San Luis Rey, California.

two types, each requiring different handling. Type I consists of seasonal flowering limoniums. This group flowers mainly in early to late summer, depending on the variety. Plants must be well established and exposed to cool temperatures to initiate flowers. If the plants are not established, they will not produce flowers, even under low temperatures. Also, even if plants are mature, without a cold period they continue growing vegetatively. Plants planted in spring to summer develop sufficiently to withstand the cool temperature treatment for flowering the following summer.

Type II limoniums are free-flowering. This group has growth that is temperature-dependent. The degree of free flowering differs among species and varieties. Varieties in this group flower throughout the year if temperature and light are appropriate.

CULTURE

Temperature and light. Although growing-temperature ranges differ among varieties, limonium is quite comfortable with 68 to 77F (20 to 25C). Type II, free-flowering varieties, flower throughout the year depending on temperature and light.

High light intensity may be substituted for some varieties, such as Saint Pierre or Beltlaard. These varieties will flower in winter with high light intensity even if night temperatures do not reach the required minimum temperature. These varieties usually need a minimum temperature of 60 to 65F (16 to 18C) for flowering in winter. Type I, seasonal-flowering limoniums, require a cold period, or natural winter, to induce flowering.

Soil. Sandy or sandy clay soil is preferred, but any well-drained soil is suitable. Good drainage is essential for limonium cultivation. Soil pH of 6.5 and an EC of 0.5 are ideal.

Fertilization. Limonium varieties do not require much nutrition. Moreover, excess fertilization is not only costly but may result in undesirably tall crop height, weak stems, and flower abortion. Following is a guide for basic fertilizing for Type I, seasonal-flowering limonium: 1.2 kg each of N, P, and K per 100 m² (34.5 oz per yd²). Mixed fertilizer or slow-release fertilizer: 300 to 400 kg of organic manure per 100 m² (553 to 737 pounds per 100 yd²), depending on soil conditions. For Type II, free-flowering hybrids, such as Saint Pierre and Beltlaard, use N:P:K at a ratio of 2:1.6:3 as basic fertilizer. Try compound fertilizer 14-12-9 at 8 kg per 100 m² (14.75 pounds per 100 yd²). Use potassium sulfate, 2 kg per 100 m² (3.7 pounds per 100 yd²).

Watering. Generally, a good water supply is important to establish plants during vegetative growth. Once plants begin producing flower stems, sharply reduce water. Mature limoniums require very little water, especially free-flowering hybrid limoniums that will grow tall. Excess watering elongates stems and makes them very weak. The exception to this is the Emille family and Charm Blue, which take much more water. Freely watering these varieties helps the flower stems grow taller.

Harvest. Greenhouse cultivation is recommended for the highest quality stems and continuous production of free-flowering types. Yet, all limoniums can be grown outdoors, except for a few tender perennials that you must keep over the minimum temperature required. Covering is essential to protect flowers from rain damage at harvest. Most varieties are ready to harvest when 70 to 80% of the flowers on the stem are open.

VARIETIES

L. hybrid Saint Pierre and Beltlaard (Type II, free-flowering) are interspecific hybrids of *L. caspium* × *L. latifolium*. Both free-flowering varieties inherited good characteristics, such as hardiness and strong stems. These two cultivars have the darkest purple-blue flowers among similar hybrids. Because both hybrids produce long stems, it is important to supply low levels of fertilizer and water. While the plants are growing, good watering is essential for plant establishment, but once they start producing flower stems, manage the plants dry. High temperatures and high nitrogen can cause flower abortion. In warm climates harvest stems in the early summer and autumn, and rest the plants in midsummer by pruning flower stems.

L. altaica Emille, Pink Emille, and Lavender Emille (Type II, free-flowering) are easy to grow in different climates. This group takes more fertilizer and water than Saint Pierre or Beltlaard. Additional fertilizer applications increase stem length. In warm climates this family produces the first flush in early summer and a second flush in autumn, resting during high summer temperatures. In climates where summers are cool, plants flower continuously.

L. perezii Violet (Type II, free-flowering) has been selected for its deep color, earliness, and high production. *L. perezii* is a tender perennial; keep temperatures over 41F (5C). Because this species is susceptible to *Sclerotinia*, soil disinfections and fungicide sprays are recommended. Manage plants dry. Use raised growing beds for good drainage. Keep fertilizer levels low, as excess nitrogen softens plants and makes them more susceptible to disease.

L. hybrid Lemon Star (Type II, free-flowering) is the first yellow interspecific hybrid, a cross between *L. aurea* and *L. sinensis*. A tender perennial, it requires a minimum winter temperature of 41F (5C). Lemon Star is very productive. Once the plants are established, they actually produce too many stems, making harvest difficult. Thin stems regularly, leaving seven to eight good stems. Lemon Star is also susceptible to soil diseases, such as *Phytopthora*, *Glomerella*, and *Rhizoctonia*. Before planting, disinfect the soil. Fertilize with low nitrogen levels; N:P:K of 1:2:2 is recommended. Lemon Star prefers cool conditions; 50 to 77F (10 to 25C) is ideal.

L. peregrinum Ballerina Rose (Type II, free-flowering), a very nice rose limonium, was developed by the Ministry of Agriculture and Fisheries, Levin Research Centre, New Zealand. Both the calyx and the flowers are large. Most limoniums are used as filler in bouquets, but Ballerina Rose has large, striking flowers that can be the main flowers. Vase life is excellent. Even when the flowers are closed, the brilliant, large calyx stays open. This cultivar is a tender perennial, so maintain a minimum temperature of 41F (5C). Stems are ready to harvest the second year after a spring planting. Higher temperatures speed the flowering time.

L. latifolium Avignon (Type I, seasonal-flowering) is a white-flowered *latifolium*. Just like *L. latifolium*, Avignon resists heat, cold, and disease. Both the flower and the calyx are white. The white color remains even after the flowers are closed.

L. hybrid Charm Blue (Type I, seasonal-flowering) is a hybrid obtained from seedlings crossed between *L. latifolium* and *L. gmelinii*. Charm Blue stands out for its purple calyx, which remains colorful even after the flowers have closed. Plant height reaches 31 inches (80 cm) for the first flowering and can go to 39 inches (100 cm) the following year. Supply plenty of water for Charm Blue, as water shortage results in poor plant establishment and, eventually, shorter stems. When flower stalks begin to elongate, continue to give water. This helps the flower stems grow taller.

The selected *L. dumosum* Tres Bien (Type I, seasonal-flowering) grows as tall as 24 to 28 inches (60 to 70 cm) in the greenhouse. The pink flowers are much brighter than those from conventional *L. dumosum*. The neat and attractive white calyx of the panicle remains after the rose blossoms disappear.

ANNUAL STATICE (*L. SINUATA*) FROM SEED

Provide a cool treatment for seedlings—50 to 55F (10 to 13C), for five to eight weeks beginning in the cotyledon stage—to encourage flowering. As day length increases, the length of cool treatment can be reduced. Transplant young plants to fields in the spring when days are 70 to 75F (21 to 24C), but before nights rise above 55 to 60F (13 to 16C). Use long days (mum lighting) on actively growing plants to promote earlier and greater flowering. Southern growers will harvest three to five months after sowing, while northern growers will harvest four to six months after sowing. Sowings every six to eight weeks in warm winter areas provide continuous cropping. Northern growers should not plant outdoors until the danger of frost is past. Flowers are long lasting (two weeks) in just water. Pack and ship only *dry* stems.

The Fortress series is well known as a cut flower. The mix includes blue, white, yellow, rose purple, and dark blue. Oriental Blue, with its rich, deep, dark blue coloring, is well known as a cut flower. The QIS group works in the field or the greenhouse.

LISIANTHUS (SEE *EUSTOMA*)

LOBELIA

by Jim Nau
Ball Horticultural Company
West Chicago, Illinois

❖ *Annual (Lobelia erinus). 1,000,000+ seeds/oz (35,273/g). Germination, see discussion below.*

Although more at home in a moderate climate such as England's, lobelias are used throughout the United States for their brilliant blue effect in combination boxes, pots, and hanging baskets (fig. 1). With the development of many delicate-toned varieties, lobelias have gained in popularity and are highly valued plants in shady or semishady areas.

Culturally, they are not difficult. Being a bit slow-growing, lobelias should be started not later than February 1 for flowering pot plants by Memorial Day. Their seed is very fine, so don't cover. They stand transplanting satisfactorily and do well in any ordinary garden soil. They do not tolerate high temperatures, but if you can give them some shade, they will stand heat much better.

Germinate lobelia at 75 to 80F (24 to 27C), then drop temperatures to 70F (21C) when the radicle has emerged. As leaves expand, decrease temperatures to 65F

FIG. 1. The special blue and lavender colors of lobelia flowers really shine in cool season coastal climates. Lobelias make great window box or hanging basket crops.

(18C). Hold mature seedlings, if necessary, at 60F (16C). (Exception: Heavenly does best at 50F, 10C). Multiple sow three seeds or more per plug or pack cell. Multiseed pellets, with six and seven seeds per pellet, are commonly used today.

Allow eight to nine weeks for green packs, 11 to 12 for flowering plants. For 4-inch (10-cm) pots, lobelias require 12 to 13 weeks to fill the container and bush out. As for 10-inch (25-cm) hanging baskets, allow 14 to 15 weeks and use eight to 12 clumps of seedlings. Clumps of seedlings are a good way to transplant lobelia. Like alyssum, lobelia seedlings are very fine and hard to handle. When transplanting by hand, it is easier to grab hold of clumps of seedlings, with six to 12 individual plants per clump, and transplant these rather than individual seedlings. The baskets will fill out faster, as well.

A popular lobelia variety is Crystal Palace—very dwarf, deep blue, with dark, bronze-green foliage. The new Riveria series—in Blue Eyes, Blue Splash, Marine Blue, Midnight Blue, Sky Blue, and White—is early and compact, growing 3 to 4 inches (8 to 10 cm) tall. The Rapid series—in Blue, Violet Blue, or White—blooms in 14 weeks and shows good heat tolerance for lobelia.

There is a class of trailing lobelia known as pendula that works out well for trailing over porch planters and for hanging baskets. Outstanding is the variety Sapphire, deep blue with white eye and light green foliage. The Fountain Series is available in three separate colors, including blue, lilac, and white. Plants make excellent baskets and fill out the containers quickly. The new Regatta Series—in Blue Splash, Lilac, Marine Blue, Midnight Blue, Rose, Sky Blue, and White—offers separate, bright colors seven to 10 days earlier than Fountain.

LOBULARIA (SWEET ALYSSUM)

L

by Jim Nau
Ball Horticultural Company
West Chicago, Illinois

❖ *Annual* **(Lobularia maritima)***. 90,000 seeds/oz (3,175/g). Germinates in eight to 10 days at 78 to 82F (25 to 28C). Seed should be left exposed to light during germination.*

Commonly called sweet alyssum, the softly scented blooms flower most profusely under cool, 60 to 68F (16 to 20C), night temperatures. It is recommended for use as either edging or border plants or in hanging baskets mixed with other annuals (fig. 1).

Sow direct to the final container, using eight to 15 seeds in each of the individual cells of a cell pack. Try for between five to 10 seedlings per one cell. For transplanting, seedlings are large enough to handle 20 to 25 days after sowing. Grow on at 50 to 55F (10 to 13C) night temperatures until ready to sell. Salable green packs are ready in seven weeks, with flowering packs ready eight to nine weeks after sowing. For 4-inch (10-cm) pots or 10-inch (25-cm) hanging baskets, allow 10 to 11 or 12 to 14 weeks, respectively.

In the southern United States, allow seven weeks for packs with color and nine weeks for flowering 4-inch (10-cm) pots. Plant in late summer to midwinter for plants that will flower as late as June. Plant in full sun to partial shade and space 8 inches (20 cm) apart to fill in.

Note: If you are growing the newer alyssum varieties, be aware that they have higher fertilizer requirements than the old varieties. Don't run them cold, hungry, and dry!

As either a garden or a landscape plant, sweet alys-

FIG. 1. Lavender, purple, and white blossoms grace the popular Easter Bonnet alyssum. No bedding plant grower's greenhouse is complete without the sweet, springlike fragrance of alyssum.

sum performs better in spring, early summer, or late summer plantings than it does during the heat of summer. However, quite often it is the only white-flowering annual that has the vigor to perform during the hottest part of the year in the northern United States. Though they may heat-stall (go out of bloom) temporarily in August, white alyssum varieties will flower again once the night temperatures cool off. Furthermore, today's newer varieties have increased heat tolerance.

The standard white-flowering variety today is New Carpet of Snow. A pure white variety up to 6 inches (15 cm) tall, it has reliable performance from year to year. Snow Crystal has the largest individual flower size of any white variety in the marketplace. It, too, is no taller than 6 inches. For uniform habit and excellent outdoor performance, the Wonderland series is ideal. The three separate colors are well matched in crop time and habit, with the aptly named Wonderland Deep Rose among the best rose flower colors on the market. The rejuvenated Easter Bonnet series includes a mix of deep violet, deep rose, deep pink, and lavender. Pastel Carpet, also a pastel alyssum mix, includes the same colors plus buff yellow. For an unusual twist to alyssum, try Apricot, a light salmon-peach color that is excellent under cooler climate conditions.

LOTUS

by John S. Rader
Proven Winners
Encinitas, California

❖ *Tender perennial (***Lotus berthelotii** *and* **Lotus maculatus***). Vegetatively propagated; growers produce pots and hanging baskets from purchased rooted cuttings.*

The lotus has been used in the greenhouse and nursery trades for nearly 10 years. Most consumers who see it for the first time react, "That's really different!" or "It looks like it has Christmas tree leaves!" The needlelike foliage is soft to the touch, and the bird's beaklike blossoms resemble tiny torches.

There are two species grown in the trade: a red-flowered variety, *Lotus berthelotii* Parrot's Beak, and an orange-yellow-flowered variety, *Lotus maculatus* Gold Flash. Both species were native to a specific group of cliffs in the Canary Islands. However, both are now extinct in the wild, surviving only in cultivation. The first of the two varieties to come on the U.S. market was the red-flowered Parrot's Beak. Silver, needlelike leaves with bright carmine-red blossoms like parrots' beaks make this plant appealing. Gold Flash was introduced into the United States in 1991. It became an instant hit. It had a new look—jade-green, flattened needlelike foliage with bright orange-yellow blooms.

Many growers of Parrot's Beak have difficulty producing a consistent bloom every spring, settling instead for the cultivar's attractive silvery green foliage. Gold Flash, on the other hand, not only flowers earlier but with greater consistency and frequency. Parrot's Beak will flower once per season, while Gold Flash will repeat-bloom. Blooms begin to appear on Gold Flash as early as December in mild winter climates, and mature plants will continue to cycle into bloom as long as night temperatures stay below 65F (18C). In blooming trials in southern California, Gold Flash demonstrated its free-flowering ability by flowering heavily in six distinct cycles from December until mid-May.

The lotus will tolerate some cold or frost. Both varieties will withstand periodic nights of 28F (−2C). At the other extreme, they will tolerate periods of moderately hot, dry weather but will not flower until nights cool.

Lotuses are ideal for hanging baskets, window boxes, mixed plantings, and ground covers.

CULTURAL TIPS

Since lotuses grow best during cooler times of the year, a good time to plant is from late fall to midwinter (October through January). In climates where temperatures do not regularly dip below freezing, Gold Flash can be grown without protection in full sun. In colder climates it is best grown in a greenhouse with no shading (5,000 to 8,000 footcandles 54 to 86 klux), cool nights (40 to 50F, 4 to 10C), and good air movement.

Plant three liners per 10-inch (25-cm) basket or one liner per 6-inch (15-cm) pot. Soft-pinch at planting; this pinch is usually enough. Maintain temperatures from 50 to 60F (10 to 16C) immediately following potting for one month to allow the liners to root out in the containers.

After liners are established, temperatures may be lowered. Lotuses are best finished with night temperatures of 40 to 50F (4 to 10C) and day temperatures of 65 to 75F (18 to 24C). The lower temperatures in these ranges (i.e., a 40F, 4C, night temperature) will produce slower plant growth but a more attractive finish.

Flower initiation on *L. maculatus* and *L. berthelotii* is mainly temperature-dependent. A period of vernalization (cold temperature exposure) is required. While maintaining high light levels, expose well-established plants to 40 to 45F (4 to 7C) night temperatures for four to six weeks. This will trigger flowering, which will continue as long as night temperatures are consistently below 65F (18C). Short days do not appear to be a significant influence on flowering.

Use loose, well-drained, soilless potting media. Maintain a slightly acid pH, around 5.0 to 6.0. Fertilize lotuses regularly at low levels (200 to 250 ppm N). High nitrogen will produce unsightly, soft plants with few flowers. The lotus is also salt-sensitive, so do not allow soluble salts to accumulate.

Growth regulators are not needed as long as light intensities are high and night temperatures are low. However, older plants can become very large. Cut back these plants to a third of the original marketable size to stimulate new, compact growth.

Aphids have been observed on lotuses. Most insecticides registered for aphids are effective without phytotoxicity. Avoid using any sprays during bloom period, however, as they can damage flowers. This is especially true of the insecticidal-type soaps or oils.

Gold Flash is prone to fungal diseases if light levels are low and irrigation inconsistent. *Botrytis*, *Pythium*, and *Rhizoctonia* are usually the pathogens involved. High light and good air movement will prevent these fungi.

Consistent watering is vital for successful lotuses. An irregular irrigation schedule of drying and overwatering will lead to needle drop and even stem dieback.

Lotuses grown properly are always an instant early-season seller. The key is to grow them cold.

LUPINUS (LUPINE)

by Jim Nau
Ball Horticultural Company
West Chicago, Illinois

❖ *Perennial (**Lupinus polyphyllus**). 1,000 seeds/oz (35/g). Germinates in 6 to 12 days at 65 to 75F (18 to 24C). Seed should be covered during germination.*

Sowings of lupine made in winter or early spring will not flower the same season from seed. To increase germination, some recommend alternating day and night temperatures of 80/70F (27/21C)—effective for seed one year or older. Overnight soaking or scarifying the outer coat will increase stand. If winter or spring sowings are needed, use a deep container (nothing smaller than 32 cells per 22-inch, 56-cm flat; 18 cells is better) or a 3- to 4-inch (8- to 10-cm) pot, with one plant per container. For green packs allow 10 to 12 weeks at 50 to 55F (10 to 13C) nights.

Keep one point in mind: lupine roots are fragile and once restricted will give rise to inferior plants and flowers. When the plants are ready to transplant, do not delay! To ensure flowering pots in the spring, sow seed in October and overwinter pots in a cold frame. Pot to 1-gallon containers in early February. Sell plants green in April and May. Plants should begin flowering for gardeners in June.

The Russell strain is the most common variety that is available as both seed and as transplants. Plants grow to 3 feet (91 cm) tall with flowers to 1 inch (2.5 cm) wide. Separate colors are available and are sold under such names as My Castle, a brick-red-flowering variety; Noble Maiden, with white flowers; and The Pages, which has carmine-rose flowers. There are other colors in the series.

Among dwarf varieties, consider using the Gallery series, which has five separate flower colors plus a mixture. Plants grow to 20 inches (51 cm) tall and make excellent quart or gallon containers for spring sales.

Lupines do no tolerate hot, humid summers, so they are best treated as annuals in these areas.

L

LYCOPERSICON (TOMATO)

by Lelion D. Elledge, Jr.
Rosebrook Nursery
Oklahoma City, Oklahoma

❖ *Vegetable* (**Lycopersicon lycopersicum**). *9,000 to 11,000 seeds/oz (317 to 388/g). Germinates in seven to 14 days at 70 to 75F (21 to 24C).*

As bedding plants, tomatoes are quite easy to produce (fig. 1). They grow fastest at 60F (16C), with ample water and fertilizer. However, growers generally prefer to keep tomato plants hungry to restrict tall, spindly growth. Tomato plants grow most quickly (and gangly) when they're warm, wet, and well fed, but they are not hurt by wide extremes in temperatures. For best results, grow the plants cool, dry, and hungry, or combine cool days with warm nights.

FIG. 1. Tomatoes are the best-selling vegetable bedding plant. Be sure to have plenty on hand during the spring.

Tomatoes are frequently grown in 3-inch (8-cm) Jiffy pots and cell packs for bedding plant sales, although many growers also produce tomato plants in 4- and 6-inch (10- and 15-cm) pots and even larger containers and tubs. Larger plants are usually staked and may even be sold in flower with green fruit. Seeding can be staggered to produce as many as three or four crops during the spring.

Tomato plants respond best in full sun when grown as bedding plants. Home gardeners will limit their fruit production if they place tomato plants in full or partial shade because the resulting plants will have many leaves and little fruit. In areas where sunlight is limited, early varieties, such as Early Girl, are recommended.

Many home gardeners grow tomatoes successfully on their patios or back porches (fig. 2) in urns or tubs up to 15 inches (38 cm) in diameter. Good, healthy plants bearing heavy yields of fruit can be grown this way. In fact, there are several new varieties bred especially for container use, such as the recently introduced Husky series.

Breeders have made great strides developing F₁ hybrids in the last few years. These new, specially bred varieties resist many of the most virulent diseases spread by nematodes, such as *Verticillium, Fusarium*, and others. Disease resistance increases the plants' yields and allows uninterrupted production. In addition, these plants are bred to produce higher yields, with larger fruit.

Tomato varieties normally fall into one of two categories: determinate and indeterminate. Home gardeners usually choose indeterminate varieties, which have a longer production window and a longer season for enjoyment. Indeterminate varieties tend to grow tall and require staking. Determinate varieties usually have a narrower production window and grow well when they are caged. Most commercial growers prefer determinate varieties, which allow fewer pickings but yield more fruit each time.

FIG. 2. For something different, offer cherry tomatoes in hanging baskets. The variety shown: Micro-Tom.

MARIGOLD (SEE *TAGETES*)

MARGUERITE DAISY (SEE *ARGERANTHEMUM*)

MATTHIOLA (STOCK)

STOCK AS A CUT FLOWER
by Will Healy
Ball Horticultural Company
West Chicago, Illinois

❖ *Annual* (**Mathiola incana**). *18,000 to 19,000 seeds/oz (635 to 670 seeds/g). Germinates in 14 days at 65 to 75F (18 to 24C). Cover seed lightly.*

Columnar, or nonbranching, stocks at one time were a major greenhouse cut flower crop in the midwestern, eastern, and southern United States. They were introduced into this country in the 1930s by George J. Ball. Today the cut flower crop is largely produced outdoors in California and Arizona (fig. 1).

Several new forms are interesting as pot and bedding plants. The Midget series is very early, dwarf (8 to 10 inches, 15 to 25 cm), and fine for bedding pack (pony and jumbo 6) and 4-inch (10-cm) pot crops. Midget is available in red, rose, violet, and white. Violet finishes 10 days later than the others in the series. It's possible to eliminate the singles by keeping only the plants with serrated leaves when selection is done at an early stage. The plain leaf types will be single and may be discarded if you wish.

Stock flowering is inhibited at high temperatures and, generally, it fails to flower past May or June in northern summers (except trysomics). Optimum temperature for blooming is about 50F (10C) nights, although the plant can survive at

FIG. 1. Stock is one of the best loved, old-time formal cut flowers. Its strong clove scent transforms a cut flower bouquet. Stock is also a fine bedding plant in cool and coastal climates.

temperatures as low as freezing. Sowings can be made from August through February to bloom from October through March in mild climates. In mild-winter areas the Midgets will bloom continuously from November to April.

While more day-length sensitivity, the Cinderella series is fantastic for pot production and produces more double-flowering plants with a sweet clove scent. Be sure to get buds set prior to September for winter and spring sales. Grow Cinderella at 50 to 52F (10 to 11C) to initiate flowers, then grow on cool. Allow 12 to 13 weeks for flowering packs. Midwinter sowings are best. September sowings require long days throughout the winter to produce uniformly flowering plants. Cinderella is available in a full range of colors, including carmine, dark blue, lavender, pink, and white.

Plants in the Harmony series, also dwarf, grow from 9 to 12 inches (23 to 30 cm) tall. Even though nonselectable, they produce the same number of doubles as selectable types. Harmony is best for pot production.

Trysomic dwarf double is a seven-week strain widely used for bedding plants in some sections of the country, especially California. Growing to a height of 12 inches (30 cm), it is the only strain that will flower under high temperatures. Plants will usually throw a central spike in May, then produce numerous side shoots that will flower later. This variety can be flowered in the pack. Seed should be sown in February. As soon as seedlings become established, move them to a cold house. Chiefly grown as a color mixture, the strain is also available in separate colors.

STOCK AS A CUT FLOWER

Stock is a true biennial, requiring a period of low temperature for flower-bud differentiation. Vegmo Plant B.V., Rijsenhout, the Netherlands, offers the following cultural tips for success with stock as a cut flower.

Stock is grown in areas of cool temperatures. It can withstand light frost but is killed at temperatures several degrees below freezing. It is grown for greenhouse flowering from January to June north of 35° latitude (i.e., north of Bakersfield, California; Oklahoma City, Oklahoma; or Charlotte, North Carolina). Stock may also be grown farther south at high altitudes or other cool regions. It is grown in the field for late winter and spring cut flowers in California. Farther north along the Pacific coast, and in some isolated spots from New York to Canada, a summer field crop is grown successfully.

Several types of stock are in the trade, but the columnar (nonbranching) type is the most popular. It is grown almost to the exclusion of other types for cut flowers. Developed in the Netherlands by Vegmo Plants, the Vegmo series stock are fully double. Because plants are selected at the seedling stage, they're available only as plugs. The series is available in 14 colors, from the pale pink of Daphne to the brick red of Siberia.

The Cheerful series provides 95% double flowers. Plants set buds without a cold period. August sowings produce harvestable cuts in November, even in warm fall regions. Cheerful is recommended for greenhouse production only. The Column series is the best choice for field production. Plants grow 2½ to 3 feet (76 to 91 cm) tall, with 50 to 60% of plants bearing double blooms.

General culture. Stock may be started from seed or plugs. Plant out at a density of 54 to 59 plants per square yard (64 to 70 m²). Nonbranching plants grown in the field are sown in rows 18 inches (46 cm) apart, and later thinned to about 3 inches (8 cm) between plants in the row. The thinning operation consists of removing the small plants; this also increases the percentage of doubles. When grown in the greenhouse, plugs are planted or thinned to 3 by 6 inches (8 by 15 cm).

Start by watering the young plants overhead to obtain uniform moisture in the soil. When the plants have filled in, water underneath with drip irrigation.

Before planting, it is important to do a soil analysis. Use liquid feed when watering, rather than adding a large amount of fertilizer before planting. Potassium sulphate, when used as a liquid fertilizer, gives the best results. Wash liquid fertilizers off the foliage to prevent scorching.

During and after flower induction, fertilize with potassium sulphate to improve stem quality. Depending on water quality, water with an EC level twice that of the regular water. For example, if your water has an EC of 1.0, add 2.0 EC from KNO_3, for a total solution EC of 3.0.

Soil. *Matthiola* grows best in a well-drained soil with a pH of 5.5 to 6.5. When grown outdoors, incorporate 10-10-10 prior to planting, depending on soil tests. Add additional potassium if needed.

Matthiola is considered a moderate feeder and will produce inferior-quality cut flowers if given excessive nitrogen. Constant-feed at 150 to 200 ppm during early stages of growth. Reduce the feed as flowers mature. Light, sandy soils require 200 to 300 ppm because of more leaching.

Stock is among the most sensitive of the florist crops to potash deficiency. This appears as a dying of the lower leaves from the tips and margins inward, progressing up the plant nearly to the top. Leaf loss is most serious at flowering.

SCHEDULING

Winter and spring planting. Stock can be planted from the November 15 until May 15. During this period it is possible to grow it under natural day-length conditions. January and February plantings can be harvested in 14 weeks; March, April, and May plantings, eight to 10 weeks.

Summer planting. Stock is sensitive to high temperatures. When average temperatures are higher than 65F (18C), flower initiation will be inhibited. The plants will continue to grow vegetatively until the temperature drops below 65F (18C). Normally, flower buds are initiated three to four weeks after planting.

To get good results from plantings between May 15 and August 15, it is necessary to keep the young plants actively growing by keeping the greenhouse as cool as possible, such as with fan-and-pad cooling. Summer plantings can be harvested in 10 to 14 weeks.

Fall planting (northern Europe). To obtain good results from plantings between August 15 and November 15, it is necessary to use HID lighting three to five weeks after planting (400 watts/m²) for 15 to 16 hours a day. Without HID lights, spikes become gappy and stretched. Stock planted from October to December takes approximately 17 weeks to harvest.

Temperature. For the first 10 days after planting, it is necessary to keep temperatures in the range of 58 to 60F (14 to 16C). After plant establishment the ideal temperature range is 52 to 58F (12 to 14C), but the average temperature should not go above 65F (18C).

Plants of proper size for bud formation require 21 days at 50 to 60F (10 to 16C) to initiate buds on all plants under natural winter conditions. Plants given this treatment developed buds after the temperature was increased to about 60F (16C).

Buds form after the appearance of the 15th leaf, if conditions are favorable. Stock grown at temperatures below 60F (16C) develop smooth-margined leaves. If plants are grown at a minimum of 65F (18C), the leaves that appear after the first 15 are lobed. Plants grown at high temperatures, having developed lobed leaves, will again produce linear leaves and buds when exposed to low temperatures.

Flowering is hastened by two weeks in winter by increasing day length (mum lighting) after low temperature has caused buds to form.

Harvest. Harvest when seven to 10 flowers are open. Stems may be cut, or simply pull out the whole plant and cut off the roots later. Remove bottom leaves prior to placing in preservative solution. Stems treated with STS have longer vase life.

PROBLEMS

Cultural factors. Short plants occur if the temperature is low when seedlings are developing. Seeds planted after December usually produce short plants. Blind plants, those that fail to form flower buds, are caused by too short a period of temperatures below 60F (16C).

Cut flowers fail to take up water when plants are low in carbohydrates, which is a result of low light intensity or high temperatures. Plants with extremely hard stems, due to age or having been grown in dry soil, often fail to take up water.

Insect and disease. Flowers are streaked or blotched with light color when plants are infected with a virus, which is spread by several kinds of aphids. Several viruses common in cruciferous plants are known to cause breaking (light areas) in the flowers. Other insects which commonly attack stock include leaf rollers and thrips.

Grayish foliage, wilting of plants, and rotting at the soil line are symptoms of root-rot or wire stem (*Rhizoctonia* spp.). Soil sterilization is the only control. *Phytophthora cryptogea* causes similar injuries, but the roots are soft rotten rather than dry rotten as with *Rhizoctonia*.

Bacterial blight (*Phytomonaincanne*) results in sudden wilting and collapse of one or both cotyledons, with the base of the cotyledon appearing water-soaked. On old plants the disease appears as dark, water-soaked areas around the leaf scars. The main stem is soft and water-soaked.

MELON (SEE *CURCUBITS*)

MIMULUS (MONKEY FLOWER)

by Jim Nau
Ball Horticultural Company
West Chicago, Illinois

❖ *Annual* (Mimulus × hybridus). *624,000 seeds/oz (22,010 seed/g).
Germinates in five to seven days at 60 to 70F (16C to 21C). Leave the
seed exposed to light during germination.*

Blooming in bold, bright colors, mimulus flowers hold up well, but they require long days of at least 13 hours light to appear. For sowings prior to February 15 in the northern United States, grow under a mum-light, long-day setup. Transplant from sowing trays or plug flats directly to the final containers. We recommend this crop for areas with cool summers when it is to be grown in the garden. This is the climate that is found in the Pacific coastal regions and selected areas of the Atlantic seaboard states. In other areas, mimulus plants can be sold in 4- or 6-inch (10- or 15-cm) pots for spring and early summer color.

For green packs allow seven to eight weeks, and up to 10 weeks for full-blooming cell packs. Four-inch (10-cm) pots take 11 weeks, and 10-inch (25-cm) hanging baskets take up to 12 weeks to fill out. Add another two weeks for baskets to be more mounded and full in appearance. When transplanting seedlings, use one or two per 4-inch pot and up to eight for each 10-inch basket.

The Calypso and Velvet series are the predominant varieties in the trade. Since there is little demand for this material due to our climate, only a limited number of varieties are available in the United States. Calypso grows to about 16 to 18 inches (41 to 48 cm) tall, with solid and bicolor flowers to 2 inches (5 cm) across. Bicolor flowers are often spotted with large crimson blotches. Mystic grows 10 to 12 inches (25 to 30 cm) tall. Bicolor orange, yellow, scarlet, wine, and ivory flowers have a "monkey face" pattern.

NARCISSUS (DAFFODIL AND PAPERWHITE)

PAPERWHITE NARCISSUS
by A.A. De Hertogh
North Carolina State University
Raleigh

❖ *Perennial* (**Narcissus** *sp.*). *Vegetatively propagated by offset bulbs.*
Growers purchase bulbs for cut flower and pot plant production.

Hardy daffodils truly are the essence of spring! Their bright golden yellow color as cuts or potted plants is everywhere in February and March.

❖ DAFFODILS

POTTED DAFFODILS

See table 1 for specific temperature schedules for potted bulb forcing. Forcing temperatures in the greenhouse are 60 to 62F (16 to 17C). Optimum stem length for pot daffodils ranges from 8 to 14 inches (20 to 36 cm). Be sure to check on the response of individual varieties. A good rule of thumb: When scheduling, plan for flowers a full week earlier than sale date. For example, schedule daffodils to flower on February 7 for Valentine's Day sales.

Here are height control suggestions from Gus De Hertogh of North Carolina State University. Apply a 1,000-ppm (6.4 oz/gal or 25 ml/l) to 2,000-ppm (12.8 oz/gal or 50 ml/l) foliar spray of Florel (ethephon) when leaves or floral stalks are 3 to 4 inches (8 to 10 cm) long. Foliage should be dry. If required, use a second application two to three days later. Do not apply if the flower bud is visible! The concentration and number of applications vary with cultivars and flowering periods. Bulbs should have received proper cold week treatment for the given flowering period prior to Florel application. Do not wet foliage for 12 hours after treatment. Florel should be applied in a well-ventilated, 60 to 65F (16 to 18C) greenhouse.

❖ PAPERWHITE NARCISSUS

Paperwhite bulbs for cut flower and pot plant forcing in the United States and Canada are produced in Israel. The marketing season normally extends from mid-November to April. Bulb sizes available range from 13/14 to 16/up cm in

TABLE 1 POT AND CUT DAFFODIL FORCING SCHEDULES

Sales season	Dates bulbs should arrive	Planting dates	Rooting room temperatures	Greenhouse forcing time & temperatures
Potted daffodils				
Christmas	Late Aug. (store at 48F)	Oct. 1.	48F to Dec. 1. 41F until Jan. 1, then 32 to 35F.	Bring into greenhouse three to four weeks prior to sale. Force at 60 to 62F.
Valentine's Day	Late Sept.	Early Oct.	48F until Dec. 1, then 32 to 35F.	Same as above.
March and April flowering	Early Nov.	Mid-Nov.	Same as for Valentine's Day. Hold at 32 to 35F until time for greenhouse forcing (two weeks prior to sale).	Same as above.
Cut daffodils				
Valentine's Day	Mid-Sept.	Late Sept.	48F until Dec. 1. 41F until Jan. 1. 32 to 35F until bulbs are brought into the greenhouse for forcing.	Bring into greenhouse three to four weeks prior to sale. Force at 55F.
March and April flowering	Early Oct.	Oct. 10 to 30.	Same as above.	Same as above.

circumference, with the most consistent results being obtained with 15/16 and 16/up cm bulbs. The cultivars available and programming information are in table 2. All have very fragrant flowers.

U.S. and Canadian flower-bulb wholesalers should have the Israeli-grown paperwhites shipped from Israel at 77 to 86F (25 to 30C). Upon arrival, they should be stored under well-ventilated conditions at 77 to 86F until either shoots or roots begin to grow out of the bulb. Then store the bulbs at 35F (2C). In general, it is advisable not to place paperwhite bulbs at 35F before November 1. Prior to shipping to the forcers, the bulbs must be placed at 63F (17C) for two to three weeks. This would be true regardless of whether the bulbs were previously stored at the warm 77 to 86F or low 35F temperatures.

The forcer should be prepared to plant the bulbs immediately on arrival. Prior to planting, check bulbs carefully. Discard any damaged bulbs. If bulbs must be stored, place them at 63F (17C) under well-ventilated conditions, but only for a few days. Take care not to damage the shoot if it has emerged from the nose of the bulb.

CUT FLOWERS

Use well-drained, sterile growing media with a pH of 6.0 to 7.0. Plant in 4-inch-(10-cm-) deep flats or trays, using 35 to 50 bulbs per large flat. Paperwhites can be planted bulb to bulb. The flats or trays do not need to be spaced out on the forcing benches.

Use a medium light-intensity (5,000 footcandles, 54 klux), well-ventilated greenhouse. No fertilization is required for paperwhites in the greenhouse. Keep growing media moist.

Use 60 to 63F (16 to 17C) night temperatures. Lower temperatures can be used, but the plants will take longer to reach the market stage. Normally, this is two to five weeks (see table 2).

Normally, diseases and insects are not a problem. However, the forcer should always look for *Botrytis* and aphids.

Cut flowers when the first floret is fully colored. If the flowers must be stored, place them upright and dry at 32 to 35F (0 to 2C). Advise wholesalers and retailers to market paperwhites when the first flower is fully colored. When this is done, the consumer will receive maximum satisfaction.

POTTED PLANTS

Use well-drained, sterile growing media with a pH of 6.0 to 7.0. Plant three bulbs in a 5-inch (13-cm) standard pot, four to five bulbs in a 6-inch (15-cm) standard pot, or seven to nine bulbs in an 8-inch (20-cm) pan. Grow plants pot to pot on the greenhouse bench. Keep the growing media moist.

N

TABLE 2   PAPERWHITE NARCISSUS FORCING PROGRAM

Variety	Planting date	Bud date	Dates to bud (avg)
Galil: all white, pot plants			
	Oct. 15–20	Nov. 25–30	40
	Nov. 15–20	Dec. 20–25	35
	Dec. 15–20	Jan. 12–17	28
	Jan. 15–20	Feb. 5–10	20
	Feb. 15–20	Mar. 1–5	15
Sheleg: all white, cut flowers			
	Oct. 15–20	Nov. 25–30	40
	Nov. 15–20	Dec. 20–25	33
	Dec. 15–20	Jan. 12–17	30
	Jan. 15–20	Feb. 5–10	30
	Feb. 15–20	Mar. 1–5	28
	Mar. 15–20	Apr. 10–15	25
Ziva: all white, cut and potted			
	Oct. 10–20	Nov. 12–17	33
	Nov. 10–20	Dec. 6–11	26
	Dec. 10–20	Dec. 30–Jan. 4	20
	Jan. 10–20	Jan. 25–30	15
	Feb. 10–20	Feb. 20–Mar. 3	10

Use a medium light-intensity (5,000 f.c., 54 klux), well-ventilated greenhouse. No fertilization is required for paperwhites in the greenhouse. Use 60 to 63F (16 to 17C) night temperatures. Lower temperatures can be used, but the plants will take longer to reach the market stage. Normally, this is two to five weeks (see table 2).

Normally, diseases and insects are not a problem. However, the forcer should always look for *Botrytis* and aphids.

To reduce excessive elongation of the flower stalk and leaves of Ziva, these plants can be sprayed to runoff with 2,000 ppm ethephon (Florel) when the shoots are 4 inches (10 cm) long. A 2,000-ppm Florel solution is 1 pint in 2½ gallons, and this treats about 500 6-inch (15-cm) pots. It is important that the foliage be dry at the time of treatment. Also, do not wet the foliage for 12 hours after treatment. Late afternoon is the best time to spray. This treatment can, however, delay flowering by two to four days.

Plants should be marketed when shoots are 8 to 10 inches (20 to 25 cm) tall and the flowers are visible. Do not wait until they begin to show color. If the plants must be stored, then store at 41F (5C). Advise wholesalers and retailers to market paperwhites when they are in the bud stage of development. Homeowners should keep the plants in the coolest area of the home in order to obtain maximum satisfaction. In USDA Zones 8 to 11, advise consumers that these bulbs can be planted outside after they have finished flowering. They will acclimatize outdoors in these areas.

NEMESIA

❖ *Annual (***Nemesia strumosa suttonii***). 90,000 seeds/oz (3,175/g).*
Germinates in 10 to 14 days at 65F (18C). Cover seed after sowing.
Temperatures above 65F will inhibit germination.

There are two distinct classes of this showy annual: Grandiflora and Nana Compacta. Grandiflora, the taller strain, will grow to about 2 feet (61 cm).

Of much greater value are the Nana Compacta types. Their dwarf habit (10 inches/25 cm in the greenhouse) makes them popular to use in combinations. For this purpose, they should be sown in January or February for April and May flowering in 2¼- or 3-inch (6- or 8-cm) pots.

While available in separate colors, *Nemesia* is chiefly used as a mixture. The Carnival Mixture, an excellent strain that is more heat-resistant and remains in flower several weeks longer than other strains, is the "traditional" *Nemesia*. For separate colors, try KLM, a blue-and-white bicolor; Mellow, a burnt red-and-white bicolor; or Orange Prince.

Decorticated, or cleaned, *Nemesia* seed is available. This allows for many more seeds per ounce and much improved germination. Sow seed direct to final containers, as *Nemesia* does not like transplanting. Grow plants cool, 55 to 60F (13 to 16C).

NICOTIANA (FLOWERING TOBACCO)

by Jim Nau
Ball Horticultural Company
West Chicago, Illinois

❖ *Annual (***Nicotiana alta***). 250,000 seeds/oz (8,818/g). Germinates in 10 to 15*
days at 70 to 75F (21 to 24C). Leave the seed exposed during germination.

Keep germination temperatures uniform for nicotianas. Sell green in smaller packs, such as 48s or 72s, or sell in bloom in large packs or pots. If plants become restricted, they may remain too small to perform well in the garden. Among the most distinctive garden annuals available, nicotianas look sharp in 4-inch (10-cm) pots and in tubs. Plants are self-cleaning and require no deadheading in the garden, which makes them ideal for landscaping (fig. 1). For green packs, allow six to eight weeks; up to 10 weeks for flowering packs. For 4-inch (10-cm) pots, use one plant per pot and give 10 to 11 weeks to finish off in color.

If growing in spring and following these recommendations for crop time, growers should have no problem flowering newer varieties. Note that nicotianas are long-day plants and flower more evenly when given day lengths of 10 hours or more. If scheduling nicotianas for fall sales, it is suggested that they be grown under long-day incandescent lights (mum light set-up) for flowering dates of late September until February. When grown cool, plants may develop white spots on the young foliage.

The Domino and Nicki series are ideal for pack production and work well in garden plantings. The free-flowering blooms remain open all day. Domino White and Nicki Red offer especially good overall performances as pot, landscape, and bedding plants.

For a wide-ranging color selection in the new varieties, choose the Starship or Metro series. Starship is extra early, while Metro

FIG. 1. Nicotiana has come bursting onto the scene with plants that flower in pleasing pastel colors in the pack or pot for the grower and provide months of satisfaction for home gardeners.

will bloom with Domino. For something special—especially in 4-inch (10-cm) or combination pots—try Havana Appleblossom. Its pastel pink and white flowers are one reason it won a Fleuroselect Gold Medal in European trials.

ORNAMENTAL CABBAGE (SEE *BRASSICA*)

OSTEOSPERMUM (AFRICAN DAISY)

❖ *Annual (Osteospermum ecklonis). Vegetatively propagated. Growers produce 4- or 6-inch (10- or 15-cm) pots or hanging baskets from purchased rooted cuttings.*

Osteospermum is one of the new kids on the block for spring bedding plant sales. Plants make excellent 4- or 6-inch (10- or 15-cm) pots as bedding plants, in combination bowls, in hanging baskets, or as flowering pot plants. Most of today's varieties have been developed by the Danes (fig.1).

FIG. 1. The new osteospermums originated in Denmark. With their large, bold flowers, they're sure to be a hit. The variety in the photo: Sunscape Daisy Lusaka.

Jan Hall, Paul Ecke Ranch, Encinitas, California, shares the following information.

Use one cutting per 4- or 6-inch pot, three cuttings per 10- or 14-inch (25- or 36-cm) basket. Finishing time from planting is 12 to 15 weeks. Plants are ready for sale when they show first flowers and have visible buds for subsequent blooms.

Pot cuttings into well-drained media on receipt. Establish strong roots by starting the cuttings warm, at 68 to 70F (20 to 21C) nights, for about two weeks, until roots reach the edges of the growing containers.

Pinch plants two to three weeks after planting, leaving six nodes above the soil line. New shoots will develop from this point. After pinching, move pots to bright conditions with cool night temperatures (45 to 55F, 7 to 13C) for four weeks to initiate flowers. In large pots or baskets, keep the plants at warm temperatures for another week or so to allow shoot development. Once flowers are initiated, temperatures can be raised to 50 to 55F (10 to 13C) nights. Constant temperatures below 50F can delay the time to flower but will produce more flower buds.

B-Nine at 2,500 ppm after the first pinch can be used to control height. Avoid using Cycocel, which causes leaf yellowing. Warm daytime temperatures, which promote faster growth, increase the need for height control.

Insects, such as aphids, thrips, and caterpillars, and the disease *Botrytis* may attack *Osteospermum*.

Retail growers can tell customers to cut plants back when flowering stops in the heat of the summer. This will control height and spur plants to generate new growth in time for cool fall nights and another spectacular flower display.

Try the Sunscape series for a complete range of bright colors. Cape Daisy comes in seven separate colors—from white to deep purple, yellow, and spoon types. Plants are mounded and generally show an early flowering response. Cape Daisy is well suited for pots or hanging basket production. The Sunny series comes in five separate colors. Three are spreading types, best suited for hanging basket production, while two are mounded types, good for pots or hanging baskets.

PAEONIA (PEONY)

❖ *Perennial (*Paeonia officinalis × P. lactiflora *hybrids). Vegetatively propagated. Growers produce 1-gallon or larger containers from purchased tubers.*

One of the old-fashioned perennial flowers for the home garden, peonies are long-lived plants that can survive from generation to generation of the families around whose homes they are planted. Preferring cold winters so as to break dormancy, peonies come in a limited range of colors, predominantly white, rose, pink, and shades in between. Plants flower in May and are hardy from USDA Zones 4 to 7. To be fair to the genus *Paeonia,* it should be understood that there are a number of different species available to the trade in this class, and that this limited information in no way touches on the diversity of the line.

Peonies are propagated by tubers that are dug in the fall, divided, and then replanted for sales the next spring. Tubers dug in late winter and early spring can flower during the same year as dug, but not with the impact that fall-dug tubers provide. Choose tubers with three to five "eyes" and plant them into gallon containers, leaving the tips of the eyes exposed. If dug in September, there will be some root development belowground, though the foliage will stay dormant until next spring.

Most growers buy in tubers and pot them during the fall, then hold plants in a cold greenhouse or cold frame until March. Plants will sprout and be ready for sale in the spring. Tubers may also be planted in the winter for spring sales. Grow on at 45F (7C) to get the plants established in the pot. Peonies require at least five weeks of weather 40F (4C) or below to break dormancy and develop flower buds.

PANSY (SEE *VIOLA* × *WITTROCKIANA*)

PAPAVER (POPPY)

by Jim Nau
Ball Horticultural Company
West Chicago, Illinois

❖ *Perennial (*Papaver orientale *and* P. nudicaule*). 95,000 seeds/oz (3,351/g). Germinates in seven to 14 days at 65 to 75F (18 to 24C). Leave the seed exposed to light during germination.*

Papaver nudicaule, the Icelandic or Arctic poppy, is treated as an annual or tender perennial throughout North America since it will flower in 15 to 17 weeks when grown in 4-inch (10-cm) pots at 50 to 55F (10 to 13C). Flowers are single and measure from 3 to 5 inches (8 to 13 cm) across. They do not tolerate the hot summers of the central and southern United States, performing their best as container plants with morning sun and afternoon shade. Flower colors include white, cream, scarlet, bronze, orange, pink, and yellow. Most often the varieties are sold in mixes rather than separate colors.

The most popular variety to the trade in recent years is Champagne Bubbles, an F_1 mixture of colors on strong, bushy plants. Sparkling Bubbles is very similar except that it is open-pollinated and takes several days longer to flower than Champagne Bubbles.

The Oriental poppies *(P. orientale)* are true perennials characterized by large orange or scarlet blossoms with a prominent black mark in the center of the flower. Flowers are single, measure to 5 or 6 inches (13 or 15 cm) across, and are displayed in either May or early June. The flowering time is relatively short, and the plants will go dormant during the summer. Be sure to mark the spot where they grew in the spring because by late August little is left to let you know where to plant your fall bulbs and other autumn-planted perennials.

As a cut flower, harvest as the bud opens but before it fully expands. Next, sear the end of the stem to prevent "bleeding" of the white latex and place the stem into water.

Seed-propagated varieties are fine, though shades can vary between plants. Of particular value is Allegro, a compact, scarlet-red variety to 18 inches (46 cm) tall. Plants require two years from seed to flower, but they are salable green in packs 10 to 12 weeks after sowing. Beauty of Livermore is a deep red, while Brilliant has bright scarlet flowers. Oriental Mix includes red, scarlet, pink, and salmon.

PELARGONIUM DOMESTICUM (REGAL GERANIUM)

by Drew Effron
Oglevee Ltd.
Connellsville, Pennsylvania

❖ *Annual* **(Pelargonium domesticum)**. *Vegetatively propagated. Most growers purchase prefinished plants from specialists.*

Prebudded Regal geraniums can be flowered year-round if you maintain night temperatures of 58F (14C) or less, which improves flowering and prevents bud

abortion. Temperature, light level, and photoperiod are critical to producing high-
quality plants (fig. 1).

Crop time is dependent on cultivar,
forcing environment, and whether the
plants are pinched. Nonpinched plants
should flower in eight to 10 weeks, while
pinched plants require 10 to 12 weeks.

Day-length control is critical for
Regals from September 15 through April
1. Sixteen to 18 hours of continuous light
must be supplied during this time until
flower buds show color. Additional lighting, supplying a minimum of 10 footcandles (108 lux) at plant level, can be supplied by stringing a line of 100-watt
incandescent bulbs at 4-foot (1.2-m) centers and 6 feet (1.8 m) above the benches.
This will supply sufficient light for two 4-
foot benches with an 18-inch (46-cm)
aisle between them. Supplemental light

FIG. 1. Regal geraniums make an elegant show as a pot plant. The variety
shown: Vavra Peggy.

during periods of low natural light can help maintain finished plant flower
quality. Natural daylight for maximum plant quality should not exceed 4,000
f.c. (43 klux).

Use a light, well-drained, porous potting medium with a pH adjusted to
5.5 to 6.0. Many commercial soilless mixes can be used successfully, as can
media high in organic material, such as a 1:1:1 soil:peat:perlite mix. Avoid
mixes with vermiculite, due to soil compaction. Pot the plants in 4½-, 5-, 6-,
or 7-inch (11-, 13-, 15-, or 18-cm) azalea pots, depending on the variety and
the forcing program.

When potting, set plants slightly deeper than the original soil level. Irrigate
three times, spread one hour from each other, on the day of potting. Over the
next three days, irrigate once daily at midday. Subsequent irrigation can be determined by weather conditions and growth stage. Regals are very sensitive to drying out, and media must be kept uniformly moist but not wet during the remainder of the growing cycle.

A constant liquid feed fertilization program of 150 to 200 ppm will be sufficient for most cultivars, environments, and geographic areas. To provide 150
ppm N, mix 9 ounces of calcium nitrate and 4½ ounces of potassium nitrate in
100 gallons of water (270 and 135 g per 400 l). Phosphorus and magnesium
should be incorporated into the media before planting to provide the correct

nutritional balance. Use regular media and foliar tests to determine changes in your fertilization type and schedule.

Flowering performance is greatly influenced by growing temperatures. Do not exceed maximum night temperatures of 58F (14C) and day temperatures of 68F (20C). Substantially cooler temperatures will increase crop time, while higher temperatures can yield stretched plants with fewer or even aborted buds. Proper ventilation is necessary throughout growth and is especially critical as plants begin to flower.

Plant spacing is based on pot size. For 6-inch (15-cm) pots, set plants at one plant per square foot (10 or 11 plants per m²). Four-inch (10-cm) pots can be spaced at 2½ plants per square foot (27 per m²).

No growth regulators are specifically labeled for regal geraniums. Unpublished research has shown effective height control when plants are sprayed with 3,000 ppm active ingredient (a.i.) of Cycocel (1:40 dilution), applied to glisten on foliage 17 days after potting.

PELARGONIUM X HORTORUM (ZONAL GERANIUM)

CUTTING GERANIUMS
by Bob Oglevee
Oglevee Ltd.
Connellsville, Pennsylvania

XANTHOMONAS DESTROYS
by Dr. Mike Klopmeyer
Ball FloraPlant
West Chicago, Illinois

GROWING GERANIUMS FROM SEED
by Jim Nau
Ball Horticultural Company
West Chicago, Illinois

P

❖ *Tender perennial* (**Pelargonium × hortorum**). *6,000 seeds/oz (212/g). Seed germinates in seven to 10 days at 70 to 75F (21 to 24C). Cover seed lightly for germination.*

❖ CUTTING GERANIUMS

Geraniums are the single most popular bedding plant in the United States today. The typical pot of red geraniums placed on the front stoop and the plants decorating the ground around a mailbox post are slices of Americana that can be seen from coast to coast (fig. 1).

FIG. 1. Cutting geraniums are available in a range of colors and plant shapes. In the photo (*clockwise from upper left*): Red Hots, Evening Glow, Cherry, Morning Mist, Isabell, and Angel.

Geraniums are most commonly sold as bedding plants in the spring. Plants are typically started from either rooted or unrooted cuttings, commonly called "zonal" geraniums, or from plug seedlings. Many growers produce both zonal and seed geraniums, although zonals put on a larger flower show in the pot and produce the plants most consumers envision when they think of geraniums. Because of disease problems—*Xanthomonas* bacterial blight, especially—most growers leave propagation to specialists who maintain disease-indexed mother stock.

In the United States growers produced $201.6 million in geraniums in 1995, according to the United States Department of Agriculture *Floriculture Crops Report*, accounting for nearly 20% of all bedding plants sales. About 70% of all geraniums are produced in pots—from 4-inch to 6-inch (10- to 15-cm) and patio containers—with zonals taking 70% of all geranium pot sales and a good share of hanging basket production, too. Seed geraniums dominate flat sales, valued at $36 million and 4 million units. Conservatively, Americans purchase about 200 million geranium plants each year—enough for each household to buy two!

According to a Professional Plant Growers Association survey, cutting geraniums account for about 10% of total bedding plant production. But cutting geraniums did not always hold such an important position in bedding plant production. During the 1950s and 1960s, geranium production declined because the crops were slow to turn over and losses to systemic diseases were large. The geranium's current position can be attributed to culture virus indexing (VI) and production of disease-free cuttings.

As the geranium market has grown, so has the variety of uses. Previously, consumers usually purchased red geraniums to be planted on Memorial Day in the cemetery. Today consumers want geraniums in a variety of colors, in both flowers and foliage, that perform in either full sun or full shade. They also want varieties that suit large, open gardens, planter boxes, windowsills, and hanging baskets (fig. 2). The 1980s and early 1990s saw new varieties with compact, dark leaves as well as reduced crop time and increased flowering.

The floribundas (Stardom geraniums), with their free-flowering habit, spreading growth habit, and excellent heat tolerance, are as at home in the landscape as they are in baskets and containers.

Culture indexing to control fungal and bacterial disease. Major geranium crop losses common 20 years ago were due to various bacterial and fungal diseases—especially bacterial blight *(Xanthomonas pelargonii)* and *Verticillium* wilt *(Verticillium albo-atrum)*—that plug the conductive tissues and make the translocation of water and nutrients nearly impossible. Because there are no chemical protectants or cures for these diseases, once a crop is infected, the disease cannot be eradicated. Hence, the losses became a part of growing. Growing plants infected with these bacteria and fungi meant managing the disease instead of the plant through low temperature, fertility, and water, which in turn slowed crop turnover.

FIG. 2. Here's a slice of Americana: a color bowl of zonal geraniums with a dracaena spike in the center.

With the advent of culture indexing, these bacterial and fungal diseases have been controlled. Briefly, culture-indexed geraniums are the product of a laboratory procedure that allows trained personnel to visually check for the presence of systemic diseases and select cuttings that do not have the diseases. To ensure that bacterial and fungal diseases do not escape detection, the indexing, or testing, procedure is repeated for three consecutive generations. The plant is then placed in a specialized greenhouse called the nucleus house. As a further check, the nucleus stock is periodically culture-indexed, and no plants are held for longer than one year.

Culture indexing allows for selection of those cuttings not infected, but it does not alter the genetic structure of the geranium. Therefore, culture-indexed plants have no change in their resistance or susceptibility to fungal or bacterial diseases. If any of these diseases are introduced during production, the plants will become infected. Indexed stock should never be mixed with nonindexed stock, and plants should not be held over from year to year. It is critical to realize that no matter how well a crop is managed, there is always the possibility of reinfection. If this happens, all the advantages of using culture-indexed stock will be lost.

P

For the best results, a grower must follow strict sanitation procedures. Before planting stock plants, the greenhouse should be clean and free from weeds, pests, and diseased plant material. If the house was previously used, sterilize the entire greenhouse. Make steam sterilization a priority treatment. Steaming should be done with moist media at a temperature of 180F (82C) for 30 minutes. This can be done by a continuous steamer on a potting machine or in bulk vaults. For items that cannot be steamed, use chemicals such as a hospital disinfectant or commercial bleach. Remove any debris, such as dead plant material, especially under raised benches.

Treat all soil with either steam or chemicals. The chance of recontamination will be reduced if all the soil in a given greenhouse area is treated. The floors should be sterilized with a formaldehyde solution before dumping treated soil on them. Containers need to be new, steam-sterilized, or soaked in a disinfectant (60 minutes in a 10% hospital disinfectant solution or 30 minutes in a 10% bleach solution).

Ground or raised pot benches need to be steam-sterilized or chemically treated. Wooden benches can be sprayed or painted with a brand of copper naphthenate, such as Cuprinol. Creosote should never be used in the greenhouse due to its toxicity to plants.

Any tool or material that comes directly or indirectly into contact with geraniums should be sterilized. Automatic watering systems and growing implements (such as shovels) should be soaked in a disinfectant as described above. All watering hoses should be soaked in a disinfectant. These hoses should be hung up so the nozzles never touch the ground. Do not use knives to take cuttings; manual breaking is preferred. When knives must be used, keep several soaking in a disinfectant. The cutting knife should be changed every 10 minutes for a clean knife that has been soaking in the disinfectant.

The clean geranium has given the grower the ability to manage the growth of the plant, as opposed to managing the disease, which results in faster crop turnovers. A careful grower should not lose any geraniums. The use of culture-indexed plants in conjunction with advanced cultural practices has provided the industry with better performing plants that are more profitable to grow.

Virus indexing. With the fungal and bacterial diseases under control, it was apparent the next major limiting factor was viral diseases. Viral diseases are much different from bacterial and fungal diseases, due to the nature of viruses. Fungal and bacterial diseases are characterized by severe crop losses, but viral diseases usually affect the quality and the overall performance of the plant. Chlorotic spots and vein clearing may indicate the presence of virus, while other symptoms, not as obvious, are reduced plant vigor, fewer and smaller flower heads, stunting, poor plant habit, and poor rooting. Unless a plant without virus is grown beside the plant with virus for comparison, these symptoms can go unnoticed. Plants may not show any of these symptoms and still carry a virus that may lead to further infection in the greenhouse.

As in the case of bacterial and fungal diseases, there are no chemical treatments to control or eliminate a virus. Virus can only be controlled using a systematic approach of heat treatment, meristem-tip culture, and virus indexing. Like culture indexing, virus indexing is a laboratory procedure to test for the presence of the pathogens. Briefly, the process starts with a culture-indexed geranium that is heat-treated for three weeks at 100F (38C) during a 16-hour day and 95F (35C) during the night. This helps to reduce virus levels in the plant. The meristem tip, a small, 0.5- to 1.0-mm cutting, is removed and propagated in test tubes in laboratory tissue culture under sterile conditions. The resulting plant must then be virus-indexed or tested for the presence of specific viruses. Heat treatment and meristem-tip culture do not guarantee virus removal but only aid in removing virus.

The differences between virus-indexed material and nonindexed material is dramatic. Plants are more vigorous and break more freely. They have more blooms because there are more florets per bloom, and the bloom lasts longer. These plants tend to bloom sooner and are more uniform in blooming time, size, and overall quality. This increases the percentage of high-quality plants, which command a higher price, and leads to faster crop turnovers. It all adds up to higher profits for the grower.

Virus-indexed plants are not immune to virus. Strict sanitation is an important step to prevent reinfection. Moreover, many viral diseases are transmitted by insects, such as aphids. It is extremely important to follow a preventative maintenance program for these pests.

SEVERAL POINTS ON PROPAGATION

The ideal cutting for propagation is a 2- to 3-inch (5- to 8-cm) terminal with an active growing tip and no physical damage. Terminal cuttings are preferred because they finish two weeks earlier than eye or heel cuttings. For disease control, break cuttings off, whenever possible. It is important to break cuttings evenly, leaving no jagged edges. When clean breaks are not possible and knives must be used, they should be soaked in a disinfectant and changed every 10 minutes. Before sticking the cuttings, the bottom inch (2.5 cm) of the cuttings should be cleaned: never leave any petioles or stipules which will be below the soil level.

Some researchers recommend rooting hormones, while others point out that rooting hormones provide little value to cutting propagation. Rooting hormones are probably unnecessary on fast-rooting varieties without fertility imbalances. However, if there is a fertility imbalance or if rooting a slow cultivar, rooting hormone seems to be of value. Rooting hormones, such as indolebutyric acid (IBA), have been shown to improve the uniformity of root development and to speed up root initiation, especially on slow-rooting cultivars, such as Precious, Sunbelt Coral, and most of the ivy geraniums.

IBA can be applied by using commercial dust or solution. If the liquid form of IBA is used, take care not to use more than a 0.10% solution, or damage will result. A fungicide such as Benlate (10%) or Captan (7%) can be added to the rooting duster or solution to help control basal rots caused by various bacteria and fungi. There are unpublished reports that Captan has also been shown to increase rooting. It is important to observe all pesticide laws when using these compounds. Rooting mixtures should always be dusted or sprayed on the basal end of the cutting. For disease control purposes, never dip the cuttings into a powder or solution.

Two stages. There are two stages in the propagation of geranium cuttings: root initiation and development, and a growth stage. Both stages are important to develop high-quality cuttings, but root initiation and development is probably the most critical. The first objective of propagation is to quickly put roots on a cutting to relieve the dehydration of the cutting. When a cutting is taken, the natural water supply to the leaves from the roots is eliminated, but the leaves continue to lose water, or transpire.

The transpiration rate must be reduced to keep the cutting alive. This is usually accomplished by using mist. Mist allows the grower to use soft, succulent cuttings, which root faster but tend to lose water more rapidly than hard cuttings. Mist is an effective propagation tool that lowers the transpiration rate of the cuttings by raising the relative humidity around the leaves. Mist also lowers the leaf and air temperatures. The cooling is so effective that leaf temperature is often 10 to 15F (6 to 8C) lower than air temperature.

The net effect is that geraniums can be propagated under relatively high levels of light, which increase the growth rate of the cutting. This higher light intensity increases the photosynthetic rate, ensuring that the cutting has the necessary carbohydrates for root initiation and development. A cutting propagated under heavy shade without mist may lack the food necessary for rapid growth because the respiration rate is higher than the photosynthetic rate.

The optimum mist program will depend on light, humidity, temperature, cultivar, and the cutting's age. Ideally, the grower should try to maintain a thin layer of moisture on the leaves in the day (fig. 3). Mist periodically at night to relieve the water stress caused by evaporation. Care must be taken not to over-mist, as this may leach the nutrients from the foliage, especially after the cutting is rooted.

FIG. 3. John Hardiman, Hardimans' Inc., Council Bluffs, Iowa, shows some very well done zonal geraniums.

Another key factor. Proper temperature control is also important to initiating root development. The grower should provide bottom or soil heating to maintain a soil temperature of 65 to 70F (18 to 21C). Air temperature should be 70F during the day, with night temperatures maintained at 62F (17C). Air temperatures above or below these recommended levels will have a negative effect on the cutting's development. Temperature above 75F (24C) promotes bud development in advance of root development, which increases the transpiration rate, while air temperature below the recommended levels lowers the cutting's growth rate.

Under ideal conditions and depending on the cultivar, callus should form on the basal end of the cutting after five days. Roots should be developed within 10 to 14 days of sticking the cutting. At this stage cuttings for finishing should be placed in 4-inch (10-cm) pots. Using lightly rooted cuttings will minimize transplant shock. More vegetative growth will be required if these cuttings are to be shipped.

The objective of the growth stage of propagation is to promote the development of vegetative growth. Cuttings at this stage are self-sufficient and should not require mist. Temperature is an important variable in the growth stage, and the temperature in the propagating area should remain above the temperature in the stock area. Day temperature should be at least 70F (21C), and the recommended night temperature is 65F (18C) or higher. These rooted cuttings should be spaced at 20 to 25 per square foot (22 to 27/m^2), and placed on a constant-feed program of 200 to 300 ppm nitrogen. Carbon dioxide can be used to increase the vegetative growth of the cuttings. The growth stage should take approximately three weeks in order to prepare the cuttings for sale.

GREENHOUSE INSECT AND DISEASE PROBLEMS

There are several pest and disease problems that the grower must be aware of in order to manage the environment. *Botrytis* blight, a disease caused by the air-borne fungus *Botrytis cinerea,* lives on aging tissue, such as old leaves, blooms, and debris. Under the right environmental conditions, *Botrytis* can attack and damage young, soft, succulent tissue. The fungus produces spores that are carried through the air, on splashing water, and on cuttings. Once on the tissue, the spores attack the plant if high humidity and free water are found in the greenhouse. When cutting stubs are infected, the disease can progress several inches down a previously healthy stem.

Botrytis is always present in the greenhouse environment. The best way to control *Botrytis* is to make conditions less favorable for its growth and development. Remove flowers when they are in the bud stage. All dead and infected plants, parts, leaves, and blooms must be cleaned up and removed from around the plants and under the benches. This reduces the sources of infection and therefore the load of spores present in the greenhouse. Because *Botrytis* spores require moisture to germinate, holding the relative humidity below the dew point will reduce the number of infections. Relative humidity can be lowered by venting on dry days or venting while heating on very humid days. Weekly applications of Exotherm Termil, Daconil 2787, Chipco 26019, Ornalin, or Benlate are good preventative measures. However, Benlate should be used with caution; prolonged use of Benlate at recommended rates has been shown to inhibit the rooting of cuttings in certain cultivars, such as Pink Camelia, Pink Fiat, and Springtime Irene.

Pests. Geraniums are attacked by a number of greenhouse pests. Pests that are particularly bothersome to geraniums include aphids, whiteflies, and spider mites. These pests are a problem because of the damage they cause and their potential to carry, and reinfect plants with, various bacterial, fungal, and viral pathogens. The most effective method to control insect damage is following a carefully managed spray program. Control weeds throughout the greenhouse because they serve as an ideal reservoir for spider mites and aphids. During the warmer months of the year, screen side vents to prevent the entry of insects such as aphids and whiteflies.

Aphids injure plants by piercing and sucking the sap from the plant, which causes the leaves to curl. This affects the appearance and the performance of the plant. Aphids are commonly found in groups on the stems and leaves. The female aphid is capable of producing 50 daughters, each maturing within a week under average conditions! Aphids can also introduce certain viruses to geraniums.

Whiteflies can be a problem on geraniums. The adult whitefly is a tiny, white, mothlike insect that can fly short distances in the greenhouses. Whitefly adults and nymphs feed by piercing and sucking on the underside of the plant's leaves. Whitefly damage can be seen in the form of a stunted, yellow plant. The whitefly completes one generation from egg to larva to adult in about 30 days. Many recommended chemicals are aimed at only one phase of the insect's life cycle, so plan spraying schedules accordingly.

Western flower thrips has become an increasingly problematic insect in all greenhouse crops, including geraniums. Thrips is a very small (2 mm or less), threadlike insect which poses a dual threat to geraniums. It is a known vector of TSWV and INSV. Also, it causes crippling damage to new growth and flowers, reducing the plant's value and marketability. While the best control for thrips is exclusion, careful monitoring and early application of insecticides will reduce potentially devastating effects.

Two-spotted spider mites, nearly invisible to the naked eye, are persistent pests with geraniums. The adults and nymphs suck sap from the plant, resulting in mottled, bleached-appearing foliage. If a plant is heavily infested, its photosynthetic rate is decreased, resulting in minimal new growth. Under average greenhouse conditions, the spider mite requires about 11 days to develop from egg to adult. As with whitefly, the recommended chemicals for spider mite attack only certain stages of the insect's life cycle. Take this into consideration when setting up spray schedules.

Because chemical registration varies between states and countries, it is impossible to recommend a general spray program here. Contact your local extension agent or university for a list of chemicals registered for use in your area.

NUTRITION

The grower should test plant tissue and media bimonthly to monitor the crop. This can be done by saturated media extract or by sending samples to the state or local university for testing. Perform foliar analysis each month. Take samples for testing by harvesting the last or newly mature leaves.

Nutrient-deficiency symptoms are often easy to detect just by sight. Nitrogen deficiency symptoms are less vigorous plant growth and a light green color on new growth. Phosphorous deficiency is characterized by small leaves and large blotches on leaf margins. A curling of the margins may occur. Potassium deficiency shows in leaves with interveinal chlorosis. As plants age the leaves become dull green with reddish spots. Calcium deficiency shows when leaves are dull looking with interveinal chlorosis; leaves can be small with no additional plant growth. Magnesium deficiency shows in reduced plant size and small leaves. Leaf margins curl down; leaves also turn reddish in color.

FAST-CROPPING 4-INCH GERANIUMS

To profitably finish geraniums, you must turn over as many geraniums as possible from a given area in the shortest time possible. Using culture- and virus-indexed geraniums with new fast-cropping techniques allows the grower to finish a 4-inch (10-cm) product in less than six weeks from a 2-inch (5-cm) plant. Furthermore, breeding and selection have produced varieties that finish pot to pot, maximizing bench space. Pot to pot varieties in the red category include Red Hots, Kim, Sassy Dark Red, Ritz, and Red Satisfaction. The pink cultivars that can be grown this way are Melody, Laura, Fox, Peaches, First Kiss, Love Song, Precious, and Veronica. Of the fancy leaf cultivars (Brocade series), Wilhelm Langguth does well pot to pot.

The fast-cropping technique requires that media be tested and adjusted to proper pH (5.5 to 6.2), salts, and fertility levels before the crop is planted, because there is no time to adjust afterwards and still grow a good crop. A slightly higher fertility regime of 300 ppm nitrogen should be used with three out of every four irrigations. This higher fertility promotes more compact plant growth with short internodes and dark green foliage. High nitrogen and soft growth do not delay flowering of geraniums.

The plants should be pushed in order to maximize their potential. Plants should not be pinched, because this would delay flowering. Plant height can be regulated with Cycocel, A-Rest, or B-Nine. Cycocel is recommended over A-Rest and B-Nine because it provides more uniform results over a broad range of cultivars. Florel should never be used to manage the height of a finished crop because it aborts flower buds for four to six weeks, thus delaying the crop. Water should not be withheld to control plant height because this would slow plant growth and lengthen finishing time.

Before spraying with Cycocel, be sure the geraniums are moist and well fertilized to prevent plant damage. The geraniums should be sprayed on a cloudy, cool day or early in the morning. When spraying with Cycocel, use a 1,500 ppm solution. This is equivalent to a 1:80 dilution. Do not use a spreader sticker with the Cycocel solution.

Cycocel should be applied 17 to 21 days after planting, when the axillary shoots are a quarter to a half inch (0.6 to 1.3 cm) in length. At this time there should be approximately seven sets of leaves on the plant. The leaves should be sprayed only to glisten, without any runoff. A 1,500 ppm Cycocel spray of a quarter to a third of a gallon should cover 100 square feet (1.0 to 1.3l per 10 m^2). If you see Cycocel running to the center of the leaf during application, the foliage can be syringed. This will reduce but not completely eliminate damage to the plant. Even if sprayed properly, yellowing may appear on the leaves in seven days, but this should disappear in three weeks, with the possible exception of one or two bottom leaves. If too much Cycocel was applied or if the conditions were not right for spraying, yellowing of the leaves will appear in three to four days, with physical damage to the plant being the end result. A grower who is unfamiliar with Cycocel, or who just wants to be cautious, can use a Cycocel spray of half strength (750 ppm) with two applications seven days apart.

Some varieties may require two 1,500-ppm applications of Cycocel to control plant height. The second application should be made 10 days after the first application. The following cultivars could be included: Sincerity, Yours Truly, Sarah, Pink Expectations, Medallion Dark Red, Aurora, and the Sunbelt varieties.

Gibberellic acid has been reported to increase flower size and life on geraniums. Treated blooms have been reported to last seven to 10 days longer than blooms of nontreated plants. A grower wishing to try gibberellic acid should begin with a 1- to 2-ppm solution. Plants should be sprayed to glisten after two or three florets have opened. Because little work has been done with gibberellic acid conduct, perform a small trial before deciding to use gibberellic acid on a large block of plants.

Direct-stick program for the mass market. Mass marketing to chain stores is a growing segment of our market. For this reason many growers use an unrooted cutting direct-stick program. Sterilized, soilless media must be prepared. The media must be well drained and have a pH of 5.6 to 6.2 for optimum growth. Containers should be spaced pot to pot on the bench. Purchase clean stock cuttings and stick directly in the pots. If cuttings must be stored, store at 55F (13C) or lower (unrooted cuttings can be safely stored at 40F, 4C). Do not store cuttings for more than 24 hours. Make sure media is watered in before sticking so cuttings can be firmly placed in pots. Do not place crops overhead to minimize disease problems. Maintain 68F (20C) media temperature. Also, use mist to relieve stress on the cuttings for the first seven to 10 days. Apply enough mist to prevent stress or wilting. Irrigate only lightly until roots appear, as wetting foliage would encourage *Botrytis*.

Cuttings should be rooted to the edge of the pot in 14 days. After cuttings are rooted, begin a constant liquid feed program of 150 to 200 ppm N-P-K. Night temperatures should run 65F (18C). CO_2 can be injected into the greenhouse at 800 to 1,000 ppm for an 8-hour period of the day or until the greenhouse temperature reaches 80F, when you should vent and turn off the CO_2.

Growth regulators can be used on the crop. Compact varieties require no application; varieties described as medium need one application. Growth regulators, such as B-Nine, Bonzi, A-Rest, and Cycocel, can be used. Check with your extension agent on which are registered in your area, and always read and follow label directions.

❖ BEWARE: *XANTHOMONAS* DESTROYS

As a geranium producer, can you afford to have even a handful of dying geraniums on your bench? To lose an entire crop? To threaten your income and drastically reduce your profits for the coming season? That's exactly what can happen if *Xanthomonas* bacterial blight reaches your benches.

Because there is absolutely no chemical cure for bacterial blight on geraniums, it's imperative that every person involved in geranium production review the basics of recognizing and preventing this devastating disease before it strikes.

Just what is *Xanthomonas*? *Xanthomonas* is a plant pathogen that causes bacterial blight in geraniums. Formally known as *Xanthomonas campestris* pv. *pelargonii* (Xcp), this bacterium infects only members of the genera *Pelargonium* and *Geranium*, including zonal geraniums (vegetative and seed), ivy geraniums (vegetative and seed), regal geraniums (Martha Washington), which normally act as symptomless carriers, scented geraniums (*Pelargonium* spp.), and perennial geraniums (*Geranium* spp.)

What should you look for? Watch your entire geranium crop for two major symptoms of *Xanthomonas*: leaf spotting and wilting of the leaves, branches, or entire plant. Symptoms may vary, depending on the geranium variety, the plant culture used, and the environmental conditions in the greenhouse. Bacteria splashing onto the leaves will cause small, round leaf spots—usually an eighth of an inch (0.3 cm) in diameter or smaller. The spots, dark brown to black, have an oily or watersoaked appearance that may worsen in warmer weather. They may appear singly or within a wedge of brown tissue.

Xanthomonas can also cause wilting. Yellow areas appear in a V-shape, with the wider part of the V on the leaf margin and the point on a vein. As the tissue is killed by the bacteria, this area will wilt, turning brown and dry. The bacteria then moves on to the stems, branches, and possibly the entire plant.

Is it really *Xanthomonas*? It is possible that other problems can cause similar symptoms. If you suspect *Xanthomonas* has infected your geraniums but are not certain, refer to a reputable color-photo reference book, like the *Ball Field Guide to Diseases of Greenhouse Ornamentals*. If you still have the least amount of uncertainty, it is vital that disease diagnosis be conducted *immediately* by a reputable university or industrial plant diagnostic laboratory.

How does *Xanthomonas* spread? Cuttings taken from infected stock plants are the primary means of spreading *Xanthomonas*. While the stock plants may not show the signs of bacterial blight, the cuttings may contain a small number of bacterial cells. The bacteria can also move quickly from plant to plant by splashing from overhead irrigation and by handling during pinching or cleaning plants. The temperatures for optimal development of *Xanthomonas* are 70 to 80F (21 to 27C). Temperatures above or below this range for an extended period of time will slow the development of symptoms.

How do you get rid of *Xanthomonas*? *There is NO chemical control that is 100% effective in eliminating this devastating pathogen.* Geraniums have no immune system, nor is there an effective antibiotic or chemical that will kill every last bacterial cell. While chemicals may reduce bacterial populations on or in geranium plants, they will never kill all of them. As soon as the suppressing effects of the chemical diminish, the bacteria will proliferate again.

If *Xanthomonas* is present, you must eliminate it! The only way to get back into the geranium business is to start over clean:

1. Destroy all geraniums on your property, including stock plants, by burning, burying, or disposing of them in your local landfill. Don't compost the infected plants on your property, as the bacteria can survive for up to six months in soil and plant debris.

2. Remove and dispose of all plant debris, pots, and soils associated with geraniums.

3. Disinfect all nonporous surfaces (benches, walls, walkways, and head house or packing areas) with a quaternary ammonia compound, such as Green Shield.

4. Finally, and very importantly, buy in new culture-indexed plants for the next season. Clean, healthy plants are key to your geranium success.

Are your other crops safe? Yes! Xcp *cannot* attack or successfully survive for long periods of time upon other floriculture crops like mums, poinsettias, New Guinea impatiens, and petunias.

❖ GROWING GERANIUMS FROM SEED

Hybrid seed geraniums—millions of them—are a staple for mass market sales in the spring. Landscapers also prefer seed geraniums for mass plantings. Seed geraniums may be grown in flats with 18 plants, in 4- or 6-inch (10- or 15-cm) pots, or in hanging baskets.

Culture. Geranium seed germinates readily in seven to 10 days at temperatures of 70 to 75F (21 to 24C). Cover seed lightly with media. Maintain high humidity around the seed during germination with mist. If growing in open flats, transplant seedlings when the first true leaves appear. Grow on at 60 to 65F (16 to 18C).

Provide supplemental light to plugs immediately following germination to speed flowering. Salable flowering packs or 4-inch (10-cm) pots take 13 to 15 weeks, while hanging baskets in flower take 17 to 18 weeks.

As with zonal geraniums, seed geraniums may be treated with Cycocel during production for height control. Begin applications 30 to 40 days after sowing. Treat two times at 1,500 ppm. For more uniform growth, treat plants three to four times at 750 ppm, with sprays timed about one week apart. Using lower rates more frequently results in stockier, well-branched plants. DIF—using cool night-warm day temperature regimes or providing an early morning temperature drop—works well in geranium height control. Do not pinch seed geraniums, as flowering would be delayed.

Because of the long crop time, most growers buy in seed geranium plugs. From a 288-plug, allow nine to 10 weeks to finish in 4-inch (10-cm) pots.

P

TABLE 1   PELARGONIUM VARIETIES BY BREEDER
(MOST POPULAR VARIETIES INDICATED WITH ASTERISK)

Name	Color	Comments
Ball FloraPlant, West Chicago, Illinois		
Designer series (medium-green foliage)		
Red	red	Medium vigorous; distinct zone; intense color; excellent landscape performance
Bright Red	red	Vigorous; distinct zone; non-burning red; excellent landscape performance
*Dark Red	dark red	Medium vigorous; distinct zone; excellent landscape performance
Rose	rose	Vigorous; distinct zone; excellent landscape performance
Salmon Rose	salmon rose	Vigorous; distinct zone; excellent landscape performance
Bright Scarlet	scarlet	Vigorous; excellent landscape performance
Scarlet	scarlet	Vigorous; distinct zone; excellent landscape performance
Coral	coral	Medium vigorous; excellent landscape performance
*Salmon	salmon	Medium vigorous; distinct zone; flowers are fuchsia rose with shades of salmon; excellent landscape performance; fills beds in quickly
Light Salmon	pastel salmon	Medium; distinct zone; fade resistant blooms; excellent landscape performance
White	pure white	Medium vigorous; excellent landscape performance
Light Pink	light pink	Vigorous; distinct zone; excellent landscape performance
Pink Pearl	light pink	Vigorous; distinct zone; excellent landscape performance
*Hot Pink	hot pink	Medium vigorous; excellent landscape performance
Pink Parfait	pink	Vigorous; distinct freckled flowers; excellent landscape performance
*Bright Lilac	dark lavender	Vigorous; excellent landscape performance
Lilac Chiffon	light lavender	Medium vigorous; excellent landscape performance
Purple Rose	purple	Medium; flowers are purple with rose and orange tinges, excellent landscape performance
*Starburst Red	novelty	Vigorous; distinct zone; unique striped flowers of red, salmon and white; excellent landscape performance
Showcase series (dark green foliage)		
*Red	red	Medium habit
Bright Red	bright red	Medium; free-flowering, non-burning blooms
Scarlet	scarlet	Slightly more vigorous than many original Showcase varieties, distinct zone
Bright Coral	coral	Medium
Dark Salmon	dark salmon	Compact medium; distinct zone; upright plants; floriferous
Salmon	salmon	Medium; distinct zone; outstanding well-branched habit
Light Salmon	pastel salmon	Vigorous; blooms do not fade in high temperatures

TABLE 1   PELARGONIUM VARIETIES BY BREEDER   (CONTINUED)

Name	Color	Comments
White	non-blushing white	Medium vigorous
Pink	medium pink	Medium; distinct zone
Pink Heart	pink	Compact medium habit; single flower; unique pink freckled flowers; grow warm to obtain good size
Pink Parfait	pink	Medium; unique pink freckled flowers

Dummen Geraniums, North Kingsville, Ohio
Pinnacle series

Name	Color	Comments
Blue	lavender	Medium flower; green foliage; slight zone
Cardinal Red	rose red	Medium flower; green leaves; slight zone
Cherry Red	cherry red	Medium flower; green foliage
Coral	coral	Medium flower; green foliage; intense zone
Crimson Red	deep red	Medium flower; green foliage; slight zone

Fischer Geraniums USA, Boulder, Colorado
Compact varieties

Name	Color	Comments
*Alba	white	Medium-early habit; medium green leaves;
Champion	orange	Early; dark green leaves
El Dorado	lilac pink	Early; dark green leaves
Explosive	dark pink	Early; dark green leaves
Omega	dark salmon	Early; dark green leaves
Rio	rose/red	Early; dark green leaves
Tiffany	hot pink	Early; medium green leaves

Medium-sized varieties

Name	Color	Comments
*Atlantis 96	purple	Medium early; dark green leaves
Blues	pink	Medium early; medium green leaves
Boogy	dark orange red	Early; dark green leaves
Brasil	lilac pink	Early; dark green leaves
Dark Polka	purple red	Medium early; dark green leaves
Montevideo	salmon	Very early; dark green leaves
Dolce Vita	light salmon	Early; medium green leaves

P

TABLE 1 PELARGONIUM VARIETIES BY BREEDER (CONTINUED)

Name	Color	Comments
Gloria	orange	Early; medium green leaves
*Samba	dark red	Early; medium green leaves
Lotus	white	Medium early; dark green leaves
Madison	salmon	Early; medium green leaves
*Bravo	dark pink	Medium early; dark green leaves
Noblesse	salmon	Early; medium green leaves
*Tango	red	Medium early; dark green leaves
Medium to vigorous-sized varieties		
Casanova	light rose	Early; medium green leaves
Diabolo	dark red	Medium early; medium green leaves
Grand Prix	scarlet	Medium early; medium green leaves
*Kardino	purple red	Medium early; medium green leaves
Magic	purple red	Medium early; medium green leaves
*Schoene Helena	salmon	Medium early, medium green leaves

Goldsmith Plants Inc., Gilroy, California
Americana series (medium-green foliage)

Name	Color	Comments
*Cherry Rose	vibrant magenta	Excellent heat performance; early to flower; good for fast cropping
Coral	clear bright coral	Excellent heat performance; early to flower; pure coral color, good for fast cropping
Dark Red	true dark red	True red color; does not fade; medium vigorous habit
Deep Rose	deep blue rose	Excellent heat performance; unique eye
*Light Pink	pastel pink	Excellent heat performance; unique pastel color; strong plant habit
Light Salmon	pastel salmon	Excellent heat performance; outstanding leaf zoning
*Pink	clear rose pink	Excellent heat performance; early to flower; good for fast cropping
*Red	bright red	Excellent heat performance; most widely used variety
Rose	blue rose	True rose color; good garden performance
Rose Splash	rose with red eye	Outstanding heat performance; unique color; single flower; large heads; very floriferous
*Salmon	warm pastel salmon	Leaf zoning; early to flower; true salmon color; good for fast cropping
Violet	fluorescent violet	Spectacular color; early to flower
White	pure white	Early to flower; floriferous; good for fast cropping

TABLE 1 PELARGONIUM VARIETIES BY BREEDER (CONTINUED)

Name	Color	Comments
White Splash	white with pink eye	Unique flower color; early to flower; excellent for patio gardening
Eclipse series (dark green foliage)		
Light Lavender	light lavender	Dark leaf foliage; compact growth habit; unique color to dark leaf class; long shelf life
Light Salmon	pastel salmon	Good heat tolerance; limited growth regulators required; can be grown at high density; long shelf life
*Red	strong, true red	Excellent heat performance; leaf zoning
Salmon Orange	salmon orange	Good heat tolerance; limited growth regulators required; can be grown at high density; long shelf life
White	bright white	Good heat tolerance; limited growth regulators required; can be grown at high density; long shelf life

Oglevee Ltd, Connellsville, Pennsylvania
Scarlet zonal varieties

Medallion Dark Red	dark red	Vigorous habit; very good heat tolerance; medium early; medium zone
*PAC Sassy Dark Red	burgundy	Compact; very good heat tolerance; early; no zone
*PAC Kim	scarlet	Medium habit; very good heat tolerance; early; no zone
Lollipop	coral	Compact; good heat tolerance; early; no zone
*Red Hots	scarlet	Compact; very good heat tolerance; early; no zone
Ritz	scarlet	Medium habit; very good heat tolerance; early; no zone
Sincerity	scarlet	Vigorous; very good heat tolerance; late; dark zone
Starburst Red	white-streaked salmon red	Compact; good heat tolerance; early; medium zone
*Yours Truly	scarlet	Medium habit; very good heat tolerance; medium flowering; medium zone

Pink zonal varieties

Bubble Gum	dark pink	Compact; good heat tolerance; early; no zone
Patriot Bright Pink	clear candy pink	Medium habit; good heat tolerance; early; no zone
Patriot Light Pink	apple blossom pink	Medium habit; good heat tolerance; early; no zone
Patriot Salmon Blush	salmon pink	Medium habit; very good heat tolerance; early; medium zone
Sweet Dreams	salmon pink	Medium habit; good heat tolerance; early; no zone
Cotton Candy	salmon pink	Medium habit; good heat tolerance; early; no zone

P

TABLE 1 PELARGONIUM VARIETIES BY BREEDER (CONTINUED)

Name	Color	Comments
*PAC Evening Glow	dark salmon	Medium habit; good heat tolerance; early; medium zone
PAC First Kiss	pink/magenta center	Medium habit; good heat tolerance; early; no zone
PAC Love Song	bicolor pink/ magenta	Medium habit; medium heat tolerance; medium flowering; no zone
*PAC Melody	dark candy pink	Medium habit; very good heat tolerance; early; no zone
Morning Mist	light salmon	Vigorous; very good heat tolerance; medium flowering; slight zone
Peaches	peach accented salmon	Medium habit; good heat tolerance; early; no zone
*PAC Pink Expectations	salmon pink	Medium habit; very good heat tolerance; early; medium zone
Sarah	rose pink	Vigorous habit; very good heat tolerance; medium flowering; no zone
Sunbelt Coral	coral	Vigorous; very good heat tolerance; late; no zone
Purple zonal varieties		
PAC Lavender Lady	dark lavender	Compact; medium heat tolerance; early; no zone
Olympia	lavender pink	Medium habit; good heat tolerance; early; no zone
Aurora	dark lavender pink	Vigorous; very good heat tolerance; medium flowering; dark zone
*PAC Fox	dark magenta	Medium habit; good heat tolerance; early; no zone
PAC Laura	lavender pink	Medium habit; good heat tolerance; early; no zone
PAC Precious	light orchid	Medium habit; good heat tolerance; early; no zone
*PAC Veronica	magenta	Medium habit; very good heat tolerance; early; slight zone
Brocade zonal variety		
Wilhelm Langguth	scarlet	Medium habit; good heat tolerance; late; no zone
White zonal varieties		
Angel	white	Vigorous; good heat tolerance; early; no zone
*PAC North Star	white	Medium habit; good heat tolerance; early; no zone
Raspbery Ice	white/pink	Compact; good heat tolerance; early; no zone
Floribunda varieties		
Maureen	orange/white	Compact; medium green leaves
Angela	light pink	Compact; medium green leaves
Elizabeth	red	Medium habit; medium dark leaves
Gypsy	lavender	Vigorous; medium dark leaves
Julia	coral	Medium habit; medium green leaves
Lucille	coral	Vigorous; medium green leaves
Marilyn	candy pink	Medium habit; medium dark leaves

Petal shattering. One of the biggest complaints about seed geraniums is petal shattering: flowers literally fall apart during shipping and on the retail shelf. Treating with silver thiosulfate (STS) prevents petal shattering and makes plants with a nicer point-of-sale presentation. Use a commercially available preparation.

Shattering can also be controlled by shipping plants just as they begin to show color. Also, avoid shipping situations that allow ethylene buildup, such as tight shipping boxes and tight truck bodies, and make sure plants are well watered before shipping to minimize stress.

Varieties. There are four basic types of seed geraniums available today: pack types, pot types, landscape types, and tetraploids (fig. 4). Like their vegetatively propagated cousins, tetraploids have four sets of chromosomes. Flower size rivals that of zonals, as does plant habit. Crop time is longer than other seed varieties, about one to three weeks longer. Tetra Scarlet has extra-large, 5-inch (13-cm) flower heads and performs well in 4-inch (10-cm) pots. Flowering is in 14 to 15 weeks. Freckles, a Rio (zonal) look-alike, has distinctive rose spots on each petal. Crop time is 15 to 16 weeks.

FIG. 4. Geraniums are almost unbeatable in window planters as this planting at Ball FloraPlant, West Chicago, Illinois, shows.

Pack geraniums include the well-known Elite series, to which the earliness of any seed geranium is compared. Red Elite, an industry icon, is a part of most seed geranium programs. The Elite series is available in a number of separate colors. Crop time is 14 weeks. Cameo (pale pink), Lone Ranger (deep red), and Neon Rose (pink and rose) are other varieties well suited to pack production.

For 4-inch (10-cm) pot production, try the Orbit, Ringo 2000, or Glamour series. All three series come in a wide color range, make top-quality plants, are Cycocel-responsive, and provide good flower presentation.

For outstanding landscape performance, the Pinto series is hard to beat. Every seed geranium program should include pots of Picasso, with its electric pink and violet flowers. While crop time is 16 weeks and other series boast of "Picasso-like colors," no one has yet to duplicate it. Avid gardeners know Picasso by name.

PELARGONIUM PELTATUM (IVY GERANIUM)

❖ *Annual (* **Pelargonium peltatum***). 6,000 seeds per oz (212 per g).
Germinates in five to 10 days at 70 to 75F (21 to 24C). Cover seed for
germination. Most ivy geraniums are vegetatively propagated due to limited selection of seed types.*

Ivy geraniums are synonymous with summer in northern Europe, where nearly
every balcony, front stoop, and patio has
a potted ivy geranium. Because of their
trailing habit, they are often called balcony geraniums (fig. 1).

Most ivy geraniums are propagated by
cuttings. Vegetative forms come in a wide
array of bright and pastel colors, with single and fully double flowers. A hanging
basket filled with an ivy geranium is a
Mother's Day standard. Bob Oglevee,
Oglevee Ltd., Connellsville, Pennsylvania,
offers the following cultural guidelines for
vegetatively propagated ivies, while Jim
Nau, Ball Horticultural Co., West Chicago, Illinois, offers guidelines for seed-
propagated ivies.

FIG. 1. Ivy geraniums make an excellent
hanging basket.

VEGETATIVE IVIES

Ivy geranium varieties can be separated into three categories depending on their
light tolerance. Varieties that must be grown under low light of 2,000 to 2,500 footcandles (22 to 27 klux) are Sybil Holmes and Simone. Other varieties—such as
Nichole, Beauty of Eastbourne, Nannette, and Balcon Royale—can be grown with
2,500 to 3,000 f.c. (27 to 32 klux). The third category is the most light-tolerant and
can be grown with light levels of 3,000 to 3,500 f.c. (32 to 38 klux) and includes
Salmon Queen, Princess Balcon, Cornell, Peppermint Candy, and King of Balcon.

Providing light levels that are too high results in small plants with tiny,
cupped leaves, small flowers, and some burning. You may also notice aborted
buds and florets, due not so much to high light but to high air and leaf temperatures. If leaf and air temperatures can be kept below 80F (27C), high light does
not affect ivy geraniums. Because leaf temperature is so important, ivy geraniums
should never be hung close to the glass, as this would only multiply the problem.

Ivy geraniums may be cultured in a soilless media in 4-, 6-, or 8-inch (10-, 15-, or 20-cm) pots or, more traditionally, in 10- and 14-inch (25- and 36-cm) hanging baskets. Use from three to five cuttings to make full baskets. Follow the nutritional guidelines provided for zonal geraniums. Cycocel sprays will be effective in controlling plant spread; follow the guidelines set forth for zonal geraniums here, too.

Ivy geraniums are also susceptible to edema, a physiological problem caused by environmental conditions. On a cloudy day when the soil is warm and moist but the air is cool and moist, the plant will absorb water rapidly, but it will lose very little water through its leaves. The cells of the plant, especially on the undersides of the leaves, swell with the excess water and form blisters that eventually burst. These broken cells later harden and turn brown, with a corky appearance.

Ivy geraniums require more attention than zonal geraniums with respect to edema. Steps to prevent and control edema must be followed through finishing. Nichole, White Nichole, Cornell, Beauty of Eastbourne, and Salmon Queen are the more edema-resistant varieties. Although edema is not caused by an insect, spider mite damage can resemble edema, so be careful in problem identification. Ivy geraniums, especially Sybil Holmes, are very susceptible to spider mites, by the way.

Zonal geraniums can receive up to 5,000 f.c. (54 klux) of light before the foliage and flowers start to burn. However, ivy geraniums must be grown in a greenhouse with only 2,500 f.c. (27 klux) of light (definitely not higher than 3,500 f.c., 38 klux). Media pH should be between 5.0 and 5.5. If using a peat-lite medium, maintain high levels of nitrogen and iron. Proper irrigation is also very important. Water plants only in the morning and remove all saucers from hanging baskets.

SEED-GROWN IVIES

Growing ivy geraniums from seed was first made possible with Summer Showers, a single-flowering mix of bright and pastel colors introduced in the mid-1980s. This Fleuroselect award-winning variety makes spectacular hanging baskets and works well as a ground cover in many regions, but it is available only as a mix. Plants are salable in hanging baskets in 18 weeks. Growers who wish to sell separate colors can plant one plug per 4-inch (10-cm) pot.

More recently, the Tornado series of ivy geraniums was introduced. As the bedding plant world's first variety developed through the advanced biotechnology of embryo rescue, it has received a lot of attention, and with good reason: it produces spectacular plants. Available in two separate colors, White and Lilac, the Tornados are salable in hanging baskets in 14 weeks. No ivy geraniums are as floriferous as the Tornados. Plant growth is naturally compact, reducing the need for growth regulators. Flowers are highlighted with deep, rich, crimson veins running the length of the petals. As an added bonus, the leaves are citrus scented.

PEONY (SEE *PAEONIA*)

PEPPER, ORNAMENTAL (SEE *CAPSICUM,* ORNAMENTAL)

PEPPER, VEGETABLE (SEE *CAPSICUM,* VEGETABLE PEPPER)

PERIWINKLE (SEE *CATHARANTHUS*)

PETUNIA (ANNUAL)

by Brian Corr
PanAmerican Seed Co.
West Chicago, Illinois

❖ *Annual* (Petunia × hybrida). *245,000 to 285,000 seeds/oz (8,642 to 10,053/g). Germinates in 10 to 12 days at 75 to 78F (24 to 26C). Do not cover the seed upon sowing.*

PLUG PRODUCTION

Sow seed into well-drained, disease-free media with a pH of 5.5 to 6.0 and electrical conductivity about 0.75 mmhos/cm (1:2 extract). The seed does not need to be covered.

Germinate seed at 75 to 78F (24 to 26C). Reduce temperatures to 68 to 75 F (20 to 24C) at cotyledon emergence. When true leaves are formed, reduce temperatures to 65 to 70F (18 to 21C). Plugs may be held at 60 to 65F (16 to 18C) from maturity until transplant (fig. 1).

Supplemental light during germination (100 footcandles,

FIG. 1. Dave Linder, Linder's Greenhouse Inc., St. Paul, Minnesota, displays Primetime Lavender petunia. Dave operates two substantial production ranges and a fine retail outlet in the Twin Cities.

1,076 lux) will improve germination uniformity and seedling quality. Seedlings must receive higher light levels immediately after germination to avoid elongation. After germination maintain light levels between 1,000 and 2,500 f.c. (11 and 323 klux). As seedlings mature, light levels may be increased up to 5,000 f.c. (581 klux) if temperature can be controlled.

Maintain 100% relative humidity until cotyledons emerge. Humidity can be reduced gradually to approximately 50% as plugs mature (fig. 2).

Fertilize with 50 ppm N from 15-0-15 as soon as radicles emerge. When cotyledons expand, increase fertilization to 100 to 150 ppm N. Use 20-10-20 with every other fertilization if growth slows. Maintain media electrical conductivity between 1.0 and 1.5 mmhos/cm (1:2 extract).

Control plug growth first by environment, nutrition, and irrigation management, then, if needed, with chemical plant growth regulators. Minimize ammonium-form nitrogen fertilizer to avoid elongation of seedlings. Temperature differential (DIF) can also be used to minimize height.

FIG. 2. Paul Cavicchio, Cavicchio Greenhouses, Sudbury, Massachusetts, holds a fine example of a 6-inch petunia grown outdoors. Note that plants produced under these cool temperatures are well branched and stocky.

If chemical plant growth regulation is necessary, test B-Nine, Bonzi, or A-Rest sprays. Test B-Nine at 2,500 to 5,000 ppm, Bonzi, at 6 to 15 ppm, or A-Rest at 26 to 132 ppm. Use the lower end of the range for the first application, and a higher rate if a second application is necessary.

GROWING ON TO FINISH

Maintain night temperatures of 55 to 60F (13 to 16C) and day temperatures of 60 to 65F (16 to 18C). Note: Petunias may be hardened off at maturity by gradually lowering temperature. If properly acclimated, petunias can tolerate temperatures down to freezing. Temperatures above 75F (24C) will significantly decrease branching and cause unwanted stem elongation. Maintain light levels as high as possible while maintaining moderate temperatures.

Use well-drained, disease-free soilless media with a medium initial nutrient charge and a pH of 5.5 to 6.3. Fertilize at every other irrigation with 15-0-15 alternating with 20-10-20 at 150 ppm nitrogen. Maintain media electrical conductivity around 1.0 mmhos/cm (using 1:2 extraction).

Effective height control of petunias can be accomplished with environmental manipulation. Once plants are established after transplant, they can be allowed to

FIG. 3. Try petunias in color bowls and hanging baskets in addition to regular cell packs. The variety shown: Flame Carpet.

wilt prior to irrigation to provide some height control (fig. 3 and 4). Height can also be controlled by withholding fertilizer, especially phosphorous and ammonium-form nitrogen. Petunias are responsive to day-night temperature differential (DIF), being shorter with a negative DIF. B-Nine, Bonzi, A-Rest, and Sumagic are effective for petunia height control, although these compounds are not labeled for this use in all locations. Always follow label recommendations. Test sprays of B-Nine at 2,500 to 5,000 ppm, Bonzi at 15 to 50 ppm, or Sumagic at 10 to 30 ppm. Do not apply plant growth—regulating compounds after buds are visible.

Allow four to six weeks from sowing to transplant and 11 to 13 weeks from transplant until flowering.

COMMON PROBLEMS

Petunias can be attacked by aphids, thrips, and whiteflies. One disease affecting petunias is pythium root rot. Symptoms are soft, brown, mushy roots. To control, drench with Subdue, Banrot, Truban, or a similar compound. *Botrytis* blight causes tan or brown leaf or flower spots followed by gray, fuzzy spores. Maintain good air circulation and reduce humidity during production. Remove dead or injured leaves or flowers. Apply preventative sprays, such as Chipco 26019, Ornalin, or Daconil, or smoke with Exotherm Termil.

Iron deficiency can be a problem. Low iron levels or pH above 6.8 will cause interveinal chlorosis in upper foliage. Certain colors within a mix are more sensitive to iron

FIG. 4. The bright-edged white rims of picotee flower patterning is always a hit for consumers looking for novelty. The variety shown: Merlin Picotee Mix.

deficiency and will show the problem, while other colors will not show symptoms. If the pH is above 6.5, lower the pH by adding iron sulfate. Apply 3 to 5 ounces of iron sulfate per 100 gallons (85 to 141 g per 400 l) as a soil drench. Wash the foliage with clear water after application. If the pH is below 6.2, apply chelated iron. To increase iron content without changing pH, apply chelated iron (12% iron) as a soil drench at 4 ounces per 100 gallons (113 g per 400 l).

Iron/magnesium toxicity is caused by extremely low pH. Toxicity is indicated by brown or tan lesions on the foliage. Switch to a base-forming fertilizer, such as 15-0-15. If symptoms do not improve or if the pH is below 5.5, irrigate the crop with a hydrated lime solution. To increase soil pH, apply 12 ounces of hydrated lime per 100 gallons of water as a soil drench (340 g per 400 l). Do not apply if the media ammonium level is above 10 ppm (1:2 extraction).

Boron deficiency can be caused if pH is high and boron concentration in the fertilizer solution is low. Boron deficiency is indicated by distorted foliage, tip dieback, and proliferation of side shoots below the meristem. Maintain pH between 5.5 and 6.3. Use a boron supplement, if necessary. As a supplement, apply borax (0.5 ounce per l00 gallons, or 8 g per 400 l) or Solubor (0.25 ounce per 100 gallons, 7 g per 400 l). Apply one or two times during production. If borax is used, dissolve in hot water.

VARIETIES

Single grandifloras. This is the most popular of the petunia classes on the market. Characterized by large flowers of 3½ to 5 inches (9 to 13 cm) across, the single grandifloras have dominated the market since White Cascade was introduced in the early 1950s. Today nearly all the varieties offered are F_1 hybrids of either frilled or rounded petal edges, in a wide range of colors. The latest development is the advent of veined petals, where the flower color is predominantly salmon, pink, coral, or of an orchid shade, with a veined petal surface that extends down into the throat. Sugar Daddy was the first variety on the market to exhibit this characteristic in this form, but the latest trend has been to breed more-compact plants rather than the taller plants of yesteryear. However, such taller plants often give superior hanging basket performance and still have value in the marketplace.

In the southern United States, plants are hardy in areas with mild or no winters, doing well in full sun. A hard frost will kill petunias. Sown in January for pack sales in mid-March, the plants will flower until August. Space 12 to 15 inches (30 to 38 cm) apart in the garden.

There are a number of excellent grandiflora varieties on the market today: Dreams, Falcon, Flash, SuperCascade, and Ultra are the leaders. The new class of spreading petunias from seed—Purple Wave, the All-America Selections award winner, and Pink Wave—are excellent for pots, baskets, and patio containers. Both are favorites for landscapers because of their spreading habits. Pink Wave is more restrained, with better branching. Apply long days to speed flowering with mum lighting techniques.

Single multiflora and single floribunda. No class of petunias has had more work and development on new varieties than multifloras. This class of petunias was once considered unpopular because of its small flower size, but has come a long way. Since the advent of the floribunda variety Summer Madness in 1983, a great many varieties have been introduced. Floribundas offer the disease tolerance of multifloras, but with the 3-inch (8-cm) flower size that is expected on medium-sized grandifloras.

Varieties like the Carpet, Celebrity, and Primetime series are of particular merit due to their compact performance in the garden. The Carpet series is getting more and more following in the southern United States as a variety that stands up well without lodging (falling apart) in the late season. The Celebrity series is noted for its larger than average flower size. Primetime's wide color range makes it an excellent choice. Finally, if you prefer smaller flower size, try the Fantasy series. Flowers of 1½ to 2 inches (4 to 5 cm) on early blooming plants in a broad range of colors are the advantages of this series.

PETUNIA (TRAILING)

by John S. Rader
Proven Winners
Encinitas, California

❖ *Annual* (**Petunia** × **hybrida** × **P. pendula**). *Vegetatively propagated. Due to high virus susceptibility, it is important to purchase young plants from a reputable company with a culture-indexing program.*

Trailing petunias are an entirely new breed of petunia. Initially developed by Suntory Ltd. in cooperation with Keisei Rose Co., both companies based in Japan, the new breed was labeled "Surfinia." Taxonomically speaking, Surfinia is the result of an initial cross of the common annual *Petunia* × *hybrida* with a South American native perennial pendula type (fig. 1). By crossing the colorful, floriferous annuals with this vigorous species, the resulting progeny are vigorous, everblooming petunia plants that bloom heavily, grow rapidly, and are much more vital than their annual parent. The varieties selected are those that have inherited growth characteristics of their wild parent with the aesthetics of their domesticated ancestor. In the United States, Supertunias were the first trailing petunia series to be widely mar-

keted. They were joined by Surfinia as supply channels increased.

CLEAN STOCK

Since most types of trailing petunias are
propagated vegetatively, it is *critical* to start
with virus-indexed material. Trailing petunias are prone to virus. If a propagator does
not renew his mother block from "elite," or
virus-indexed, clonal stock annually, the odds
are very high that a virus can develop in the
carried-over mother block. Trailing petunias
are susceptible to potato "Y" virus, tobacco
mosaic virus, tomato spotted wilt virus, and
cucumber mosaic virus. There are probably
others that have not yet been identified. Virus
is not something for growers to fear if they
are purchasing their lining-out stock from a
reputable source with an elite stock program.
In addition, growers should never hold over
previous year's plant material for cuttings or
a second year of sales. Such plants are
extremely susceptible to virus, and the risk is
not worth any economic savings.

FIG. 1. Surfinia, the granddaddy
of today's vegetative petunias, puts
on a spectacular show in hanging
baskets, patio planters, window
boxes, or in ground beds. It's practically indestructible.

Growers of trailing petunias in the United Kingdom learned a very expensive
lesson on the value of clean stock. In 1994 about 5 million trailing petunias were
dumped due to widespread infection of tobacco mosaic virus and potato "Y" virus.
Due to the exploding demand, suppliers elected to maintain their mother blocks a
second season to increase supply faster. Unknown to the growers, the held-over
stock was infected. The viruses soon spread to the new elite stock. The entire inci-
dent received broad publicity—all the way to the consumer press, with headlines
blaring the message of diseased plants for sale. The value of clean stock is now
understood from grower to consumer.

CULTURAL INFORMATION

Propagation. Unlike annual petunias, trailing petunias are propagated vegetatively by cuttings. These new strains of petunias produce very little viable
seed. They are also not true to type when grown from seed. All Proven
Winners varieties, Supertunia or Surfinia, have been maintained clean in asep-
tic conditions in tissue culture. They are especially vigorous and as disease-

free as possible. Soft tip cuttings root best on bottom heat with overhead mist. Growers should be advised, however, that all varieties of trailing petunias on the market are either patented or in the process of having a patent obtained. Propagation is by license only.

Planting. Due to their fast growth, plant trailing petunias later than most crops. These petunias are versatile. They may be grown in baskets, tubs, combination pots, or 4-inch (10-cm) pots. Plant one plant to a 6-inch (15-cm) basket; three to four plants to an 8- to 10-inch (20- to 25-cm) basket. Finish time for a basket is about six weeks. When grown in as 4-inchers (for eventual consumer transplanting to garden or container), plant one plant per 4-inch pot and expect about a four-week finish time. Plant in a well-drained, organic potting mix which has a pH of 5.0 to 6.0. Take care not to plant any deeper than the liner soil level. One pinch one to two weeks after planting is normally all the pinching that will be required. Pinching is not recommended to finish a 4-inch pot. Full sun is essential for optimum finish. If shaded, these vigorous plants will appear stretched, and flowering will be reduced. Most varieties are day-length sensitive. Flowering can be significantly reduced by shorter days.

Florel can be used instead of a pinch, if used during long days. It can also help maintain more uniform growth and flowering. Care should be taken not to use Florel late in the fall or early spring, as blooming can be delayed.

Temperature. Optimum growth temperatures are 75 to 83F (24 to 28C) days, with nights between 55 to 60F (13 and 16C). However, trailing petunias are not so sensitive that they cannot be grown at temperatures outside of this range. This means that successful growth can be achieved during hot summer days and nights, as well as cooler winter days in mild climates. As far as cold tolerance, these new petunias will handle an occasional cold night around 25 to 28F (–4 to –2C). During hot summer days, most trailing petunias will tolerate 95 to 100F (35 to 38C).

Water. Although these new breeds grow fast and can become very large, their water requirement is surprisingly low. They are best maintained on the dry side. In areas where water is high in salts, leach periodically.

Fertilizer. Here is perhaps the most challenging cultural requirement. Since trailing petunias are so fast-growing, it is difficult to keep up with them nutritionally. The general rule is to double the average feed program. More specifically, if a constant-feed program is in place, fertilizer should be applied at 300 to 350 ppm N, 100 to 150 ppm P, and 200 to 250 ppm K. If feed is applied periodically, rates should be 400 to 450 ppm N, 200 to 250 ppm P, and 300 to 350 ppm K. A periodic dose of Fe, applied at 5 to 10 ppm once or twice during the production phase, will enhance growth and finish quality. A third fertilization program that works combines a slow-release fertilizer and an average constant-feed program, such as 250 ppm N, 75 ppm P, and 150 ppm K. The slow-release fertilizer should be high in nitrogen and applied at the recommended rate.

To emphasize how hungry these new petunias can be, the grounds supervisor of The Timbers, an office building in Solana Beach, California, mixed four times the recommended amount of Osmocote with the potting mix when planting trailing petunias. The results were vigorous, healthy green growth with heavy blooms and no burn! It is not recommended, however, that growers quadruple any fertilizer rate.

Pest control. There are no serious pest problems for most growers. A standard preventative spray program will lessen the likelihood of whiteflies, which can infest neglected, poorly maintained plants. Also, watch for aphids and mites. Growers in areas where bud worms can be a problem should spray preventatively with *Bacillus thuringiensis* spray. Bud worms can strip every bloom off of even a heavily blooming plant in very short time. In parts of the country where slugs and snails can be a problem, appropriate metaldehyde sprays or baits should be applied.

As far as disease problems go, it is very important to start with vigorous, virus-free liners. Although not overly susceptible to disease, some fungal pathogens can attack trailing petunias. In most cases, such infections are grower-induced by overwatering, underfeeding, growing in low light conditions, or planting liners too deep. *Pythium* appears to be the most frequently occurring fungus. A drench with a broad-spectrum fungicide combination, such as Chipco 26019 at 16 ounces per 100 gallons (479 g per 400 l) and Subdue at 2 ounces per 100 gallons (60 g per 400 l) at planting time will add a little extra insurance.

CULTURAL TIPS FROM THE GROWER TO THE CONSUMER

Since trailing petunias are a new plant for garden centers and consumers, it is important that customers feel confident they can succeed with them, especially when trailing petunias command a premium price. They need to understand that there are distinct differences between these new trailing, vegetative types and the standard annual seed petunias. Banners, labels, and cultural information should be supplied at the point of purchase.

Tell consumers the following: In the ground, plant one 4-inch container per square yard (about one 10-cm pot/m²) of sunny garden space. Trailing petunias planted in beds will fill in rapidly, producing a spreading carpet in about a month. Supertunias prefer a well-drained soil high in organic matter. In fact, an organic mulch, such as a bark or wood shavings, makes a good surface for shoots to spread on. Rooting along the spreading, vinelike branches is uncommon, but a single branch can spread 4 to 6 feet (1.2 to 1.8 m) from the original plant. These petunias do not like to have excessive moisture on their foliage.

Whether planting in the ground, baskets, planters, or whatever, the largest challenge for home gardeners is to keep up with the plants nutritionally. Trailing petunias grow so fast and are so floriferous that they need to be fertilized often and heavily. Home gardeners should feed with a good all-purpose fertilizer high in nitrogen and iron. The same general rule for growers also applies to consumers, that is, to double the recommended fertilizer rate. The grower may want to apply a slow-release fertilizer just prior to shipping or sale to provide some ensurance of feed after sale. Again, it is important to double the rate.

Consumers should also be advised regarding the water requirements of trailing petunias, which can be deceptive in their moisture needs. It is easy for the average gardener to assume that because these plants get so large and grow so fast, they use large amounts of water. Actually, the opposite is true. Constantly wet, saturated soils will render trailing petunias weak and chlorotic, especially when they are planted in the garden. Again, a well-drained soil is important.

A huge benefit for consumers is that there is no need to clean or disbud these new types of trailing petunias to maintain active, flowering growth. They are indeterminate in their growth and flowering habits. Spent blooms shrivel away, while at the same time new blooms are emerging to take their place. Trailing petunias respond well to cutting back. If plants become overgrown or unsightly, simply cut back and reduce watering.

There are numerous new varieties still yet to come on the market: upright pot types, small flowering miniatures, bicolors, bushy types, and the list goes on. Trailing petunias are here to stay. The key for the grower is to always buy from a reputable clean-source propagator. Such a propagator is one who has gone to the time and expense of putting a variety through the tissue culture cleanup and establishment processes, having selected the very best varieties with which the grower can succeed.

PHLOX

❖ *Annual* (**Phlox drummondii**). *14,000 seeds/oz (494/g). Germinates in 10 to 15 days at 60 to 65F (16 to 18C). Dark.*

❖ *Perennial* (**P. paniculata**). *2,000 to 2,500 seeds/oz (71 to 88/g). Germinates in 15 to 25 days at 60F (16C) once seed has been pretreated with a moist, chilling period.*

ANNUAL PHLOX

A very showy, colorful annual, phlox probably requires a little more attention than common annuals such as petunias when planted to outdoor beds. Allow 10 to 11 weeks from sowing for flowering packs, 13 to 14 weeks for 4-inch (10-cm) pots. Sow direct to the final container for best results, as plants do not like transplanting. Otherwise, grow them as plugs under cool, dark, and dry conditions.

There are both tall and dwarf classes of phlox, known as Grandiflora (15 to 18 inches, 38 to 46 cm) and Nana Compacta (6 to 8 inches, 15 to 20 cm), respectively. The former is not as widely used as the dwarf form, which is most popular and of greatest commercial value.

FIG. 1. Shown here in Vaughan's bedding plant field trials, Globe Mix phlox makes a good, low growing bedding plant for full sun.

The most widely used dwarf forms are Globe Mixture, Petticoat Mixture, and Twinkle Mixture. Globe Mixture has an almost perfectly round, ball-like growth habit and an exceptionally free-flowering characteristic (fig. 1). It branches out beautifully from the base of the plant, in marked contrast to the other types of annual phlox, and has a good range of showy colors. Twinkle Mixture, an early bloomer, produces an abundant amount of dainty, starred flowers with pointed petals. Petticoat Mixture is a variety of star-shaped flowers in a bright range of colors on plants to 8 inches (20 cm) tall.

PERENNIAL PHLOX

Then there is the whole class of perennial phloxes. Briefly, the most important group is the summer phlox, also known as garden phlox (*P. decussata, P. paniculata, P. maculata,* and more). There are dozens of brilliant reds, purples, salmons, and vari-colored varieties in this group, propagated by divisions and root cuttings. Grown from seed, plants will not flower the first year. Divisions purchased from commercial propagators can be potted in the later winter or early spring and sold seven to nine weeks later. Grow at 55F (13C) nights. They like a fairly well-enriched soil and should be kept fairly moist. Plant out on 18-inch (46-cm) centers. Their normal flowering season is mid-July to September. Beltsville Beauty is a blend of *P. paniculata* types under one name. Seed should be exposed to moist, freezing temperatures for several weeks for best germination. Perennial phloxes are hardy from zones 4 to 8. They perform best in warm and dry areas. Plants will not tolerate a long duration of hot and humid weather—they are extremely susceptible to powdery mildew.

P

For rock garden work, the dwarf perennial *Phlox subulata* makes a brilliant showing. It is division-propagated, flowers in May and June, and makes a mat of bright green foliage throughout the season. There are a dozen or so other choice varieties, varying from white through bright pink, red, and crimson. All of this *P. subulata* group are procumbent, 4 to 6 inches (10 to 15 cm) tall, and are used for carpet plantings and rockeries. For a full flower show, make sure purchased liners or divisions potted in the winter have been properly vernalized.

POINSETTIA (SEE *EUPHORBIA PULCHERRIMA*)

POPPY (SEE *PAPAVER*)

PORTULACA (MOSS ROSE)

by Jim Nau
Ball Horticultural Company
West Chicago, Illinois

❖ *Annual (**Portulaca grandiflora**). 280,000 seeds/oz (9,877 seed/g). Multiple sow three seed or more per plug or pack cell. Germinates in seven to 10 days at 75 to 80F (24 to 26C).*

Portulacas are excellent flowering plants for areas with poor soil and full sun. These plants work well in rock walls, as well, but will not tolerate poorly drained soil. Flower colors range from yellow, white, cream, fuchsia, scarlet, all the way to violet. Flowers are double in appearance and can develop to 2½ inches (6 cm) across. The greatest area for advancement in this class are those selections whose blooms stay open even under cloudy conditions. At present only a limited number of varieties do so, and even these don't remain fully open.

For green packs in the North, allow 10 to 11 weeks, and 13 weeks for flowering cell packs of three to five seedlings per cell. Southern growers can flower packs in eight to nine weeks. To finish 48-cell packs from plugs, allow five weeks. Most often portulaca is sown directly to the final cell pack, like alyssum, with at least eight to 10 seeds used per cell. Thinning is not necessary unless filling cells devoid of any seedlings. For something a little different, offer portulaca in 10-inch (25-cm) hanging baskets, allowing 13 to 14 weeks to flower and become salable.

The large, double blooms of Sundance Mixture and similar varieties are usually the quickest to flower and stay open. The Sundial series (fig. 1), an F₁ also blooms early (10 to 14 days earlier) and offers a good range of separate colors plus a mix. Double Mixture and Calypso Mixture display wide ranges of colors, but their flowers close up if light is limited.

In a different species closer to the true purslane type, Wildfire Mixture is the earliest variety to flower in baskets and the most vigorous garden performer. The single flowers bloom in rose, yellow, and white. However, the flowers open and close more erratically than other types. Furthermore, Wildfire Mixture can reseed itself and become a nuisance.

FIG. 1. This reliable full-sun performer is a champion. The variety shown: Sundial.

PRIMULA (PRIMROSE)

by Ed Markham, horticulturist
Kent, Washington

❖ *Annual and perennial* **(Primula acaulis).** *25,000 to 33,000 seeds/oz. (875 to 1,155 seeds/g).* **P. polyanthus.** *34,000 seeds/oz. (1,190 seeds/g).* **P. malacoides.** *280,000 seeds/oz. (9,800 seeds/g). Seed germinates in 21 to 28 days at 60 to 65F (16 to 18C). Do not cover seed.* **P. obconica.** *190,000 seeds/oz. (6,650 seeds/g). Seed germinates in 10 to 20 days at 68F (20C). Do not cover seed.*

The primrose family, Primulaceae, consists of several genera (fig. 1). The family's plants characteristically have one or a cluster of stems with flowers having five-lobed calyxes and corollas. Interestingly, the family has only one other genus of major commercial importance, namely *Cyclamen*. The genus *Primula* has about 600 species, only about six of which are of commercial importance. The majority of primulas were natives of China. For those looking for a complete account of all of the primula species, the book *Primula* by John Richards is available from Timber Press.

PRIMULA ACAULIS

Within the past 20 years, there has been a major switch from the old Pacific Giant *polyanthus* primula, bearing flowers on stems 8 to 12 inches (20 to 30 cm) tall, to the *acaulis* type, which is virtually stemless, thus bearing its flowers on much dwarfer and compact plants. Extensive breeding work continues, bringing a great number of colors and types to the trade, generally marketed in 4- to 6-inch (10- to 15-cm) and combination pots.

FIG. 1. Bright, colorful, and fragrant primula is a sure sign of spring at Molbak's, Woodinville, Washington. The variety shown: Danova.

Primula germination requires close attention to narrow tolerances of temperature, light, and moisture. These conditions must be carefully monitored. At plug Stage 1, the seed must be in a 100%-moisture environment in a well-aerated media. Do not cover the seed, because it needs light for germination. Temperatures must be within 3F (–16C) of a norm of 62F (17C). Any variance from these conditions can cause poor germination of this relatively high-cost seed. At plug Stages 2 and 3, frequent but light watering is important for maintenance of good aeration. Proper moisture is essential at this critical time. Because of the critical requirements for primula seedlings, consider buying in plugs from a primula specialist.

At transplanting, about eight weeks after sowing, do not cover the growing tip. The media should have a pH of 5.4 to 6.2. The fertilizer should have a high potash (K) level. Use of a 13-2-13 is recommended, along with special attention to regular soil tests, especially for pH and EC.

Following transplanting, grow on to the 10-leaf stage at temperatures ideally in the 50- to 60F (10- to 16C) range. Shade the growing-on range for cooling and to hold light levels to around 3,000 footcandles (32 klux). Primula responds very well to DIF to help avoid stretched and leafy plants. During the darker months, maintain cool temps of 35 to 40F (2 to 4C) and grow under good light levels and ventilation.

Diseases generally are a result of poor environmental control. Avoid soft growth, maintain good ventilation, and keep the foliage dry. Insects are rarely a problem. Strawberry root maggots love primula roots in the garden, though.

Extensive and very competitive breeding has brought many new colors and types to *P. acaulis*. Check your catalogue for varieties recommended for early to late blooming. Recent additions to large floret types are Danova and Hethor.

Lovely is a smaller-flowered early bloomer. Improved Pageant has a great color range. Also consider the newer introductions like Quantum, Lucento, and Fama. Always include a bit of Finesse for its classy white "thread" bordering the florets. Most of the labeling refers to garden use, but it is likely that most of the plants will land on the table or windowsill during cold, wet, and dark days. Good idea! Later varieties—and now the very hardy Wanda types (such as Wanda Supreme) with bronze-red foliage and wider color range—are great for garden use. In any case, always remember that the straight colors are easiest to handle from start to market, with their limitless choices and combinations.

PRIMULA POLYANTHUS

The cultural needs of *P. polyanthus* are like those for *P. acaulis*. Their market is more likely to be for garden use because their taller-stemmed habit makes them more suitable and prominent for this purpose. Imaginative landscapers often use *P. acaulis* in the foreground with *P. polyanthus* in the background.

P. polyanthus has not had the breeding attention of *P. acaulis*, though there are now a few shorter-stemmed *P. polyanthus* types, such as the newer Rainbow or Concorde.

PRIMULA MALACOIDES

Known as the fairy primrose because of its delicate whorls of mostly pastel colors, *P. malacoides* is truly a sleeper in the field of winter pot plants, as well as that of early outdoor use. *P. malacoides* is enhanced by the recent addition of F_1 hybrids that germinate readily under conditions like those for *P. acaulis*. Like all primulas, grow them in a well-aerated soil, with 6.0 pH optimum. Grow on at 55 to 60F (13 to 16C) and drop to 50F (10C) for high-quality finish in 4-inch (10-cm) multiplanted or for great combinations with ferns and bulbs.

Use a fertilizer low in nitrogen and high in potash. Avoid salt buildup with continuous soil analysis.

The new F_1 Prima opens many possibilities for pots. It is also great for late winter in a protected window box or patio planter. Try it. You and your customers will be pleasantly rewarded (fig. 2).

P

PRIMULA OBCONICA

In the past, growing *P. obconica* was often an unpleasant experience. The advent of new "primin-free" strains, however, have eliminated problems with skin reactions, though it may take some time to convince those who have suffered the rash. *P. obconica* is a warmer-temperature, five- to seven-month crop that does not require the cool temperatures of most of the other primulas (fig. 3). It makes upright trusses of colorful, large blooms for great 5- or 6-inch (13- or 15-cm) pot plants. It also does very well in the early spring cool garden.

Germinate at 65F (18C). Transplant in about eight weeks. Grow on at 60 to 65F (16 to 18C), then quality finish at 50 to 55F (10 to 13C). Avoid salt buildup, and do not allow plants to dry to wilt, because of possible leaf margin burn. Nutrition requirements are similar to those of other primulas.

Primin-free *P. obconia* varieties like Libre open up a whole new market.

FIG. 2. Growers often produce potted primula for bedding plant sale outdoors in unheated hoop houses. This crop was at Turkey Creek Farms, Houston, Texas.

FIG. 3. *Primula obconica* is produced as a flowering pot plant for winter and spring sales. The variety shown: Juno Red Picotee.

HARDY PRIMULAS

P.vialii, P. denticulata, P. japonica, P. auricula, and others offer possibilities as perennials in this era of perennial popularity. Check with your seed salesman for unique and spectacular blooms to help fill another niche with primulas.

Finally, it must be said that there has been too much 4-inch (10-cm) commodity syndrome with the primula crop. Primulas have enormous variety in size, use, combination, and color potential, limited only by the imagination employed during production and marketing.

PROTEA

by Elizabeth C. Gerritsen
Salinas, California

❖ *Perennial shrubs* (**Protea** *spp.*). *Vegetatively propagated. Primarily used as a cut flower.*

Protea is native to areas of Australia, nearby New Caledonia, South Africa, and South and Central America. Fortunately, planting proteas in the rich microclimates of California and Hawaii have proven successful, as well. The Proteaceae family includes many well-known genera, such as *Banksia, Leucospermum, Leucodendron,* and the very popular *Grevillea* and *Protea.* Proteas are often mistaken for tropical flowers because of their exotic qualities. Culture is best suited to a Mediterranean-like climate with mild temperatures, low humidity, and high light. However, proteas can tolerate extreme temperature fluctuations, although it will be short-lived in such a situation.

Proteas grow well on a hillside or mound in well-draining soil. They require deep, weekly watering while establishing. Once established, they grow nicely with minimum care. Soil should drain well and be slightly acidic (pH of 5.0), and remember to never add phosphates.

The well-adjusted protea plant will often bloom after two years, but it can take as long as four for a substantial bloom.

The floral industry is beginning to see the merits of proteas as cut flowers. Proteas blend very nicely with traditional and tropical flower varieties and have a very long shelf life if handled properly. Remember to keep buckets clean and the flowers cool with good air circulation. Black spots can develop on protea foliage because of exceptionally wet field conditions or postcutting mishandling.

When pruning proteas, cut back sharply, and the flower will develop on the new wood. The plant is hardy, and in its natural environment it is burnt to the ground by wildfires, only to reappear the next season.

The primary season for planting proteas is autumn, and secondarily spring.

Besides being fairly drought-tolerant, proteas have very few pest damage problems. Even though the blooms make lovely homes for a variety of insects, there is surprisingly little damage caused by them. To remove insects, immerse the cut flower in a bucket of clean water and shake gently—good news for those who like to handle pesticides as little as possible.

Weeding should be done between the plants, but just cut the weeds aboveground so as not to disturb the protea's root structure.

PYRETHRUM (PAINTED DAISY)

❖ *Perennial* (**Chrysanthemum coccineum,** *though it may also be listed as*
  C. roseum, Pyrethrum coccineum, *or* **P. roseum**). *18,000 seeds/oz,*
  (635 g). Germinates in two to three weeks at 60 to 70F (16 to 21C).

Pyrethrum coccineum or *P. roseum*—the painted daisy—is among the most valuable and best known of this easily grown perennial. Pyrethrums include a number of useful, long-stemmed, cut flower varieties that bloom heavily in June with
a few scattered flowers during the rest of summer. In England, named varieties
that are propagated by divisions are very popular. The climate there enables them
to be marketed on heavy 2- to 3-inch stems. Some growers have tried this division-propagated stock, but found they die out shortly, while seed-grown plants
will do well for three to four years. Double strains have been developed, and
while they are improvements upon the single form, the strength of at least two-
year-old plants is necessary to produce double flowers; the first year nearly all
will be singles. There are separate-color varieties available, but the Double Mix
strain is most popular. Another English strain known as Robinson's Mixture produces large, single, attractive flowers.

   For green packs for spring sales allow 10 to 12 weeks when grown at 55 to
60F (13 to 16C).

REGAL GERANIUM (SEE *PELARGONIUM DOMESTICUM*)

RHODODENDRON (FLORIST AZALEA)

by Roy A. Larson
Department of Horticultural Science
North Carolina State University
Raleigh

❖ *Shrubs,* (**Rhododendron obtusum, Rhododendron simsii**). *Vegetatively*
  propagated from cuttings. Most growers buy in prefinished plants and
  force them into flower for fall, winter, and spring pot plant sales.

At one time florist azaleas and poinsettias were sold in similar quantities in
florist retail shops for Christmas. When bedding plant growers began adding
more poinsettias to their inventories so they could make more efficient use of
their greenhouse facilities and retain their labor forces, the volume of poinsettias

FIG. 1. Young azalea liners at Blackwell Nurseries Inc. in Semmes, Alabama.

grown in the United States increased greatly, but azalea sales did not change drastically. While poinsettias require about a four-month crop time from propagating to flowering, the azalea could require two to three years, depending on the plant size desired. The interval between the timed or final pinch and flowering could be a minimum of 26 weeks. Expenses incurred in production do limit the sales of azaleas in the mass market, but sales through such channels are increasing. It is not unusual, however, to see that the price of azaleas is higher than for almost any other flowering potted plant.

Many growers once propagated their own cuttings and kept the plants on their premises for about three years before the plants were given the final pinch. The final pinch often was timed so the plants would be in flower for a specific holiday. Many of these growers were not located in geographic areas whose climates were conducive to rapid growth. Now many greenhouse firms purchase plants from producers in more favorable climates, who can more efficiently and quickly grow the plants to the desired sizes (figs. 1 and 2). One now can purchase azalea plants that are vegetative; or that have received their final pinch and have dormant flower buds; or whose dormancy has been overcome, and all the purchaser has to do is force the plants.

Each added procedure increases the cost of the product, and it is wise for buyers to compare costs to determine which stage of development is best for their businesses. For example, a dormant, budded plant will cost less than one that has been precooled by the propagator. Either natural or

FIG. 2. Production of braided azalea trees at Yoder Brothers, Alva, Florida.

refrigerated cooling will be necessary for dormant plants, though, so the dormant plants might not be the best plants to purchase.

Production of azalea liners now is done in large numbers in areas such as southern Florida, Mobile, Alabama, and Oregon. New York still ranks high in the annual azalea statistics, primarily because one large greenhouse firm has facilities in Florida, South Carolina, and Long Island.

CULTIVARS

There are many azalea cultivars available to nurserymen and flower growers. People who purchase florist azaleas generally are looking for characteristics that are quite different from purchases for the landscape. Hardiness is not a major trait required for the florist azalea, as many flowering plants are sold in regions where the climate prohibits use of evergreen azaleas outdoors. In areas where azaleas are popular in the landscape, however, such as in the Southeast and the Pacific Coast, the suitability of a cultivar for both indoors and outdoors could be a fringe benefit of the purchase. Color, form, and uniformity of flowering are major characteristics that a forcer must consider.

Red flowers still are most popular, though their percentage of sales has declined with the changes that have occurred in interior decoration. Pink and variegated flowers have gained in popularity, and white-flowered cultivars probably have remained quite stable in sales. Color preferences vary from season to season, from one region to another, and among ethnic groups.

Customers also have at least six choices in flower forms. Flowers can be single, semidouble, or double, and there are hose-in-hose forms (flowers whose sepals resemble petals in color and size) of these three types. Some stamens develop petal-like forms, giving the appearance of semidouble or double flowers.

Some very nice but slow-growing azalea cultivars have declined in popularity because the grower usually cannot get higher prices for such cultivars, compared to those that break and grow vigorously after a pinch. Examples of slow-growing cultivars are Leopold Astrid, California Sunset, and Albert and Elizabeth. New cultivars continue to be developed. Some cultivars are the results of breeding programs, such as the eight cultivars recently introduced by Yoder Brothers: Athena, Champagne (fig. 3), Cherish, Goddess, Mystic, Party Favor, Remembrance, and Sachet. Many new cultivars are sports of existing cultivars: Coral Dogwood, Variegated Dogwood, and Nancy Marie are all sports of the original Dogwood. New cultivars also are being introduced from Europe, and in recent years the Vogel series has been an addition to the selection available in the United States.

Emphasis once was placed on the origin of cultivars and whether they were Rutherfordianas, Pericats, Belgian indicas, Kurumes, or belonged in other groups. Such information is interesting, particularly to the specialist. One can refer to definitive azalea references, such as Fred Galle's book *Azaleas* (Timber Press), if such information is desired.

ENVIRONMENTAL FACTORS

A knowledge of the effects of environmental factors is important at all stages of azalea production and is especially crucial near the conclusion of the crop, when so much has already been invested in it. Failure to apply water when it is needed, inadequate or excessive heat and light, and insufficient ventilation can nullify in a day what it has taken months to achieve.

FIG. 3. The variety Champagne, a new rhododendron introduction.

Water. It is difficult to rate which environmental factor is most important, but in some areas of the country, water would be the most important consideration. Water quality and quantity must both be considered. If one waits too long to irrigate, the medium, usually very high in organic matter, can pull away from the sidewalls of the container; the water, when applied, will just run down the inside of the pot. Water will still be deficient. Media such as peat moss are very difficult to moisten, once becoming dry, and several applications might be needed before the peat moss gets moist. On the other hand, some of the most serious plant pathogens affecting azaleas, such as *Phytophthora cinnamomi* (azalea decline or little leaf disease), are water molds, so excessive amounts of water could predispose the plants to devastating disease problems.

Different watering systems have been devised to reduce labor costs of manual watering and to prevent excessive water runoff. Sprinkler irrigation frequently is used, as many plants can be watered at one time, but much of the water falls between the pots, and foliar disease organisms can be spread by splashing water. Drip tube irrigation is being installed by some growers to address these problems. Plants are placed on plastic, so the water does not just run into the ground.

Water quality is a serious problem in some regions of the country, and the expense of correcting the problem can be high. Salts, such as chlorine, sodium, sulfate, iron, boron, and carbonates, should be monitored if poor water quality is encountered.

R

Light. Producers of azalea liners are generally located in regions where light intensity is high, and a material such as saran might be needed to reduce the intensity and the consequent high leaf temperature. Some producers might cool off the plants by applying moisture overhead, as the evaporating water has a cooling effect. The cooling is only temporary, however, and the danger of spreading disease organisms with splashing water already has been mentioned.

Azalea growers often load trucks with the plants in the field, then ship the plants to their customers, who force them. The trucks usually are refrigerated, so when the plants arrive at the customer's greenhouse, the medium is quite cool. If the plants are placed under bright light upon arrival, such as in an unshaded greenhouse, the plants will lose moisture through transpiration faster than the water will be absorbed. Leaf burn and abscission often are the consequences of such treatment.

Azaleas respond to day length, but not as dramatically as do chrysanthemums or poinsettias. Vegetative growth will be stimulated under long days, while short days will promote flower bud initiation and early development. Control of photoperiod is needed if year-round flowering of azaleas is desired on a precise schedule. For many years we have extended the day length to 16 hours or used mum lighting from 10 P.M. to 2 A.M. to provide long days from September 1 to March 31. We also have used black cloth to provide short days from April 1 to August 31. A nine- or 10-hour day length is a practical short-day treatment to fit in with the workday.

Light intensity also is an important factor during the precooling of azaleas to break flower bud dormancy. Research at the University of Florida has shown that light intensities as high as 120 footcandles (1 klux) in a cooler that is maintained at a maximum temperature of 50F (10C) will result in better plants and faster forcing than if the light intensity is down as low as 5 f.c. (54 lux) (fig. 4). A daily 12-hour lighting schedule is most common. Incandescent light bulbs are the cheapest to install, but they do add heat to the cooler. We have used fluorescent light tubes in half of our cooler, and the leaves closest to the tubes developed a purplish color.

FIG. 4. Trials at the University of Florida, Gainesville, show that forcing was accelerated if light intensity was 120 footcandles in the cooler (*right*) rather than 30 footcandles (*left*).

Light quality and its effects on azalea growth and flowering have not been extensively studied, but it now is known that crowding plants has a pronounced effect on light quality, as leaves overlapping other leaves act as filters. Shoots on plants exposed to the far-red end of the spectrum will be more elongated than those exposed to the red band, as the leaves absorb the far-red light. Lateral branching also will be reduced when far-red light is absorbed, so plant shape and size are adversely affected by crowding.

Temperature. A night temperature of 65F (18C) would be satisfactory for most of the duration of an azalea crop. Lateral shoot development will be adequate at such a temperature, and flower bud initiation and early development will be satisfactory. Once flower buds reach a certain stage of development, they will cease further development and become dormant. Temperatures ranging from 35 to 50F (2 to 10C) or applications of gibberellic acid will be needed to overcome that dormancy so forcing can proceed. In some areas of the country, 35 to 50F (2 to 10C) can be provided under natural conditions in the fall months, and refrigerated storage is not needed for that particular period. Natural cooling is not as precise or predictable as refrigerated storage, however.

Refrigerated storage units can be purchased, or they can be designed and constructed by the grower. Some growers purchase used refrigerator boxes removed from trucks, or they might even obtain a refrigerated railroad car. The grower should be aware of the special needs of an azalea cooler. Lights must be installed if the temperatures are to be maintained at 45 to 50F (7 to 10C), or leaf drop will occur. Plants precooled at the warmer temperatures and placed under lights will require watering, preferably twice a week, again to avoid leaf drop. A relative humidity of 80 to 90% is also recommended. The cold storage for Kurume azaleas is four weeks, and it is six weeks (1,000 hours) for most other types. Some cultivars, such as those in the Vogel series, will flower quite uniformly even without cooling, but flowering will be more accurately controlled with precooling.

Natural cooling is least expensive and least controlled. It can be used satisfactorily for spring flowering, such as for the Easter season and Mother's Day, in most parts of the United States. Plants must be protected from freezing (35F, 2C, would be a safe minimum temperature). "Holding" greenhouses should be shaded to prevent high leaf temperatures and water loss during the day when the root balls are still cold.

Media. This factor is very important but it primarily is under the control of the propagators and liner producers, rather than under the control of the forcer. Frequently, the plants will be purchased in the final container, or the plants might be shifted to containers only slightly larger, if a timed or final pinch is to be made by the forcer. Very little medium is required to add volume during trans-

R

planting. Acid peat moss is still preferred by many growers, but cost or unavailability might require a substitute. Shredded pine fibers or pine bark humus are popular alternatives in the southeast, while other regions utilize organic matter that is readily inexpensive and available. The medium should be acidic to ensure vigorous root systems and to minimize potential nutrient deficiencies. A good medium is also free of pathogens, insects, and weed seed.

Pinching. This operation will not be needed if the plants are at the desired size when purchased. The grower will have to pinch the plants at least once if larger plants are wanted. Pinching can also be used as the base point in timed flowering. It usually takes at least 26 weeks from when the plants are pinched to when they are salable as flowering plants, and that's under controlled conditions. Natural precooling could prolong the production time.

More lateral shoots will initiate and develop if plants are given a soft pinch since more nodes are retained on each shoot. The final pinch can be a laborious and costly task because there are usually many shoots on each plant. Machines and chemicals have been used to eliminate most of the hand labor, but some major azalea producers have gone back to using battery-powered clippers or even pruning shears. Efforts are made to shape the plants at the same time that apical dominance is being overcome. Machines usually give the plants a standard shape and miss some of the youngest shoot tips. These shoots will elongate and extend beyond the plant canopy. Chemical pinching agents, such as Atrimmec and Off-Shoot-O, will affect the young shoots, but they cannot eliminate unwanted shoots or reduce the length of shoots. These pinching agents should be given consideration if manual pinching seems to be too big a task since, in just a few hours, plants can be chemically pinched, which could take weeks to do manually. The chemicals must be carefully prepared and applied, however, to provide the positive results desired and to avoid phytotoxicity. Rates and suggested time and method of application are indicated on the labels and should be followed.

Breaking flower bud dormancy. The characteristic of azaleas to reach a certain stage of flower bud development and then become dormant is protective, as it prevents flowering just as freezing temperatures arrive in the fall. It is a hurdle for azalea forcers, however, and cultivar selection, cool temperatures, or gibberellic acid are the only ways to clear it.

Cultivar selection is not the most viable option at this time. Hopefully, breeding programs will give us cultivars that do not become dormant. The difficulty with current cultivars that do not seem to go dormant is that they often flower when the forcer might not want them to be in flower, or only a few flowers may be open at one time. Irregular flowering is not a desirable trait when one is trying to produce high-quality florist azaleas.

The influence of temperature on the breaking of dormancy is discussed in the "Temperature" section. The need to provide some mechanism to provide cool temperatures to break dormancy can be costly and often limits the quantity of plants that can be forced into flower at one time. Cooler space becomes the limiting factor. Labor is also required to move plants into and out of the storage area.

Gibberellic acid. For 40 years we have known that gibberellic acid can be used to break azalea flower bud dormancy, but it wasn't until the summer of 1995 that the chemical received EPA label clearance for that purpose.

The following GibGro 4LS recommendations closely coincide with our research findings over a 30-year period.

> Differences in responsiveness may vary from one cultivar to another, or from one set of growing conditions to another, or from one cultural management system to another. Therefore, prior to widespread use, a small number of plants from each cultivar under a specific set of growing and management conditions should be tested to verify desired efficacy.

> Spray plants to run-off. The actual spray application rate will vary, depending on plant size and spacing density. Thorough spray coverage is essential for uniform flowering. Do not apply after flower buds show color.

> PARTIAL SUBSTITUTION OF COLD (Three Sprays):

> A representative spray application rate that has been proven effective for 6" (15 cm) potted plants spaced at a density of 1 per square foot is 1 gallon/200 square feet.

> GUIDE: Apply three sprays of 250 to 500 ppm a.i. At weekly intervals after three to four weeks of chilling.

> TOTAL SUBSTITUTION OF COLD:

> GUIDE: Apply four to six sprays at 1,000 ppm a.i. at weekly intervals. Plants must be at Stage 5 of floral development (style elongated and open) before first spray is applied.

Application rates and recommended water volume are shown in a table on the GibGro 4LS label.

PESTS

If one is only forcing the plants, there is a relatively short time when pests can become a problem, if one buys healthy, insect-free plants. There are some insects and diseases that everyone engaged in azalea production should know, however.

Insects. Infestations of insects are less likely to occur when one has azalea plants on the premises for only a few months, compared to the two-year period experienced in the past. Most liner producers do a conscientious job of controlling insects. These facts do not mean that insects are never a problem for the forcer, however, as they can attack the plants at any stage of development.

Azalea leafminer can be a serious pest. Ironically, it is a more serious pest when plants are grown in the greenhouse than when plants are outdoors, and it can be found in the greenhouse at any time of the year. Leaf tips turn brown, so the injury is very conspicuous, particularly when just about every leaf on every plant is injured.

Spider mites can be troublesome. Leaves that turn bronze in color frequently reveal that spider mites are present. Webbing also occurs, giving the plants a "trashy" appearance. Mites develop resistance to pesticides that are used too often, so the same form of miticide should not be used again and again. Syringing the foliage is one method some growers use to help control the pest, but know that some foliar diseases are spread by splashing water. Removing weeds near the area where the azaleas are grown should also reduce invasion by mites.

Azalea lace bug is very damaging on outdoor azaleas but is seldom encountered in the greenhouse. Heavy infestations can make foliage look white.

Diseases. The longer the crop duration, the greater the chance for exposure to disease, so the major risk is with the propagator and liner producer, not with the forcer. Azaleas are subject to some pathogens at almost any stage of development, and some disease organisms that strike during forcing need to be discussed.

Cylindrocladium blight and root rot can attack azalea plants from the propagation bench right through the forcing period. All parts of the plants seem to be subject to attack by the pathogen, *Cylindrocladium scoparium*. The disease progresses very quickly. Avoidance is the best policy, but appropriate fungicides can be used if the pathogen becomes established. One should try to avoid subjecting plants to splashing water, as spores from diseased plants are spread that way.

Phytophthora cinnamomi (azalea decline or little leaf disease) is a much slower-developing pathogen, but its consequences can be deadly. The organism is a water mold, so plants grown in media that are poorly drained can be likely victims. Clean stock and a satisfactory root media decrease the chances for the disease. There are helpful fungicides available, as well.

Powdery mildew can affect azalea foliage under certain environmental conditions. Cool storage temperatures and the high relative humidities recommended can be ideal conditions for the development of the powdery white growth on the leaves. Sulfur is one effective chemical; other fungicides are also effective.

Nematodes and azaleas once were almost synonyms in some regions of the country, but pathogen-free media and effective chemicals have lessened the occurrence and importance of nematodes.

Plants suspected of being afflicted with a disease should be diagnosed by qualified plant pathologists. Then appropriate control methods should be practiced.

Physiological disorders

Not all azalea problems are caused by insects or plant pathogens, and these disorders are termed *abiotic* or *physiological disorders*. In some instances the problem can be a consequence of some action taken or not taken months prior to the appearance of the disorder.

Failure of plants to flower. Perhaps the most frequent cause of failure to flower is not allowing sufficient time for flower bud initiation and early development after the plants have received their final pinch. Warm temperatures (at least 65F, 18C at night) are required to stimulate lateral shoot development, and the same temperatures are conducive for flower initiation. At least 14 weeks are needed from the pinch date to when dormancy-breaking treatments are started. Uneven flowering will occur if this schedule isn't followed, if flowering occurs at all.

Failure to flower or uneven flowering can also occur if dormancy breaking treatments are inadequate. The requirements to break dormancy were mentioned earlier.

Leaf drop. Water stress is a frequent cause of leaf drop, and this can occur at any stage of the crop. It can also occur if the light intensity in refrigerated storage is inadequate. Exposure to ethylene in coolers that previously had contained apples can cause leaf drop, but this disorder is quite cultivar dependent. High soluble salts, root rot, or any other cause of root damage can cause the leaves to fall off.

Bypass shoots. Shoots that surround the flower bud or flower are termed *bypass shoots*. These shoots, in sufficient numbers, could cause flower buds to abort. The bypass shoots can be so conspicuous that they detract from the beauty of the plant, even if they don't cause bud abortion. Bypass shoots will be very troublesome if too long a delay occurs from when the plants are pinched to when the dormancy-breaking treatments are started.

Manual removal of the bypass shoots was at one time the only way to handle the problem, as none of the growth regulators seemed to have an impact on the development of the shoots. Researchers at the University of Florida did show that 100 ppm of Bonzi, applied seven weeks before the plants went into the cooler, would cause the bypass shoots to be so short that they wouldn't be noticed.

Year-round flowering

The florist azalea can be timed with precision for flowering at any time of the year. Manipulation of temperature and day length is the key to scheduling. Refrigerated storage is necessary if one wishes to practice year-round flowering, and the ability to light the plants for long days or pull black cloth for short days makes the schedule more feasible. Day-length control was discussed in the section on light.

Azaleas for a niche market. Though azaleas have not been popular items in mass market outlets, the situation is changing, especially for plants grown in 4- to 6-inch (10- to 15-cm) pots. It is difficult for a medium-sized or small grower to compete with the large operations, so some unique products would be a logical solution. Braided azalea trees, shown in fig. 2, and pyramidal azaleas are being offered by some growers. Azalea hanging baskets are prized plants, requiring two to three years to complete, but one can go to the other extreme and produce mini-azaleas in pots ranging from two to four inches (5 to 10 cm) in size and requiring a production period of about eight months (fig. 5).

Care and handling. Azaleas sold at the right stage of development and given adequate water and light in the interior site should remain attractive for at least one month; under optimum conditions a longevity of six weeks is possible. Azaleas should be marketed when 25 to 30% of the flowers are open, unless a spectacular display of flowers is wanted for a special occasion. Satisfactory conditions would be a temperature not exceeding 70F (21C), a relative humidity of at least 50%, and avoidance of water stress. It is very hard for a consumer to rewet a media that has dried out, and it might be necessary for the consumer to submerge the root ball in a pail of water to moisten it.

Azaleas that are relatively hardy outdoors in the area where the flowering plants are sold give the customer the opportunity to enjoy the beauty of the plant indoors, then in the home landscape. Plants that flower during the winter months cannot be placed outdoors until the danger of frost is over, and consumers should be warned of the consequences of exposing the plants to harsh conditions.

FIG. 5. Mini azaleas are a low-priced alternative for consumers. The variety shown: Paloma.

ROSA (ROSE)

by Gary Pellett
Poulsen Roser Pacific, Inc.
Central Point, Oregon

Ron Ferguson and Keith Zary
Jackson & Perkins Roses
Bear Creek Gardens, Inc.
Medford, Oregon

❖ *Shrub (**Rosa** spp.). Vegetatively propagated. Growers begin crops using bought-in plants. Generally, plants are budded or grafted, except for mini roses grown as flowering pot plants.*

GROWING GARDEN ROSES FOR CUT FLOWER PRODUCTION

Roses grow in most areas of the world. The genus *Rosa* is widely adaptable and can be encouraged to grow almost anywhere. In climates as diverse as Germany, Holland, Italy, Kenya, Mexico, Brazil, California, and the Midwestern regions of the United States, there are thousands of acres of roses produced for cut flower sales (fig. 1).

There are, however, certain ideal conditions for growing garden roses as cut flowers. In general, the best quality is produced with cool nights coupled with sunny, warm days. Ideal spring, summer, and fall temperatures for outdoor rose production are minimum night temperatures in the range of 45 to 65F (7 to 18C) and maximum daytime temperatures in the range of 75 to 90F (24 to 32C). There are considerable variety differences in cold temperature response. Some varieties will still grow well as cold as 40F (4C) at night, while others need 50F (10C) or above to form well-shaped flower heads.

FIG. 1. Len Busch Roses, Plymouth, Minnesota, are substantial rose growers. Pat Busch (*right*) and grower Pat Etzel show off a sample of their wares.

The best climates for outdoor cut flower production are those with moderate to low humidity and which receive limited rainfall (less than 30 inches, 76 cm, yearly) during the growing season. The timing of rainfall is important. Winter rains while plants are dormant are not as problematic as summer rains while plants are in bloom. In the production regions named above, irrigation is generally required.

R

In areas with higher rainfall (greater than 30 inches, 76 cm, yearly), roses will also grow well, but special attention must be paid to control disease. Continual high humidity presents challenges in powdery mildew control. Still, persons evaluating production areas should keep in mind that significant outdoor rose production exists in such rain-prone areas as Germany, Holland, and Jamaica, to name a few.

In areas where strong winds are experienced during the growing season, provide windbreaks. Many simple systems are effective. Trees, bamboo lattice, snow fencing, and commercial plastic fabrics are commonly used.

In certain high-light and -temperature areas, flower and foliage quality are enhanced by producing under shade cloth. Generally, stem length is improved, flower bud size increases, flower colors fade less, foliage expansion (leaf size) is enhanced, and foliage gloss is improved. The grower should conduct trials prior to committing to one shade density for the entire plantation. Densities in use range from 15 to 50% shade. White shade has been used successfully in some warmer areas. Shade also can aid in screening out some insects. Some producers along the Mediterranean coast use shade cloth and/or plastic that can be pulled over the crop or removed, as climate conditions warrant.

Site selection. In most regions roses are best grown in full sun, although, as mentioned, certain areas warrant the use of shade cloth part of the year. Plants will tolerate some shading from trees or buildings, for instance, but best production can be obtained by having the plantation where plants will receive a full day of sun. If choice of site selection is between full morning sun or full afternoon sun, uninterrupted morning sun is preferred. Avoid planting roses close to any areas where trees, shrubs, or grass might compete for soil moisture or nutrition intended for roses.

In selecting a site for roses, make air circulation a prerequisite. Freely circulating air will promote the rapid drying of foliage and help control fungal diseases. If planting on a slope, locate roses toward the top of the rise. This will encourage free air movement. Low-lying areas can be more susceptible to frosts.

Additionally, consider locating the production site away from sources of dust, which might cause unsightly deposits on leaves or blooms. Although it is possible to rinse foliage, it is best to avoid extremely dusty sites.

Deer, rabbits, and other animals may love roses more than people do. If deer or rabbits are present, usually the only effective way to avoid plant damage is by fencing. The authors' experience with alternative controls is that they don't work over time. Although not absolutely required, electricity and telephone capability at the production site is helpful. In particular, electrical irrigation and spraying equipment are timesaving devices. Irrigation can be more easily automated when electricity is available. Depending on the size of the operation, a 220-volt, three-phase installation may be the most economical over time.

The site should have a good all-weather road that can accommodate entry by a semitrailer. Generally, such items as rose bushes and production supplies are often delivered by semitrailer, making access an important detail. The plans for marketing and selling should be considered when evaluating a potential growing site. Location and market access often determine whether it is to be a retail or a wholesale operation. If sales are to be direct to the public, location of the sales area is critical. Having the production site adjacent to the marketing site is often a plus, since it allows the customer to confirm visually that the product is locally grown. If the operation is oriented toward wholesale sales, the site location needs to allow direct delivery to the wholesaler by the grower or by an express freight handler.

The quality and quantity of available irrigation water is critical to success in areas where irrigation is required.

Soil preparation. Rule 1 of cut rose fertilization: Always use a soil analysis and recommendations from a laboratory experienced in floricultural crops to guide fertilization rates and practices. Also, know your water quality and adjust fertilization and irrigation accordingly. Rule 2 of cut rose fertilization: See Rule 1. Soil and water tests are key factors to success.

Roses will thrive in many types of soil, from pure sand to heavy clays, if soil preparation and irrigation is conducted properly. Prior to adding soil amendments, any supplemental drainage requirements, such as underground tiles or plastic drains, should be addressed. Ideal soil is deep, well-drained, and fertile.

If soil is sandy, gravelly, or with low water-holding capacity or cation exchange capacity, amendments should be chosen that improve water and nutrient retention. Peat moss, coco peat, fine bark, sawdust, many types of compost, and vermiculite would be beneficial in improving rose growth in these types of soils. If soil is high in clay or silt, generally, consider soil amendments that will improve soil aeration and drainage. Medium-ground pine bark, rice hulls, straw, coarsely ground corncobs, composted peanut hulls, coarse washed sand, perlite, and volcanic scoria are just a few of the materials used worldwide.

If possible, the soil should be worked to a depth of 18 to 24 inches (46 to 61 cm). In addition to any of the above soil amendments, well-composted manure can be considered as a soil amendment and fertilizer. It should be well composted and added in judicious amounts, since excessive amounts can and will cause problems, such as excessive soil salinity. Let your soil lab be your guide.

Another approach is to not grow in the soil at all but to force the roses in containers. In Italy some of the best outdoor roses in the world are grown by producers who plant two to three plants per 20-liter (5.3-gallon) container. The containers are filled with volcanic pumice, vermiculite, or a combination of both. Irrigation is accomplished by drip tubes. Containers are set out in rows on nursery ground cloth for weed control. The grower in an area of extremely low winter temperatures would need to consider additional winter protection for such a production system.

Always base a fertilizer program on soil analysis done by a competent soil laboratory with experience in floricultural soil fertility management. This will reveal potential problems that can be alleviated before planting. For example, roses grow best in soils that are slightly acidic; ideal soil pH for roses should run 6.2 to 6.5. The lab can also recommend preplant soil amendments and any fertilizers that should be incorporated prior to planting.

Good drainage is a necessity. If your soil does not have good drainage, consult your local county extension agent or soil specialist for recommendations. In many soils, ripping (deep cultivation) prior to planting improves root performance and plant health. If possible, attempt to cultivate 12 to 18 inches (30 to 46 cm) deep prior to planting roses. Drain tiles or pipe may also aid in improving drainage on difficult sites.

It may also be appropriate to consider soil fumigation. This largely depends upon what crops have been planted previously in the location and what fumigants are allowed in your area. Generally, broad-spectrum soil fumigants that control fungal diseases, weeds, soil insects, and nematodes are best for roses. Contact your extension service specialist or your rose specialist for details, since regulations vary with location.

Variety selection. Correct variety selection is a key factor in the success of an operation (table 1). Not just any garden rose variety will work, since the variety must be able to perform as a cut flower after harvest. This means the flowers must have acceptable vase life and shipability, attractive foliage, and acceptable color under commercial postharvest conditions. Some of the criteria for variety selection include demand for specific colors in your market, stem length, disease resistance, productivity, fragrance, and vase life.

The suggested color percentage of the plantation is only 35 to 50% red, which is significantly lower than greenhouse cut rose operations. Pinks, yellows, bicolors, and blends are all in demand. Since most greenhouse cut roses have little or no fragrance, growers should make fragrance an important criterion in variety selection. However, in general, extremely fragrant roses often have very soft petals, which can cause a less than ideal vase life.

Building rose plants for production. The following types of dormant rose plants can initiate an outdoor cut flower production program: two-year rose plants, started-eye rose plants (one year), June-budded rose plants, and bench-grafted plants. The larger two-year plants have more stored energy, start with more vigor, and will be quicker into cut flower production than the other, smaller plants.

For maximum performance, cut flower rose plants should be purchased from a specialist rose propagator. During purchase keep in mind that the propagator

TABLE 1   GARDEN ROSES FOR CUT FLOWERS

Variety	Color	Fragrance	Other
Red			
Legend	Nonfading R		Excellent form
Olympiad	Bright R	Little	Very long-stemmed hybrid tea
Oklahoma	Black-R	Yes	Hybrid tea
Tuxedo	Dark R	Some	Hybrid tea
American Pride	Dark R	None	Hybrid tea; good vase life
Kardinal	Excellent bright R		Long vase life; very resistant to mildew
Dallas	Bright R		Large, premium-quality blooms; good vase life; excellent mildew resistance
Black-red novelty			
Taboo	Black-R	Yes	Long-stemmed hybrid tea from Europe; very interesting; worth testing
White			
Honor	Clean W	Some	Large flower; very long stem
White Masterpiece	W	Some	Very large flower
Pristine	W with Pi tinge		Elegant
Sheer Bliss	W with Pi center	Yes	
Crystalline	W fringed with Pi		Hybrid tea; excellent vase life and shipability
Fountain Square	W with Pi tinge		Sport of Pristine; upright grower; straight stems
Pink			
Ivory Tower	Very light Pi	Yes	
Fragrant Memory	Medium Pi	Very much	
Touch of Class	Coral-Pi blend	Little	
First Prize	Deep rose-Pi	Little	
Diadem	Clear Pi		Spray rose from Germany; hardy outdoor var.; should be cut when buds barely show color; pinch out terminal bud when pea size to cause lateral buds to form sprays.
Apricot			
Brandy	Deep Ap	Some	
Medallion	Light Ap	Some	Very large blooms
Orange-red			
Tropicana	Light O	Some	
Fragrant Cloud	Light O-R	Extreme	Good substance
Marina	Bright O	Slight	Floribunda; good vase life and shipability
Artistry	Coral O		Very vigorous; long stems
Lavender			
Sterling Silver	Lav	Yes	
Blue Ribbon	Deep Lav	Yes	
Yellow			
Oregold	Deep lemon	Some	
Gold Medal	Golden Y bud	Some	
Sunbright	Bright Y	Some	
Magic Lantern	O-R to copper	Some	Sport of Gold Medal; same as GM except for deeper color
Blends			
Voodoo	Y-peach-O blend	Some	

Note: Most of these roses are long-stemmed hybrid teas, even when not explicitly described so.

R

must be licensed for the varieties being purchased. The plants should be produced under a virus-indexing program. Also, varieties should be budded on a suitable understock for the climate and soil of the region of production. Many understocks are used worldwide; a few to be considered are *Rosa Dr. Huey*, *R. manetti*, *R. indica major*, and *R. multiflora*.

Planting. Roses purchased for outdoor cut rose production should be planted in rows. The following spacing is commonly used in the industry:

Between rows	39 inches (100 cm)
Spacing between plants in row	8 to 12 inches (20 to 30 cm)
Plant density	14,000 to 20,000 per acre (35,000 to 50,000/ha)

Overall, the look will be one of a hedge of roses as one looks down a planted row. It is advisable to use a string line to ensure straight rows.

If the field is dry, it should be irrigated approximately one week prior to planting in order to have good soil in a workable condition but with good soil moisture.

High quality roses are produced in many planting scenarios—raised rows, raised beds, flat plantings, and even planting in the bottom of the furrow. Experience, soil type, drainage, and irrigation method all should be considered when choosing the planting method. One successful way of preparing the ground beds for planting is to simply trench or plow out a row of desired length in the prepared soil using a small tractor. If the soil is a little dry, the open row should be irrigated before planting.

When your roses arrive, open the boxes to inspect the contents and then store in a cool, shaded place or in refrigeration. It is best if roses are planted immediately. If the rose plants must be stored prior to planting, they should be stored at 34 to 36F (1 to 2C) and left sealed in their original shipping cartons. At these temperatures, dormant rose plants may be stored for two or three weeks. Do not store with fruits or vegetables or any other sources of ethylene.

Hydration and humidity. Many growers hydrate dormant plants by immersing them in water several hours prior to planting. Plant roses one at a time, ensuring the roots are spread out uniformly and fully covered with soil. *At all costs, avoid allowing the plants to dehydrate!* Plant the rose so the bud union is at or just above the soil line. Irrigate the planted roses immediately.

After planting, it is advisable (especially under dry and warm conditions) to mound pine bark around each plant or lightly cover the plants with clean straw to help protect the rose from dehydration, direct sun, or drying winds. After two or three weeks, this bark or straw can be spread out around the base of the plants to serve as a mulch.

To encourage maximum bud emergence, it is advisable to maintain high relative humidity around the newly planted roses. If the weather is rainy and cool, this will occur naturally. However, if drier conditions exist, the grower must syringe, sprinkle, or mist the plants briefly during the middle portion of the day. At this phase of starting the plants, it is critical that the grower take care not to overwater (waterlog) the soil. Too much water reduces soil oxygen, which is required for healthy root emergence and growth. The grower should focus on syringing the plants during the middle portion of the day to reduce stress until sufficient root emergence has occurred to support top growth. If there is any doubt, check the roots and soil moisture.

As is the case with most things, even overdoing syringing can have its down-side. Diminish syringing during the later portion of the afternoon in order to allow the plant foliage to dry off during the evening and night. If not, the grower is allowing conditions for rapid disease development. Once good root growth has begun, taper off midday syringing to allow further root development.

Directing growth. After the roses have been planted and have begun to grow, the grower begins to direct the growth in order to build up the basal shoots to form a chassis to carry future production. This building begins several weeks after the plants have sprouted. To ensure that roots have grown enough to compensate for vigorous top growth, the first flush or two of flowers should not be harvested.

As the eyes on the original canes force and begin to produce new canes, several options are available for the grower to build the plants for future production. The traditional method is to pinch back the new growth. As canes grow and form a flower head, cut back each new cane to the second five-leaflet leaf from above the point of shoot emergence. This can be done when the flower bud is about the size of a pea. This pinch will force one or two eyes below the pinch to break and force new canes. These stems can be harvested as cut flowers. In the case of smaller plants or varieties, the above procedure can be duplicated.

Irrigation. Roses require abundant irrigation to produce quality flower stems. If water quality is unknown, a water analysis will reveal potential problems. Once again, use only laboratories experienced in rose culture. In particular, excesses in sodium, boron, electrical conductivity, and pH in the irrigation water are of importance.

There are regions where commercial roses can be grown without irrigation, but generally speaking, it is necessary to be able to irrigate. There are a few main methods to accomplish irrigation under commercial conditions. In row-flood irrigation, areas between rows are flooded with water. This way uses more water than drip irrigation, but it has an advantage over sprinklers in that it does not wet the foliage. With drip irrigation, water is applied through a drip tube and emit-ted in limited quantities to the root system. Soluble fertilizer can be applied by injection. Using sprinkler (overhead) irrigation, water is applied uniformly to an

R

entire field area or via micro sprinklers in the bed areas only. As a rule of thumb, approximately 1 inch (2.5 cm) of water should be applied per week. It is an efficient irrigation method. However, the water quality must be excellent to avoid foliage residue, for best growth, and to enhance foliage quality, generally. Acidified water would not be desirable.

As always, the penetration of the irrigation water should be checked by using a soil probe. The soil should be wet to a depth of 12 to 18 inches (30 to 46 cm). Winter soil moisture is also important. Plants should not be allowed to dehydrate in the winter.

Fertilization. Generally speaking, roses require moderate to high fertilization levels. The best pH range for roses is 6.2 to 6.5. Soil salts should be kept in the moderate to low range for best yield. All fertilizer programs need to be based on soil analysis by a qualified laboratory.

Several methods or combinations of methods can be used to fertilize rose plants for cut flower production. There are many kinds of organic fertilizers. They generally need to be tilled into the soil and become slowly available. Composted manure is an excellent fertilizer, soil amendment, and source of minor elements.

In general, established roses require dry granular fertilization in the spring, in late June, and in mid-August. As always, the amount and type should be based upon a soil analysis. However, as a general rule, each fertilizer application using a dry granular fertilizer should amount to about 1 to 2 ounces (28–57 g) per plant per feeding of 10-10-10 or 10-15-10 fertilizer. This fertilizer is put into a band about 6 to 8 inches (15 to 20 cm) from the base of each plant. Simply bury it below the soil surface and irrigate. Minor elements can also be applied, as indicated in the soil analysis, by using dry fertilizers.

Constant liquid feed is a very efficient method of fertilization. Drip systems, in addition to reducing water usage, have the advantage of allowing fertilizer to be injected into the irrigation lines with the objective of giving a small amount of fertilizer with each irrigation. Fertilizer injection should be discussed with the irrigation supplier when the installation is being designed. If feeding through a drip system, most growers will feed at 100 to 150 ppm N-P-K for two irrigations, then irrigate with clear water, then repeat.

Disease and insect control. The types of diseases and insects that will attack a commercial rose plantation vary due to location, time of year, adjacent crops, and weather. The most common diseases of outdoor cut roses are powdery mildew *(Sphaerotheca pannosa)*, downy mildew *(Peronospora sparsa)*, Botrytis *(Botrytis spp.)*, root rot *(Pythium, Phytophthora, Fusarium, or Rhizoctonia)*, and nematodes *(Meloidogyne* spp. principally*)*. The *Compendium of Rose Diseases* (APS Press, The American Phytopathological Society, 3340 Pilot Knob Road, St. Paul, MN 55121) offers complete descriptions and color photos.

The most common insect pests of outdoor cut roses are red spider mites, aphids, thrips, and *Lepidoptera* larva.

There is an array of products from organic to inorganic to be used in pest and disease control. Due to local or federal regulations, a review of pest control options should be undertaken with the local or regional extension horticulture specialist or local chemical supplier. One note of caution: Many products cause foliar phytoxicity on roses. If a producer is not experienced with the reaction of the particular variety to a spray, a thorough testing must be made to determine if there will be problems.

Cutting and pruning. Most garden roses used for cut flower production should be cut when the calyx loosens and reflexes and the outer petals have begun to unfold. Generally, flowers are cut at the second five-leaflet leaf above the previous cut, slightly above the node (point of leaf attachment). If extra length is desired, stems may be undercut (below the last cut) or cut to the first five-leaflet leaf. Undercutting must be done infrequently, or not enough foliage would be left on the plant to ensure rapid and productive growth of the next crop.

Most commercial operations use "cut and hold" shears, which allow the individual harvester to carry the harvested stems in one arm while cutting with the other.

Pruning is generally done in late winter. For most varieties, prune to strong canes about 18 inches (46 cm) above the bud union.

Postharvest care and grading. Roses are generally cut morning and evening. This practice may not be necessary in field production during cooler times of the production cycle, but during the warmest times of the year, it may be necessary if best-quality blooms are to be harvested.

Roses, when cut from the plant, should be placed immediately in a hydrating solution, such as water acidified (using citric acid) to a pH of 3.5. For best results, the stems should be put in buckets in the field. The buckets should be collected and moved to a refrigerator unit at 34F (1C) to continue the hydration process. After a few hours, when the stems are again full of water, they can be graded, bunched, and then returned to cold storage. Buckets used for harvest and cold storage should be clean. After grading, water used in buckets for storing roses should be treated with chlorine to keep bacteria from developing. Using chlorine at the rate of 100 ppm will eliminate bacteria from rose buckets. Sanitation is important to good flower quality.

R

Roses generally are graded according to length of stem. Most growers grade in increments of 4 inches (10 cm). Standard grades are:

- Short   10 to 14 inches (25 to 36 cm)
- Medium   14 to 18 (36 to 46)
- Long   18 to 22 (46 to 56)
- Extra long   22 to 26 (56 to 66)
- Fancy   26 to 30 (66 to 76)
- Extra fancy   30 or more (76 up)

All broken, crooked, or bent stems are culled, as are those roses with poor head shape, bent necks, or disease.

Packing and shipping. After grading, most cut roses are bunched in bundles of 12 to 25 stems. Roses are often bunched in a round pack, then wrapped with a commercial plastic or cellophane product. For large-headed varieties, many growers use a spiral pack, and others nest roses, so that 12 flowers are arranged below the other 13. The wrap must never be tight, as roses grow during storage and shipping. If packed too tight, the growing heads will actually bruise each other or snap off. Also, extend the wrap two inches (5 cm) above the flower heads in the bunch in order to help avoid bruising the tops of the flowers. Label all bunches by grade and variety. After wrapping, stems are usually tied together with rubber bands, string, or twist ties.

The grower close to the final market often delivers in water and refrigerated trucks. If selling at roadside markets or through local florists, the grower should stress that the flowers are locally produced and that they are garden roses. If being shipped long distances, roses are packed in insulated, standard-size boxes, cleated in, and often iced. Express overnight service, bus lines, air freight, or refrigerated trucks are common modes of transport.

Information sources. Frequently, we are asked to recommend sources of information for cut flower producers. While the industry has a shortage of practical information on cut rose production for commercial use, following are two excellent books:

- *Roses: A Manual of Greenhouse Rose Production,* Roses Inc., P.O. Box 99, Haslett, MI 48840.

- *A Compendium of Rose Diseases*, The American Phytopathological Society, 3340 Pilot Knob Road, St. Paul, MN 55121.

Additionally, for guidance on some of the basics of the greenhouse and horticulture industry, there are several useful books.

- *The Greenhouse Environment*, John Wiley & Sons, 605 Third Avenue, New York, NY 10016.

- *The Farm Chemical Handbook*, Meister Publishing, Willoughby, OH 44094.

- *The Commercial Greenhouse*, Delmar Press, Albany, NY 12205.

- *Commercial Greenhouse Operation & Management*, Reston Publishing Co., Reston, VA 22090.

It is highly recommended that all rose growers join Roses Inc., a trade association for greenhouse growers of fresh cut roses: Roses Inc., 1152 Haslett Road, PLO Box 99, Haslett, MI 48840, phone 800/968-7673, 517/339-9544; fax 516/339-3760. It has an excellent bulletin on *Care and Handling of Fresh Cut Roses.* Roses Inc., also conducts very educational annual meetings and tours and distributes rose research information to its members.

CUT ROSES IN THE GREENHOUSE

Much of the preceding information for growing garden roses as cut flowers also applies to traditional greenhouse rose culture in ground beds.

Prepare soil based on soil analysis from a soil-testing laboratory familiar with floriculture production. Sterilize or pasteurize soil prior to planting. Adjust soil pH to 6.0 to 6.5.

After planting, pinch plants at four weeks. Depending on light levels, variety, and temperature, it is sometimes recommended to pinch a second time five to six weeks after the first pinch. Time to flower from the pinch depends on the variety and the time of year. Generally allow 42 to 56 days, approximately six to eight weeks.

The total time to cutting first flowers from planting can vary from 70 to 90 days (10 to 13 weeks) in a one-pinch program to 108 to 127 days (15 to 18 weeks) in a two-pinch program.

Building the plants. One way to build up newly planted roses, or to rebuild reserves in established plants, is a method known as deheading and deshooting. The technique is labor-intensive, but it does develop basal shoots, which aid in the future production of cut flowers.

When growers dehead, they generally follow one of two options. They remove all developed flower heads that show color by snapping them off immediately below the flower bud (i.e., at the peduncle). Growers should dehead all stems that reach this phase during the program. Alternatively, they can allow all flowers in a bed to completely flower out and drop their petals. At this time the flower head is usually removed as described above, snapping flowers off at the base (at the peduncle).

ONE DUTCH ROSE GROWER STEPS OFF THE "MORE PRODUCTION IS BETTER" TREADMILL

Do you want to make more money? Then increase production volume through new varieties, higher efficiency, or expansion. So has gone the thinking among Dutch rose growers for years.

But in the end it is not how many stems you produce that matters; it is how much money you keep by the end of the day. Unfortunately, for too many growers, the only clear road they see leading to higher profits is more and more production volume. For these growers, volume is equal to success.

Dutch rose grower Bill Steenks is gambling that he'll be able to earn more money by growing fewer stems of better quality and earning more per stem. Ironically, his motivation to change his production comes from low market prices caused by the "more production is better" thinking. "If we grow roses and make a lot of money, then there is no impetus to change. But when prices are low, then you look at how you can change to make more money," he says.

Breaking traditions. Steenks Rozen's 5-acre (2-ha) glasshouse in Baarlo on the Netherlands' eastern border is a $2.9 million (5.5 million Dfl) investment to pursue his theory. The range, turnkey glass with 10-foot- (3-m-) tall gutters, overhead curtain systems, and total energy and supplemental lighting, also uses the knotwilleke production system (bent stem technique). The technique bends lateral stems down to increase the amount of leaves making a photosynthetic base for the rose plant. Because greenhouses using the technique have aisles filled with foliage, it looks more like "jungle" culture than the standard prim, woody look of typical rose growing (fig. 2).

How many stems should you bend down? There are no set guidelines. Stems with buds that look like they will be of poor quality are bent down, for example. Bill says there must be a balance: bending down too many leaves will take energy away from flowering.

Bill is also breaking with tradition by growing plants in gutters on raised benches and using coco medium as his substrate. The guideline he uses for knowing how many stems to bend down: to have enough foliage covering aisles and the space between rows of plants so that it is totally dark under the growing bench.

FIG. 2. Many California cut rose producers are switching from ground beds to hydroponics systems using the bent stem technique from Holland. The greenhouse is Dramm & Echter, Encinitas, California. The variety shown: Kardinal.

Production. The new 2-hectare range has 160,000 yellow rose plants. About one-fourth are Frisco, and the rest are split between Surprise (it changes from apricot orange to rose), Eskimo (white), and Sasha (red). Anticipated annual production is 8 million stems. While Frisco is the leading high producer in the Netherlands and the rose most people think of when they think of high-volume Dutch rose production, Bill is taking a different tack with it. "With Frisco you can get 500 to 600 stems per meter^2 per year [465 to 557 per 10 ft^2] and sell them for 0.27 Dfl ($0.146) or 0.28 Dfl ($0.151) per stem. I will do it at 400 stems per meter^2 per year [372 per 10 ft^2] and get 0.45 Dfl ($0.243) or 0.50 (Dfl) ($0.270) per stem because my roses are high quality.

"Until now the 500- to 600-stem Frisco growers were making a good profit. There was a big market for that type of product at that price. But now production levels are really high, so prices will go down even more."

Bill says that the "more production is better" thinking is prevalent in the Netherlands because of the auction system. "Thinking in yields/numbers/kgs is because of the auction. But there comes a point when that thinking goes wrong. You must produce something the market will pay *more* for. That's how you serve the market, not by dumping."

Culture. Bill's rose culture is on raised benches, all reflective white metal. Rose plants grow in plastic bags of coco fiber—six plants per bag, and eight per square meter (7.4 per 10 ft^2).

The coco fiber has a lower pH, about 4.8 to 5.3, which is one of the main reasons Bill thinks they are growing so well. As many growers have learned, rock wool culture of roses has a pH of 5.5 to 6.3—at a lower pH, the rock wool would break down. Irrigation is drip, and because the coco bags sit in troughs on benches, water is recirculated easily. Supplemental lights can supply 4,000 watts per square meter (372/ft^2) for 18 hours a day.

All plants at Steenks Rozen are on natal brier rootstock. In Bill's own trials, he found that natal brier gives 15 to 20% higher yields, better quality, and longer stems. "The rootstock area is important for the future of all rose production—to experiment with new rootstocks to get longer, better quality stems." He adds, "Natal brier is doing well, but if we really begin to select and trial rootstock, it can go higher."

Competing against imports. When questioned about rose imports into the Netherlands, he holds strong views, atypical for a Dutch grower. "You cannot stop the imports. . . . Everyone is growing roses," he says. "The way to fight for your market is to produce better roses or cheaper roses, but you cannot simply close the market.

"I think what will happen in the Netherlands is that if we want to survive [competition from Africa], we must get a brand name, and the best growers must do that together." Vase life, flower opening, and scent: these are the marketing angles to pursue.

Deshooting follows deheading. Rose plants will respond to deheading by initiating numerous lateral shoots. These new lateral shoots need to be removed as they develop. This usually involves two to three passes per week through the growing bed. The program will not work unless this program is performed continuously. Normally, the only shoots removed are those that initiate on what would have been cut flower stems. Do not remove lower or ground shoots during this operation, since the purpose of the program, after all, is to encourage basal development. Dehead and deshoot for six to eight weeks. Afterwards remove upper stems as if they were cut flower stems.

Troubleshooting leaf drop. There are numerous reasons why rose plants may suddenly drop old leaves. Most of these are environment and medium related.

Common atmospheric characteristics inducing leaf drop include excessive humidity; changes in temperature, especially low night temperatures; and changes in light level, especially overcast or other low-light conditions. A rose plant may have leaf drop from ethylene exposure or may suffer phytotoxicity from a pesticide or from sulfur burning.

Undesirable medium qualities, such as excess salinity or a low nitrogen level, can also be responsible for leaf drop. Low soil pH can cause manganese or aluminum toxicity or magnesium or calcium deficiency, but high pH can also be very unfavorable. If the problem seems to be high bicarbonates in the soil or irrigation water, use water with less than 122 ppm (2 meq/l) of bicarbonates or acidify the irrigation water.

A rose plant could lose roots for various reasons, which might lead to leaf drop. Nematodes may be causing trouble. Downy mildew is a more noticeable problem. Finally, rose leaves do not live forever, so there could be leaf senescence from simple old age.

Spray rose pointers. Spray roses have been an exciting addition to cut rose production the past few years. Because sprays are used heavily in bouquets and florists' garden arrangements, the color mix is heavy in pinks and pastels (Evelien, Porcelina, Princess).

Research at Jackson & Perkins, Bear Creek Gardens, offers the following growing tips for spray roses in the greenhouse. Keep plants low. The best quality and production occur when plants are maintained under 48 inches (120 cm) to the top of the crop. Accomplish this by cutting the harvest hard at the first five-leaf or lower. Plants are generally quick into production, especially from a started eye. It is not necessary to build the plant a long time. Most of the varieties are relatively short-cycled.

Remove the central bud when it reaches pea size. Check plants twice weekly for this. Cutting can begin when the second flower begins to crack open. Excessive lateral branches (more than five to six laterals) should be removed early enough so as not to leave large scars on stems and to promote more uniform flowering in the remaining buds.

POT ROSES

Pot roses have been grown for decades as a commercial crop (fig. 3). Perhaps the first large-scale production of pot roses were the grafted varieties of Mother's Day roses that were potted and pinched for holiday sales.

Growers and consumers look to varieties with ease of culture, high bloom counts, good colors, flowers with classical rosebud shape, year-round availability, and above all, excellent shelf life (table 2). There are a number of rose hybridizers that specialize in commercial pot rose varieties worldwide; the principal hybridizing companies are Poulsen, DeRuiter, Meilland, and Jackson & Perkins. Currently, the Poulsen varieties dominate the North American and European industry.

As varieties and cultural techniques have improved, more pot roses are being forced from plants produced from cuttings instead of budded stock. Additionally, varieties for larger pot sizes are now being offered on the market as own-root products.

A tremendous expansion in pot rose production has occurred worldwide in the past five to 10 years. The four principal producing countries, ranked in order of units, are Denmark, Holland, United States, and Canada. It is estimated that there are approximately 45 million pot roses produced yearly in these four leading countries.

FIG. 3. Mini roses, even in tiny 2- and 3-inch pots have become the rage for supermarket sales. These plants, branded Pinocchio, were grown by the nursery Gartneriet Rosa, Denmark, and are packaged to sell two pots at a time.

Improved varieties and culture. The meteoric increase in pot rose production and demand has come about from several factors. With uniform cultural

R

TABLE 2   SUGGESTED POT ROSE COLOR MIXES

Sales season	Red %	White %	Yellow %	Pink %	Coral, Salmon %
General	20–25	15–20	15	35	10–15
Christmas	75–85	15–25			
Valentine's Day	50	10–15		20–30	10–15
Easter and Secretary's Day	20–30	15–25	10–15	15–25	10–15
Mother's Day	40	10–15	15–20	20–25	10–15

requirements, bloom counts, and improved shelf life, the varieties are highly specialized and the result of intense selection by hybridizers. Just as important is the greater knowledge that research and production specialists have garnered to understand the reactions of the rose plant under certain conditions.

The last decade has seen great improvements in greenhouses, environmental controls, and automated equipment, allowing exact environmental control and precise timing of cultural activities. Among these improvements have been exact temperature and humidity control, high-pressure fog for improved rooting and ebb-and-flood irrigation. HID lights allow year-round production. There is now automated spacing equipment, as well as automated shearing equipment for mechanized pinching. Growth regulators now are available to effectively manage the finished height and greatly improve the foliage quality.

Market and consumer interest. Researchers in both Europe and North America have worked to understand the rose's requirements in the marketing and consumer environment. Demand has shifted from spring- and holiday-dominated sales to an almost year-round program. The consumer can enjoy good duration of flowering in the home, and later replant it into the garden almost year-round. Few floral products offer this advantage, which will aid in further increasing demand.

The combination of all these factors—specialized varieties, better culture, and versatility in usage—has led to pot roses being available to a wider consumer audience throughout the entire year. As retailers, buyers, and consumers have more positive experiences with pot roses, demand will continue to increase.

Market demand in the United States and Canada is principally in 4- and 6-inch (10- and 15-cm) pots, which parallels other greenhouse potted flowering crops. Currently, the principal channel in distribution is through supermarkets and chain stores, although as more 6-inch products are on the market, florist use is expanding. Demand for colors seems to be about 50% red and 50% other colors. Buyers look for pot roses with high bloom counts and abundant lateral sprays. The high flower counts also allow retailers or growers to remove any spent blooms while the pots continue to flower in the retail environment.

Production overview. Pot roses are perhaps the most demanding flowering pot plant to produce (table 3). Propagation of roses requires specialized equipment and knowledge, as well as a steady supply of high-quality cuttings. Required equipment includes HID lighting, a fog propagation system, mechanized mowing equipment, bottom and top heat, and modern environmental controls.

Because of the specialization and precision needed, it is recommended that growers chose a specialist propagator that offers rooted liners for potting

TABLE 3 GROWER'S GUIDE TO FINISHING POT ROSES

Item	Transplant pot size[a] inches	cm	Shear[c] week[d]	Finish Target season[f]	wk[d]	Spacing inches	cm	Growth time After fin. spacing	Total wks[d]
40-cavity liner	4	10	2	Winter	5	6½	17	4–6	9–11
				Spring	3			4–5	7–8
32-cavity liner	5–6	13–15	3	Winter	6	10–12	25–30	6	12[g]
				Spring	4			6	10[g]
14-inch (36-cm) stem-grafted tree rose	6	15	4[e]	Winter	12	10	25	6	18[h]
				Spring	8			6	14[h]
4-inch (10-cm) prefinished pot	N.a.[b]		Don't	Winter	2	6½	17	4	6
				Spring	1			4	5

[a] In all cases, one transplant per pot. Also, the initial spacing is equal to the pot size.
[b] Not applicable.
[c] To promote branching, cut back plant to 2-3 inches (5-8 cm) above the pot top.
[d] After potting.
[e] A grafted tree rose is a short-cycled product that is sheared lightly once prior to delivery to grower.
[f] Winter finish time refers to crop finishing up to and including Easter. Spring finish time applies to any crop from Mother's Day through November 1. Use spring finish time year-round if plants are grown under HID lights with at least 450 footcandles (5 klux).
[g] Grow out times for 32-cell roses and hanging roses are estimates only.
[h] Production times for 10-inch (25-cm) miniature tree roses are based on grow out at contract growing facilities in Salt Lake City, Utah.

up. Pot roses are also available as prefinished 4-inch (10-cm) material. This possibility is especially interesting for the grower with a tight schedule or without good winter light.

Miniature pot rose culture. Forcing of pot roses from rooted liners or prefinished material requires focus on certain cultural requirements.

Media: Most well-drained, commercial peat-perlite media mixes are acceptable. Sterilization is suggested. Optimum pH range is 5.5 to 6.7. European-style mixes use a high percentage of peat moss.

Place the liner at its soil level at transplant. Use a postplant preventive fungicide drench (Banrot or Chipco/Subdue).

Light: Use unshaded greenhouses in winter in all areas. Glass is best. Except for high-light areas, use unshaded greenhouses in fall and spring. HID lighting is highly recommended for Valentine's Day crops grown from liners for most of the United States and Canada. Pot rose specialists use HID year-round to obtain consistently high quality.

In most cases, use shade in summer months to produce the best foliage and flower quality. Shade young plants when light exceeds 3,000 footcandles (32

R

klux). From Week 4 or 5 until finish, shade plants when radiation exceeds 4,100 f.c. (44 klux).

Temperature and humidity: Ideal temperatures are 60 to 65F (16 to 18C) nights and 70 to 75F (21 to 24C) days, although pot roses will grow well at 80F (27C). Start the crop at the higher range of temperatures and gradually reduce temperatures approximately two weeks after final spacing. This will increase crop time, but it will improve shelf life. Avoid wide fluctuations in temperature throughout the entire crop.

For best mildew and *Botrytis* control, use overhead steam or hot water pipes for heating. Horizontal airflow is beneficial in the control of *Botrytis* and powdery mildew.

Never allow pot roses to dry or to stand in water for extended time periods. Dehumidify the greenhouse, if necessary, to keep night humidity below 90 percent.

Spacing: For 4-inch (10-cm) pot production, space pot-tight until pinch, then space on 6- to 7-inch (15- to 18-cm) centers, yielding three to four plants per square foot (31 to 42 plants/m^2). If a grower notices excessive leaf yellowing on lower leaves at finish time, the timing of the initial spacing should be moved up.

Fertilization and salinity: Use a continuous-fertilization program. Mini pot roses are heavy feeders. In a constant liquid feed program, nitrogen should be approximately 160 to 220 parts per million (ppm), phosphate (P$_2$O$_5$) at 60 ppm, and potash (K$_2$O) at 160 ppm, plus trace elements. As with all potted crops, manage salinity carefully. The EC range should be 1.7 to 2.1 millimhos/cm.

Growth regulators: Depending upon the variety and pot size, plant quality is normally enhanced by a growth regulator. Bonzi (paclobutrazol), where registered, is the most commonly used.

Insects and diseases. The major insect pests include aphids, spider mites, and western flower thrips. Control aphids with Orthene, mites with Pentac or Avid, and western flower thrips with Avid, Mavrik, and Orthene.

The major pot rose diseases are powdery mildew, downy mildew, and *Botrytis*. For control, keep foliage dry and maintain proper night temperature and humidity (below 90 to 95%). If powdery mildew occurs, control with Chipco, triforine, or sulfur burners. Downy mildew generally is a potential problem in the spring and fall.

Specific chemicals or brand names are mentioned here (and elsewhere) for information only. Follow current label directions for use and application rates. The legal use of all chemicals varies by region.

GROWING CONTAINER ROSES FOR GARDEN CENTER SALES

The use of bare-root rose plants for planting into containers is at an all time high (fig. 4). Many nurseries around the country have been producing container roses for years; however, with increased demand and interest, some are growing container roses for the first time.

Generally speaking, the following types of dormant rose plants are used to containerize and force into bloom for garden center sales:

* Two-year-old budded rose plants. Common understocks in North America are *R. Dr. Huey* and *R. multiflora*.

* One-year, own-root, field-grown rose plants.

* Rooted rose liners.

* Tree (standard) roses, budded on understock. They vary from one to three years in age.

There are a multitude of types of roses, generally classified by their growth character and/or the flower type: hybrid teas, floribundas, hedge roses, climbers, ground cover roses, and more. The grower should review with potential customers their demands and place orders early for planting stock. Orders should generally be placed during the prior spring or summer for best availability.

Most plants in North America are grown in California, Arizona, and Texas. Harvests begin as early as October and continue through January, although timing can be subject to weather. Be sure that the supplier is offering virus-indexed material from fumigated fields. Due to ongoing programs by the various state agricultural departments, growers should anticipate that either they or their customers will be inspected for rose virus. Additionally, virus-clean plants will grow with more vigor and better performance.

Preplant handling and preparation. Upon receiving your roses, open the cartons to inspect the condition of the plants.

Storage of dormant plants can present a challenge. Often a grower doesn't have proper refrigeration facilities or sufficient space for holding rosebushes for extended periods. Dormant rose plants are extremely sensitive to ethylene, which will cause a delayed start or even death to bare-root rose plants. The gas penetrates almost instantly through the boxes and plastic, so don't think that sealed boxes are a protection from the gas. Refrigerator facilities should be ventilated prior to use and not contain any fruit or vegetables while roses are in storage. Optimum storage temperature for dormant rose bushes is 34F (1C).

If you are going to place the planted containers directly into a warm environment (70F, 21C), it is recommended to "sweat" the plants before planting. This is accomplished by first opening the box, spraying plain water over the plants in the carton, then closing the plastic liner as tightly as possible and closing the box. Stack the closed cartons in storage at 60 to 70F (16 to 21C). The high humidity and warm temperature will help the plants begin to force. Check plants daily until the eyes have forced shoots to one-eighth to one-fourth inch (3 to 6 mm) in length.

R

When the containerized plants are to be placed directly into a cold growing environment (40 F, 5C), it is best not to sweat the plants. This way the plants can initiate rooting at a slower rate, prior to onset of top growth. Frost control may be required. Sprinkling with plain water while temperatures are below freezing is effective. Also, a heat source or covering with a blanket will aid in frost protection.

Prior to planting, it is recommended to soak the roots overnight to hydrate the plants.

A wide range of container sizes are used; however, 2-gallon and 3-gallon sizes are the most common. To reduce plant desiccation, the canes should be pruned to 6 inches (15 cm) above the bud union. If you have sweated the eyes out, cut the canes after the plants are in the containers. Otherwise, the eyes would be broken off during handling, and the plants would have more chance of drying out. Spray a fungicide over the freshly cut canes as soon as possible. Chipco 26019 (iprodione) has been shown to be effective in preventing cane dieback from *Botrytis*. Root pruning is not recommended, unless root size does not allow the plant to fit into container.

Remember: it is extremely important not to allow the plants to dry out during the planting operation!

Container media and planting. The planting mix should be well drained and contain a high amount of organic matter (approximately 80%). Calcium and phosphorus should be added to the soil mix before potting. Use superphosphate 0-45-0 at 1 to 2 pounds per cubic yard (593 to 1187 g/m³). Depending on the pH, add limestone at 1 to 2 pounds per cubic yard or gypsum at 1 pound per cubic yard (limestone will raise pH; gypsum has no effect on pH). The ideal pH for roses is 6.0 to 6.5.

There are many formulas or mixes for growing media. Often, the mix used is determined by the raw materials that are available locally and desirable because of low prices.

Composted bark will lighten the mix. Other composts can be used, but they should be trialed prior to commercial use. Also, check pH and salts.

When planting the rose, the bud union (swollen area) should be at or slightly above the soil line so that water can collect and soak into the soil. Do not overfill the container.

The canned roses are now ready to place into the growing area. Since many nurseries are using hoop houses, this discussion is in accord with this technique. Spacing the plants is done as either final or pot to pot. The best plants will be produced on open spacing with 12- to 16-inch (30- to 40-cm) centers.

Water plants in thoroughly. Watering by hand twice is usually best to ensure full saturation of the entire container. If labor is not available, overhead sprinkling will have to be used. Overhead sprinkling should operate until water runs from the weep holes in the bottoms of the containers. After the plants are watered in and the soil is wet, apply a soil drench using Subdue 2E. A spray of Chipco 26019 over the fresh-cut canes will help prevent disease.

Growing on. Initially, humidity should be kept high during the day. This can be accomplished by periodic spraying. Usually, this is started in midmorning and stopped by 3:00 to 4:00 P.M. This is very important and can make the difference in how well the plants break and leaf out. Always be alert for hot sunny days following a cloudy spell, when temperatures can climb rapidly, damaging the tender new foliage. Proper venting and humidity control is the solution.

FIG. 4. Thousands of retail growers and garden centers purchase dormant bareroot roses and pot them up for growing on. Home gardeners love them. The grower is Al's Fruit and Shrub, Woodburn, Oregon.

Media should not be allowed to dry out during the first two weeks of the program. Irrigation during the breaking-out stage should be at least weekly, even if spraying is being done for humidity. Fertilizer can be applied in the second to third week. New hair roots are sensitive to excessive fertilizer. A dry or liquid application of 20-0-20 will not be too hot, yet it will allow enough nitrogen to stimulate growth. If applying by hand, a tablespoon of dry fertilizer per pot will work. Liquid can be applied as a drench or foliar overhead spray. The soil should be wet before applying either form.

Many growers use slow-release fertilizers. Common formulations are 18-9-9, 14-14-14, and 19-6-12. With this type of feed, nutrients will be continuously available to the plant. It is suggested that micronutrients be added to the fertilizer program to improve overall quality. Normally, micronutrient application commences at Week 4.

Problems. Failure to leaf out is the most common problem with canned roses. It is generally due to desiccation or drying out of the canes and the roots during planting or during the initial weeks after planting. Following the outlined procedures for handling and planting helps to eliminate this problem.

As the plants leaf out, use preventative sprays and environmental management to control diseases and pests. The most important foliar diseases of container roses are downy mildew *(Peronospora sparsa)*, powdery mildew *(Sphaerotheca pannosa)*, and *Botrytis (Botrytis cinerea)*. Begin spraying weekly with Dithane M-45 and Alliete when plants have foliage. The most common insect pests on container roses are aphids *(Macrosiphum* spp.*)*, worms *(Lepidoptera* spp.), thrips *(Frankliniella* spp.*)* and red spider mites *(Tetranychus* spp.*)*.

RUDBECKIA (GLORIOSA DAISY, BLACK-EYED SUSAN)

by Jim Nau
Ball Horticultural Company
West Chicago, Illinois

❖ *Perennial* (**Rudbeckia hirta**). *25,000 to 80,000 seeds/oz (882 to 2,822/g). Germinates in five to 10 days at 70F (21C). The seed can be lightly covered during germination.*

Rudbeckias are excellent perennials for well-drained areas in full-sun locations (fig. 1). Gloriosa daisies are often treated as annuals in the Midwest or in any area where the autumn can be cold and wet. They do not appreciate overly moist conditions, and the dwarfer varieties often die from foliar diseases brought on by excessive moisture in combination with high humidity, such as powdery mildew. Plants flower readily from seed the first season, and varieties like Marmalade, Goldilocks, Becky Mix, and Toto will even flower well in pots or packs.

FIG. 1. Rudbeckia is a reliable perennial—producing hundreds of bright golden yellow flowers during summer months.

For flowering 4-inch (10-cm) pots in May, sow in January and use one plant per pot. For the taller varieties, allow 11 to 15 weeks for green packs.

Among varieties, both Marmalade and Goldilocks are dwarf selections to only 15 inches (38 cm) tall in the garden. Marmalade does have problems with powdery mildew, as noted above, but provides color from planting until August. Flowers are single, to 3 inches (8 cm) across, in a golden yellow to light orange color.

Goldilocks, flowering 10 days earlier, is a semidouble- to double-flowering variety to 12 inches (30 cm) tall in the garden. Flowers range from 3 to 4 inches (8 to 10 cm) across, and the variety is excellent in packs or pots.

Becky Mix is a simple mix with two predominant flower colors—orange and orange-red. Occasionally, a golden yellow color pops up. Becky is a pot plant variety going only 10 inches (25 cm) tall in the pot, 12 to 14 inches (30 to 36 cm) tall in the garden. Flowers are single and 2½ to 3½ inches (6 to 9 cm) across.

Indian Summer is an All-America Selections award winner with huge, 5- to 9-inch (13- to 23-cm), single to semidouble, golden orange flowers on plants 30 to 36 inches (76 to 91 cm) tall.

In the taller varieties, Double Gold is of particular value for cutting or as a background plant in the garden. Double Gold has double flowers up to 6 inches (15 cm) across in a stable color of golden orange.

SAINTPAULIA (AFRICAN VIOLET)

by Arnold W. Fischer
Arnold Fischer Greenhouses
Fallbrook, California

❖ *Annual (***Saintpaulia ionantha***). Vegetatively propagated from leaf cuttings. Large growers self-propagate, smaller growers purchase liners from specialists.*

African violets never go out of style. These long-lasting houseplants can be cropped year-round. The African violet crop can be timed and, grown efficiently, brings a good return.

Varieties differ in colors, shapes, and sizes. Most mass production varieties are compact and miniature, with fast, uniform growing habits and good flower shows (fig. 1). Standard varieties are generally compact for 4-inch (10-cm) pots. Miniature varieties are designed for 2-inch (5-cm) pots. For special uses, or to distinguish plants from competitor varieties, there are more variety groups on the market: super compact varieties for 3-inch (8-cm) pots, maxi varieties for 6-inch (15-cm) pots, and medi varieties for 2½-inch (6-cm) pots. All varieties should have long-lasting, upright flowers, marketed only after comprehensive trials.

To successfully grow African violets, the proper environment must be maintained: the pH and salt levels have to be right; plantlets have to be

S

healthy and must come from a good propagator's stock; and the best bench space and greenhouse for production should be available.

GETTING OFF TO A GOOD START

After potting, keep the media moist. The plants should be kept in a warm, humid microclimate with media temperature of 75F (24C) for 25 to 30 days, depending on growth performance, especially in the wintertime. Lower temperatures will prolong crop time. An under-bench heating system is recommended to keep a uniform media temperature. Irrigation should be from overhead, with the water temperature equaling room temperature, for about three to four weeks after potting, until spacing.

FIG. 1. African violets have decorated millions of kitchen windowsills all over the United States and are a mainstay of supermarket floral departments from coast to coast.

LATER CULTURE

Right after spacing, apply a good overhead irrigation, then begin with capillary mat irrigation. Check weekly for dry pots to ensure the capillary system is working properly. Maintain air temperature at 69 to 70F (21 to 22C) day and night. To prevent *Botrytis*, humidity has to be controlled. Plants can be grown not only on a capillary irrigation mat, but also with an ebb-and-flood or trough watering system. Supply fertilizer at every irrigation. Early morning irrigation is recommended. If you continue to water from overhead, water temperature must be about room temperature (70F, 21C) to prevent the occurrence of white ring spots on the foliage.

CROP TIME

From a potted liner to a blooming African violet takes, in general, 10 to 12 weeks. Miniatures flower in eight to 10 weeks. The crop time depends largely on the growing environment—with high tech greenhouses, plan on production time of eight to 10 weeks for standard varieties. Even in the wintertime with a good heat retention system, it is possible to use a warmer night temperature (75F, 24C) than day temperature (70F, 21C), cool days and warm nights. Under such growing temperatures, plants do not stretch as much—they remain compact.

TABLE 1 AFRICAN VIOLET PRODUCTION SCHEDULE

Sales target		Order	Potting time	
Holiday	Date	liners (mo)	Standard variety	Miniature variety
Thanksgiving	Nov. 22–28	July	mid-Sept.	late Sept.
Christmas	Dec. 25	Aug.	late Sept. to early Oct.	mid-Oct.
New Year's Day	Jan. 1	Aug.–Sept.	early–mid-Oct.	mid–late Oct.
Valentine's Day	Feb. 14	Sept.–Oct.	late Nov.	mid-Dec.
Easter	Mar. 23–Apr. 25	Oct.	mid–late Dec.	late Dec.
Mother's Day	May 8–14	Oct.–Nov.	late Feb. to early March	early–mid-March

LIGHTING AND SHADING

African violets do not like direct sunlight. Their environment should preferably supply them with about 1,500 footcandles (16 klux) of diffused light. In summertime this light level can be obtained by shading. The greenhouse ought to have at least one movable shading system on the inside, combined with a permanent summer outside shade. Artificial lighting in wintertime is not required but can be helpful. Supplementary light from HID or Gro-Lux lamps (900 f.c., 10 klux), can benefit newly planted pots during December and January. Artificial lighting is a must for double-deck benching.

MEDIA AND FERTILIZER

The medium must have high air capacity in addition to being a good reservoir of water and a buffer for holding nutrients. Violets perform best in a light, well-drained, peat-lite mix or soil-based media. Media should contain a minimum amount of nutrients but must be low in total soluble salts. A pH of 6.0 to 7.0 and salts below 1,200 milligrams per cubic yard ensure a good growing environment. To predict the availability of a fertilizer, it must be measurable. A high-percentage fertilizer has fewer filler salts; therefore, smaller amounts are needed. Never use organic fertilizer! The total salts in the media determine plant growth and the requirements of liquid feeding. A light liquid feed program is recommended. The nutrition equivalent should be 1 part N, 0.8 P, 1.2 K (potash may be as high as 1.25 or 1.5); potash improves flower quality.

Use caution not to overfertilize. It is always possible to adjust fertilization unless the leaves become thick and brittle from overfertilization. Leaves should always be flexible and soft. African violets are very sensitive to high salts. Test media and water for nutrients, pH, and salts regularly. Carbon dioxide can be helpful in the wintertime. Between November and February, 400 to 600 ppm is enough.

S

Preventing Diseases and Pests

Basics for preventing disease include: keeping clean greenhouses and surroundings; using the best source for plantlets; starting with healthy plants, potting them on time, growing them in the right environment and in good facilities, and applying preventive spraying. Prevention rather than cure is easier to manage. Preventive sprays should be used only as needed. Some pesticides can act as growth regulators, and vigorous growth can be jeopardized by some specific chemicals.

The most common pests are thrips and foliar nematodes. *Attention:* It is important to know the life cycle of insects, especially thrips. If infestation has already occurred and the crop is in bloom, plants are no longer salable. A spray program for thrips must be at three- to four-day intervals for at least six weeks to match the insect's life cycle. Various chemicals can be helpful. Another pest, cyclamen mites (only seen with a microscope), look like thrips but are easier to manage.

Powdery mildew most commonly occurs in the spring when cold air is coming through the vents. It is easy to handle if you act quickly after first traces. *Phytophthora* is the most common fungus disease. Contamination generally comes from unclean areas and soil. Just as important is to buy clean, young plants. Apply media drench fungicides for control. *Botrytis* could also be a problem.

Conclusion

Seed propagated African violet liners have recently become available from specialist propagators. Follow crop times as listed in table 1 for vegetatively propagated liners. The series is known as Memento and comes in 12 colors.

To successfully produce African violets year-round, do not underestimate the requirements of proper environment, culture, and timing. Cultural instructions for finished plant production of African violets should be used as guidelines only. They need to be tailored to the production methods and capability of each individual greenhouse operation.

Salvia

by Jim Nau
Ball Horticultural Company
West Chicago, Illinois

❖ *Annual (Salvia splendens). 7,500 seeds/oz (265/g). Germinates in 12 to 15 days at 75 to 78F (24 to 26C). Seed should be left exposed to light during germination.*

❖ *Tender perennial (***Salvia farinacea***). 24,000 seeds/oz (847/g). Germinates in 12 to 15 days at 75 to 78F (24 to 26C). Seed should be left exposed to light during germination.*

❖ *Perennial (***Salvia** × **superba,** *also known as* **S. nemorosa, S. sylvestris** *var.* **superba***). 25,000 seeds/oz (872/g). Germinates in four to eight days at 72F (22C). Do not cover seed after sowing.*

All salvias are excellent in 4- and 6-inch (10- and 15-cm) containers (fig. 1) or cell packs and are of special importance in the landscape and home garden. *S. splendens,* including such varieties as Red Hot Sally and the Carabiniere series, is characterized as a group by flower colors in bold red, white, lilac, salmon, wine (burgundy), and blue. Flowers are held on upright spikes, and the plants fill in readily on 12-inch (30-cm) centers in the garden (fig. 2).

S. *farinacea* is available in either pastel blue or off-white, so it works best in settings with other pastel colors or contrasting bold colors. Plants have a finer leaf and grow taller than the *splendens* varieties. *S. farinacea* types make excellent dried cut flowers and background plants for the flower garden. Treated as an annual in the northern United States and as a tender perennial in the South, this class is an overlooked variety that does well in warm, dry locations. It is especially suited to landscape sales.

FIG. 1. Salvia is an excellent crop for 4-inch pots. The variety shown: Cleopatra Scarlet.

PRODUCTION

If growing salvia plugs, be careful of high salts during early production stages, especially for *S. farinacea*. To control growth, use Bonzi, A-Rest, B-Nine, or DIF. Do not hold plugs, as holding reduces rooting out after transplant. Plugs take five to six weeks from sowing.

For flowering packs of *S. splendens,* allow nine to 11 weeks from sowing for the more dwarf varieties; sell taller strains, such as America and Bonfire, green in the pack seven to eight weeks after sowing. Four-inch (10-cm) pots require 11

S

FIG. 2. For something unusual, try salvia in one of the bright salvia color mixes. This planting at the Park Seed trials in Greenville, South Carolina, looked like a multicolored jewel box.

to 13 weeks for the smaller strains to flower. *S. farinacea* requires eight to nine weeks for salable green packs and up to 16 weeks for flowering 4-inch pots. *S. farinacea* does not flower well in the cell pack, though it can be grown for this market. It does better going to the garden from a 4- or 6-inch (10- or 15-cm) pot.

In the southern United States, allow eight to nine weeks for flowering packs or dwarf *splendens* varieties; sell taller types green. Space *splendens* 10 to 12 inches (25 to 30 cm) apart in full sun to partial shade in the garden. Planted from March to May, it will flower until November, though blooming sporadically in July and August. Not a hardy crop, *splendens* cannot tolerate frost.

ANNUAL VARIETIES

In varieties, for free-flowering performance, look to Red Hot Sally, Fuego, and Vista, all among the earliest and most compact salvias. Red Hot Sally is a slightly darker shade of red. Scarlet Queen, comparable to Fuego, and Scarlet King, comparable to Red Hot Sally, are good pack performers with outstanding garden performance. Other excellent choices are the Empire or Sizzler series, which are made up of separate colors. Of particular value is the Empire Lilac, a light purple–flowering variety. For taller background accents, try Flare. Flare is 10 to 14 days later to flower, but it provides a long season of color, right up until frost.

In the *farinacea* types, the Victoria series is excellent all around, providing free-flowering performance from planting until weather extremes take away from it. For something smaller, try Rhea, which grows about 4 inches (10 cm) shorter than the Victorias. The new variety Strata is an All-America Selections award winner with graceful silvery white and mid-blue florets. Plants grow to 12 to 14 inches (30 to 36 cm) high. Reference is also a two-toned *farinacea* type.

PERENNIAL SALVIA

S. × superba is a tried and true performer in the garden. Plants grow 15 to 24 inches (38 to 61 cm) tall, with a spread nearly equal. Flowers are violet-blue, sometimes rose, and appear as spikes, like on other salvias. Plants are hardy from Zones 4 to 8 and flower naturally from May to June. Trimming plants back encourages a second flowering flush in July and August.

Allow nine to 11 weeks from sowing for salable green cell packs, finished at 50 to 55F (10 to 13C) nights. Plants will flower reliably the first year from seed, even from sowings as late as February. Seed-grown *S. superba* is a variable crop—plants may or may not be uniform. The variety Blue Queen, or Stratford Blue (the same plant), has fewer off-types from seed.

Other varieties include East Friesland, which flowers from June to August, with deep violet-purple flowers on plants 18 to 24 inches (46 to 61 cm) tall. Also available is May Night, with dark violet-purple flowers on 20-inch (51-cm) plants. Both of these varieties come true only from vegetatively propagated plants. Seed sold under these names will not come true.

SATIN FLOWER (SEE *CLARKIA*)

SCABIOSA (PINCUSHION FLOWER)

by Jim Nau
Ball Horticultural Company
West Chicago, Illinois

❖ *Annual (S. atropurpurea). 4,500 seeds/oz (159/g). Germinates in 12 days at 70F (21C). Perennial (S. caucasica). 2,400 seeds/oz (85/g). Germinates in 10 to 18 days at 65 to 70F (18 to 21C).*

Often called pincushion flower, the annual form of scabiosa is another outdoor summer-cutting item that produces reasonably long stems, even under trying outdoor conditions. Scabiosa should be placed out as early as soil will permit. Fall sowing under favorable conditions is even better. Annual scabiosa also does well forced for spring flowering in a cool house (50F, 10C), but in a warm house or during warm weather, the naturally slender stems draw up rather weak. For May flowering the seed should be sown in March and benched in April. For cutting purposes on a small scale, the Imperial Giants Mix is a good strain.

The perennial form, like the annual, is a cool-temperature plant. While both suffer in excessive summer heat, the perennials are inclined to die out if it gets too trying. Probably for this reason, they are not widely used in the Midwest, but they are the choice where summers are more moderate. Allow 10 to 12 weeks for salable, green 32-cell packs from sowing. Grow cool in 55F (13C) or lower nights. House Hybrids is a mix of blue- and white-flowering varieties growing 2 to 3 feet (61 to 91 cm) tall. Fama, more uniform, is available in blue, and Perfecta is lilac or white (Perfecta White).

SCAEVOLA

by John S. Rader
Proven Winners
Encinitas, California

❖ *Annual (Scaevola aemula). Vegetatively propagated. Growers produce pots and baskets from purchased rooted cuttings.*

With its unusual five-petaled fan flowers, this Australian genus exists in over 60 species and can also be found throughout many islands of the South Pacific. Named for the Roman hero Mucius Scaevola, who demonstrated his bravery (or stupidity) by burning off his hand, its flowers do in fact resemble a hand! In Hawaii there are two species, one from the mountains, one from the beach. Folklore has it that the plants are separated lovers. When they are reunited, the two half-flowers will make a whole.

Scaevola plants have been increasing in popularity and application for about 10 years. The first variety to be introduced on the market was *Scaevola aemula* Mauve Clusters, a small-flowered, compact variety that initially was grown as a hanging basket plant. Recently, growers have found it is best used as a ground cover in mild-winter areas. Mauve Clusters has demonstrated superior performance in hot, dry areas, such as the Central Valley of California and the deserts of Nevada and Arizona. The small fan flowers give their best showing during the warm, long days of summer.

In 1988 Kientzler KG of Gensingen, Germany, rocked the flowering plant industry in Europe with its introduction of *Scaevola aemula* Blue Wonder (fig. 1). Blue Wonder has a robust growth habit and more and larger flowers than Mauve Clusters.

Blue Wonder became an instant hit in the U.S. in 1990. Its rapid popularity and resulting high demand was the motivation for the formation of the Proven Winners propagation and marketing group. Blue Wonder has since been replaced with New Wonder, which is superior to it in all categories. Trials have shown it to be more compact, heavier blooming, and with larger flowers.

FIG. 1. Blue is a tough color for flowers, but not when you're talking about scaevola. If you're looking to add more blue crops to your mix, try New Wonder scaevola in hanging baskets for a spectacular show!

CULTURAL TIPS

The best time to plant New Wonder is February to May in all areas of the U.S. However, this cultivar can flower all year and may be grown in warmer climates year-round.

Plant three liners per 8- to 10-inch basket for a 12- to 14-week finish. Leaving two to three nodes, pinch once at planting and again in the same way, three weeks later. If necessary, pinch a third time to adjust shape. New Wonder can also be grown as a 6-inch (15-cm) pot plant with one liner. Follow the 8- to 10-inch basket pinching schedule. For bedding purposes, plant one liner in a 4-inch (10-cm) pot with one pinch. Finish time for a 4-inch is about six to seven weeks.

New Wonder performs best with a soil pH of 5.0 to 5.5. A light, well-drained medium is essential. Constant feed with nitrogen at approximately 250 to 300 ppm and potassium at 170 ppm is recommended. Periodically supplement with iron at 5 ppm; otherwise, average micronutrient amounts are required.

Scaevola can develop phosphate toxicity, which can be recognized by yellow-red chlorosis in older leaves and bleached yellow chlorosis in young leaves. In extreme phosphate toxicity cases, leaves will turn necrotic and shrivel. Avoid fertilizers that contain high phosphorous. A constant-feed program through drip irrigation with periodic leaching is best to reduce phosphate buildup.

Scaevola plants in soilless, organic potting media typically have a high water requirement, especially in warm, bright-light conditions. Never allow a scaevola plant to wilt once blooming begins. Flowering can be detrimentally affected. On the other hand, if planted in the ground as a bedding plant, New Wonder is drought-tolerant.

New Wonder can be grown in a wide range of temperatures. Grow outdoors or in the greenhouse with night temperatures from 55 to 65F (13 to 18C). For outdoor container growers, this cultivar has tolerated nights as cold as 28F (-2C). *Scaevola* is generally very heat-tolerant. New Wonder has favorably withstood

S

temperatures well over 100F (38C). The best light exposure is full sun, 5,000 to 9,000 footcandles (54 to 97 klux). Any shade will result in soft, spindly growth and a delay in flowering.

When growers experience difficulties with *Scaevola,* it is usually due to insufficient nitrogen feed or low light levels. Keep nitrogen and light levels high. In addition, night temperatures should be maintained at 65F (16C) or below. Warm nights can make plants stretch, especially if daytime light levels are low. Grow late crops outdoors in full sun and not in the greenhouse.

Florel has been helpful in controlling height and producing good branching. Apply about two to three weeks after the first pinch. It may not be necessary to hand-pinch if Florel is used. Use at the 500-ppm rate. Bonzi is very effective in height control but has little effect on branching. Exact rates seem to vary from one grower's experience to another's. Thus, it is difficult to recommend any rate at this time.

If propagated from elite (virus-indexed) mother stock, and average cultural sanitation practices are employed, New Wonder is not particularly susceptible to plant disease. Insect problems are usually minimal. Leaf miners and whiteflies appear to be the most aggressive pests. Thrips can infest the flowers but do not seem to affect flower development. A preventative insect control program should be implemented. This cultivar does not appear to be sensitive to properly applied chemical sprays. Flowers can be damaged by pesticide sprays with active ingredients that are dissolved in petroleum-based solvents, however. The same is true for insecticidal soaps.

MORE VARIETIES

A second-generation, small-flowering variety has been developed that is superior to the original Mauve Clusters. Scaevola hybrid Petite Wonder is a bluer, more heavily flowered selection. The small fan flowers are held up above deep green foliage, resulting in a much more attractive finished plant than Mauve Clusters. Following the cultural practices as for New Wonder will likewise result in success with Petite Wonder.

New varieties are being tested and bred. The value of this genus is only beginning to be realized. With the wide popularity of Wonder-type scaevola, breeders are coming out with new selections with many different growth habits and colors. The grower can look forward to having multiple choices from continually improved selections to offer the customer.

SCHIZANTHUS (POOR MAN'S ORCHID)

by Will Healy
Ball Horticultural Company
West Chicago, Illinois

❖ *Annual (***Schizanthus** × **wisetonensis***). 45,000 seed/oz (1,587/g). Germinates in seven to 14 days at 70 to 75F (21 to 24C). Do not cover seeds. However, light inhibits germination, so place seed trays in total darkness.*

Schizanthus is not a new crop, but today's varieties are very different from those of the past, with a short, compact habit and shatter-resistant flowers (fig. 1). Older types often were tall and open, with flowers that shattered easily. It has attractive, pastel-colored flowers that give it the common name "poor man's orchid." The foliage is light green and fernlike.

An excellent crop for winter and spring, *Schizanthus* not only grows well at cool temperatures but actually performs best then.

FIG. 1. New breeding is reviving sales of this old-time pot plant and garden favorite, schizanthus.

GERMINATION AND SEEDLING GROWTH

Sow seed in plug trays and cover very lightly or not at all. Fertilize with nitrate-form fertilizer for more compact plugs than if using a fertilizer with a significant amount of ammonium nitrogen. Fertilize with 100 ppm nitrogen during plug production. The optimum germination temperature has not been identified, but 70 to 75F (21 to 24C) is acceptable. Plugs should be transplanted as soon as they can be pulled from the tray, before they are rootbound.

CROP CULTURE

Temperature and timing. Growth is affected by temperature, with flowering fastest at higher temperatures but having the most compact and uniform habit at lower temperatures. Grown at 47F (8C), the plants require 165 days to flower, while

S

those grown at 65F (18C) flower in 90 days. Plants grown at the lower temperatures were shorter with a better growth habit. An acceptable plant can be grown at either temperature, so the grower must weigh the cost of fuel against the cost of longer crop time. An average temperature of 60F (16C) is a good compromise, with days to flower increased to 100, yet with very acceptable plant habit.

Besides temperature, light intensity also affects days to flower. At 53F (12C) *Schizanthus* requires 143 days to flower at an average of 816 footcandles (9 klux), but only 125 days at an average of 1,084 f.c. (12 klux). Photoperiod does not appear to influence flowering.

Today's schizanthuses do not require pinching to produce uniform plants. Pinched plants do have a somewhat more uniform habit, but days to flower is increased slightly.

Nutrition. While no controlled experiments have been conducted, *Schizanthus* appears to be a moderate feeder. Fertilization with a balanced fertilizer containing primarily nitrate nitrogen at 200 ppm has been successful.

Foliage is naturally light green, and deficiency of any mineral nutrient will accentuate this undesirable trait. Although no controlled studies have been conducted, it is believed that magnesium sulfate (Epsom salts) at 1 pound per 100 gallons (479 g per 400 l) applied as a drench just prior to flowering will improve foliage color.

Pests. No significant insect pests have been identified on schizanthus, and no phytotoxicity has been observed from any of the pesticides used. Monitor for thrips, whiteflies, and aphids, taking appropriate control measures if insects are observed.

Schizanthus is relatively disease-free. The most consistent problem is anthracnose-like leaf spot on the lower foliage. Plants grown on ebb-flood benches have little or no leaf spot, while plants that are overhead-watered show symptoms. Sprays with Chipco 26019 appear to reduce the severity of the disease. *Botrytis* can also be a problem during periods of high humidity. A general recommendation for disease prevention is to avoid overhead watering, maintain ample air movement, and water early in the day to allow the foliage to dry.

Height control. Today's new varieties of schizanthuses are naturally short and should be in proportion to their containers if grown well. They respond to applications of A-Rest, B-Nine, Bonzi, and Sumagic. Two sprays with 2,500 ppm B-Nine will result in shorter plants and darker green foliage, with little or no delay in flowering, although B-Nine is not registered for use on schizanthus plants.

Avoiding high temperatures, providing adequate light (less than 1,000 f.c., 11 klux), and spacing the plants so the foliage does not overlap is usually sufficient to prevent excessively tall plants without chemical regulators.

Postproduction handling. *Schizanthus* has excellent pot-production longevity, with today's new varieties lasting two weeks in a home or office with minimal care. These schizanthuses can withstand three days of shipping in closed boxes and remain salable, although shipping does reduce the expected longevity of the plant for the consumer by a few days.

Varieties. The Royal Pierrot series is breaking ground in making this an easy-to-produce pot plant. The series' first color, Lavender, offers genetic compactness and plant uniformity, finishing is just 14 weeks from seed to flower in a 6-inch (15-cm) pot. Royal Pierrot is also great for 4- or 4½-inch (10- or 11-cm) pots.

SCHLUMBERGERA (HOLIDAY CACTUS)

by Thomas H. Boyle
University of Massachusetts
Amherst

❖ *Tropical, woody perennial,* **Schlumbergera × buckleyi** *(holiday cactus or Christmas cactus);* **Schlumbergera truncata** *(Thanksgiving cactus, zygocactus, or holiday cactus). Vegetatively propagated. Growers take their own cuttings or purchase cuttings from specialists.*

The botanical names for Christmas cactus and Thanksgiving cactus have been revised several times, which has left horticulturists (and taxonomists) bewildered about the correct names for these plants. The popular houseplant known as Christmas cactus is actually an interspecific hybrid of *S. truncata* and *S. russelliana* that originated about 150 years ago in England. The currently accepted botanical name for Christmas cactus is *S. × buckleyi*—the "×" indicates that it is an interspecific hybrid. Christmas cactus has purplish brown anthers, ribbed ovaries, and leaf segments with rounded margins. It flowers near Christmas in the northern hemisphere, hence the common name.

In contrast, Thanksgiving cactus *(S. truncata)* has yellow anthers, rounded ovaries with no ribs, and segments with prominent teeth on the margins. It flowers about six weeks earlier than Christmas cactus. Today's commercial varieties are either "pure" Thanksgiving cacti or complex hybrids derived from crossing Thanksgiving cactus with Christmas cactus. The complex hybrids exhibit characteristics that are intermediate between the two parents. Many growers refer to Christmas cactus, Thanksgiving cactus, and their complex hybrids by an all-inclusive name: holiday cactus (fig. 1).

S

VARIETIES

Table 1 lists some holiday cactus varieties that are grown commercially. Varieties differ in the following traits: plant habit (pendulous, semipendulous, or erect); growth rate; degree of branching; temperature tolerance; size and shape of segments and flowers; flower color; and time of flowering under natural photoperiods. Growers should select varieties that perform well under their environmental and cultural conditions and satisfy consumer preferences. Plant size, flower color, and number of flowers per container are the main criteria that consumers use to evaluate holiday cactus. Many holiday cactus varieties are patented. Growers in the United States must sign license agreements in order to propagate them.

FIG. 1. No Christmas pot plant display at retail is complete without a few pots of holiday cactus. This easy-to-grow crop will reflower for consumers year after year.

GROWTH FROM CUTTINGS TO TRANSPLANTING

Plants are propagated by rooting mature, single-segment cuttings obtained from vegetative stock plants. Remove cuttings from stock plants by twisting the segment 180 degrees and pulling it upward. Take cuttings from the top of the stock plants, and use only the

TABLE 1   COMMERCIAL VARIETIES OF HOLIDAY CACTUS

Color	Varieties[a]
Pink, magenta, or violet	Alexis, Amanda, Barbara, Camilla, Christmas Charm, Christmas Magic II, Dark Sonja, Eva, Holiday Splendor, Isabelle, Lavender Doll, Lavender Doll II, Madisto, Madonga, Madrilane, Masanga, Naomi, Nicole, Rocket, Sarah, Sonja, Thor-Louise, Yantra
Scarlet or red	Claudia, Dark Marie, Kris Kringle, Linda, Madeleine, Marie (Maria[a]), Noris, Red Radiance, Starbrite, Thor-Alise, Zaraika
Salmon or orange	Christmas Cheer, Christmas Fantasy, Frida, Ilona, Madsolme, Malibu, Peach Parfait, Santa Cruz, Sleigh Bells, Twilight Tangerine
Yellow	Cambridge, Christmas Flame, Gold Charm
White	Bridgeport, Gina, Jaffa (Sanne[a]), Madelone, Thor-Britta, White Christmas

[a] Synonymous names are given in parentheses.

mature terminal and subterminal segments for propagation. Avoid taking cuttings near the base of stock plants because they are more likely to be contaminated with soilborne pathogens. Collect cuttings in clean, pathogen-free containers.

Unrooted cuttings can be stored for up to three months at 50 to 59F (10 to 15C) and 85 to 95% relative humidity. Propagate cuttings in cell packs or 1½- to 2-inch (4- to 5-cm) pots, using two to four cuttings per cell or pot. Propagation media should be free of pathogens and well-drained. Keep the temperature of the propagation media at 70 to 75F (21 to 24C) during rooting. Segments will root equally well using intermittent mist, high-humidity tents, or periodic hand-watering, as long as the media remain moist and warm. Conditions during propagation are highly conducive to the spread of soilborne pathogens. Strict sanitation will reduce diseases and minimize the need for fungicides.

Propagate cuttings between November and March for sales the following November and December. Cuttings propagated during naturally short days (SD) from early September until late April should be given long-day (LD) photoperiods to promote vegetative growth. LD photoperiods can be provided using "night-break" lighting from 10 P.M. to 2 A.M. at 5 to 10 footcandles (108 klux) minimum light intensity at plant level.

A newly propagated cutting will often produce only one new shoot, resulting in a sparsely branched plant. If new shoots are pinched off at about six to eight weeks after sticking cuttings, then multiple shoots (two to four) will develop. Pinching is a laborious task but will result in a fuller, higher quality plant.

Cuttings propagated between November and March will be ready for transplanting in April, May, or June. Use one cell per 3½- to 4-inch (9- to 10-cm) pot, three cells per 5- to 6-inch (13- to 15-cm) pot, four cells per 6-inch (15-cm) hanging basket, and eight to 10 cells per 8-inch (20-cm) hanging basket. Plants can be grown at pot-tight spacing during most or all of the growing period.

CULTURE

Use a growing medium that is well-drained, pathogen-free, and adjusted to a pH of 5.7 to 6.5. Many growers use a commercially formulated soilless mix that is composed primarily of sphagnum peat.

The frequency of irrigation will vary, depending on the type of growing medium, plant establishment, and environmental conditions. Well-established plants may need to be irrigated every two to three days in sunny, warm weather, or every five to eight days in cool, cloudy weather. Recently potted plants should be irrigated less frequently than well-established plants. In general, holiday cactus will tolerate underwatering better than overwatering. Saturation of the growing media for prolonged periods will predispose root systems to attack by soilborne disease organisms.

High-quality plants can be produced using a balanced N-P-K fertilizer with micronutrients and applying 150 to 200 ppm N at each watering. Some growers use calcium nitrate and potassium nitrate to supply about 180 ppm N, 390 ppm K, and 53 ppm Ca at each watering, also applying a balanced N-P-K fertilizer with micronutrients once a month at 150 ppm N. High-quality plants can be produced using either nitrate or ammonium as the nitrogen source. Plants grown in soilless media will benefit from periodic applications of magnesium sulfate (Epsom salts) at 20 ounces per 100 gallons (600 g per 400 l), giving about 150 ppm Mg. Leaching with plain water is done as necessary to prevent high soluble salt levels. Begin fertilization as soon as roots develop on newly propagated cuttings.

The pH of growing media should be kept above 5.5. Plants will take up high amounts of iron and manganese when the pH drops below 5.5, leading to serious plant damage. Growing medium samples should be analyzed at regular intervals to monitor pH and soluble salts. Avoid using fertilizers with strongly acid reactions.

Under optimum growing conditions, holiday cactus will produce one tier of growth every six to eight weeks. Greenhouse temperatures should be maintained at 62 to 65F (17 to 18C) nights and 68 to 72F (20 to 22C) days. Provide ventilation above 74F (23C). Maintain light intensity at 1,500 to 3,000 f.c. (16 to 32 klux) on a year-round basis. Strong sunlight or high temperatures (greater than 90F, 32C) can cause the new growth to become chlorotic on some cultivars (see the summer chlorosis section).

By early September the plants should be (1) two to three segments long in 3-inch (8-cm) pots; (2) three to four segments long in 3½-inch (9-cm) pots; or (3) four to five segments long in 4½-inch (11-cm) pots. Plants that are taller than desired may be shortened by "leveling." Plants are leveled by having their terminal segments twisted off. Plants should be leveled during the first week of starting SD for flower induction. When correctly done, leveling will yield heavily budded plants that are uniform in height.

FLOWERING

Flowering requirements. Flowering of holiday cactus is controlled by photoperiod and temperature. When temperatures are below about 59F (15C), flower initiation will occur under any photoperiod, including continuous irradiation. When temperatures are between 60 and 75F (16 and 24C), plants will initiate flowers under SD but will remain vegetative under LD. Thus, holiday cactus is an SD plant when grown at temperatures ranging from 60 to 75F. The *critical day length* (the photoperiod separating SD from LD responses) is between 12½ and 14 hours for plants grown at 64 to 65F (18C) nights and 70 to 72F (21 to 22C) days. Growers can use *natural flowering* or *controlled flowering* for producing holiday cactus.

Natural flowering. Flower buds will initiate on holiday cactus when the natural photoperiod becomes shorter than the critical day length. To be exact, the long nights trigger flowering, not the short days. In the northern United States, natural flowering will occur primarily in mid-November when plants are grown under natural day lengths and 62 to 65F (17 to 18C) nights (table 2). The plants will flower earlier, though, if night temperatures dip into the 50s (10 to 16C) during August and early September. In the southern United States, natural flowering will occur mainly in early to mid-December.

As a method, natural flowering has two main disadvantages. Each variety has a limited flowering period; thus, several varieties are needed to provide a continuum of flowering plants. Also, crop scheduling is difficult because the exact time of flowering varies from year to year (table 2).

Controlled flowering. Holiday cactus can be scheduled on a year-round basis by using controlled flowering. This method requires accurate control of temperature and photoperiod. Maintain greenhouse temperatures at 62 to 65F (17 to 18C) during the night and 68 to 72F (20 to 22C) during the day. During natural LD (late April to early September in the northern United States), flowering can be induced by reducing the day length to eight or nine hours daily (giving 16 or 15 hours of continuous darkness, respectively). It is important to maintain SD conditions on a daily basis for at least three weeks. The grower should take proper precautions to prevent high temperatures under black cloth; poor or uneven budset may occur if the temperature exceeds 75F (24C) during SD.

In holiday cactus the rate of flower bud development is influenced by the average daily temperature (ADT). Flower bud development is accelerated as the ADT increases from 59F to 75F, 15 to 24C (table 3). The information in table 3

TABLE 2   NATURAL FLOWERING TIMES OF HOLIDAY CACTUS VARIETIES

Cultivar	Mean date of flowering[a]				Range (days)
	1988	1989	1990	1991	
Christmas Charm	Nov. 12	Nov. 17	Nov. 7	Nov. 15	10
Christmas Fantasy	Nov. 18	Nov. 24	Nov. 12	Nov. 18	12
Gold Charm	Nov. 14	Nov. 25	Nov. 18	Nov. 17	11
Lavender Doll	Nov. 21	Nov. 25	Nov. 18	Nov. 21	7
Lavender Doll II	Nov. 27	Dec. 5	Nov. 30	Nov. 25	10
Rocket	Nov. 14	Nov. 18	Nov. 5	Nov. 11	9
White Christmas	Nov. 15	Nov. 24	Nov. 14	Nov. 19	10
Mean of all seven cultivars	Nov. 17	Nov. 20	Nov. 15	Nov. 18	

Note: Trialed over a four-year period at the University of Massachusetts, Amherst. Greenhouse temperatures were maintained at 65F (18C) nights and 70–72F (21–22C) days.
[a] First flower open. Each mean represents five to eight pots.

TABLE 3   TEMPERATURE, BUD LENGTH, AND FLOWERING OF HOLIDAY CACTUS

Bud length (mm)[a]	Days to flowering			
	59F	65F	70F	75F
1	70	50	39	32
2	58	41	32	26
3	51	37	28	23
4	47	33	26	21
5	43	31	24	19
6	40	28	22	18
7	37	27	21	17
8	35	25	19	16
9	33	24	18	15
10	31	22	17	14
11	30	21	17	14
12	28	20	16	13
13	27	19	15	12
14	26	18	14	12
15	25	18	14	11
16	24	17	13	11
17	23	16	13	10
18	22	15	12	10
19	21	15	12	9
20	20	14	11	9
25	16	12	9	7
30	13	9	7	6
40	8	6	5	4
50	5	3	3	2
60	2	1	1	1

Source: Data from Lange, N., & R. Heins. 1992. How to schedule Thanksgiving cactus . . . and optimize flower number. *Greenhouse Grower* 10(9):62–64.

Note: Camilla, Dark Marie, and Madisto varieties under four different average daily temperatures. 59F, 65F, 70F, 75F = 15C, 18C, 21C, 24C, respectively.

[a] One mm = 0.039 inch; 1 inch = 25.40 mm.

can be used to schedule the time of flowering. Measure the lengths of the largest buds on several plants and calculate the average bud length. Find the number in column 1 that corresponds to average bud length, then read across to determine how many days are required for the buds to open. For example, if the largest buds averaged 8 mm in length, approximately 35, 25, 19, or 16 days would be required for these buds to open if the ADT were kept at 59, 65, 70, or 75F (15, 18, 21, or 24C), respectively.

Delaying flowering with long days. Growers in the northern United States often find that some holiday cactus varieties flower too early for Christmas sales when plants are grown under natural photoperiods. Flowering can be delayed by maintaining plants under LD starting in the first week of September. Use "night-break" lighting from 10 P.M. to 2 A.M. (about 5 to 10 f.c. at top of plants) to prevent flowering under natural SD. Incandescent lamps are most commonly used for providing LD, but high-pressure sodium lamps are also effective. Temperatures must be maintained above 60F (16C) and preferably 62 to 65F (17 to 18C) during LD photoperiods. Lighting from sunset until about 10 P.M. will also keep plants vegetative, but night-break lighting is more cost-effective.

Irrigation during flower induction. Water stress during flower induction will not enhance budset and may actually reduce it. Keep the growing medium evenly moist during flower initiation and development.

Growth regulators. Benzyladenine (BA) has been shown to increase branching of vegetative plants and increase the number of flower buds on reproductive plants. To increase the number of flower buds, apply 50 to 100 ppm BA as a spray when pinpoint buds are visible (natural flowering) or about 10 to 12 days after starting SD photoperiods (controlled flowering).

Flower and bud drop in holiday cactus can be induced by environmental stress (water, light, temperature) or exposure to ethylene. Silver thiosulphate (STS) has been shown to substantially reduce flower bud drop of holiday cactus. Apply STS (200 ppm) as a spray when flower buds are one-fourth to one-half inch (6 mm to 1 cm) long. STS is not registered for use on holiday cactus. Use a spreader-sticker with BA or STS sprays on holiday cactus.

Shipping. Open flowers of holiday cactus are easily damaged during sleeving and shipping. Therefore, plants are usually shipped in the bud stage in order to minimize injury. The optimum stage for shipping depends on the postproduction environment. Plants should be shipped in the "large bud" stage (one to two days from opening) if the customer plans to display the plants in an environment with low light (less than 300 f.c., or 3 klux) and low humidity, such as a supermarket. However, plants can be shipped in the "small bud" stage (10 to 14 days from opening) or "medium bud" stage (four to seven days from opening) if the customer plans to display the plants in an environment with high light (1,500 to 3,000 f.c., 16 to 32 klux) and high humidity, such as a greenhouse.

PROBLEMS

Diseases. Holiday cactus is subject to several diseases. *Fusarium oxysporum* is a fungus that causes root or stem rots. Infected segments produce reddish orange sunken spots and then abscise. Orange spores develop in the lesions and are spread easily by water or air. Stems topple over when the basal segment becomes infected. Banrot (40WP and 8G) and Cleary's 3336-F are registered for control of *Fusarium* on holiday cactus.

Phytophthora parasitica and *Pythium aphanidermatum* are fungi that also incite root or stem rots. *Phytophthora* root and stem rot is characterized by necrotic stem lesions with faded reddish borders, gray-green discoloration of the stems, and segment abscission. *Pythium* root and stem rot is similar but segment abscission is rare. Banrot (40WP and 8G) is registered for control of *Phytophthora* and *Pythium* on holiday cactus.

Bipolaris cactivora (formerly *Drechslera cactivora*) is a fungus that causes stem rot. Symptoms include blackened, sunken lesions up to one-half inch (1 cm) in diameter. Black spores develop in the lesions, giving them a fuzzy appearance. Infected segments commonly abscise. Daconil is effective in controlling bipolaris stem rot on holiday cactus but is not labeled for this crop.

Erwinia carotovora is a bacterium that causes soft rot in many cacti, including holiday cactus. The initial symptom is usually a blackened, wet, slimy lesion that develops on the basal segment and progresses upward in the shoot. Plants wilt, collapse, and usually die. Bacteria are spread by splashing water. Since this disease is caused by a bacterium, fungicides will not control the disease. Discard any plants that are suspected of having bacterial soft rot.

Insect pests. Fungus gnats *(Bradysia* spp.) and flower thrips *(Frankliniella* spp.) are the two major insect pests of holiday cactus. Adult fungus gnats do not cause injury, but the larvae feed on roots and stem tissue. Feeding damage may lead to infection by disease-causing organisms. Fungus gnats can be controlled by avoiding overwatering, controlling algae growth, and applying pesticides that are registered for this pest. Flower thrips are tiny (one twenty-fifth inch, or 10 mm in length), slender insects that feed on immature segments and flower buds. Feeding causes growth distortion, flecking on fully expanded petals, and bud drop. Both the adults and immature stages are injurious to plants. Heavy infestations may cause severe economic losses. Reduce thrips populations by screening greenhouse vents, discarding infested plants, and applying registered pesticides.

Summer chlorosis. Prolonged exposure to high light intensities or high temperatures (greater than 90F, 32C) can cause new growth to become chlorotic (yellow) on some varieties. Symptoms are particularly noticeable during summers that are warmer than usual. Among the 43 holiday cactus cultivars that were trialed at the University of Massachusetts in 1991 and 1992, Amanda, Dark Sonja, Jaffa, and Sonja became very chlorotic; Barbara, Frida, and Isabelle became moderately chlorotic; and the other cultivars exhibited little or no summer chlorosis. Cultivars that become chlorotic in summer will often green up on arrival of cooler temperatures, but some plants may still remain chlorotic at flowering. Growers should select varieties that show little or no summer chlorosis if their greenhouses exceed 90F for extended periods or if they plan to induce flowering during summer months using artificial SD (black cloth).

SEDUM

by Jim Nau
Ball Horticultural Company
West Chicago, Illinois

❖ *Perennials* (**Sedum acre, S. spurium,** *and* **S.** × *Autumn Joy). 400,000 seeds/oz (14,109/g). Fresh seed germinates in eight to 14 days at 72F (22C). Seed should be left exposed to light during germination.*

S. acre is a small-flowered sedum variety with flowers to one-half inch (1 cm) across. Flowers are single and yellow in color. Plants are 2 to 4 inches (5 to 10 cm) tall and are smaller overall in habit than *S. spurium.*

S. spurium is more invasive than the previous variety, with bronze-red foliage that appears more intense as the evenings cool in autumn. Flowers are bright rose in appearance and up to one-half inch (1 cm) across. Flowers are single and held on foliage 6 inches (15 cm) tall. *S. spurium* has broad leaves, rather than the needlelike foliage of *S. acre.*

Both species are hardy to USDA Zones 3 through 9. *S. acre* flowers in June, while *S. spurium* flowers in July and August.

Division, cuttings, and seed are the common forms of propagation. Sowings made in winter or spring of either variety can be sold green in the packs 11 to 14 weeks later. *S. acre* will not flower the same season from seed and needs to be overwintered in a quart or gallon container for flowering the first year. *S. spurium* will flower the same season from seed, usually in late July or August. However, the best bloom from seed comes the second season.

Among varieties, *S. acre* is sold under the name Golden Carpet, and the most uniform of the selections are those vegetatively propagated. In *S. spurium,* the species is often sold under this name or *S. spurium coccineum,* and both work well in the perennial garden. *S.* × Autumn Joy (also called Indian Chief and Herbstfreude) is often listed as a cultivar of *S. spectabile.* Autumn Joy has become a staple for fall garden center sales, with its bright pink flowers that appear in August and September on mounded plants to 2 feet (61 cm). Flowers turn bronze with the first frost. Autumn Joy can only be propagated vegetatively. Take stem cuttings or divide the plants in the spring or fall.

S

SENECIO (CINERARIA)

by David W. Niklas
Clackamas Greenhouses, Inc.
Aurora, Oregon

❖ *Annual (Senecio cruentus). 75,000 to 150,000 seeds/oz (2,646 to 5,291 seeds/g). Germinates in 10 to 14 days at 70 to 75F (21 to 24C). Leave seed exposed to light for germination.*

Cinerarias are colorful pot plants that, when properly grown as compact, well-spaced plants, can be a profitable addition to winter and early spring sales (fig. 1). Demand is high from January through April, especially for mass market sales at Valentine's Day and Easter. Cinerarias are traditionally produced in 5- or 6-inch (13- or 15-cm) pots, but new compact varieties make excellent 4-inch (10-cm) pots that can be highly profitable. Cinerarias come in a wide range of colors. Some strains have predominantly solid colors, while others have a high percentage of white "eyes." Generally, a high percentage of bicolors is desirable to retailers.

FIG. 1. Cineraria is a great cool season pot crop in cold climates and an excellent window box crop for mild, coastal climates. The variety shown: Venus Blue Shades.

The following schedule is for January to February bloom in the mild winters of the Pacific Northwest. Desired blooming dates and local temperatures during the growing season will necessitate altering the schedule. After flower initiation, a procedure described later, cineraria bloom dates can be controlled by manipulating forcing temperature. *Cineraria* can tolerate temperatures from 40 to 68F (4 to 20C), although temperatures above 63F (17C) should be avoided. At lower temperatures, growth is greatly retarded, and care must be taken to avoid overwatering.

Sowing time for 6-inch (15-cm) plants is in early August, while 4-inch (10-cm) plants can be sown two weeks later. Germinate at 70 to 75F (21 to 24C), but transfer to 65F (18C) after true leaves develop. Seed is small, so select a well-prepared, pasteurized medium for sowing. Do not cover the seed with medium, but take steps to maintain high humidity.

As soon as plants are established, transplant to larger cells. This will avoid problems with damping off. Since cinerarias are leafy plants, choose at least a 2-inch (5-cm) cell. I prefer to use 50-cells or larger to avoid stretching problems. At this stage, temperatures of 60 to 62F (16 to 17C) will promote good vegetative growth. Transplanting usually occurs in mid-September. Plants remain in the 50-cells for six to eight weeks, or until between October 15 and 30. An application of B-Nine at 0.25% applied as soon as plants are well established will reduce stretch and improve quality.

Between October 15 and 30, shift plants to the final containers. Keep temperature at 60 to 62F (16 to 17C) at night for two weeks to promote rooting, then drop the temperature to between 45 and 55F (7 and 13C) for a minimum of four weeks to initiate buds. During this cool period, keep plants pot-tight. An additional spray or sprays of B-Nine at 0.25% should be used to control stretch, if necessary. This will depend on pot size grown and variety selected.

In early December space the plants and raise temperatures to between 60 and 65F (16 and 18C). Early-flowering varieties will be eight to 10 weeks from sale. Lower average temperatures delay flowering, but plants will exhibit less vegetative growth, resulting in more compact plants with less reliance on growth regulators.

During December, January, and early February, cinerarias can take all the light available. In March and April, shade should be applied to reduce water stress. A sunny day in late January or early February can cause even well-watered cinerarias to wilt. Cinerarias are light feeders, and too much nitrogen will produce leafy plants that take up too much bench space to be profitable. Be careful not to let the pH rise too high, as plants would develop chlorosis due to iron deficiency.

Cinerarias are subject to the big three insect pests: aphids, whiteflies, and thrips. Careful monitoring and a well-thought-out spray rotation are essential for control of all three. Tomato spotted wilt virus, spread by thrips, has been reported on cineraria, so control of thrips is essential. *Verticillium* can be a problem in the early stages of the crop, so good sanitation must be maintained. *Botrytis* and powdery mildew can occur at flowering, but good air movement and cultural practices prevent these problems. *Pythium* root rot often occurs if plants are overwatered.

The following varieties are good for 5- or 6-inch (13- or 15-cm) production: Cindy, Dwarf Erfurt, Tourette, Improved Festival, Sonnet, and Venus. For 4-inch (10-cm) plants you can use the following: Amigo, Cindy, Starlet, Dwarf Erfurt, and Cupid. Cindy and Dwarf Erfurt require increased B-Nine applications.

S

SENECIO   (DUSTY MILLER)

by Jim Nau
Ball Horticultural Company
West Chicago, Illinois

❖ *Tender perennial (for various genera and species, see below).*
  Germinates in 10 to 15 days at 72 to 75F (22 to 24C). Do not cover the
  seed upon sowing.

"Dusty miller" is actually a common name for a number of plants that have gray, woolly foliage. These plants are perennials in warm winter areas but are treated as annuals north of Zone 8, especially if winter protection is not provided. The yellow flowers borne on these plants the second season after sowing are relatively unimportant since it is the foliage that is sold in the U.S. bedding plant trade. However, dusty miller also makes excellent dried foliage plants that work well in floral arrangements. The flowers are of merit because they make a unique contrast with the foliage and provide ample color in June and July in the southern United States.

FIG. 1. The silvery gray-green foliage of dusty miller makes it an excellent choice for combination pots and in the garden. These plants are off to a good start in a Styrofoam pack, typical of British bedding plant production. The grower is Bypass Nurseries, England.

In cropping, allow 11 to 12 weeks for all varieties to be sold without flowers in a cell pack (fig. 1). A 4-inch (10-cm) pot will take up to 14 weeks to fill out, using one plant per pot. If the plants get away from you, B-Nine is registered for use on this crop.

In the southern United States, allow eight to 10 weeks for foliage sales in packs and 11 to 12 weeks for sales in 4-inch (10-cm) pots, one plant per pot. Dusty miller is ideal for landscape and container plantings in the South and over-winters best in the milder regions.

The genus, species, and seed count per ounce varies from variety to variety. Also note that all the heights listed below are based on the foliage in the garden without flowers. Silverdust (*Senecio cineraria*, 50,000 to 100,000 seeds/oz,

1,764 to 3,527/g) is one of the most popular varieties on the market today. Its deeply notched leaves are covered with a woolly mat of gray on plants to 8 inches (20 cm) tall. Cirrus is also listed under *S. cineraria* but differs from Silverdust by not having as much of a gray cast to the foliage. Its leaves are the least notched of any of the dusty miller varieties in this discussion. Cirrus grows to 10 inches (25 cm) tall.

Diamond (*Cineraria maritima,* also *Centaurea candidissima,* 50,000 seeds/oz, 1,764/g) is the most vigorous and has the largest leaf of the dusty millers. It grows to 15 inches (38 cm) tall and fills in well, making it a top choice for landscapers to use in mass plantings. These plants appear very similar to Silverdust except that they are larger overall, and with a slightly less gray cast than Silverdust. Silver Lace (*Chrysanthemum ptarmiciflorum*, 200,000 seeds/oz, 7,055/g) is one of the dwarf varieties available. It differs from other varieties in its leaves, which are twice notched, giving a lacy appearance to the plant. It grows to 10 inches (25 cm) tall, and its foliage is a lighter shade of gray than the other varieties here.

SHASTA DAISY (SEE *LEUCANTHEMUM*)

SINNINGIA (GLOXINIA)

❖ *Annual (*Sinningia speciosa*). 800,000 seeds per oz (28,000 per g). Germinates in 14 to 21 days at 65 to 70F (18 to 21C). Leave seed exposed to light during germination.*

Gloxinia has increased in popularity so much that many growers consider it to be a major crop. The wide variety of colors and types include double and single varieties, standard solid colors that run predominantly red, pink, deep purple, lavender, and white, as well as many shades of two-tones with white centers or rims (fig. 1).

Gloxinias can be grown from tubers or seed. Tubers are available only in midwinter, so this limits their flowering period to the early spring months. However, seedlings are available all year long. Most growers purchase gloxinias as seedlings from specialty growers instead of sowing seed themselves.

Growers choosing to start with seed must use much care and have a warm, moist house with 65 to 70F (18 to 21C) nights. The seed is very fine and should be planted in light media, with very little, if any, covering over the seed. The tiny seed takes two to three weeks to germinate and will be ready for first transplanting in about six weeks. Second transplanting goes directly into 6-inch (15-cm) pots about

S

FIG. 1. Gloxinia, with its stately, oversized flowers, is a staple among pot plants for spring sales. The variety shown: Glory.

four weeks later, making total crop time from seed to spring flowering about six months.

Gloxinia seedlings in small pots can be purchased any week of the year from specialty growers. Many growers raise them as a year-round crop without any special requirements. They do require a reasonably warm night temperature of 65F (18C) with at least 75F (24C) days to grow properly, but no night lighting, black shade, pinching, or disbudding is needed for normal growth.

Bought-in seedlings should be unpacked as soon as received, placed in trays in a warm greenhouse, watered lightly to help acclimate them, and potted when convenient in a few days. Do not leave gloxinia seedlings too long in small pots as they would become stunted rapidly and would flower prematurely with much smaller plants and only a few buds. In fact, any type of shock to the gloxinia seedlings during the first few weeks can produce stunted, inferior plants that bloom early with only a few flowers. Too much fertilizer, high light, high temperature, or loss of root system causes premature flowering.

GLOXINIA FINISHING

Starting with small gloxinia seedlings, the following procedures will produce good finished plants in 10 to 14 weeks, depending on the season. Gloxinia grows much faster in the long, warm days of summer.

Plants should be potted deep, one-quarter to one-half inch (6 to 13 mm) from the crown of plant, in a loose potting soil containing plenty of peat and some soil amendments—such as perlite, vermiculite, calcine clay, or coarse sand—for good aeration. Better grades of peat-lite mixes can also be used. Heavy soil mixes with poor aeration will result in poor root development and stunted foliage.

Set the seedling well down into the soil to stabilize the plant, leaving only the four uppermost leaves above the soil level. Gently tap the pot to level the soil, but do not pack it around the plant, as gloxinias like a very loose, open soil mix. Next, the plant should be watered in lightly with a good dual-purpose fungicide, such as Banrot or Subdue. This will eliminate most disease problems that might show up much later in the crop.

Gloxinias are moderate feeders. You will obtain excellent results using weekly feedings, alternating 2 pounds per 100 gallons (960 g per 400 l) of a 15-16-17 peat-lite formula with plain calcium nitrate at the same strength. If desired, slow-release fertilizers, such as Osmocote 14-14-14, may be incorporated into media, but they must be used at one-fourth the recommended strength. Be careful not to use any fertilizer with excessive amounts of phosphate or urea (ammonia), as gloxinia reacts poorly to both. Remember, 20-20-20 has a very high amount of both and is not recommended.

Greenhouse temperatures should range from 65F (18C) nights to 75F (24C) days for best results. In northern climates tempered water is recommended for gloxinias. Water below 50F (10C) can cause injury to foliage and root systems.

Optimum light intensity for gloxinia is about 2,000 to 2,500 footcandles (22 to 27 klux). Excessive light, above 3,000 f.c. (32 klux), will cause yellow, blotched foliage, hard growth, or small, irregular, light brown spots on leaves.

To produce a choice gloxinia, B-Nine can be used 12 to 16 days after potting. This will shorten the main stem and leaf petioles, resulting in a sturdy, well-shaped plant. The suggested B-Nine rate on gloxinias is only one-third that used on mums—approximately 0.10%. This can be made by dissolving B-Nine SP at the rate of 2 teaspoons per gallon of water. A second application may be used seven to 10 days later for plants grown under low-light conditions (under 2,000 f.c.). Bonzi has shown promise recently, particularly in reducing flower stem stretch during warm periods of the year. Use one application of Bonzi at 1 ounce per gallon (30 g per 4 l) of water as a light foliar spray (only) when buds start to grow above the foliage (three to four weeks before flowering). This controls leaf expansion and stretch of primary flower stalks.

For growing temperatures below 65F (18C), add two weeks to the above schedule. Do not allow temperatures to fall below 60F (16C). Gloxinias are delayed by low day or night temperatures and dark, overcast winter days. Auxiliary lights, such as fluorescent or HID lights that produce 200 f.c. (2 klux) or more at bench level, used in the daylight periods and extended into the night (8 A.M. to 10 P.M.), will speed up winter growth by several weeks.

TABLE 1   GLOXINIA CROP PRODUCTION PERIODS

Season	Months	Length (weeks)
Winter	December–March	13–14
Spring	April–May	12
Summer	June–September	10
Fall	October–November	11–12

INSECTS AND DISEASE

Cyclamen mites, too small to be seen without a magnifier, may be detected by stiffening and discolored, reddish brown center leaves. Pentac or Kelthane will kill cyclamen mites as well as other mites. Army worms or loopers can be controlled with Mavrik, Dipel, or Resmethrin aerosol.

Thrips attack the growing tips of small plants and can cause the leaves to grow out deformed with cuts, elongated holes, or ragged edges. Because of increased thrips problems, including some strains of virus (tomato spotted wilt virus) that can be carried by thrips, the suggested method of control is to alternately spray several different insecticides, such as Orthene or Avid. During periods of heavy infestation, spraying twice a week may be necessary to control any influx of thrips that could come from outside or from freshly hatched adults out of the soil pupae stage.

Immediately after potting, apply a good fungicide drench. A second application may be applied four weeks later for complete disease control to last the entire production time. A light foliar rinse should be applied after fungicide drench to eliminate possible injury to foliage from residue.

VARIETIES

There are two basic types of gloxinias available as seedlings—the regular florist grade strains, which are ideal for 6- to 6½-inch (15- to 17-cm) pots, and now the fast-growing Super Compact strain, developed for 4- to 5-inch (10- to 13-cm) pots, which are so ideal for the mass market. They are a much faster crop, flowering in six to nine weeks on the bench, when they can be sleeved and boxed for shipping. Super Compact strains will flower three to four weeks ahead of the standard types, can be grown on 9-inch (23-cm) centers, and produce two-and-one-half to three times as many plants in the same space during the course of one year.

The Glory series, with single flowers, is available in separate colors for cropping year-round. Additionally, the Avanti series features flexible leaves that bend, allowing plants to be sleeved and boxed. Avanti is available in nine separate colors.

SNAPDRAGON (SEE *ANTIRRHINUM*)

SOLIDAGO (GOLDENROD)

by Jeff McGrew
McGrew Horticultural Products and Services
Mount Vernon, Washington

❖ *Perennial (***Solidago** × **hybrida***). 220,000 seeds/oz (7,760 seeds/g).*
Germinates in five to eight days at 70F (21C).

Solidago, or goldenrod, is native to much of the world. A hardy perennial in the garden, goldenrod has also become an excellent commercial cut flower through modern breeding and selection.

Species solidago can be easily grown from seed. Grow seedlings under long days with night temperatures of 50 to 60F (10 to 16C). It is also vegetatively propagated. For reliable plant performance and successive flowering, most commercial cut flower growers chose to buy in vegetatively propagated cuttings of hybrids such as Tara and Yellow Submarine. Tara, a golden yellow, has plume-shaped flowers, while Yellow Submarine has medium yellow, open-spreading flowers. Both are well suited for fresh or dried production, and both are mostly day-length neutral, in some regions, blooming naturally two to three times during the year. Yellow Submarine is a vigorous grower, producing 24- to 36-inch (61- to 91-cm) plants.

Plant at a density of 1½ cuttings per square foot (16/m²) in low-light growing areas, two cuttings per square foot in high-light growing areas.

CULTURE

Provide media with medium texture, excellent drainage, and low salt levels. When plants are actively growing during long-day periods, they are low to moderate feeders. During bud initiation or short days, reduce nitrogen. Clear water during the last week is recommended. High soil salt levels can negatively affect production and stem quality. EC levels of 0.75 to 1.5 are optimal.

The soil should be kept moist during vegetative growing periods. When short days are started, a *gradual* drying-down period helps bud setting and toning. During the finishing, a flagging (slight wilting) of leaves between waterings is commonly practiced.

Solidago is a full-sun crop. Use long days (greater than 16 hours) to maintain vegetative growth for winter and spring crops. When shoots lengthen to 18 inches (46 cm) or so, begin short days to induce flowering. It flowers naturally in late summer. Night temperatures of 50 to 60F (10 to 16C) are acceptable during long days. These cool temperatures help keep the plants growing vegetatively. During bud initiation (short days), raise night temperatures to 62 to 65F

S

(17 to 18C) to speed this process. Also, higher night temperatures give a better and more complete bud set (more flowers per stem).

Diseases and insects. Powdery mildew can be a problem when temperatures are cool and moderate humidity is present. Maintain good air movement and keep foliage dry. Rust can cover goldenrod in the late summer. Pine trees are the host, so plant far away from pine tree stands. Chemical controls are not available.

GREENHOUSE FORCING

By using mum lighting and black cloth techniques, rooted cuttings can be planted year-round in most areas of the U.S. One layer of support wire is required.

Plant rooted cuttings and do not pinch for two to three weeks. A solid root system is critical for quality production. Pinch after two to three weeks, leaving two or three sets of leaves on the plant. Pinching is optional, some growers preferring not to pinch. The harvest then acts as the pinch. This pinch will generate two to three top shoots, and additional ground shoots will begin emerging from around the root system. When these newly developed shoots reach about 15 to 18 inches (38 to 46 cm) total average plant height, begin the short day cycle (bud initiation period). If cyclic mum lighting is used, a total day length of 16 to 17 hours is recommended. This vegetative period takes from five to eight weeks, depending on growing temperatures and cultivar.

The short day cycle (black cloth period) should include no more than 12 hours of total light for optimum flower initiation. This short day period takes from four to five weeks, if night temperatures of 60 to 65F (16 to 18C) are maintained. Total crop time generally takes from 10 to 16 weeks, depending on growing temperatures, cultivar, and the flush the plants are being forced from. Generally, growers obtain three to five flushes per plant before production stock is discarded.

When all flowering stems are harvested from a planted area, cut the plants back *totally* to the ground, fully removing partially cut stems, stubble, and old wood. If old wood is left above the ground, it will produce prematurely budded stems of no value. Total removal of all old stems stimulates the root system to produce underground shoots. These underground shoots will develop into the quality stems and flowers for the next flush or growing period.

When flowering stems have swollen (the buds beginning to show color), the short day period is completed. During harvest, apply long days again to ensure that ground shoots which begin developing during harvest will maintain their vegetative growing state. The grower can generally expect 3½ to 4½ flower flushes per year.

Outdoor growing

Natural-season flowering will occur in August to September (only once per year), depending on growing conditions and the variety being grown. No long-day extension or short-day manipulation is required for this flowering period. *Solidago* will overwinter in most areas. After overwintering, the field crop should be cut back (totally to the ground) sometime in May or early June. The ground shoots will develop as previously described and bloom again in fall.

In some temperate growing areas, as in southern California and Florida, a second flowering crop harvested in the spring may be possible. Natural warm night temperatures will determine if this is possible.

Harvest flowers when they are half open. Store and ship dry. Good postharvest practices will give shelf life of seven to 10 days. The harvest period itself usually lasts seven to 10 days, first to last.

SQUASH (SEE *CURCUBITS*)

STATICE (SEE *LIMONIUM*)

STOCK (SEE *MATTHIOLA*)

STRAWBERRY (SEE *FRAGARIA*)

STRELITZIA (BIRD OF PARADISE)

❖ *Tropical* (Strelitzia *spp.*). *Vegetatively propagated. Grown as a tropical cut flower or flowering tropical foliage plant in warm climates.*

The common orange-and-blue bird of paradise flower *(Strelitzia reginae)* has been used in the floral industry as an unusual tropical cut for years. *S. nicolai* produces similar flowers in white and does well as an interiorscape foliage plant. It has large, shiny leaves and grows in a planar fashion, similar to a compass tree. Flowering usually occurs when the plant is roughly 10 feet (3 m) tall and five years old. It will bloom in an interiorscape that has at least 400 footcandles (4 klux) of daylight.

Strelitzia nicolai and *S. reginae* can be grown from seed, but they take seven years to flower. In Hawaii division of the rhizomes is common. Stage 4 liners from tissue culture are often used to begin plants in 6- or 8-inch (15- or 20-cm)

S

pots, two seedlings per pot. Crop time to a finished 10-inch (25-cm) pot will be 14 to 18 months.

Spacing should be pot tight until plants reach 12 to 14 inches (30 to 36 cm). They can then be potted up to a 10-inch pot, two plants per pot, and finished spacing should be 18 to 24 inches (46 to 61 cm).

Any well-drained common potting mix with high organic matter is good. The pH should be between 5.5 and 6.5. In good light these plants have high water consumption, but they should not be kept wet. Shade cloth denser than 30% should not be used, and temperatures are best kept between 65 and 90F (18 and 32C).

Strelitzias do not like high soluble salts. A slow-release granular (12:6:8) can be used as a topdressing, or a light nutrient solution can be applied under drip irrigation.

Scale and mealybugs can sometimes be problems, and some fungal diseases occur. There are several well-known pesticides and broad-spectrum fungicides labeled for strelitzia.

It is important to acclimatize these plants for shipping. Increase shade, cut back water, and leach well two to three weeks before shipping. Temperatures during shipment should range from 55 to 80F (13 to 27C).

Due to the plant's slow growth rate and structure, careful handling is critical during shipping to prevent broken or damaged leaves. The plant simply will not "fill in" a missing leaf. Therefore, any void will remain permanently and can cause serious disfigurement.

Generally, finished plants in 10-inch (25-cm) pots are 3 feet (91 cm) tall, in 14-inch (36-cm) pots 5 to 6 feet (1.5 to 1.8 m) tall, and 17-inch (43-cm) pots 7 to 8 feet (2.1 to 2.4 m) tall. Strelitzias can be field-grown as specimen plants in sizes 10 feet (3 m) and up. They should be root-pruned six months before digging. After digging place each plant in a 28- to 60-inch (0.7- to 1.5-m) container of potting mix, irrigate well, and place in sun again until rooted to the sides. To acclimatize, place the plants in a shadehouse 45 to 60 days before shipping, cut back on water, and leach well.

STREPTOCARPUS (CAPE PRIMROSE)

by Bob Michael
Oglevee, Ltd.
Connellsville, Pennsylvania

❖ *Annual (***Streptocarpus** × **hybridus***). 992,250 seeds/oz (35,000/g). Germinate at 75F (24C). Leave seed exposed to light. Grown as a flowering potted plant. Propagation is primarily vegetative, although recently, newer, more tolerant seed varieties have become available.*

Streptocarpus is an easy-to-grow flowering pot crop requiring conditions similar to those of gloxinia and African violet (fig. 1). A member of the Gesneriaceae family, *Streptocarpus* has deep green leaves, which contrast with vibrant flower colors ranging from soft pastels to hot pinks and purples. It is adaptable to year-round growing and, under the proper conditions, will flower almost continuously. This multiuse crop is ideal for a windowsill with filtered, indirect light. It is also used as an annual,

FIG. 1. Thalia streptocarpus, a FloraStar award winner, makes a showy pot plant.

planted outside in a shaded setting, which gets light only early in the morning or very late in the afternoon.

Newer hybrids of *Streptocarpus*, or cape primrose, are making commercial production both easier and quicker. Since the introduction of Concorde, the first F_1 *Streptocarpus*, newer hybrids have been introduced that offer a fuller and longer color range and are more tolerant of a wide range of greenhouse environmental conditions. Flowering plants can be produced from a seed in about four to five months in the summer or five to six months during the winter. From a rooted liner cutting, they take seven to eight weeks in the summer or 10 to 14 weeks in the winter.

PROPAGATION

Vegetative. *Streptocarpus* can be produced by three techniques involving leaf sections. One technique is to use leaf tip sections about 3 inches long and 1½ inches wide (8 and 4 cm). Stick the bottom half of the leaf section into the propagation medium. A second technique involves removing the top and bottom portion of the leaf just on either side of the midvein. Turn the leaf over and make quarter-inch (0.6-cm) incisions 1 inch (2.5 cm) apart on the midvein and stick it lengthwise in the propagating media. The third and most common technique involves cutting the leaf in half just above and below the midvein and sticking these two leaf sections lengthwise in the propagating medium.

The medium for leaf section propagation can be straight vermiculite; one part peat, one part perlite, and one part vermiculite; or straight peat. The peat should be of the highest quality; blonde peat is preferred. The medium should be

S

kept moist at all times, but not oversaturated, and its temperature should be between 70 and 75F (21 and 24C). Once the leaf sections are stuck, an initial fungicide drench is required. Do not mist the leaf sections; this would promote rot and *Botrytis*. Start feeding these leaf sections 10 days after propagation at 100 ppm with 20-10-20. Feed every other irrigation; *Streptocarpus* is very sensitive to high salts. Plantlets will start to emerge and can be transplanted in about three months into a cell pack.

Seed. *Streptocarpus* seed is very fine, 35,000 seeds per gram. Sow the seeds thinly on the surface of premoistened peat and place the rooting tray on a heated bench at 75F (24C). Do not cover seed with peat; use glass or plastic to cover your rooting tray to keep it from drying out. Remove the covering as soon as the seeds start to germinate. Do not place the rooting trays in direct light. Three to four weeks after sowing, young seedlings can be pricked out of the seed tray and transplanted into cell packs. These young seedlings are each characterized by one large single leaf, which is prominent throughout the life of the crop. Place the seedlings in a glass greenhouse at 65 to 70F (18 to 21C) until new growth is visible, then drop the temperature to 60 to 65F (16 to 18C). Keep the young plants moist but not wet, and protect them from direct sunlight; shading is necessary to reduce the light intensity to 750 to 1,000 footcandles (8 to 11 klux). These young plants are ready to transplant into their final pot size once the flower buds are visible.

Basic crop culture

Whether *Streptocarpus* is produced from seed or rooted liner cuttings, the following cultural information is the same. Transplant the young plants into a balanced peat-perlite mix that has a high degree of soil porosity and drainage. Some soilless mixes may require the use of supplemental fritted trace elements. A soil pH of 5.6 is preferred; however, a range of 5.2 to 5.6 is acceptable. *Streptocarpus* can be planted in a variety of pot sizes, from a 4-inch (10-cm) azalea with one plant per pot, to a 6½-inch (17-cm) pot with two plants, to an 8-inch (20-cm) hanging basket with three plants, or a 10-inch (25-cm) basket with four plants.

Start feeding transplants one week after potting. *Streptocarpus* needs high nutritional levels for the best plant and flower quality. However, it is sensitive to high salts. Maintain salt levels below 100 mhos (10^{-5}). Feed at a rate of 100 to 125 ppm N and K using 15-15-15 or 20-10-20. Allow a 10% drip-through to help avoid salt buildup. Leaching is important to reduce salt buildup and should be done every third irrigation. If watering overhead, cold water (water that is 10F, −12C, cooler than the air temperature) can cause leaf spotting. Avoid spotting by using subirrigation, drip tubes, or tempering the water temperature. Let the media dry out between waterings to avoid root and crown rot.

Streptocarpus grows best at 1,000 to 1,200 f.c. (11 to 13 klux) of light. At lower temperatures, they will withstand higher light intensities of up to 1,500 f.c. (16 klux), and at higher temperatures during the summer, light intensity less than 1,000 f.c. is necessary.

The recommended temperature range for optimal growth is 65F (18C) nights and 75F (24C) days; begin venting at 70F (21C). Growing temperatures from 50 to 60F (10 to 16C) will not affect the quality, but they will increase crop time. Temperatures from 80 to 90F (27 to 32C) will produce quicker vegetative growth, but they will reduce the plant and flower quality.

Streptocarpus performs well in humid conditions, but do not allow any standing water on the leaves or flowers, especially at night. This would spot both the leaves and the flowers, reducing their salability. Horizontal airflow is extremely helpful in drying off the plants as well as reducing the incidence of *Botrytis*.

Bud removal for the first four weeks will generate more vegetative growth early on, making the finished product fuller and more salable. Bud removal will also even out the flowering of the crop so that a higher percentage of the crop is marketable on one ship date.

The main insect pest problems of *Streptocarpus* are fungus gnats (especially on the seedlings), aphids, whiteflies, thrips, mealybugs, and cyclamen mites. The main diseases are crown rot and *Botrytis*. Proper management of the greenhouse environment, correct watering techniques, insect-scouting the greenhouse crops, and weed control both in and outside the greenhouse will reduce the occurrence of these pests and diseases.

VARIETIES

Seed varieties available are the Weismoor hybrids, with colors ranging from pink, white, blue, and lilac to pink and red-fringed. Flower size is 4 to 5 inches (10 to 13 cm). Concord Mix is early to flower, three weeks earlier than Weismoor. Concord flowers freely and has a good range of pastel shades. Delta Blue has a growth habit similar to Concord's but is two weeks earlier to flower. Royal Mix is two weeks later to flower than Concord and has a blend of rich velvety purple, deep pink, and red.

Rooted cuttings are available in the Olympus series, which is early to bloom. Olympus flowers are held well above the foliage and appear continuously. The leaves are small enough for 4- to 5½-inch (10- to 14-cm) pot production and sleeve easily. Colors range from bicolor dark blues and purples to pure whites. Bavarian Belle is another series propagated from cuttings that features larger flowers and smaller, more narrow leaves, compared to seed varieties. These are well suited for pot sizes ranging from 4- to 10-inch (10- to 25-cm) hanging baskets. Colors range from rich purple to pastel blue, soft pinks, and shades of deep

S

reds. There is also an unnamed series containing the varieties Helena, Alicia, Samantha, and Michelle. This series is noted for its short leaves, smaller than most other series, compact growth habit, rapid flower response, and floriferousness. The individual flowers are smaller than those of other series, but what it lacks in flower size, it more than makes up for in sheer volume of flowers. Colors range from white with lavender-pink centers to soft pink to red to purple. Recommended pot sizes are 4-inch (10-cm) azalea to a 6-inch (15-cm) hanging basket (three cuttings); 8-inch (20-cm) hanging baskets (four cuttings); and 10-inch (25-cm) hanging baskets (five cuttings).

SUNFLOWER (SEE *HELIANTHUS*)

SUTERA (BACOPA)

❖ *Tender perennial (***Sutera cordata***). Vegetatively propagated. Growers produce pots or baskets from purchased rooted cuttings.*

Bacopa makes an excellent addition to combination planters or as a hanging basket item. Bacopa also makes a tremendous ground cover. The cultural information that follows is from Proven Winners.

Plant rooted cuttings from January to March for spring finish. Use three to four liners per 8- to 10-inch (20- to 25-cm) pot; one to two liners per 4- to 6-inch (10- to 15-cm) pot. Allow 12 to 14 weeks for finishing. Use a well-drained, peat-perlite medium with pH slightly acid at 6.0 to 6.5. Pinch once at planting.

Grow plants at moderate to high light levels, 3,500 to 8,000 footcandles (38 to 86 klux). Maintain 65 to 75F (18 to 24C) day temperatures, 50 to 60F (10 to 16C) at night. Bacopa will tolerate cold temperatures to 30 to 32F (–1 to 0C) and high temperatures to 85 to 90F (29 to 32C). Maintain plants slightly on the dry side, but never to the point of wilting. Avoid wetting foliage. Use 200 to 250 ppm N, 65 to 75 ppm P, and 125 to 165 ppm K constant liquid feed. Otherwise, apply a periodic feed of 300 to 400 ppm N, 100 to 150 ppm P, and 200 to 300 ppm K.

Watch for whiteflies. Don't use insecticidal soaps, which could cause foliage burn. Bacopa is somewhat susceptible to *Pythium* in damp, low-light situations. Drench with a fungicide, such as Subdue, for *Pythium* at planting.

SWEET PEA (SEE *LATHYRUS*)

TAGETES (MARIGOLD)

by Jim Nau
Ball Horticultural Company
West Chicago, Illinois

❖ *Annual. 9,000 seeds/oz (317/g). Germinates in seven days at 72 to 75F (22 to 24C). Cover the seed lightly after sowing. African or American marigolds (*Tagetes erecta*); French marigolds (*Tagetes patula*); Triploid marigolds (*Tagetes erecta × patula*).*

One of the premier plants of all annuals, marigolds perform well in dry as well as moist conditions and display color all season long. No annual garden is complete without the bright colors that marigolds add (fig. 1). They are excellent in borders, as cut flowers, in the landscape, and in any other setting which needs long-term color.

Marigolds are available as either single, semidouble, or fully double flowers, in yellow, orange, and gold. Red and crimson are available in

FIG. 1. Stately, formal elegance is the best way to describe American marigold flowers. Newer breeding has enabled growers to produce them in packs for sale with flowers. The variety shown: Perfection Gold.

the triploids and French types, though these colors are absent in African varieties. However, there are white-flowering plants as well among Africans. In double-flowering varieties, there are crested doubles, where the blooms appear mounded and full, or an anemone form, which is a flat, wide flower whose center is recessed.

Flowers range in size from 1 inch (2.5 cm) on French varieties to as broad as 5 inches (13 cm) on some of the African varieties. Plants range in height from 6 inches (15 cm) to 3 feet (91 cm) on plants that fill in well when spaced out 12 inches (30 cm) apart in the garden. Marigolds have a strong scent, which is one of their key faults. As cut flowers, strip the foliage on lower stems before shipping to decrease the scent.

Production note: Marigolds are susceptible to micronutrient toxicity, especially to iron and manganese, at low pH. Be sure to keep the pH of growing

media above 6.3 throughout production. You can manage pH with nitrate fertilizers (14-0-14, 15-0-15), or begin with a high pH charge in media. As pH goes
down, iron uptake increases, which causes yellow spots on leaves—iron toxicity.
There's also variety sensitivity. The Boys do not show symptoms, while African
marigolds are the most sensitive.

FRENCH MARIGOLDS

Crop times differ from variety to variety, but many will flower in eight to 11
weeks from sowing seed. Some varieties flower in only seven weeks after a
spring sowing. For green packs, allow six or seven weeks depending on variety.
For blooming cell packs, allow eight to 11 weeks, and up to 13 weeks for 4-inch
(10-cm) pot sales, with one plant per pot. For those preferring baskets, allow 11
to 14 weeks for five or six plants per 10-inch (25-cm) basket.

In the southern United States, allow seven to nine weeks for flowering packs,
10 to 11 weeks for flowering 4-inch (10-cm) pots. Plant to the garden around the
end of May, after all danger of frost has passed, and again in August for flowering until frost. Space 8 inches (20 cm) apart in full sun to partial shade. Plants
may heat-stall during the hottest parts of summer.

Among double-crested flowering varieties, plants grow 6 or 8 inches (15 to
20 cm) high, and on occasion, up to 12 inches (30 cm) tall. The most dwarf variety on the market is the Little Devil series, which has 1-inch (2.5-cm) blooms in
five separate colors on plants to 8 inches (20 cm) tall.

Little Devil Fire, a top performer, has maroon petals with a gold border on a
double-flowering, crested type bloom. The Boy, Janie, and Little Hero series are
slightly taller, growing up to 10 inches (25 cm) tall. Janie Tangerine is an excellent variety with deep orange flowers on mounded plants. Yellow Boy is still one
of the top-selling yellow-flowering varieties in the U.S. trade today. Unlike many
yellow-colored flower varieties on the market, Yellow Boy is a true yellow
(bright yellow) as opposed to a golden yellow.

Finally, the Bonanza and Hero series, each with a number of strong performers, feature the largest flower size of any French marigolds on the market.
The Bonanza series ranges from 10 to 12 inches (25 to 30 cm) tall, with flowers that grow to 2 inches (5 cm) or more across. Of the six varieties available,
Bonanza Yellow, Bonanza Flame, and Bonanza Orange are three of the series'
best overall performers. Hero's flowers range up to 2½ inches (6 cm) across on
plants to 12 inches (30 cm) tall. Hero Red is a red-and-yellow bicolor with a red-
flecked yellow crest. Hero Flame is similar to Bonanza Flame except it has a
larger flower overall, and Hero Orange is a deep orange that rivals Bonanza
Orange as being the deepest orange available.

Of fully double varieties, there is only one series of any merit, the Gates.
Golden Gate is a large-flowered variety with fully double blooms in mahogany

with gold edges (fig. 2). The full double-ness of the variety comes out in 4-inch (10-cm) pots and in the garden. Small cell packs tend to restrict the plant and decrease overall performance. Once planted to the garden, Golden Gate will bloom fully double. Other Gates include Orange, Yellow, and Garden, a mix.

Among double-flowering anemone, or Queen types, the best varieties are the Safari and the Aurora series. Safari Queen, named for her ancestor, Queen Sophia, has russet red flowers edged with golden bronze. Performance surpasses that of the well-known Queen Sophia. In the Aurora series, Aurora Yellow Fire is an unusual variety with golden yellow petals that have a deep crimson blotch at the base of each petal. Auroras are available in five additional colors and a mix.

In single-flowering marigolds, the Disco series is the key variety in the U.S. trade. Single-flowering marigolds are not as tolerant of heat and humidity as the double-flowering varieties. They perform best in British gardens and areas of North America that have similar conditions.

FIG. 2. The large, fully double flowers of Golden Gate marigolds really shine in patio pots. As this fine pot demonstrates, marigolds offer a lot of flower power for home gardeners.

AFRICAN MARIGOLDS

African varieties (often called American marigolds) flower earlier under short days. For sowings after February 15, start with short days upon germination and continue for 14 days, maintaining darkness from 5 P.M. to 8 A.M. Short days can actually be started from the time of sowing on, using inverted standard flats to cover the germination trays.

Although African marigolds require up to two more weeks to flower than their French and triploid counterparts, a short-day treatment helps produce a smaller overall plant habit, plus earlier and more uniform blooming. Allow nine to 10 weeks for green packs and 11 to 12 for flowering cell packs. For flowering 4-inch (10-cm) pots, allow 12 to 13 weeks, using one plant per pot. In the southern United States, allow 10 to 11 weeks for flowering green pack sales, 11 to 12 weeks for flowering 4-inch pots.

T

One of the best varieties is All Seasons Discovery Yellow. It makes an excellent 4-inch (10-cm) pot or bedding plant, with fully double flowers to 3-inch (8-cm) across on plants to 14 inches (36 cm) tall. The Antiqua series, available in four colors, is early and uniform.

The Inca series is still one of the premier varieties for landscaping. The three separate colors of orange, yellow, and gold have the largest flower size of any double-flowering African. Flowers grow to 5 inches (13 cm) across on plants to 18 inches (46 cm) tall in the garden. Excel, Marvel, and Perfection, all from newer breeding, are trying to edge Inca out with faster crop times and better flower quality.

Among taller African varieties, the Lady or Galore series are two of the best choices for cut flower growers, and also can be used as annual hedges in the home garden. Primrose Lady, with flowers to 3 inches (8 cm) across, is the best yellow-flowering African on the market. If you are looking for hedge marigolds, try Gold Coin or Jubilee.

TRIPLOID MARIGOLDS

Triploid marigolds provide the longest lasting color of any type of marigold. They will usually stay in color through August's hot weather. Though germination is considerably lower than for French and African types, triploid marigolds have the advantage of quick finishing. Approximately 90% of the varieties bloom in just seven to eight weeks. However, since they are shy to set seed, it is sometimes difficult to get the varieties you want.

Green packs require seven to eight weeks, while flowering, salable cell packs take up to nine weeks. For 4-inch (10-cm) pots and 10-inch (25-cm) hanging baskets, allow 10 to 11 weeks. In the southern United States, allow seven to eight weeks for flowering packs, 10 to 11 weeks for flowering, 4-inch pots.

THUNBERGIA (BLACK-EYED SUSAN VINE)

❖ *Annual (Thunbergia spp.). 1,100 seeds/oz (39/g). Germinates in 12 days at 70 to 75F (21 to 24C).*

Where a splash of real color is wanted in annual climbers, porch boxes, or hanging baskets, you'll find *Thunbergia alata* hard to beat. It's easy and cheap to propagate, fast-growing, and has lots of black-eyed orange, buff, and yellow flowers (fig. 1). It is also very useful as a screening material when planted to open ground and allowed to cover a trellis or fence.

Presoak seed overnight before sowing to speed germination. Allow eight to nine weeks for flowering 4-inch (10-cm) pots that are sown direct without transplanting, since the roots are tender and resent frequent repotting. If seedlings are transplanted, move them 12 to 18 days after sowing. For 10-inch (25-cm) hanging baskets, allow 10 to 11 weeks, with six to seven seedlings per container. Plants trail tremendously and flower profusely.

The Susie series, a fairly recent introduction, consists of three distinct colors—orange, white, and yellow—with eyes, as well as a mixture that contains both eyed and solid colors. These are especially suited for hanging baskets. Alata Mix includes yellow, buff, and orange, all with black eyes.

FIG. 1. Thunbergia is an easy-to-grow hanging basket crop.

TOMATO (SEE LYCOPERSICON)

TORENIA (WISHBONE FLOWER)

❖ *Annual* (**Torenia fournieri**). *500,000 seeds/oz (17,637/g). Germinates in seven to 15 days at 70F (21C). Seed should be left exposed to light during germination and maintain 70F (21C) soil temperatures.*

Once popular as a summer pot plant, *Torenia* is chiefly used today as a summer bedding annual, where it makes a surprisingly bright showing, flowering from June until frost. It can also be used as a midwinter pot plant, at which time its bright blue, burgundy, orchid, rose, or violet blooms are most welcome. Pinching will help develop a bushy plant.

Torenia fournieri is most commonly known as the wishbone flower, named for the wishbone formed by the fusion of two stamens in each flower. A less known common name is Florida pansy, as this is a nonfading substitute for pansies, which can lose their color in southern climates. Although individual flowers resemble those of the snapdragon, a member of the same family, torenias have a subtle charm all their own.

T

Torenia is versatile. Though doing best in cool semishade, it will tolerate a very wide range of conditions—from just a few hours of morning sun to full sun in many areas, if moisture is adequate. Even the leaves are attractive, a glossy green, which become bronzed as the cool nights of fall arrive. Although it's not the first flower to bloom in the garden, it makes up by looking good all summer, right up to frost. Newer, compact cultivars make excellent 4-inch (10-cm) flowering pot plants that last well on a sunny windowsill indoors. Best of all, *Torenia* is a quick and easy greenhouse crop to produce.

PROPAGATION

Transplant bought-in seedlings to cell packs or 4-inch (10-cm) pots in three to four weeks.

For your own plugs, germinate seed in a growth chamber at 70 to 75F (21 to 24C) soil temperature. Remove from the chamber at early emergence, usually six days. Light them for 18 hours at 750 footcandles (8 klux) to hasten flowering. Fertilize in the plug tray with 50 ppm N constant liquid feed. This feeding makes an attractive, salable plug with good, dark green foliage. Transplant into cell packs or 4-inch (10-cm) pots at four to five weeks.

GROWING ON

Transplant into well-drained, sterile media. For pot production, one seedling per 3- to 4-inch (8- to 10-cm) pot is sufficient; three to four seedlings fit a 6-inch (15-cm) pot. Use a soilless mix with a pH of 6.0 to 6.2.

Maintain high light levels when growing on. Full light intensity will result in compact plants, even where temperatures are warmer than the ideal. *Torenia* is nonphotoperiodic and can be flowered year-round.

Torenias can be grown cool; 58F (14C) nights and 62F (17C) days are ideal. They can be grown at warmer temperatures in high-light areas and still remain compact, but growing warm in low light will cause stretching.

Torenias are heavy feeders. Provide constant feed of 200 ppm N from a Peat-Lite Special formulation with trace elements. Proper fertilization enhances the green leaf color for a more attractive plant at point of purchase.

Minimize the need for chemical height control by growing with high light and low DIF (difference between day and night temperatures). Bonzi has been effective at controlling height when used as a drench at 4 to 8 ppm or as a spray at 30 ppm.

Torenias may be affected by common greenhouse pests, such as whiteflies and aphids, which are usually controlled by standard spray programs. No phytotoxicities have been noted in trials. For control of powdery mildew, grow in a well-ventilated environment and watch for excess water and humidity.

VARIETIES

The Clown series of F$_1$ performs in all kinds of weather. These plants are densely branched and compact, blooming all summer long. Plants grow 8 to 10 inches (20 to 25 cm) tall. Clown is great for packs, pots, or color bowls and comes in eight colors.

TRACHELIUM (THROATWORT)

❖ *Tender perennial (Trachelium caeruleum). 2,660,000 seeds/oz (93,827/g). Germinates in five to seven days at 65 to 68F (18 to 20C). Leave seed exposed to light for germination. Keep seed and seedling trays under short days to prevent premature flowering.*

Trachelium has become a popular cut flower which is used as filler in bouquets (fig. 1). Its panicle flowers come in striking lavender and dark blue as well as creamy white and pastel pink. Most growers purchase plugs rather than propagate *Trachelium*.

Cultural information here is provided by Vegmo Plant, Rijsenhout, the Netherlands. Crop time varies from 10 weeks for July plantings to 17 weeks for November through January plantings. Plant out at a density of 54 plants per square yard (65/m^2) for single-stem production; 18 plants per square yard (20 to 24/m^2) for pinched production.

Flowering is encouraged by long days and inhibited by short days. Provide continuous 16-hour days, or use night interruption lighting (mum lighting). Light at an intensity of 15 watts/m^2. Light until harvest to avoid misshapen, pyramidal flowers. For crops planted from the end of May until early July,

FIG. 1. The delicate lacelike lavender, blue, or white flowers of trachelium burst onto the cut flower scene in the early 1990s. Ever since, it has been turning up more and more frequently as the mixed bouquet filler of choice for discriminating retailers. The variety shown: Lake Tahoe.

T

encourage vegetative growth before initiating flowers by applying short days (12 hours) during weeks two and four. The exact length of short day treatment depends on light levels and temperatures.

Tracheliums prefer cooler temperatures and therefore do not do well out-doors in the hot summers of the South or Midwest. For fall and winter crops, grow at 50 to 52F (10 to 11C) for the first three weeks, then raise night temperatures to 58F (16C) and days to 60F (16C). Spring crops can be grown at 60F nights and 62 to 68 F (17 to 20C) days. In colder climates, additional heat may be necessary to maintain 52F nights and 62F days.

Plants require average fertility. During bud formation, apply additional potassium nitrate. Be sure to rinse off the plants afterward to avoid burning. Do not use ammonium fertilizers on tracheliums. Also, tracheliums are very sensitive to high manganese levels in the soil. Use care when steaming, as steam releases manganese bound in the soil, which can result in manganese toxicity. Toxicity symptoms will show as leaf damage during the last stage of development. Steaming for four hours or less and using iron chelate prevents leaf damage. Apply iron chelate three to five weeks after planting.

Harvest the flowers when the panicles are three-quarters open. Flowers normally have a 10-day to two week vase life in water.

Varieties. Blue Wonder (blue), Merli Blue (blue for summer production), Lake Superior (dark blue and purple), Lake Avalon (pink), and Lake Powell (white) are all available as plugs.

VERBENA

❖ *Annual (Verbena × hybrida). 10,000 seeds/oz. (353/g). Germinates in 20 days at 65F (18C) dark.*

One of our old standby bedding plants, *Verbena* owes its popularity to its bright, varied colors, low-cost propagation, and a free, all-season flowering habit, even under full sun (fig. 1). There are two basic types of verbenas. The spreading type is procumbent or carpetlike, covering a fairly wide area. The upright type is free-branching but distinctively bushy in habit. Though both types tend to go in and

FIG. 1. Colorful, low-growing verbena is an important part of every bedding plant grower's "other" bedding plant assortment. The variety: Romance Violet with Eye.

out of bloom during the summer, the spreading varieties seem to be more free-flowering and less prone to cyclical flowering patterns.

Propagation is usually from seed sown in mid-February for 3-inch (8-cm) blooming pots by Memorial Day. They may be sown as late as the end of March, transplanted into packs or flats, and will still make a good showing outdoors, but not as early as when started in pots. Germination, which is not rapid, is the key to success with verbenas. Chill seed for one week before sowing. The night before sowing, water in flats and use a Banrot drench at one half teaspoon per gallon (7.8 ml per 4 l) of water. Sow seed without additional watering in. Cover flat with black plastic until germination begins.

Because of their genetic makeup, most verbena varieties are poor germinators; following these guidelines can increase germination percentages to 65% or better. Verbena seed is extremely susceptible to rot when moisture is high. Overwatering at time of sowing will reduce germination. Some growers have had better success by preparing the seed flat in the late afternoon, watering it, and then sowing the seed the following morning without any further watering. Once established, *Verbena* should be grown at about 50F (10C) nights and should be pinched back early. Crop times differ for spreading and upright types, with the spreading types generally flowering 10 to 18 days sooner than their upright counterparts.

Germinate plugs at 75 to 80F (24 to 27C). Reduce temperatures to 70F (21C) when the cotyledons expand. When true leaves develop, decrease temperatures to 65 to 70F (18 to 21C).

VARIETIES

In the spreading class, the best variety is the Romance series. Romance is basal branching, with numerous clusters of quarter-inch (64-mm) blooms in a wide range of colors. Romance grows to 10 inches tall (25 cm) and works well in containers and in the annual border. Also in the spreading verbena class is the Amour series and the Quartz series. Amour Pink is a unique color.

As for the upright verbenas, the best are the Novalis series and Trinidad, which is a vibrant rose flower color on plants to 10 inches (25 cm) tall. Novalis grows to 10 inches tall in the garden and is characterized by having basal-branching plants that are free-flowering. Novalis is a good pack performer, too.

VERBENA SPECIES FROM SEED

Verbena speciosa Imagination germinates in five to seven days at 70F (21C). Imagination blooms continuously through the summer, throwing hundreds of deep violet blue, one-half-inch (1-cm) flowers in clusters. Plants spread up to 2 feet (61 cm). Once Imagination becomes established, it is a superb, drought-tolerant landscape performer. Imagination is also great in color bowls and combination pots (fig. 2).

Sterling Star, also a *V. speciosa*, is even larger than Imagination. The plants have similar habit and form, and the lavender blue flowers appear all summer. Sterling Star is excellent for sloping hills or parks. You can "plant it and forget it."

Verbena canadensis germinates in five to 10 days at 75F (24C). A tender perennial, which should be treated as an annual in northern areas, these plants are mildew-tolerant and grow 12 to 14 inches (30 to 36 cm). The flowers are deep rose-violet.

A new type of verbena is sweeping the country: the vegetative varieties. Tapien, a leading variety, is an extremely floriferous verbena ground cover that may become a bedding plant standard.

FIG. 2. For your customers seeking something new, try Imagination *Verbena speciosa*, an All-America Selections award winner. Imagination makes a lovely hanging basket.

NEW VEGETATIVE TRAILING VERBENA

Trailing verbena is also a great addition to color bowls. Will Healy, Ball Horticultural Company, West Chicago, Illinois, offers these culture tips. Grow vegetative trailing verbena at 68 to 75F (20 to 24 C) days and 62 to 65F (17 to 18C) nights. At low temperatures, verbena foliage yellows. Plants do best in cool temperatures of 55 to 65F (13 to 18C) in full sun. Provide 4,500 footcandles (48 klux) for best growth and compact plants. When temperatures are high, excessive light causes leaf necrosis.

Use well-drained media with a pH of 6.2 to 6.5. Because plants are sensitive to high soluble salts, make sure water alkalinity is below 140 and EC is below 0.5. Verbena roots are very sensitive to water management, so avoid under or overwatering. Maintain good fertility levels (nitrogen at 150 to 250 ppm once a week) to keep plants actively growing.

Four-inch pots can be finished from one liner in five to six weeks; 6-inch pots from two liners in six to seven weeks, and 10-inch baskets from four to five liners in 10 to 12 weeks.

Verbena are day length neutral, but are affected by the total amount of accumulated light. Flowers develop after three to nine leaf pairs are initiated. Pinch plants above the fourth or fifth set of leaves (2 to 2½ inches above soil) once roots have reached the edge of the growing container. Pinch plants two to three times during production to increase branching and improve plant shape. To encourage flowering rather than vegetative growth, avoid excessive use of ammonia fertilizers and provide good light levels and proper irrigation.

Apply Cycocel at 1,500 to 2,000 ppm, five to six days after the pinch and apply every two weeks thereafter to control growth. Negative DIF (cool days/warm nights) is also effective for height control. Florel promotes plant branching; avoid applications within eight weeks of sale.

Plants are susceptible to a variety of insects and disease, so maintain good cultural practices. Avoid stem canker (*Botrytis*) by keeping media dry, but do not let media dry out completely. Thrips, whiteflies, and aphids may also be problems.

Tapien is available in bright pink and violet. Plants are fast growing and make a bright colored ground cover when planted outdoors in high light and cool temperatures.

VERONICA (SPIKE SPEEDWELL)

by Jim Nau
Ball Horticultural Company
West Chicago, Illinois

❖ *Perennial (***Veronica spicata***). 221,000 seeds/oz (7,795/g). Germinates in seven to 14 days at 65 to 75F (18 to 24C). Leave the seed exposed to light during germination.*

Veronicas are excellent summer-flowering perennials that bloom freely the first season from a winter sowing. The blue-, white-, or rose-pink-flowering spikes are borne in June and July, and the plants will flower until August. Veronicas as a group are hardy to USDA Zones 4 to 8.

Of easy culture, *V. spicata* is salable green in a pack in 10 to 13 weeks after sowing. It will flower during the summer the first year after sowing. Plants flower more dependably the second season after sowing.

In varieties, *V. spicata* is available in both seed- and vegetatively propagated material. Those that are seed-propagated will be of several different habits and heights as well as differing shades of blue. Though strong garden performers without being too vigorous, more uniform selections are available from cuttings or division.

FIG. 1. Blue Bouquet veronica is a strong garden performer and makes striking 4-inch pots.

From seed you can obtain *V. spicata* species or Blue Bouquet (fig. 1), with dark green foliage and medium blue flowers. This is the best seed variety available; sow in mid-January for flowering plants in May and June.

Vegetative varieties—all spectacular—are also available. The best known is Sunny Border Blue, a Perennial Plant Association plant of the year. Plants grow 15 to 20 inches (38 to 51 cm) tall and have deep violet-blue flowers. Blue Charm and Crater Lake are also blue, while Icicle is white, and Red Fox is deep rose pink.

Related material includes *V. repens,* which is a dwarf-flowering plant to 1½ inches (4 cm) tall and 12 to 18 inches wide (30 to 46 cm) that flowers in April and May. These plants are best used in rock gardens where the soil is allowed to dry out. Plants are salable in 10 to 12 weeks after sowing and are available in single-flowering forms only.

Finally, *V. incana* (woolly speedwell) is the most unique of the three veronica types but also the most difficult to germinate and grow. If the variety does not germinate well, place the seed into moist peat and refrigerate for 10 days to two weeks; then sow as noted above. *V. incana* can succumb to high humidity and heat, even during germination. Plants have a gray-green foliage, like dusty miller, and require 12 to 14 weeks to be salable green in the cell packs. Neither *V. repens* nor *V. incana* will flower the same season from seed.

VINCA (SEE *CATHARANTHUS*)

VINCA (COMMON PERIWINKLE, COMMON MYRTLE)

by Jim Nau
Ball Horticultural Company
West Chicago, Illinois

❖ *Perennial* (**Vinca minor**). *Vegetatively propagated from tip cuttings.*

This trailing woody vine is a popular groundcover. Evergreen and easy to maintain, the plants grow 6 to 8 inches (15 to 20 cm) tall and spread several feet (1 m or more). Flowers are lavender blue and single, contrasting nicely with the deep green, oval foliage.

Vinca vine is easily propagated from tip cuttings whenever plants are actively growing. Take 2½-inch (6-cm) cuttings, which will be salable in five to six weeks in a 135-cell tray. To even plants up, trim cuttings once before sale. Pot commercial transplants (2¼ inches, 6 cm) in the fall for spring sales in 4-inch (10-cm) pots or quarts the following spring.

Varieties include *V. minor* var. *alba,* with white flowers. Areola has leaves with cream-yellow centers. Bowles variety has deep blue flowers on plants that mound more toward the crown and have thicker, more glossy foliage. Miss Jekyll is pure white on dwarf plants with small leaves. Sterling Silver is variegated with cream-yellow veins down the center of each leaf; its flowers are blue.

Vinca major has larger leaves on plants that are fuller and more robust. Flowers are also blue and single, measuring 1 to 2 inches (2.5 to 5 cm) across. *V. major* grows faster than its relative and makes an excellent addition to hanging baskets and combination pots and bowls. Plants are hardy in Zone 7 and south.

VIOLA (JOHNNY JUMP-UP, VIOLET)

by Jim Nau
Ball Horticultural Company
West Chicago, Illinois

❖ *Perennials (* **Viola tricolor, V. cornuta, V. odorata**). *30,000 to 40,000 seeds/oz (1,058 to 1,411/g). Germinates in seven to 14 days at 65 to 70F (18 to 21C). Cover seed lightly after sowing.*

Invaluable as edging plants, these violas are also very effective when used in mass plantings to front a shrub border. Although not as large as pansies, they are more free-flowering and are available in a wide range of colors—yellow, red, purple, lavender, apricot, and white. Their culture is much the same as for pansies. Late summer or fall sowings are, if anything, even more successful because of violas' unusual hardiness. Violas will do very well in light shade, although they prefer full sun. They will thrive in any good garden soil, but in order to do their best, they should be planted in a fertile soil with a good supply of organic matter. Hardy to Zones 6 through 8, plants are best treated as annuals or short-lived perennials.

V

In general, for either spring or fall sales, allow 11 to 12 weeks for flowering cell packs when grown at 50 to 55F (10 to 13C) nights.

V. tricolor got its name from its unusual tricolor blooms and free-flowering performance. Although the flowers themselves are small, they appear in abundance. The flowers are naturally colored deep purple, purple-black, yellow, white, and/or blue. Today's varieties feature all of these in various combinations. Alpine Summer, an F_1, has true tricolor flowers of deep purple, lavender, and yellow. This plant, more prostrate than the well-known Helen Mount, has slightly smaller three-quarter-inch (2-cm) flowers. Helen Mount, the standard of this species, is known as Johnny jump-up. It is less uniform in habit and flower color than Alpine Summer. Blue Elf, also known as King Henry or Prince Henry, is vivid purple with lavender and white shading.

V. cornuta, the horned violet, is similar to *V. tricolor* in appearance and culture, but the plants look tidier and more dwarf. These violas require one to two weeks longer to flower. Two new series have livened this class with more uniform plant habit, earlier flowering, and bright colors in the past years: Jewel, an F_1 with separate colors, and the Princess series (fig. 1), also with separate colors.

V. odorata is the violet of poetry. The species is fragrant, although some varieties are without scent. Plants spread easily and are true perennials. Germination is more difficult. To ensure success, provide a moist chilling treatment. Plants can also be vegetatively propagated. Queen Charlotte, available as seed, is a fragrant violet-blue, while White Czar is pure white with deep purple whiskers. These flowers are not fragrant.

FIG. 1. The new violas—a multiflora approach to pansies—look great in flats and provide home gardeners with excellent performance. Try them for fall sales—gardeners can overwinter the plants. The variety: Princess Mix.

VIOLA × WITTROCKIANA (PANSY)

by Will Healy
Ball Horticultural Company
West Chicago, Illinois

❖ *Annual (***Viola × wittrockiana***). 20,000 seeds/oz (705/g). Germinates in seven to 10 days at 65 to 75F (18 to 24C). Cover lightly with vermiculite.*

Pansies are *the* undisputed annual for winter and spring flowering (fig. 1). In the South, pansies are planted in the fall, with flowering starting as soon as the temperatures reach the low 50s (10 to 12C). In the North pansies are planted during the fall or early spring, with flowering in late spring.

Growers should start planning their pansy programs in July. Table 1 highlights the key details for a successful pansy program.

When developing your pansy program keep in mind that there are several seed treatments available to improve the number of usable seedlings. Although raw or untreated seed produces acceptable results in the winter, most growers use primed or pregerminated seed for summer sowing (fig. 2). When raw seed is used in the summer, germination occurs in waves, which makes it difficult to manage seedling development within the tray. Pregerminated seed germinates uniformly, which results in uniform seedling development within the tray. Pregerminated seed will yield more usable seedlings per tray. Seeding accuracy quickly becomes the limiting factor in producing full trays when using pregerminated seed.

FIG. 1. During the 1990s pansies skyrocketed to the forefront of bedding plant production. Like many growers, the Yoshitomi Bros., West Linn, Oregon, grow their pansies in unheated cold frames. Pansies do best in the garden when they are planted in the fall and overwintered.

During the warm season of the year, growers use plug sizes ranging from 512 to 288 to improve germination and ensure plant survival after planting. At transplant time a rule of thumb is "As the temperature increases, the plug size should also increase to ensure survival."

V

TABLE 1 PANSY PRODUCTION SCHEDULING

Stage	Duration	Night temp (F)	Soil conditions			Fertilizer			Growth regulator
			Moisture	pH	EC[a]	Formulation	Rate (ppm)	Frequency	
Sowing to transplant, 512 plug									
1	3–7 days	65–75	Wet	5.0–5.5	0.50	No ammonia	0	None	None
2	7 days	65–75	Moist	5.5–5.8	0.50	10-0-14	50–75	1 wk	B-Nine
3	14–21 days	60–75	Medium to dry	5.5–5.8	0.75	20-10-20 14-0-14	100–150 100–150	Alt wk[b] Alt wk[b]	N-Nine or A-Rest
4	7 days	55–65	Medium	5.5–5.8	0.75	14-0-14	100–150	2 wks	A-Rest or Bonzi
Transplant to flower, 4-inch (10-cm) pot or 367 tray									
If fall	6–8 wks	65–75	Med to dry	6.0	1.0	20-10-20 14-0-14	100–150 100–200	1 time 1 wk[c]	Bonzi or Sumagic
If winter	8–10 wks	55–65	Med to dry	6.0	1.0	20-10-20 14-0-14	100–150 100–150	1 time 1 wk[c]	B-Nine

[a] mmhos
[b] Alternate between the two formulations, with one feeding per week.
[c] After onetime feeding with first formulation, feed weekly with other.

GERMINATION

Sow seed into a well-drained, disease-free medium. Use a medium with a pH of 5.5 to 5.8, low electrical conductivity, and little or no ammonium-form nitrogen. Cover seed lightly with coarse vermiculite.

For plugs, from sowing to radicle emergence maintain 70 to 75F (21 to 24C). Temperatures above 80F (27C) would reduce germination percentage and uniformity. Seed temperature is more important than air temperature, so measure soil temperature. Increase shading above the seedlings, if necessary. Primed seed is less sensitive to high temperatures and should be used during hot weather. Reduce temperature to 60 to 75F (16 to 24C) after cotyledons emerge. Most growers use chambers to accurately control the germination temperature during Stage 1.

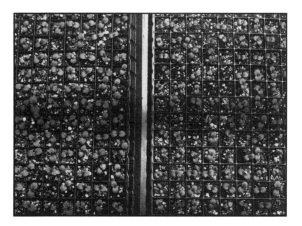

FIG. 2. One reason pansy production has boomed is seed treatments. To take advantage of fall sales, growers have to germinate seed and grow on seedlings in the summer—tough to do for cool loving pansies. Enter primed seed. When seed is primed the germination process has already begun, giving growers a leg-up on the crop. This photo shows the difference between Genesis primed pansy seed on the left and untreated raw pansy seed on the right. Notice how much more uniform the primed seed tray is.

The moisture level around the seed is critical for success. To maintain the critical moisture level, thoroughly wet the flat before sowing. Use a fog system to maintain the moisture level in the flat. The large pieces of vermiculite create a "tent" around the seed, maintaining high humidity without smothering the seed. If you use fine vermiculite, the vermiculite plates smother the seed and prevent germination!

Light is not necessary for germination, but seedlings must receive light immediately after germination to avoid elongation. Use light levels up to 5,000 footcandles (54 klux) if temperature can be controlled.

FERTILIZATION

For top-quality pansies, use a fertilization program that follows a few rules: (1) avoid urea-containing fertilizers; (2) use fertilizers containing ammonia (20-10-20) to promote soft growth, leaf expansion, and stem stretch; and (3) use fertilizers containing just nitrate (14-0-14) to promote root growth, compact habit, and well-toned plants.

V

Growers alternate between 20-10-20 and 14-0-14, depending on how the crop is developing. The initial feed for both plugs and finished crops is 14-0-14. Once roots develop, alternate with 20-10-20 to expand the leaves. As the leaves start to expand, switch back to 14-0-14 to prevent excessive stretch. During warm periods of the year, pansies require periodic supplemental boron (Solubor at a quarter ounce per 100 gallons, or 7.5 g per 400 l, or borax at twice that rate) to prevent boron deficiency.

Growth regulators

A-Rest, B-Nine, Bonzi, or Sumagic will keep plugs and finished plants compact. Excessive ammonia fertilizer and wet soils promote stem stretch, which requires additional growth regulators. When using ammonia fertilizers, remember to apply the growth regulator before fertilizing for better height control. Since growth regulators require two days or more to become effective, apply the growth regulator as the plant starts reaching the allowable size, not after it is too tall. When developing a growth regulator program, growers use B-Nine and A-Rest during the plug stage and Bonzi and Sumagic on the finished crop. Apply only as much growth regulator as is needed to control growth for two to three weeks.

Effective spray rates are B-Nine at 2,500 to 5,000 ppm; A-Rest at 16 to 33 ppm; and Bonzi at 5 to 10 ppm. Do not apply growth regulators after visible bud.

Disease control

Under certain environmental conditions, pansies experience a variety of diseases. Understand the environmental conditions that promote the disease, then manage the environment to reduce disease problems. When using chemicals to control disease outbreaks, follow the labels carefully.

Although black roots are a *Thielaviopsis* symptom, don't panic if your pansies have black roots. Pansy roots are transparent when they are wet and appear black because of the surrounding soil color. With *Thielaviopsis*, though, root hairs will often be damaged or destroyed. Foliage is often chlorotic. Reusing flats and placing plants on contaminated soil will spread the disease. Wet soil with a pH of 6.2 or higher promotes *Thielaviopsis*. When growing outdoors, use iron sulfate drenches to keep the soil pH below 6.0. Drench with Banrot or Cleary's 3336.

There are several leaf spots that produce gray, black, blue, or maroon lesions on pansy foliage. Plants that are poorly fertilized or grown under nonoptimum conditions, such as high humidity, will have more leaf spot diseases. Keeping the foliage dry going into the night is the primary control method to minimize spread of foliar diseases. Maintaining a regular fertilization program to encourage strong growth will reduce disease infections.

With yellow growing points, although most symptoms point to boron deficiency, the growing point is cream colored; boron-deficient plants have green growing points. The problem appears when plants are placed in a high-temperature environment.

When growing *Viola* outdoors, growers periodically experience *Pythium* or *Rhizoctonia* infections when excessive rains occur. The continuously saturated soil promotes the spread of root rot diseases. Drench with Subdue, Banrot, or Truban.

Nutrient deficiencies

With boron deficiency, young foliage becomes thick and puckered, yet remains green. The tip may abort, with numerous small shoots developing below the meristem. Maintain pH below 6.0 to ensure boron availability. Supplement fertilization with one or two applications of a boron source during production, especially if temperatures are above 85F (29C). As a boron supplement, apply borax (0.5 ounce per 100 gallons, 4 g per 100 l) or Solubor (0.25 ounce per 100 gallons, 2 g per 100 l). Apply one or two times during production. If borax is used, dissolve in hot water.

Iron deficiency results in interveinal chlorosis of the youngest leaves. Maintain pH below 6.0. Iron deficiency may result from root damage due to *Thielaviopsis* or *Pythium* root rot. Supplemental iron may be helpful in extreme cases. To reduce soil pH, apply 3 to 5 ounces of iron sulfate per 100 gallons (22 to 37 g per 100 l) as a soil drench. Wash the foliage with clear water after application.

Magnesium deficiency presents as interveinal chlorosis of middle foliage. Magnesium deficiency is more prevalent with high calcium levels. Supplement with Epsom salts once or twice per crop. Apply 1 to 2 pounds of Epsom salts per 100 gallons (120 g per 100 l) as a soil drench.

Varieties

There are many varieties! The biggest trend in varieties is that F_1 varieties are plainly taking over the market because of their uniformity, clear, consistent color, and superior plant performance. Mainly they fall into three broad categories, according to their flower size.

Medium-flowered series. Crystal Bowls (clear faced), Maxim (blotched), Melody (clear and blotched colors), Rally (clear and blotched colors), Sky (clear and blotched colors), and Universal Plus (clear and blotched colors). These medium-flowered series are very free flowering and tolerant of wide weather extremes. They are the staple for pack production.

Large-flowered series. Bingo (clear and blotched colors), Crown (clear faced), Delta (clear and blotched colors), Imperial (clear and blotched colors), Regal (blotched), and Roc (clear and blotched colors). Large-flowered pansies are popular for packs and are the staple for pot production. Landscapers lean heavily toward them.

Extra large-flowered series. These include the Happy Face series, Majestic Giants, and Super Majestic Giants. Extra large–flowered pansies have 3½- to 4-inch (9- to 10-cm) blossoms. They're so large they tend to flop around in poor weather. But consumers love them.

YEAR-ROUND PANSIES IN THE NORTH?

Pansies overwinter in the North! Alex Gerace, Welby Gardens, Denver, is a pansy specialist who has successfully overwintered pansies planted in beds outdoors. He recommends fall, winter, and spring pansies to Denver gardeners and suggests planting pansies on top of fall bulbs. In this way they will be enjoyed through the fall and again with the earliest breaks in the weather in early spring. Normally, they are replaced during the summer. Alex recommends that winter pansies not be covered (it tends to rot them). Also, place them in sunny locations where any snow cover will promptly melt.

YARROW (SEE *ACHILLEA*)

ZANTEDESCHIA (CALLA LILY)

by A.A. De Hertogh
North Carolina State University
Raleigh

❖ *Tender perennials (**Zantedeschia** spp.). Vegetatively propagated. Growers purchase tubers or rhizomes from commercial propagators.*

Calla lilies originated in South Africa. The primary sources of rhizomes or tubers that are forced either as cut flowers or potted plants are grown in the United States, New Zealand, Israel, and the Netherlands. Some are species types, but most are hybrids (see table 1).

Unless you have previous experience forcing callas, try them first on a small scale to determine the adaptability of the species or varieties to specific forcing and marketing conditions.

Depending on growing conditions and types used, calla lilies can be in flower almost all year. However, the major flowering season for greenhouse-grown callas is early winter through late spring. They are also grown outdoors in USDA Climatic Zones 8 to 11. The flowering rhizomes and tubers are usually

TABLE 1   ZANTEDESCHIA FOR FORCING AS CUT FLOWERS AND POTTED PLANTS

Species or cultivar	Production sources	Flower color	Days to 1st flower	Foliage color	Height[b]		Minimum ghse temp	
					Inches	Cm	F	C
Z. aethiopica	Israel, U.S.	White	70–95	Green	18–22	45–55	55	13
Z. albomaculata	Netherlands, U.S.	White	60–85	Spotted	14–18	35–45	60	16
Z. elliottiana	U.S.	Golden yellow	55–80	Spotted	14–18	35–45	60	16
Z. rehmannii	Netherlands, U.S.	Rosy pink to lavender	55–85	Green[a]	12–16	30–40	60	16
Solfatare	Netherlands	Golden yellow	55–80	Spotted	14–18	35–45	55	13
Hybrid Yellow	U.S.	Golden yellow	55–65	Spotted	16–18	40–45	60	16
Apricot Glow	New Zealand	Apricot	65–70	Green	12–14	30–35	60	16
Celeste	New Zealand	Deep pink	65–70	Spotted	14–16	35–40	60	16
Galaxy	New Zealand	Lavender	55–65	Spotted	12–14	30–35	60	16
Golden Sun	New Zealand	Dark yellow	55–65	Green	14–16	35–40	60	16
Tosca	New Zealand	Red	65–70	Green	14–16	35–40	60	16

[a] Usually narrow
[b] Average plant height without any growth regulators

sold in three sizes (diameters): 1½ to 1¾ inches, 1¾ to 2 inches and 2 to 2½ inches and, from U.S. sources, 2½ inches and up. *Z. aethiopica* and most New Zealand cultivars are available for planting in the fall, while the others are available after December. *Z. aethiopica* can be grown continuously and does not need to rest. The other callas have a rest period.

Very little specific information is available on postharvest handling of calla lilies. After harvest, they are cleaned, inspected, graded, and then given 68 to 86F (20 to 30C) for seven days. Callas are subsequently stored at 35 to 48F (2 to 9C) for the *Z. aethiopica* types and at 48 to 75F (9 to 24C) for the colored types, with the lower temperatures generally being used when sprouting occurs. A minimum of six weeks of postlifting storage is needed before the first planting can be made. Suppliers must inform forcers as to when the rhizomes or tubers were lifted.

On arrival, always check a few rhizomes or tubers to be certain they are free from serious diseases such as soft rots, physical damage, or physiological disorders like chalking. If a large amount of rhizomes or tubers have soft rot, do not try to save the crop. Report the problem to your supplier.

Provided the rhizomes or tubers have had at least six weeks of postlifting storage, forcers should be prepared to plant rhizomes or tubers as soon as they arrive. If they must be stored, place them at 41 to 65F (5 to 18C).

Z

FIG. 1. Callas make great pot plants or cut flowers. The variety: Pink Sensation.

Calla lilies require well-drained, sterilized planting media fairly high in organic matter, and a pH 6 to 6.5. Prior to planting, carefully inspect and discard any diseased rhizomes or tubers.

When callas are forced as cut flowers, they are normally grown in large pots so they can easily be moved in the greenhouse. Generally, they are planted three to an 8½-inch (22-cm) pot. It is usually best not to transfer the rhizomes or tubers from pot to pot but to continue growing them in the same pot for several years. For potted flowering plants, use one rhizome or tuber per 5-inch (13-cm) pot. Some shorter forcing types are suitable for 4-inch (10-cm) pots. When planting, cover the rhizomes or tubers with 1 inch (2.5 cm) of growing media.

When forced as potted plants, they can generally be grown pot to pot until the last two to three weeks, when they need some spacing. When grown as cut flowers in pots, some spacing is generally needed as the plants come into flower. It will depend on the growth of the foliage and the layout of the greenhouse.

Force callas in a medium-to-high light intensity (greater than 2,500 footcandles, 27 klux) greenhouse. Very low light intensities, especially as the flowers begin to color, can cause fading of some of the colored cultivars. They may need shading in the summer months. There is no known pronounced photoperiod for flowering of calla lilies; however, short photoperiods can reduce plant height, and long days (night interruption) can increase plant height.

After planting, water thoroughly. Then keep planting media only slightly moist. Do not allow the media to dry out, especially when forcing *Z. aethiopica.*

After planting, start plants at 60 to 65F (16 to 18C). Once the plants begin to sprout, grow white callas at 55F (13C) night temperature and 65 to 70F (18 to 21C) day temperature, and colored callas at 60F (16C) night temperature and 65 to 70F day temperature. Use a well-ventilated greenhouse.

Calla lilies do not require a heavy fertilization program. Use 200 ppm N of 20-20-20 one to two times per month.

Use a preventative bacterial rot *(Erwinia)* control program. *Erwinia* is the major disease of callas. Always inspect the bulbs prior to planting. To minimize leaf disease, keep the greenhouse's relative humidity low and the air moving. Occasionally there are leaf and flower blights and some insect problems, but not many.

There are two main injury disorders that can occur with calla lilies. Chalking has been reported, but the exact causes are not known—mechanical damage and sunburning have been implicated. Also, flower and leaf abnormalities can be caused by the use of gibberellic acid preplant dips.

Three plant growth regulators have shown promise for calla lilies. Many of the colored callas have shown an increase in the number of flowers produced per rhizome or tuber by using preplant GA3 or GA4 dips at 500 ppm for 10 minutes. Promalin also promotes flowering. Extensive trials are needed in this area, and none are EPA-cleared at the present time. Calla lilies will normally begin flowering about 55 to 95 days after planting.

Cut flowers one day prior to pollen shed. If they are to be stored for up to three days, place them dry at 41F (5C). Commercial floral preservatives containing sugar and a fungicide are generally advised to reduce stem splitting and rolling.

Market flowering potted plants when two to three flowers are fully colored (fig. 1). Advise consumers to keep the plants in the coolest area of the home to obtain maximum plant life. The plants do, however, need direct sunlight.

ZINNIA

❖ *Annual (**Zinnia angustifolia**). 60,000 to 78,000 seeds per oz (2,100 to 2,730 per g). Germinates in four to eight days at 70 to 72F (21 to 22C). Cover seed for germination.*

❖ *Annual (**Zinnia elegans**). 2,000 to 6,000 seeds per oz (70 to 210 per g). Germinates in three to seven days at 70 to 72F (21 to 22C). Cover seed for germination. Seed may be sown direct to the final growing container or to the field.*

Zinnias, especially *Z. elegans,* have graced gardens for generations. Their showy, heat-tolerant flowers can be found from coast to coast. The best part is they are easy to grow—for growers and gardeners. More recently, *Z. angustifolia* has made its way into the bedding plant scene as a superb landscape performer (fig. 1).

In the south allow four to five weeks for green pack sales, eight weeks for flowering 4-inch (10-cm) pots. Sell medium and tall varieties green. Zinnia that is allowed to flower in the pack is highly susceptible to attack by *Botrytis* and other diseases. Seed may be sown direct to the final container. Keep media at 70F (21C) minimum until radicle emergence. Most zinnia complaints from grow-

Z

FIG. 1. *Zinnia angustifolia* is an excellent crop for growers catering to the landscape trade and those looking to offer home gardeners an unusual plant. Virtually indestructible, plants produce literally hundreds of bright white flowers all summer long.

ers are due to trying to germinate too cool. At cool temperatures, zinnias rot in the ground. After germination maintain 60F to encourage stocky plant growth.

Grown as outdoor-produced cut flowers, zinnias can be harvested reliably from early July through frost. Sow seed into pots in March and grow cool to carry plants. Plant out in the field in late May and provide good moisture. Space plants 12 by 12 inches (30 by 30 cm). Dahlia- and cactus-flowered zinnias are preferred, although some growers include the small flowers of Lilliputs for variety.

Seed can be sown direct to the field on June 1 for flowers by late July. Provided good soil and water as needed, zinnias grow quickly. Sow seed in rows 2 feet (61 cm) apart and thin plants to 6 inches (15 cm) apart in the row. Subsequent sowings can be made through June. Disbudding greatly improves flower quality.

Among varieties to grow for cut flowers, State Fair Mix grows to 30 to 36 inches (76 to 91 cm) tall and produces flowers measuring 5 to 6 inches (13 to 15 cm) across, with broad petals in a full range of bright colors. It shows good resistance to *Alternaria* and mildew. The Ruffles series, an F_1, has five separate colors and a mix. Plants grow to 30 inches tall and produce 2½-inch (6-cm), ruffled flowers on long, stiff stems. Splendor, also an F_1, produces 4- to 5-inch (10 to 13-cm), fully double flowers with semiruffled petals. Plants grow to 2 feet (61 cm) tall; they can be spaced closer together, as they form a hedge. It is advisable to stake taller-growing cut flower varieties.

Varieties for pot and bedding production include the F_1 series Peter Pan, Dasher, Short Stuff, and Dreamland (fig. 2). All come in a wide range of separate colors and are well suited for pack or pot sales.

Open-pollinated zinnias flower later than hybrids by up to one week and show a more variable plant habit.

The newest zinnias to come on the scene, the *Z. angustifolia* types, have become especially popular for landscape plantings. Their medium-sized 1½- to 2-inch (4 to 5 cm) flowers appear in profusion on informal, mounded plants, and they thrive in the heat!

In the north allow eight weeks for green packs, nine to 10 weeks for flowering packs. For 4-inch (10-cm) pots and hanging baskets, sow seed in late January for full flowering by mid-May. Once flowering begins, plants stay in bloom as long as temperatures are warm.

The Star series comes in gold, orange, white, and a mix. Plants grow 10 to 12 inches (25 to 30 cm) tall and produce an abundance of fade-resistant, 2-inch (5-cm) single

FIG. 2. The new, dwarf zinnias, such as Dreamland Mix, are an excellent way to brighten up sunny summer flower beds.

blooms. Classic Golden Orange produces 1½-inch (4 cm) single flowers on spreading plants ideal as a ground cover.

ZONAL GERANIUM (SEE *PELARGONIUM × HORTORUM*)

Z

APPENDIX 1

Holiday Dates

	1997	1998	1999	2000	2001	2002	2003	2004
New Year's Day	Jan. 1	Jan. 1	Jan. 1	Jan. 1	Jan. 1	Jan. 1	Jan. 1	Jan. 1
St. Valentine's Day	Feb. 14	Feb. 14	Feb. 14	Feb. 14	Feb. 14	Feb. 14	Feb. 14	Feb. 14
Presidents' Day	Feb. 17	Feb. 16	Feb. 15	Feb. 21	Feb. 18	Feb. 18	Feb. 17	Feb. 16
Ash Wednesday	Feb. 12	Feb. 25	Feb. 17	Mar. 8	Feb. 28	Feb. 13	Mar. 5	Feb. 25
St. Patrick's Day	Mar. 17	Mar. 17	Mar. 17	Mar. 17	Mar. 17	Mar. 17	Mar. 17	Mar. 17
Palm Sunday	Mar. 23	Apr. 5	Mar. 28	Apr. 16	Apr. 8	Mar. 24	Apr. 13	Apr. 4
Easter Sunday	Mar. 30	Apr. 12	Apr. 4	Apr. 23	Apr. 15	Mar. 31	Apr. 20	Apr. 11
Passover	Apr. 22	Apr. 11	Apr. 1	Apr. 7	Mar. 27	Apr. 16	Apr. 5	Apr. 23
Secretary's Day	Apr. 23	Apr. 22	Apr. 21	Apr. 26	Apr. 25	Apr. 24	Apr. 23	Apr. 21
Mother's Day	May 11	May 10	May 9	May 14	May 13	May 12	May 11	May 9
Memorial Day	May 26	May 25	May 31	May 29	May 28	May 27	May 26	May 31
Father's Day	June 15	June 21	June 20	June 18	June 17	June 16	June 15	June 20
Labor Day	Sept. 1	Sept. 7	Sept. 6	Sept. 4	Sept. 3	Sept. 2	Sept. 1	Sept. 6
Grandparent's Day	Sept. 7	Sept. 13	Sept. 12	Sept. 10	Sept. 9	Sept. 8	Sept. 7	Sept. 12
Rosh Hashanah	Oct. 2	Sept. 21	Sept. 11	Sept. 29	Sept. 17	Sept. 6	Sept. 26	Sept. 15
Yom Kippur	Oct. 11	Sept. 30	Sept. 20	Oct. 8	Sept. 26	Sept. 15	Oct. 5	Sept. 24
Boss's Day	Oct. 15	Oct. 21	Oct. 20	Oct. 18	Oct. 17	Oct. 16	Oct. 15	Oct. 20
Sweetest Day	Oct. 18	Oct. 17	Oct. 16	Oct. 21	Oct. 20	Oct. 19	Oct. 18	Oct. 16
All Saints' Day	Nov. 1	Nov. 1	Nov. 1	Nov. 1	Nov. 1	Nov. 1	Nov. 1	Nov. 1
Dia de los Muertes	Nov. 1-2	Nov. 1-2	Nov. 1-2	Nov. 1-2	Nov. 1-2	Nov. 1-2	Nov. 1-2	Nov. 1-2
Election Day	Nov. 4	Nov. 3	Nov. 2	Nov. 7	Nov. 6	Nov. 5	Nov. 4	Nov. 2
Thanksgiving Day	Nov. 27	Nov. 26	Nov. 25	Nov. 23	Nov. 22	Nov. 28	Nov. 27	Nov. 25
Hanukkah	Dec. 24	Dec. 25	Dec. 4	Dec. 21	Dec. 9	Nov. 29	Dec. 20	Dec. 7
Christmas Day	Dec. 25	Dec. 25	Dec. 25	Dec. 25	Dec. 25	Dec. 25	Dec. 25	Dec. 25

APPENDIX 2

USDA PLANT HARDINESS ZONE MAP

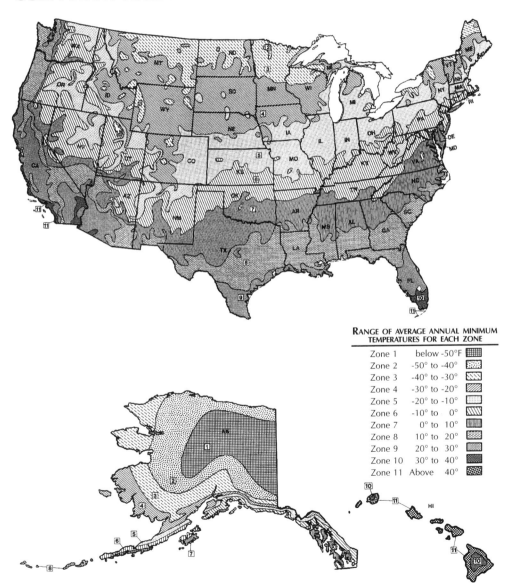

RANGE OF AVERAGE ANNUAL MINIMUM TEMPERATURES FOR EACH ZONE	
Zone 1	below -50°F
Zone 2	-50° to -40°
Zone 3	-40° to -30°
Zone 4	-30° to -20°
Zone 5	-20° to -10°
Zone 6	-10° to 0°
Zone 7	0° to 10°
Zone 8	10° to 20°
Zone 9	20° to 30°
Zone 10	30° to 40°
Zone 11	Above 40°

INDEX

A

B